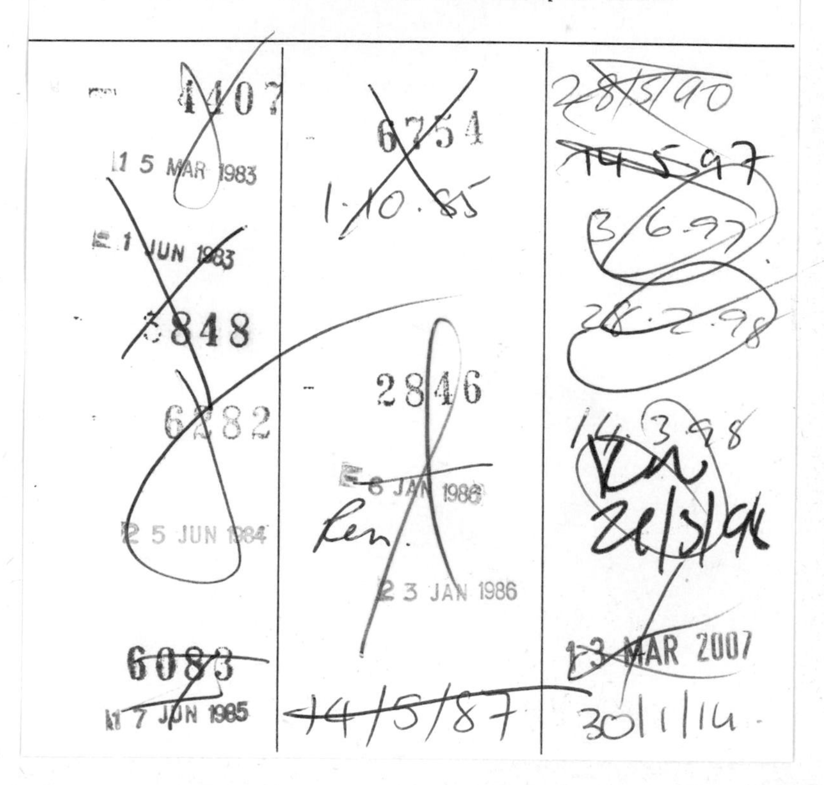

CONTROL OF PIG REPRODUCTION

Proceedings of Previous Easter Schools in Agricultural Science, published by Butterworths, London

*SOIL ZOOLOGY Edited by D. K. McL. Kevan (1955)
*THE GROWTH OF LEAVES Edited by F. L. Milthorpe (1956)
*CONTROL OF THE PLANT ENVIRONMENT Edited by J. P. Hudson (1957)
*NUTRITION OF THE LEGUMES Edited by E. G. Hallsworth (1958)
*THE MEASUREMENT OF GRASSLAND PRODUCTIVITY Edited by J. D. Ivins (1959)
*DIGESTIVE PHYSIOLOGY AND NUTRITION OF THE RUMINANT Edited by D. Lewis (1960)
*NUTRITION OF PIGS AND POULTRY Edited by J. T. Morgan and D. Lewis (1961)
*ANTIBIOTICS IN AGRICULTURE Edited by A. M. Woodbine (1962)
*THE GROWTH OF THE POTATO Edited by J. D. Ivins and F. L. Milthorpe (1963)
*EXPERIMENTAL PEDOLOGY Edited by E. G. Hallsworth and D. V. Crawford (1964)
*THE GROWTH OF CEREALS AND GRASSES Edited by F. L. Milthorpe and J. D. Ivins (1965)
*REPRODUCTION IN THE FEMALE MAMMAL Edited by G. E. Lamming and E. C. Amoroso (1967)
*GROWTH AND DEVELOPMENT OF MAMMALS Edited by G. A. Lodge and G. E. Lamming (1968)
*ROOT GROWTH Edited by W. J. Whittington (1968)
*PROTEINS AS HUMAN FOOD Edited by R. A. Lawrie (1970)
*LACTATION Edited by I. R. Falconer (1971)
*PIG PRODUCTION Edited by D. J. A. Cole (1972)
*SEED ECOLOGY Edited by W. Heydecker (1973)
HEAT LOSS FROM ANIMALS AND MAN: ASSESSMENT AND CONTROL Edited by J. L. Monteith and L. E. Mount (1974)
*MEAT Edited by D. J. A. Cole and R. A. Lawrie (1975)
*PRINCIPLES OF CATTLE PRODUCTION Edited by Henry Swan and W. H. Broster (1976)
*LIGHT AND PLANT DEVELOPMENT Edited by H. Smith (1976)
*PLANT PROTEINS Edited by G. Norton (1977)
ANTIBIOTICS AND ANTIBIOSIS IN AGRICULTURE Edited by M. Woodbine (1977)
CONTROL OF OVULATION Edited by D. B. Crighton, N. B. Haynes, G. R. Foxcroft and G. E. Lamming (1978)
POLYSACCHARIDES IN FOOD Edited by J. M. V. Blanshard and J. R. Mitchell (1979)
SEED PRODUCTION Edited by P. D. Hebblethwaite (1980)
PROTEIN DEPOSITION IN ANIMALS Edited by P. J. Buttery and D. B. Lindsay (1981)
PHYSIOLOGICAL PROCESSES LIMITING PLANT PRODUCTIVITY Edited by C. Johnson (1981)
ENVIRONMENTAL ASPECTS OF HOUSING FOR ANIMAL PRODUCTION Edited by J. A. Clark (1981)
EFFECTS OF GASEOUS AIR POLLUTION IN AGRICULTURE AND HORTICULTURE Edited by M. Unsworth and D.P. Ormrod (1982)
CHEMICAL MANIPULATION OF CROP GROWTH AND DEVELOPMENT Edited by J. S. McLaren (1982)

**These titles are now out of print but are available in microfiche editions*

Control of Pig Reproduction

D.J.A. COLE, PhD
G.R. FOXCROFT, PhD
University of Nottingham School of Agriculture

BUTTERWORTH SCIENTIFIC
London Boston Durban Singapore Sydney Toronto Wellington

First published 1982

British Library Cataloguing in Publication Data

Control of pig reproduction.
1. Swine – Reproduction – Congresses
I. Cole, D.J.A. II. Foxcroft, G.R.
636.4′08926 SF396.9

ISBN 0–408–10768–5

Typeset by Scribe Design Ltd, Gillingham, Kent
Printed and bound by Mansell Bookbinders Ltd, Witham, Essex

PREFACE

When considering the potential efficiency of modern animal production systems, it is well recognized that reproductive efficiency is invariably a major limiting factor. In contrast, in other areas of management such as growth and nutrition, optimal economic returns may be achieved at less than maximum performance. Such observations are clearly relevant to the management of commercial pig herds and our failure to achieve optimum reproductive performance lies both in an inability to correctly diagnose non-pathological conditions of infertility and in a limitation in the number of carefully evaluated techniques for controlling pig production. In both situations further improvements in practical management can only be based on a better understanding of the basic reproductive processes involved.

With these points in mind it was intended that the 34th Easter School in Agricultural Science should act as a forum to bring together current knowledge on pig reproduction and to pay particular attention to its possible application in the field. The title *Control of Pig Reproduction* can be taken to imply a consideration of both the physiological control of reproduction and the control exerted by man in the management of his animals. The choice of participants reflects this approach and it is gratifying that the conference itself succeeded in bringing together so many experts in different disciplines with the common aim of directing their knowledge towards the improvement of pig reproduction. Equally the Easter School provided an opportunity for scientists from laboratories in many parts of the world to meet personally in Nottingham; it is hoped that the lively social, as well as scientific interactions that ensued, will provide an impetus for new, as well as continuing, collaborative efforts to advance our knowledge in this area.

D.J.A. Cole
G.R. Foxcroft

ACKNOWLEDGEMENTS

It is a pleasure to acknowledge the efforts of those who presented papers and participated in the discussions of the Easter School and thus ensured its success. We are indebted to Professor E.C. Amoroso who opened the Conference and to those who acted as Chairmen: Dr D.R. Melrose, Professor B.N. Day, Dr R.B. Heap, Professor P.J. Dziuk, Dr D.B. Crighton and Dr A. Aumaitre.

The University of Nottingham wishes to express its gratitude to the following organizations whose financial assistance made possible the contributions from overseas speakers:

J. Bibby Agriculture Ltd.
BOCM Silcock Ltd.
The British Council
Colborn Nutrition Limited
Dalgety Spillers Agriculture Ltd.
Hoechst Aktiengesellschaft
Imperial Chemical Industries Limited
Intervet International B/V
Pauls and Whites Foods Ltd.
Pig Improvement Company Ltd.
RHM Agriculture Limited
Roche Products Limited
Upjohn Limited

Finally, we would like to thank all those staff and students who gave their time to help in so many ways. We would like to mention our particular appreciation to Mrs Jose Newcombe and Mrs Pam Courtman for organizing the secretariat and to Jim Tilton and Andy Paterson for arranging the social programme.

CONTENTS

I

THE MALE, MATING AND CONCEPTION

1

MALE SEXUAL DEVELOPMENT

B. COLENBRANDER, M.T. FRANKENHUIS† and C.J.G. WENSING
Department of Anatomy, Veterinary Faculty, State University of Utrecht, The Netherlands

Sexual development is an array of processes which takes place in a regular sequence. During development the body which is sexually bipotential differentiates in either a male or female direction. The first step in this development is determined when the ovum is fertilized with either a 19 X or a 19 Y sperm cell, the pig having 38 ($2n$) chromosomes. The principal

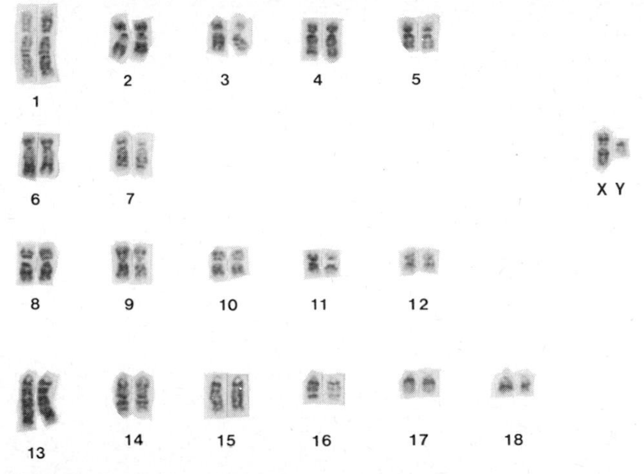

Figure 1.1 G-banded metaphase chromosomes from a cultured blood lymphocyte. The G-banding procedure included treatment with trypsin and staining with Giemsa. For a detailed description of chromosomal banding patterns in the pig see Lin *et al.* (1980). By courtesy of Dr A.A. Bosma

task of the Y chromosome is to direct the embryonic indifferent gonad to differentiate as a testis. This chromosome is rather limited in size in most mammals (*Figure 1.1*).

Irrespective of chromosome constitution, the anlage of the genital system is indifferent i.e. no morphological difference in gonadal or accessory genital structures is observed between the sexes in the first stages of embryonic development. The mammalian embryo has a tendency to

†Also of Department of Histology and Cell Biology, Medical Faculty, State University of Utrecht, The Netherlands

develop into a female. Male development is due to the intervention of major regulatory genes. One is Y-linked and specifies a plasma membrane protein (H–Y antigen). Testicular differentiation does not occur in the absence of H–Y antigen. Moreover, an X-linked gene regulates the function of the H–Y structural gene (Ohno, 1977; Wachtel and Koo, 1978). Development of the indifferent gonad into a testis is then induced. Subsequently the testis produces hormones that induce phenotypic male development. Differentiation of the male accessory genital organs is androgen-dependent, as is the sex-specific differentiation of a number of nuclei in the central nervous system (Dörner, 1976). An X-linked gene mediates the responsiveness of target cells to steroid hormones (androgens) produced by the testis (Ohno, 1977; Wachtel, 1979).

In the pig three periods of morphological and functional differentiation of the testis can be distinguished. The first period starts with gonadal differentiation at 26 days post coitum and lasts until approximately 60 days post coitum; the second is the perinatal period and the third from puberty onwards. In this review we will focus attention on the morphological and endocrine aspects of testicular development, before adulthood.

Early foetal period

The indifferent gonad can be recognized at approximately 21 days post coitum, as a bilateral proliferation of mesenchymal blastema covered by coelomic epithelium along the medial side of the mesonephros (Allen, 1904; Black and Erickson, 1968; Pelliniemi, 1975a). The surface epithelium extends as primitive cords into the gonadal blastema. The primordial germ cells, which have reached the gonad by amoeboid movements, (Jirasek, 1976) are located in the mesenchyme under the surface epithelium and in the cords. They increase in number by mitosis. At 24 days post coitum the male and female gonads are morphologically similar. An indication of testicular differentiation can be seen at the age of 26 days, when the primordial germ cells become entirely enclosed within the cords. The cords are separated from the adjacent mesenchyme by a partial discontinuous basal lamina (Pelliniemi, 1975b).

Leydig cells

Leydig cells appear at the age of 30 days. They proliferate rapidly to reach a maximal stage of development at approximately 38 days post coitum.This is followed by a substantial regression in their number and size (Ancel and Bouin, 1903; Allen, 1904; Whitehead, 1904; Moon and Hardy, 1973).

The well-differentiated Leydig cells are characterized by the presence of a considerable amount of tubular and/or vesicular smooth endoplasmic reticulum (SER). The rough endoplasmic reticulum (RER) is scanty and consists of two types: long and short profile RER. The long type is situated in close proximity to the mitochondria (van Vorstenbosch, Colenbrander and Wensing, 1981). There can be an intimate contact between Leydig cells and gap junctions are common as shown in *Figure 1.2(a)* and *(b)*.

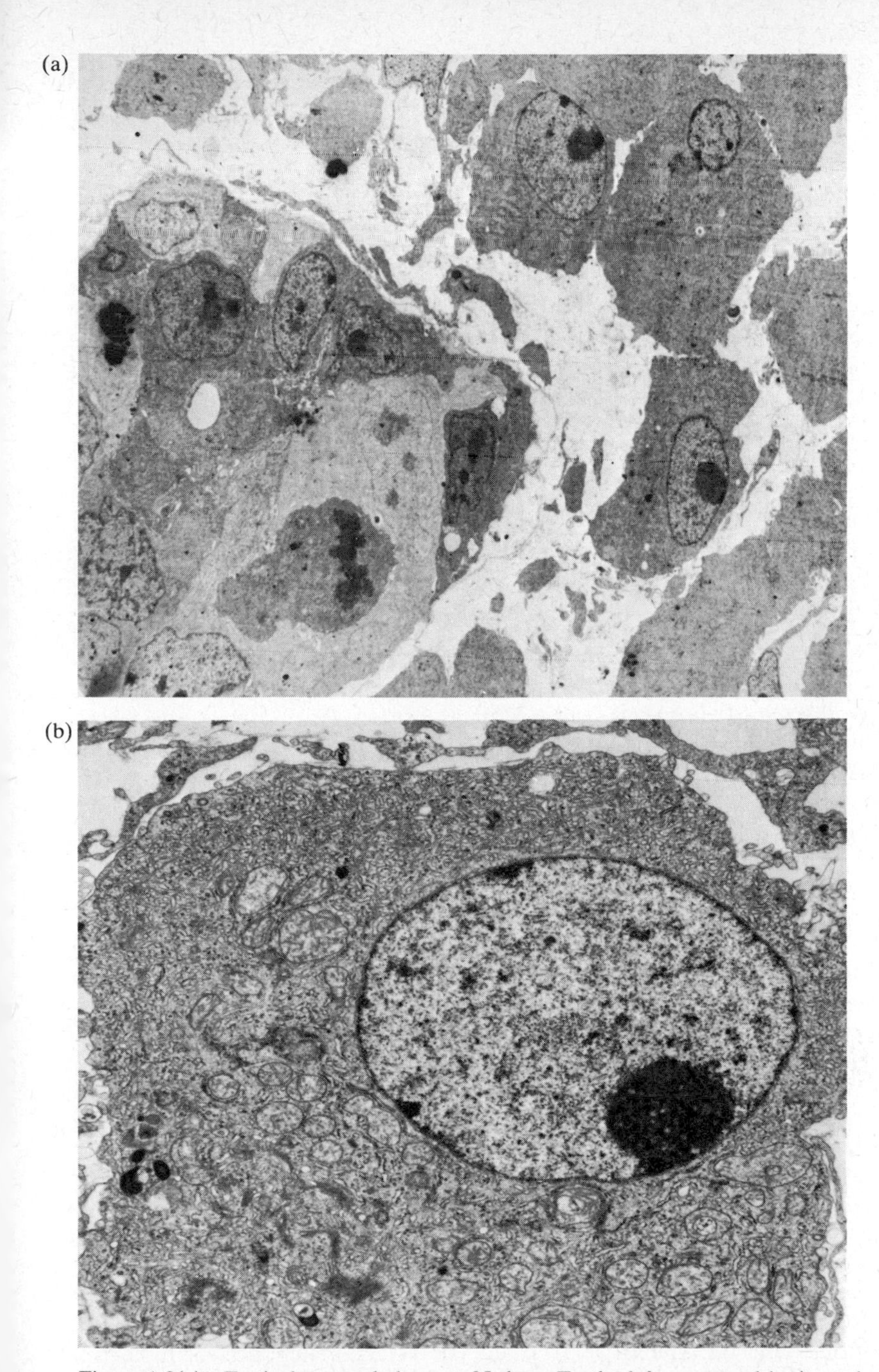

Figure 1.2(a) Testicular morphology at 35 days. To the left a sex cord is situated with a dividing germ cell. The peritubular space is empty; Leydig cells are well developed. (Magn. 2200×). (b) Leydig cell with an oval nucleus and a prominent nucleolus. Large quantities of branched tubular SER. Long profile RER is coupled with mitochondria. The mitochondria display some degenerative forms. (Magn. 8500×). By courtesy of Dr C.J.A.H.V. van Vorstenbosch

Sertoli cells

Sertoli cells or sustentacular cells are characterized by a nucleus which is irregular in size and the mitochondria are elongated (Pelliniemi, 1975b). There is, however, very little information on Sertoli cell morphology in the foetal pig after 26 days post coitum.

Germ cells

The germ cells are recognized as large round cells having a large round nucleus with finely dispersed chromatin. At 27 days the germ cells are located singly within the testicular cords and are completely surrounded by Sertoli cells (Pelliniemi, 1975b). The number of germ cells in the testis of the foetal pig gradually increases by mitotic division from 42 days onwards (van Straaten and Wensing, 1977). They do not enter meiosis until the prepubertal period (Swierstra, 1968) whereas in female foetuses meiosis starts at about 40–48 days post coitum (Black and Erickson, 1968; Mauleon, 1978).

Two factors are involved in the onset of meiosis, Meiosis Inducing Substance (MIS) and Meiosis Preventing Substance (MPS). MIS in mammals is produced in the rete from mesonephric-derived cells both in the testis and the ovary. MPS which is produced within the testicular tissue (possibly within the testicular cords) might participate in preventing meiosis in the foetal testis (Byskov, 1979).

Testicular hormones

In the last ten years much information has become available on the biochemical aspects of testicular function. The testis already produces testosterone at an early stage and this steroid can be detected at about 26 days. Maximal testicular testosterone concentrations are found at about 35 days and decline thereafter (Raeside and Sigman, 1975). A comparable pattern of changes in testosterone secretion by the foetal pig testis has been observed in organ culture (Stewart and Raeside, 1976).

The biological activity of the androgens secreted by the foetal pig testis (±30 days) has been demonstrated in co-culture experiments with rat prostates (Moon, Hardy and Raeside, 1973). Concentrations of serum testosterone are higher (~4 ng/ml) at 35 days (Ford, Christenson and Maurer, 1980) than in either the earlier or in the later part of the foetal period (Colenbrander, de Jong and Wensing, 1978; Ford, Christenson and Maurer, 1980). Although there is no doubt about androgen production in the testis, the location of androgen synthesis within the gonad is still uncertain. In adult gonads, androgens are produced by the Leydig cells. However, SER which is considered to be involved in steroid production (Christensen, 1975) could not be observed in 27-day old Leydig cells (Pelliniemi, 1975b). Enzyme histochemical activity of Δ^5-3β-hydroxysteroid dehydrogenase (3β-HSD) could not be detected in Leydig cells but only in the sex cords (Moon and Raeside, 1972). However, preliminary histochemical studies in our laboratory revealed 3β-HSD activity in Leydig cells around 35 days. Further histochemical studies at light and electronmicroscopical levels may solve the question of the site of androgen production in the early foetal period.

The onset of androgen production in the gonad is probably independent of gonadotrophic hormone stimulation. Testis differentiation and testosterone production occurs in organ culture in the absence of gonadotrophic hormones and hence is probably due to genetic programming of the steroid-producing cells. Human chorionic gonadotrophin (HCG) can stimulate testosterone production by the early foetal pig testis (Raeside and Middleton, 1979) but it remains questionable whether physiologically active extra-pituitary gonadotrophin-like compounds exist in the pig (Amoroso and Porter, 1970; Ziecik and Flint, 1980). Data on gonadotrophic hormone production by the foetal pituitary are scarce. At 55 days serum luteinizing hormone (LH) concentrations are below detectable levels even after stimulation with LH releasing hormone (Colenbrander *et al.*, 1977; Colenbrander *et al.*, 1980). Moreoever, foetal decapitation in the pig does not interfere with the morphological development of Leydig cells before 60 days (van Vorstenbosch, Colenbrander and Wensing, 1981). Thus, gonadotrophic hormone-dependent androgen secretion probably does not take place in the first phase of testicular development.

One other hormone produced by the testis of great significance for male sexual differentiation is the anti-Müllerian hormone (AMH). This protein is synthesized by foetal (and neonatal) Sertoli cells and mediates the regression of the Müllerian ducts (Jost, 1947; Blanchard and Josso, 1974). In the 27-day old pig foetus at the time of gonadal differentiation, AMH production is very low, but thereafter gradually increases. Total inhibition of rat Müllerian duct growth can be observed in co-cultures with 29 days pig testis after three days of culture (Tran, Meusy-Dessolle and Josso, 1977). The production of AMH by the Sertoli cells lasts until after birth (Tran, Meusy-Dessolle and Josso, 1981).

What is the effect of the testicular hormones on the extra-gonadal tissues? As already mentioned the AMH is responsible for the regression of the Müllerian ducts. The androgens produced by the testis are essential for the development of the male accessory sex organs and for the sexual differentiation of the brain. Development and differentiation of the male accessory sex glands and external genitalia follow testicular differentiation and are androgen-dependent. The sequence of development of somatic sex characteristics is fixed and often identical in different mammals (Jost, 1965; Price and Ortiz, 1965). Differentiation of the Wolffian duct and the anlage and development of the bulbo-urethral glands, prostate, seminal vesicles and the external genitalia occurs between 26 days, when androgen secretion starts, and 50 days (Koning, 1942; Zietschmann and Krölling, 1955; Gier and Marion, 1970). The anlage of the bulbo-urethral gland can be seen in the lateral wall of the urethral sinus at 30 days (32 mm). Development of the prostate can be observed in the lateral wall of the pelvic urethra after 40 days (57 mm). The seminal vesicles differentiate in connection with the Wolffian ducts after 50 days (88 mm) (Koning, 1942).

Sexual differentiation of the brain, which in the adult male results in tonic gonadotrophic hormone secretion and male mating behaviour; is also androgen-dependent (Dörner, 1976). Exposure of female foetuses to exogenous testosterone earlier than 90 days leads to an impaired functioning of the stimulatory oestrogen feedback mechanism (Elsaesser and Parvizi, 1979). Testosterone treatment of neonatal females does not affect this feedback system (Colenbrander, 1978). This indicates that androgen secretion during the first phase of testicular development seems to be involved in brain sex priming. More information on sex differentiation of the brain and the role of the ovary in this process is discussed in Chapter 5 by Elsaesser.

Between 65 and 95 days testicular activity seems to be reduced. The testis/body weight ratio decreases, which means that testicular growth lags behind body growth (van Straaten and Wensing, 1977). Testosterone production by the Leydig cells (Colenbrander, de Jong and Wensing, 1978; Ford, Christenson and Maurer, 1980) is low. Sertoli cells probably still produce AMH as this hormone is present at 40 days and immediately after birth, but data on the period in between are lacking (Tran, Meusy-Dessolle and Josso, 1977; 1981). The number of germ cells in the testis gradually increases (van Straaten and Wensing, 1977), but refined data on the morophology of these cells in this period are also lacking.

Another aspect of male sexual development which occurs in this period of relative inactivity of the Leydig cells is the descent of the testis. An extra-abdominal position, resulting in a lower temperature of the gonad, is essential for spermatogenesis in adulthood. The process of descent is brought about by a swelling and subsequent regression of the extra-abdominal part of the gubernaculum. This results in a movement of the testis towards the inguinal canal in the swelling phase and a descent into the scrotum during the regression phase (Wensing, 1968; Wensing, Colenbrander and van Straaten, 1980).

The process starts at approximately 60 days post coitum. The testis passes the inguinal canal at about 85 days and reaches the bottom of the scrotum shortly after birth. It has been postulated that gonadotrophic hormones regulate the process of descent (Hadziselimovic and Herzog, 1980). However, it is unlikely that gonadotrophic hormones are important as testicular descent in the decapitated foetus is undisturbed (Colenbrander *et al.*, 1979). Testosterone is also unlikely to be an important regulating factor of gubernacular outgrowth, as in this period testosterone concentrations are low. In the decapitated foetus, where Leydig cell development is disturbed after 60 days, timing of the descent is normal (Colenbrander *et al.*, 1979). However testicular hormones still seem to play an essential role in this process as foetal castration at 60 days post coitum results in arrest of gubernaculum development and consequently no testicular descent occurs (Colenbrander, Macdonald and Elsaesser, unpublished observation). As the Leydig cells are rather undifferentiated after 55 days, a factor produced by the sex cords is probably involved in gubernacular outgrowth. Testosterone might be involved in the regression of the gubernaculum, which occurs after 85 days when serum testosterone concentrations are increasing (Colenbrander, de Jong and Wensing, 1978). Also in decapitated foetuses and neonatal pig freemartins which have very low testosterone concentrations, gubernacular regression is retarded (Colenbrander and Wensing, 1975; Colenbrander *et al.*, 1979).

After 90 days, when the testis has reached the neck of the scrotum, the second phase of testicular development starts.

Perinatal period

The second (perinatal) phase of testicular development extends from 90 days post coitum until three weeks after birth. In this period there is an increase in the testis/body weight ratio. Testis weight increases between 100 days post coitum and three weeks after birth, from 86 mg to about 2 g in Landrace pigs (van Straaten and Wensing, 1977). This is mainly due to Leydig cell differentiation and proliferation.

Leydig cells

At three weeks after birth when the Leydig cells are maximally developed, they form almost 65% of the testicular volume. Two types of Leydig cells can be distinguished in the early postnatal period (van Straaten and Wensing, 1978). The intertubular cells are the first to be formed and are located between groups of coiled sex cords. They are already present, although less differentiated, in the early foetal period. The second type, the peritubular cells, are enclosed within the coils of the sex cords. They probably differentiate from peritubular mesenchymal elements (Moon and Hardy, 1973; Dierichs, Wrobel and Schilling, 1973) (*Figure 1.3(a)*). Apart from total Leydig cell volume, the cell nuclear size, cell volume and other histological features mark the differentiation of the interstitial cells (Dierichs, Wrobel and Schilling, 1973; van Straaten and Wensing, 1977). NADH-diaphorase and hydroxysteroid dehydrogenase (HSD) activity, as demonstrated histochemically (*Figures 1.3(b)* and *1.5*), is intense in the second postnatal week and declines thereafter (Moon and Raeside, 1972; van Straaten and Wensing, 1978). The histochemical activity is most prominent in the intertubular Leydig cells whereas the peritubular cells show less HSD activity. The perinatal Leydig cells are surrounded by a basement membrane and cell connections appear as gap junctions and desmosome-like structures. There is an increase in the number of mitochondria with tubular cristae and electron dense inclusions. The smooth endoplasmic reticulum is abundant and consists of branched tubular and extensive whirls of SER; the latter may be fenestrated and microfilaments diminish sharply around birth. The RER is quantitatively unimportant (*Figures 1.4(a)* and *(b)*) (Dierichs, Wrobel and Schilling, 1973; van Vorstenbosch, Colenbrander and Wensing, 1981). Leydig cell development in the late foetal and postnatal period is dependent upon gonadotrophic hormones (Morat, 1977; Colenbrander *et al.*, 1979).

After one month, regression of interstitial cells occurs and degeneration of mitochondria and dilatation of the cisternae of the smooth endoplasmic reticulum is observed. Leydig cell regression in this period even results in a decrease of the testis weight.

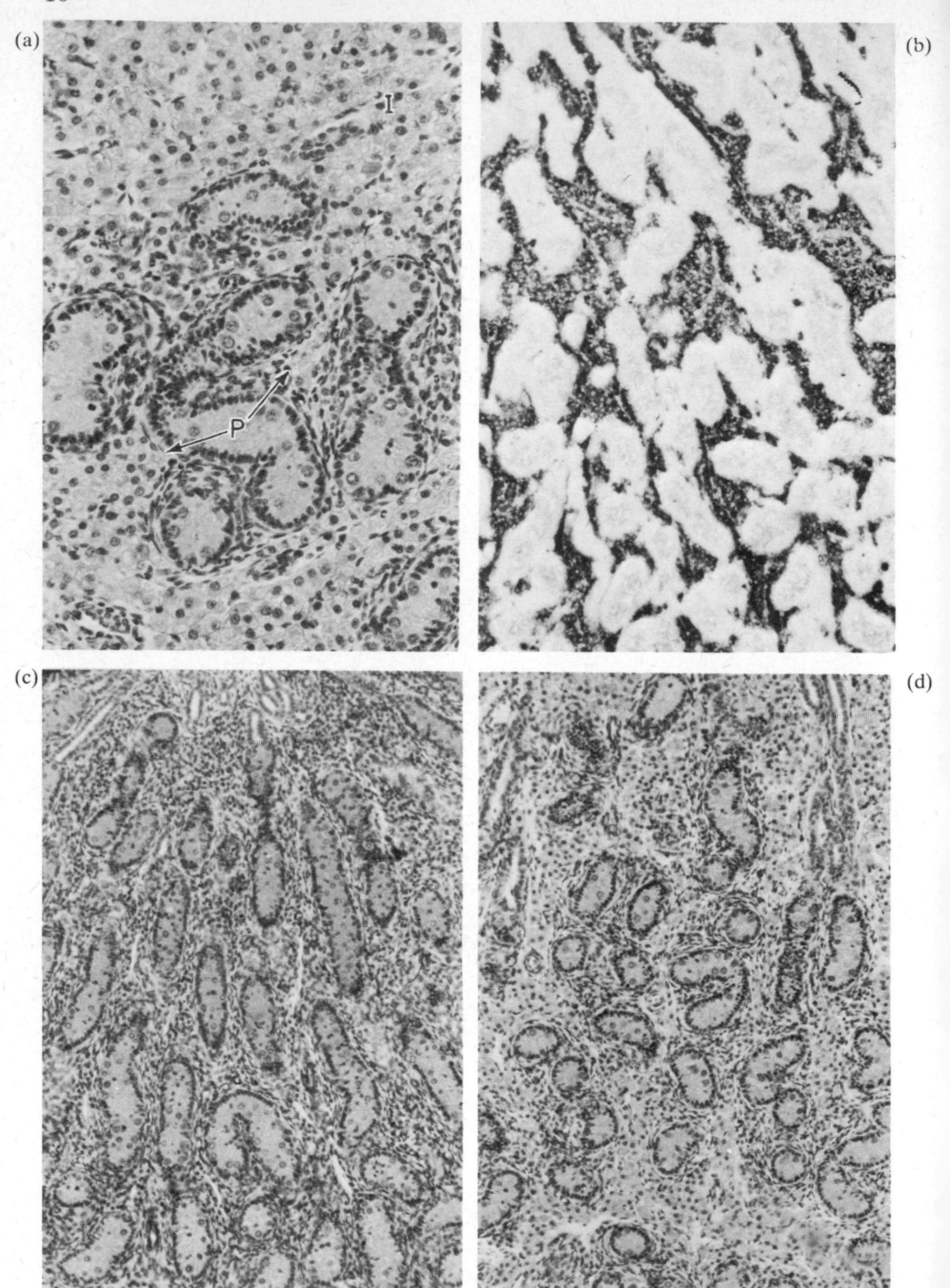

Figure 1.3 (a) Intertubular (I) and peritubular (P) Leydig cells 113 days post coitum (Bouin fixed, trichrome stained). (Magn. 189×). (b) NADH diaphorase activity of Leydig cells, 113 days post coitum. (Magn. 75×). (c)–(d) Coiling of the sex cord during development, 85 and 113 days post coitum respectively. (Bouin fixed, trichrome stained). (Magn. 96×).

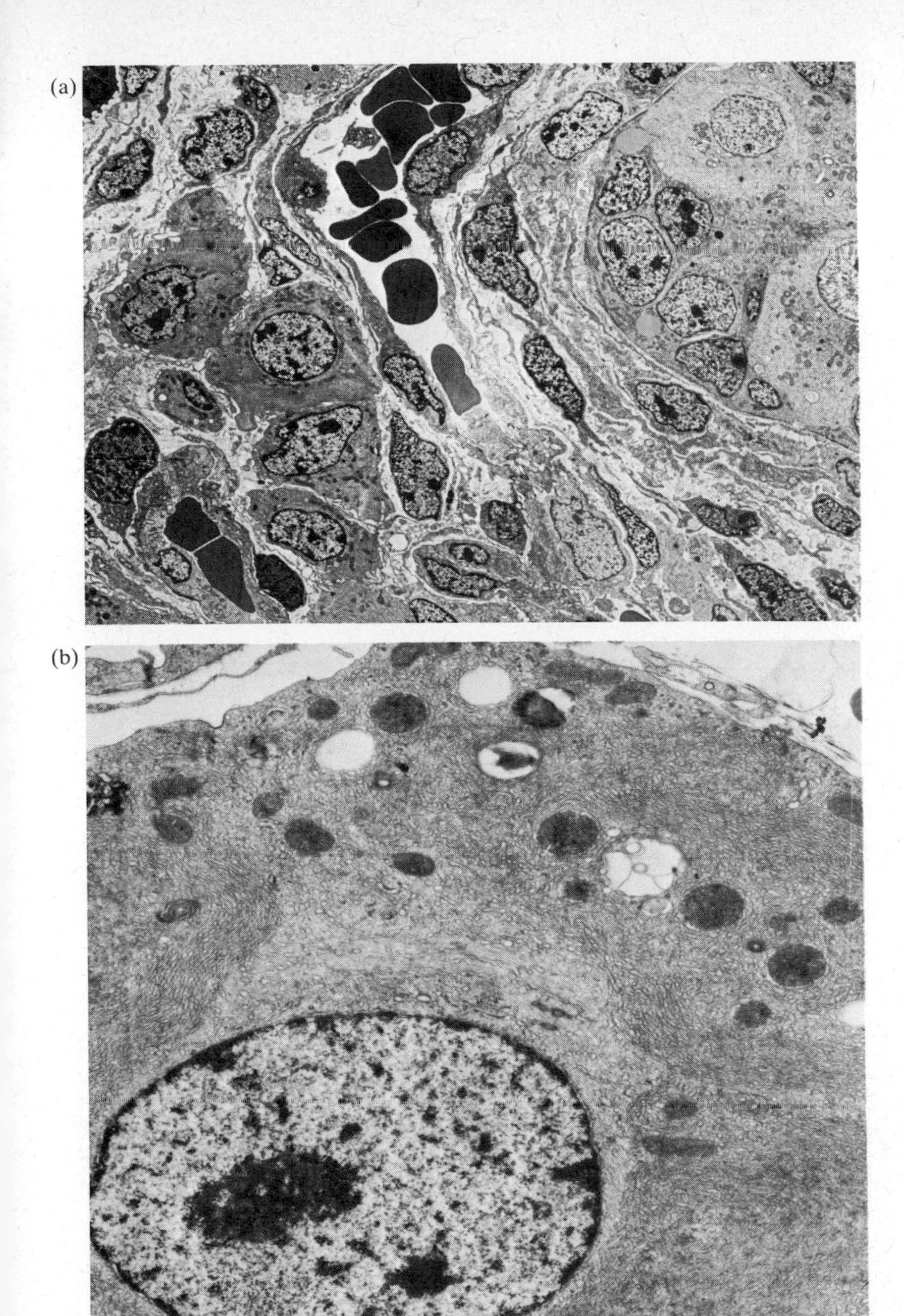

Figure 1.4(a) Testicular morphology at 113 days. To the right a sex cord is situated with germ cells, and Sertoli cells with lipid droplets. Several layers of mesenchymal cells surround the sex cord. To the left differentiated Leydig cells are located. (Magn. 1600×). (b) The Leydig cells contain conspicuous smooth endoplasmic reticulum (SER) mainly displayed in the form of whirls. Branched tubular SER is persistent in appreciable amounts. Near to the nucleus 10 nm microfilaments are present. (Magn. 10800×). By courtesy of Dr C.J.A.H.V. van Vorstenbosch

Sertoli cells

Data on the development of Sertoli cells in the second half of the foetal period are scarce although they are the main component of the sex cords (or seminiferous tubules). The diameter of the seminiferous tubules is about 60 μm and only increases after germ cell differentiation starts. However, tubular length increases markedly after 90 days post coitum (van Straaten and Wensing, 1977). The tubules which at first are rather straight, develop a coiled appearance in the perinatal period (*Figure 1.3(c)* and *(d)*). Four weeks after birth the rate of tubular length growth decreases again (van Straaten and Wensing, 1977). The mitotic activity of Sertoli cells which is high in the first month, then also decreases (Tran, Meusy-Dessolle and Josso, 1981). The blood testis barrier is still not formed at this age. At six weeks the Sertoli cells morphologically form a monomorphic cell population. They are juxtaposed on the internal edge of the seminiferous cords and on the apical side of the cells numerous extensions can be observed. The cytoplasm contains abundant SER and numerous ribosomes (Dierichs and Wrobel, 1975; Wrobel and Dierichs, 1975; Chevalier, 1978).

Germ cells

The number of germ cells per testis increases gradually in the late foetal and neonatal period. It does not seem to be influenced by the perinatal Leydig cell differentiation although there is a reduction in increase after four weeks after birth (van Straaten and Wensing, 1977). Germ cells (gonocytes) are morphologically characterized by a regular, round to oval shape and a centrally located nucleus. Concentration of cytoplasmic organelles is sparse in comparison with the Sertoli cells.

Testicular hormones

The differentiation of the testis in the perinatal period is also accompanied by an increase in steroid production. Serum testosterone concentrations which are relatively low (<0.5 ng/ml) in the second half of the foetal period rise perinatally and are maximal (~1.5 ng/ml) in the second week after birth (Meusy-Dessolle, 1974; 1975; Colenbrander, de Jong and Wensing, 1978; Ford, Christenson and Maurer, 1980) (*Figure 1.5*). This rise and decline in serum testosterone concentrations runs parallel to a similar change in testicular androgen concentration (Booth, 1975; Segal and Raeside, 1975). However, not only quantitative changes but also important qualitative changes occur. Testosterone which is the main androgen in the foetal and early neonatal period is quantitatively much less important after six weeks than, for example, androstenediol and dehydroepiandrosterone (Ruokonen and Vihko, 1974; Booth, 1975).

Less is known about hormone production by the neonatal Sertoli cell. Hormones produced by the Sertoli cells are inhibin, AMH (anti-Müllerian hormone), oestradiol and MPS (meiosis preventing substance). The presence of MPS in Sertoli cells is still hypothetical and further study will be

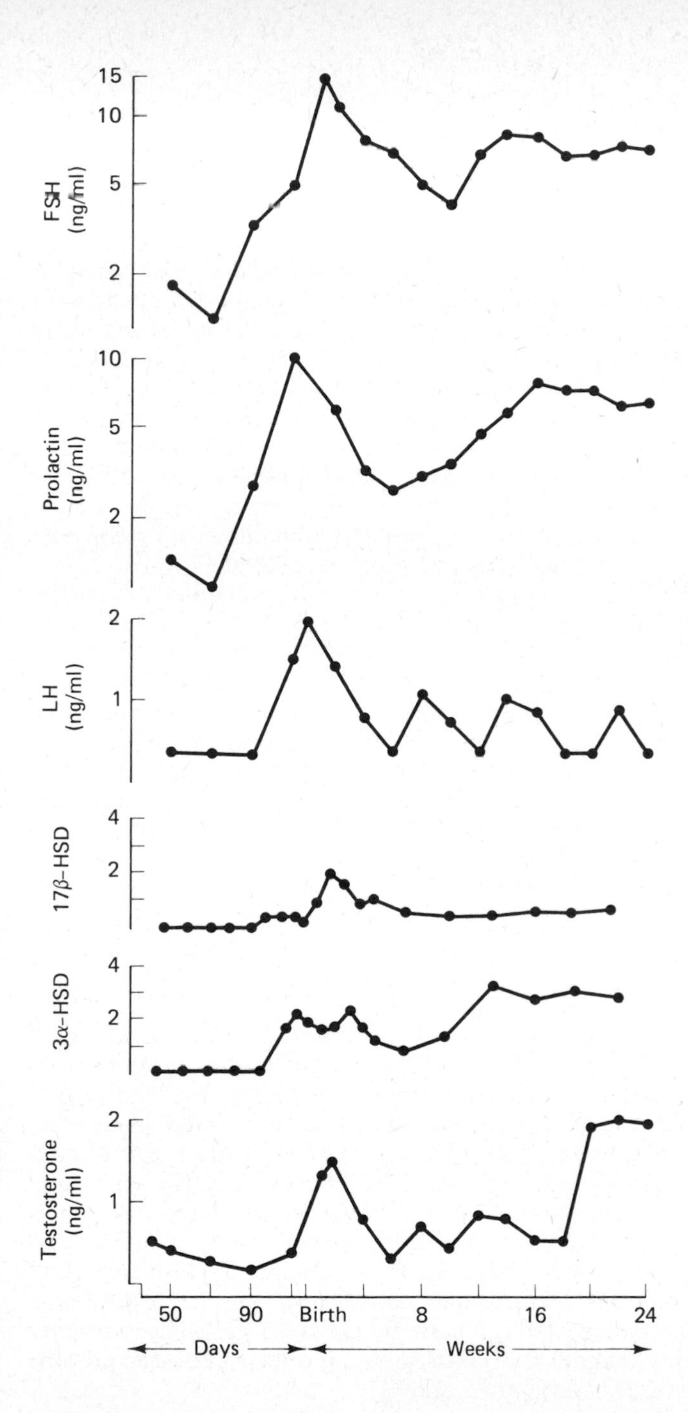

Figure 1.5 Changes in serum pituitary hormone concentrations, histochemical activity of the Leydig cells and serum testosterone concentrations during development. From Colenbrander and van der Wiel, unpublished observations, Colenbrander *et al.*, 1977; van Straaten and Wensing 1978; Colenbrander, de Jong and Wensing, 1978

needed to clarify this topic. Also no data are available in the male pig on inhibin, a non-androgenic hormone produced by the Sertoli cells, which has been demonstrated in porcine follicular fluid (Welschen *et al.*, 1977; Steinberger, 1979). This polypeptide plays a significant role in the feedback control by the testis of FSH production by the pituitary (Setchell, 1978; de Jong *et al.*, 1979) and may be responsible for a decline in serum FSH concentrations in the male pig 3–4 weeks after birth (see below).

AMH, which is already produced at 27 days, is still present in the postnatal period, although it could not be detected in the first few days of postnatal life. In the second week after birth AMH activity increases to a high level similar to that observed in the late foetal period; thereafter it slowly decreases and practically disappears after 70 days (Tran, Meusy-Dessolle and Josso, 1981). Oestradiol (E_2) is produced by the Sertoli cell, from steroid precursors produced in the Leydig cell (Dorrington and Armstrong, 1975) but may also be produced by the Leydig cell itself (Rommerts and Brinkmann, 1981). Serum E_2 concentrations are very high (1.1±0.3 ng/ml) in the late foetal period due to placental production (Bowerman, Anderson and Melampy, 1964; Macdonald *et al.*, 1979), but rather low (<0.2 ng/ml) in the neonatal and prepubertal animal. E_2 plays an important role in male sexual development in the boar. It is known to suppress Leydig cell function by decreasing the activities of 17β-hydroxylase (Brinkmann *et al.*, 1980), but also in the pig it stimulates the accessory sex glands and male sexual behaviour (Joshi and Raeside, 1973).

The protein binding of steroids has not been studied extensively but the binding of progesterone increases from 83% to 92% between 57 and 111 days (Macdonald *et al.*, 1980).

Pituitary

Pituitary hormones which are important for gonadal development are luteinizing hormone (LH), follicle stimulating hormone (FSH) and prolactin.

Morphological studies of the pituitary indicate that growth hormone-producing cells are already differentiated at 46 days and thyroid stimulating hormone (TSH)- and FSH-producing cells at 70–76 days (Liwska, 1975; 1978). Preliminary results of immunohistochemical staining for prolactin show that at 70 days, only very few cells (1.2±0.6%) stain positively; at the end of gestation many more cells (8.7±0.8%) show a positive reaction (Colenbrander *et al.*, 1982). A comparable staining pattern has been observed for LH-producing cells (Meyer, personal communication). Higher amounts of LH and FSH are detectable in the pituitary at the end of gestation when compared with younger ages (Melampy *et al.*, 1966).

Serum LH concentrations rise at the end of gestation, are high in the second week after birth and decline thereafter (*Figure 1.5*). The increase of serum LH in the foetus can be mimicked at an earlier stage by injecting LH releasing hormone, which does not raise LH levels at 50 days, induces a positive, mediocre response in some foetuses at 70 days but produces a marked increase in all foetuses at 110 days (Colenbrander *et al.*, 1980).

So, in the neonatal period LH concentrations rise and induce a parallel rise in testosterone. No feedback exists in the first weeks after birth, as castration does not result in increased LH levels (Ford and Schanbacher, 1977). However, when the animal is castrated at eight weeks, LH concentrations rise (Colenbrander *et al.*, 1977) suggesting that the feedback is probably established in the third week, when both LH and testosterone decline in control animals (*Figure 1.5*).

FSH and prolactin concentrations are rather low before 90 days post coitum. They gradually rise to maximal values around birth when the serum concentrations are 14.0±0.7 ng/ml and 10.1±1.2 ng/ml respectively; thereafter concentrations of both hormones decline (Colenbrander, van de Wiel and Wensing, 1980; Colenbrander *et al.*, 1982) (*Figure 1.5*). FSH is essential for the initiation of spermatogenesis in puberty via stimulation of Sertoli cells. It also induces an increase in androgen binding protein (ABP) production by Sertoli cells (Steinberger, 1971). There are clear indications that prolactin plays a crucial role in testicular function. Absence of prolactin in mice leads to gonadal atrophy or underdevelopment of the gonad. High concentrations of prolactin also result in sterility (Bartke *et al.*, 1978).

Pubertal period

After the second period of Leydig cell differentiation and regression, the final phase of testicular development gradually approaches. At three months the testis/body weight ratio increases again, mainly due to tubular development. This is not caused by Sertoli cell multiplication, as the mitotic activity of these cells has declined after 30 days (Tran, Meusy-Dessolle and Josso, 1981), but to the spectacular increase in the number of germ cells (van Straaten and Wensing, 1977). The Leydig cells differentiate again and increase in number after 2.5 months.

Leydig cells

After 13 weeks the Leydig cells which are enclosed by tubuli (peritubular) differentiate and maximal development is reached at about four months of age. Morphologically the Leydig cells have much in common with the perinatally differentiated Leydig cells. The SER, characteristic for steroid-producing cells, is present in large quantities and the mitochondria contain closely packed tubules. Between adjacent cells membrane fusions can be observed (Belt and Cavazos, 1967; Dierichs, Wrobel and Schilling, 1973). HSD activity increases in the third month (van Straaten and Wensing, 1978) (*Figure 1.5*) and the peritubular cells show marked 3α-HSD and 17β-HSD activity. The intertubular cells remain small and are histochemically inactive (van Straaten and Wensing, 1978). Also in other species different populations of Leydig cells are present in the testis (Rommerts and Brinkmann, 1981) and this may be due to a difference in location. Some are located in close proximity to Sertoli cells which can influence Leydig cells by specific secretion products (Aoki and Fawcett, 1978). Others are in close contact with lymph fluid or blood vessels (Fawcett, Neaves and Flores, 1973). In the fourth month androgen concentrations in the testis increase again. Testosterone, however, is quantitatively unimportant and the main steroids present are 16-androstenes (Ruokonen and Vihko, 1974; Booth, 1975; Gray *et al.*, 1971; Colenbrander, de Jong and Wensing, 1978; Florcruz and Lapwood, 1978).

Sertoli cells

Sertoli cell development slowly progresses after birth. Mitotic activity decreases after 30 days, Sertoli cell junctions appear at six weeks and the blood testis barrier is formed around 100 days and completed after 120 days (Tran, Meusy-Dessole and Josso, 1981).

In the mature testis two types of Sertoli cells characterized by the presence of a large lipid droplet are present. Sertoli cells with a light nucleus extend from the basement membrane to the lumen of the tubule and show the same fine structure as those in other mammalian species, i.e. numerous filaments and ribosomes, abundant SER and typical junctions with adjacent Sertoli cells and with germ cells. The second type of Sertoli cell has a dark nucleus and is found only in the basement region of the seminiferous epithelium. These cells contain a limited cytoplasm which contains SER and numerous filaments (Chevalier, 1978). The Sertoli cell structure and function is hormonally controlled by FSH and androgens (Hansson *et al.*, 1975; Chevalier, 1979); however their functional activity also seems to be partly regulated by the germ cells. In the rat protein synthesis and secretion by the Sertoli cells is influenced by the stage of development of the spermatids that surround the Sertoli cells (Ritzen and Boctani, 1981).

Germ cells

Spermatogenesis is initiated during the pubertal development of the testis. In the perinatal period only undifferentiated germ cells (gonocytes) are present; at 10 weeks after birth differentiation of germ cells can be observed (van Straaten and Wensing, 1977) and at three months spermatogonia and pachytene spermatocytes are present and sometimes round spermatids (stage 1–8) can also be found. After four months spermatogenesis is completed in many seminiferous tubules.

The whole spermatogenic cycle takes 8.6 days (Swierstra, 1968). The kinetics of the spermatogenic epithelium in the adult boar have been studied in whole mounts of the seminiferous tubules (Frankenhuis, de Rooy and Kramer, 1980), thus enabling spermatogonia and spermatogonial divisions to be identified. Spermatid development is comparable with spermatid development in small laboratory rodents and can be subdivided into 16 steps, characterized by changes in the development of the acrosome or the shape of the nucleus of the spermatids (*Table 1.1*). Four classes of spermatogonia can be distinguished:

(a) Undifferentiated A spermatogonia [A single (A_s), A paired (A_{pr}), A aligned (A_{al})];
(b) Differentiated A spermatogonia [A_1, A_2, A_3, A_4];

Table 1.1 CELL ASSOCIATION IN EACH OF THE TWELVE STAGES OF THE CYCLE OF THE BOAR SEMINIFEROUS EPITHELIUM

13	14	14	15	15	15	16	16				
1	2	3	4	5	6	7	8	9	10	11	12
P	P	P	P	P	P	P	P	P	P	D	M_1 M_2
A_3/A_4	A_4/In	In	In/B	B	B/PLS	PLS	L	L	Z	Z	P
		A_1	A_1	A_1	A_1	A_1	A_1	A_1/A_2	A_2	A_2/A_3	A_3
Undifferentiated type A spermatogonia (A_s, A_{pr}, A_{al})											

Each column consists of the various cell types making a cellular association. These associations succeed one another from left to right in the table. A_s, A_{pr}, A_{al}, A_1, A_2, A_3 and A_4: spermatogonia type A; In: intermediate type spermatogonia; B: spermatogonia type B; PLS: preleptotene spermatocytes; L: leptotene; Z: zygotene; P: pachytene; D: diplotene and diakinesis; M_1: first meiotic division; M_2: second meiotic devision; 1–16: steps in spermatid development.

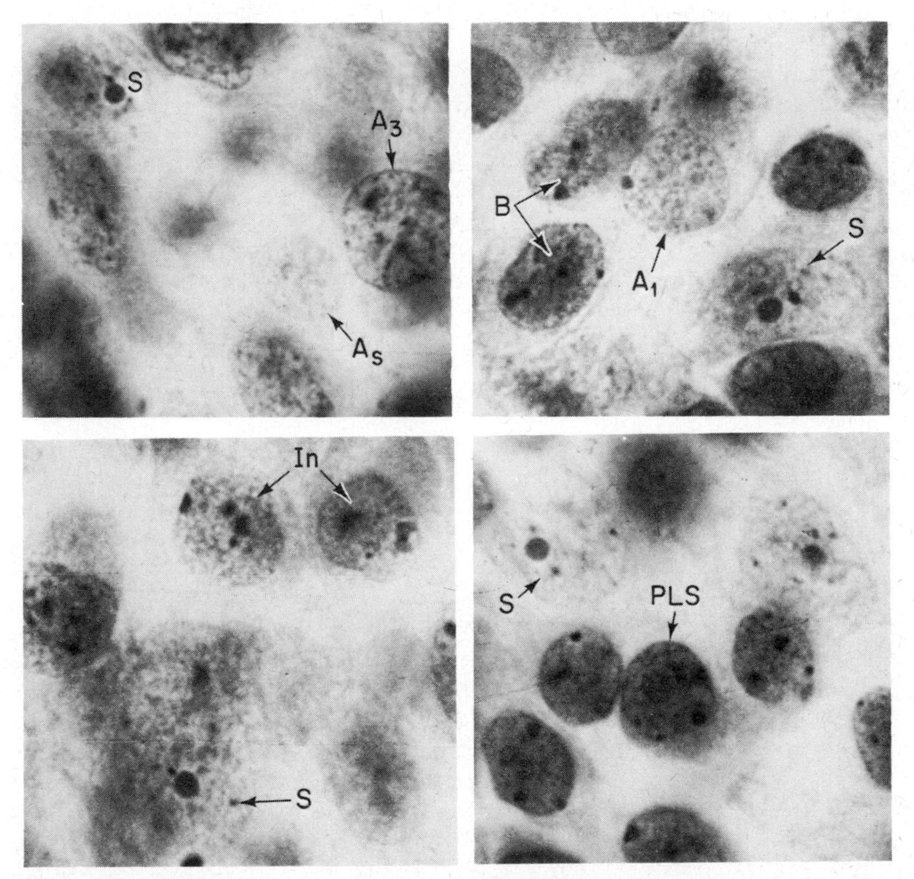

Figure 1.6 Whole mounted seminiferous tubules in the plane of focus where Sertoli cells, spermatogonia and preleptotene spermatocytes occur i.e. just under the basement membrane. The tubules were fixed in Zenker for 24 hours and stained with Harris' hemalum to identify spermatogonia and spermatogonial divisions. S : Sertoli cell; A_s, A_1, A_3 : spermatogonia type A; B : spermatogonia type B; In: intermediate type spermatogonia; PLS: preleptotene spermatocyte. (Magn. 1640×)

(c) Intermediate spermatogonia (In); and
(d) B spermatogonia (B) (*Figure 1.6*).

Undifferentiated A spermatogonia can be identified most easily in those parts of the tubules where other types of spermatogonia are in their G_2 phase, as their cell cycle is not in phase with the cell cycle of the differentiated A and intermediate spermatogonia.

In all probability the A single spermatogonia are to be regarded as the stem cells of spermatogenesis. When in this model of spermatogonial stem cell renewal an A single spermatogonium divides, it gives rise to two A paired spermatogonia, that either stick together and by the second division form A aligned spermatogonia, or migrate away from each other to become two new A single spermatogonia. The A_1 spermatogonia, the first of the differentiated A spermatogonia, arise from A aligned spermatogonia. Along the length of the tubules four waves of divisions of A spermatogonia can be subdivided into four generations (A_1, A_2, A_3, A_4). The last of these waves gives rise to intermediate spermatogonia (In), intermediate between A and B spermatogonia with respect to the amount of chromatin in the nucleus. Intermediate spermatogonia divide into type B spermatogonia and these into preleptotene spermatocytes. Thereafter the prophase of the first meiotic division starts, followed by the second meiotic division. By studying whole mounts of the seminiferous tubules it has become clear that the kinetics of the spermatogenic epithelium in the boar demonstrates a striking similarity with that in small laboratory rodents.

During testicular differentiation, which is paralleled by increased androgen secretion, the accessory sex glands also differentiate. Their weight markedly increases in the later part of puberty (Egbunike, 1979) as well as their histochemical activity (Aitken, 1960; Wrobel and Fallenbacher, 1974). Hypophysectomy or castration results in underdevelopment of the accessory sex glands and a decrease in weight and histochemical activity can be observed (Wrobel, 1969; Morat *et al.*, 1980).

The factors which determine the onset of puberty in the male still remain to be unravelled. Endocrinological changes, but also environmental factors are important (Ramaley, 1979; Egbunike, 1979). Serum concentrations of gonadotrophic hormones in the pig do not change remarkably in the pubertal period (*Figure 1.5*). However, minor changes in the balance of pituitary hormone concentrations i.e. LH, FSH, and prolactin, may play a decisive role. Another important event could be the formation of the blood testis barrier (Ramaley, 1979). Data are also scarce on the developmental pattern of gonadotrophic hormone and the steroid receptors of Leydig and Sertoli cells in the pig during development. Preliminary *in vitro* studies on the regulation of HCG receptors in the Leydig cells of immature pigs demonstrate that prostaglandin $F_{2\alpha}$ induces a decrease in the number of HCG binding sites (Haour *et al.*, 1981). Changes in the receptor population could probably also play a role in the onset of puberty.

Conclusions

In the pig three periods of testicular development can be distinguished. The first period is between 26 and 60 days of gestation, at the time of sexual differentiation. The second period occurs perinatally when the increase in testicular weight is mainly due to the increase in number of Leydig cells and their state of development. The third is during puberty when spermatogenesis starts.

Testicular development between 26 and 60 days is probably independent of gonadotrophic hormone stimulation. The Leydig cells are the most likely source of the early peak of serum testosterone, seen at 35 days. The second phase of Leydig cell development seems to be initiated by foetal pituitary activity. Rising gonadotrophic hormone concentrations in the peripheral circulation initiate Leydig cell differentiation which results in increased testicular androgen secretion. The significance of this perinatal Leydig cell differentiation is still unclear. After birth when a functional feedback system is established between the gonad and the hypothalamo–pituitary system, both gonadotrophic hormone and androgen concentrations in the peripheral circulation decline. The factors which are essential for the third period of Leydig cell differentiation are still unknown.

Sertoli cell development seems to be more continuous when compared to that of the Leydig cells. AMH, produced by the Sertoli cell, is already present at 27 days post coitum and can still be detected after birth. Just after birth elongation of the seminiferous tubules is temporarily enhanced, possibly due to high serum FSH concentrations which stimulate the Sertoli cell. No remarkable changes in gonadotrophic hormone concentrations occur between 100 and 120 days after birth when the blood testis barrier is formed.

From the early foetal period germ cell development does not seem, at least quantitatively, to be influenced by hormonal changes. The germ cells gradually increase in number by mitosis. Two months after birth differentiation of germ cells starts and meiosis occurs. At four months spermatid development is completed in many seminiferous tubules and kinetically shows a striking similarity to that in laboratory rodents. The factors which are involved in the onset of puberty still remain to be unravelled.

References

AITKEN, R.N. (1960). A histochemical study of the accessory genital glands of the boar. *J. Anat.* **94**, 130–142

ALLEN, B.M. (1904). The embryonic development of the ovary and testis of the mammals. *Am. J. Anat.* **3**, 89–146

AMOROSO, E.C. and PORTER, D.G. (1970). The endocrine functions of the placenta. In *Scientific Foundations of Obstetrics and Gynaecology*, (E.E. Phillip, J. Barnes, and M. Newton, Eds.), pp. 556–586. London, Heineman

ANCEL, P. and BOUIN, P. (1903). Histogenèse de la glande interstitiele du testicule chez le porc. *C. r. Seanc. Soc. Biol.* **55**, 1680–1682

AOKI, A. and FAWCETT, D.W. (1978). Is there a local feedback from the seminiferous tubules affecting activity of the Leydig cells? *Biol. Reprod.* **19**, 144–158

BARTKE, A., HAFIEZ, A.A., BEX, F.J. and DALTERIO, S. (1978). Hormonal interactions in regulation of androgen secretion. *Biol. Reprod.* **18**, 44–54

BELT, W.D. and CAVAZOS, L.F. (1967). Fine structure of the interstitial cells of Leydig in the boar. *Anat. Rec.* **158**, 333–350

BLACK, J.L. and ERICKSON, B.H. (1968). Oogenesis and ovarian development in the prenatal pig. *Anat. Rec.* **161**, 45–56

BLANCHARD, M.G. and JOSSO, N. (1974). Source of the Anti-Mullerian hormone synthesized by the fetal testis. Mullerian-inhibiting activity of fetal bovine Sertoli cells in tissue culture. *Pediat. Res.* **8**, 968–971

BOOTH, W.D. (1975). Changes with age in the occurrence of C_{19} steroids in the testis and submaxillary gland of the boar. *J. Reprod. Fert.* **42**, 459–472

BOWERMAN, A.M., ANDERSON, L.L. and MELAMPY, R.M. (1964). Uterinary estrogens in cycling, pregnant, ovariectomized and hysterectomized gilts. *Iowa St. J. Sci.* **38**, 437–445

BRINKMANN, A.O., LEEMBORG, F.G., ROODNAT, E.M. DE JONG, F.H. and VAN DER MOLEN, H.J. (1980). A specific action of estradiol on enzymes involved in testicular steroidogenesis. *Biol. Reprod.* **23**, 805–809

BYSKOV, A.G. (1979). Regulation of meiosis in mammals. *Anns Biol. anim. Biochim. Biophys.* **19**, 1251–1261

CHEVALIER, M. (1978). Sertoli cell ultrastructure. I. A comparative study in immature, pubescent adult and cryptorchid pigs. *Anns Biol. anim. Biochim. Biophys.* **18**, 1279–1292

CHEVALIER, M. (1979). Sertoli cell ultrastructure. II. Morphological effects of hypophysectomy in pubescent pigs. *Anns Biol. anim. Biochim. Biophys.* **19**, 583–596

CHRISTENSEN, A.K. (1975). Leydig cells. In *Handbook of Physiology; Section 7, Endocrinology; Vol. 5, Male Reproductive System,* (R.O. Greep and E.B. Astwood, Eds), pp. 57–94. Baltimore, Waverley Press Inc.

COLENBRANDER, B. (1978). Aspects of sexual differentiation in the pig. Thesis, Utrecht

COLENBRANDER, B. and WENSING, C.J.G. (1975). Studies on phenotypically female pigs with ovarian aplasia and inguinal hernia. *Proc. Koninklijke Nederlandse Akademie v. Wetenschap* **C78**, 33–46

COLENBRANDER, B., DE JONG, F.H. and WENSING, C.J.G. (1978). Changes in serum testosterone concentrations in the male pig during development. *J. Reprod. Fert.* **53**, 377–380

COLENBRANDER, B., VAN DE WIEL, D.F.M. and WENSING, C.J.G. (1980). Changes in serum FSH concentrations during fetal and prepubertal development in pigs. *Proc. VI Int. Congr. Endocrinology, Melbourne,* 228 (Abstract)

COLENBRANDER, B., KRUIP, Th.A.M., DIELEMAN, S.J. and WENSING, C.J.G. (1977). Changes in serum LH concentrations during normal and abnormal sexual development in the pig. *Biol. Reprod.* **17**, 506–513

COLENBRANDER, B., MACDONALD, A.A., PARVIZI, N. and ELSAESSER, F. (1980). Changing responsiveness of fetal pig pituitary to LH-RH. *Proc. Int. Union Physiol. Sci. Budapest* **Vol. XIV**, p. 1106 (Abstract)

COLENBRANDER, B., ROSSUM-KOK, C.M.J.E. van, STRAATEN, H.W.M. van, and WENSING, C.J.G. (1979). The effect of fetal decapitation on the testis and other endocrine organs in the pig. *Biol. Reprod.* **20**, 198–204

COLENBRANDER, B., MACDONALD, A.A., MEIJER, J.C., ELLENDORFF, F., VAN DE WIEL, D.F.M. and BEVER, M.M. (1982). Prolactin in the pig fetus (Abstract). *Eur. J. Obstet. Gynaec. Reprod. Biol.,* in press.

DE JONG, F.H., WELSCHEN, R., HERMANS, W.P., SMITH, S.D. and VAN DER MOLEN, H.J. (1979). Inhibin, follicular fluid and Sertoli cell medium. *J. Reprod. Fert., Suppl.* **26**, 47–59

DIERICHS, R., and WROBEL, K.H. (1975). Membranspezialisierungen bei Sertolizellen des Schweines. *Verh. anat. Ges.* **69**, 845–847

DIERICHS, R., WROBEL, K.H. and SCHILLING, E. (1973). Licht-, und Elektronenmikroskopische Untersuchungen an den Leydigzellen des Schweines während der postnatalen Entwicklung. *Z. Zellforsch. mikrosk. Anat.* **143**, 207–227

DORRINGTON, J.H. and ARMSTRONG, D.T. (1975). FSH stimulates estradiol-17β synthesis in cultured Sertoli cells. *Proc. natn. Acad. Sci. USA* **72**, 2677–2681

DORNER, G. (1976). Sex hormone-dependent differentiation processes. In *Hormones and Brain Differentiation,* pp. 94–127. Amsterdam, Elsevier

EGBUNIKE, G.N. (1979). Development of puberty in Large White boars in a humid tropical environment. *Acta Anat.* **104**, 400–405

ELSAESSER, F. and PARVIZI, N. (1979). Estrogen feedback in the pig. Sexual differentiation and the effect of prenatal testosterone treatment. *Biol. Reprod.* **20**, 1187–1193

FAWCETT, D.W., NEAVES, W.B. and FLORES, M.N. (1973). Comparative observations on intertubular lymphatics and the organization of the interstitial tissue of the mammalian testis. *Biol. Reprod.* **9**, 500–532

FLORCRUZ, S.V. and LAPWOOD, K.R. (1978). A longitudinal study of pubertal development in boars. *Int. J. Androl.* **1**, 317–330

FORD, J.J. and SCHANBACHER, B.D. (1977). Luteinizing hormone secretion and female Indosis behaviour in male pigs. *Endocrinology* **100**, 1033–1038

FORD, J.J., CHRISTENSON, R.K. and MAURER, R.R. (1980). Serum testosterone concentrations in embryonic and fetal pigs during sexual differentiation. *Biol. Reprod.* **23**, 583–587

FRANKENHUIS, M.T., DE ROOY, D.G. and KRAMER, M.F. ((1980). Spermatogenesis in the pig. In *Proc. 9th Int. Congr. Anim. Reprod. A.I., Madrid,* 265

GIER, H.T. and MARION, G.B. (1970). Development of mammalian testes and genital ducts. *Biol. Reprod. Suppl.* **1**, 1–23

GRAY, R.C., DAY, B.N., LASLEY, J.F. and TRIBBLE, F. (1971). Testosterone levels of boars at various ages. *J. Anim. Sci.* **33**, 124–126

HADZISELIMOVIC, F. and HERZOG, B. (1980). Etiology of testicular descent. In *Descended and Cryptorchid Testis,* (E.S.E. Hafez, Ed.), pp. 138–147. The Hague, Nijhoff

HANSSON, V., WEDDINGTON, S.C., NAESS, O. and ATTRAMADAL, A. (1975). Testicular androgen binding protein (ABP)—A parameter of Sertoli cell secretory function. In *Hormonal Control of Spermatogenesis,* (F.S. French, V. Hansson, E.M. Ritzen and S. Nayfeh, Eds.) pp. 323–336. New York, Plenum Press

HAOUR, F., MATHER, J., DRAY, F. and SAEZ, J.M. (1981). Regulation of HCG receptors by gonadotrophin in porcine Leydig cells. *Int. J. Androl.* (Abstract), in press

JIRASEK, J.E. (1976). Principles of reproductive embryology. In *Disorders of Sexual Differentiation,* (J.L. Simpson, Ed.), pp. 51–110. New York, Academic Press

JOSHI, H. and RAESIDE, J.I. (1973). Synergistic effects of testosterone and oestrogens on accessory sex glands and sexual behaviour of the boar. *J. Reprod. Fert.* **33**, 411–423

JOST, A. (1947). Recherches sur la differenciation sexuelle de l'embryon de lapin. 3. Role des gonades foetales dans la differenciation sexuelle somatique. *Archs. Anat. microsc. Morph. exp.* **36**, 271–315

JOST, A. (1965). Gonadal hormones in the sex differentiation of the mammalian fetus. In *Organogenesis,* (R.L. de Haan and H. Ursprung, Eds.), pp. 611–628. New York, Holt, Rinehart and Winston

KONING, J.N. (1942). Over de ontwikkeling van den sinus urogenitalis en de accessoire geslachtsklieren van het varken. Thesis, Utrecht

LIN, C.C., BIEDERMAN, B.M., JAMRO, H.K., HAWTHORNE, A.B. and CHURCH, R.B. (1980). Porcine chromosome identification and suggested nomenclature. *Can. J. Genet. Cytol.* **22**, 103–116

LIWSKA, J. (1975). Development of the adenohypophysis in the embryo of the domestic pig. *Folia morph.* **34**, 211–217

LIWSKA, J. (1978). Ultrastructure of the adenohypophysis in the domestic pig. Part I: Cells of the pars anterior. *Folia Histochem. Cytochem.* **16**, 307–314

MACDONALD, A.A., COLENBRANDER, B., ELSAESSER, F. and HEILHECKER, A. (1980). Progesterone production by the pig fetus and the response to stimulation by adreno-corticotrophin. *J. Endocr.* **85**, 34–35P

MACDONALD, A.A., ELSAESSER, F., PARVIZI, N., HEILHECKER, A., SMIDT, D. and ELLENDORFF, F. (1979). Progesterone, oestradiol 17β and luteinizing hormone concentrations in blood and amniotic fluid of chronically catheterised pig fetuses. *J. Endocr.* **80**, 14P

MAULEON, P. (1978). Ovarian development in young mammals. In *Control of Ovulation,* (D.B. Crighton, N.B. Haynes, G.R. Foxcroft and G.E. Lamming, Eds.), pp. 141–158. London, Butterworths

MELAMPY, R.M., HENRICHS, D.M., ANDERSON, L.L., CHEN, C.L. and SCHULTZ, J.R. (1966). Pituitary follicle stimulating hormone and luteinizing hormone concentrations in pregnant and lactating pigs. *Endocrinology* **78**, 801–810

MEUSY-DESSOLLE, N. (1974). Evolution du taux de testosterone plasmatique au cours de la vie foetale chez le porc domestique. *C.r. hebd. Séanc. acad. Sci., Paris* **278**, 1257–1260

MEUSY-DESSOLLE, N. (1975). Variations quantitatives de la testosterone plasmatique chez le porc male de la naissance à l'âge adulte. *C.r. hebd. Séanc. acad. Sci., Paris* **281**, 1875–1878

MOON, Y.S. and HARDY, M.H. (1973). The early differentiation of the testis and interstitial cells in the fetal pig and its duplication in organ cultures. *Am. J. Anat.* **138**, 253–268

MOON, Y.S. and RAESIDE, J.I. (1972). Histochemical studies on hydrosteroid dehydrogenase activity of fetal pig testes. *Biol. Reprod.* **7**, 278–287

MOON, Y.S., HARDY, M.H. and RAESIDE, J.I. (1973). Biological evidence for androgen secretion by the early fetal pig testes in organ culture. *Biol. Reprod.* **9**, 330–337

MORAT, M. (1977). Action morphogène des hormones gonadotropes sur les cellules de Leydig du verrat. I. Effects de l'hypophysectomie. *Arch. Anat. microsc. Morph. exp.* **66**, 119–142

MORAT, M., LOCATELLI, A., TORQUI, M., CHEVALIER, M., CHAMBON, M., and

DUFAURE, J.P. (1980). Effects de l'hypophysectomie puis de l'administration de la gonadotropine HCG sur le taux de testostérone et sur la structure de l'épididyme et des glandes accessoires chez le verrat. *Reprod. Nutr. Dev.* **20**, 61–76

OHNO, S. (1977). The Y-linked H–Y antigen locus and the X-linked Tfm locus as major regulatory genes of the mammalian sex determining mechanism. *J. Ster. Biochem.* **8**, 585–592

PELLINIEMI, L.J. (1975a). Ultrastructure of gonadal ridge in male and female pig embryos. *Anat. Embryol.* **147**, 19–34

PELLINIEMI, L.J. (1975b). Ultrastructure of the early ovary and testis of pig embryos. *Am. J. Anat.* **144**, 89–112

PRICE, D. and ORTIZ, E. (1965). The role of fetal androgen in sex differentiation in mammals. In *Organogenesis,* (R.L. de Haan and H. Ursprung, Eds.), pp. 629–652. New York, Holt, Rinehart and Winston

RAESIDE, J.I. and MIDDLETON, A.T. (1979). Development of testosterone secretion in the fetal pig testis. *Biol. Reprod.* **21**, 985–989

RAESIDE, J.I. and SIGMAN, J. (1975). Testosterone levels in early fetal testes of domestic pigs. *Biol. Reprod.* **13**, 318–321

RAMALEY, J.A. (1979). Development of gonadotropin regulation in the prepubertal mammal. *Biol. Reprod.* **20**, 1–31

RITZEN, E.M. and BOCTANI, C. (1981). Cyclic secretion of proteins by the rat seminiferous tubule, depending on the stage of spermatogenesis. *Int. J. Androl.* In press.

ROMMERTS, F.F.G. and BRINKMANN, A.O. (1981). Modulation of steroidogenic activities in testis Leydig cells. *Molec. Cell. Endocr.* **21**, 15–28

RUOKONEN, A. and VIHKO, R. ((1974). Steroid metabolism in testis tissue: concentrations of unconjugated and sulphated neutral steroids in boar testis. *J. Ster. Biochem.* **55**, 33–38

SEGAL, D.H. and RAESIDE, J.I. (1975). Androgens in testis and adrenal glands of the fetal pig. *J. Ster. Biochem.* **6**, 1439–1444

SETCHELL, B. (1978). Endocrinological control of the testis. In *The Mammalian Testis,* pp. 332–358. Cornell, Cornell University Press

STEINBERGER, A. (1979). Inhibin production by Sertoli cells in culture. *J. Reprod. Fert. Suppl.* **26**, 31–45

STEINBERGER, E. (1971). Hormonal control of spermatogenesis. *Physiol. Rev.* **51**, 1–22

STEWART, D.W. and RAESIDE, J.I. (1976). Testosterone secretion by the early pig testis in organ culture. *Biol. Reprod.* **15**, 25–28

STRAATEN, H.W.M. van, and WENSING, C.J.G. (1977). Histo-morphometric aspects of testicular morphogenesis in the pig. *Biol. Reprod.* **17**, 467–472

STRAATEN, H.W.M. van, and WENSING, C.J.G. (1978). Leydig cell development in the testis of the pig. *Biol. Reprod.* **18**, 86–93

SWIERSTRA, E.E. (1968). Cytology and duration of the cycle of the seminiferous epithelium of the boar; duration of the spermatozoan transit through the epididymis. *Anat. Rec.* **161**, 171–186

TRAN, D., MEUSSY-DESSOLLE, N. and JOSSO, N. (1977). Anti-Müllerian hormone is a functional marker of foetal Sertoli cells. *Nature, Lond.* **269**, 411–412

TRAN, D., MEUSY-DESSOLLE, N. and JOSSO, N. (1981). Anti-Müllerian activity and formation of the blood testis barrier in the testis of the developing pig. *Int. J. Androl.* (Abstract)

VORSTENBOSCH, C.J.A.H.V. van, COLENBRANDER, B. and WENSING, C.J.G. (1981). The influence of fetal decapitation on the ultrastructure of Leydig cells during testicular development in the pig. *Int. J. Androl.* (Abstract)

WACHTEL, S.S. (1979). H–Y antigen in the mammalian fetus. *Anns Biol. anim. Biochim. Biophys.* **19**, 1231–1237

WACHTEL, S.S. and KOO, G.C. (1978). H–Y antigen and abnormal sex differentiation. *Birth Defects* **14**, 1–7

WELSCHEN, R., HERMANS, W.P., DULLAART, J. and DE JONGH, F.H. (1977). Effects of an inhibin-like factor present in bovine and porcine follicular fluid on gonadotrophin levels in ovariectomized rats. *J. Reprod. Fert.* **50**, 129–131

WENSING, C.J.G. (1968). Testicular descent in some domestic mammals. I. Anatomical aspects of testicular descent. *Proc. Koninklijke Nederlandse Akademie v. Wetenschap* **C 71**, 423–434

WENSING, C.J.G., COLENBRANDER, B. and VAN STRAATEN, H.W.M. (1980). Normal and abnormal testicular descent in some mammals. In *Descended and Cryptorchid Testis,* (E.S.E. Hafez, Ed.), pp. 125–137. The Hague, Nijhoff

WHITEHEAD, R.H. (1904). The embryonic development of the interstitial cells of Leydig. *Am. J. Anat.* **3**, 167–182

WROBEL, K.H. (1969). Zur Feinstruktur des Samenblasenepithels beim Schwein. *Zent. VetMed.* Reihe **A 16**, 400–415

WROBEL, K.H. and DIERICHS, R. (1975). Wachstum and Differenzierung der Sertolizellen im Schweinehoden während des postnatalen Ontogenese. *Anat. Anz.* **69**, 723–724

WROBEL, K.H. and FALLENBACHER, E. (1974). Histologische und Histochemische Untersuchung zur postnatalen Ontogenese des Nebenhodens beim Schwein. *Anat. Histol. Embryol.* **3**, 85–99

ZIECIK, A.J. and FLINT, A.P.F. (1980). Gonadotrophin-like substance in pig placenta and embryonic membranes. *J. Endocr.* **85**, 25P

ZIETSCHMANN, O. and KROLLING, O. (1955). Die Entwicklung des Harngeslechtsapparates. *Lehrbuch der Entwicklungsgeschichte der Haustiere,* pp. 373–412. Berlin, P. Parey

2

TESTICULAR STEROIDS AND BOAR TAINT

W.D. BOOTH
A.R.C. Institute of Animal Physiology, Animal Research Station, Huntingdon Road, Cambridge, UK

The testis of the mature boar is a large organ weighing as much as 500 g, and occupying a conspicuous position in the scrotum. This fact may well have impressed those concerned with domestication of the pig, to the extent that they became aware of the testis being the source of determinants responsible for the development of male characteristics. Some of these characteristics were undesirable in so far as boars could develop a high libido, presenting husbandry problems, and the carcasses of older animals were often unpalatable in texture with a strong flavour and odour. Similar problems with the intact males of other meat producing animals, eventually led to the routine practice of castrating animals not required for breeding.

However, contrary to established opinion, the entire male pig at least to bacon weight, does not generally present the problems which castration was intended to eliminate. In fact there are considerable advantages to be gained both economically and gastronomically by producing meat from the young entire boar. During development, intact boars utilize their food efficiently, resulting in a more rapid growth rate and production of a leaner carcass than that of castrated boars or gilts. Although it is evident that testicular steroids are the active agents, there are few reports (Booth, 1980a) indicating which of the many steroids produced by the boar testis are involved in stimulating the development of male characteristics. The boar testis, like that of the stallion, is a prolific producer of oestrogen as well as androgens, and it is possible that oestrogens may act synergistically with androgens to produce both anabolic and androgenic effects.

A considerable body of evidence has accrued on the significance of one major group of steroids produced by the boar testis, namely the 16-unsaturated C_{19} steroids (16-androstenes) (Gower, 1972; Claus, 1979; Booth, 1980b). Interest in these musk-smelling compounds was aroused initially when it was suggested that they may be responsible for the well-known taint or 'off odour' in the carcasses of mature boars as well as being involved in chemical communication (Sink, 1967). Subsequently, Patterson (1968a,b) reported that, indeed, 16-androstenes are the major compounds responsible for boar taint. Furthermore these compounds when presented in aerosol form to oestrous pigs, facilitate the induction of the mating stance (Melrose, Reed and Patterson, 1971; Reed, Melrose and Patterson, 1974), and thus provide one of the first demonstrations of a mammalian pheromone which has been identified chemically.

It is the aim of this review to give a detailed account of testicular steroids in the boar, with particular reference to the biosynthesis, metabolism and physiological role of androgens, oestrogens and 16-androstenes. A primary objective is to emphasize that the boar testis is not only a versatile and prolific steroid-producing organ, but also the knowledge gained through endocrine studies on this organ is important to the more practical aspects of reproduction and meat production in the pig.

Androgens

TESTICULAR PRODUCTION

Parkes (1966) noted that the testes of certain Chaeromorpha including the pig contain abundant interstitial tissue, and suggested that this could be related to a capacity of the boar testis to produce large quantities of steroid. Evidence for this suggestion already existed since Prelog and Ruzicka (1944) had isolated large amounts of the musk-smelling 16-androstenes, 3α-androstenol (5α-androst-16-en-3α-ol) and 3β-androstenol (5α-androst-16-en-3β-ol) from pig testes. However, Lindner (1960) was the first to undertake a detailed study of testicular androgens in large farm animals including the pig. Lindner (1961) determined testosterone and androstenedione in the testis and spermatic vein blood of boars, and found a close correlation between the content of steroid in the testis and blood; the greatest amounts of steroid were found in mature boars. Large amounts of 17-ketosteroid existing primarily as the sulphate conjugate of dehydroepiandrosterone (DHA) (Gower, Harrison and Heap, 1970), were found in boar urine (Huis in't Veld, Louwerens and Reilingh, 1964; Lunaas and Velle, 1965; Raeside, 1965), and indicated that most of the 17-ketosteroids in boar urine originate in the testis. Later Liptrap and Raeside (1970) measured DHA and oestrogen in boar urine to evaluate the endocrine function of the cryptorchid testis. Although the excretion of these steroids was comparable to that in normal boars, Liptrap and Raeside (1971) subsequently showed that the cryptorchid testis was refractory to human chorionic gonadotrophin (HCG) until returned to the scrotum (Liptrap and Raeside, 1972).

The biosynthesis of testicular steroids in mammals has been reviewed by Setchell (1978). In the testis, testosterone is synthesized from pregnenolone through two major pathways, the so called 4-ene and 5-ene pathways, and in the boar evidence suggests that the 5-ene pathway is particularly important (see *Figure 2.1*). Baulieu, Fabre-Jung and Huis in't Veld (1967) found larger quantities of DHA sulphate and testosterone than unconjugated DHA and androstenedione in the testis and spermatic vein blood of the boar, and recently Setchell *et al.* (personal communication) confirmed these findings for spermatic vein blood, and in addition found even higher concentrations of steroid, particularly DHA sulphate, in testicular lymph. Raeside and Howells (1971) also isolated 5-androstenediol (5-androstene-3β,17β-diol) sulphate from spermatic vein blood. Ruokonen and Vihko (1974) provided a detailed account of steroid

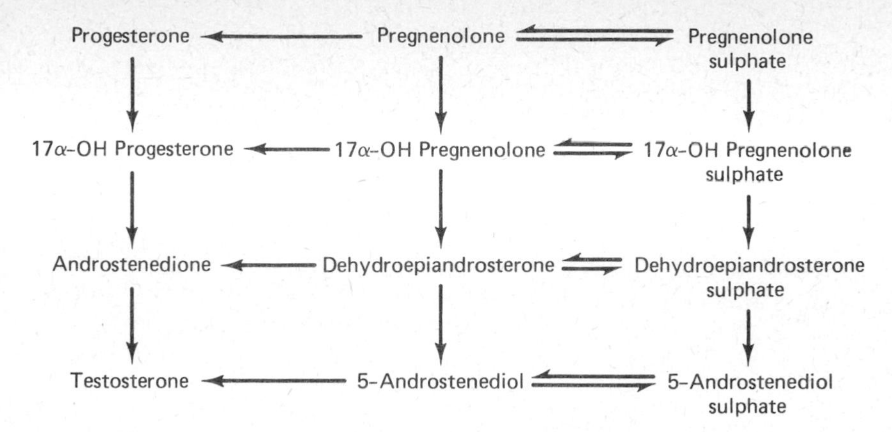

Figure 2.1 Biosynthesis of androgens from C_{21} steroids in boar testis.

sulphates in the boar testis and as DHA, 5-androstenediol and 3β-androstanediol (5α-androstane-3β,17β-diol) were the main C_{19} steroids present as monosulphates, speculated that the 5-ene pathway for testosterone synthesis was probably important in the boar. Furthermore the preponderance of steroid sulphates suggested that these might act as intermediates or regulators of testosterone synthesis (Roberts *et al.*, 1964; Notation and Ungar, 1969; Payne and Jaffe, 1970). Booth (1975) also found that DHA and 5-androstenediol (free and sulphated) were quantitatively more important than testosterone or androstenedione in boar testicular extracts and particularly in those of mature animals (see *Figure 2.2*). A similar pattern for androgens was also found in the testicular tissue of intersex pigs (Booth and Polge, 1976). Previous speculation concerning

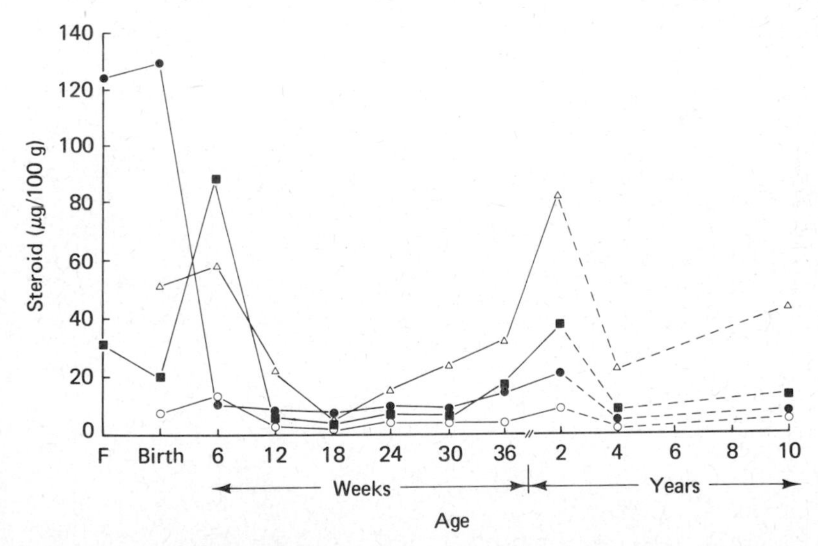

Figure 2.2 Age-related changes in the concentration of unconjugated androgens in boar testis. Values between 84 days gestation (F) and two years are means, values for individual boars older than two years are related by a broken line. ● testosterone; ○ androstenedione; ■ DHA; △ 5-androstenediol. From Booth (1975)

the role of sulphated androgen intermediates in testosterone synthesis was substantiated when Ruokonen (1978) incubated minces of boar testis with both unlabelled and radioactively labelled DHA sulphate; 5-androstenediol sulphate was the main metabolite, but free 5-androstenediol, testosterone and androstenedione were also formed. Related to this was the finding that 3β-hydroxy,5-ene steroid sulphates were formed from pregnenolone sulphate with the sulphate moiety remaining intact (Gasparini, Hochberg and Lieberman, 1976). A key enzyme in steroid-producing tissues converting 5-ene to 4-ene steroids is 3β-hydroxy,5-ene steroid dehydrogenase (4-ene/5-ene isomerase). Dufour and Raeside (1968) demonstrated histochemically the presence of this enzyme in mature boar testis with free DHA and 5-androstenediol as substrates. Testosterone 5-androstenediol and oestradiol-17β were suitable substrates for the histochemical demonstration of 17β-hydroxysteroid dehydrogenase in boar testis, and these studies were extended to foetal pigs (Moon and Raeside, 1972). More detailed studies have been carried out on steroid interconversions in various preparations of boar testis. Inano and Tamaoki (1975) studied the cofactor requirements for the interconversion of androstenedione and testosterone by 17β-hydroxysteroid dehydrogenase, which is distributed evenly between agranular and granular microsomes (Cooke and Gower, 1977), and Shimizu (1978) found that pregnenolone was converted by the microsomal fraction to 5-androstene-3β,17α-diol.

Testosterone is the predominant androgen in the testis of the foetal pig (Segal and Raeside, 1975; Booth, 1975; see *Figure 2.2*), and significant fluctuations in its occurrence are found during foetal life. Testosterone production by the foetal testis is elaborated at the time of sexual differentiation of the gonad (day 26 onwards) (Pelliniemi, 1976) as indicated by the levels of steroid in the testis (Raeside and Sigman, 1975), *in vitro* production by the testis (Stewart and Raeside, 1976), levels in the umbilical artery and amniotic fluid (Ford, Christenson and Maurer, 1980) and amounts in peripheral serum (Colenbrander, de Jong and Wensing, 1978). By the time sexual differentiation of the genital system has been initiated (day 40) there is a fall in testosterone production until about 100 days of gestation, when levels in the testis (Segal and Raeside, 1975) and plasma (Colenbrander, de Jong and Wensing, 1978) increase to term. After parturition there is wide individual variation in the production of testosterone and androstenedione (Hoffmann, Claus and Karg, 1970; Gray *et al.* 1971; Elsaesser, Konig and Smidt, 1972), which has been confirmed and extended to other steroids by Booth (1975). However despite this, patterns of steroid production are apparent between birth and maturity. Steroid determinations on the testis (Elsaesser, Konig and Smidt, 1972; Booth, 1975) and peripheral blood (Elsaesser *et al.*, 1976; Colenbrander, de Jong and Wensing, 1978; Tan and Raeside, 1980) have shown that relatively high levels of androgen are present during the postnatal period (see *Figure 2.2*), decreasing during early puberty before increasing again between late puberty and maturity. In the peripheral blood of the adult boar, testosterone is usually >2 ng/ml. However much higher levels of DHA sulphate, 16–175 ng/ml, (Booth, 1980a; Tan and Raeside, 1980; Setchell *et al.*, personal communication) and 5-androstenediol sulphate (Booth, 1980a) are present.

Ford and Schanbacher (1977) found elevated levels of plasma luteinizing hormone (LH) during the first three weeks of postnatal life, but in other studies age-related changes in androgen production were not paralleled by similar changes in LH secretion (Elsaesser *et al.*, 1976; Lapwood and Florcruz, 1978), although during puberty LH showed a highly pulsatile mode of secretion. Reports of a diurnal variation in androgen secretion are contradictory. The data of Claus and Gimenez (1977) and Claus and Hoffmann (1980), indicate that the highest levels of testosterone in spermatic vein and peripheral plasma occur during the afternoon and after midnight. However, Sanford *et al.* (1976) and Lapwood and Florcruz (1978) did not find a diurnal variation in either testosterone or LH secretion at any age; the latter workers stress that the contradictory findings in relation to a diurnal variation in hormone secretion may be due to differences in controlling for environmental factors such as light and stress. This aspect has been studied to some extent by Liptrap and Raeside (1975) and Pitzel *et al.* (1980). These investigators have shown that in both Yorkshire boars and miniature pigs respectively, the administration of either corticosteroids or ACTH caused a rise in plasma testosterone (60–90 minutes post injection) which is followed by a decrease (6–12 hours post injection). No diurnal variation was found for testosterone following ACTH, but marked diurnal changes were still found for corticosteroids (Pitzel *et al.*, 1980). It is postulated that the gradual decrease in LH after ACTH treatment is due to the negative feedback of the initial increase in testosterone; however the way corticosteroids influence testosterone secretion is not known. Ellendorff *et al.* (1975) and Liptrap and Raeside (1978) have shown that copulation results in an increase in plasma testosterone, and the continued presence of an oestrous pig or exposure to an aggressive boar leads to similar endocrine changes. Since Liptrap and Raeside (1978) found plasma corticosteroids increased rapidly before testosterone levels rose after exposure of their boars to other pigs, this adds further support for a role of corticosteroids in mediating testosterone release in certain behavioural situations.

Elsaesser *et al.* (1976) have measured 5α-dihydrotestosterone and progesterone in the peripheral plasma of miniature boars; progesterone is mainly testicular in origin since castration lowers the plasma concentration. The plasma levels of 5α-dihydrotestosterone are a fifth to a tenth that of testosterone, and although the Leydig cells could be a direct source of 5α-dihydrotestosterone (Morat and Courty, 1979), extragonadal sources are possible as is the case in the male rabbit (Booth and Jones, 1980).

Oestrogens

TESTICULAR PRODUCTION

Among the earliest evidence for oestrogens in the boar was the determination of large amounts of these steroids in urine (see review by Velle, 1966) with oestrone as the predominant oestrogen, followed by oestradiol and oestriol (Busch and Ittrich, 1968; see also *Table 2.1*). The testis seems to be the major source of oestrogens, since oestrone and oestradiol-17β have been isolated from the testis (Velle, 1958a), castration results in a marked fall in urinary oestrogens, and gonadotrophin stimulation causes a significant increase in urinary oestrogen excretion (Lunaas and Velle, 1965; Raeside, 1965). Oestrogen excretion is similar to androgen excretion in so far as it shows wide individual variation (Velle, 1958b), and considerable daily variation (Raeside, 1965).

The pathways for the biosynthesis of oestrogen in the boar testis have not been elucidated, but it is likely that androstenedione and testosterone are obligatory intermediates (Engel, 1973). Furthermore since DHA and its sulphate are precursors of oestrogen in those tissues capable of converting 5-ene C_{19} steroids to 4-ene C_{19} steroids (Andersen and Lieberman, 1980), the boar testis would appear to be well equipped for oestrogen synthesis.

Recently oestrogens have been measured in peripheral plasma, spermatic vein blood, testicular lymph and rete testis fluid of boars (see *Table 2.1*). In keeping with earlier findings for oestrogen in urine, oestrone is the

Table 2.1 COMPOSITE DATA ON THE OCCURRENCE OF OESTROGENS IN THE ADULT BOAR

	Urine (mg/24h)	*Peripheral blood* (ng/ml)	*Spermatic vein blood* (ng/ml)	*Testicular lymph* (ng/ml)	*Rete testis fluid* (ng/ml)
Unconjugated oestrogens:					
Total		~0.20[b]			
		0.06–0.25[c]	0.08–2.37[c]		
		0.06[e]	0.30[e]	0.05[e]	0.08[e]
		0.04–0.53[d]			
Oestrone		0.07[d]			
Oestradiol		0.03[d]			
Conjugated oestrogens:					
Total		0.75–4.00[c]	0.78–7.58[c]		
		10.1[e]	29.2[e]	284[e]	2.45[e]
		8.41–198[d]			
Oestrone	1.10–26.8[a]	20.1[d]			
Oestradiol	0.11–5.29[a]	8.50[d]			

Data taken from the following sources:
[a]Raeside (1965);
[b]Velle (1976);
[c]Claus and Hoffmann (1980);
[d]Booth (1980);
[e]Setchell *et al.* (personal communication).

predominant oestrogen in plasma, particularly as the sulphate. It is noteworthy that unconjugated oestrogen in boar plasma is higher than that in the plasma of oestrous gilts (Henricks, Guthrie and Handlin, 1972). The results of Booth (1980a) indicate that some of the oestrogen in boar plasma may originate from peripheral aromatization of androgens, since higher levels of both unconjugated and sulphate conjugated oestrogen were found in the plasma of castrated boars receiving testosterone and 5-androstenediol, than in controls receiving oil only. Claus and Hoffmann (1980) found that plasma oestrogens increased from puberty to maturity, and in response to HCG; conjugated oestrogen also showed a diurnal variation similar to testosterone.

Target organ responses to androgens and oestrogens

The testis is both a producer and target organ for steroids. Androgens produced primarily in the Leydig cells play a vital role in supporting spermatogenesis (Setchell, 1978). Richards and Neville (1973) found that the seminiferous tubules of the rat converted DHA to 5-androstenediol and this may occur in the boar, since the highest levels of 5-androstenediol in relation to DHA are found in the boar testis after the onset of spermatogenesis (see *Figure 2.2*). In the rat testis, oestrogen may be directly involved in the regulation of androgen synthesis in addition to an indirect negative feedback effect on pituitary LH secretion (Ford and Schanbacher, 1977; Kalla *et al.*, 1980). Similarly, the high oestrogen levels in boar testis, particularly in the lymph (see *Table 2.1*) may also have a local effect in the gonad, modulating its potential to produce equally large quantities of C_{19} steroid.

In studies involving the administration of oestrogen to intact boars, it is not clear whether the effects on androgen production are direct, or through the modulation of LH release. Echternkamp *et al.* (1969) found that 96 mg of diethylstilboestrol implanted into young boars had no significant effect on plasma androgen activity or on the weight of the testes and seminal vesicles, but the weight of the prostate and boar odour were reduced. Schilling and Lafrenz (1977) found injections of 5 and 10 mg oestradiol valerionate to boar piglets on days 8 and 25 of life, increased plasma androgen levels and the development of interstitial tissue when the pigs were 5 months old; however degenerative changes were present in the spermatogenic epithelium, in keeping with known deleterious effects of exogenous oestrogen on the testis (Kaur and Mangat, 1980). Some of these findings are difficult to interpret, the more so since different oestrogens, treatment regimes, age of animal and physiological indices were investigated. This aspect is reinforced in studies on the castrated boar aimed at elucidating the role of androgens and oestrogens on accessory organ function and sexual behaviour. Joshi and Raeside (1973) castrated boars around 12 months of age which had been trained to mount a dummy, and then gave them a series of intramuscular injections for 6-week periods of testosterone alone, or alternating with diethylstilboestrol, oestrone or oestradiol-17β in combination with testosterone. Whereas testosterone alone had only a slight effect on restoring the volume of ejaculates and libido, oestrogen in combination with testosterone significantly enhanced both the secretory activity of the accessory organs and libido; these findings indicated a synergistic role for oestrogens with androgens. On the other hand, Booth (1980a) did not find a synergistic effect of oestrone with androgens on the development of accessory organs in castrated boars (*Table 2.2*). Since the ratio between androgen and oestrogen, and the total dose of steroid were similar to that used by Joshi and Raeside (1973), one explanation for a certain lack of agreement between the two studies could be the gross difference in the age of castration before steroid treatment. Perhaps therefore the boar is somewhat refractory to oestrogen (Linde, Einarsson and Gustafsson, 1975), or alternatively the fact that most of the circulating oestrogen is sulphated could mean that the steroid is in a relatively inactive form. However preliminary results from a study in progress indicate that the prostate and seminal vesicles of the boar have the capacity to convert ^{3}H-labelled oestrone sulphate *in vitro* to free oestrone and oestradiol-17β. One of the aims of the investigation by Booth (1980a) and a subsequent unpublished study using 1 mg/5 kg 5-androstenediol, was achieved by the demonstration that this steroid has significant androgenic activity in the boar, particularly with regard to stimulating the seminal vesicles (*Table 2.2*). This finding is supported by recent findings in the author's laboratory showing that the prostate and seminal vesicles have the capacity *in vitro* to convert ^{3}H-labelled 5-androstenediol to testosterone, 5α-dihydrotestosterone and 5α-androstanediols.

Table 2.2 RESPONSE OF ACCESSORY ORGANS (AT 38 WEEKS OF AGE) IN BOARS CASTRATED PREPUBERTALLY AND RECEIVING TWICE WEEKLY INJECTIONS OF STEROIDS IN OIL BETWEEN 14 AND 38 WEEKS OF AGE

Steroid treatments	*No. of pigs*	*Gland weights* (g)			*Seminal vesicles*	
		Prostate	*Seminal vesicle*	*Bulbo-urethral*	*Fructose* (mg)	*Zinc* (mg)
A	3	9.20±2.62*	38.0±5.59	70.9±10.9	62.5±24.3	5.55±1.32
B	3	5.40±0.57	32.8±3.43	49.0±4.86	34.1±12.7	6.13±1.20
C	3	7.17±1.37	28.9±3.31	56.9±5.63	46.7±31.2	2.69±0.57
D	3	4.20±1.50*	64.2±8.86†	40.7±7.10	104±40.7	17.7±1.54†
E	3	0.56±0.03	0.94±0.11	5.29±0.14	0.053±0.017	0.035±0.005
F	4	11.8±1.27	211±5.51	122±2.60	69.4±23.0	60.5±11.3

Steroid treatments: A, testosterone (2 mg/5 kg) + oestrone (1 mg/5 kg); B, 5-androstenediol (2 mg/5 kg) + oestrone (1 mg/5 kg); C, testosterone (2 mg/5 kg); D, 5-androstenediol (2 mg/5 kg); E, untreated castrated pigs; F, untreated intact boars.
Data expressed as the mean ±S.E.M.
*Significantly different from each other, $P<0.05$.
†Significantly different from treatments A–C, $P<0.05$.
From Booth (1980a)

Since DHA and 5-androstenediol sulphates are major steroids in boar plasma, studies on the fate of these conjugates are particularly important for a greater understanding of androgen action in the male pig. In this regard Joshi (1971) found that DHA sulphate supplements the effect of testosterone on accessory organs, and preliminary evidence (Booth, unpublished) has been obtained for the conversion of ^{3}H-labelled DHA and 5-androstenediol sulphates to the corresponding free steroids in *in vitro* incubations with minces of the prostate and seminal vesicles. Since androgens and oestrogens are present in testicular fluid (Setchell *et al.*, personal communication) it remains to be demonstrated what effects these might have on the epididymis.

There is a sexual dimorphism in the submaxillary salivary gland of the pig (Booth, Hay and Dott, 1973), and indirect evidence indicates that testicular steroids are involved. Booth (1977) demonstrated *in vitro* that the submaxillary gland of the boar responds like a typical androgen target organ by converting testosterone to 5α-dihydrotestosterone and 5α-androstanediols. Furthermore, 5α-dihydrotestosterone is present in greater amounts than testosterone in the submaxillary glands of boars, but not in female pigs (Booth, 1972). Flood (1973) and Booth, Hay and Dott (1973) showed histochemically that pig submaxillary glands contain steroid metabolizing enzymes, but the latter workers using DHA as substrate, also showed that the activity of 3β-hydroxy-5-ene steroid dehydrogenase was greatest in the serous cells of boar glands. The finding that free DHA and 5-androstenediol are converted to testosterone and 5α-reduced steroids in the submaxillary gland of the boar, suggests that these steroids may act as prohormones adding to the total androgen output of the testis. The sexual dimorphism in the submaxillary gland of the pig seems to be related to the accumulation of the musk-smelling 16-androstenes in the gland of the boar, and their release into saliva as pheromones (Booth, 1980b).

Joshi and Raeside (1973) found that oestrogen enhanced the effect of testosterone on libido in castrated boars. This finding agrees with the concept that the effects of aromatizable androgens on sexual behaviour are mediated by oestrogens (Callard, Petro and Ryan, 1978). On the other hand oestrogen alone induces female sexual behaviour in castrated pigs (Ford and Schanbacher, 1977), thus indicating that 5α-reduction of testosterone is also required for the expression of male behaviour in the pig as in sheep (Parrott, 1978). It is noteworthy that 5-androstenediol did not support copulatory behaviour in castrated boars as well as testosterone (Booth, 1980a), a finding in keeping with studies in the rat (Morali *et al.*, 1974) and one explanation for this is that only some of the 5-androstenediol is converted to testosterone for aromatization. Other effects of testicular steroids at the cerebral level in the boar, have been studied by Parvizi *et al.* (1977) who showed that implants of 5α-dihydrotestosterone, testosterone and other steroids into the mediobasal hypothalamus or amygdala, affected both the stimulatory and inhibitory mechanisms of LH release.

Much of the work on the anabolic effects of androgens and oestrogens in the pig have involved the use of synthetic steroids, whereupon it has been demonstrated that oestrogens have a pronounced growth-promoting effect which is enhanced by androgen (Fowler, 1976). However, the use of synthetic steroids leads to concern over their possible effects on human health, since they may accumulate in the carcasses. Until convenient methods are available for monitoring levels of synthetic steroids in carcasses (Kroes *et al.*, 1976), it has been suggested that the use of naturally occurring steroids as anabolic agents might overcome the problem of depot effects (Velle, 1976). Rossouw, Skinner and Kemm (1971) injected pigs from weaning to porker weight with androstenedione and observed a growth-promoting effect of the steroid. The effect was greatest in intact boars, suggesting other testicular steroids were involved. In a recent study by the author, castrated boars received subcutaneous injections of 5-androstenediol (1 mg/kg, twice weekly) from 12–40 weeks of age. The pigs were slaughtered at 40 weeks of age, and the Meat and Livestock Commission carried out a detailed carcass evaluation. As found in a previous study (see *Table 2.2*), 5-androstenediol had a pronounced androgenic effect on the accessory organs, but no statistically significant anabolic effects were found. There was, however, a trend for some carcass characteristics of the steroid-treated pigs to be intermediate between those of intact and castrated boars (see *Figure 2.3* and *Table 2.3*). Further studies of this nature are needed involving larger groups of animals.

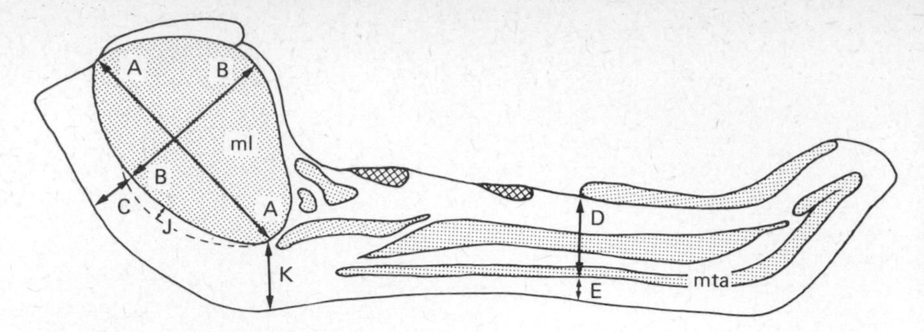

Figure 2.3 A possible anabolic role for 5-androstenediol in the boar. Measurements were taken on the cut surface of the forequarter at the head of the last rib. Values (given in *Table 2.3*) are the means for five castrated boars receiving 5-androstenediol (1 mg/kg) between 12 and 40 weeks of age, and five castrated boars and three intact boars receiving oil injections only. Although there were no statistically significant differences between the groups, there was a trend towards greater leanness in the intact boars, and to a lesser extent in steroid-treated pigs compared with untreated controls. ml—muscle longissimus, mta—muscle transversus abdominis. (Booth unpublished, the author is grateful for the carcass analyses carried out by the Meat and Livestock Commission).

Table 2.3 RELATIVE DISTRIBUTION OF MUSCLE AND FAT ON CUT SURFACE OF FORE-QUARTER AT THE HEAD OF THE LAST RIB IN INTACT BOARS, CASTRATED BOARS AND CASTRATED BOARS TREATED WITH 5α-ANDROSTENEDIOL

Measurements[(a)]	*Intact boar* (mm)	*Castrate + 5α-Androstenediol* (mm)	*Castrate* (mm)
A	92.0	89.8	90.8
B	56.0	52.6	51.6
C	18.0	19.4	20.2
D	26.3	25.4	23.0
E	8.33	9.60	10.0
J	4.67	4.60	6.00
K	25.7	24.8	26.8

[(a)]For definition of A–K see *Figure 2.3*.

16-Androstenes

TESTICULAR PRODUCTION

The 16-androstenes are quantitatively probably the most abundant steroids produced by the pig testes (Prelog and Ruzicka, 1944; Claus, 1970; Ruokonen and Vihko, 1974; Booth, 1975; Booth and Polge, 1976; see *Table 2.4*). It was generally assumed that these musk-smelling steroids occurred in only a few species such as the pig and man as metabolites of androgens. However, a series of investigations by Gower and co-workers (see review by Gower, 1972) indicated that androgens are unlikely to be significant precursors of 16-androstenes in pig testes. Gower and Ahmad (1967) showed *in vitro* that boar testes convert pregnenolone to give high yields of 3β-androstenol (10–15%) and some 3α-androstenol (1–2%).

Table 2.4 THE AMOUNT OF 16-ANDROSTENES IN TESTICULAR TISSUE AND SUBMAXILLARY GLANDS OF BOARS AND INTERSEX PIGS (μg/GLAND/ANIMAL)

	Boars		*Intersex pigs*[a]		
	Immature (12 wk) (Pooled tissue, 4 pigs)	*Mature (2 yr) (Mean, 2 pigs)*	*Ovary (left) Ovotestis (right)*	*Ovotestis (left) Testis (right)*	*Bilateral testes*
Total wt of testes/animal (g)	19.2	970	23.5	28.5	171
3α-Androstenol	8.90	2791	33.6	112	1672
3β-Androstenol	1.49	12 608	154	460	2482
3β-Androstadienol	0.65	269	1.11	8.44	Trace
5α-Androstenone	0.23	180	0.60	11.2	75.2
Androstadienone	ND	13.8	Trace	1.87	22.9
Total wt of submaxillary glands/animal (g)	11.7	276	44.3	70.1	64.0
3α-Androstenol	2.25	10 942	67.3	381	729
3β-Androstenol	0.59	268	19.0	48.5	133
3β-Androstadienol	Trace	18.7	Trace	6.73	27.0
5α-Androstenone	1.00	1266	4.16	12.3	11.3

[a]Data for three pigs showing different degrees of masculinization. ND, None detected. The data has not been corrected for extraction losses (% recovery 20–30%).
After Booth (1975); Booth and Polge (1976)

Progesterone also acts as a substrate (Ahmad and Gower, 1968) but not 17β-hydroxyprogesterone, DHA, testosterone, testosterone acetate or 16-dehydroprogesterone, indicating that 16-androstenes are formed from C_{21} steroid precursors before 17β-hydroxylation and side-chain cleavage (see *Figure 2.4*). Subsequently, Katkov and Gower (1970) and Brophy and Gower (1972) showed that 3β-androstadienol (5,16-androstadien-3β-ol) and androstadienone (4,16-androstadien-3-one) are important intermediates between pregnenolone and 5α-reduced 16-androstenes. The enzyme

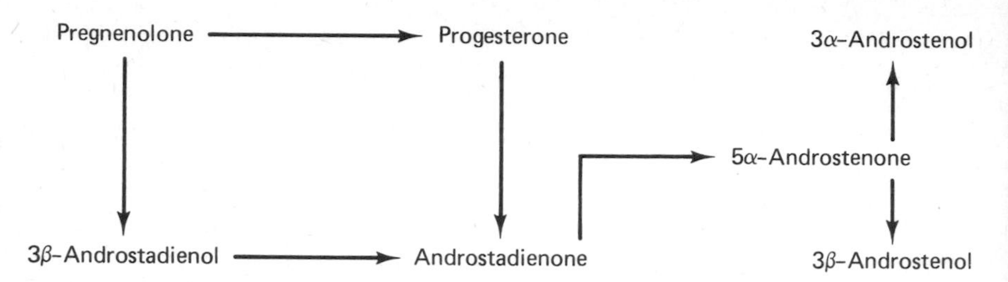

Figure 2.4 Biosynthesis of 16-androstenes from C_{21} steroids in boar testis.

system responsible for the conversion of pregnenolone to 3β-androstadienol is referred to as 'andien-β synthetase'; however the steps between pregnenolone and 3β-androstadienol have not been defined. Andien-β synthetase resides in the microsomal fraction of the testes (Gower and Loke, 1971; Shimizu and Nakada, 1976), and predominantly (66%) in the agranular components (Cooke and Gower, 1977). Mason, Park and Boyd (1979) using a microsomal preparation from immature pig testes in incubations with ^{14}C-labelled pregnenolone, found 3β-hydroxypregn-5,16-dien-3-one in addition to 3β-androstadienol and 17α-hydroxypregnenolone and concluded that this C_{21} steroid might be an intermediate in 16-androstene synthesis, at least in young pigs.

After an infusion of ^{3}H-labelled pregnenolone into the spermatic artery of a mature boar, Saat *et al.* (1972) found radioactively labelled 3α/3β-androstenols, primarily as sulphates, in the testes and spermatic vein blood. Ruokonen and Vihko (1974) also isolated these steroid sulphates from boar testes. Recently sulpho-conjugation of 3β-androstenol has been found in porcine liver (Fish, Cooke and Gower, 1980), but 3β-androstenol predominates as the glucuronide in urine (Gower, Harrison and Heap, 1970).

Claus, Hoffmann and Karg (1971) using gas–liquid chromatography showed that both the major boar taint steroid 5α-androstenone (5α-androst-16-en-3-one) and testosterone in blood plasma of boars increased with age and the amounts of 5α-androstenone (6.0–22.3 ng/ml) were greater than those in females (0.8–2.0 ng/ml) or castrated boars (1.3–2.7 ng/ml). Similar results for 5α-androstenone were obtained by Andresen (1974) and Claus and Hoffmann (1980) using a radioimmunoassay; however a diurnal variation was found by Claus and Gimenez (1977) and Claus and Hoffmann (1980), but not by Andresen (1975a). Andresen (1975a), Carlstrom *et al.* (1975) and Claus and Hoffmann (1980) found a biphasic increase in plasma 5α-androstenone after HCG treatment; the first peak was reached within 3 hours, and the second larger peak by 28–30 hours. Similar levels of unconjugated 3α-androstenol to unstimulated 5α-androstenone were measured in boar plasma (Bicknell and Gower, 1976), but lower levels (0.24–0.77 ng/ml) of the steroid alcohol compared with the ketone were found in the plasma of female and castrated pigs.

ACCUMULATION IN PERIPHERAL TISSUES

The presence of 16-androstenes in peripheral tissues of the pig is of interest for two main reasons. First, these compounds are among the few mammalian pheromones identified chemically, and this has therefore resulted in a greater interest in studies on olfaction in the pig (see reviews by Claus, 1979; Booth, 1980b). Secondly man's olfactory sense is sensitive to these compounds (Kloek, 1961) and many people dislike their odour, particularly when associated with pig meat (boar taint).

Patterson (1968a) isolated 3α-androstenol from the submaxillary glands of boars, but not male castrates or females. This finding was confirmed in more extensive studies by Booth (cited by Gower, 1972), Katkov, Booth and Gower (1972) and Booth (1975), (see also *Table 2.4*). High concentrations of 16-androstenes were also found in the submaxillary glands of intersex pigs (Booth and Polge, 1976; and *Table 2.4*). Although there are reports that 16-androstenes are present in the parotid glands of boars (Claus, 1970; Claus, Hoffmann and Karg, 1971), it seems that the submaxillary gland is primarily involved in concentrating these odorous steroids (Patterson, 1968a; Katkov, Booth and Gower, 1972). Furthermore, recent work in the author's laboratory, using polyacrylamide gel

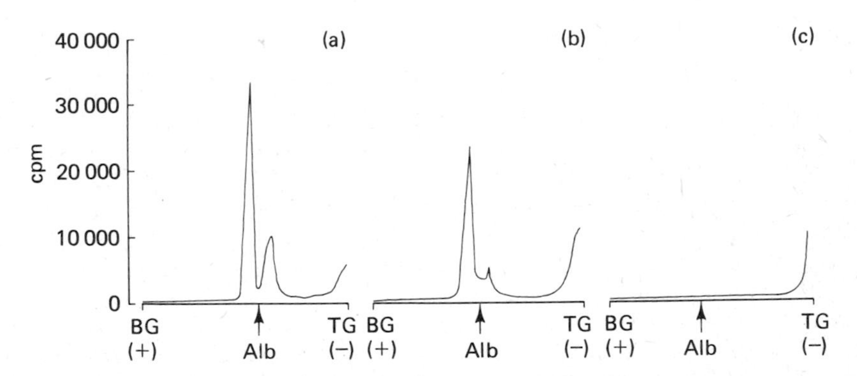

Figure 2.5 Binding of (5α,6α-^{3}H) 5α-androstenone (cpm) (Isocommerz, Dresden) to high molecular weight components in salivary gland cytosols and saliva of mature boar. Samples (80 μl) of 1 in 20 dilution of submaxillary gland (a) and parotid gland (c) cytosols prepared in a 10 mM Tris based buffer containing 1% propanediol, and 1 in 10 dilution of mixed saliva (b) were incubated overnight at 4 °C with ^{3}H-labelled 5α-androstenone and run on 7% polyacrylamide slab gel electrophoresis essentially after Davis (1964). Gel slices (2 mm) were immersed in a toluene-based scintillation medium overnight before counting. BG, bottom of gel; TG, top of separating gel; Alb., albumin. Protein concentrations in (a) 62 μg, (b) 20 μg, (c) 43 μg. From Booth (unpublished)

electrophoresis has demonstrated that proteins strongly binding 5α-androstenone are present in the saliva and cytosols of porcine submaxillary glands but not of the parotid glands (see *Figure 2.5*). These findings, and the observation that removal of the submaxillary glands in boars reduces their ability to induce the mating stance in oestrous pigs (Perry *et al.*, 1980), is further evidence indicating that the boar submaxillary gland has a special role in eliminating the pheromonal steroids. Since 3α-androstenol is the predominant 16-androstene in boar saliva (Booth, 1980b) and submaxillary gland, the binding of this steroid in these media is under investigation. In blood plasma, the amounts of 3α-androstenol (Bicknell and Gower, 1976) and 5α-androstenone (Claus and Hoffmann, 1980) are similar, therefore the increased amounts of 3α-androstenol relative to 5α-androstenone in the submaxillary gland and saliva are probably due to conversion of the ketone to the alcohol by the very active 3α-hydroxysteroid dehydrogenase in the salivary gland (Katkov, Booth and Gower, 1972; Booth, 1977). A similar ratio of 3α-androstenol to 5α-androstenone is found in the apocrine sweat glands of the boar (Stinson and Patterson, 1972), but it is not known if this ratio depends on 3α-hydroxysteroid dehydrogenase in the sweat glands.

One practical consequence of the study of 16-androstenes as pheromones is the commercial availability of aerosols containing 5α-androstenone to facilitate the detection of oestrus in pigs housed in the absence of a boar. This application was established on the basis of the work carried out by Melrose, Reed and Patterson (1971) and Reed, Melrose and Patterson (1974), and reviewed in relation to other aspects of olfaction and reproduction in the pig (Booth, 1980b).

The first isolation of 5α-androstenone as the major compound responsible for taint in boar fat, was achieved by Patterson (1968b). Subsequently Claus (1970), and Claus, Hoffmann and Karg (1971) reported that 1.03–7.49 μg/g of 5α-androstenone were present in the fat of postpubertal boars and Beery and Sink (1971) also demonstrated the presence of 3α-androstenol in boar fat. Kaufman, Ritter and Schubert (1976) obtained similar values to Claus, Hoffmann and Karg (1971) for 5α-androstenone in boar fat, and values <0.1 μg/g in the fat of castrated boars and females. Andresen (1975b) established a radioimmunoassay for 5α-androstenone in fat, and confirmed the data of earlier work. Malmfors and Andresen (1975) found that the concentration of 5α-androstenone in boar fat was positively correlated ($n = 0.51$–0.54, $P<0.001$) with the intensity of boar taint determined either by the soldering iron technique (Jarmoluk, Martin and Fredeen, 1970), or heating fat samples to 180 °C. The concentration of 5α-androstenone in fat was not related to that in plasma ($n = 0.36$), but after HCG treatment (Malmfors *et al.*, 1976), 5α-androstenone increased in both plasma and fat. In a study using more animals, Lundstrom *et al.* (1978) found high correlations between backfat and plasma levels of 5α-androstenone after HCG treatment.

Artificial control of boar taint production

Since castration not only removes the undesirable taint steroids, but also the main source of anabolic steroids, several approaches have been investigated to selectively reduce the production of 16-androstenes, but maintain the production of anabolic steroids in intact boars.

INHIBITION OF 16-ANDROSTENE SYNTHESIS

Synthetic compounds such as diethylstilboestrol (Ecktenkamp *et al.*, 1969) and 19-norethisterone acetate (Rommel, Otto and Blödow, 1975) when administered to intact boars, reduce boar taint in the carcasses, which is likely to be due to a decrease in 16-androstene synthesis associated with the observed atrophy of Leydig cells. However, as mentioned earlier (p.33), the use of synthetic hormones involves a health risk due to possible depot effects. An important consequence of studies on the biosynthesis of 16-androstenes in boar testes, would be to find substances which selectively inhibit enzymes responsible for their synthesis, without affecting the synthesis of androgens; however this aim has not been achieved. Brophy and Gower (1974) found that 5α-pregnane-3,20-dione, and a series of 17β-derivatives of testosterone (Kaufman and Schubert, 1980) inhibited both 16-androstene and androgen biosynthesis.

IMMUNIZATION

Specific antibodies can be raised against steroids after they have been conjugated chemically with proteins such as bovine serum albumin, and this phenomenon has permitted the development of radioimmunoassays. As a method for eliminating boar taint, boars were injected repeatedly with 5α-androstenone linked to bovine serum albumin (Claus, 1975). The levels of 5α-androstenone in fat were reduced in the five boars studied, but limitations such as site and frequency of injection, and age at the start of injections were realized. More boars have since been immunized against 5α-androstenone by Patterson (personal communication), and the results are encouraging. In boars slaughtered at 90 kg, the amounts of 5α-androstenone were reduced considerably in blood (46%, n = 3) and fat (72%, n = 12).

GENETIC SELECTION

Comments have been made that boar taint is more apparent in some breeds than others e.g. Landrace and Pietrain>Large White. Breeding studies on Landrace pigs (Jonsson and Andresen, 1979; Willeke *et al.*, 1980) have shown that 'high' and 'low' lines arise for 5α-androstenone and testosterone. As the low line effect seems to be due to a delay in puberty, this could result in an undesirable reduction in anabolic effects.

ENVIRONMENTAL FACTORS

Claus (1977) reported that between August and January, the natural mating period of the wild boar, higher levels of 5α-androstenone were present in the fat of domestic boars during their second year of life. Patterson (personal communication) also found the highest levels of 5α-androstenone in the fat (0.62 μg/g) of a group of Landrace boars (87 kg) during the natural mating period, but in other combinations of sex and breed (Large White × Landrace) × Large White, the highest levels of 5α-androstenone were found outside the natural mating period. In all groups the levels of taint steroid (~0.5 μg/g) indicated that intact boars $<$ 90 kg can be marketed for meat throughout the year. Evidence from a study by Bonneau and Desmoulin (1980) suggests that social conditions affect the levels of 5α-androstenone in boars, with higher amounts in grouped pigs particularly of mixed sexes above 80 kg when sexual activity would be increased at puberty. Bonneau and Desmoulin (1980) and Patterson (personal communication) found a wide individual variation in 5α-androstenone levels (0.21–2.55 μg/g) at 95 kg; perhaps some of this variation was due to dominance hierarchies being established, an aspect needing further investigation.

Taint and boar meat

A survey carried out by Rhodes (1971) indicated that bacon produced from boars (108 kg) was as acceptable as bacon from gilts of similar weight, to over 350 people in 125 households and $<$ 1% of consumers found boar bacon less acceptable on the basis of odour. A second survey (Rhodes, 1972) involving 419 households consisting of 1560 persons, showed that pork joints from carcasses of 24-week old boars were as acceptable as joints from gilts of the same age.

Most of the results from fundamental studies on testicular steroids in the boar, have been obtained during the time boar meat has been introduced to the British market. Generally the results of the endocrinologist support the finding that immature boars at porker weight are producing minimum quantities of 5α-androstenone. However, as the boar matures to over 100 kg, the risk of taint increases with the increased testicular output of 16-androstenes during puberty. This aspect is of particular concern to countries outside the UK who would like to introduce boar meat, but where traditionally pigs are often slaughtered at heavier weights. The knowledge that intact boars are a desirable source of meat, if the risk of taint could be overcome, has led to the investigation of methods which might be practicable for monitoring taint in individual carcasses at the slaughter house. A suitable method must be quick so as not to interfere with the rapid throughput of modern slaughterhouses, and furthermore must be sensitive and reliable. The soldering iron technique of Jarmoluk, Martin and Fredeen (1970) is quick but subjective, dependent upon the considerable variability in human olfactory acuity. By combining this technique with fat biopsies taken from living pigs (Lundstrom, Malmfors and Hansson, 1973), pigs can be monitored before slaughter. Quantitative assessment of 5α-androstenone by radioimmunoassay, although sensitive, is elaborate and time-consuming. Recently, however, Andresen (1979) has investigated the possibility of a rapid radioimmunoassay for 5α-androstenone, involving the absorption of fat onto filter paper before direct assay without solvent extraction; the practical use of this method is being assessed. Another possible method for assessing boar taint is that based on the findings of Forlund, Lundstrom and Andresen (1980). A positive correlation was found between the size of the bulbourethral glands and 5α-androstenone in fat (n = 0.56–0.75), but a correlation of only 0.28–0.34 was found between gland weights and boar taint assessed by the soldering iron method, a finding in keeping with the opinion that boar taint is not entirely due to 16-androsterone steroids. Indeed, Lundstrom *et al.* (1980) have indicated that skatole and indole in boar carcasses can enhance boar taint above that produced by 5α-androstenone alone; this adds a new dimension to the subject of boar taint, which is beyond the scope of this review.

Conclusions

The essential theme of this review has been to emphasize that the boar testis produces a variety of steroids. Of these, three groups of biologically significant steroids namely androgens, oestrogens and 16-androstenes, have been discussed. Studies on these groups of steroids in the boar have shown that this animal should be ideal for future work aimed at answering some of the questions posed by the endocrinologist e.g. is there a role for steroid sulphates other than simply products for excretion, to what extent do prohormones add to the total effect of androgens and oestrogens in target tissues, and what is the role for oestrogen in the male? Already there is considerable evidence to show that in the boar, there is a synergism between the endocrine function of the testis and the production and release of the pheromonal 16-androstene steroids, and this has led to problems with attempts to artificially dissociate these two functions in relation to boar meat production. However, with the knowledge that the risk of boar taint is minimal in young boars, it seems that the market for boar meat should continue to expand, particularly if immunization against 5α-androstenone becomes practicable as a means of ensuring that precocious boars do not have tainted carcasses.

Acknowledgements

I wish to thank Dr B.P. Setchell and his colleagues and Dr R.L.S. Patterson for allowing me to include some of their unpublished data.

References

AHMAD, N. and GOWER, D.B. (1968). The biosynthesis of some androst-16-enes from C_{21} and C_{19} steroids in boar testicular and adrenal tissue. *Biochem. J.* **108**, 233–241

ANDERSEN, N.G. and LIEBERMAN, S. (1980). C_{19} steroidal precursors of estrogens. *Endocrinology* **106**, 13–18

ANDRESEN, O. (1974). Development of a radioimmunoassay for 5α-androst-16-en-3-one in pig peripheral plasma. *Acta Endocr.* **76**, 377–387

ANDRESEN, O. (1975a). 5α-androstenone in peripheral plasma of pigs, diurnal variation in boars, effects of intravenous HCG administration and castration. *Acta Endocr.* **78**, 385–391

ANDRESEN, O. (1975b). A radioimmunoassay for 5α-androst-16-en-3-one in porcine adipose tissue. *Acta Endocr.* **79**, 619–624

ANDRESEN, O. (1979). A rapid radioimmunological evaluation of the androstenone content in boar fat. *Acta vet. scand.* **20**, 343–350

BAULIEU, E.E., FABRE-JUNG, I. and HUIS IN'T VELD, L.G. (1967). Dehydroepiandrosterone sulfate: a secretory product of the boar testis. *Endocrinology* **81**, 34–38

BEERY, K.E. and SINK, J.D. (1971). Isolation and identification of 3α-hydroxy-5α-androst-16-ene and 5α-androst-16-en-3-one from porcine adipose tissue. *J. Endocr.* **51**, 223–224

BICKNELL, D.C. and GOWER, D.B. (1976). The development and application of a radioimmunoassay for 5α-androst-16-en-3α-ol in plasma. *J. Ster. Biochem.* **7**, 451–455

BONNEAU, M. and DESMOULIN, B. (1980). Evolution de la teneur en androstenone du tissu adipeux dorsal chez le porc male entier de type Large White: variations selon les conditions d'elevage. *Reprod. Nutr. Dev.* **20**, 1429–1437

BOOTH, W.D. (1972). The occurrence of testosterone and 5α-dihydrotestosterone in the submaxillary salivary gland of the boar. *J. Endocr.* **55**, 119–125

BOOTH, W.D. (1975). Changes with age in the occurrence of C_{19} steroids in the testis and submaxillary gland of the boar. *J. Reprod. Fert.* **42**, 459–472

BOOTH, W.D. (1977). Metabolism of androgens *in vitro* by the submaxillary salivary gland of the mature domestic boar. *J. Endocr.* **75**, 145–154

BOOTH, W.D. (1980a). A study of some major testicular steroids in the pig in relation to their effect on the development of male characteristics in the prepubertally castrated boar. *J. Reprod. Fert.* **59**, 155–162

BOOTH, W.D. (1980b). Endocrine and exocrine factors in the reproductive behaviour of the pig. *Symp. Zool. Soc. Lond.* **45**, 289–311

BOOTH, W.D. and JONES, R. (1980). The extragonadal origin of 5α-reduced androgens in the peripheral circulation of the adult male rabbit. *Int. J. Androl.* **3**, 692–702

BOOTH, W.D. and POLGE, C. (1976). The occurrence of C_{19} steroids in testicular tissue and submaxillary glands of intersex pigs in relation to morphological characteristics. *J. Reprod. Fert.* **46**, 115–121

BOOTH, W.D., HAY, M.F. and DOTT, H.M. (1973). Sexual dimorphism in the submaxillary gland of the pig. *J. Reprod. Fert.* **33**, 163–166

BROPHY, P.J. and GOWER, D.B. (1972). 16-unsaturated C_{19} 3-oxo-steroids as metabolic intermediates in boar testis. *Biochem. J.* **128**, 945–952

BROPHY, P.J. and GOWER, D.B. (1974). Studies on the inhibition by 5α-pregnane-3,20-dione of the biosynthesis of 16-androstenes and dehydroepiandrosterone in boar testis preparations. *Biochim. Biophys. Acta* **360**, 252–259

BUSCH, W. and ITTRICH, G. (1968). Untersuchungen über die Ostrogenausscheidung beim Eber. *Endokrinologie* **53**, 100–105

CALLARD, G.V., PETRO, Z. and RYAN, K.J. (1978). Phylogenetic distribution of aromatase and other androgen-converting enzymes in the central nervous system. *Endocrinology* **103**, 2283–2290

CARLSTROM, K., MALMFORS, B., LUNDSTROM, K., EDQUIST, L.E. and GAHNE, B. (1975). The effect of HCG on blood plasma levels of 5α-androstenone and testosterone in the boar. *Swed. J. agric. Res.* **5**, 15–21

CLAUS, R. (1970). Bestimmung von Testosteron und 5α-androst-en-3-on, einen Ebergeruchsstoff, bei Sweinen. Dr. agr. thesis. Technical High School, Munich

CLAUS, R. (1975). Neutralization of pheromones by antisera in pigs. In *Immunization with Hormones in Reproductive Research*, (E. Nieschlag, Ed.), pp. 189–197. Amsterdam, North Holland Publishing Company

CLAUS, R. (1979). Pheromone bei Säugetieren unter besonderer Berlicksichtigung des Ebergeruchsstoffes und seiner Beziehung zu anderen Hodensteroiden. *Adv. Anim. Nutr.* **10**, 1–136

CLAUS, R. and GIMENEZ, T. (1977). Diurnal rhythm of 5α-androst-16-en-3-one and testosterone in peripheral plasma of boars. *Acta endocr.* **84**, 200–206

CLAUS, R. and HOFFMANN, B. (1980). Oestrogens, compared to other steroids of testicular origin in blood plasma of boars. *Acta endocr.* **94**, 404–411

CLAUS, R., HOFFMANN, B. and KARG, H. (1971). Determination of 5α-androst-16-en-3-one, a boar taint steroid in pigs, with reference to relationships to testosterone. *J. Anim. Sci.* **33**, 1293–1297

COLENBRANDER, B., DE JONG, F.H. and WENSING, C.J.G. (1978). Changes in serum testosterone concentrations in the male pig during development. *J. Reprod. Fert.* **53**, 377–380

COOKE, G.M. and GOWER, D.B. (1977). The submicrosomal distribution in rat and boar testis of some enzymes involved in androgen and 16-androstene biosynthesis. *Biochim. Biophys. Acta* **498**, 265–271

DAVIS, B.J. (1964). Disc electrophoresis. Method and application to human serum proteins. *Ann. N.Y. Acad. Sci.* **121**, 404–427

DUFOUR, J. and RAESIDE, J.I. (1968). Histochemical demonstration of Δ^5-3β- and 17β-hydroxysteroid dehydrogenases in the testes of the boar. *J. Reprod. Fert.* **16**, 123–124

ECHTERNKAMP, S.E., TEAGUE, H.S., PLIMPTON, R.F. and GRIFO, A.P. (1969). Glandular development, hormonal response and boar odor and flavor intensity of untreated and diethylstilbestrol implanted boars. *J. Anim. Sci.* **28**, 653–658

ELLENDORFF, F., PARVIZI, N., POMERANTZ, D.K., HARTJEN, A., KÖNIG, A., SMIDT, D. and ELSAESSER, F. (1975). Plasma luteinizing hormone and testosterone in the adult pig: 24 hour fluctuations and the effect of copulation. *J. Endocr.* **67**, 403–410

ELSAESSER, F., KÖNIG, A. and SMIDT, D. (1972). Der testosteron und androstendiongehalt im eberhoden in abhängigkeit vom alter. *Acta endocr.* **69**, 553–566

ELSAESSER, F., ELLENDORFF, F., POMERANTZ, D.K., PARVIZI, N. and SMIDT, D. (1976). Plasma levels of luteinizing hormone, progesterone, testosterone and 5α-dihydrotestosterone in male and female pigs during sexual maturation. *J. Endocr.* **68**, 347–348

ENGEL, L.L. (1973). The biosynthesis of oestrogens. In *Handbook of*

Physiology, 2, Sect. 7, Part 1, (R.O. Greep and S.R. Geiger, Eds.), p. 467. Washington, American Physiological Society

FISH, D.E., COOKE, G.M. and GOWER, D.B. (1980). Investigation into the sulphoconjugation of 5α-androst-16-en-3β-o1 by porcine liver. *FEBS Lett.* **117**, 28–32

FLOOD, P.F. (1973). Histochemical localization of hydroxysteroid dehydrogenases in the maxillary glands of pigs. *J. Reprod. Fert.* **32**, 125–127

FORD, J.J. and SCHANBACHER, B.D. (1977). Luteinizing hormone secretion and female lordosis behaviour in male pigs. *Endocrinology* **100**, 1033–1038

FORD, J.J., CHRISTENSEN, R.K. and MAURER, R.R. (1980). Serum testosterone concentrations in embryonic and fetal pigs during sexual maturation. *Biol. Reprod.* **23**, 583–587

FORLUND, D.M., LUNDSTROM, K. and ANDRESEN, O. (1980). Relationship between androstenone content in fat, intensity of boar taint and size of accessory sex glands in boars. *Nord. VetMed.* **32**, 201–206

FOWLER, V.R. (1976). Some aspects of the use of anabolic steroids in pigs. In *Anabolic Agents in Animal Production*, (F.C. Lu and J. Rendel, Eds.), p.109. Stuttgart, Georg Thieme

GASPARINI, F.J., HOCHBERG, R.B. and LIEBERMAN, S. (1976). Biosynthesis of steroid sulfates by the boar testes. *Biochemistry* **15**, 3969–3975

GOWER, D.B. (1972). 16-unsaturated C_{19} steroids. A review of their chemistry, biochemistry and possible physiological role. *J. Ster. Biochem.* **3**, 45–103

GOWER, D.B. and AHMAD, N. (1967). Studies on the biosynthesis of 16-dehydrosteroids. The metabolism of (4-^{14}C) pregnenolone by boar adrenal and testis *in vitro*. *Biochem. J.* **104**, 550–556

GOWER, D.B. and LOKE, K.H. (1971). Studies on the subcellular location and stability of the enzyme system involved in the biosynthesis of 5,16-androstadien-3β-o1 from 3β-hydroxy-5-pregnen-20-one (pregnenolone). *Biochim. Biophys. Acta* **250**, 614–616

GOWER, D.B., HARRISON, F.A. and HEAP, R.B. (1970). The identification of C_{19} 16-unsaturated steroids and estimation of 17-oxosteroids in boar spermatic vein plasma and urine. *J. Endocr.* **47**, 357–368

GRAY, R.C., DAY, B.N., LASLEY, J.F. and TRIBBLE, L.F. (1971). Testosterone levels of boars at various ages. *J. Anim. Sci.* **33**, 124–126

HENRICKS, D.M., GUTHRIE, H.D. and HANDLIN, D.L. (1972). Plasma oestrogen, progesterone and luteinizing hormone levels during the oestrous cycle in pigs. *Biol. Reprod.* **6**, 210–218

HOFFMANN, B., CLAUS, R. and KARG, H. (1970). Bestimmung von Testosteron im peripheren blut von Schwein und rind mit einer Doppelisotopenderivat Verdünnungsmethode. *Acta endocr.* **64**, 377–384

HUIS IN'T VELD, L.G., LOUWERENS, B. and REILINGH, W. (1964). The origin of urinary dehydroepiandrosterone in boars. *Acta Endocr.* **46**, 185–196

INANO, H. amd TAMAOKI, B. (1975). Relationship between steroids and pyridine nucleotides in the oxido-reduction catalyzed by the 17β-hydroxysteroid dehydrogenase purified from the porcine testicular microsomal fraction. *Eur. J. Biochem.* **53**, 319–326

JARMOLUK, L., MARTIN, A.H. and FREDEEN, H.T. (1970). Detection of taint (sex odour) in pork. *Can. J. Anim. Sci.* **50**, 750–752

JONSSON, P. and ANDRESEN, O. (1979). Experience during two generations of within lines boar performance testing, using 5α-androst-16-ene-3-one (5α-androstenone) and an olfactory judgement of boar taint. *Annls Genet. Select. anim.* **11**, 241–250

JOSHI, H.S. (1971). Synergistic effects of testosterone and oestrogens on male accessory glands of castrated rats and boars. PhD Thesis. University of Guelph

JOSHI, H.S. and RAESIDE, J.I. (1973). Synergistic effects of testosterone and oestrogens on accessory sex glands and sexual behaviour of the boar. *J. Reprod. Fert.* **33**, 411–423

KALLA, N.R., NISULA, B.C., MENARD, R. and LORIAUX, D.L. (1980). The effect of estradiol on testicular testosterone biosynthesis. *Endocrinology* **106**, 35–39

KATKOV, T. and GOWER, D.B. (1970). The biosynthesis of androst-16-enes in boar testis tissue. *Biochem. J.* **117**, 533–538

KATKOV, T., BOOTH, W.D. and GOWER, D.B. (1972). The metabolism of 16-androstenes in boar salivary glands. *Biochim. Biophys. Acta* **270**, 546–556

KAUFMAN, G. and SCHUBERT, K. (1980). Inhibition of 16-androstene biosynthesis in boar testis preparations by known and new steroids. *J. Ster. Biochem.* **13**, 351–358

KAUFMAN, G., RITTER, F. and SCHUBERT, K. (1976). Quantitative determination of the boar taint substance 5α-androst-16-en-3-one in fat. *J. Ster. Biochem.* **7**, 593–597

KAUR, C. and MANGAT, H.K. (1980). Effects of estradiol dipropionate on the biochemical composition of testis and accessory sex organs of adult rats. *Andrologia* **12**, 373–378

KLOEK, J. (1961). The smell of some steroid sex-hormones and their metabolites. Reflections and experiments concerning the significance of smell for the mutual relation of the sexes. *Folia psychiat. neurol. neurochir. neerl.* **64**, 309–344

KROES, R., HUIS IN'T VELD, L.G., SCHULLER, P.L. and STEPHANY, R.W. (1976). Methods for controlling the application of anabolics in farm animals. In *Anabolic Agents in Animal Production*, (F.C. Lu and J. Rendel, Eds.), p.192. Stuttgart, Georg Thieme

LAPWOOD, K.R. and FLORCRUZ, S.V. (1978). Luteinizing hormone and testosterone secretory profiles of boars: effects of stage of sexual maturation. *Theriogenology* **10**, 293–306

LINDE, C., EINARSSON, S. and GUSTAFSSON, B. (1975). The effect of exogenous administration of oestrogens on the function of the epididymis and the accessory sex glands in the boar. *Acta vet. scand.* **16**, 456–464

LINDNER, H.R. (1960). Testicular endocrine function in domestic animals. PhD Thesis. Cambridge University

LINDNER, H.R. (1961). Androgens and related compounds in the spermatic vein blood of domestic animals. *J. Endocr.* **23**, 171–178

LIPTRAP, R.M. and RAESIDE, J.I. (1970). Urinary steroid excretion in cryptorchidism in the pig. *J. Reprod. Fert.* **21**, 293–301

LIPTRAP, R.M. and RAESIDE, J.I. (1971). Urinary steroid excretion in response to endogenous and exogenous gonadotrophin stimulation of cryptorchid testes in the pig. *J. Reprod. Fert.* **25**, 55–60

LIPTRAP, R.M. and RAESIDE, J.I. (1972). Effect of gonadotrophin stimulation on urinary steroid excretion after relocation of normal and cryptorchid testes in the boar. *J. Reprod. Fert.* **30**, 465–467

LIPTRAP, R.M. and RAESIDE, J.I. (1975). Increase in plasma testosterone concentration after injection of adrenocorticotrophin into the boar. *J. Endocr.* **66**, 123–131

LIPTRAP, R.M. and RAESIDE, J.I. (1978). A relationship between plasma concentrations of testosterone and corticosteroids during sexual and aggressive behaviour in the boar. *J. Endocr.* **76**, 75–85

LUNAAS, T. and VELLE, W. (1965). The effect of gonadotropins and synthetic gestagens on testicular steroid secretion in swine. *Acta endocr. Suppl.* **100**, 41

LUNDSTRÖM, K., MALMFORS, B. and HANSSON, I. (1973). A simple biopsy technique for obtaining fat and muscle samples from pigs. *Swed. J. agric. Res.* **3**, 211–213

LUNDSTRÖM, K., MALMFORS, B., HANSSON, I., EDQVIST, L.E. and GAHNE, B. (1978). 5α-androstenone and testosterone in boars. Early testing with HCG, sexual stimulation and diurnal variation. *Swed. J. agric. Res.* **8**, 171–180

LUNDSTRÖM, K., HANNSON, K.-E., FJELKNER-MODIG, S. and PERSSON, J. (1980). Skatole—another contributor to boar taint. *Meeting of Meat Research Workers, Colorado Springs, USA* Abstract F13

MALMFORS, B. and ANDRESEN, O. (1975). Relationship between boar taint intensity and concentration of 5α-androst-16-en-3-one in boar peripheral plasma and back fat. *Acta agric. scand.* **25**, 92–96

MALMFORS, B., LUNDSTRÖM, K., HANSSON, I. and GAHNE, B. (1976). The effect of HCG and LH-RH on 5α-androstenone levels in plasma and adipose tissue of boars. *Swed. J. agric. Res.* **6**, 73–79

MASON, J.I., PARK, R.J. and BOYD, G.S. (1979). A novel pathway of androst-16-ene biosynthesis in immature pig testis microsomal fractions. *Biochem. Soc. Trans.* **7**, 641–643

MELROSE, D.R., REED, H.C.B. and PATTERSON, R.L.S. (1971). Androgen steroids associated with boar odour as an aid to the detection of oestrus in pig artificial insemination. *Br. vet. J.* **127**, 497–501

MOON, Y.S. and RAESIDE, J.I. (1972). Histochemical studies on hydroxysteroid dehydrogenase activity of fetal pig gonads. *Biol. Reprod.* **7**, 278–287

MORALI, G., LARSSON, K., PEREZ-PALACIOS, G. and BEYER, C. (1974). Testosterone, androstenedione, and androstenediol: effects on the initiation of mating behaviour of inexperienced castrated rats. *Horm. Behav.* **5**, 103–110

MORAT, M. and COURTY, Y. (1979). Dosage simultane, par radioimmunologie, de l'androstenedione, de la testosterone et de la dihydrotestosterone. Application a l'etude du fonctionnement des cellules de Leydig. *C. r. Seanc. Soc. biol., Paris* **173**, 1070–1077

NOTATION, A.D. and UNGAR, F. (1969). Regulation of rat testis steroid sulfatase. A kinetic study. *Biochemistry* **8**, 501–506

PARKES, A.S. (1966). The testes of certain Chaeromorpha. *Symp. Zool. Soc. Lond.* **15**, 141–154

PARROTT, R.F. (1978). Courtship and copulation in prepubertally castrated male sheep (wethers) treated with 17β-estradiol, aromatizable androgens, or dihydrotestosterone. *Horm. Behav.* **11**, 20–27

PARVIZI, N., ELSAESSER, F., SMIDT, D. and ELLENDORFF, F. (1977). Effects of intracerebral implantation, microinjection, and peripheral application of sexual steroids on plasma luteinizing hormone levels in the male miniature pig. *Endocrinology* **101**, 1078–1087

PATTERSON, R.L.S (1968a). Identification of 3α-hydroxy-5α-androst-16-ene as the musk odour component of boar submaxillary salivary gland and its relationship to the sex odour taint in pork meat. *J. Sci. Fd Agric.* **19**, 434–438

PATTERSON, R.L.S. (1968b). 5α-androst-16-ene-3-one:- compound responsible for taint in boar fat. *J. Sci. Fd Agric.* **19**, 31–38

PAYNE, A.H. amd JAFFE, R.B. (1970). Comparative roles of dehydroepiandrosterone sulfate and androstenediol sulfate as precursors of testicular androgens. *Endocrinology* **87**, 316–322

PELLINIEMI, L.J. (1976). Ultrastructure of the indifferent gonad in male and female pig embryos. *Tiss. Cell* **8**, 163–174

PERRY, G.C., PATTERSON, R.L.S., MacFIE, H.J.H. and STINSON, C.G. (1980). Pig courtship behaviour: pheromonal property of androstene steroids in male submaxillary secretion. *Anim. Prod.* **31**, 191–199

PITZEL, L., HARTIG, A., HOLTZ, W. and KÖNIG, A. (1980). Plasma corticosteroid, testosterone and LH levels after ACTH injection in male pigs. *Acta Endocr. Suppl.* **234**, 35–36

PRELOG, V. and RUZICKA, L. (1944). Über zwei moschusartig riechende Steroide aus Schweinetestes-Extracten. *Helv. Chim. Acta* **27**, 61–66

RAESIDE, J.I. (1965). Urinary excretion of dehydroepiandrosterone and oestrogens by the boar. *Acta endocr.* **50**, 611–620

RAESIDE, J.I. and HOWELLS, G.A. (1971). The isolation and identification of androstenediol sulfate from spermatic vein blood and testes of the boar. *Can. J. Biochem.* **49**, 80–84

RAESIDE, J.I. and SIGMAN, D.M. (1975). Testosterone levels in early fetal testes of domestic pigs. *Biol. Reprod.* **13**, 318–321

REED, H.C.B., MELROSE, D.R. and PATTERSON, R.L.S. (1974). Androgen steroids as an aid to the detection of oestrus in pig artificial insemination. *Br. vet. J.* **130**, 61–67

RHODES, D.N. (1971). Consumer testing of bacon from boar and gilt pigs. *J. Sci. Fd Agric.* **22**, 485–490

RHODES, D.N. (1972). Consumer testing of pork from boar and gilt pigs. *J. Sci. Fd Agric.* **23**, 1483–1491

RICHARDS, G. and NEVILLE, A. (1973). Androgen metabolism in rat interstitial tissue and seminiferous tubules. *Nature, Lond.* **244**, 359–361

ROBERTS, K.D., BANDI, L., CALVIN, H.I., DRUCKER, W.D. and LIEBERMAN, S. (1964). Evidence that steroid sulfates serve as biosynthetic intermediates. IV. Conversion of cholesterol sulfate *in vivo* to urinary C_{19} and C_{21} steroid sulfates. *Biochemistry* **3**, 1983–1988

ROMMEL, P., OTTO, E. and BLÖDOW, G. (1975). Orientierungsversuche zur medikamentellen Beseitigung des Geschlechtsgeruches beim Eber. 3. Mitt: Intramuskulare Applikation von 19-Norathisteronazetat. *Mh. VetMed* **30**, 292–297

ROSSOUW, A.F., SKINNER, J.D. and KEMM, E.H. (1971). The effect of androstenedione on growth, carcass composition and reproductive development of porkers. *S. Afr. J. Anim. Sci.* **1**, 85–89

RUOKONEN, A. (1978). Steroid metabolism in testis tissue: the metabolism of pregnenolone, pregnenolone sulfate, dehydroepiandrosterone and dehydroepiandrosterone sulfate in human and boar testes *in vitro*. *J. Ster. Biochem.* **9**, 939–946

RUOKONEN, A. and VIHKO, R. (1974). Steroid metabolism in testis tissue: concentrations of unconjugated and sulfated neutral steroids in boar testis. *J. Ster. Biochem.* **5**, 33–38

SAAT, Y.A., GOWER, D.B., HARRISON, F.A. and HEAP, R.B. (1972). Studies on the biosynthesis *in vivo* and excretion of 16-unsaturated C_{19} steroids in the boar. *Biochem. J.* **129**, 657–663

SANFORD, L.M., SWIERSTRA, E.E., PALMER, W.M. and HOWLAND, B.E. (1976). The profile of luteinizing hormone and testosterone secretion in the boar. *VIII Int. Congr. Anim. Reprod. A.I., Kracow* **3**, 96–99

SCHILLING, E. and LAFRENZ, R. (1977). Östrogengaben an männliche Saugferkel und die Wirkung auf Hodenentwicklung und Androgenproduktion zur Zeit der Geschlechtsreife. *Zuchthygiene* **12**, 145–151

SEGAL, D.H. and RAESIDE, J.I. (1975). Androgens in testes and adrenal glands of the fetal pig. *J. Ster. Biochem.* **6**, 1439–1444

SETCHELL, B.P. (1978). Endocrinology of the testis. In *The Mammalian Testis,* (C.A. Finn, Ed.), pp. 109–180. London, Paul Elek

SHIMIZU, K. (1978). Formation of 5-[17β-^{2}H] androst-3β,17α-diol from 3β-hydroxy-5-[17,21,21,21-^{2}H] pregnen-20-one by the microsomal fraction of boar testis. *J. biol. Chem.* **253**, 4237–4241

SHIMIZU, K. and NAKADA, F. (1976). Formation of [17-^{2}H] androsta-5,16-dien-3β-o1 from [17,21,21,21-^{2}H] pregnenolone by the microsomal fraction of boar testis. *Biochim. Biophys. Acta* **450**, 441–449

SINK, J.D. (1967). Theoretical aspects of sex odor in swine. *J. Theoret. Biol.* **17**, 174–180

STEWART, D.W. and RAESIDE, J.I. (1976). Testosterone secretion by the early fetal pig testes in organ culture. *Biol. Reprod.* **15**, 25–28

STINSON, G.C. and PATTERSON, R.L.S. (1972). C_{19}-Δ^{16} steroids in boar sweat glands. *Br. vet. J.* **128**, xli–xlii

TAN, H.S. and RAESIDE, J.I. (1980). Developmental patterns of plasma dehydroepiandrosterone sulfate and testosterone in male pigs. *Anim. Reprod. Sci.* **3**, 73–81

VELLE, W. (1958a). Investigations on naturally occurring estrogens in ruminants and pigs. Thesis. Oslo

VELLE, W. (1958b). Further studies on urinary oestrogen excretion by the boar. *Acta endocr.* **29**, 395–400

VELLE, W. (1966). Urinary oestrogens in the male. *J. Reprod. Fert.* **12**, 65–73

VELLE, W. (1976). Endogenous anabolic agents in farm animals. In *Anabolic Agents in Animal Production*, (F.C. Lu and J. Rendel, Eds.), p. 159. Stuttgart, Georg Thieme

WILLEKE, H., CLAUS, R., PIRCHNER, F. and ALSING, W. (1980). A selection experiment against 5α-androst-16-en-3-one, the boar taint steroid, in adipose tissue of boars. *Z. Tierzücht. Zücht. Biol.* **97**, 86–94

3

INTERRELATIONSHIPS BETWEEN SPERMATOZOA, THE FEMALE REPRODUCTIVE TRACT, AND THE EGG INVESTMENTS

R.H.F. HUNTER
School of Agriculture, University of Edinburgh, Edinburgh, UK

This chapter describes some principal physiological events during transport of spermatozoa in the female reproductive tract from the time of ejaculation until penetration of the egg investments at the start of fertilization. The objective is to review our current understanding of the relevant processes and to highlight areas requiring further research; it is also hoped to contrast certain aspects of sperm transport in pigs with those in ruminants. Although much of the material discusses processes not obviously susceptible to manipulation, topics related to the successful practice of artificial insemination are emphasized.

Previous reviews of sperm transport in the female reproductive tract of pigs include those of Hunter (1973a; 1975a,b; 1980), Polge (1978) and Einarsson (1980), and a comprehensive coverage of the early literature is found in these works. The volume edited by Hafez and Thibault (1975) compares mechanisms of sperm transport and storage in a wide range of vertebrates, whilst fertilization itself is described in pigs by Thibault (1959), Hancock (1961), Baker, Dziuk and Norton (1967), Hunter (1972a, 1974), Szollosi and Hunter (1973; 1978), Baker and Polge (1976) and Polge (1978).

Deposition of semen and uterine transit

The specific stimulus provoking ejaculation in the boar is the interlocking of the glans penis in the cervical ridges of an oestrous gilt or sow. Due to the influence of ovarian steroid hormones, these muscular ridges become taut and oedematous at oestrus (Burger, 1952; Smith and Nalbandov, 1958), and their arrangement is such that the lumen of the cervix tapers towards the internal os, enabling the glans penis to be gripped by the female tissues (*Figure 3.1*). During thrusting of the penis through the anterior vagina and into the cervical canal, there is also a twisting component to the penile shaft, causing the spiral folds of the glans to interdigitate tightly with the cervical wall. The pressure derived from this 'lock' leads to ejaculation, a situation which is mimicked during manual collection of boar semen.

As a result of these anatomical specializations and because of the large volume of the boar ejaculate (150–500 ml), semen makes only passing contact with the innermost portion of the cervical canal before entering the body of the uterus (*Figure 3.1*). Deposition is therefore effectively *intrauterine* and, in the context of sperm transport to the site of fertilization, little further consideration needs to be given to the structure of the cervix. An initial distribution of semen in the uterus may be achieved due to the force of ejaculation and the volume of fluid involved, assuming that little or no leakage occurs into the vagina. However, contractions of the myometrium are vigorous during oestrus (Corner, 1923; Keye, 1923), and

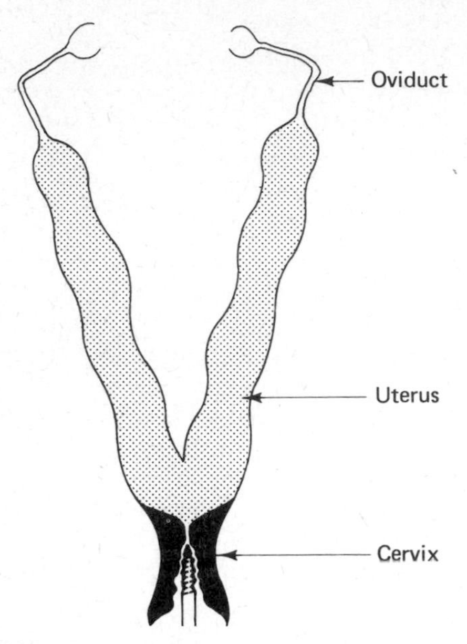

Figure 3.1 Diagram to illustrate the manner in which the spirally-arranged folds of the glans penis interlock with the prominent muscular ridges of the cervix in oestrous pigs. As a result of these specializations, ejaculation is effectively into the uterus and the large volume of semen distends the two cornua by the end of the protracted period of mating.

these should assist transport of the male secretions and, in the pre-ovulatory interval, redistribution of fluid between the two horns. Reflex release of oxytocin would enhance uterine contractions (see Knifton, 1962) and a stimulatory rôle of the semen itself should be considered. Although very low concentrations of prostaglandins E_2 and $F_{2\alpha}$ were reported in boar semen (Hunter, 1973a), an increased concentration of prostaglandin $F_{2\alpha}$ was found in the uterine venous blood of a gilt sampled 15 minutes after mating when the uterus was still full of semen (*Figure 3.2*). This result was not endorsed in six other animals sampled 35–60 minutes after mating, but the notion that semen may activate a local release of smooth muscle stimulants is worth examining further.

One other aspect concerning myometrial contractions should be mentioned at this stage. Unilateral fertilization, that is the fertilization of eggs in only one Fallopian tube with failure of spermatozoa to reach eggs in the contralateral tube, is a frequent sequel to the delayed insemination of pigs (Kvasnitsky, 1959; Hancock, 1962; Hunter, 1967). Since uterine contractions are programmed by the balance of ovarian steroid hormones, and because ovarian steroid secretion will have changed in favour of progesterone under conditions of delayed insemination, an altered pattern of uterine contractions may explain this failure of sperm transport and fertilization. If contractile waves of low frequency occlude one horn of the uterus during insemination, then the semen would be delivered unilaterally. Also the more relaxed condition of the cervix leads to leakage of semen with a belated mating or insemination, and with insufficient fluid remaining in the uterus to enable a redistribution between the two horns, this may also be a cause of diminished fertility.

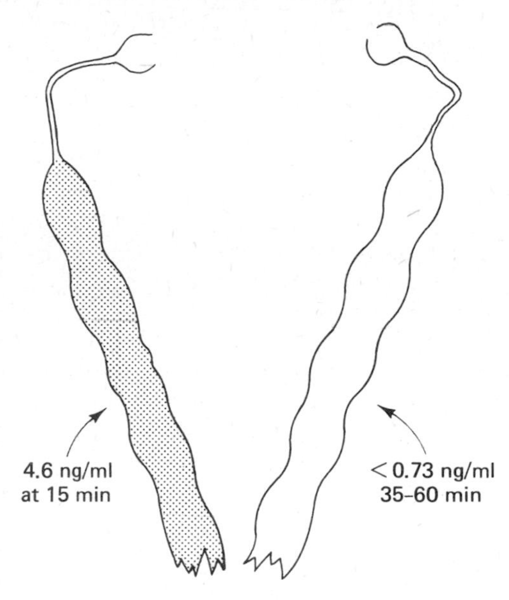

Prostaglandins in boar semen (ng/ml)

	E_2	$F_{2\alpha}$
Collection	<0.27	<0.34
Uterus	<1.1	<2.8

Figure 3.2 Diagrammatic representation of the uterine horns in a group of gilts to indicate the concentration of prostaglandin $F_{2\alpha}$ in uterine venous blood at various intervals after mating. The concentration for the left horn (4.6 ng/ml) represents the situation 15 minutes after the end of mating when the lumen is still distended by the ejaculate. The concentration shown against the right horn (<0.73 ng/ml) represents the highest concentration in a series of six gilts sampled 35–60 minutes after mating when the bulk of the ejaculate was no longer grossly detectable. From Hunter and Poyser (unpublished). These figures are compared in the lower part of the diagram with the concentration of prostaglandins in freshly-ejaculated semen and in semen aspirated from the uterus shortly after mating (see Hunter, 1973a).

During the protracted period of coitus (e.g. 5–15 minutes), semen is not emitted as a homogeneous fluid. If the ejaculate is collected as a series of fractions, these will consist first of watery and gel pre-sperm secretions, then a sperm-rich fraction, followed by a post-sperm and gelatinous fraction (Rodolfo, 1934a; McKenzie, Miller and Bauguess, 1938). Such a sequence may take 3–5 minutes and can be considered as one full wave of ejaculation. After renewed thrusting and re-establishment of the cervical lock, a second and indeed third wave of ejaculation may follow. Whether the fractions have a special significance in the transport of spermatozoa is uncertain, but they may favour displacement of the sperm-rich fraction to the region of the utero-tubal junctions (McKenzie, Miller and Bauguess, 1938; Du Mesnil du Buisson and Dauzier, 1955a). In any event, under conditions of natural mating performed in the pre-ovulatory interval, the uterine horns should be fully distended with semen by the time mating is finished, a situation that has been verified by laparotomy and at autopsy. The gelatinous material of the ejaculate acts as a cervical plug for a variable period after mating, thereby preventing immediate leakage of the uterine contents.

The total number of spermatozoa deposited may vary from $4–8 \times 10^{10}$ cells, with the concentration of spermatozoa in the whole ejaculate being $1–3 \times 10^8$ cells/ml (Rodolfo, 1934b; McKenzie, Miller and Bauguess, 1938; Polge, 1956). A concentration of this order, or even higher if there is preferential transport of the sperm-rich fraction, will therefore bathe the utero-tubal junction by the completion of mating with a mature boar; uterine transport of semen should not therefore be a physiological problem. Accordingly, discussion of the process of sperm transport after mating now needs to focus on the rôle of the tract between the utero-tubal junction and the site of fertilization. Diminished fertility may arise, however, after insemination of a reduced volume of diluted semen due to ineffective transport of spermatozoa to the utero-tubal junction. The use of frozen-thawed boar semen may further exacerbate the problem, as will insemination too early in oestrus.

Sperm transport through the utero-tubal junction

As far as the transport of spermatozoa is concerned, the utero-tubal junction of pigs functions, in several respects, in a manner similar to the cervix of ruminants. Before developing this analogy, the morphology of the utero-tubal junction must be briefly described. As shown in *Figure 3.3* the longitudinal folds of the isthmus extend caudally into the tip of the uterine horn, where they are arranged as a series of polypoid processes. These have a prominent mucosa which is sensitive to the balance of ovarian steroids and during the period of oestrus the polypoid processes swell due to oedema in the mucosa. This swelling is particularly pronounced in the pre-ovulatory phase of oestrus (*Figure 3.3*), but begins to subside within 8–12 hours of the time of ovulation. The processes therefore form a powerful barrier in the tract during the first 40–48 hours of oestrus, and are the principal physical obstacle to sperm transport. Although a series of quite recent studies has examined the rôle of the utero-tubal junction in relation to sperm transport in pigs (Hunter, 1972b; 1973a; 1973b; 1975b; 1981; Hunter and Hall, 1974a,b), the anatomy of this region first received attention in the classical studies of Andersen (1928) and Lee (1928).

The utero-tubal junction of oestrous pigs is not patent to the passage of fluids from the uterus when tested by injection into the uterine lumen. Indeed, as observed by Lee as long ago as 1928, the wall of the uterus will

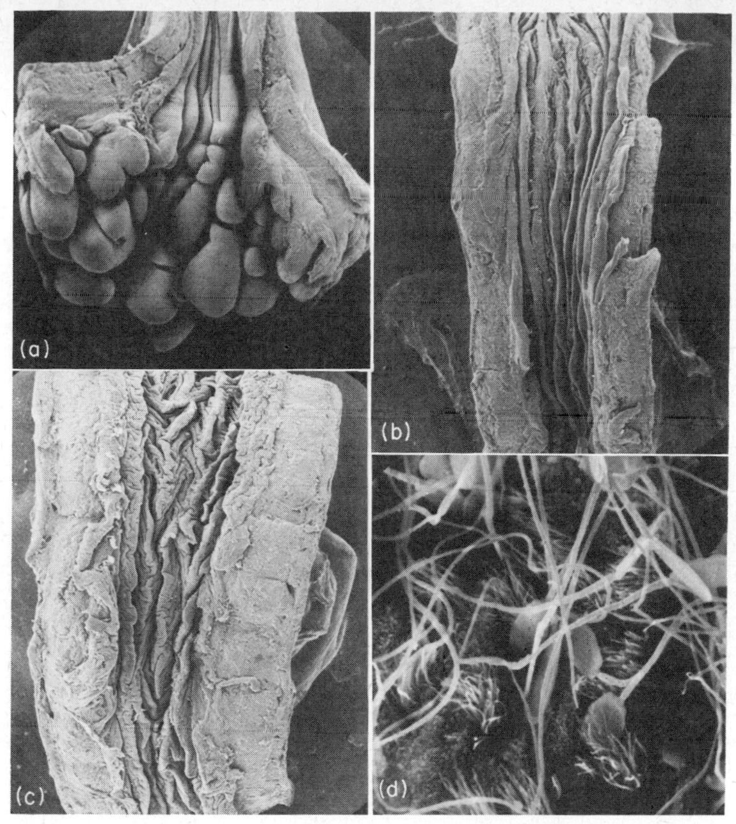

Figure 3.3 Scanning electron micrographs of the utero-tubal junction and isthmus of pigs during the period of oestrus. (a) The utero-tubal junction showing the highly swollen polypoid processes that extend rostrally into the isthmus as longitudinal folds. The oedematous polypoid processes act as a major barrier to seminal plasma and immotile spermatozoa in the pre-ovulatory phase of oestrus. (b) and (c) To show the thick muscular wall in the lower portion of the isthmus which, together with the swollen longitudinal ridges and folds, functions to regulate the ascent of spermatozoa. (d) Spermatozoa on the surface of a furrow between the terminal folds of the isthmus. Many of the sperm heads appear to be engaged by their tips in the epithelial surface and occasionally in contact with cilia. From Fléchon and Hunter (1981)

rupture before fluid can be forced through the junction into the Fallopian tube. Although these remarks concern the potential for bulk passage of fluids, there is no evidence from insemination of radio-opaque material that even minor quantities of whole semen would cross the utero-tubal junction (Polge, 1978). This report endorses an earlier failure to trace constituents of seminal plasma in the Fallopian tubes at oestrus by biochemical methods (Mann, Polge and Rowson, 1956). Taken together, these findings strongly suggest that whole semen does *not* enter the tubes of pre-ovulatory animals in detectable amounts after mating, and infers that spermatozoa gain the lumen of the isthmus by virtue of their own motility. Both these comments bring out the analogy between the rôle of the utero-tubal junction in pigs and that of the cervix in ruminants.

Further evidence that the utero-tubal junction divests the ascending spermatozoa of any gross association with seminal plasma was reported by Hunter and Hall (1974a). Using measurement of capacitation time as a sensitive physiological indicator, they noted that spermatozoa passing through the utero-tubal junction had a consistent advantage of approximately two hours compared with those instilled directly into the tubes. This observation was strengthened by the two hours delay in capacitation time after introduction of 0.02–0.1 ml aliquots of cell-free seminal plasma into the tubal isthmus (Hunter and Hall, 1974a). Once more, sperm motility was considered essential for negotiation of the utero-tubal junction prior to ovulation, and those studies claiming passage of dead cells have been criticized either as being unphysiological (e.g. Baker and Degen, 1972) or possibly performed late in oestrus when the utero-tubal junction is no longer fully oedematous and fertilization should have occurred (e.g. First *et al.*, 1968a). Moreover, if dead spermatozoa can freely negotiate the utero-tubal junction prior to ovulation, then perhaps the same facility should be extended to polymorphonuclear leucocytes and bacteria present in the uterine lumen after mating. There is no evidence that this is the case, nor did a study of the tubal isthmus by scanning electron microscopy reveal such cells (Fléchon and Hunter, 1981).

Despite all this evidence in favour of the utero-tubal junction acting to prevent passage of whole semen, polymorphs and bacteria in the preovulatory phase of oestrus, one recent report has claimed that boar seminal plasma does enter the tubes after insemination on the second day of oestrus (Einarsson *et al.*, 1980). However, the method of homogenizing Fallopian tube tissues in order to detect radiolabelled compounds suspended in the inseminate was unable to distinguish between radioactivity in the lumen and that in the wall, as might have been brought about by local vascular or lymphatic transport. Even so, muscular movements associated with the region of the utero-tubal junction might explain the observations of Einarsson *et al.*, but the failure of sperm capacitation and fertilization in the presence of seminal plasma would then need to be considered.

Sperm ascent up to and through the utero-tubal junction is relatively rapid when compared with the situation in ruminants. Whilst spermatozoa have been observed in the ampulla within 15–30 minutes of mating or artificial insemination (Burger, 1952; Ito, Kudo and Niwa, 1959; First *et al.*, 1968b), the interval required for spermatozoa to traverse the utero-tubal junction in numbers sufficient to fertilize the eggs has been calculated from post-coital separation of the isthmus from the utero-tubal junction (Hunter and Hall, 1974b; Hunter, 1981). The conclusion here was that a population of spermatozoa sufficient to give *maximum* fertilization is established in the tubes within 1–2 hours of mating. Because this rate of formation of adequate sperm reservoirs in the tubes corresponds closely with the time course of seminal plasma elimination from the uterus, one explanation for the exceptional volume of the boar ejaculate can be offered. The intrauterine deposition of 150–500 ml of fluid may have evolved to ensure successful passage of whole semen to the utero-tubal junction of all oestrous females in a breeding herd (i.e. gilts and parous sows) so that spermatozoa are present at the junctions for a period of time sufficient to enable formation of reservoirs in the caudal isthmus of each tube. This argument is supported by the fact that surgical insemination of as little as 0.05 ml semen directly into the caudal isthmus leads to fertilization (Polge, Salamon and Wilmut, 1970; Hunter, 1973b). Therefore, from the point of view of the future development of artificial insemination, a technique that permits deposition of a few millilitres of whole semen at the top of each uterine horn against the utero-tubal junction should be fruitful, and should also enable a much wider propagation of each boar ejaculate.

Whilst the utero-tubal junction has been suggested as a reservoir for spermatozoa (Du Mesnil du Buisson and Dauzier, 1955a; Rigby, 1966), it has been argued from several points of view that the physiological reservoir is more likely to be established in the lowermost portions of the isthmus (Hunter and Léglise, 1971; Hunter, 1972b, 1973a); the evidence from experiments involving post-coital transection of the tract would support this argument (Hunter and Hall, 1974b; Hunter, 1981). Nonetheless, there are glandular openings within the processes of the utero-tubal junction (Fléchon and Hunter, 1981), so short-term storage of spermatozoa might be expected in this region.

Sperm regulation and release by the isthmus

The tactical problem after completion of mating and distension of the uterus with semen is to secure transport of an appropriately-sized population of viable spermatozoa from the utero-tubal junction to the site of

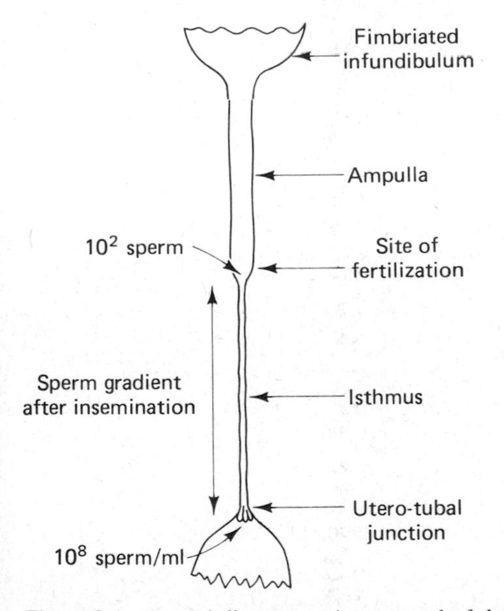

Figure 3.4 A semi-diagrammatic portrayal of the Fallopian tube to illustrate the extremely steep sperm gradient that must be obtained between the utero-tubal junction and the site of fertilization at the ampullary-isthmic junction if polyspermy is to be avoided. The figures indicate the concentrations of spermatozoa that would bathe the utero-tubal junction at the completion of mating and the total number of spermatozoa that might achieve the site of fertilization *at any one time.*

fertilization at the ampullary-isthmic junction. However, in the light of the concentration of spermatozoa bathing the utero-tubal junction at the end of mating, the rôle of the isthmus must be seen more as a *regulator* of sperm transport than as a section of the tract *facilitating* transport of cells. This is illustrated in *Figure 3.4*: a gradient in sperm concentration decreasing from 10^8 cells/ml semen to an approximate total of 10^2 cells *at any one time* must be achieved between the utero-tubal junction and the site of fertilization, a distance of less than 10–12 cm in most gilts and sows. If the number of capacitated spermatozoa in the vicinity of the freshly ovulated eggs were to increase above such a figure, then the risk of polyspermic fertilization would be high. Polyspermy entails penetration of more than one spermatozoon into the egg cytoplasm, a lethal condition in mammals (*Figure 3.5*).

The precise manner in which the isthmus forms a gradient in sperm concentration is unknown. Even so, the lumen of this portion of the tube is extremely constricted in the pre-ovulatory period, for the longitudinal

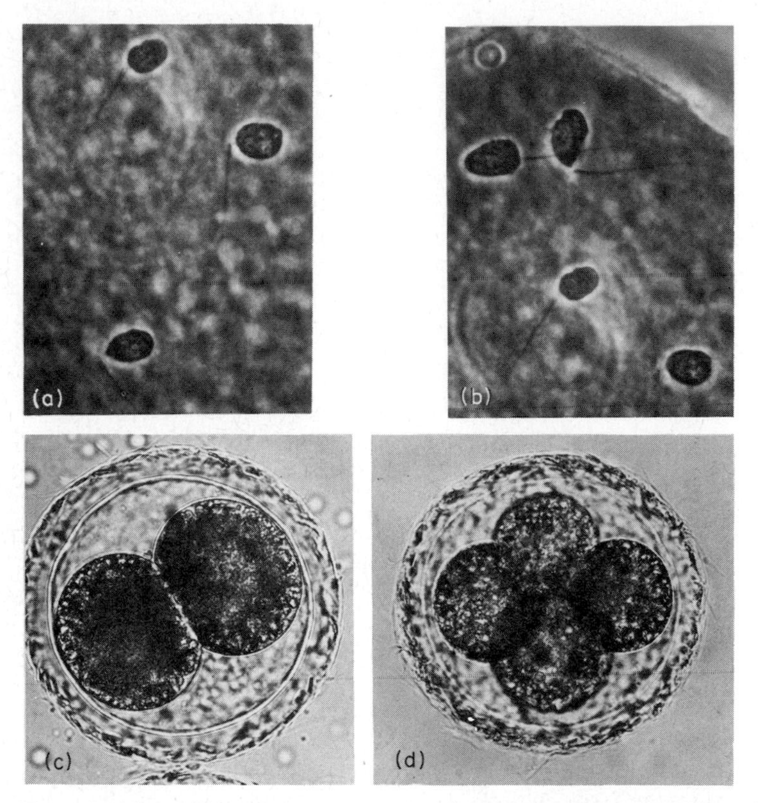

Figure 3.5(a) and (b) Whole-mount preparations of pig eggs fixed in 25% acetic-alcohol and stained with 0.5% aceto-orcein to show polyspermic penetration of the vitellus. The similar stages of transformation of the sperm heads suggest closely synchronous (and recent) penetration of the egg membranes. (c) and (d) Whole-mount preparations of living pig embryos at the 2- and 4-celled stages of development to show the very large numbers of sperm heads trapped in the zona pellucida as a result of the block to polyspermy in the innermost part of this membrane. More than 200 spermatozoa are within the zona of the 4-celled embryo.

folds protrude inwards and are oedematous. Physical reduction in the patency of the tube therefore serves to regulate the number of spermatozoa passing to the site of fertilization. Although these comments concentrate on swelling in the mucosa, isthmic patency is also controlled through its powerful layers of circular and longitudinal muscles. Two aspects of myosalpingeal function are of interest (reviewed by Hunter, 1977; 1980). First, the α-adrenergic receptors in the isthmic muscle are known to be activated under oestrogen dominance, leading to contraction, whereas β-adrenergic receptors are potentiated by progesterone secretion, leading to relaxation. Second, prostaglandins of the F series in the myosalpinx may act to cause contraction with oestrogen dominance, whereas those of the E series may promote relaxation under progesterone dominance. Thus, the influence of pre-ovulatory oestrogen secretion ensures regulation of sperm ascent by enhancing contraction of the myosalpinx, a situation that changes close to the time of ovulation.

Instead of a smoothly decreasing sperm gradient along the length of the isthmus, there is evidence that spermatozoa are largely restricted to the caudal tip (1–2 cm) of the isthmus during much of the pre-ovulatory interval (Hunter, 1981). Release of spermatozoa from this reservoir commences shortly before the time of ovulation, probably under the influence of a local transfer of ovarian follicular hormones to the wall of the isthmus. Pre-ovulatory secretion of progesterone may be involved, but a local vascular or lymphatic transfer of follicular relaxin and prostaglandins should also be considered. Whatever the nature of the mechanisms, such ovarian regulation of the patency of the isthmus would explain why peri- or post-ovulatory mating can lead to a more rapid transport of spermatozoa to the site of fertilization than mating early in oestrus (Du Mesnil du Buisson and Dauzier, 1955b; Hunter, 1981).

Discussion of sperm gradients in the isthmus would not be complete without stressing the deleterious influence of removing or overcoming such gradients. Three experimental approaches have all led to the condition of polyspermy, inferring that the gradient is no longer functional and that the egg membranes have been confronted simultaneously by more than one competent spermatozoon. First was the approach of surgical resection of most of the isthmus followed by end-to-end anastomosis of the remaining portions of the tube (Hunter and Léglise, 1971). Second was the reduction in oedema and myosalpingeal contraction obtained by microinjections of a solution of progesterone under the serosal layer of the isthmus and utero-tubal junction (Hunter, 1972b). Third was the instillation of excessive numbers of spermatozoa into the tube by introducing a 'rounded' insemination needle through the utero-tubal junction and into the isthmus (Polge, Salamon and Wilmut, 1970; Hunter, 1973b).

Apart from their intrinsic motility, sperm transport in the isthmus is assisted by waves of muscular activity and by the influence of cilia. Both have been examined in post-mortem preparations of pig material (Gaddum-Rosse and Blandau, 1973; 1976; Blandau and Gaddum-Rosse, 1974) and in tissues obtained at laparotomy (Fléchon and Hunter, 1981); cilial activity and waves of contraction appear to progress from the utero-tubal junction to the site of fertilization. Again, these activities are under endocrine control and although of an ad-ovarian direction at the time of ovulation, they are reversed within 48 hours when the embryos or unfertilized eggs are descending to the uterus. In fact, the phase of accelerated sperm transport close to the time of ovulation is probably associated more with myosalpingeal activity than with the beat of tubal cilia or sperm flagella. On the other hand, contractions and cilial beat in the ampulla are programmed in an ab-ovarian direction towards the site of fertilization, which explains why an appropriately timed intraperitoneal insemination may lead to fertilization (Hunter, 1978).

Capacitation and sperm interactions with the egg investments

Ovulation occurs approximately 40 hours after the onset of oestrus, and eggs released on to the densely ciliated surface of the fimbriated infundibulum are propelled to the site of fertilization within 45 minutes, still invested in cumulus cells (Hunter, 1974). These cells surrounding individual eggs aggregate very soon after ovulation so that the eggs shed by each ovary reach the ampullary-isthmic junction assembled within a plug of cumulus cells (Hancock, 1961). Such a plug may persist for several hours after ovulation in unmated animals, but spermatozoa promote its rapid dissolution, presumably by means of acrosomal enzymes. Whilst the endocrine events associated with ovulation may stimulate a release of spermatozoa from the isthmus (see above, p.57), initial contact of the follicular contents with the tubal epithelium could also provoke a final phase of sperm transport to the ampullary-isthmic junction. This is not to infer that the follicular fluid passes down the tube, but because it is rich in hormones such as prostaglandins of the E and F series and relaxin, transient contact with the epithelium could influence the function of the isthmus.

Capacitation of boar spermatozoa requires some 2–3 hours after mating or intracervical insemination (Hunter and Dziuk, 1968). By contrast, if aliquots of whole semen are introduced directly into the Fallopian tubes, the process of capacitation is delayed by approximately two hours. The uterus and tubes therefore act synergistically to accelerate achievement of the capacitated state, primarily by removing elements of the seminal plasma during sperm passage through the utero-tubal junction (Hunter and Hall, 1974a). Contact of spermatozoa with the egg investments or, alternatively, with the microenvironment around denuded eggs may be necessary for the completion of capacitation and the concomitant increase in sperm motility.

Swelling of the acrosome is thought to be an immediate sequel to capacitation of boar spermatozoa, and precedes the vesiculation reaction between the plasma and outer acrosomal membranes (Szollosi and Hunter, 1978). Despite the intense progressive motility of capacitated spermatozoa, hyaluronidase is considered necessary to depolymerize a passage between the follicular cells (see Austin, 1961). Again, dogma holds that the acrosome reaction is necessary before hyaluronidase can be released from viable spermatozoa, but Szollosi and Hunter (1978) inferred escape of this enzyme from the swollen acrosome *before* actual commencement of vesiculation. This would enable the fertilizing sperm to reach the surface of the zona before exhibiting the acrosome reaction. Constituents of the zona

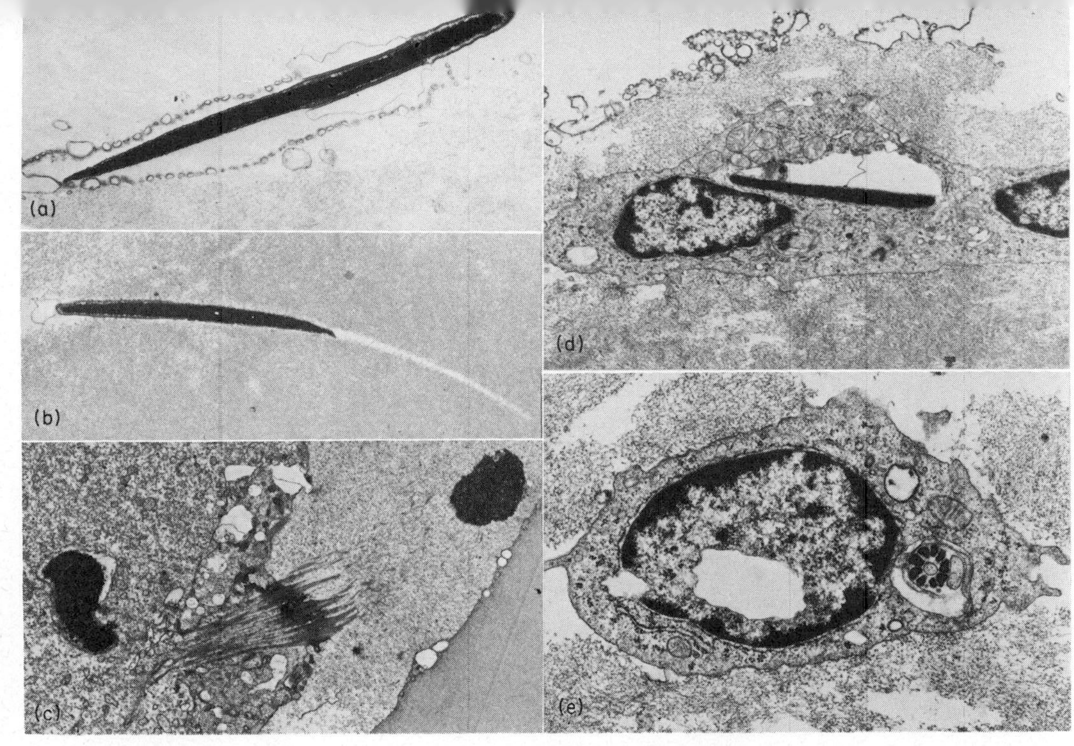

Figure 3.6 Electron micrographs of boar spermatozoa, pig eggs and cells of the cumulus oophorus. (a) Spermatozoon on the surface of the zona pellucida showing the extensive vesiculation between the outer acrosomal and plasma membranes that constitutes the acrosome reaction. (b) Spermatozoon in the substance of the zona pellucida showing the distinct penetration pathway that is inferred to be digested by lytic enzymes of the inner acrosomal membrane. Note the spongy outer and more compact inner regions of the zona pellucida. (c) Section through the peripheral region of an egg showing the microtubules and mid-body of the spindle, and completion of the second meiotic division shortly after activation by the fertilizing sperm. Note the absence of organelles in the cytoplasm of the second polar body, and that the chromatin in this vestigial body has not attracted elements of a nuclear envelope. (d) and (e) Follicle cells of the corona radiata still intimately associated with the zona pellucida and containing engulfed spermatozoa. Note the transverse section through the sperm tail in (e). From Szollosi and Hunter (1973; 1978)

pellucida or indeed factors emanating from the egg cytoplasm might therefore trigger the acrosome reaction, which would be a means of conserving acrosomal enzymes until the egg itself was confronted. Boar spermatozoa penetrate the zona under the dual influence of incisive progressive motility and that of lytic enzymes. Whilst ultrastructural studies have failed to reveal the sperm penetration filament reported by Dickmann and Dziuk (1964), a pathway can be detected in the substance of the zona after passage of the fertilizing sperm (Szollosi and Hunter, 1973). Spermatozoa not penetrating the zona pellucida may be incorporated by cells of the corona radiata (*Figure 3.6*).

One of the components of egg activation initiated by membrane fusion of the fertilizing sperm with the egg plasma membrane is the cortical reaction whereby the contents of the cortical granules are released into the perivitelline space (Austin, 1956; Szollosi, 1967; Fléchon, 1970). The manner in which this reaction causes a block to polyspermy in the zona pellucida still requires clarification, but there is no obvious structural change in this accessory membrane. The block is located in the innermost portion of the zona pellucida, for the heads of accessory spermatozoa continue to penetrate into the zona of newly fertilized eggs (Thibault, 1959; Hancock, 1961). Despite the stability of the block to polyspermy, the outermost portion of the zona remains permeable to spermatozoa for up to two days after initial penetration, a situation which explains the massive increase in zona sperm numbers as the egg descends through the isthmus. In fact, 2- and 4-cell eggs may each contain in excess of 200 spermatozoa (*Figure 3.5*), and yet only one sperm has crossed the perivitelline space to enter the cytoplasm. Spermatozoa in the isthmus not entering the zona pellucida or being incorporated by follicular cells may be transported back into the uterus at the time of embryo passage (Hunter, 1978), but whether this accounts for the whole of the residual tubal population remains uncertain, as does its ultimate fate.

Acknowledgements

Published and unpublished studies of the author referred to in the text were supported by grants from the Agricultural Research Council, the Wellcome Trust and the Ford Foundation. Drs J.E. Fléchon, N.L. Poyser and D. Szollosi kindly gave permission to publish photographs and data arising from collaborative projects.

References

ANDERSEN, D.H. (1928). Comparative anatomy of the tubo-uterine junction. Histology and physiology in the sow. *Am. J. Anat.* **42**, 255–305

AUSTIN, C.R. (1956). Cortical granules in hamster eggs. *Exp. Cell Res.* **10**, 533–540

AUSTIN, C.R. (1961). *The Mammalian Egg*. Oxford, Blackwell Scientific Publications

BAKER, R.D. and DEGEN, A.A. (1972). Transport of live and dead boar spermatozoa within the reproductive tract of gilts. *J. Reprod. Fert.* **28**, 369–377

BAKER, R.D., DZIUK, P.J. and NORTON, H.W. (1967). Polar body and pronucleus formation in the pig egg. *J. exp. Zool.* **164**, 491–498

BAKER, R.D. and POLGE, C. (1976). Fertilization in swine and cattle. *Can. J. Anim. Sci.* **56**, 105–119

BLANDAU, R.J. and GADDUM-ROSSE, P. (1974). Mechanism of sperm transport in pig oviducts. *Fert. Steril.* **25**, 61–67

BURGER, J.F. (1952). Sex physiology of pigs. *Onderstepoort J. vet. Res. Suppl.* No. 2

CORNER, G.W. (1923). Cyclic variation in uterine and tubal contraction waves. *Am. J. Anat.* **32**, 345–351

DICKMANN, Z. and DZIUK, P.J. (1964). Sperm penetration of the zona pellucida of the pig egg. *J. exp. Biol.* **41**, 603–608

DU MESNIL DU BUISSON, F. and DAUZIER, L. (1955a). Distribution et résorption du sperme dans le tractus génital de la truie: survie des spermatozoides. *Annls Endocr.* **16**, 413–422

DU MESNIL DU BUISSON, F. and DAUZIER, L. (1955b). La remontée des spermatozoides du verrat dans le tractus génital de la truie en oestrus. *C. r. Séanc. Soc. Biol.* **149**, 76–79

EINARSSON, S. (1980). Site, transport and fate of inseminated semen. *Proc. 9th Int. Congr. Anim. Reprod. A.I.* **1**, 147–158

EINARSSON, S., JONES, B., LARSSON, K. and VIRING, S. (1980). Distribution of small- and medium-sized molecules within the genital tract of artificially inseminated gilts. *J. Reprod. Fert.* **59**, 453–457

FIRST, N.L., SHORT, R.E., PETERS, J.B. and STRATMAN, F.W. (1968a). Transport and loss of boar spermatozoa in the reproductive tract of the sow. *J. Anim. Sci.* **27**, 1037–1040

FIRST, N.L., SHORT, R.E., PETERS, J.B. and STRATMAN, F.W. (1968b). Transport of boar spermatozoa in estrual and luteal sows. *J. Anim. Sci.* **27**, 1032–1036

FLÉCHON, J.E. (1970). Nature glycoprotéique des granules corticaux de l'oeuf de lapine. *J. Microsc.* **9**, 221–242

FLÉCHON, J.E. and HUNTER, R.H.F. (1981). Distribution of spermatozoa in the utero-tubal junction and isthmus of pigs, and their relationship with the luminal epithelium after mating: a scanning electron microscope study. *Tissue Cell* **13**, 127–139

GADDUM-ROSSE, P. and BLANDAU, R.J. (1973). *In vitro* studies on ciliary activity within the oviducts of the rabbit and pig. *Am. J. Anat.* **136**, 91–104

GADDUM-ROSSE, P. and BLANDAU, R.J. (1976). Comparative observations on ciliary currents in mammalian oviducts. *Biol. Reprod.* **14**, 605–609

HAFEZ, E.S.E. and THIBAULT, C.G., Eds. (1975). *The Biology of Spermatozoa.* Basel, Karger

HANCOCK, J.L. (1961). Fertilisation in the pig. *J. Reprod. Fert.* **2**, 307–331

HANCOCK, J.L. (1962). Fertilisation in farm animals. *Anim. Breed. Abstr.* **30**, 285–310

HUNTER, R.H.F. (1967). The effects of delayed insemination on fertilisation and early cleavage in the pig. *J. Reprod. Fert.* **13**, 133–147

HUNTER, R.H.F. (1972a). Fertilisation in the pig: sequence of nuclear and cytoplasmic events. *J. Reprod. Fert.* **29**, 395–406

HUNTER, R.H.F. (1972b). Local action of progesterone leading to polyspermic fertilisation in pigs. *J. Reprod. Fert.* **31**, 433–444

HUNTER, R.H.F. (1973a). Transport, migration and survival of spermatozoa in the female genital tract: species with intra-uterine deposition of semen. In *Sperm Transport, Survival and Fertilising Ability*, (E.S.E. Hafez and C. Thibault, Eds.). Paris, Colloques de l'Institut National de la Santé et de la Recherche Médicale

HUNTER, R.H.F. (1973b). Polyspermic fertilisation in pigs after tubal deposition of excessive numbers of spermatozoa. *J. exp. Zool.* **183**, 57–64

HUNTER, R.H.F. (1974). Chronological and cytological details of fertilisation and early embryonic development in the domestic pig, *Sus scrofa. Anat. Rec.* **178**, 169–186

HUNTER, R.H.F. (1975a). Physiological aspects of sperm transport in the domestic pig, *Sus scrofa.* I Semem deposition and cell transport. *Br. vet. J.* **131**, 565–573

HUNTER, R.H.F. (1975b). Physiological aspects of sperm transport in the domestic pig, *Sus scrofa.* I. Semen deposition and cell transport. *Br. vet. J.* **131**, 681–690

HUNTER, R.H.F. (1977). Function and malfunction of the Fallopian tubes in relation to gametes, embryos and hormones. *Eur. J. Obstet. Gynaec. Reprod. Biol.* **7**, 267–283

HUNTER, R.H.F. (1978). Intraperitoneal insemination, sperm transport and capacitation in the pig. *Anim. Reprod. Sci.* **1**, 167–179

HUNTER, R.H.F. (1980). *Physiology and Technology of Reproduction in Female Domestic Animals.* London and New York, Academic Press

HUNTER, R.H.F. (1981). Sperm transport and reservoirs in the pig oviduct in relation to the time of ovulation. *J. Reprod. Fert.* **63**, 109–117

HUNTER, R.H.F. and DZIUK, P.J. (1968). Sperm penetration of pig eggs in relation to the timing of ovulation and insemination. *J. Reprod. Fert.* **15**, 199–208

HUNTER, R.H.F. and HALL, J.P. (1974a). Capacitation of boar spermatozoa: synergism between uterine and tubal environments. *J. exp. Zool.* **188**, 203–214

HUNTER, R.H.F. and HALL, J.P. (1974b). Capacitation of boar spermatozoa: the influence of post-coital separation of the uterus and Fallopian tubes. *Anat. Rec.* **180**, 597–604

HUNTER, R.H.F. and LÉGLISE, P.C. (1971). Polyspermic fertilisation following tubal surgery in pigs, with particular reference to the role of the isthmus. *J. Reprod. Fert.* **24**, 233–246

ITO, S., KUDO, A. and NIWA, T. (1959). Studies on the normal oestrus in swine with special reference to proper time for service. *Annls Zootech., Série D, Suppl.* 105–107

KEYE, J.D. (1923). Periodic variations in spontaneous contractions of uterine muscle, in relation to the oestrous cycle and early pregnancy. *Bull. Johns Hopkins Hosp.* **34**, 60–63

KNIFTON, A. (1962). The response of the pig uterus to oxytocin at different stages in the oestrous cycle. *J. Pharmac. Pharmacol., Suppl.* **14**, 42T–43T

KVASNITSKY, A.V. (1959). Méthode fractionnée d'insemination artificielle des truies. *Annls Zootech., Série D, Suppl.* 43–58

LEE, F.C. (1928). The tubo-uterine junction in various animals. *Bull. Johns Hopkins Hosp.* **42**, 335–357

McKENZIE, F.F., MILLER, J.C. and BAUGUESS, L.C. (1938). The reproductive organs and semen of the boar. *Res. Bull. Mo. Exp. Stn*, No. 279

MANN, T., POLGE, C. and ROWSON, L.E.A. (1956). Participation of seminal plasma during the passage of spermatozoa in the female reproductive tract of the pig and horse. *J. Endocr.* **13**, 133–140

POLGE, C. (1956). Artificial insemination in pigs. *Vet. Rec.* **68**, 62–76

POLGE, C. (1978). Fertilisation in the pig and horse. *J. Reprod. Fert.* **54**, 461–470

POLGE, C., SALAMON, S. and WILMUT, I. (1970). Fertilising capacity of frozen boar semen following surgical insemination. *Vet. Rec.* **87**, 424–428

RIGBY, J.P. (1966). The persistence of spermatozoa at the utero-tubal junction of the sow. *J. Reprod. Fert.* **11**, 153–155

RODOLFO, A. (1934a). The physiology of reproduction in swine. II. Some observations on mating. *Philipp. J. Sci.* **55**, 13–18

RODOLFO, A. (1934b). The physiology of reproduction in swine. I. The semen of boars under different intensivenesses of mating. *Philipp. J. Sci.* **53**, 183–203

SMITH, J.C. and NALBANDOV, A.V. (1958). The rôle of hormones in the relaxation of the uterine portion of the cervix in swine. *Am. J. vet. Res.* **19**, 15–18

SZOLLOSI, D. (1967). Development of cortical granules and the cortical reaction in rat and hamster eggs. *Anat. Rec.* **159**, 431–446

SZOLLOSI, D. and HUNTER, R.H.F. (1973). Ultrastructural aspects of fertilisation in the domestic pig: sperm penetration and pronucleus formation. *J. Anat.* **116**, 181–206

SZOLLOSI, D. and HUNTER, R.H.F. (1978). The nature and occurrence of the acrosome reaction in spermatozoa of the domestic pig, *Sus scrofa*. *J. Anat.* **127**, 33–41

THIBAULT, C. (1959). Analyse de la fécondation de l'oeuf de la truie après accouplement ou insémination artificielle. *Annls Zootech., Série D, Suppl.* 165–177

4

ARTIFICIAL INSEMINATION

H.C.B. REED
Meat and Livestock Commission, Pig Breeding Centre, Thorpe Willoughby, Selby, North Yorkshire, UK

The techniques and application of artificial insemination (AI) have changed considerably since it was first advocated as a practical proposition by the Japanese workers, Ito *et al.* (1948) over thirty years ago. Although Polge (1956) described the techniques necessary for the field application of AI in some detail no real development of commercial services took place until Rutgers (1966) reported a marked increase in the use of AI in Holland (120 000 inseminations per year) following a bad outbreak of foot and mouth disease in 1962. In the last five years there has been a marked increase in the development of field AI services in Holland and certain other countries (Denmark and the Democratic and Federal Republics of Germany).

Functions of AI

It is important that any review of AI should include, if only briefly, some reference to the current reasons for AI development, which are reflected in the AI services provided, and also the future needs of the pig industry. These reasons, which are listed below, vary in importance from country to country.

Genetic improvement of livestock

In all species AI enables superior tested sires to be used to a much greater extent than would be possible with natural mating. In the case of pigs the primary function of the AI service is (a) to provide a wider variety of new genes for herds at 'nucleus' level thus enabling greater selection intensity to be practised, and (b) to produce replacement breeding stock and, in certain situations, slaughter generation stock in commercial herds.

Maintenance of high health status

If herds are to make genetic progress fresh genetic material has to be introduced but this immediately increases the risk of introducing disease into the herd. However, this risk can be minimized, if not eliminated, by the use of AI (Reed, 1978; Larsen *et al.*, 1978; Pursel *et al.*, 1980b).

Adoption of batch farrowing procedures

Use of batch weaning, a necessity in large herds, means batch serving of sows with periods of heavy use of boars. Supplementation of boar services with AI can prevent overuse of boars thus enabling services to be restricted to four or five per week for each boar. AI could have an important role in conjunction with oestrous synchronization if the use of compounds such as the progestagen allyl trenbolone (RU-2267, altrenogest) described by Martinat-Botte *et al.* (1980) and Pursel *et al.* (1980a) were to give acceptable results under commercial conditions.

Reduction of service costs

In countries with high density pig populations, notably Holland and Denmark, AI centres have been able to offer a service at a lower cost than for natural service (Willems, personal communication).

International exchange of genetic material

The high cost of transporting live animals long distances by air, coupled with the problem of health certification, have stimulated a limited but increasingly important use of AI. Liquid semen has been used in certain situations but where the semen has to be quarantined to meet health regulations or is likely to be exposed to extremes in ambient temperature, the use of frozen semen can be justified in spite of the higher costs and lowered conception rates associated with present procedures (L.A. Johnson, 1980).

Current AI usage

The development of AI in Europe has been reviewed by Willems (1977) and more recently on a world wide basis by Bonnadonna and Succi (1980) and Iritani (1980). Detailed figures for the use of frozen semen are also given in Iritani's paper. These reports together with those from more recent data are summarized in *Table 4.1*. Of the 34 countries for which figures are quoted, the largest use of AI is in the USSR (2.4 million animals), the German Democratic Republic (1.02 million) and China (970 000). However, AI is now also used extensively in Western Europe (1.66 million). These reports indicate that over 7 million animals are bred by AI annually; however, less than 20 000 inseminations are carried out with frozen semen.

Table 4.1 ESTIMATED WORLD PIG AI USAGE

Country	*Year*	*Total no. females served*	*Total no. inseminated*	*No. inseminated with frozen semen*
Australia[(a)]	1977/78	?	3000	Experimental
Austria[(a)]	1977/78	405742	50312	Experimental
Belgium[(a)]	1977/78	626000	11894	Experimental
Brazil[(b)]	1977	?	5950	?
Bulgaria[(b)]	1977	?	145193	?
Canada[(c)]	1978	?	15697	7000
China[(d)]	?	23000000	970000	Experimental
Czechoslovakia[(a)]	1977/78	525605	170000	Experimental
Denmark[(e)]	1979/80	1560000	525000	Experimental
Eire[(f)]	1979/80	?	3000	Experimental
Finland[(a)]	1977/78	122700	67945	Experimental
France[(a)]	1977/78	2400000	80000	800
Germany (DR)[(a)]	1977/78	1195900	1016515	Experimental
Germany (FR)[(a)]	1977/78	2309873	241034	Experimental
Great Britain	1979/80	1641600	55532	Experimental
Greece[(a)]	1977/78	600000	Exptl.	Experimental
Hungary[(a)]	1977/78	713000	232321	560
Japan[(a)]	1977/78	1850000	48000	Experimental
Korea[(a)]	1977/78	399000	78530	Experimental
Malaysia[(b)]	1977	?	3896	?
Netherlands[(g)]	1979/80	?	485922	Experimental
Norway[(a)]	1977/78	134000	65300	Experimental
Peru[(a)]	1977/78	352250	Exptl.	Experimental
Philippines[(a)]	1977/78	62500	62500	—
Poland[(b)]	1977	?	150431	?
Singapore[(a)]	1977/78	9370	7357	Experimental
Spain[(a)]	1977/78	1212000	35000	1000
Sweden[(h)]	1978/79	510000	20375	Experimental
Switzerland[(a)]	1977/78	?	17725	600
Thailand[(a)]	1977/78	2989751	6711	Experimental
Turkey[(b)]	1977	?	429361	?
USA[(a)]	1977/78	15000000	30000	9000
USSR[(b)]	1977	?	2400000	?
Yugoslavia[(a)]	1977/78	1406000	90696	Experimental

Source: [(a)]Iritani (1980); [(b)]Bonadonna and Succi (1980); [(c)]28th Report on AI in Canada (1978); [(d)]Peilieu *et al.* (1980); [(e)]Svineavl og produktion i Danmark (1980); [(f)]Smith, D.J., personal communication; [(g)]Willems, personal communication; [(h)]Johnson, E. (1980).

Type of AI service provided

Inseminator service

Initially virtually all inseminations outside Eastern Europe were carried out by technicians operating from AI centres or sub-centres, many of whom were simultaneously providing a service for cattle. In Western Europe the inseminator service is still the major one in use particularly in those countries with dense pig populations such as Denmark and Holland (Willems, 1977). Inseminations are increasingly being carried out mainly by specialist pig inseminators although half the countries have received support from the cattle AI services (Willems, 1977). However, in some countries inseminations are also carried out by self-employed inseminators.

Semen delivery service (SDS)

Since the early report of a Semen Delivery Service (SDS) in Britain in the middle 1960s (Melrose, Reed and Pratt, 1968), there has been a steady but slow expansion of this type of service under which semen is sent, on request, direct to producers who carry out their own inseminations. The advantages of such a service are:

(a) Liquid semen from a range of superior performance tested boars from all AI centres in a country can be made available on a nationwide basis instead of being restricted to the area covered by the centre's Inseminator Service.
(b) Two inseminations can be carried out at the optimum time in the oestrous period instead of one insemination at the time the inseminator happens to visit the farm. In such situations efficient producers can often achieve higher fertility results with inseminations carried out by themselves than with those performed by a trained inseminator from the AI centre.
(c) The risk of introducing disease into the herd is reduced because there is no movement of vehicles or personnel between farms as is the case with an Inseminator Service.
(d) The service is cheaper to operate because the potential for semen sales is nationwide and it eliminates the cost of labour and transport incurred by inseminators travelling to farms.

On the other hand, the difficulties in operating such a service cannot be overlooked, namely:

(a) Reliable transport facilities are essential to ensure that liquid semen reaches its destination within 24 hours.
(b) Users of the service need to be given adequate training. In some countries (Federal Republic of Germany, Denmark and Holland) AI centres stipulate that the producer must attend a short training course before being allowed to receive semen. This is not a requirement in the UK but one-day training courses are provided by certain centres.
(c) Producers lacking confidence can be reluctant to use the Semen Delivery Service but this can be overcome by training courses and demonstrations of the AI techniques.
(d) Arrangements for monitoring AI results are advisable to provide a means of assessing how effectively the service is being used by different herds.

'On farm' semen collection/insemination programmes

AI has been used on a 'within farm' basis for many years on the large state farms and combines in Eastern Europe but in recent years there has also been increased use by the large farms and breeding companies in other countries. Good fertility results can be obtained and little additional technical expertise is required in those situations where raw semen containing high spermatozoal numbers is used shortly after collection.

However, where it becomes necessary to dilute and store semen for some hours before use, some laboratory facilities are required which may be difficult for individual farms or breeding companies to justify financially. This problem has been overcome in Spain and the USA where AI centres provide diluent for organizations carrying out their own 'within farm' AI programmes.

Review of the technical aspects of AI in the light of field developments

HEALTH AND MANAGEMENT OF BOARS FOR AI

Health considerations

It is clear from earlier remarks that increased priority must be placed on the health status of the AI boar stud and on the conditions under which semen is collected, processed and distributed, by attention to the following factors:

(a) Screening of boars before entry and during a 28-day isolation period for those diseases for which there is a reliable diagnostic test supported by a satisfactory herd history. Such a list would include foot and mouth disease, swine vesicular disease (SVD), tuberculosis, brucellosis, leptospirosis, Aujeszky's disease and transmissible gastroenteritis (TGE). Screening of this nature is only possible when routine vaccination against the disease concerned is not practised.
(b) Introduction of boars into an isolation compound for one month prior to entry into the main stud and keeping the boars in the main stud strictly segregated from all other livestock.
(c) Maintenance of high standards of hygiene at semen collection.
(d) Addition of antibiotics to diluted semen to control non-specific bacteria which may be present in semen and which are sensitive to penicillin and streptomycin.

Methods of housing

Comparatively little attention has been paid to the specific needs of AI boar studs. The general principles covering the housing of sows are, of course, relevant to the boar but in those countries which experience high ambient temperatures it is particularly important that boars are kept in insulated housing as Wettemann *et al.* (1976) have shown that boars subjected to elevated temperatures (above 31 °C) showed impaired reproductive performance. More recently, Wettemann and Desjardins (1979) have suggested that the lowered male reproductive performance following heat stress is, in part, related to alterations in testicular endocrine function and in spermatid maturation.

Normally boars are kept singly but some AI centres have, for economy reasons, found it quite practical to keep two boars per pen. *Figure 4.1* shows a purpose-built house for 80 boars individually penned on a large AI centre in the Federal Republic of Germany.

Figure 4.1 Purpose-built house for 80 individually penned boars on a large AI Centre in the Federal Republic of Germany. By courtesy of Dr R. Hahn

Boar training

The great majority of boars can normally be trained for use in AI without difficulty as indicated by data for 394 boars which entered the two Meat and Livestock Commission (MLC) Pig Breeding Centres between 1964 and 1980 (*Table 4.2*). The percentage of boars successfully trained for AI under 10 months of age at entry (92%) was significantly higher ($P<0.001$) than boars 10–18 months at entry (70%). There was also a significant difference ($P<0.001$) in the percentage of 10–18 month old boars (70%) trained compared with the seven boars over 18 months of age at entry (0%). Significant differences between breeds were also noted, Large White boars being better than Landrace ($P<0.005$ for boars under 10 months of age at

Table 4.2 ANALYSIS OF BOARS TRAINED FOR AI AT DIFFERENT AGES AT ENTRY INTO AI CENTRE (1964–1980)

Breed	*Boars under 10 months at entry*			*Boars 10–18 months at entry*			*Boars over 18 months at entry*		
	Total no.	*No. trained*	*%*	*Total no.*	*No. trained*	*%*	*Total no.*	*No. trained*	*%*
Large White	171	167	98	9	9	100	1	–	0
Landrace	135	122	90	22	14	64	5	–	0
Hampshire	21	14	67	2	–	0	1	–	0
Other breeds (Welsh, Lacombe, BSB and Duroc)	23	20	87	4	3	75	–	–	–
TOTAL	350	323	92	37	26	70	7	–	0

Significance of differences in boar training: Boars (<10 months) versus boars (10–18 months), $P<0.001$; Boars (10–18 months) versus boars (>18 months), $P<0.001$; Large White boars (<10 months) versus Landrace boars (<10 months), $P<0.005$; Large White boars (10–18 months) versus Landrace boars (10–18 months), N.S.; Hampshire boars versus other breeds of boar, $P<0.05$.
From Meat and Livestock Commission.

Table 4.3 INCIDENCE OF BOARS WHICH MOUNTED AND EJACULATED AT FIRST INTRODUCTION TO DUMMY SOW (1964–1980)

Breed	*All boars trained for which data available*			*Trained boars, which were subsequently culled because of low libido*		
	Total no. trained	*No. trained at first attempt*	*%*	*Total no. trained*	*No. trained at first attempt*	*%*
Large White	161	113	70	5	3	60
Landrace	96	68	71	17[a]	9	53
Hampshire	13	4	31	2	–	0
Other breeds (Welsh, Lacombe, BSB and Duroc)	23	10	43	4	–	0
TOTAL	293	195	66	28	12	43

[a]Excludes six Landrace boars culled for low libido for which detailed training data not available.
Significance of differences in boars which mounted and ejaculated at first introduction to dummy sow:
Large White and Landrace v. all other breeds, $P<0.001$
Boars culled for low libido v. boars culled for other reasons, $P<0.05$.
From Meat and Livestock Commission.

entry). The Hampshire breed was significantly more difficult to train than the other breeds ($P<0.05$ for boars under 10 months of age at entry). MLC data in *Table 4.3* show that when boars were introduced to the dummy sow for the first time 195 boars (66%) mounted and ejaculated successfully. The Large White and Landrace boars were significantly better than the other breeds (Hampshire, Welsh, Lacombe, British Saddleback (BSB) and Duroc) in this respect ($P<0.001$). Of the trained boars, which were subsequently culled because of low libido, only 12 (43%) mounted and ejaculated at the first introduction to the dummy sow compared with 195 (66%) eventually culled for reasons other than low libido ($P<0.05$).

Collection technique

It is generally accepted that the most satisfactory method for collecting semen is by fixation of the penis by the hand method using a dummy sow as a teaser. However, the shape and size of dummy sows used varies widely as also does the semen collection site. Some AI centres favour a separate collection pen while others use a portable dummy sow in the boar's pen.

Collection frequency and semen production

This subject has been reviewed by du Mesnil du Buisson and Paquignon (1977). In our own experience the optimum collection frequency seems to be once every four or five days although du Mesnil du Buisson and Paquignon have suggested that semen collection can be increased to more than once or twice a week without markedly affecting the amount of spermatozoa collected from each boar. However, Swierstra and Dyck (1976) found that when ejaculates collected at 72 and 24 hour intervals were compared, the former ejaculates contained three times as many spermatozoa. It is clear that there is a considerable variation between boars in the number of spermatozoa produced, which is thought to be influenced mainly by the size of the testes (du Mesnil du Buisson and Paquignon, 1977). Hemsworth and Galloway (1979) have reported that spermatozoa numbers in the sperm-rich fraction of the ejaculate can be significantly increased in the short term by allowing the boar to have a false mount and thereafter restraining it in an adjacent pen for two minutes before collecting the semen.

SEMEN PRESERVATION

Liquid semen

Diluents in current use are listed in *Table 4.4* and recent comparative fertility results, where available, are summarized in *Table 4.5*. Egg yolk/glucose, one of the original diluents to be used in the field, is now only

Table 4.4 DILUENTS CURRENTLY USED FOR STORING BOAR SPERMATOZOA IN THE LIQUID STATE

Diluent	*Reference*
Egg yolk/Glucose	Aamdal and Hogset (1957)
IVT (Illinois Variable Temperature)	VanDemark and Sharma (1957); Modified for pigs by du Mesnil du Buisson and Jondet (1961)
Kiev (also known as Plishko, Varohm, Merck, GCHC, Guelph or EDTA)	Plishko (1965)
Kharkov (also known as Trilon B)	Serdiuk (1968)
BL-1	Pursel, Johnson and Schulman (1973)
SCK7 (Modified IVT)	Developed by Walls Meat Co. Ltd. (1975) (Walters, personal communication)
Zorlesco	Gottarai, Brunel and Zanelli (1980)

used in Belgium (Willems, 1977). IVT (Illinois Variable Temperature) diluent which was used extensively in Europe in the sixties and early seventies has been largely superseded by Kiev diluent. It can be seen in *Table 4.5* that both diluents give very similar fertility results (Larsson, Swensson and Wass, 1979; Kuiper and de Haas, 1980; Reed, MLC unpublished data) but the Kiev diluent is much simpler to prepare and, for this reason, is more widely used by commercial AI centres. There appears to be little difference in fertility between first and second day semen but L.A. Johnson *et al.* (1980) reported that third day semen gave a significantly lower farrowing rate. BL-1 diluent, which was first described by Pursel, Johnson and Schulman (1973), has been reported to give significantly lower fertility results than Kiev diluent (Paquignon *et al.*, 1980b; L.A. Johnson *et al.*, 1980). Where 1–3 day old Kiev diluted semen has been compared with 4–6 day old SCK7 diluted semen, better fertility results have been obtained with Kiev diluent (Swensson, 1977; Paquignon *et al.*, 1980b) although somewhat more encouraging results were reported for SCK7 diluent by Fischer-Pereira Cunha (1979). However this diluent has only been used to a very limited extent and there is no published evidence

Table 4.5 COMPARISON OF FERTILITY RESULTS OBTAINED WITH LIQUID SEMEN DILUENTS

Reference	*Diluent*	*Semen age*	*No. animals*	*Success rate (%)*	*Average no. pigs born*
Larsson *et al.* (1979)	IVT	0–12 hours	158	63.3[a]	11.4
		13–36 hours	455	64.8[a]	11.4
		37–60 hours	201	63.7[a]	11.9
		Total	814	64.3[a]	
	EDTA Glucose (Kiev)	0–12 hours	155	64.5[a]	11.0
		13–36 hours	421	60.8[a]	11.6
		37–60 hours	189	65.1[a]	11.2
		Total	765	62.6[a]	
Kuiper and de Haas (1980)	IVT	1st and 2nd day	1113	78.1[a]	10.92
	Kiev	1st and 2nd day	921	79.2[a]	10.84
Reed (1981)[c]	IVT	1–4 days	744	70.8[a]	9.9
	EDTA (Kiev)	1–4 days	694	71.6[a]	9.7
Paquignon *et al.* (1980b)		1st day	42	69.1[a]	10.2
		2nd day	88	76.2[a]	9.6
	Guelph (Kiev)	3rd day	48	68.8[a]	11.7
		4th day	48	81.3[a]	8.8
		5th day	18	61.2[a]	7.8
		Total	178	72.5[a]	10.3
		1st day	38	63.2[a]	11.4
	BL-1	2nd day	78	70.5[a]	10.8
		3rd day	35	60.0[a]	11.0
		Total	151	66.2[a]	11.0
Johnson *et al.* (1980)	Kiev	1–3 days	1280	69.3[a]	10.1
	BL-1	1–3 days	1283	60.5[a]	9.8
		1st day	825	70.2[a]	10.4
	Kiev & BL-1	2nd day	891	65.9[a]	9.8
		3rd day	847	58.7[a]	9.5
Paquignon *et al.* (1980b)		1st day	27	66.7[b]	
	BL-1	2nd day	51	70.6[b]	
		3rd day	16	81.2[b]	
		Total	94	71.3[b]	
		4th day	27	81.5[b]	
		5th day	27	48.2[b]	
	SCK7	6th day	31	61.3[b]	
		7th day	19	63.2[b]	
		Total	104	63.5[b]	
Fischer-Pereira Cunha (1979)		1st day	189	62.96[a]	10.3
	Kiev	2nd day	201	64.18[a]	9.9
		Total	390	63.69[a]	10.1
		1st day	111	69.36[a]	9.9
		2nd day	121	66.11[a]	9.1
	SCK7	3rd day	110	57.27[a]	9.7
		4th day	129	68.99[a]	9.6
		5th day	91	59.34[a]	8.4
		Total	562	64.59[a]	9.4
Swensson (1977)	IVT	1–3 days	124	65.3[a]	10.9
	SCK7	5–6 days	116	44.8[a]	9.1
	IVT	1–3 days	504	88.3[b]	
		4th day	143	83[b]	
	SCK7	5th day	170	82[b]	
		6th day	189	80[b]	

[a]Percentage farrowing rate.
[b]Non-return at 54 days.
[c]MLC unpublished data.

to demonstrate its superiority over other diluents in terms of fertility results with semen older than three days.

In a recent preliminary report Gottarai, Brunel and Zanelli (1980) gave details of a new diluent which could be used for storing semen for 8–12 days but no fertility data were given to support this claim. On the basis of the above reports and experience with the MLC AI services, Kiev would seem to be the diluent of choice for use under commercial conditions. Semen extended in this diluent can be used for up to three days with acceptable fertility results and use of semen stored for up to five or six days can be justified in export situations or where nucleus breeders need to nominate particular boars even though the use of such semen is likely to be associated with lower fertility.

Frozen semen

Numerous reports have been published since Polge, Salamon and Wilmut (1970) first obtained successful fertility results with frozen semen. Various methods developed by different groups of workers have been summarized by Graham, Crabo and Pace (1978), Larsson (1978), Pursel (1979) and L.A. Johnson, (1980). The main steps in the different procedures described are somewhat similar in that they all involve a period of equilibration, concentration by centrifugation, addition to diluent and glycerol at some stage prior to freezing in pellets or straws. Recent fertility results obtained by different workers for the main methods described were summarized by L.A. Johnson, (1980) as shown in *Table 4.6*. Reasonable

Table 4.6 SUMMARY OF FERTILITY RESULTS OBTAINED WITH DIFFERENT IMPROVED FREEZING PROCEDURES

Freezing procedure	*No. animals*	*% pregnant*	*Litter size*
Beltsville method	378	53	8.2
French procedure	293	58	8.8
German method	805	65	8.5

From Johnson, L.A. (1980)

results have been obtained with all the methods listed, but the Beltsville method (Pursel and Johnson, 1975) seems to have been favoured by the commercial AI organizations using frozen semen under field conditions in the USA, Canada and Spain. The processing and freezing procedures are summarized below:

Equilibration at 20° for 2 hours
↓
Removal of seminal plasma by centrifugation
↓
Addition of BF5 diluent
↓
Cool to 5 °C
↓
Addition of further BF5 diluent and glycerol
↓
Freeze 0.2 ml pellets on dry ice
↓
Liquid nitrogen

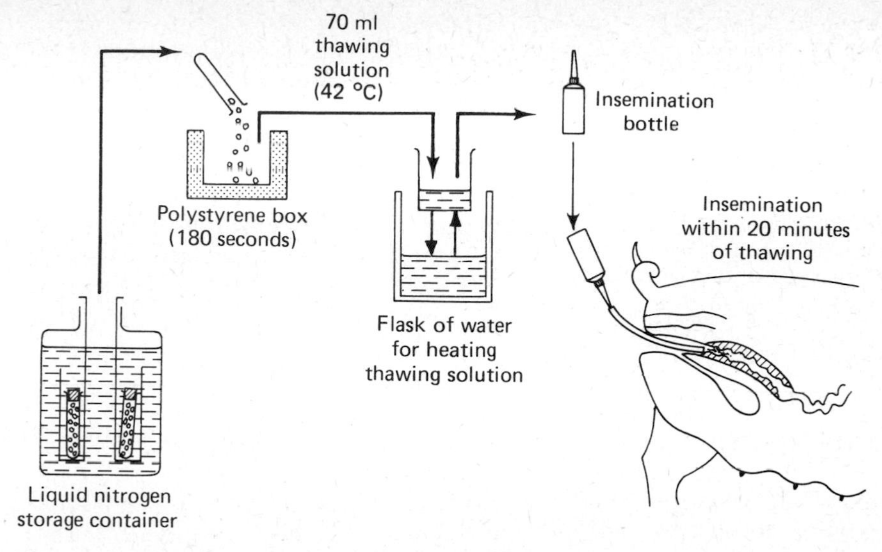

Figure 4.2 Thawing procedure (based on Beltsville technique)

Pursel, Schulman and Johnson (1978) showed that Orvus ES Paste, which is one of the components of BF5 diluent, has a beneficial effect on semen preservation. The thawing procedure which involves the addition of 70 ml Beltsville thaw solution at 42 °C, is summarized in *Figure 4.2*. The Beltsville procedure has also been used in Britain for imported and exported semen. Interim results for these inseminations together with other field results for which Meat and Livestock Commission has been responsible are summarized in *Table 4.7*. It is of interest to note that there is comparatively little difference in fertility between pigs inseminated by full-time inseminators and those by producers doing their own inseminations. The results are also in agreement with the conception rate of 39% (and 8.8 pigs born alive) (321 animals) obtained by the Animal Breeding Research Organisation, Edinburgh with frozen semen imported from

Table 4.7 FERTILITY RESULTS OBTAINED FROM FROZEN SEMEN 1975–1980 (a) IN THE UK; (b) EXPORTED

Category	*No. breeders*	*No. pigs inseminated*	*No. pregnant*	*Conception rate (%)*	*Average no. pigs born*
(a) MLC SEMEN USED IN UK					
Pigs inseminated on farms by MLC inseminators	8	67	30	45	6.1
Pigs inseminated on farms by breeders	4	34	14	41	6.9
(b) EXPORTED/IMPORTED SEMEN					
Pigs inseminated with MLC exported semen by overseas breeders	5	160	54	34	6.6
Pigs inseminated with semen imported from USA by UK breeders	7	178	53	30	7.4

North America (Will, personal communication). They also confirm the findings of Johnson *et al.* (1980) who found that conception rates and mean number of pigs born alive achieved with frozen semen (47% and 7.1 respectively) were considerably lower than those obtained with liquid semen (79% and 9.9) from the same boars. Some of the factors influencing conception rates of pigs inseminated with frozen semen under commercial conditions have been evaluated by Paquignon *et al.* (1980a).

The reports previously listed, coupled with our own experience, have shown that frozen semen is unlikely to be used on a widespread commercial scale in the foreseeable future because of the following shortcomings:

(a) The high spermatozoal dose (5×10^9 spermatozoa per insemination) needed to achieve reasonable fertility under field conditions markedly increases the cost of providing a frozen semen service compared with that for liquid semen.
(b) Laboratory assessments of semen quality are of very limited value in predicting a boar's potential fertility.
(c) There is a considerable variation in the freezing ability between boars.
(d) The semen processing procedure is complicated and time-consuming compared with the procedures used for liquid semen.
(e) Producers carrying out their own inseminations need to have ready access to liquid nitrogen supplies.
(f) The timing of insemination(s) in relation to ovulation and hence oestrus detection is very much more critical if acceptable fertility results are to be obtained (Larsen, 1976).
(g) Average conception rates and litter sizes achieved with frozen semen are lower than those obtained with liquid semen.

OESTRUS DETECTION PROCEDURES

A positive response to the back pressure test in the absence of the boar has, for many years, been regarded as the optimum method for identifying the correct time for a single insemination within the oestrus. The difficulty of eliciting a positive response in all animals to the back pressure test in the absence of the boar has led to the development of synthetic boar odour aerosol to elicit the physical symptoms of oestrus in such animals (Melrose, Reed and Patterson, 1971; Reed, Melrose and Patterson, 1974). These aerosol packs† have been extensively used in those countries with small herds or where wide use of an Inseminator Service has been made.

Schilling and Rostel (1964) reported that the lowest pH values of vaginal mucus secretion were found in mid-oestrus. An instrument‡ relying on this principle has been marketed by a commercial organization who claim that it can be used to accurately pinpoint the optimum time for insemination. However, as no data appear to have been published on the use of this instrument it is not possible to evaluate the effectiveness of the technique as an aid in timing inseminations correctly.

†'Boar Mate', Antec International Ltd., Sudbury, England.
‡'Walsmeta', Masterbreeders (Livestock Development) Ltd., Basingstoke, England

Under Semen Delivery Service conditions the use of a boar for heat detection is the method of choice because semen must be ordered sufficiently early to reach the farm to enable the inseminations to be carried out at the correct time. Thereafter two inseminations can be carried out at an interval of 8–16 hours without further recourse to oestrus detection.

Table 4.8 RECOMMENDED PROCEDURE FOR OESTRUS DETECTION

Type of AI service	*Method of oestrus detection*	
	Boar available	*No boar available*
'Inseminator' Service (One insemination/oestrus)	Boar to detect oestrus onset. Back pressure in absence of boar to time insemination.	Back Pressure Test + Boar Odour Aerosol
'Semen Delivery' Service (Two inseminations/oestrus)	Boar to detect oestrus onset.	Back Pressure Test + Boar Odour Aerosol

The most reliable routine for oestrus detection depends, therefore, on the availability of a boar and recommendations for use in relation to the type of AI service available and frequency of oestrus checking as indicated in *Table 4.8.*

INSEMINATION TIMING

Polge (1969) stated that the principal factor governing fertility and litter size was the relationship between time of insemination (or mating) and the time of ovulation during oestrus. He reported that ovulation occurred between 36 and 50 hours after the onset of heat and also pointed out that injection of human chorionic gonadotrophin (HCG) at around the onset of oestrus would result in ovulation 40–42 hours later. Since then Niswender, Reichert and Zimmerman (1970) have shown that the LH peak coincided with the observed onset of oestrus as measured by boar acceptance. However, recent work by Foxcroft (personal communication) suggests that

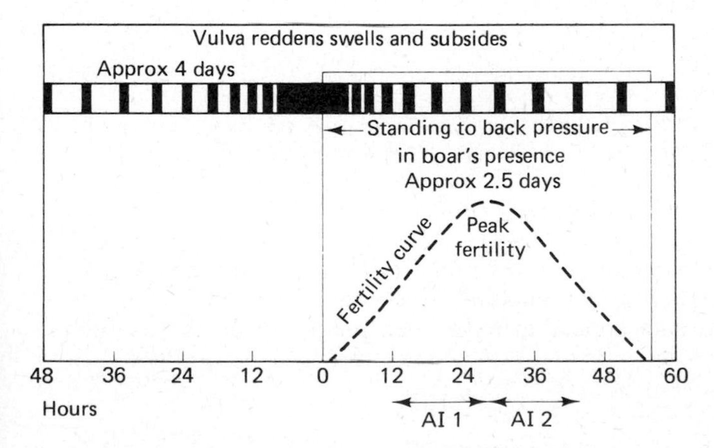

Figure 4.3 Recommended times for insemination

the LH surge might vary from as much as 12 hours before to 12 hours after the onset of boar acceptance. This would seem to suggest that, where possible, there is a strong case for carrying out more than one insemination in each oestrus, even though Boender (1966) has shown that inseminations carried out very early and very late in oestrus result in low fertility. *Figure 4.3* shows the timing recommended to producers using the MLC Semen Delivery Service. In practice this is governed by the method of semen dispatch to the farm as indicated in *Table 4.9*. It should be noted that the interval between inseminations for a postal dispatch is likely to be only 7–8 hours (09.00 and 16.30 hours on day 2) compared with 16 hours for a rail dispatch (16.30 hours on day 1 and 09.00 hours on day 2) because postal

Table 4.9 TIMES CURRENTLY RECOMMENDED FOR INSEMINATION OF PIGS IN HERDS USING MLC SEMEN DELIVERY SERVICE IN BRITAIN

Animal first stands to boar	*Method of semen dispatch*	*First insemination*	*Second insemination*
Morning Check (Day 1)	Rail Collected from AI Centre	p.m. Day 1	a.m. Day 2
	Post	a.m. Day 2	p.m. Day 2
Afternoon Check (Day 1)	Collected from AI Centre	p.m. Day 1	a.m. Day 2
	Rail Post	a.m. Day 2	p.m. Day 2

Table 4.10 FERTILITY RESULTS RELATED TO METHOD OF SEMEN DISPATCH AND AGE OF SEMEN AT DISPATCH (SDS 1977–79)

Dispatch method	*Semen age at dispatch*	*No. known results*	*No. pregnant*	*Conception rate (%)*	*Average no. pigs born*
Collected from AI Centre	1st day	1902	1507	79	10.4
	2nd day	253	199	79	10.1
	3rd day	13	11	85	11.8
Rail	1st day	11 107	8164	74	10.1
	2nd day	1805	1287	71	10.2
	3rd day	52	34	65	9.6
Post	1st day	9598	6708	70	9.6
	2nd day	415	267	64	9.7
	3rd day	–	–	–	–

Statistical analysis based upon data for semen ages at dispatch Days 1 and 2 only.
Significance of differences in % success rate:
'Dispatch Day 1' versus 'Dispatch Day 2', $P<0.01$
'Semen collected' versus 'Dispatch by rail', $P<0.001$
'Dispatch by rail' versus 'Dispatch by post', $P<0.001$
From Meat and Livestock Commission.

dispatches take 24 hours to reach their destination compared with arrival the same day for the other two dispatch categories. Data for the MLC Semen Delivery Service are shown in *Table 4.10*. Although there were significant differences for semen dispatched by the different methods ($P<0.001$ for semen collected versus rail dispatch and $P<0.001$ for rail dispatch versus postal dispatch), it cannot be concluded that insemination timing was the sole reason for the lower fertility because fertility levels also seem to be influenced by semen age at dispatch ($P<0.01$ for day 1 versus day 2 semen) and also possibly by month of the year (see *Figure 4.6*). There is also a need to determine the influence of 'within herd' effects

Table 4.11 COMPARISON OF FERTILITY RESULTS FOR SOWS INSEMINATED TWICE AND THREE TIMES IN EACH OESTRUS (INTERIM RESULTS)

Duration of oestrus (hours)	*No. inseminations per oestrus*	*No. sows*	*No. unknown results*	*No. known results*	*No. pregnant*	*Conception rate (%)*	*No. pigs born*	*No. litters*	*Average no. pigs born*
>40	Triple AI (Trial)	25	–	25	25	100	295	24	12.3
	Double AI (Control)	25	–	25	23	92	237	22	10.8
<40	Double (Non trial AI)	82	2	80	66	83	659	65	10.1
Total		132	2	130	114	88	1191	111	10.7

Significance of differences in % conception rate:
- Triple AI versus Double AI (>40 hr oestrus) N.S.
- Triple AI versus Double AI N.S.

Significance of differences in litter size:
- Triple AI versus Double AI (>40 hr oestrus) N.S.
- Triple AI versus Double AI $P<0.01$

From Meat and Livestock Commission

because different herds not only use different methods of semen dispatch and oestrus detection but also sometimes anticipate the onset of oestrus and do not necessarily use the semen on the same day as its arrival.

Under field conditions in Britain some producers mate animals, which are in oestrus for three days or more, a third time. *Table 4.11* shows the interim results of a small MLC trial carried out on one farm with very good records where alternate sows in oestrus for more than 30–40 hours were inseminated twice or three times at intervals varying from 8–16 hours. Fertility results are also given for sows in oestrus less than 40 hours, which were artificially inseminated over the same period. No significant differences in conception rate or litter size were observed between the two sow groups in oestrus for over 40 hours but if all sows are included, those inseminated three times had a significantly better litter size ($P<0.01$) than sows inseminated only twice during oestrus. Further data are required before definite conclusions can be drawn but these results are in agreement with those found by Tilton (personal communication). Although Hunter (1975) has reported that an adequate sperm reservoir can be maintained in the utero-tubal junction for at least 24 hours, it is possible that an adequate supply of spermatozoa cannot be maintained for the same period under commercial AI conditions using limited spermatozoal numbers ($1–3 \times 10^9$) compared with natural mating conditions involving higher spermatozoal numbers.

There is also evidence that mixed semen from different boars results in better fertility (Pacova and Dupal, 1978) but this is unlikely to be acceptable to producers using AI to produce replacement breeding stock. However this procedure could have some application in large commercial herds making extensive use of AI for the production of slaughter generation pigs.

INSEMINATION TECHNIQUE

The spermatozoal dose and inseminate volume used for liquid semen under commercial conditions varies from 1×10^9 to 3×10^9 and 50 to 150 ml respectively. In some countries additional diluent is added to the diluted semen prior to insemination. It is now also usual for semen to be transported in disposable polythene bottles with spouted tops, the end of which is cut off prior to being attached to the catheter. Different countries tend to favour different types of catheter but the version most widely used is the spiral catheter evolved by Melrose and O'Hagan (1961). It is particularly favoured by producers carrying out their own inseminations because they can readily recognize when it is in the correct position for insemination from its characteristic 'locking' action in the posterior folds of the cervix. Attempts are being made to produce a disposable spiral catheter at an economic price.

FERTILITY LEVELS

Conception rates reported for AI services from different countries vary widely (60–90%) depending on the criteria used for measuring fertility, type of service provided, management and size of herd, number of inseminations per oestrus and sow parity. The detailed recording and monitoring carried out in herds using the MLC Semen Delivery Service has allowed an assessment of the factors that appear to influence field fertility results. The main factors are summarized in *Tables 4.12* to *4.14* and *Figures 4.4* to *4.6*.

In *Figure 4.4* it can be seen that there was a marked improvement in the average conception rate achieved during the first seven years of the service

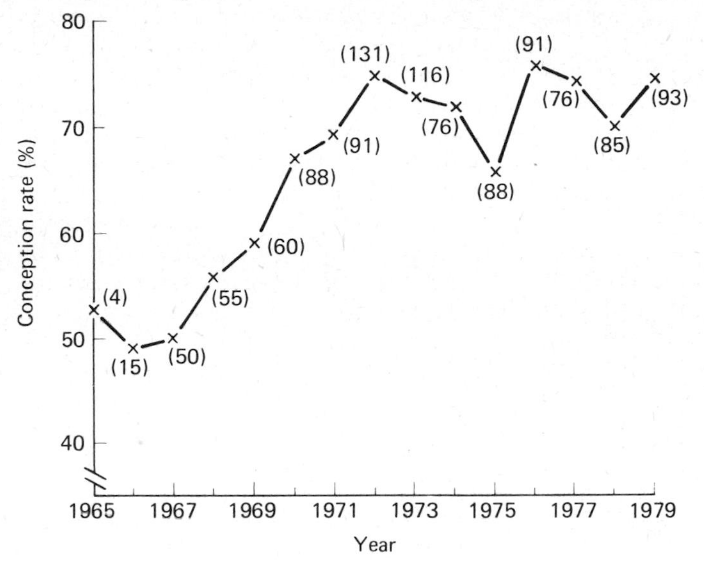

Figure 4.4 Mean annual percentage conception rate for Semen Delivery Service 1965–1979 (known results in hundreds). From Meat and Livestock Commission

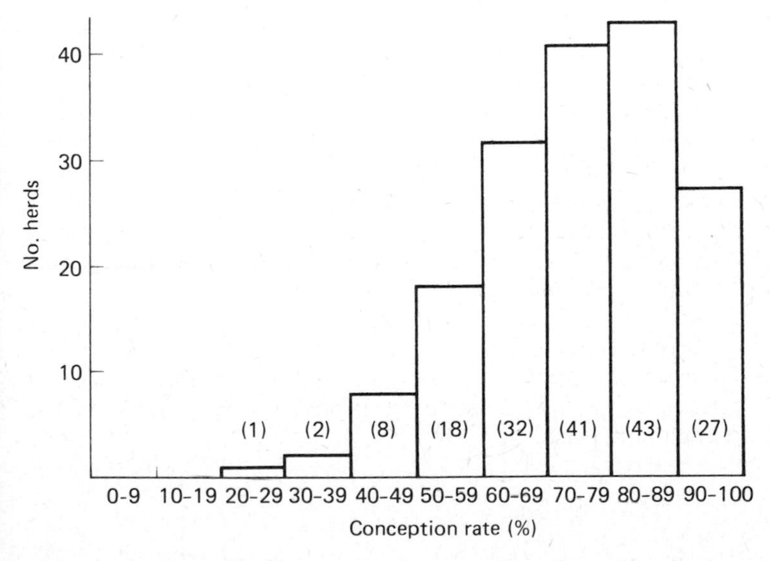

Figure 4.5 Variation in annual herd percentage conception rate levels (based on herds with over 10 known results for SDS 1979–1980). From Meat and Livestock Commission

Table 4.12 HERD CATEGORIES USING THE MLC SEMEN DELIVERY SERVICE (1979–1980)

Herd category	*No. herds*	*No. semen doses dispatched*	*No. known results*	*No. pregnant*	*Conception rate (%)*	*Average no. pigs born*
Nucleus and reserve nucleus herds	51	1098	550	369	67	10.2
Breeding companies	25	590	363	251	69	9.9
Commercial herds	935	13171	7848	5927	76	10.2
TOTAL	1011	14859	8761	6547	75	10.2

Difference in % conception rate for commercial herds versus other herd categories is significant (P<0.001)
From Meat and Livestock Commission.

Table 4.13 FERTILITY RELATED TO SEMEN USAGE (SDS 1979–80)

Herd annual semen usage (doses)	*No. herds*	*No. doses dispatched*	*No. known results*	*No. pregnant*	*Conception rate* (%)	*Average no. pigs born*
1–4	517	1151	686	458	67	10.2
5–9	195	1260	710	494	70	10.4
10–19	141	1906	1133	813	72	10.2
20–49	102	3204	1763	1287	73	10.0
50–99	32	2204	1346	1036	77	10.2
100–249	16	2555	1394	1087	78	10.5
250 or over	8	2579	1729	1372	79	9.7
TOTAL	1011	14859	8761	6547	75	10.2

Increasing % conception rate with usage is highly significant (P<0.001).
From Meat and Livestock Commission.

Table 4.14 FERTILITY RELATED TO HERD SIZE (SDS 1979–80)

Herd size (sows/gilts of breeding age)	*No. herds*	*No. known results*	*No. pregnant*	*Conception rate* (%)	*Average no. pigs born*
1–4	234	425	248	58	10.5
5–19	116	324	215	66	10.7
20–49	138	760	571	75	10.4
50–99	160	1304	999	77	10.3
100–249	203	2713	2053	76	10.4
249–499	40	924	689	75	10.1
500 or over	19	2063	1588	77	9.5
Herd size unknown	101	248	184	74	10.0
TOTAL	1011	8761	6547	75	10.2

Herds with under 20 sows/gilts have significantly lower % conception rate than herds with over 20 sows/gilts (P<0.001).
From Meat and Livestock Commission.

Table 4.15 FERTILITY RELATED TO PARITY (SDS 1979–80)

Parity	*No. known results*	*No. pregnant*	*Conception rate* (%)	*Average no. pigs born*
Sows	5532	4375	79	10.4
Gilts	1136	787	69	9.0

Sows have significantly higher % conception rates than gilts (P<0.001)
From Meat and Livestock Commission.

(1965–72) but little improvement has been obtained in the last seven years. *Table 4.12* shows that commercial herds, which accounted for 89% of the semen demand, achieved better farrowing rates than nucleus and breeding company herds ($P<0.001$). The distribution of herd fertility levels for those herds with more than 10 known results is shown in histogram form in *Figure 4.5*. It can be seen that a high proportion of herds achieve conception rates in excess of 70%. Herd conception rates, as one would expect, are related to AI usage (*Table 4.13*). The 56 herds (6%) using over 50 doses of semen per annum account for 49% of the semen uptake—an important factor when considering priorities for field advisory work. *Table 4.14* shows that herds with under 20 sows and gilts of breeding age had significantly lower fertility than herds with over 20 animals of breeding age ($P<0.001$). In Britain producers carrying out their own inseminations are sometimes deterred from using artificial insemination on maiden gilts because the technique is said to be more difficult and associated with lower fertility than with sows. MLC data (*Table 4.15*) show that the average conception rate and total number of pigs born was 79% and 10.4 pigs for sows and 69% and 9.0 pigs for gilts respectively. Although the differences were significant ($P<0.001$), producers are still justified in using a semen delivery service for maiden gilts particularly if the objective is to produce replacement breeding stock.

It has already been mentioned that high ambient temperatures can have an adverse effect on boar fertility. In addition to this factor, semen used in the Semen Delivery Service can be in transit for up to 24 hours and stored for up to another 48 hours under varying conditons on farms. Even in temperate countries such as Britain the dispatch of semen by public transport can result in semen being exposed to a wide range of environmental temperatures even though it is packed in expanded polystyrene boxes. Liquid semen diluents in current use such as IVT and Kiev do not give ideal protection to semen under wide variations in ambient temperature. Analysis of ten years MLC data for the Semen Delivery Service showed that there is a small but real underlying seasonal trend in mean conception rate and litter size numbers although there is a considerable

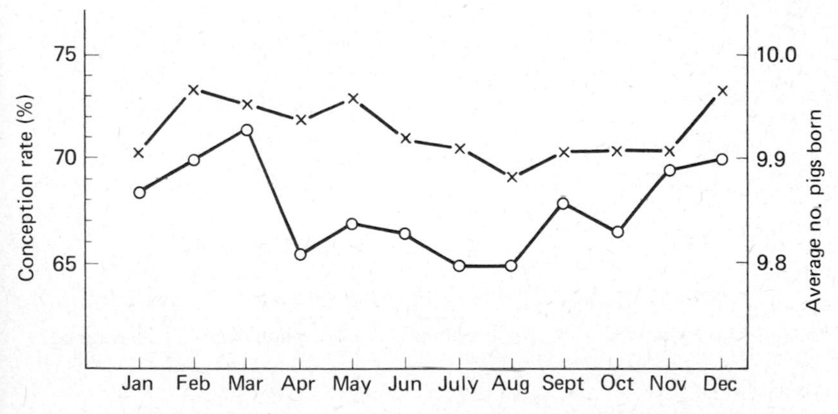

Figure 4.6 Mean seasonal variation in percentage conception rate (×——×) and average number of pigs born (○——○) for SDS 1970–1979 inclusive (92 000 results). From Meat and Livestock Commission

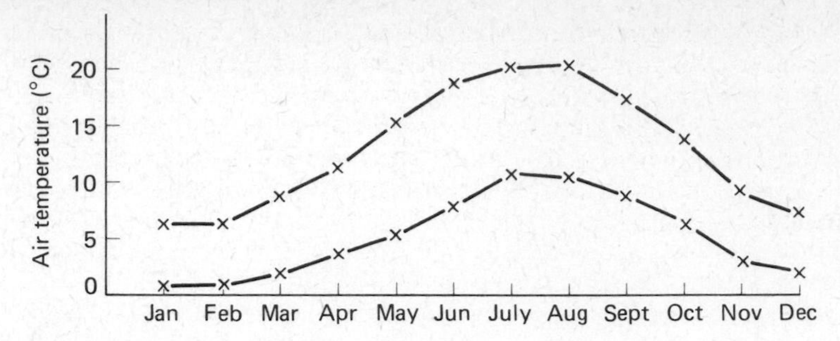

Figure 4.7 Average daily mean maximum and minimum air temperatures recorded by Meteorological Office at Cawood, near Selby, 1970–1979 inclusive

variation between individual years. In general there is a tendency for conception rates to decrease as temperature increases in the summer months and to increase as temperature decreases in the winter months (*Figure 4.6*). However the range of average daily mean maximum and minimum air temperatures recorded near the MLC Centre was small (*Figure 4.7*) compared with countries experiencing greater extremes in climate.

TRAINING, ADVISORY AND MONITORING PROCEDURES NECESSARY FOR PRODUCERS USING A SEMEN DELIVERY SERVICE

In those situations where AI organizations provide Semen Delivery Service facilities for producers, adequate education, training in techniques and monitoring facilities must also be available to ensure satisfactory fertility results irrespective of whether liquid or frozen semen is used. Three important aspects are discussed below.

Provision of explanatory literature

All prospective users should be provided with clear instructions covering the administrative and technical procedures involved to enable them to decide if they wish to use the service available.

Provision of training facilities

The type of training provided varies. Some organizations make it a condition that producers undergo a short training course usually of one day (North America and Denmark). Other organizations prefer producers to use the Inseminator Service before using the Semen Delivery Service (Holland and Bavaria, Federal Republic of Germany). In Britain, the MLC have organized one-day group training courses, given individual 'on farm' training for the large herds or provided training sessions at AI centres, but course attendance is not a pre-requirement for use of the service. Results from the MLC Semen Delivery Service would suggest that 'on farm' training or attendance at a one-day training course is quite sufficient for the producer of average ability to acquire the necessary confidence and experience to use AI satisfactorily. However it is essential that farm personnel, who will do the actual technical work, receive the training rather than the owner. The use of frozen semen does not require a longer training period but strict attention to detail especially the timing of the inseminations is essential.

Monitoring AI fertility results

Monitoring fertility results in herds making regular use of the service is essential. The availability of computer facilities permits immediate access to the appropriate data on a regular basis. A permanent herd history record for regular AI users should also be maintained at the centre. In the MLC service a report is completed at the time of the first herd visit to show herd size and herd breeding policy, personnel involved, heat detection, natural service and insemination procedures. A confidential assessment of the herd's capabilities for the use of AI and recommendations, if any, are also recorded. This record is updated at subsequent visits which, in the case of regular users (over 20 doses per annum), is made about once per year.

Some AI centres in Britain and Holland consider it is also advisable that progeny of AI boars be monitored for developmental abnormalities. Where the incidence of an abnormality becomes unacceptably high the boar is culled. Atresia ani, scrotal hernias and congenital splay legs are examples of the more common abnormalities observed. Such monitoring procedures can, however, be carried out more effectively through an Inseminator Service where AI personnel regularly visit the herds of AI users, than through a Semen Delivery Service.

Future developments

The high costs of the inseminators' time and travelling coupled with health considerations have favoured the development of a Semen Delivery Service under British conditions and more recently in certain European countries. However in some countries (USA) the lack of suitable transport facilities has ruled out the possibility of any widespread use of a Semen Delivery Service unless frozen semen can be used. In areas of dense pig population where mileage travelled per insemination is low, an Inseminator Service with a single insemination is used routinely, but in some situations technicians inseminate at the time of a visit and leave a second dose for the producer to inseminate later, thereby giving the fertility advantage from a double insemination. This practice would become more attractive if disposable insemination equipment were inexpensive and readily available. However specific guidance could be given to all producers on oestrus detection frequency and the optimum number and timing of inseminations particularly in large herds with intensively managed sows where accurate oestrus detection can be difficult. The advice given has often to be varied according to such individual herd circumstances.

Further improvements in semen storage techniques would enable a more efficient AI service to be provided for producers. In the case of liquid semen, the Semen Delivery Service would have a much wider appeal if diluted semen could be used for up to seven days without any drop in fertility. Semen could then be dispatched on a weekly basis thereby reducing the cost and possible inconvenience of semen transport.

It has already been established that there is a use for frozen semen in import/export situations in spite of the high costs associated with current techniques. These costs would be considerably reduced if the spermatozoa

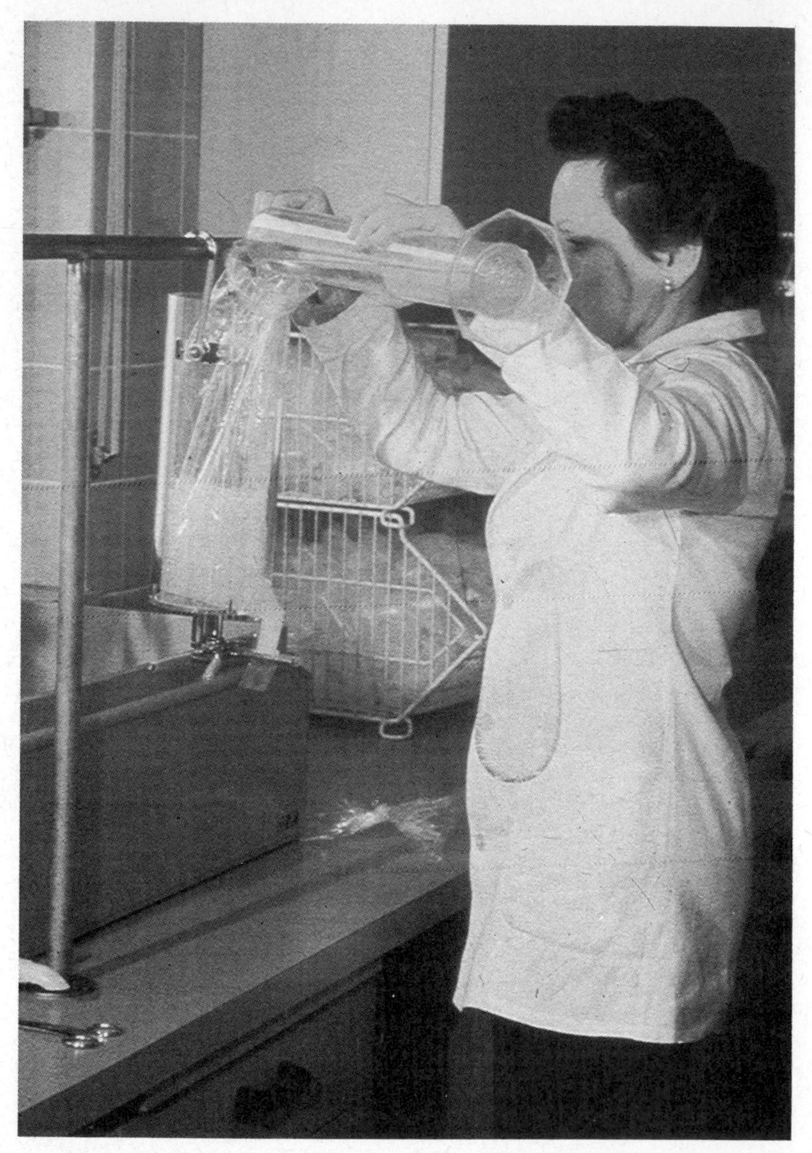

Figure 4.8 Dilution of boar semen with Kiev extender in plastic bag also used for the collection of semen. By courtesy of Dr R. Hahn

numbers/dose could be considerably reduced without lowering fertility and if poor 'freezer' boars could be quickly identified from laboratory assessments of semen quality.

The marked expansion of AI in certain countries to the stage where a limited number of AI centres are now using more than 100 000 doses/year has provided a stimulus to streamline various AI centre procedures, especially semen collection, dilution and dispatch techniques. A good example of this trend is at the Neustadt AI centre in the Bavarian area of Germany where diluent is added to semen in the same plastic bag in which the semen is collected (*Figure 4.8*). More attention will also have to be paid to the design of large scale boar accommodation and overall design of AI centres in view of the increasing importance of health. In the latter respect AI centres in most countries are already subject to certain government AI regulations. In addition to using separate staff, health control would also be improved if premises used by field technicians were separate from the boar stud and the facilities used for collection and dilution of the semen. Close control of AI centres in this way may help to stimulate AI demand when the European Economic Community introduces the swine fever eradication programme in mid-1981. Independent quarantine facilities for semen will also become necessary if the demand for frozen semen should develop. However the main factor influencing the type and extent of AI services provided in the future will undoubtedly be the cost of providing such services.

References

AAMDAL, J. and HOGSET, I. (1957). Artificial insemination in swine. *J. Am. vet. med. Ass.* **131**, 59–64

BOENDER, J. (1966). The development of AI in pigs in the Netherlands and the storage of boar semen. *Wld Rev. Anim. Prod.*, **II** (*Special Issue*), 29–44

BONADONNA, T. and SUCCI, G. (1980). Artificial insemination in the world. *Proc. 9th. Int. Congr. Anim. Reprod. A.I., Madrid*, pp. 1–15

DU MESNIL DU BUISSON, F. and JONDET, R. (1961). Utilisation du CO_2 dans l'insemination porcine. *Proc. 4th Int. Congr. Anim. Reprod., The Hague* **4**, 822–827

DU MESNIL DU BUISSON, F. and PAQUIGNON, M. (1977). Production du sperme de verrat—utilisation par insemination artificielle. *Proc. 28th Eur. Ass. Anim. Prod., Brussels*, P/3.01

FISCHER-PEREIRA CUNHA, R.A. (1979). [Comparative trials on artificial insemination in the pig with particular reference to the SCK7 diluent]. Thesis. Freie Universitat Berlin German Federal Republic. (Cited by *Pig News and Information* (1980) **1**, 276)

GOTTARAI, L., BRUNEL, L. and ZANELLI, L. (1980). New dilution media for artificial insemination in pigs. *Proc. 9th Int. Congr. Anim. Reprod. A.I., Madrid* **3**, 275

GRAHAM, E.F., CRABO, B.G. and PACE, M.M. (1978). Current status of semen preservation in the ram, boar and stallion. *J. Anim. Sci.* **47**, *Suppl. II*, 80–119

HEMSWORTH, P.H. and GALLOWAY, D.B. (1979). The effect of sexual stimulation on the sperm output of the domestic boar. *Anim. Reprod. Sci.* **2**, 387–394

HUNTER, R.F. (1975). Physiological aspects of sperm transport in the domestic pig, *Sus scrofa*. II. Regulation, survival and fate of cells. *Br. vet. J.* **131**, 681–689

IRITANI, A. (1980). Problems of freezing spermatozoa of different species. *Proc. 9th Int. Congr. Anim. Reprod. A.I., Madrid* **1**, 115–132

ITO, S., NIWA, T., KUDO, A. and MIZUHO, A. (1948). Studies on the artificial insemination in swine. *Res. Bull. Chiba zootech. Exp. Stn* **55**, 1–74

JOHNSON, E. (1979). Svinsemin i Denmark. *Svinskotsel* **69**, 20

JOHNSON, E. (1980). [Pig AI in 1979. Good conception rates and an increase in the number of sows inseminated]. *Svinskotsel* **70**, 36–37. (Cited by *Anim. Breed. Abstr.* (1980). **48**, 813)

JOHNSON, L.A. (1980). Artificial insemination of swine: fertility with frozen boar semen. *Proc. 5th Int. Pig Vet. Soc., Copenhagen*, 37

JOHNSON, L.A., AALBERS, J.G., WILLEMS, C.M.T. and RADEMAKER, J.H.M. (1980). Fertility of boar semen stored in BL-1 and Kiev extenders at 18 °C for three days. *Proc. 5th Int. Pig Vet. Soc., Copenhagen*, 33

JOHNSON, L.A., AALBERS, J.G., WILLEMS, C.M.T. and SYBESMA, W. (1979). Effectiveness of fresh and frozen boar semen under practical conditions. *J. Anim. Sci.* **49, ***Suppl.* **1**, 306

KUIPER, C.J. and DE HAAS, A.J. (1980). Some aspects of artificial insemination in an industrial breeding and production program. *Proc. 5th Int. Pig Vet. Soc., Copenhagen*, p.36

LARSEN, R.E., HURTGEN, J.P., HILLEY, H.D. and LEMAN, A.D. (1978). Diseases transmissible with artificial insemination. *Proc. 19th Ann. George A. Young Conf., Lincoln, Nebraska,* 59–69

LARSSON, K. (1978). Current research on the deep freezing of boar semen. *Wld Rev. Anim. Prod.* **14**, 59–64

LARSSON, K., SWENSSON, T. and WASS, K. (1979). A field trial on the fertility of liquid boar semen after utilization of two different diluents. *Nord. VetMed.* **31**, 337–338

MARTINAT-BOTTE, F., BARITEAU, F., MAULEON, P. and SCHEID, J.P. (1980). Oestrus control in gilts with a progestagen treatment (RU2267). *Proc. 9th Int. Congr. Anim. Reprod. A.I., Madrid,* **3**, 111

MELROSE, D.R. and O'HAGAN, C. (1961). Investigations into the technique of insemination in the pig. *Proc. IVth Int. Congr. Anim. Reprod. A.I., The Hague* **4**, 855–859

MELROSE, D.R., REED, H.C.B. and PATTERSON, R.L.S. (1971). Androgen steroids associated with boar odour as an aid to the detection of oestrus in pig artificial insemination. *Br. vet. J.* **127**, 497–502

MELROSE, D.R., REED, H.C.B. and PRATT, J.H. (1968). Developments in the use of pig artificial insemination by the farmer. *Proc. 6th Int. Congr. Anim. Reprod. A.I., Paris* **2**, 1087–1089

NISWENDER, G.D., REICHERT, L.E. and ZIMMERMAN, D.R. (1970). Radioimmunoassay of serum levels of luteinizing hormone throughout the estrous cycle in pigs. *Endocrinology* **87**, 576–580

PACOVA, J. and DUPAL, J. (1978). [The effect of heterospermy on conception rate and fertility of inseminated sows and gilts]. *Zivocisna vyroba* **23**, 735–741 (Cited by *Pig News and Information* (1980) **1**, 374)

PAQUIGNON, M., BUSSIÈRE, J., BARITEAU, F. and COUROT, M. (1980a). Effectiveness of frozen boar semen under practical conditions of artificial insemination. *Theriogenology* **14**, 217–226

PAQUIGNON, M., BUSSIÈRE, J., BARITEAU, F., LE MAIGNAN DE KERANGAT, G. and COUROT, M. (1980b). Efficacite des dilueurs Guelph et SCK7 pour la conservation prolongée à l'état liquide du sperme de verrat. *Journées de la recherche porcine en France,* **12**, pp. 157–160. Paris, L'Institut Technique du Porc

PEILIEU, C., TUNG, W., AN, M., YANG, Y.S. and LIANG, K.Y. (1980). Recent development of animal reproduction and A.I. in China. *Proc. 9th Int. Congr. Anim. Reprod. A.I., Madrid* **3**, 409

PLISHKO, N.T. (1965). [A method of prolonging the viability and fertilising ability of boar spermatozoa]. *Svinovodstvo* **19**, 37–41. (Cited by *Anim. Breed. Abstr.* (1966) **34**, 89)

POLGE, C. (1956). Artificial insemination in pigs. *Vet. Rec.* **68**, 62–76

POLGE, C. (1969). Advances in reproductive physiology in pigs. *J. Austr. Inst. agric. Sci.* **35**, 147–153

POLGE, C., SALAMAN, S. and WILMUT, I. (1970). Fertilising capacity of frozen boar semen following surgical insemination. *Vet. Rec.* **87**, 424–428

PURSEL, V.G. (1979). Advances in preservation of swine spermatozoa. In *Animal Reproduction, BARC. Symposium No. 3*, (H. Hawk, Ed.), pp. 145–157. Montclair, New Jersey, Allanheld, Osmun & Co. Ltd

PURSEL, V.G. and JOHNSON, L.A. (1975). Freezing of boar spermatozoa: Fertilising capacity with concentrated semen and a new thawing procedure. *J. Anim. Sci.* **40**, 99–102

PURSEL, V.G., JOHNSON, L.A. and SCHULMAN, L.L. (1973). Fertilizing capacity of boar semen stored at 15 °C. *J. Anim. Sci.* **37**, 532–535

PURSEL, V.G., SCHULMAN, L.L. and JOHNSON, L.A. (1978). Effect of Orvus ES Paste on acrosome morphology, motility and fertilizing capacity of frozen-thawed boar semen. *J. Anim. Sci.* **47**, 198–202

PURSEL, V.G., ELLIOTT, D.O., NEWMAN, C.W. and STAIGMILLER, R.B. (1980a). Synchronization of estrus in swine with allyl trenbolone and subsequent fertility with frozen semen. *Proc. 9th Int. Congr. Anim. Reprod. A.I., Madrid* **3**, 113

PURSEL, V.G., McVICAR, J.W., GEORGE, A.E. and WATERS, H.A. (1980b). Guideline for international exchange of swine semen and embryos. *Proc. 9th Int. Congr. Anim. Reprod. A.I., Madrid* **2**, 301–308

REED, H.C.B. (1978). Genetic movement and health. *Proc. Pig Vet. Soc.* **3**, 45–55

REED, H.C.B. MELROSE, D.R. and PATTERSON, R.L.S. (1974). Androgen steroids as an aid to the detection of oestrus in pig artificial insemination. *Br. vet. J.* **130**, 61–67

RUTGERS, H. (1966). Organisation and results of artificial insemination in pigs in the Netherlands. *Wld Rev. Anim. Prod.* **II** (*Special Issue*) 55–63

SCHILLING, E. and ROSTEL, W. (1964). Investigating methods of diagnosing oestrus in the sow. *Dt. tierärztl. Wschr.* **71**, 429–436. (Cited by *Anim. Breed. Abstr.* (1965). **33**, 453)

SERDIUK, S.I. (1968). Boar sperm storage: medium and method. *Proc. 6th Int. Congr. Anim. Reprod. A.I., Paris* **2**, 1155–1157

SWENSSON, T. (1977). [Experiments with a new diluent for boar semen]. *Svinskotsel* **67**, 29. (Cited by *Anim. Breed. Abstr.* (1977). **45**, 683

SWIERSTRA, E.E. and DYCK, G.W. (1976). Influence of the boar and ejaculation frequency on pregnancy rate and embryonic survival in swine. *J. Anim. Sci.* **42**, 455–460

VANDEMARK, N.L. and SHARMA, U.D. (1957). Preliminary fertility results from the preservation of bovine semen at room temperature. *J. Dairy Sci.* **40**, 438–439

WETTEMANN, R.P. and DESJARDINS, C. (1979). Testicular function in boars exposed to elevated ambient temperature. *Biol. Reprod.* **20**, 235–241

WETTEMANN, R.P., WELLS, M.E., OMTVEDT, I.T., POPE, C.E. and TURMAN, E.J. (1976). Influence of elevated ambient temperature on reproductive performance of boars. *J. Anim Sci.* **42**, 664–669

WILLEMS, C.M. (1977). Development of A.I. in pigs in Europe. *Proc. 28th Eur. Ass. Anim. Prod., Brussels*, P/3.03/I

II

REPRODUCTIVE DEVELOPMENT IN THE GILT AND SOW

5

ENDOCRINE CONTROL OF SEXUAL MATURATION IN THE FEMALE PIG AND SEXUAL DIFFERENTIATION OF THE STIMULATORY OESTROGEN FEEDBACK MECHANISM

F. ELSAESSER
Institut für Tierzucht und Tierverhalten Mariensee (FAL), 3057 Neustadt 1, Federal Republic of Germany

Sexual maturation, culminating in puberty and fertility, is a complex process which may involve maturational changes at all levels of the central nervous system–hypothalamo–pituitary–gonadal axis. It is the purpose of this review, based on published and unpublished work from our own laboratory, as well as the work of others, to summarize our present knowledge of the control of sexual maturation in the female pig. Although several other pituitary hormones and adrenal function may be relevant to the process of sexual maturation, because of the restricted availability of such data in the pig only the control of gonadotrophin secretion and ovarian function will be considered. Where information necessary for the understanding of sexual maturation in the pig is lacking, references will be made to work in other species. However, no attempt is made to integrate these data into a general concept of sexual maturation.

The description of the ontogeny of ovarian morphology and steroidogenesis will be followed by a consideration of the developmental pattern of gonadotrophin secretion, attempting to correlate them with the maturation of ovarian function and changes in the responsiveness of the ovary to gonadotrophins. Another aspect to be discussed is the responsiveness of the pituitary to luteinizing hormone (LH) releasing hormone at different stages of development. Special attention will be paid to the ontogeny of negative and positive gonadal steroid feedback control of gonadotrophin secretion as well as to the sexual differentiation of the LH surge mechanism.

Development of ovarian function and gonadal steroid levels

As shown by Allen (1904) the porcine foetal gonad becomes differentiated at about 31–32 days of gestation to an ovary which contains egg nests. In a recent study Oxender *et al.* (1979) have examined ovarian development during the foetal period and postnatally for up to 90 days of age (*Figure 5.1*). The percentage of egg nests observed in the ovary decreased as foetal age increased and egg nests were seldom observed in ovaries from pigs 20 or more days after birth. Starting at about 60–70 days of gestation primordial follicles became the dominant oogonic structure, accounting for

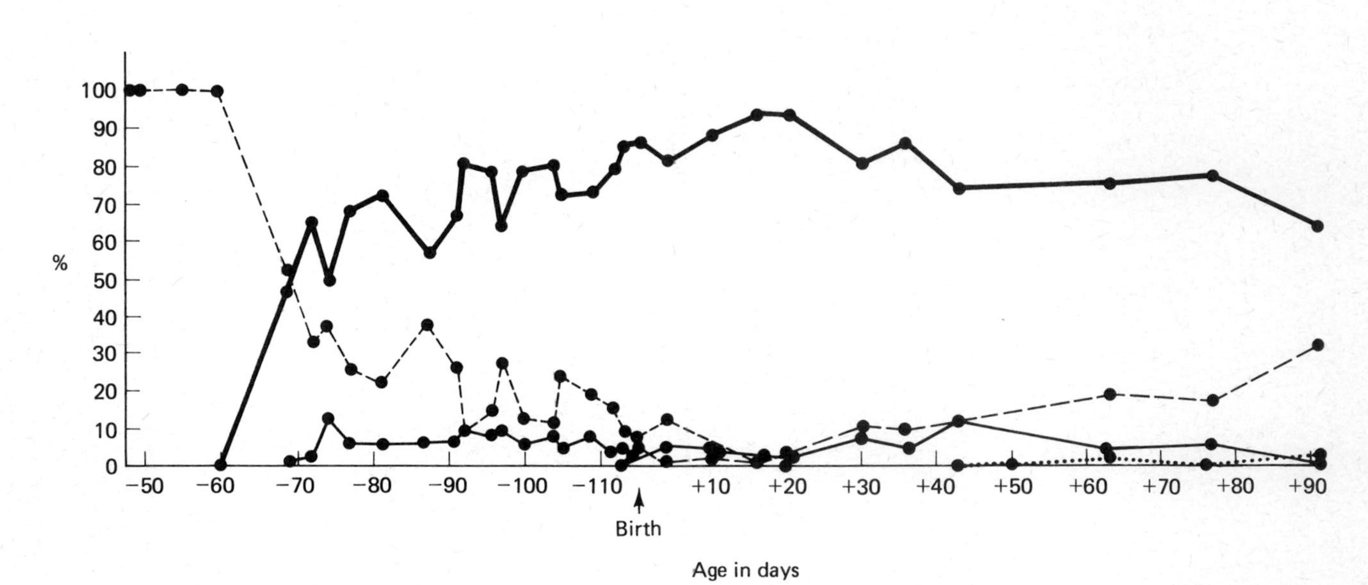

Figure 5.1 The number of oogonic structures in ovarian section from pigs 49 days post coitum to 92 days post partum graphed as a percentage of the total. ▬ Primordial follicles, —— primary follicles, --- secondary follicles, ······ tertiary follicles. Courtesy of Oxender *et al.* (1979)

about 80% of the ovarian follicles until 90 days after birth. The first primary follicles appeared in ovarian sections from foetuses at 70 days of gestation, whereas secondary follicles were initially observed perinatally and by 90 days postnatally accounted for nearly 30% of the total oogonic structures. Tertiary follicles were never observed in pigs younger than 60 days of age and one or more tertiary follicles were found in only a few of the ovarian sections from pigs of 60–90 days of age.

Although it is not known exactly when ovarian steroidogenic ability develops in the pig, it appears that the ovary matures later than the testis in this respect. Histochemical studies in which the activity of two key enzymes for steroidogenesis were measured (Oxender *et al.*, 1979) revealed that from 49 days of gestation to at least 90 days after birth neither 3β-hydroxysteroid dehydrogenase nor 17β-hydroxysteroid dehydrogenase activity was detectable in the ovaries.

The foetus is no doubt exposed to high levels of oestradiol and progesterone and the levels of both steroids further increase towards the end of gestation. However, available data suggest that the placenta is the major site of production (Barnes, Comline and Silver, 1974; Elsaesser *et al.*, 1976; Macdonald *et al.*, 1979; Macdonald *et al.*, 1980; Choong and Raeside, 1974). The physiological significance of these high sex steroid levels in the foetal circulation remains to be determined and it is unknown whether they exhibit feedback effects on gonadotrophin release.

Within six days of birth, plasma oestradiol declines to undetectable levels (≤20 pg/ml) and is still undetectable at 60 and 160 days of age (Elsaesser and Foxcroft, 1978; Elsaesser and Parvizi, 1979). Recent observations (Stickney, 1982) seem to indicate a small increase in concentrations of plasma oestradiol, however, from about 6 pg/ml to 16 pg/ml between 150 and 210 days of age.

Plasma progesterone levels seem to follow a similar pattern to that described for oestradiol-17β. Immediately after birth, levels decline and remain low throughout the prepubertal period (Elsaesser *et al.*, 1976; Elsaesser, Parvizi and Ellendorff, 1978), although occasionally higher levels are observed. Again, information on the immediate prepubertal period is lacking.

At this point it is appropriate to recall that the level of a given hormone is determined by its secretion rate, its interconversion from precursors and its metabolic clearance. In the gilt there is evidence to suggest that the metabolism of oestradiol changes with maturity. Identical doses of oestradiol benzoate/kg body weight produced higher concentrations of oestradiol in the plasma of gilts at 60 days of age than in piglets 6 days old and the highest levels were recorded in gilts at 160 days of age (Elsaesser and Foxcroft, 1978; Elsaesser and Parvizi, 1979). In a recent collaborative study (Elsaesser, Stickney and Foxcroft, 1982) using an isotope infusion technique, the metabolic clearance rate of oestradiol/kg body weight was found to be higher in immature than in peripubertal gilts (*Table 5.1*).

If the rate of inactivation of oestradiol increases more slowly with age than does body weight and therefore blood volume, constant levels of oestradiol can be achieved without the necessity for the ovary to increase its oestradiol production. Thus, in addition to any direct activation of

Table 5.1 METABOLIC CLEARANCE RATE (MCR) AND PRODUCTION RATE (PR) OF OESTRADIOL-17β IN GILTS ($\bar{x}$ ± SEM)

Age (days)	*Body weight* (kg)	*n*	*Infusion material*	*MCR* (ml/min)	*MCR* (ml/min/kg)	*Conversion rate* $E_1/E_2 \times 100$ (%)	*% binding of oestradiol in plasma*	*PR* (ng/min)
60	19.7±1.2	5	Oestradiol	1173±311[a][b]	62.6±18.8[a]	33.3±4.2[a]		37.2±15.5
			^{3}H-Oestradiol	2133±274	116.0±14.5[a][b]	31.6±3.7[b]	79.8±1.4	62.5±15.6[a]
160	66.0±2.9	4	Oestradiol	2938±679[a]	47.9±10.4	14.3±3.3[a]		108.1±33.0
			^{3}H-Oestradiol	3027±340[b]	48.5±4.8[b]	15.9±2.1[b]	81.3±1.1	111.5±7.9[a]

Values with the same superscript are significantly different from each other (P at least ≤ 0.05)
Courtesy of Elsaesser, Stickney and Foxcroft (1982)

ovarian steroidogenesis, a decrease in the metabolic clearance rate facilitates the elevation of gonadal steroid levels during puberty.

Developmental patterns of luteinizing hormone (LH), follicle stimulating hormone (FSH) and prolactin in the pituitary and plasma and the control of ovarian development

In view of the well-known stimulatory effects of gonadotrophins on the ovary in the mature animal, the question arises whether there is any direct evidence that the foetal pituitary secretes gonadotrophins and if so, is there any evidence to suggest that follicular development is controlled by the pituitary? Prior to day 80 of gestation gonadotrophin activity in the foetal pituitary is low or undetectable as shown by bioassay (Smith and Dortzbach, 1929; Melampy *et al.*, 1966) and more recently by the histochemical studies of Liwska (1975). Thereafter gonadotrophic activity (Smith and Dortzbach, 1929) or LH concentrations in the anterior pituitary (Melampy *et al.* 1966) increase. Cell types probably equivalent to FSH secreting cells have been described in foetuses at 70–79 days of gestation and were observed sporadically at day 51 of gestation (Liwska, 1978).

This pattern of gonadotrophic activity in the pituitary corresponds well with the changes of plasma LH and FSH concentrations during foetal life (*Figure 5.2*). Circulating LH concentrations before day 80 of gestation are low or undetectable (Elsaesser *et al.*, 1976; Colenbrander, *et al.*, 1977). During the last few weeks of foetal life LH levels increase, probably a few days earlier in females than in males, and remain elevated for some time after birth (Colenbrander *et al.*, 1977; Elsaesser, Parvizi and Ellendorff, 1978). Somewhat lower LH concentrations are found during the last three weeks of gestation in chronically catheterized pig foetuses (Macdonald *et al.*, 1979).

Serum FSH levels in the foetus behave similarly to the pattern described for LH. Before day 80 of gestation serum FSH concentrations are low. After 80 days a sharp increase in serum FSH concentrations occurs in female foetuses, but a much slower rise occurs in male foetuses. The levels remain relatively constant after birth (Colenbrander, van de Wiel and Wensing, 1980).

To date the pattern of LH and FSH secretion has not been described in detail throughout the whole period from birth to puberty and information from the period immediately before first ovulation is especially lacking. During the first five weeks of life, average LH concentrations decrease in Landrace pigs as well as in miniature pigs. Although miniature pigs mature younger than Landrace pigs, with first oestrus occurring at about 15–20 weeks of age, the mean LH levels appear to remain unchanged during the period between 5 and 24 weeks of age (Colenbrander *et al.*, 1977; Elsaesser *et al.*, 1976; Elsaesser, Parvizi and Ellendorff, 1978).

The developmental pattern of FSH release differs from that of LH. After birth plasma FSH gradually rises to a maximum at about 10 weeks of age and then appears to decline slightly towards puberty (Colenbrander, van de Wiel and Wensing, 1980).

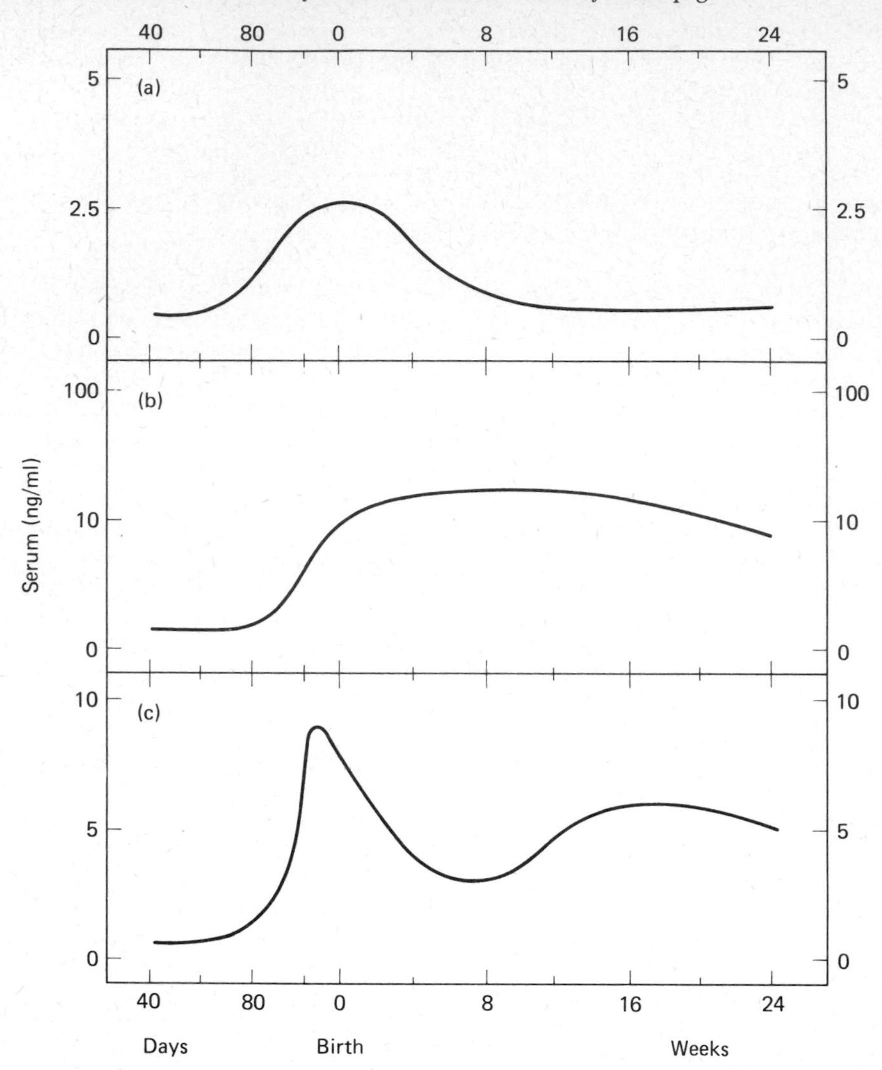

Figure 5.2 Serum levels of (a) LH, (b) FSH and (c) prolactin during sexual development in female pigs. (a) after Elsaesser *et al.* (1976) and Colenbrander *et al.* (1977); (b) after Colenbrander *et al.* (1980); (c) after Colenbrander *et al.* (1981)

Prolactin levels are elevated at birth, decline to a minimum when the gilt is five weeks of age and then increase again (Colenbrander *et al.*, 1981). The physiological importance of this pattern remains speculative, since a definite role for prolactin has thus far not been described in the pig.

As mentioned before, neither changes in hydroxysteroid dehydrogenase activity nor in follicular morphology occur which might be correlated with the period of greater secretion of LH or FSH near the time of birth. This contrasts to the male pig, in which the prenatal rise in serum concentration of FSH is paralleled by marked growth in the length of the seminiferous tubules and an increase in activity of Leydig cells (Colenbrander, van de

Wiel and Wensing, 1980). Further support for the conclusion that early ovarian development, but not testes development, is independent of normal functioning of the pituitary is provided by the observation that foetal decapitation does not interfere with normal germ cell development in the foetus (Colenbrander, Wensing and van Rossum Koh, 1981).

Later phases of follicular growth are, however, regulated by gonadotrophin secretion. The development of tertiary follicles or antral follicles which are first present at eight weeks after birth (Mauleon, 1964; Oxender *et al.*, 1979) may represent a stage when follicles become sensitive to gonadotrophins. Support for the idea that follicular development becomes dependent on gonadotrophins at this stage of development emerges from the observation of the failure of gonadotrophins to cause increased follicular development in gilts of five weeks of age. The critical age at which increased follicular development and ovulations can be induced by exogenous gonadotrophin treatment appears to be about nine weeks (Casida, 1934; Kather and Smidt, 1975; Oxender *et al.*, 1979) and will be discussed elsewhere in this book (Paterson, Chapter 7).

The mechanism by which the pituitary gains control over follicular development in the pig remains to be determined. However, it seems likely that this process is related to the development of gonadotrophin receptors in the ovary, as indicated by studies in the female rat (Siebers *et al.*, 1977). The continuous exposure of the ovary to comparatively high levels of FSH may play a role in this context.

In the mature sow, as in other species, the secretion of LH is pulsatile or episodic in nature and the amplitude and frequency of the LH episodes change during the oestrous cycle (Elsaesser and Parvizi, 1977; Foxcroft, 1978). In view of data from Foster and Ryan (1979) and Wildt, Marshall and Knobil (1980), which stress the importance of the frequency of LH pulses for the onset of puberty, possible maturational changes in the frequency-amplitude characteristics of episodic LH secretion in the gilt are of great interest. In the chronically catheterized foetus, minor fluctuations in LH levels occur during the last week of foetal life (Macdonald *et al.*, 1979). When pigs at 17 days of age are monitored by blood samples taken at two hour intervals, the amplitude of LH discharges appears to increase. This finding might explain the elevated average LH levels at this time (Colenbrander *et al.*, 1977). In gilts at 9–10 weeks of age episodes of LH release occur spontaneously with a mean frequency of 1.3 peaks/hour, but this frequency could be reduced to zero by oestradiol-17β treatment (*Figure 5.3*; Foxcroft, Pomerantz and Nalbandov, 1975). When profiles of LH levels were based on samples taken at 10 minute intervals, there was a gradual increase in the frequency of LH episodes from prepuberty to late puberty (Stickney and Foxcroft, personal communcation), but corresponding data for the period immediately preceding first ovulation are lacking.

It is possible that the postulated increase in frequency of the LH pulses (Foster and Ryan, 1979) occurs in the gilt only during the night period. In gilts of 160 days of age, a significantly greater proportion of raised LH concentrations occurred during the night than during daylight but this difference did not exist in gilts of 60 days of age (*Table 5.2*; Elsaesser and Foxcroft, 1978). Thus a development in nocturnal LH release appears similar to that of the human, in which sleep-related increments in the

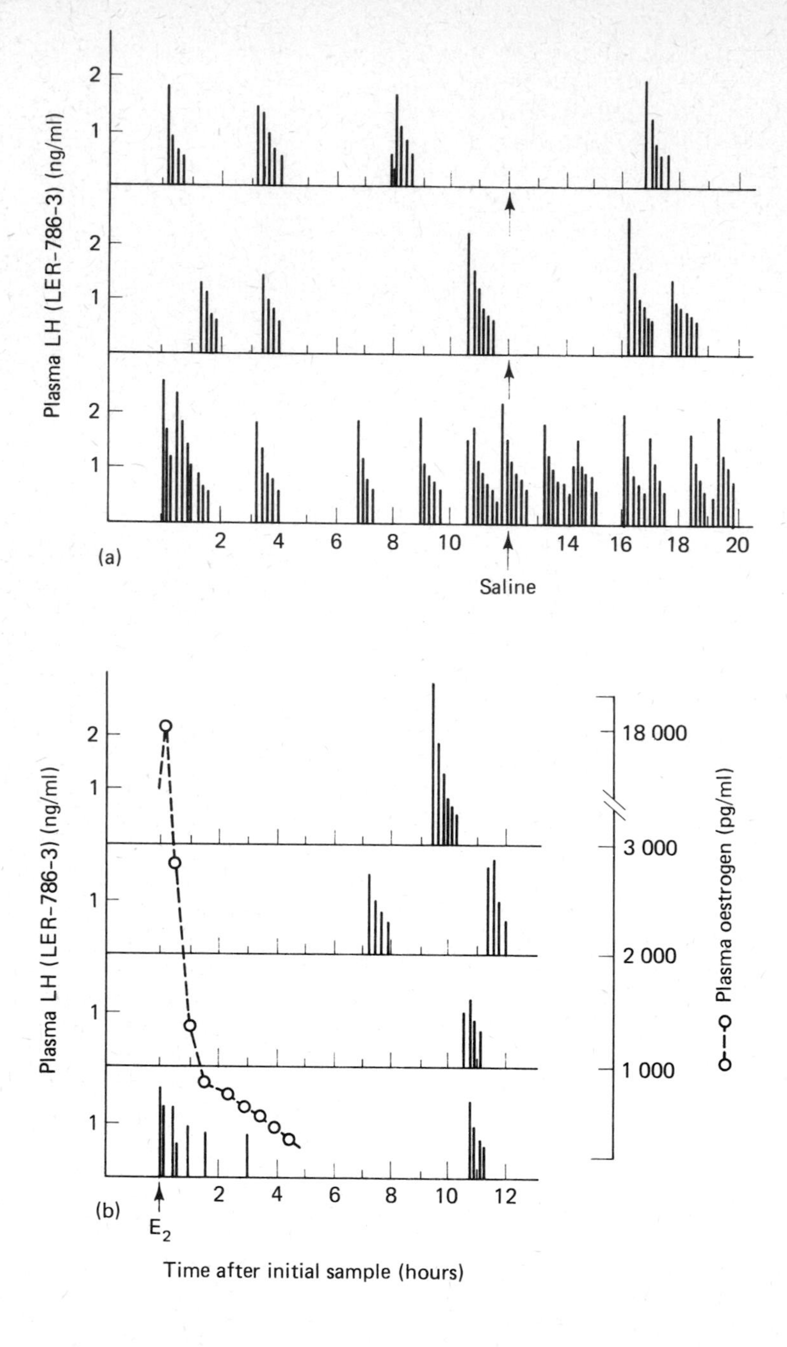

Figure 5.3 (a) Spontaneous episodic release of LH in three untreated prepubertal female pigs. Saline injection given 12 hours after initial sample. Samples were drawn every 10 minutes for LH assay. LH levels not represented by vertical bars were less than 0.7 ng LH/ml. (b) Spontaneous LH release in the 12 hour period following intravenous injection of 5.0 μg oestradiol-17β. Levels of peripheral oestrogens indicated were the means for the same four animals. By courtesy of Foxcroft, Pomerantz and Nalbandov (1975)

Table 5.2 χ^2 ANALYSIS (WITH YEATE'S CORRECTION) OF ELEVATIONS IN LH LEVELS FROM INFANTILE AND PERIPUBERTAL GILTS

Age (days)	*Nos. LH estimates* $\geq$ *1 ng/ml plasma* per 72 hours	*per 3 days*	*per 3 nights*
60 (n = 12)	82	44	38
160 (n = 12)	100	33[a]	67[a]

Blood samples were taken at 4 hour intervals for 72 hours.
[a] $P \leq 0.001$, $\chi^2 = 23.8$
From Elsaesser and Foxcroft (1978)

secretion of gonadotrophin have been observed in pubertal children (Weitzman *et al.*, 1975). It may be assumed that the nocturnal increase in number and/or amplitude of LH episodes is associated with the onset of, or an increase in, ovarian oestradiol secretion.

The responsiveness of the pituitary

The factors responsible for the pattern of plasma gonadotrophin secretion during development in the pig are still poorly understood. The evidence presented, that the ontogeny of LH secretion starts during foetal life, raises the question as to whether the hypothalamus controls pituitary function prior to birth. *In vivo* the pituitary of the 55 day old foetus is unable to respond to a single injection of LH releasing hormone. However, between 70 and 100 days of gestation the proportion of animals responding with a rise in concentrations of LH in the umbilical artery increases and all foetuses older than 100 days are capable of releasing LH with no apparent sex differences (Colenbrander *et al.*, 1980).

Preliminary data obtained *in vitro* demonstrate a similar ontogeny of pituitary responsiveness although in these studies a sex difference was apparent (*Figure 5.4*). Monolayer cultures of female foetal anterior pituitary cells exposed for four hours to LH releasing hormone had only a slightly increased LH release at 60 days of gestation in the presence of the maximal doses of LH releasing hormone (10^{-8} M). A typical dose response curve was obtained from pituitaries recovered from foetuses at 80 days of gestation and in cultures from female foetuses at 105 days of gestation both basal LH release and maximal response doubled (*Figure 5.4*). In both age groups basal and maximal LH release from male pituitary cells and the LH releasing hormone concentration required to produce a half maximal stimulation (ED_{50}) of LH release, was lower compared with female pituitary cells (Elsaesser, Bruhn, Parvizi and Heilhecker, unpublished observations). These findings correspond well with the above-mentioned earlier increment of LH levels in female compared with male foetuses and possibly relate to differences in plasma testosterone concentrations.

Evidence to suggest that the foetal hypothalamus is able to release sufficient amounts of LH releasing hormone to enhance pituitary LH secretion is derived from the response to electrical or electrochemical stimulation (Bruhn, Parvizi and Ellendorff, 1981) and will be discussed in detail by Ellendorff and Parvizi in Chapter 9.

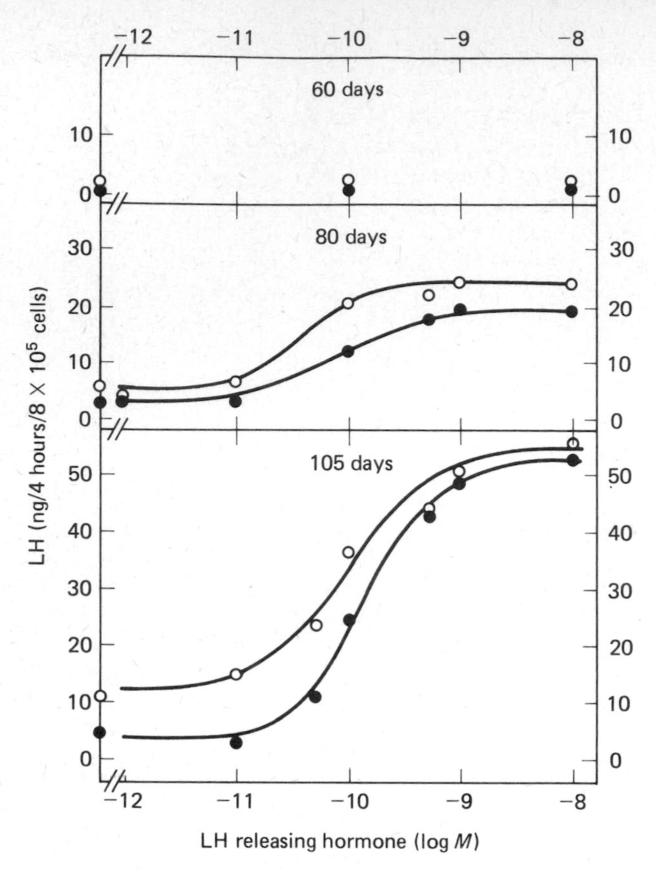

Figure 5.4 Effect of increasing concentrations of LH releasing hormone on LH release by porcine anterior pituitary cells from male (●) and female (○) foetuses. Results are presented as means of data obtained from quadruple monolayer culture dishes. By courtesy of Elsaesser, Bruhn, Parvizi and Heilhecker, unpublished observations

The postnatal development of pituitary responsiveness to LH releasing hormone has not yet been evaluated in the gilt. Currently we are studying this problem by establishing dose response curves to LH releasing hormone in pituitary monolayer cultures and the effect of pretreatment with gonadal steroids. From preliminary results we know that pituitary cells from 60 day old gilts produce basal and maximal LH releases about 20 times greater than those observed from cells of foetuses at 105 days of gestation (Bruhn, Parvizi and Elsaesser, unpublished observations).

Maturation of negative feedback of gonadal steroids

Developmental changes in the pattern of gonadotrophin secretion may at least in part reflect maturational changes in the feedback mechanisms of the central nervous system–pituitary–gonadal axis. Byrnes and Meyer (1951) and Ramirez and McCann (1965) postulated a maturational change in the negative feedback action of gonadal steroids. This hypothesis, called the gonadostat theory, has been favoured as an explanation for the mechanisms involved in the onset of puberty. This theory of sexual maturation states that the threshold for negative feedback in the hypothalamo–pituitary unit, the 'gonadostat', increases with age and thereby the sensitivity to the inhibiting effect of sex steroids progressively decreases. This results in increased secretion of gonadotrophin.

Two types of experiments have been performed to demonstrate changes in negative feedback of gonadal steroids in the female pig. One type examines the response to removal of the gonads and the other examines the effect of replacement therapy with gonadal steroids on plasma LH levels.

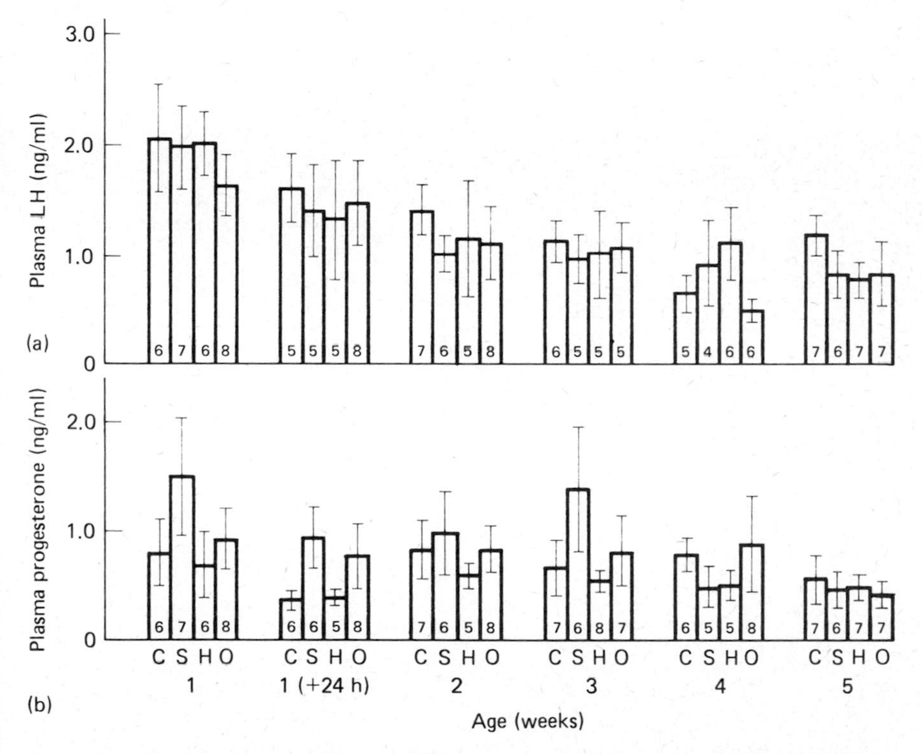

Figure 5.5 Concentrations of (a) LH and (b) progesterone in the plasma of immature female miniature pigs at various ages. Results are means ±S.E.M.; the numbers of pigs are given at the base of each column. C—controls; S—sham-operated; H—hemi-ovariectomized; O—ovariectomized (all operations were performed at one week of age). By courtesy of Elsaesser, Parvizi and Ellendorff (1978)

Levels of LH in plasma do not increase during the four weeks following gonadectomy at one week of age in either sex (*Figure 5.5*, Elsaesser, Parvizi and Ellendorff, 1978). An explanation for the failure of LH to increase after ovariectomy is the possibility that the ovary is producing no steroids. Indeed, the plasma levels of progesterone are generally low and unaffected by treatment and, as noted before, it may be assumed that the ovary is unable to produce oestrogen at this age (Oxender *et al.*, 1979). The finding that LH decreases between one and five weeks of age in ovariectomized as well as intact animals, indicates that the ovary does not play an important role in the decrease of plasma LH levels occurring at this period of development (Elsaesser, Parvizi and Ellendorff, 1978). While the response to ovariectomy at eight weeks of age has not been tested, the data of Colenbrander *et al.* (1977) are consistent with the view that at this age the hypothalamo–pituitary unit of the male pig has developed its competence to detect and to respond to the removal of the gonads. Increased levels of LH in freemartin pigs (gonads absent) suggest that at eight weeks of age LH release in the female also comes under the control of gonadal secretion. Gilts ovariectomized at either 60 or 130 days of age have

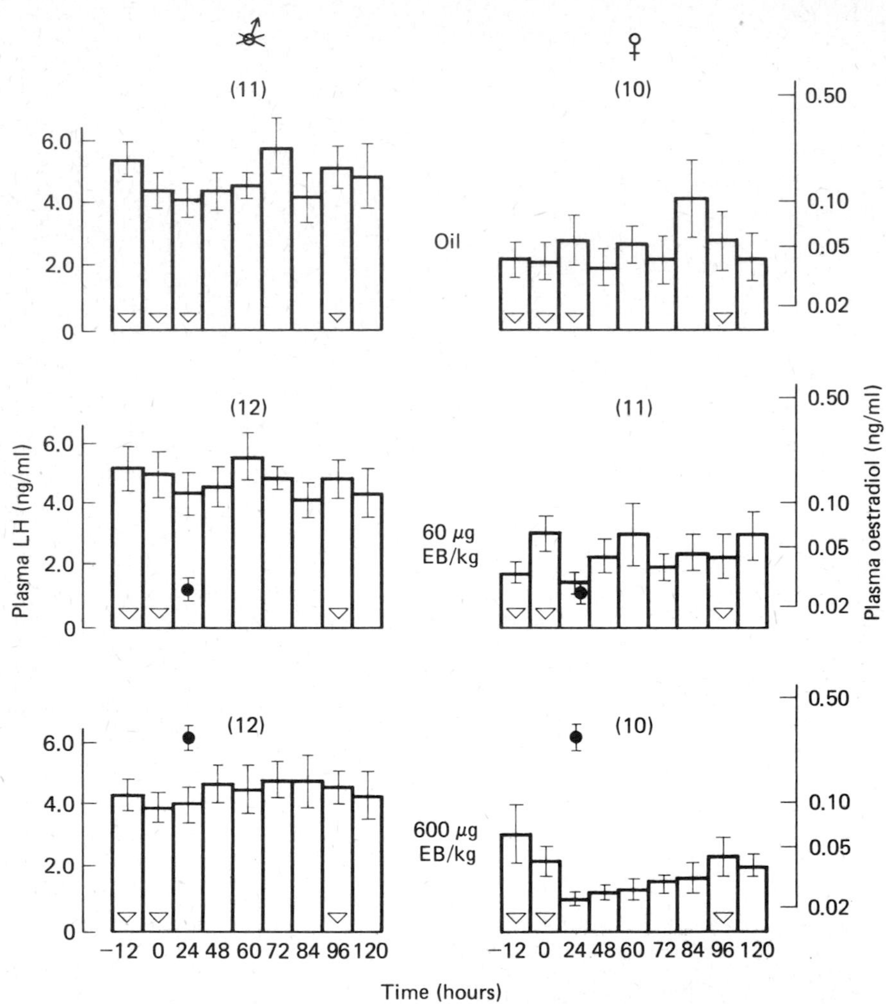

Figure 5.6 Concentrations of LH (open bars) and oestradiol-17β (●, logarithmic scale) in the plasma of orchidectomized male and intact female pigs after treatment with sesame oil or oestradiol benzoate (EB) at six days of age. ▽—plasma concentration of oestradiol-17β≤20 pg/ml. Results are means ±S.E.M.; number of pigs are shown in parentheses. By courtesy of Elsaesser and Parvizi (1979)

elevated LH and FSH levels at 160 days of age (Foxcroft, Stickney and Elsaesser, unpublished observations).

From the effect of exogenous oestradiol on plasma LH it appears that the central nervous system–pituitary unit of the newborn female pig is able to recognize and to respond to changes in concentrations of oestradiol-17β. The negative oestrogen feedback matures earlier in the female than in the male (*Figure 5.6*; Elsaesser and Parvizi, 1979). However, negative feedback responses can only be evoked by a sustained increase in the concentration of oestradiol-17β in the plasma. This is concluded from the finding that the concentration of LH in plasma of female pigs at six days of age was significantly suppressed 24 hours after the administration of 600 μg oestradiol benzoate/kg body weight whereas treatment with 60 μg oestradiol benzoate/kg body weight did not affect LH.

As gilts mature, the effectiveness of oestrogen in suppressing the secretion of LH seems to be enhanced. For example, intravenous injections of oestradiol or the subcutaneous implanation of oestradiol in silastic capsules, suppress LH levels in 9–10 week old gilts. These studies also demonstrated that the negative feedback action of oestradiol is exerted through inhibition of spontaneous episodic LH release (Pomerantz, Foxcroft and Nalbandov, 1975; Foxcroft, Pomerantz and Nalbandov, 1975).

The hypothalamus as well as the anterior pituitary may be a site of oestradiol negative feedback action in the immature gilt and nuclear and cytoplasmic receptors for oestradiol and progesterone have been demonstrated in both the hypothalamus and the pituitary. No difference in the number of binding sites was found in gilts between one month and 5.5 months of age. However, it is claimed that the capacity of cytoplasmic binding sites for oestradiol in mature cycling pigs is 15-fold higher than in prepubertal pigs (Diekman and Anderson, 1979).

There are no studies which clearly indicate a decrease in the feedback sensitivity of gonadal steroids at puberty in the gilt. There is evidence that the negative feedback action of testosterone declines during sexual maturation in the boar. In the absence of any dramatic changes in levels of LH, this decline in feedback sensitivity to testosterone may be a compensatory, rather than an initiating mechanism, necessary for the maintenance of LH release in the presence of raised levels of testosterone during puberty (Elsaesser, Parvizi and Ellendorff, 1978). Subtle changes in the amplitude and/or frequency of episodic LH release, however, cannot be excluded.

Development of the stimulatory oestrogen feedback mechanism

Expression of the stimulatory oestrogen feedback mechanism is essential for the onset of cyclic ovarian activity, since it triggers the surge release of LH which in turn induces ovulation.

The capability of the hypothalamo–pituitary unit to respond to an increase in circulating oestrogen levels with a surge of LH develops gradually as the gilt matures. As mentioned before, a single intramuscular injection of 60 or 600 μg oestradiol benzoate/kg body weight will not elicit an LH discharge in gilts at seven days of age (Elsaesser and Parvizi, 1979). At 14 days of age, female miniature pigs, given a single dose of oestradiol benzoate respond with a significant increase in the levels of LH in plasma 60–72 hours later (*Figure 5.7*; Elsaesser, Parvizi and Ellendorff, 1978). The concentration of oestradiol in plasma must be raised both over a certain period of time and over a certain threshold level to evoke positive response (Yamaji *et al.*, 1971), e.g. an intramuscular injection of 6 μg oestradiol benzoate/kg body weight (Elsaesser, unpublished) does not increase the levels of LH in the plasma of immature miniature gilts.

These observations do not imply that the control system which responds to oestrogen is fully mature in the miniature pig at 14 days of age. In fact, it is evident from a comparison of the characteristics of oestrogen-induced LH surges in 60 and 160 day old domestic gilts, that several maturational changes occur (*Figure 5.8*; Elsaesser and Foxcroft, 1978). In 60 day old

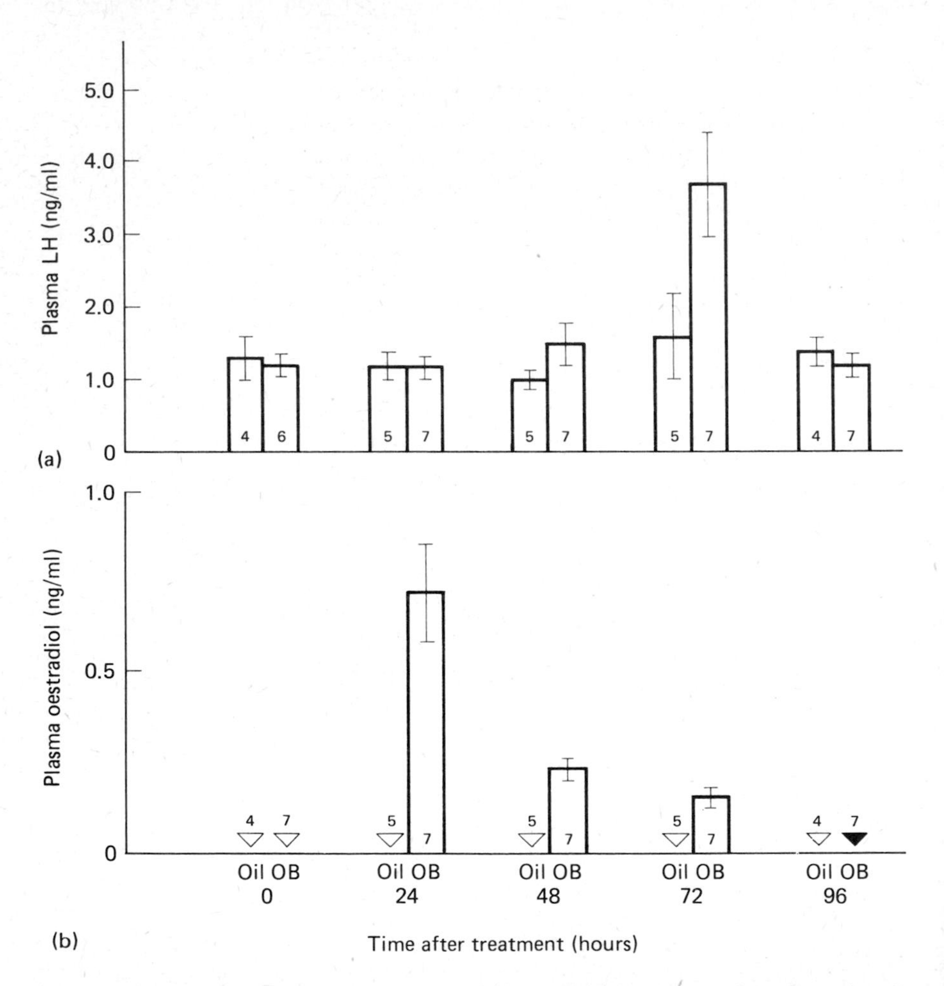

Figure 5.7 Concentrations of (a) LH and (b) oestradiol in plasma of newborn female miniature pigs after treatment with sesame oil or oestradiol benzoate (OB, 0.6 mg/kg body weight) at two weeks of age. ▽—Oestradiol concentration ⩽20 pg/ml plasma; ▼—oestradiol concentration less than 100 pg/ml plasma. Results are means ±S.E.M.; numbers of pigs are shown at the base of each column. By courtesy of Elsaesser, Parvizi and Ellendorff (1978)

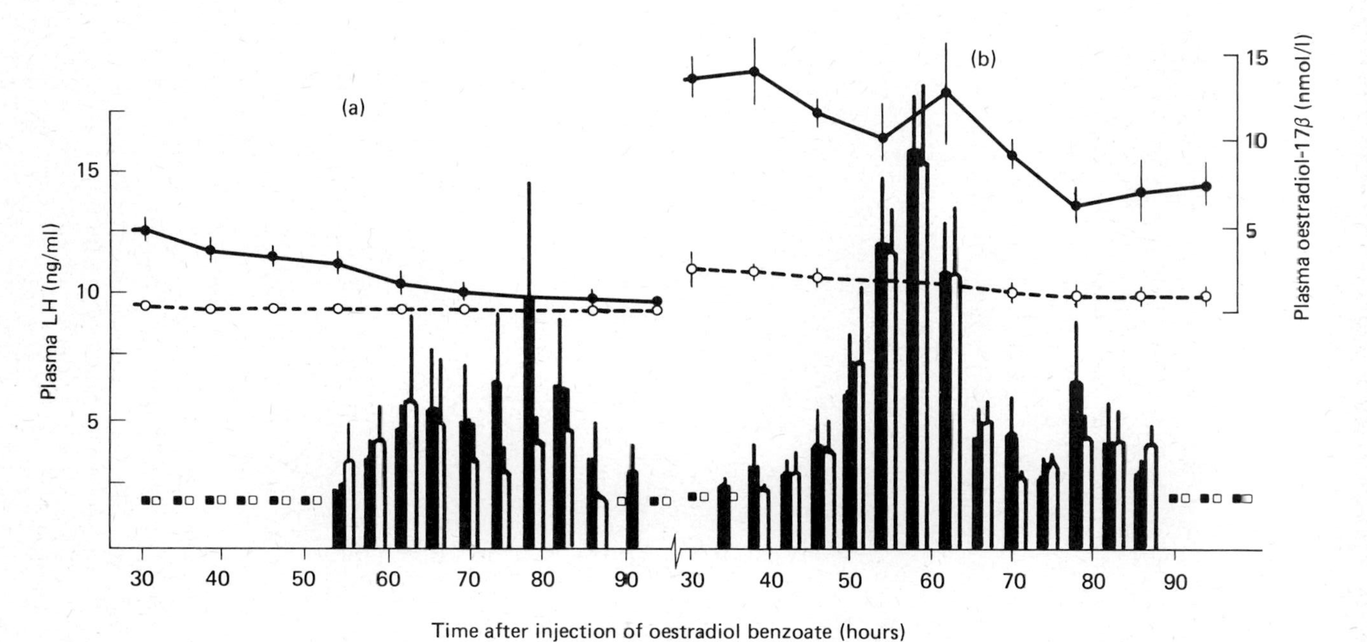

Figure 5.8 Plasma concentrations (means ±S.E.M.) of LH (histogram bars) and oestradiol-17β (graph lines) in (a) 60-day old and (b) 161-day old gilts treated with 60 μg oestradiol benzoate/kg (broken lines, open bars) or 600 μg oestradiol benzoate/kg (solid lines, solid bars). □■—Concentration of LH below the limit of detection of the assay. By courtesy of Elsaesser and Foxcroft (1978)

gilts the pattern of LH release subsequent to the injection of oestradiol benzoate is more variable and the mean concentrations of LH are lower than in gilts 160 days of age. The pattern of LH release is also less well defined in the younger gilts. It appears that the dose of oestradiol benzoate modifies the profile of the LH surge in younger gilts. Another major maturational change relates to the overall mean times to both the onset of the LH discharge and to the peak level of LH, which are greater in gilts at 60 days of age.

A notable feature of the stimulatory oestrogen feedback action in the immature gilt is the tendency for biphasic and sometimes even triphasic surges with a nocturnal preference (*Figure 5.9*). This phenomenon has

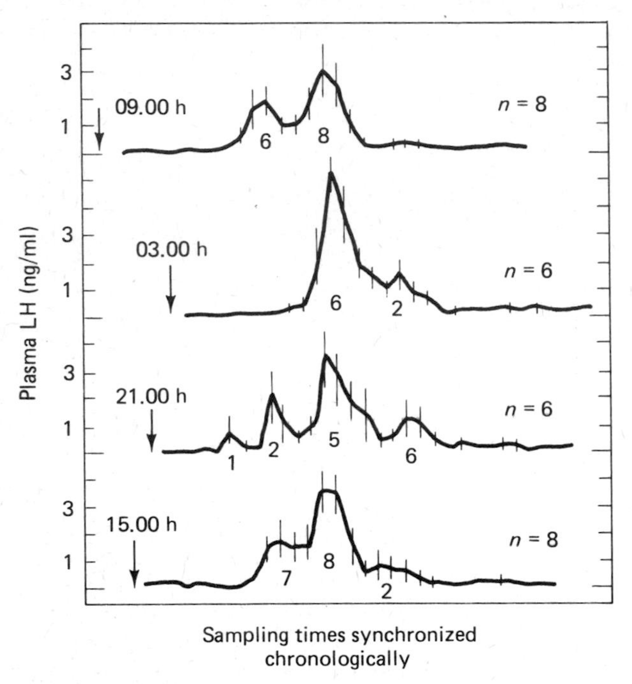

Figure 5.9 Plasma concentrations (means ±S.E.M.) of LH in gilts (80 kg) treated with oestradiol benzoate (20 μg/kg) at different times of the day (↓). From Foxcroft and Piontek (personal communication)

recently been studied in more detail (Foxcroft and Piontek, personal communications) by comparing the gonadotrophin response to the administration of oestradiol benzoate at different times of the day. Both the time to surges, and the number of surges, varied with the time of injection and phases of LH release occurred preferentially during the hours of darkness. A single large LH surge showing the characteristics of the LH surge in mature females was induced when oestradiol was given at 03.00 hours. The authors suggest that a diurnal/nocturnal influence interacts with the LH surge mechanism in the immature gilt. It is of interest that even in the mature cyclic sow the majority of spontaneous surges of LH may be associated with the night and early morning periods (Parvizi *et al.*, 1976); however, following exogenous oestradiol treatment in weaned mature sows (Edwards, 1980) as well as during late lactation (Elsaesser and Parvizi, 1980), no such synchronization of LH surges to the nocturnal period exists and a single LH discharge is usual, as is the case in the immature lamb (Foster and Karsch, 1975). Thus it seems that either further maturation of the control system which governs the LH discharge occurs, or that the cyclic changes in ovarian hormones present in the mature female predispose the hypothalamo–pituitary unit to a single discharge of LH.

It is not known why the LH surge mechanism is unable to operate soon after birth, nor is it clear what determines its gradual maturation. That the medial preoptic–anterior hypothalamic area may form an essential part of the mechanisms that control ovulation in the pig, has been shown by the induction of cystic ovaries in pigs with lesions in this area (Döcke and Busch, 1974). It is unlikely that the pituitary of the neonate is unable to release LH in a short pulse, because our previous studies suggest that the pituitary of the newborn male and female pig responds to LH releasing hormone in the same way as the adult (Elsaesser *et al.*, 1974). However, although it is possible that the pituitary in these animals is not able to release LH in a surge-like manner similar to the preovulatory surge, this has not been studied so far in the pig. The recent aforementioned finding that the number of cytoplasmic and nuclear oestradiol receptors in the hypothalamus and in the pituitary does not change in the gilt between one and 5.5 months of age (Diekman and Anderson, 1979) would argue against the assumption that changes in specific oestradiol binding sites are involved in the development of oestrogen-induced LH secretion.

With respect to possible causes of the maturation of the stimulatory oestrogen feedback, preliminary unpublished results of Foxcroft, Stickney and Elsaesser suggest that oestrogens are not involved in this process up to 60 days of age, as indicated by the failure to hasten the maturity of the LH surge response to oestrogen in immature gilts, which had been implanted with silastic capsules containing small amounts of oestradiol. In groups of gilts in which the LH response was evaluated at 160 days no further maturation of the positive feedback mechanism occurred in females ovariectomized at 60 days of age. However, substantial maturation was observed in animals either ovariectomized at 130 days or in gilts given implants of oestradiol following ovariectomy at 60 days of age. Ovarian oestrogen appears therefore to have a role in the maturation of the LH surge mechanism.

On the basis of available information, therefore, the LH surge mechanism may be considered essentially mature some time before the onset of ovarian cyclicity. The conclusion can then be drawn that the timing of the first cyclic discharge during puberty is not limited by the capability of the central nervous–pituitary system to respond to the positive feedback action of oestradiol. Rather it appears that before puberty the ovary is unable to release a signal in the form of an oestradiol surge.

SEXUAL DIFFERENTIATION OF THE STIMULATORY OESTROGEN FEEDBACK MECHANISM AND THE EFFECT OF PRENATAL TESTOSTERONE TREATMENT

One interesting feature of the stimulatory oestrogen feedback mechanism is its sexual dimorphism in various species, which is a consequence of testicular androgen secretion during a critical period of foetal or neonatal life (Barraclough, 1966; Gorski, 1973).

In the miniature pig, the LH surge control system is sexually differentiated as early as 14 days of age. Unlike the newborn female (miniature) pig, the newborn male pig is unable to release LH in response to oestrogen (Elsaesser, Parvizi and Ellendorff, 1978). That this sex difference is not due to testicular secretion at the time of oestradiol benzoate treatment is shown by the lack of an LH discharge in 14 day old or 160 day old male castrates. However, subsequent to the negative feedback action of oestradiol, LH levels in 160 day old castrated boars rise to pretreatment levels at 72 hours, although the plasma levels of oestradiol are still elevated (*Figure 5.10*, Elsaesser and Parvizi, 1979). A similar observation has been made by

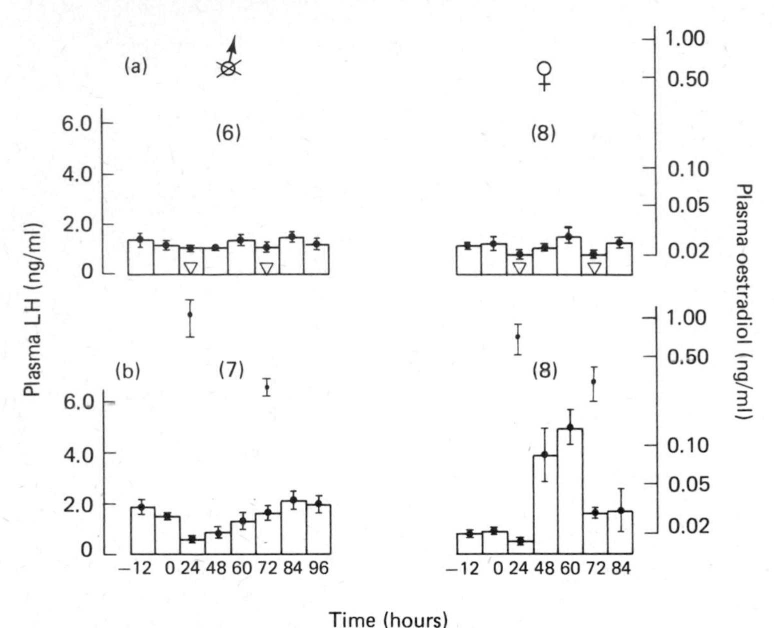

Figure 5.10 Concentrations of LH (open bars) and oestradiol-17β (●, logarithmic scale) in the plasma of orchidectomized male and intact female pigs after treatment with (a) sesame oil, or (b) oestradiol benzoate (60 μg/kg) at 160 days of age. ▽—plasma concentration of oestradiol-17β ≤20 pg/ml. Results are means ±S.E.M.; numbers of pigs are shown in parentheses. By courtesy of Elsaesser and Parvizi (1979)

Ford and Schanbacher (1977) following daily treatment of castrated boars with oestradiol benzoate, suggesting the possibility that the sexual dimorphism is quantitative rather than qualitative. In this context it is of interest that castrated boars display clear sexual receptive behaviour when treated with oestradiol benzoate.

The findings that the foetal testis increases its secretion of testosterone during differentiation of the internal and external genitalia between days 30 and 50 of foetal life (Colenbrander, de Jong and Wensing, 1978; Raeside and Middleton, 1979; Ford, Christenson and Maurer, 1980) and that neonatal treatment with testosterone is ineffective in inducing an ovulatory sterility (Zimbelman and Lauderdale, 1973), raise the possibility that a sexual differentiation of the stimulatory oestrogen feedback system occurs early in foetal life.

So far no definite answer can be given to the question of whether the concept of androgen-dependent sexual differentiation of the LH control system is valid for the pig. In prepubertal female offspring exposed to testosterone propionate via their mother (intramuscular injection on three occasions separated by two-day intervals, 5 mg/kg on days 30, 50 or 70 of foetal life) the positive response of LH to oestradiol benzoate was impaired (*Figure 5.11*) and by 250 days the weight of the ovaries and the uterus was reduced and the number of animals that had ovulated was depressed. Gilts treated with testosterone propionate later during foetal life did not differ from controls in these respects (Elsaesser and Parvizi, 1979).

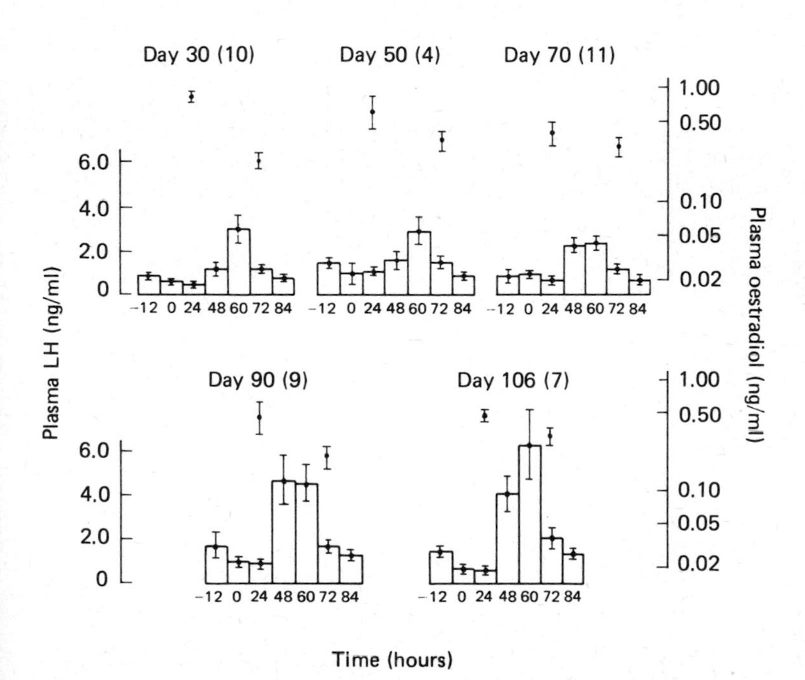

Figure 5.11 Concentrations of LH (open bars) and oestradiol-17β (●, logarithmic scale) in the plasma of intact female pigs after treatment with oestradiol benzoate (60 μg/kg) at 160 days of age. The mothers of these animals had been treated with testosterone propionate on days 30, 50, 70, 90 or 106 of pregnancy. Results are means ±S.E.M.; numbers of pigs are shown in parentheses. By courtesy of Elsaesser and Parvizi (1979)

The effects of transuterine intrafoetal testosterone treatment support the view that the pig foetus is well protected against the masculinizing effects of maternally administered androgens and that complete suppression of the LH surge mechanism as in the genetic male might be a transient effect.

Clitoral enlargement was observed in females which had been treated intrafoetally with testosterone (20 mg/foetus) on day 40 or 50 of gestation. This treatment also blocked the stimulatory oestrogen feedback mechanism prepubertally and at the time of puberty oestrous behaviour occurred at irregular intervals. Later in life oestrous cycles became more regular, and plasma progesterone profiles as well as the observation of corpora lutea or corpora albicans at laparoscopy indicated the occurrence of ovulation. In some animals failure to ovulate was suggested by elevated LH levels and the occurrence of follicular cysts. Treatment with testosterone on day 30 of gestation resulted in phenotypically masculinized females with empty scrotal sacs and a penis; the LH control mechanism, however, was not disturbed and at about 300 days of age progesterone profiles indicated normal ovarian activity. This was later confirmed by examining the ovaries at laparoscopy or laparotomy (Elsaesser, Ellendorff and Parvizi, unpublished observations).

Although it is tempting to speculate that the critical period for sexual differentiation of the control of surge LH release occurs in the pig during the second trimester of gestation, the inability to achieve total defeminization of the LH control system, including suppression of regular cycles and ovulation, does not rule out the possibility that the observed effects are pharmacological findings. The notion mentioned above, that ovarian secretion might play a role in the development of the stimulatory oestrogen feedback action in the female, has led us (Colenbrander, Parvizi and Elsaesser, unpublished observations) to compare the competence of male and female pigs which have been castrated early in life and pretreated with low amounts of oestradiol for 10 days to discharge LH. Whereas the LH surge mechanism can be activated by a similar treatment in the orchidectomized rhesus monkey (Karsch, Dierschke and Knobil, 1973), LH levels in the castrated boar rose only slightly, but significantly above pretreatment levels. In gilts ovariectomized at 20 days of age the magnitude of the LH discharge was smaller compared with gilts ovariectomized at 150 days of age. It remains to be determined whether ovariectomy even earlier in life totally abolishes the stimulatory oestrogen feedback action. More direct evidence that ovarian secretions are involved in the sexual differentiation of the LH surge mechanism is difficult to obtain in the pig, since attempts to transplant ovaries to orchidectomized piglets have so far failed.

Conclusion

It is evident from this review that sexual maturation is associated with changes at all levels of the central nervous system–pituitary–ovarian axis. Early ovarian development appears to be independent of gonadotrophic control, while later phases of ovarian development are regulated by gonadotrophin secretion. The anterior pituitary is active long before birth as indicated by the rise in titres of gonadotrophin and changes in its responsiveness to LH releasing hormone *in vivo* and *in vitro*. Negative feedback control of LH release by ovarian secretions is absent during the first weeks after birth and appears to develop between 5 and 8 weeks of age. The stimulatory oestrogen feedback LH discharge mechanism gradually develops and can be considered essentially mature shortly before the onset of puberty.

Although it is still premature to explain the pubertal process in the female pig, it appears that the onset of puberty is neither limited by the functioning of the stimulatory oestrogen feedback mechanism, nor by the pituitary or the ovary, as each can be activated prepubertally by treatment with appropriate hormones. Evidence in the pig and in other species suggests that puberty is brought about by a reduction in the intrinsic CNS inhibitory mechanism and/or a decrease in the negative feedback action of gonadal steroids, resulting in stimulation of pulsatile LH releasing hormone release and consequently augmentation of episodic LH secretion. This in turn stimulates ovarian function. In addition to the activation of ovarian steroidogenesis, a maturational decrease in the weight-corrected metabolic clearance rate of oestradiol appears to contribute to the elevation of oestradiol levels and thus to the constitution of an effective oestrogen feedback signal, which triggers the first cyclic discharge of LH.

Acknowledgements

The work was supported by grants from the German Research Foundation (DFG).

References

ALLEN, B.M. (1904). The embryonic development of the ovary and testis of the mammals. *Am. J. Anat.* **3**, 89–144

BARNES, R.J., COMLINE, R.S. and SILVER, M. (1974). Foetal and maternal plasma progesterone concentrations in the sow. *J. Endocr.* **62**, 419–420

BARRACLOUGH, C.A. (1966). Modifications in the CNS regulation of reproduction after exposure of prepubertal rats to steroid hormones. *Recent Prog. Horm. Res.* **22**, 503–539

BRUHN, TH., PARVIZI, N. and ELLENDORFF, F. (1981). Ability of the fetal hypothalamus to alter LH-secretion in response to electrical and electrochemical stimulation. *Acta endocr.* **96**, *Suppl.* **240**, 46

BYRNES, W.W. and MEYER, R.K. (1951). The inhibition of gonadotrophic hormone secretion by physiological doses of estrogen. *Endocrinology* **48**, 133

CASIDA, L.E. (1934). Prepubertal development of the pig ovary and its relationship to stimulation with gonadotrophic hormones. *Anat. Rec.* **61**, 389–396

CHOONG, C.H. and RAESIDE, J.I. (1974). Chemical determination of oestrogen distribution in the foetus and placenta of the domestic pig. *Acta endocr.* **77**, 171–185

COLENBRANDER, B., de JONG, F.H. and WENSING, C.J.G. (1978). Changes in serum testosterone concentrations in the male pig during development. *J. Reprod. Fert.* **53**, 377–380

COLENBRANDER, B., van de WIEL, D.F.M. and WENSING, C.J.G. (1980). Changes in serum FSH concentrations during fetal and pre-pubertal development in pigs. *VIth Int. Congr. Endocr.* Abstract No. 37

COLENBRANDER, B., WENSING, C.J.G. and van ROSSUM KOH, C.M.J.E. (1981). The decapitated pig fetus as a model for the study of the morphological and functional development of some endocrine organs. *Acta morph. neerl.-scand.* (Abstract, in press).

COLENBRANDER, B., KRUIP, TH.A.M., DIELEMAN, S.J. and WENSING, C.J.G. (1977). Changes in serum LH concentrations during normal and abnormal sexual development in the pig. *Biol. Reprod.* **17**, 506–513

COLENBRANDER, B., MacDONALD, A.A., PARVIZI, N. and ELSAESSER, F. (1980). Changing responsiveness of fetal pig pituitary to LHRH. *XXVII Int. Congr. Physiol. Sci.*, Abstract No. 1106

COLENBRANDER, B., MacDONALD, A.A., MEIJER, J.C., ELLENDORFF, F., van de WIEL, D.F.M. and BEVERS, M.M. (1981). Prolactin in the pig fetus. *Eur. J. Obstet. Gynaec. Reprod. Biol.* (Abstract, in press).

DIEKMAN, M.A. and ANDERSON, L.L. (1979). Quantitation of nuclear and cytoplasmic receptors for estradiol-17β and progesterone in the pituitary, hypothalamus, and uterus in the gilt during prepubertal development. *Proc. 71st Ann. Meet. Am. Soc. Anim. Sci., University of Arizona, Tucson,* pp. 290–291

DÖCKE, F. and BUSCH, W. (1974). Evidence for anterior hypothalamic control of cyclic gonadotrophin secretion in female pigs. *Endocrinology* **63**, 415–421

EDWARDS, S. (1980). Reproductive physiology of the post parturient sow. PhD Thesis. University of Nottingham

ELSAESSER, F. and FOXCROFT, G.R. (1978). Maturational changes in the characteristics of oestrogen-induced surges of luteinizing hormone in immature domestic gilts. *J. Endocr.* **78**, 455–456

ELSAESSER, F. and PARVIZI, N. (1977). Prepubertal active immunization against gonadal steroids: effect on estrus, ovulation and the oscillatory pattern of plasma LH and progesterone in the female pig. *Acta endocr.* **84**, *Suppl.* **208**, 107–108

ELSAESSER, F. and PARVIZI, N. (1979). Estrogen feedback in the pig: sexual differentiation and the effect of prenatal testosterone treatment. *Biol. Reprod.* **20**, 1187–1193

ELSAESSER, F. and PARVIZI, N. (1980). Partial recovery of the stimulatory oestrogen feedback action on LH release during late lactation in the pig. *J. Reprod. Fert.* **59**, 63–67

ELSAESSER, F., PARVIZI, N. and ELLENDORFF, F. (1978). Steroid feedback on luteinizing hormone secretion during sexual maturation in the pig. *J. Endocr.* **78**, 329–342

ELSAESSER, F., STICKNEY, K. and FOXCROFT, G.R. (1982). A comparison of metabolic clearance rates of oestradiol-17β in immature and peripubertal female pigs and possible implications for the onset of puberty. *Acta endocr.* (in press)

ELSAESSER, F., ELLENDORFF, F., PARVIZI, N. and KONIG, A. (1974). Response of the pituitary and testes to LHRH in the neonatal miniature pig. *Acta endocr. Suppl.* **184**, 29

ELSAESSER, F., ELLENDORFF, F., POMERANTZ, D.K., PARVIZI, N. and SMIDT, D. (1976). Plasma levels of luteinizing hormone, progesterone, testosterone and 5α-dihydrotestosterone in male and female pigs during sexual maturation. *J. Endocr.* **68**, 347–348

FORD, J.J. and SCHANBACHER, B.D. (1977). Luteinizing hormone secretion and female lordosis behaviour in male pigs. *Endocrinology* **100**, 1033–1038

FORD, J.J., CHRISTENSON, R.K. and MAURER, R.R. (1980). Serum testosterone concentrations in embryonic and fetal pigs during sexual differentiation. *Biol. Reprod.* **23**, 583–587

FOSTER, D.L. and KARSCH, F.J. (1975). Development of the mechanism regulating the preovulatory surge of luteinizing hormone in sheep. *Endocrinology* **97**, 1205–1209

FOSTER, D.L. and RYAN, K.D. (1979). Endocrine mechanisms governing transition into adulthood: a marked decrease in inhibitory feedback action of estradiol on tonic secretion of luteinizing hormone in lamb during puberty. *Endocrinology* **105**, 896–904

FOXCROFT, G.R. (1978). The development of pituitary gland function. In *Control of Ovulation* (D.B. Crighton, G.R. Foxcroft, N.B. Haynes and G.E. Lamming, Eds.), pp. 117–138. London, Butterworths

FOXCROFT, G.R., POMERANTZ, D.K. and NALBANDOV, A.V. (1975). Effects of estradiol-17β on LH-RH/FSH-RH-induced, and spontaneous, LH release in prepubertal female pigs. *Endocrinology* **96**, 551–557

GORSKI, R.A. (1973). Perinatal effects of sex steroids on brain development and function. *Prog. Brain Res.* **39**, 149–163

KARSCH, F.J., DIERSCHKE, D.J. and KNOBIL, E. (1973). Sexual differentiation of pituitary function: apparent difference between primates and rodents. *Science* **179**, 484–486

KATHER, L. and SMIDT, D. (1975). Vergleichende Untersuchungen zur ovariellen Reaktion infantiler weiblicher Schweine der Deutschen Landrasse und des Göttinger Miniaturschweines auf gonadotrope Stimulierung. *Zuchthygiene* **10**, 10–15

LIWSKA, J. (1975). Development of the adenohypophysis in the embryo of the domestic pig. *Folia morph.* **34**, 211–217

LIWSKA, J. (1978). Ultrastructure of the adenohypophysis in the domestic pig (*Sus scrofa domestica*). Part I: Cells of the pars anterior. *Folia histochem. cytochem.* **16**, 307–314

MACDONALD, A.A., ELSAESSER, F., PARVIZI, N., HEILHECKER, A., SMIDT, D. and ELLENDORFF, F. (1979). Progesterone, oestradiol-17β and luteinizing hormone concentrations in the blood and amniotic fluid of chronically catheterized pig foetuses. *J. Endocr.* **80**, 14P

MACDONALD, A.A., COLENBRANDER, B., ELSAESSER, F. and HEILHECKER, A. (1980). Progesterone production by the pig fetus and the response to stimulation by adrenocorticotrophin. *J. Endocr.* **85**, 34–35

MAULEON, P. (1964). Deroulement de l'ovogenèse comparé chez differents mammiferes domestiques. *Proc. IVth Int. Congr. Anim. Reprod. A.I., The Hague,* **2**, 348–354

MELAMPY, R.M., HENRICKS, D.M. ANDERSON, L.L. CHEN, C.L. and SCHULTZ, J.R. (1966). Pituitary follicle-stimulating hormone and luteinizing hormone concentrations in pregnant and lactating pigs. *Endocrinology* **78**, 801–804

OXENDER, W.D., COLENBRANDER, B., van de WEIL, D.F.M. and WENSING, C.J.G. (1979). Ovarian development in fetal and prepubertal pigs. *Biol. Reprod.* **21**, 715–721

PARVIZI, N., ELSAESSER, F., SMIDT, D. and ELLENDORFF, F. (1976). Plasma luteinizing hormone and progesterone in the adult female pig during the oestrous cycle, late pregnancy and lactation and after ovariectomy and pentobarbitone treatment. *J. Endocr.* **69**, 193–203

POMERANTZ, D.K., FOXCROFT, G.R. and NALBANDOV, A.V. (1975). Acute and chronic estradiol-17β inhibition of LH release in prepubertal female pigs: time course and site of action. *Endocrinology* **96**, 558–563

RAESIDE, J.I. and MIDDLETON, A.T. (1979). Development of testosterone secretion in the fetal pig testis. *Biol. Reprod.* **21**, 985–989

RAMIREZ, V.D. and McCANN, S.M. (1965). Inhibitory effect of testosterone on luteinizing hormone secretion in immature and adult rats. *Endocrinology* **76**, 412–417

SIEBERS, J.W., PETERS, F., ZENZES, M.T. SCHMIDTKE, J. and ENGEL, W. (1977). Binding of human chorionic gonadotrophin to rat ovary during development. *J. Endocr.* **73**, 491–496

SMITH, P.E. and DORTZBACH, C. (1929). The first appearance in anterior pituitary of the developing pig foetus of detectable amounts of the hormones stimulating ovarian maturity and general body growth. *Anat. Rec.* **43**, 277–297

STICKNEY, K. (1982). The physiology of oestrogen-induced puberty in the gilt. PhD Thesis, University of Nottingham

WEITZMANN, E.D., BOYAR, R.M., KAPEN, S. and HELLMAN, L. (1975). The relationship of sleep and sleep stages to neuroendocrine secretion and biological rhythms in man. *Recent Prog. Horm. Res.* **31**, 399–441

WILDT, L., MARSHALL, G. and KNOBIL, E. (1980). Experimental induction of puberty in the infantile female rhesus monkey. *Science* **207**, 1373–1375

YAMAJI, T., DIERSCHKE, D.J., HOTCHKISS, J., BHATTACHARYA, A.N., SURVE, A.H. and KNOBIL, E. (1971). Estrogen induction of LH release in the rhesus monkey. *Endocrinology* **89**, 1034–1041

ZIMBELMAN, R.G. and LAUDERDALE, J.W. (1973). Failure of prepartum or neonatal steroid injections to cause infertility in heifers, gilts and bitches. *Biol. Reprod.* **8**, 388–391

6

FACTORS AFFECTING THE NATURAL ATTAINMENT OF PUBERTY IN THE GILT

P.E. HUGHES
Department of Animal Physiology and Nutrition, University of Leeds, UK

Puberty attainment in the gilt represents the onset of reproductive capability, since the first behavioural oestrus normally coincides with the pubertal ovulation. In the wild pig this first oestrus would occur in the late autumn when the gilt is approximately eight months of age (Signoret, 1980). However, in the domestic gilt the aim should be to stimulate the precocious attainment of puberty, the gilt being first mated, at second oestrus, at about 200 days of age.

It is both desirable and necessary to induce early puberty attainment in the gilt as she is a costly non-productive animal until the initiation of the first pregnancy. Thus, a review of the major factors influencing puberty attainment in the gilt is of value in identifying those measures which may be taken to reduce pubertal age and hence minimize the cost of introducing replacement gilts.

Factors influencing puberty attainment

AGE, LIVEWEIGHT AND RATE OF GROWTH

Several recent reviews have considered the relationship between puberty attainment and the age, liveweight and growth rate of the gilt (Hughes and Varley, 1980; Kirkwood, 1980). In each case the three factors were considered together as they are intimately related. In view of this it is difficult to dissociate the effects of each individual factor. Furthermore, in addition to having possible influences on puberty attainment themselves, they also reflect the effects of both genotype and nutrition. Thus, rather than review the effects of chronological age and livewieght on puberty attainment *per se* it is more useful to consider which of the two provides the more accurate reflection of the stage of development of the gilt. This measure of physiological development (often referred to as the physiological age) is used to assess the degree of maturity of the hypothalamic–pituitary–ovarian axis, and hence the ability of the gilt to respond to puberty stimulation.

Many reports suggest that gilts normally attain puberty at approximately 200–210 days of age (Duncan and Lodge, 1960; Anderson and Melampy, 1972). However, since the range in age at puberty can be as wide as 102 days (Aherne *et al.*, 1976) to 350 days and above (Brooks and Smith, 1980) the average figure appears to be of little value. Equally, gilt liveweight at puberty is extremely variable, reported pubertal weights varying from 55 kg (Aherne *et al.*, 1976) to over 120 kg (Hughes and Cole, 1975).

Despite this variability it has been reported repeatedly that chronological age is a more accurate indicator of degree of development than is liveweight (Robertson *et al.*, 1951a; Robertson *et al.*, 1951b; Duncan and Lodge, 1960; Hughes and Cole, 1976). It appears that the variability in gilt response to external stimuli (particularly boar contact) is greater at a constant liveweight than at a constant chronological age (Hughes and Cole, 1976).

There can be little doubt that many of the above data reflect the influences of many intrinsic and extrinsic factors, such as genotype, environment and boar contact, on puberty attainment. Nevertheless, they serve to illustrate the shortcomings of using either chronological age or liveweight as an indicator of a gilt's stage of development. Despite this, the consensus of opinion would appear to favour chronological age as the best available indicator of the physiological age of a gilt. Indeed, most studies on the effects of imposed stimuli on puberty attainment in the gilt have used a constant chronological age as the starting point for the stimulation. If the consensus of opinion is correct then the implication is that, under conventional feeding systems, the influence of nutrition on puberty attainment is minimal—otherwise the liveweight of the gilt would provide the better measure of physiological development. However, although this latter point remains in doubt (see below) it is reasonable to suggest that there exists a lower liveweight threshold below which gilts will not achieve puberty (Dickerson, Gresham and McCance, 1964).

If liveweight does exert little influence on puberty attainment it might also be supposed that the rate of liveweight increase, or growth rate, would also have a minimal effect on the timing of puberty. Many reports do suggest that this is the case in the gilt (Burger, 1952; Haines, Warnick and Wallace, 1959; Gossett and Sorensen, 1959; Sorensen, Thomas and Gossett, 1961; Goode, Warnick and Wallace, 1965), although this is not necessarily true in all species (Brody, 1945; Monteiro and Falconer, 1966). Despite this conclusion Reutzel and Sumption (1968) have demonstrated that a positive genetic relationship exists between age and puberty and daily liveweight gain in the gilt. These data imply that, while genotype may influence puberty attainment, variations in growth rate of animals of similar genotype will have little effect on puberty.

NUTRITION

Whilst it has been shown that both liveweight and growth rate may, under normal conditions, exert little influence on puberty attainment in the gilt, it must be emphasized that this conclusion is rather tenuous. Since both factors may reflect the nutrient supply to the animal it may be easier to reach more positive conclusions through studying the effects of variations in feed supply on puberty attainment. Thus, the relationships between the attainment of puberty and both plane of feeding and dietary composition are considered here.

Plane of feeding

Many early experiments to elucidate nutritional effects on reproduction in the pig used full- and limited-feeding systems as the basis for the studies. This means that the dietary regimes compared contained different levels of both energy and protein. Therefore, this work must be considered separately from the more recent investigations of energy and protein effects which have employed isonitrogenous and isoenergetic diets respectively.

The results available from these studies are both variable and conflicting. Several workers have reported that full feeding of prepubertal gilts results in the early attainment of puberty when compared with their restricted-fed counterparts (Burger, 1952; Zimmerman *et al.*, 1960; Goode, Warnick and Wallace, 1965; Friend, 1974; 1976). However, other results suggest that puberty attainment is either unaffected by plane of feeding (Robertson *et al.*, 1951b; Christian and Nofziger, 1952; Lodge and MacPherson, 1961, Pay and Davies, 1973) or is delayed by full feeding (Self, Grummer and Casida, 1955; Hafez, 1960; MacPherson, Hovell and Jones, 1977). These results are summarized in *Table 6.1.*

Table 6.1 THE EFFECTS OF PLANE OF NUTRITION ON PUBERTY ATTAINMENT IN GILTS

Age at puberty (days)		*Source*
High plane diet	*Low plane diet*	
198	203	Robertson *et al.* (1951b)
188	235	Burger (1952)
170	167	Christian and Nofziger (1952)
223	208	Self *et al.* (1955)
195	217	Haines *et al.* (1959)
212	182	Hafez (1960)
195	205	Zimmerman *et al.* (1960)
178	176	Lodge and MacPherson (1961)
262	292	Goode *et al.* (1965)
189	187	Pay and Davies (1973)
184	201	Friend (1974)
173	194	Friend (1976)
203	186	MacPherson *et al.* (1977)

The fact that significant results have been obtained to show that high plane feeding can both stimulate and inhibit puberty attainment suggests that other factors may also have been involved to confound the results. Indeed, both breed and seasonal effects are apparent in several of the experiments (e.g. Zimmerman *et al.*, 1960). More importantly, many of the experiments used boars for oestrus detection, hence introducing a further stimulatory influence (e.g. Robertson *et al.*, 1951b; Christian and Nofziger, 1952; Self, Grummer and Casida, 1955; Haines, Warnick and Wallace, 1959; Zimmerman *et al.*, 1960; Goode, Warnick and Wallace, 1965; Pay and Davies, 1973; MacPherson, Hovell and Jones, 1977). Since breed, season and boar contact are all known to influence puberty attainment, the interpretation of the above results must be questioned. Until such time as the interactions between plane of nutrition and these other stimulatory influences are understood no definite conclusion can be reached on the effects of feed level on puberty attainment in gilts.

Energy intake

The relationship between energy intake and puberty attainment in the gilt has been extensively reviewed by Anderson and Melampy (1972) and, more recently, by Hughes and Varley (1980). It was apparent in 1972 that no definite relationship between the two factors could be established on the basis of the available data. Indeed, Anderson and Melampy (1972) concluded from 14 experiments that restricted energy intake delayed puberty by an average of 16 days in nine experiments, whereas the restricted diet hastened the onset of puberty by 11 days in five other experiments (see *Table 6.2*).

Table 6.2 THE EFFECTS OF ENERGY INTAKE ON PUBERTY ATTAINMENT IN GILTS

Number of trials	*ME intake* (MJ/day)		*Pubertal age* (days)		*Pubertal bodyweight* (kg)	
	Restricted	*Full*	*Restricted*	*Full*	*Restricted*	*Full*
9	23.2	37.5	217	201	74	91
5	25.2	37.5	201	212	74	94

From Anderson and Melampy (1972).

Unfortunately, more recent studies do not clarify the situation. In fact, in four experiments studying the effects of energy intake on puberty attainment in the gilt, three concluded that no influence was apparent (Aherne *et al.*, 1976; Friend, 1977; Friend *et al.*, 1979) whereas the fourth suggested that a high energy intake resulted in a delayed onset of puberty (Etienne and Legault, 1974).

Once again these results are undoubtedly confounded by other stimuli, such as boar contact (Gossett and Sorensen, 1959; Sorensen, Thomas and Gossett, 1961; Friend, 1977; Friend *et al.*, 1979). Equally, factors such as degree of energy restriction, gilt age at the start of restriction, and dietary composition may have influenced the results. Hughes and Varley (1980) suggested that the restricted energy intakes used in many of the above studies were, in fact, adequate for an acceptable rate of growth. On this basis it may be expected that little effect would be seen on puberty attainment until a more severe level of restriction was imposed. Indeed, when Burger (1952) severely restricted energy intake (to 50% of *ad libitum*) puberty attainment was delayed by 46 days.

It seems that little progress has been made since Anderson and Melampy reviewed the situation in 1972. Nevertheless, it is reasonable to assume that puberty attainment will not be adversely affected by the levels of dietary energy being fed in practice, providing these contain sufficient energy to ensure an adequate level of liveweight growth.

Fat intake

Little information is available on the effects of dietary fat on puberty attainment in the gilt. It is generally assumed that fat will exert an influence only through its contribution to the total energy content of the diet. However, Witz and Beeson (1951) have reported that puberty attainment is delayed in gilts fed fat-free diets. What remains to be established is whether this effect was due to dietary energy level or to a specific fat requirement by the gilt.

Protein intake

So far it has been assumed that the variations in gilt age at puberty as a result of plane of feeding were due to variations in energy intake. However, since protein intake would also vary with plane of feeding the effects of protein intake and aminoacid balance on puberty attainment must also be considered.

Several reports show that pubertal age increases when gilts are fed a protein-deficient diet (Davidson, 1930; Baker, 1959; Adams *et al.*, 1960; Cunningham *et al.*, 1974; Wahlstrom and Libal, 1977). On the other hand, minor variations in protein intake, in diets containing conventional levels of protein, do not appear to affect puberty attainment significantly (Friend, 1977; Friend *et al.*, 1979).

Fowler and Robertson (1954) reported that gilts fed protein of animal origin reached puberty significantly earlier than counterparts fed plant protein. This suggests that aminoacid imbalances in the diet may influence puberty attainment. Indeed, more recent evidence substantiates this, dietary supplementation with lysine and methionine reducing gilt age at puberty (Larrson, Nilsson and Olsson, 1966; Friend, 1973).

Vitamin and mineral intake

The roles of vitamins and minerals in pig reproduction generally, and puberty attainment in particular, have received very little attention. However, it does seem probable that severe restrictions of many vitamins and minerals, including vitamin A (Hughes, 1934), vitamin B_{12} (Johnsen, Moustgaard and Højgaard-Olsen, 1952) and manganese (Plumlee *et al.*, 1956), may delay sexual development.

Overall, it has been suggested that nutrient supply, when sufficient for commercially acceptable growth rates, is unlikely to have any significant influence on puberty attainment in the gilt (Tassell, 1967; Brooks and Cole, 1974).

GENETICS

Ramirez (1973) states that 'the ontogeny of puberty is genetically determined for each species, but the normal rate of maturation can be altered by the environment and this can modify the stage of development'. This implies that genotype will exert a considerable influence on puberty attainment in gilts, although this influence may be masked by environmental stimuli.

Several workers have detected breed differences in age at puberty in the pig (Phillips and Zeller, 1943; Self, Grummer and Casida, 1955; Etienne and Legault, 1974; Christenson and Ford, 1979). However, the variability in pubertal age within a breed is considerable, this being cited as a reason why other workers have failed to find significant breed differences (Robertson *et al.*, 1951a,b; Warnick *et al.*, 1951; Zimmerman *et al.*, 1960) – see *Table 6.3*. This within-breed variability in pubertal age may reflect the effects of other stimulatory influences (e.g. boar contact). Additionally, Christenson and Ford (1979) have concluded that significant breed–environment interactions may also exist, these only serving to further confound the situation. Thus, as in previous sections, the conflicting nature of the data prohibits the drawing of useful conclusions.

Table 6.3 THE EFFECTS OF BREED ON PUBERTY ATTAINMENT IN THE GILT

Source	*Breed used*	*Age at puberty* (days)
Robertson *et al.* (1951a)	Chester White	203
	Poland China	198
Robertson *et al.* (1951b)	Chester White	201
	Poland China	204
Warnick *et al.* (1951)		
Trial 1	Chester White	236
	Yorkshire	243
Trial 2	Chester White	259
	Yorkshire	249
Self *et al.* (1955)		
Trial 1	Chester White	225
	Poland China	206
Trial 2	Chester White	211
	Poland China	216
Zimmerman *et al.* (1960)	Chester White (restricted-fed)	198
	Poland China (restricted-fed)	192
	Chester White (full-fed)	206
	Poland China (full-fed)	204
Dyck (1971)	Yorkshire	199
	Lacombe	197
Christenson and Ford (1979)	Landrace	173
	Large White	211
	Hampshire	207
	Duroc	224
	Yorkshire	221

In contrast, heterotic effects on puberty attainment in the gilt have been repeatedly and consistently demonstrated. The data presented in *Table 6.4* clearly show that the crossbred gilt attains puberty earlier than its purebred counterpart. Such heterotic effects are not, however, confined to crosses between breeds. Foote *et al.* (1956), studying Chester White, Yorkshire and Chester White × Yorkshire gilts, reported that mean age at puberty was significantly reduced in line-cross gilts when compared with inbred gilts (194 versus 228 days respectively).

Most reproductive traits that show heterosis tend to be of low heritability. Indeed, Reutzel and Sumption (1968) have reported that the heritability of pubertal age is low or zero, when assessed on a paternal half-sib

Table 6.4 THE DIFFERENCE IN PUBERTAL AGE BETWEEN PUREBRED AND CROSSBRED GILTS

Source	*Breed of sire*	*Breed of dam*	*Difference in pubertal age between crossbred gilts and their purebred counterparts (days)*
Warnick *et al.* (1951)	Trial 1:		
	Chester White	Yorkshire	—4
	Trial 2:		
	Chester White	Yorkshire	—3
Foote *et al.* (1956)	Poland China	Duroc	—11
	Poland China	Yorkshire	—32
	Duroc	Yorkshire	—36
Short *et al.* (1963)	Duroc	Yorkshire	—10
	Yorkshire	Duroc	—5
Clark *et al.* (1970)	Poland China	Yorkshire	—24
	Yorkshire	Poland China	—6

analysis. However, these workers did obtain a heritability estimate of 0.49 when a daughter–dam regression was used. Such data may reflect a dam effect on the pubertal age of offspring. Other workers have also noted family effects (Burger, 1952) and, particularly, sire effects on gilt age at puberty (Reddy, Lasley and Mayer, 1958; Short, 1963; Hughes and Cole, 1975). Hence, while a reliable figure for the heritability of pubertal age is not available, there would appear to be some genetic basis for selection of this trait.

CLIMATIC ENVIRONMENT

Although the domestic pig does not have a specific breeding season, it is derived from a seasonally breeding animal and might therefore be expected to show variations in fertility with time of year.

Many reports suggest that season of birth exerts a marked effect on puberty attainment in the gilt. Unfortunately, many of these data are conflicting (*Table 6.5*), some reports indicating that autumn-born gilts reach puberty earlier than their spring-born counterparts (Wiggins, Casida and Grummer, 1950; Gossett and Sorensen, 1959; Mavrogenis and Robison, 1976) while others indicate the reverse (Sorensen, Thomas and

Table 6.5 THE EFFECTS OF SEASON OF BIRTH ON PUBERTY ATTAINMENT IN THE GILT

Source	*No. of gilts*	*Age at puberty* (days)	
		Spring-born	*Autumn-born*
Gossett and Sorensen (1959)	52	216	206
Zimmerman *et al.* (1960)			
Trial 1	71	196	208
Trial 2	60	203	190
Sorensen *et al.* (1961)	98	210	208
Mavrogenis and Robison (1976)	153	237	202

Gossett, 1961; Scanlon and Krishnamurthy, 1974). Other studies have either failed to observe seasonal differences at all (Wise and Robertson, 1953; Self, Grummer and Casida, 1955; Christenson and Ford, 1979), found significant differences in one year but not the next (Warnick *et al.*, 1951), or found significant differences in only one of two breeds (Zimmerman *et al.*, 1960) Such inconsistent results must, once again, be reflections of stimulatory influences other than those associated with season. Thus, although it might be expected that spring-born gilts would mature earlier, as did their wild ancestors (Signoret, 1980), there is little evidence to substantiate this. Additionally, it is not clear that puberty attainment in the wild gilt is in response to seasonal changes, since boar contact is also initiated in the autumn.

Table 6.6 THE EFFECTS OF PHOTOPERIOD ON PUBERTY ATTAINMENT IN THE GILT

Source	*No. of gilts*	*Lighting regime*	*Age at puberty* (days)
Dufour and Bernard (1968)	–	complete darkness	208
	–	natural photoperiod	219
Hacker *et al.* (1974)	16	complete darkness	222
	16	12 hours light:12 hours dark	183
Hacker *et al.* (1976)	16	6 hours light: 18 hours dark	232
	16	18 hours light: 6 hours dark	190
Hacker *et al.* (1979)	6	complete darkness	200
	6	natural photoperiod (9.0–10.8 hours light per day)	165
	6	18 hours light: 6 hours dark	175
Ntunde *et al.* (1979)	12	complete darkness	193
	12	natural photoperiod (9.0–10.8 hours light per day)	176
	12	18 hours light: 6 hours dark	177

Seasonal differences will reflect changes in both photoperiod and temperature. It is therefore worthwile to consider the effects of these two factors in isolation. Most studies (*Table 6.6*) on photoperiod demonstrate that increasing daylength advances the onset of puberty in gilts (Martinat *et al.*, 1970; Surmuhin and Ceremnyh, 1970; Surmuhin *et al.*, 1970; Hacker, King and Bearss, 1974; Hacker, King and Smith, 1976; Hacker *et al.*, 1979; Ntunde, Hacker and King, 1979). The one conflicting report is that of Dufour and Bernard (1968) who found that gilts reared in complete darkness reached puberty earlier than those reared in normal daylength. While this result agrees with data from other species (e.g. Zacharias and Wurtman, 1964), the rearing of gilts in complete darkness is an abnormal situation that is likely to yield abnormal results. Thus, it may be concluded that long daylengths will be conducive to the early maturation of gilts. The actual optimum light:dark schedule has not been elucidated, although the available data (Klotchkov, Klotchkova, Kim and Belyaev, 1971; Hacker, King and Smith, 1976: Hacker *et al.*, 1979; Ntunde, Hacker and King, 1979) does indicate that 17–18 hours light/day should be sufficient.

The effects of temperature on puberty attainment in gilts are less well documented. High environmental temperatures have been reported to exert an adverse influence on several aspects of reproduction in pigs

(Corteel, Signoret and Du Mesnil du Buisson, 1964; Edwards *et al.*, 1968; Teague, Roller and Griffo, 1968; Love, 1978), with limited evidence indicating that this may also be true for puberty attainment (Schmidt and Bretschneider, 1954; Stickney *et al.*, 1978). Conversely, Dyck (1974) has observed that the rearing of gilts under conditions of low environmental temperatures has no adverse influence on the attainment of puberty. It may be concluded that moderately low environmental temperatures will facilitate the early attainment of puberty in the gilt. However, it is also worth applying a note of caution, since work with other species has indicated that both very high and very low temperatures may delay the attainment of sexual maturity (Mandl and Zuckerman, 1952; Donovan and van der Werff ten Bosch, 1965).

When the data on photoperiod and temperature are combined to provide seasonal effects it is not surprising that variations in pubertal age are not consistent. Spring-born gilts will be stimulated to reach puberty early by the increasing daylength, but will be inhibited by high environmental temperatures. Conversely, autumn-born gilts will be stimulated by lower temperatures but inhibited by short daylengths. Thus, the actual location of the experimental site (together with the environmental control of the building, where applicable), and hence the actual temperatures and photoperiod experienced by the gilts, will, to a large extent, determine the response obtained.

SOCIAL ENVIRONMENT

It has long been known that 'stress' factors may both stimulate and inhibit puberty attainment in rodents, the response being dependent on the type and level of 'stress' applied (Duckett, Varon and Christian, 1963; Morton, Denenberg and Zarrow, 1963). In the gilt there is also limited evidence to indicate that some forms of 'stress' may stimulate the early attainment of puberty.

Table 6.7 THE EFFECTS OF STOCKING DENSITY ON PUBERTY ATTAINMENT IN THE GILT

Source	*No. of gilts*	*No. of gilts/pen*	*Age at puberty* (days)
Jensen *et al.* (1970)[a]	–	1	260
		5–7	262
		6–10	261
Robison (1974)	15	1	225
	30	30	216
Mavrogenis and Robison (1976)	50	1	222
	103	30	207
Ford and Teague (1978)[b]	32	8	217
	96	12	199
Christenson and Ford (1979)			
Trial 1	31	8	206
	48	24	207
Trial 2	32	8	247
	46	24	253

[a]Data averaged from three trials
[b]Data from three trials where space restriction was also varied.

Several workers have reported that the crowding of gilts will stimulate early sexual maturation (Robison, 1974; Mavrogenis and Robison, 1976), although other workers have found no effect (Jensen *et al.*, 1967, 1970; Christenson and Ford, 1979). These data are summarized in *Table 6.7.* What is not clear at present is whether crowding effects, if they exist, are mediated through space restriction (a form of 'stress') or through the presence of many other females (Ford and Teague, 1978).

Undoubtedly, the best known 'stress' factor in pigs is that of transportation. Many workers have reported that transporting prepubertal gilts will result in the precocious attainment of puberty (Du Mesnil du Buisson and Signoret, 1961; Paredis, 1961; Signoret, 1970). However, this stimulus may not be effective until the gilt is very near puberty, since Signoret (1972) reported that only 26.5% of gilts weighing 90–110 kg responded to transportation. When this effect is apparent it is clearly independent of boar stimulation since the peak of oestrus occurrence is unaltered when boar exposure is delayed for 10 days after transportation (Signoret, 1970) as shown in *Figure 6.1*.

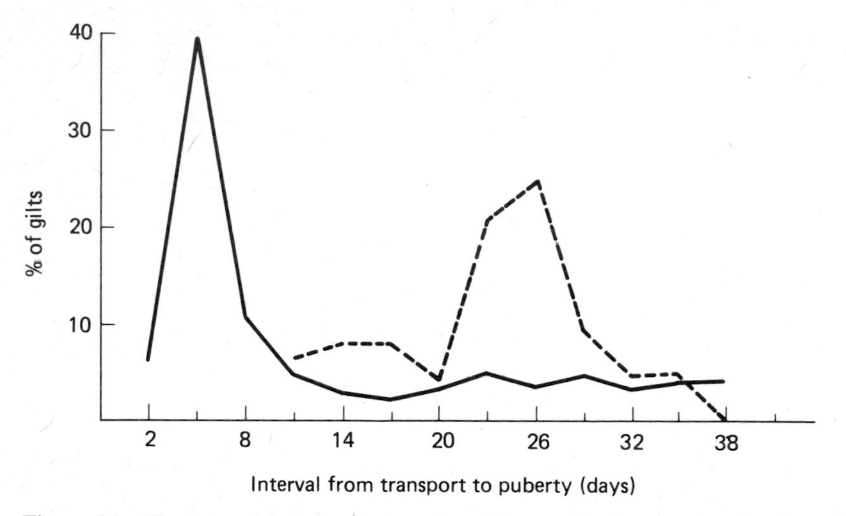

Figure 6.1 The interval between transport and the onset of oestrus in gilts. Dotted line indicates results when presentation to the boar is delayed for 10 days. From Signoret (1970)

The most recent studies in this area have concentrated on the interactions of mixing strange gilts, transportation and relocation, these being tested in the presence or absence of boar contact. The data on these 'management stresses', summarized in *Table 6.8* indicate that all these 'stress' factors may have stimulatory effects on puberty attainment in gilts. However, the results presented in *Table 6.8* also indicate that boar contact is a more potent form of puberty stimulation than is any form of 'stress'. Finally, it is worth noting that boar contact itself may also include a 'stress' component (Kirkwood, Forbes and Hughes, 1981).

BOAR CONTACT

It has already been pointed out that contact with a boar will induce puberty attainment in the prepubertal gilt. Indeed, this fact has been repeatedly

Table 6.8 THE EFFECT OF 'MANAGEMENT STRESS', WITH OR WITHOUT BOAR CONTACT, ON PUBERTY ATTAINMENT IN THE GILT

Source	*No. of gilts*	*'Management stress'*			*Boar contact*	*Treatment age* (days)	*Age at puberty* (days)
		Mixing[a]	*Transport*	*Relocation*			
Bourn *et al.* (1974)	—	−	−	−	+	135 or 165 days	167
	34	+	+	−	+	135 or 165 days	161
Zimmerman *et al.* (1974)	68	−	+	+	−	125, 150 or 175 days	204
	68	−	+	+	+	125, 150 or 175 days	181
Bourn *et al.* (1976)	44	+	+	+	−	135 or 165 days	193
	44	+	+	+	+	135 or 165 days	176
Scheimann *et al.* (1976)	—	+	−	−	−	−	(0%)[b]
		+	+	−	−		(26%)
		−	+	+	+		(79%)
		+	+	+	+		(88%)
Zimmerman *et al.* (1976)	25	+	−	−	−	165 days	(7%)[c]
	25	+	+	−	−	165 days	(8%)
	25	+	+	+	−	165 days	(28%)
	25	+	+	+	+	165 days	(87%)

[a]Mixing of strange gilts.
[b]Percentage of gilts reaching puberty within 17 days of the start of treatment.
[c]Percentage of gilts reaching puberty within 10 days of the start of treatment.

Table 6.9 THE EFFECTS OF GILT AGE AT FIRST BOAR CONTACT ON PUBERTAL AGE AND THE INTERVAL FROM FIRST BOAR CONTACT TO PUBERTY

Source	*Gilt age at boar introduction* (days)	*Interval from first boar contact to puberty* (days)	*Age at puberty* (days)
Zimmerman *et al.* (1969)	103	67	170
Zimmerman *et al.* (1974)	125	47	172
Kirkwood and Hughes (1979)	125	54	179
Zimmerman *et al.* (1969)	126	31	157
Kirkwood and Hughes (1979)	132	42	174
Bourn *et al.* (1974)	135	21	156
Hughes and Cole (1976)	135	34	169
Bourn *et al.* (1976)	135	40	175
Kirkwood and Hughes (1979)	139	27	166
Kirkwood and Hughes (1979)	146	21	167
Zimmerman *et al.* (1974)	150	32	182
Kirkwood and Hughes (1979)	153	28	181
Brooks and Smith (1980)	160	11	171
Paterson and Lindsay (1980)	160	15	175
Hughes and Cole (1976)	160	18	178
Kirkwood and Hughes (1979)	160	19	179
Brooks *et al.* (1969)	163	7	170
Brooks and Cole (1970)	165	7	172
Bourn *et al.* (1974)	165	8	173
Bourn *et al.* (1976)	165	13	178
Kirkwood and Hughes (1979)	167	12	179
Brooks and Cole (1969)	171	21	192
Kirkwood and Hughes (1979)	174	24	198
Zimmerman *et al.* (1974)	175	15	190
Kirkwood and Hughes (1979)	181	16	197
Brooks and Cole (1973)	183	16	199
Brooks and Cole (1973)	188	14	202
Hughes and Cole (1976)	190	15	205
Brooks and Smith (1980)	200	16	216
Brooks and Cole (1973)	215	10	225

demonstrated by many workers (Brooks and Cole, 1969; 1970; Brooks, Pattinson and Cole, 1969; Robison, 1974; Hughes and Cole, 1976; 1978; Mavrogenis and Robison, 1976; Thompson and Savage, 1978; Kirkwood and Hughes, 1979).

The major factor controlling the efficiency of boar contact as a puberty stimulus is the age of the gilt at the time of boar introduction. This is clearly shown in *Table 6.9* and *Figures 6.2* and *6.3*. When boar contact is initiated at very young gilt ages (3–4 months), pubertal response is minimal and sexual development may possibly be delayed (Zimmerman, Carlson and Nippert, 1969; Brooks and Cole, 1969; Doroshkov, 1974; Hughes and Cole, 1976). This is considered to reflect the relatively young physiological ages of these gilts at the time the stimulus is introduced. Consequently, it has been suggested by Brooks and Cole (1969) that the young gilt may become habituated to the boar stimulus at a stage in development when she is too young to respond. Thus, when the gilt reaches an age when a pubertal response to boar contact might be expected (see below), it is not forthcoming and puberty occurs at a similar age to that of unstimulated gilts.

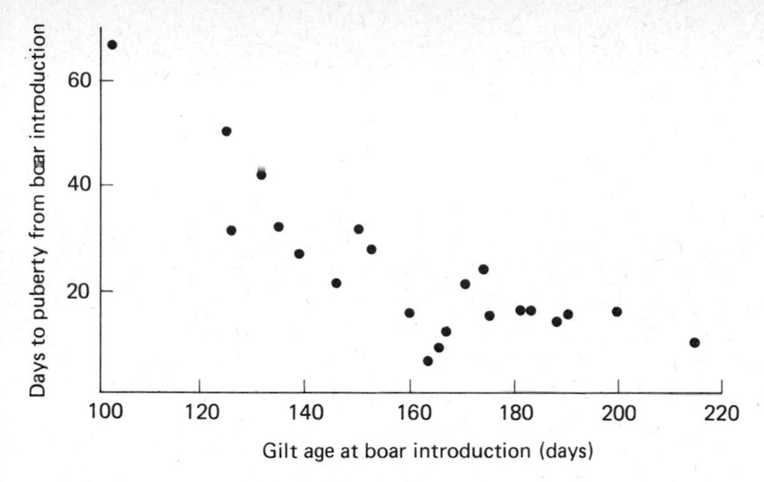

Figure 6.2 The relationship between gilt age at boar introduction and the interval from first boar contact to puberty (data relate to *Table 6.9*)

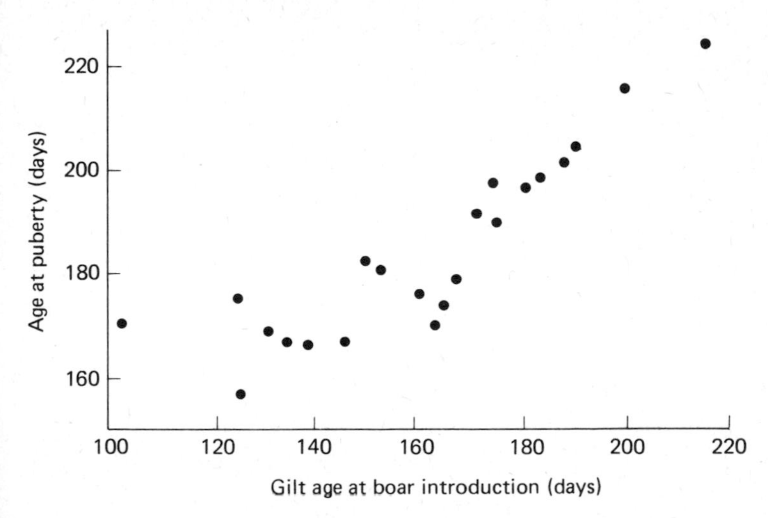

Figure 6.3 The relationship between gilt age at boar introduction and age at puberty (data relate to *Table 6.9*)

Conversely, when boar introduction is delayed until the immediate prepubertal period (six months of age and above) the response is again limited but for a different reason. When boar contact is initiated at these late gilt ages puberty attainment is stimulated, the interval from first boar contact to puberty being similar to that seen in gilts first stimulated at 160–165 days of age. However, by virtue of the relatively old ages of the gilts at boar introduction, the actual pubertal ages of these gilts are not much reduced below those of unstimulated animals.

While both very young and relatively old gilts demonstrate limited responses to boar contact, gilts first exposed to a boar at 150–170 days of age display a significant precocity in sexual development. Indeed, when boar introduction occurs at gilt ages in the region of 160 days, both the interval from first boar contact to puberty and gilt age at puberty are minimized (*Figures 6.2 and 6.3*). Additionally, initiating boar contact at this age results in maximum synchronization of the pubertal oestrus (Brooks and Cole, 1970; Hughes and Cole, 1976).

A report by Brooks and Cole (1970) suggested that gilt response to boar introduction at about 160 days of age could be improved by employing a rotation of boars rather than contact with a single boar. However, several confounding factors present in this experiment prohibited the automatic acceptance of this result. Indeed, more recent studies, albeit using limited numbers of animals, have failed to substantiate the earlier claim. The results of these recent experiments (Kirkwood, 1980; Kirkwood and Hughes, 1980a) indicate that neither a rotation of boars nor continuous boar contact improve gilt response above that obtained with single boar contact for 30 minutes/day (*Table 6.10*).

Table 6.10 THE EFFECTS OF 30 MINUTES/DAY BOAR CONTACT, CONTINUOUS BOAR CONTACT AND ROTATIONAL BOAR CONTACT ON PUBERTY ATTAINMENT IN THE GILT

Source	*No. of gilts/ treatment*	*Gilt age at puberty* (days)		
		30 min/day	*continuous*	*rotational*
Kirkwood and Hughes (1980a)	10	199	195	—
Kirkwood (1980)	6	187	183	194

All boar contact was initiated at 165 days of age.

The actual mechanism by which boar contact stimulates puberty attainment in gilts has only recently been investigated. It has long been known that, in other species, this 'male effect' is mediated via a primer pheromone (Vandenbergh, 1969), this, in turn, stimulating an endogenous LH (luteinizing hormone) release in the recipient female (Bronson and Desjardins, 1974). In the pig it was unclear whether the male effect was due to a pheromone, or whether other boar stimuli such as 'stress', auditory and visual cues were involved. A recent experiment carried out by Kirkwood, Forbes and Hughes (1981) set out to investigate this problem by considering the ability of boars to stimulate puberty attainment in anosmic gilts. The results of this trial show that boar contact is ineffective in stimulating precocious puberty attainment in gilts that are unable to perceive pheromones (*Table 6.11*). This implies that the 'boar effect' is mediated via a primer pheromone. It should, however, be recognized that other boar-originating stimuli may act synergistically with the primer pheromone(s) to produce the complete 'boar effect'.

Table 6.11 THE INFLUENCE OF BOAR CONTACT[a] ON PUBERTY ATTAINMENT IN INTACT AND OLFACTORY BULBECTOMIZED GILTS

Treatment	*Interval from first boar contact to puberty* (days)	*Age at puberty* (days)
Control—no boar contact or surgery	74	234
Boar exposure—olfactory bulbectomized	70	230
Boar exposure—sham-operated	44	204
Boar exposure—no surgery	48	208

[a]From 160 days of age.
From Kirkwood, Forbes and Hughes (1981).

The nature of the boar pheromone responsible for puberty stimulation has yet to be identified. The major free pheromones produced by the boar originate in the submaxillary salivary gland, and this is considered to be the most probable source of the puberty-accelerating pheromone (Booth, 1980; Kirkwood and Hughes, 1980b; Kirkwood, Forbes and Hughes, 1981). Indeed, it has been suggested that the active pheromone responsible for puberty induction is 3α-androstenol, but this claim remains to be substantiated.

Most reproductively active pheromones produced by the boar are secreted at very low levels in the young boar, and do not rise to adult levels until the boar approaches one year of age (Booth, 1975). It is, therefore, necessary to consider whether the young postpubertal boar is capable of inducing precocious puberty attainment in the gilt. The one small trial that has been conducted on this subject (*Table 6.12*) indicated that the boar

Table 6.12 THE EFFECT OF EXPOSURE TO BOARS[(a)] OF DIFFERENT AGES ON PUBERTY ATTAINMENT IN THE GILT

Treatment	*Interval from first boar contact to puberty* (days)	*Age at puberty* (days)
Control—no boar exposure	39	203
Exposure to 6.5-month old boar	42	206
Exposure to 11-month old boar	18	182
Exposure to 24-month old boar	19	182

[(a)]30 minute daily exposure from an average gilt age of 164 days.
From Kirkwood and Hughes (1981).

may not be able to stimulate puberty attainment in the gilt until he is approximately 9–10 months of age (Kirkwood and Hughes, 1981). Thus, it may be suggested that older boars are used to induce puberty attainment in replacement gilts. Finally, since entire litter brothers would not be stimulatory to their sisters, it seems unlikely that mixed-sex rearing would result in an habituation effect in the gilt—a conclusion supported by the recent work of Nathan and Cole (1981).

Conclusions

It is apparent from the foregoing discussion that the rate of sexual development in gilts can be modified by many factors. However, it is equally clear that there has been a lack of definitive research to identify the influences of individual stimulatory/inhibitory factors on puberty attainment. The problem arises, in the main, as a result of using boars to detect the occurrence of the pubertal oestrus. Since it has already been demonstrated that boar contact is the single most effective puberty stimulus, it is difficult to assess the true influence of other factors being examined in such experiments. Whilst the boar contact may be applied over all treatments, and hence may be considered as a standard part of the experiment, it will certainly mask the influences of the applied variables (e.g. nutrition, genotype or climatic environment) and may also interact with them to confound interpretation. In view of these problems, it has proved necessary to apply a note of caution when interpreting the majority of available data on factors influencing puberty attainment. Indeed, it is suggested that future experiments in this area should, with the exception of those studies aimed at elucidating aspects of the 'boar effect', avoid the use of boars for oestrus detection. It is recognized that such a move would make the detection of oestrus difficult but, equally, it would allow easier interpretation of results. Under such conditions oestrus detection may be carried out using one or more of three available methods. These are:

(a) visual observations of vulval and behavioural changes,
(b) regular blood sampling and the measurement of plasma progesterone concentrations, and
(c) slaughter followed by ovarian examination for the presence of corpora lutea and corpora albicantia.

This system, using all three methods, has been applied to good effect by Hughes and Cole (1978), and has proved to be as effective (if not as simple) a method of oestrus detection as the use of a boar.

Finally, despite the reservations expressed above, it may be concluded that boar contact is the single most effective natural means of stimulating precocious puberty attainment in the gilt. Response to this stimulus may, however, be augmented by the additional imposition of 'social stress'. Under commercial conditions nutritional and climatic influences on sexual development should be minimal in temperate regions, although both are likely to exert adverse influences under sub-tropical and tropical conditions. Lastly, genotype differences may prove to be useful in a selection programme in the future, but, for the present, there is insufficient information available to justify such a programme.

References

ADAMS, C.R., BECKER, D.E., TERRILL, S.W., NORTON, M.W. and JENSEN, A.H. (1960). Rate of ovulation and implantation in swine as affected by dietary protein. *J. Anim. Sci.* **19**, 1245 (Abstract)

AHERNE, F.X., CHRISTOPHERSON, R.J., THOMPSON, J.R. and HARDIN, R.T. (1976). Factors affecting the onset of puberty, postweaning oestrus and blood hormone levels of Lacombe gilts. *Can. J. Anim. Sci.* **56**, 681–692

ANDERSON, L.L. and MELAMPY, R.M. (1972). Factors affecting ovulation rate in the pig. In *Pig Production* (D.J.A. Cole, Ed.), pp. 329–366. London Butterworths

BAKER, B. (1959). Effects of changing levels of nutrition on reproduction of gilts. *J. Anim. Sci.* **18**, 1160 (Abstract)

BOOTH, W.D. (1975). Changes with age in the occurrence of C_{19} steroids in the testis and submaxillary gland of the boar. *J. Reprod. Fert.* **42**, 459–472

BOOTH, W.D. (1980). A study of some major testicular steroids in the pig in relation to their effect on the development of male characteristics in the pre-pubertally castrated boar. *J. Reprod. Fert.* **59**, 155–162

BOURN, P., CARLSON, R., LANTZ, B. and ZIMMERMAN, D.R. (1974). Age at puberty in gilts as influenced by age at boar exposure and transport. *J. Anim. Sci.* **39**, 987 (Abstract)

BOURN, P., KINSEY, R., CARLSON, R. and ZIMMERMAN, D.A. (1976). Puberty in gilts as influenced by boar exposure and "transport phenomenon". *J. Anim. Sci.* **41**, 344 (Abstract)

BRODY, S. (1945). *Bioenergetics and Growth*. New York, Reinhold

BRONSON, F.H. and DESJARDINS, C. (1974). Circulating concentrations of FSH, LH, oestradiol and progesterone associated with acute male-induced puberty in mice. *Endocrinology* **94**, 1658–1668

BROOKS, P.H. and COLE, D.J.A. (1969). Effect of boar presence on the age at puberty of gilts. *Rep. Sch. Agric. Univ. Nott.* 74–77

BROOKS, P.H. and COLE, D.J.A. (1970). Effect of the presence of a boar on attainment of puberty in gilts. *J. Reprod. Fert.* **23**, 435–440

BROOKS, P.H. and COLE, D.J.A. (1973). Meat production from gilts which have farrowed. 1. Reproductive performance and food conversion efficiency. *Anim. Prod.* **17**, 305–315

BROOKS, P.H. and COLE, D.J.A. (1974). The effect of nutrition during the growing period and the oestrous cycle on the reproductive performance of the pig. *Livest. Prod. Sci.* **1**, 7–20

BROOKS, P.H., PATTINSON, M.A. and COLE, D.J.A. (1969). Reproduction in the young gilt. *Rep. Sch. Agric. Univ. Nott.* 65–67

BROOKS, P.H. and SMITH, D.A. (1980). The effect of mating age on the reproductive performance, food utilisation and liveweight change of the female pig. *Livest. Prod. Sci.* **7**, 67–78

BURGER, J.F. (1952). Sex physiology of pigs. *Onderstepoort J. vet. Res., Suppl.* **2**, 41–52

CHRISTENSON, R.K. and FORD, J.J. (1979). Puberty and estrus in confinement-reared gilts. *J. Anim. Sci.* **49**, 743–751

CHRISTIAN, R.E. and NOFZIGER, J.C. (1952). Puberty and other reproductive phenomena in gilts as affected by plane of nutrition. *J. Anim. Sci.* **11**, 789 (Abstract)

CLARK, J.R., FIRST, N.L., CHAPMAN, A.B. and CASIDA, L.E. (1970) Age at puberty in four genetic groups of swine. *J. Anim. Sci.* **31**, 1032 (Abstract)

CORTEEL, J.M., SIGNORET, J.P. and DU MESNIL DU BUISSON, F. (1964). Seasonal variations in the reproduction in the sow and factors favouring temporary anoestrus. *5th Int. Congr. Anim. Reprod. A.I., Trento*, **3**, 536–540

CUNNINGHAM, P.J., NABER, C.H., ZIMMERMAN, D.R. and PEO, E.R. (1974). Influence of nutritional regime on age at puberty in gilts. *J. Anim. Sci.* **39**, 63–67

DAVIDSON, H.R. (1930). Reproductive disturbances caused by feeding protein-deficient and calcium-deficient rations to breeding pigs. *J. agric. Sci., Camb.* **20**, 233–239

DICKERSON, J.W.T., GRESHAM, G.A. and McCANCE, R.A. (1964). The effect of undernutrition and rehabilitation on the development of the reproductive organs: pigs. *J. Endocr.* **29**, 111–118

DONOVAN, B.T. and VAN DER WERFF TEN BOSH, J.J. (1965). *Physiology of Puberty*. Monographs of the Physiological Society No. 15, London, Arnold

DOROSHKOV. V.B. (1974). Stimulatory effect of males on reproductive performance of female pigs. *Anim. Breed Abstr.* **42**, 3830

DUCKETT, G.E., VARON, H.H. and CHRISTIAN, J.J. (1963). Effects of adrenal androgens on parabiotic mice. *Endocrinology* **72**, 403–409

DUFOUR, J. and BERNARD, C. (1968). Effect of light on the development of market pigs and breeding gilts. *Can. J. Anim. Sci.* **48**, 425–430

DU MESNIL DU BUISSON, F. and SIGNORET, J.P. (1961). Influence des facteurs externes sur le declenchement de la puberte chez la truie. *Ann. Zootech.* **11**, 53–59

DUNCAN, D.L. and LODGE, G.A. (1960). Diet in relation to reproduction and viability of the young. *Comm. Bur. Anim. Nutr. Tech. Comm.* No. 21

DYCK, G.W. (1971). Puberty, post-weaning estrus and estrus cycle length in Yorkshire and Lacombe swine. *Can. J. Anim. Sci.* **51**, 135–140

DYCK, G.W. (1974). Effects of a cold environment and growth rate on reproductive efficiency in gilts. *Can. J. Anim. Sci.* **54**, 287–292

EDWARDS, R.L., OMTVEDT, I.T., TURMAN, E.J., STEPHENS, D.F. and MAHONEY, G.W.A. (1968). Reproductive performance of gilts following heat stress prior to breeding and in early gestation. *J. Anim. Sci.* **27**, 1634–1637

ETIENNE, M. and LEGAULT, C. (1974). Effects of breed and diet on sexual precocity in the sow. *Journées de la recherche porcine en France*, pp. 52–62. Paris, L'Institut Technique du Porc

FOOTE, W.C., WALDORF, D.P., CHAPMAN, A.B. SELF, H.L., GRUMMER, R.H. and CASIDA, L.E. (1956). Age at puberty of gilts produced by different systems *J. Anim. Sci.* **15**, 959–969

FORD, J.J. and TEAGUE, H.S. (1978). Effect of floor space restriction on age at puberty in gilts and on performance of farrows and gilts. *J. Anim. Sci.* **47**, 828–832

FOWLER, S.H. and ROBERTSON, E.L. (1954). Some effects of source of protein and an antibiotic on reproductive performance of gilts. *J. Anim. Sci.* **13**, 949–954

FRIEND, D.W. (1973). Influence of dietary amino acids on the age at puberty of Yorkshire gilts. *J. Anim. Sci.* **37**, 701–707

FRIEND, D.W. (1974). Pubertal age and composition of uterus in gilts. *J. Anim. Sci.* **39**, 975 (Abstract)

FRIEND, D.W. (1976). Nutritional effects on age at puberty and plasma amino acid level in Yorkshire gilts and on chemical composition, nucleic acid, fatty acid and hydroxy proline content of the uterus. *J. Anim. Sci.* **43**, 404–412

FRIEND, D.W. (1977). Effect of dietary energy and protein on age and weight at puberty of gilts. *J. Anim. Sci.* **44**, 601–607

FRIEND, D.W., LARMOND, E., WOLYNETZ, M.S. and PRICE, K.R. (1979). Piglet and pork production from gilts bred at puberty: chemical composition of the carcass and assessment of meat quality. *J. Anim. Sci.* **49**, 330–341

GOODE, L., WARNICK, A.C. and WALLACE, H.D. (1965). Effect of dietary energy levels upon reproduction and the relation of endometrial phosphatase activity to embryo survival in gilts. *J. Anim. Sci.* **24**, 959–963

GOSSETT, J.W. and SORENSEN, A.M. (1959). The effect of two levels of energy and seasons on reproductive phenomena in gilts. *J. Anim. Sci.* **18**, 40–47

HACKER, R.R., KING, G.J. and BEARSS, W.H. (1974). Effects of complete darkness on growth and reproduction in gilts. *J. Anim. Sci.* **39**, 155 (Abstract)

HACKER, R.R., KING, G.J., NTUNDE, B.N. and NARENDRAN, R. (1979). Plasma oestrogen, progesterone and other reproductive responses of gilts to photoperiod. *J. Reprod. Fert.* **57**, 447–451

HACKER, R.R., KING, G.J. and SMITH, V.G. (1976). Effects of 6 and 18 hour light on reproduction in gilts. *J. Anim. Sci.* **43**, 228 (Abstract)

HAFEZ, E.S.E. (1960). Nutrition in relation to reproduction in sows. *J. agric. Sci., Camb.* **54**, 170–178

HAINES, C.E., WARNICK, A.C. and WALLACE, H.D. (1959). The effects of two levels of energy intake on reproductive phenomena in Duroc Jersey gilts. *J. Anim. Sci.* **18**, 347–354

HUGHES, E.H. (1934). Some effects of vitamin A-deficient diets on reproduction of sows. *Agric. Res.* **49**, 943–953

HUGHES, P.E. and COLE, D.J.A. (1975). Reproduction in the gilt. 1. Influence of age and weight at puberty on ovulation rate and embryo survival in the gilt. *Anim. Prod.* **21**, 183–189

HUGHES, P.E. and COLE, D.J.A. (1976). Reproduction in the gilt. 2. Influence of gilt age at boar introduction on the attainment of puberty. *Anim. Prod.* **23**, 89–94

HUGHES, P.E. and COLE, D.J.A. (1978). Reproduction in the gilt. 3. The effect of exogenous oestrogen on the attainment of puberty and subsequent reproductive performance. *Anim. Prod.* **27**, 11–20

HUGHES, P.E. and VARLEY, M.A. (1980). *Reproduction in the Pig.* London, Butterworths

JENSEN, A.H., GEHRING, M.M., BECKER, D.E. and HARMON, B.G. (1967). Effect of space allowance and tethering on pre- and post-pubertal behaviour of swine. *J. Anim. Sci.* **26**, 1467 (Abstract)

JENSEN, A.H., YEN, J.T., GEHRING, M.M., BAKER, D.H., BECKER, D.E. and HARMON, B.G. (1970). Effects of space restriction and management on pre- and post-pubertal response of female swine. *J. Anim. Sci.* **31**, 745–750

JOHNSEN, H.H.K., MOUSTGAARD, J. and HOJGAARD-OLSEN, N. (1952). The significance of vitamin B_{12} for the fertility of sows and gilts. *Anim. Breed Abstr.* **23**, 784

KIRKWOOD, R.N. (1980). Puberty in the gilt. PhD Thesis. University of Leeds

KIRKWOOD, R.N., FORBES, J.M. and HUGHES, P.E. (1981). Influence of boar contact on attainment of puberty in gilts after removal of the olfactory bulbs. *J. Reprod. Fert.* **61**, 193–196

KIRKWOOD, R.N. and HUGHES, P.E. (1979). The influence of age at first boar contact on puberty attainment in the gilt. *Anim. Prod.* **29**, 231–238

KIRKWOOD, R.N. and HUGHES, P.E. (1980a). A note on the efficacy of continuous vs limited boar exposure on puberty attainment in the gilt. *Anim. Prod.* **31**, 205–207

KIRKWOOD, R.N. and HUGHES, P.E. (1980b). A note on the influence of 'boar effect' component stimuli on puberty attainment in the gilt. *Anim. Prod.* **31**, 209–211

KIRKWOOD, R.N. and HUGHES, P.E. (1981). A note on the influence of boar age on its ability to advance puberty in gilts. *Anim. Prod.* **32**, 211–214

KLOTCHKOV, D.V., KLOTCHKOVA, A. Ya., KIM, A.A. and BELYAEV, D.K. (1971). The influence of photoperiodic conditions on fertility in gilts. *10th Int. Cong. Anim. Prod., Paris*, pp. 1–8

LARSSON, S., NILSSON, T. and OLSSON, B. (1966). Some aspects of the metabolic effect of amino acid supplementation of pig diets. *Acta vet. scand.* **7**, 47–53

LODGE, G.A. and MACPHERSON, R.M. (1961). Level of feeding during early life and the subsequent reproductive performance of sows. *Anim. Prod.* **3**, 19–28

LOVE, R.J. (1978). Definition of a seasonal infertility problem in pigs. *Vet. Rec.* **103**, 443–446

MACPHERSON, R.M., HOVELL, F.D. De B. and JONES, A.S. (1977). Performance of sows first mated at puberty or second or third oestrus and carcass assessment of once bred gilts. *Anim. Prod.* **24**, 333–342

MANDL, A.M. and ZUCKERMAN, S. (1952). Factors influencing the onset of puberty in albino rats. *J. Endocr.* **8**, 357–364

MARTINAT, F., LEGAULT, C., DU MESNIL DU BUISSON, F., OLLIVIER, L. and SIGNORET, J.P. (1970). Retardation of puberty in the sow. In *Journées de la recherche porcine en France,* pp. 47–54. Paris, L'Institut Technique du Porc

MAVROGENIS, A.P. and ROBISON, O.W. (1976). Factors affecting puberty in swine. *J. Anim. Sci.* **42**, 1251–1255

MONTEIRO, L.S. and FALCONER, D.S. (1966). Compensatory growth and sexual maturity in mice. *Anim. Prod.* **8**, 179–192

MORTON, J.R.C., DENENBERG, V.H. and ZARROW, M.X. (1963). Modification of sexual development through stimulation in infancy. *Endocrinology* **72**, 439–445

NATHAN, S. and COLE, D.J.A. (1981). Boar contact during rearing and attainment of puberty in the gilt. *Anim. Prod.* **32**, 370 (Abstract)

NTUNDE, B.N., HACKER, R.R. and KING, G.J. (1979). Influence of photoperiod on growth, puberty and plasma LH levels in gilts. *J. Anim. Sci.* **48**, 1401–1406

PAREDIS, F. (1961). Investigations on fertility and artificial insemination in the pig. *Anim. Breed Abstr.* **30**, 2697

PATERSON, A.M. and LINDSAY, D.R. (1980). Induction of puberty in gilts. 1. The effects of rearing conditions on reproductive performance and response to mature boars after early puberty. *Anim. Prod.* **31**, 291–297

PAY, M.G. and DAVIES, T.E. (1973). Growth, food consumption and litter production of female pigs mated at puberty and low body weights. *Anim. Prod.* **17**, 85–91

PHILLIPS, R.W. and ZELLER, J.H. (1943). Sexual development in small and large types of swine. *Anat. Rec.* **85**, 387–400

PLUMLEE, M.D., THRASHER, D.M., BEESON, W.W., ANDREWS, F.N. and PARKER, H.E. (1956). The effects of manganese deficiency upon the growth, development and reproduction of swine. *J. Anim. Sci.* **15**, 352–367

RAMIREZ, V.D. (1973). Endocrinology of puberty. In *Handbook of Physiology* (S.R. Geiger, Ed.), *Section 7, Endocrinology, Vol. II, Part 1*, pp. 1–28. Washington, D.C., American Physiological Society

REDDY, V.B., LASLEY, J.F. and MAYER, D.T. (1958). Genetic aspects of reproduction in swine. *Res. Bull. Mo. agric. Exp. Stn.* No. 666

REUTZEL, L.F. and SUMPTION, L.J. (1968). Genetic and phenotypic relationships involving age at puberty and growth rate in gilts. *J. Anim. Sci.* **27**, 27–30

ROBERTSON, G.L., CASIDA, L.E., GRUMMER, R.H. and CHAPMAN, A.B. (1951a). Some feeding and management factors affecting age at puberty and related phenomena in Chester White and Poland China gilts. *J. Anim. Sci.* **10**, 841–866

ROBERTSON, G.L., GRUMMER, R.H., CASIDA, L.E. and CHAPMAN, A.B. (1951b). Age at puberty and related phenomena in outbred Chester White and Poland China gilts. *J. Anim. Sci.* **10**, 647–656

ROBISON, O.W. (1974). Effects of boar presence and group size on age at puberty in gilts. *J. Anim. Sci.* **39**, 224 (Abstract)

SCANLON, P.F. and KRISHNAMURTHY, S. (1974). Puberty attainment in slaughter weight gilts in relation to month examined. *J. Anim. Sci.* **39**, 160 (Abstract)

SCHEIMANN, C.A., ENGLAND, D.C. and KENNICK, W.H. (1976). Initiating estrus in pre-pubertal confinement gilts. *J. Anim. Sci.* **43**, 210 (Abstract)

SCHMIDT, K. and BRETSCHNEIDER, W. (1954). The outward course of the sexual cycle in the sow. *Anim. Breed. Abstr.* **32**, 1046

SELF, H.L., GRUMMER, R.H. and CASIDA, L.E. (1955). The effects of various sequences of full and limited feeding on the reproductive phenomena in Chester White and Poland China gilts. *J. Anim. Sci.* **14**, 573–592

SHORT, R.E. (1963). Influence of heterosis and plane of nutrition on reproductive phenomena in gilts. M.S. Thesis. University of Nebraska

SHORT, R.E., ZIMMERMAN, D.R. and SUMPTION, L.J. (1963). Heterotic influence on reproductive performance in swine. *J. Anim. Sci.* **22**, 868 (Abstract)

SIGNORET, J.P. (1970). Reproductive behaviour of pigs. *J. Reprod. Fert. Suppl.* **11**, 105–117

SIGNORET, J.P. (1972). The mating behaviour of the sow. In *Pig Production* (D.J.A. Cole, Ed.), pp. 295–314. London, Butterworths

SIGNORET, J.P. (1980). Mating behaviour of the pig. In *Reviews in Rural Science, IV. Behaviour*, (M. Wodzicka-Tomazewska, T.N. Edey and J.J. Lynch, Eds.), pp. 75–78, Australia, Unversity of New England

SORENSEN, A.M., THOMAS, W.B. and GOSSETT, J.W. (1961). A further study on the influence of level of energy intake and season on reproductive performance of gilts. *J. Anim. Sci.* **20**, 347–359

STICKNEY, K., FOXCROFT, G.R., GARSIDE, D.A. and MORTIMER, M.J. (1978). Aspects of oestrogen-induced puberty in the gilt. *Anim. Prod.* **26**, 388–389 (Abstract)

SURMUHIN, A.F. and CEREMNYH, V.D. (1970). The effect of light on the development of the reproductive organs of gilts. *Anim. Breed. Abstr.* **39**, 3687

SURMUHIN, A.F., CEREMNYH, V.D., TIMOFEEV, V.P. and POZNIKOVA, A.A. (1970). Development of gilts subjected to different light regimes. *Anim. Breed. Abstr.* **39**, 3602

TASSELL, R. (1967). The effects of diet on reproduction in pigs, sheep and cattle. *Br. vet. J.* **123**, 170–176

TEAGUE, H.S., ROLLER, W.L. and GRIFFO, A.P. Jr. (1968). Influence of high temperature and humidity on the reproductive performance of swine. *J. Anim. Sci.* **27**, 408–411

THOMPSON, L.H. and SAVAGE, J.S. (1978). Age at puberty and ovulation rate in confinement as influenced by exposure to a boar. *J. Anim. Sci.* **47**, 1141–1144

VANDENBERGH, J.G. (1969). Male odour accelerates female maturation in mice. *Endocrinology* **84**, 658–660

WAHLSTROM, R.C. and LIBAL, G.W. (1977). Effect of dietary protein during growth and gestation on development and reproductive performance of gilts. *J. Anim. Sci.* **45**, 94–99

WARNICK, A.C., WIGGINS, E.L., CASIDA, L.E., GRUMMER, R.H. and CHAPMAN, A.B. (1951). Variation in puberty phenomena in inbred gilts. *J. Anim. Sci.* **10**, 479–493

WIGGINS, E.L., CASIDA, L.E. and GRUMMER, R.H. (1950). The effect of season of birth on sexual development in gilts. *J. Anim. Sci.* **9**, 277–280

WISE, F.S. and ROBERTSON, G.L. (1953). Some effects of sexual age on reproductive performance in gilts. *J. Anim. Sci.* **12**, 957 (Abstract)

WITZ, W.M. and BEESON, W.M. (1951). The physiological effects of a fat deficient diet on the pig. *J. Anim. Sci.* **10**, 957–959

ZACHARIAS, L. and WURTMAN, R.J. (1964). Blindness: its relation to age of menarche. *Science* **144**, 1154–1155

ZIMMERMAN, D.R., BOURN, P. and DONOVAN, D. (1976). Effect of "transport phenomenon" stimuli and boar exposure on puberty in gilts. *J. Anim. Sci.* **42**, 1362 (Abstract)

ZIMMERMAN, D.R., CARLSON, R. and LANTZ, B. (1974). The influence of exposure to the boar and movement on pubertal development in the gilt. *J. Anim. Sci.* **39**, 230 (Abstract)

ZIMMERMAN, D.R., CARLSON, R. and NIPPERT, L. (1969). Age at puberty in gilts as affected by daily heat checks with a boar. *J. Anim. Sci.* **29**, 203 (Abstract)

ZIMMERMAN, D.R., SPIES, H.G., RIGOR, E.M., SELF H.L. and CASIDA, L.E. (1960). Effects of restricted feeding, crossbreeding and season of birth on age at puberty of swine. *J. Anim. Sci.* **19**, 687–694

7

THE CONTROLLED INDUCTION OF PUBERTY

A.M. PATERSON
A.R.C. Group for Hormones and Reproduction in Farm Animals, University of Nottingham School of Agriculture, Sutton Bonington, Loughborough, Leicestershire, UK

Puberty, defined as the time at which ovulation and oestrus first occur in association with normal luteal function, usually takes place at about 200 days of age in the gilt (Duncan and Lodge, 1960). As the replacement gilt is reared solely for reproductive purposes, a delay in the attainment of puberty means a delay in the commencement of productive life. With the recent trend towards intensive confinement systems of pig production it has become increasingly desirable to be able to control the onset of puberty in the gilt. As discussed in Chapter 6, age at puberty can be affected by many factors including nutrition, genotype and the environment. Although there are numerous reports in the literature concerning environmental or physiological conditions which may either hasten or retard puberty in the gilt, few include studies of the associated hormonal changes. Therefore in reviewing those methods used in an attempt to control the onset of puberty in the gilt, particular emphasis is given to the endocrinology and mode of action of the technique involved.

The endogenous control of puberty

An understanding of the endogenous control of puberty, which is central to attempts to control puberty, has been hampered by a lack of basic research on this topic in the pig. As has been pointed out in Chapter 5, it is still not possible to explain fully the pubertal process in the pig in terms of the neuroendocrine mechanisms involved. However, some important points are clear and bear consideration at this stage.

The hypothalamic–pituitary unit is essentially intact at birth. The hypothalamus contains gonadotrophin releasing hormones and the pituitary is capable of releasing gonadotrophins in response to synthetic releasing hormones (Debeljuk, Arimura and Schally, 1972; Foster, Cook and Nalbandov, 1972). Recent work in the sheep (Ryan and Foster, 1980) has shown that as puberty is approached the frequency of episodic luteinizing hormone (LH) release increases to about one episode per hour. This is believed to provide the trophic stimulus for final follicular development, which in turn increases oestradiol production and leads to the induction of the first pre-ovulatory LH surge as the positive feedback mechanism becomes operative. These workers were successful in inducing precocious puberty in ewe lambs by mimicking the increased LH pulse pattern with exogenous LH. Such work has not been reported in the pig but the frequency of pulsatile LH release in the gilt does increase during development (see Chapter 5); the model for the initiation of puberty as proposed for the lamb may therefore also be applicable to the gilt.

The mechanisms governing the changes in gonadotrophin secretion during development have been recently reviewed (Levasseur, 1977; Foxcroft, 1978; Foster, 1980; Elsaesser, Chapter 5) and these mechanisms will not be considered in detail in this review. In general, two theories have evolved to explain the increase in gonadotrophin secretion as puberty is approached and conflict exists as to which theory best explains the onset of puberty, although they are certainly not mutually exclusive. The first involves the concept of differential feedback sensitivity (Ramirez and McCann, 1963; Grumbach *et al.*, 1974) and has become known as the classic 'gonadostat' hypothesis. This postulates that the hypothalamus is highly sensitive to negative feedback by low levels of circulating oestrogens, which suppress the secretion of releasing hormones and hence the gonadotrophins. As puberty is approached there is a change in the set point of the hypothalamus which becomes less sensitive to negative feedback, allowing secretion of gonadotrophins to increase. The second theory proposes that there is specific inhibition of the secretion of releasing hormones from the hypothalamus by the central nervous system (Davidson, 1974; Levasseur, 1977). The cortex, limbic system and pineal gland are all believed to have some influence on the hypothalamus in a balanced inhibition of the secretion of releasing hormones (Gorski, 1974). These extra-hypothalamic structures presumably monitor factors in the internal and external environment and mediate their influence on gonadotrophin secretion via neuroanatomical pathways.

Induction of puberty with the male

The introduction of the mature male to the prepubertal female can advance and synchronize puberty in a number of species. This phenomenon was first described in the mouse by Whitten (1956a,b) and has since been reported in rats (Cooper and Haynes, 1967), deer-mice (Bronson and Marsden, 1964), sheep (Dýrmundsson and Lees, 1972) and pigs (du Mesnil du Buisson and Signoret, 1962; Brooks and Cole, 1969; 1970). The results obtained from using boars to induce puberty have been reviewed by Hughes (Chapter 6) and will not be repeated here.

In mice it is the odour of the male's urine which is the major exteroceptive factor involved in the stimulation of the female (Dominic, 1965; Bronson and Whitten, 1968). The physical presence of the male is not essential as females can be stimulated by transferring bedding soiled with male urine to their cage (Parkes and Bruce, 1962; Vandenbergh, 1969; 1975), but it must be changed frequently to ensure a response. Bronson and Maruniak (1975) found that while urine alone stimulated a uterine weight response, physical contact with an intact male or urine plus physical contact with a castrated male produced a much stronger response. They concluded that the main stimulatory effect was pheromonal but tactile cues were also important.

Wilson and Bossert (1963) divided pheromones into two main types—signaller pheromones which produce a behavioural response and primer pheromones which cause a physiological change. In the pig the source of the pheromones which stimulate puberty is not well defined. When Kinsey *et al.* (1976) exposed gilts to a range of treatments including boar urine odour and physical contact with boars, they found that boar contact advanced puberty but the odour of their urine did not. Kirkwood and Hughes (1980) were able to advance puberty by exposing gilts to the odour of a vacant boar pen, but the administration of the urinary pheromone 5α-androst-16-en-3-one (Δ^{16}) which has signaller pheromone properties (Reed, Melrose and Patterson, 1974) had no effect on the age of gilts at puberty. This suggests that the primer pheromones which stimulate puberty in the gilt are not of urinary origin. Boar saliva is known to contain a reproductively-active pheromone complex in which the major active compound is 3α-androstenol (Gower, 1972; Booth, 1975; 1977). Having studied the urinary pheromones and the mating behaviour of pigs, Kirkwood, Forbes and Hughes (1981) have postulated that the source of the primer pheromone in the boar is the submaxillary salivary gland.

The physiological mechanism by which the male stimulates puberty in the female mouse has been studied by Bronson and his co-workers at Austin, Texas. Their work has established that the introduction of the male causes a rise in the basal LH concentrations of the female (Bronson and Desjardins, 1974; Bronson and Maruniak, 1976). This rise takes place within 30 minutes of the introduction of the male and the ovary responds to the higher concentration of LH by secreting oestradiol. Circulating levels of oestradiol are elevated in six to twelve hours and remain elevated for up to 48 hours leading to the initiation of positive feedback as LH, FSH and progesterone concentrations increase in a pattern which mimics that seen in the proestrus phase of the adult cyclic mouse and ovulation takes place.

It has been suggested that the period of elevated oestrogen is essential for the final maturation of the positive feedback system in the hypothalamus, which facilitates the release of gonadotrophins in response to oestrogen stimulation. In this context, Bronson (1975) and Bronson and Maruniak (1976) have postulated that the effect of the male is exclusively on basal levels of LH and oestrogen and that the secretion of the other hormones are altered indirectly by the action of oestradiol.

Few studies on the hormonal events associated with the onset of puberty in the gilt after exposure to boars have been published. Recent work at the University of Missouri (Paterson, Cantley and Day, 1980; Esbenshade *et al.*, 1982) has characterized some of the changes taking place at this time but the picture is by no means yet complete. Blood samples were taken twice daily from nine gilts around the time of relocation and exposure to mature boars. *Figure 7.1* shows the mean hormone levels in these gilts from day —6 to day 6, with day 0 being the first day of the pubertal oestrus. Plasma oestradiol-17β (E_2) began increasing from day —5 and reached peak levels during proestrus, on day —3 ($n = 5$) or day —2 ($n = 4$). Plasma LH levels prior to day —1.5 and after day 1.5 fluctuated between 0.5 and 1 ng/ml and were considered basal. The pre-ovulatory surge of LH was measured on day —1 ($n = 4$) or day 0 ($n = 5$). Progesterone concentrations were less than 0.3 ng/ml until after oestrus and then they rose to about

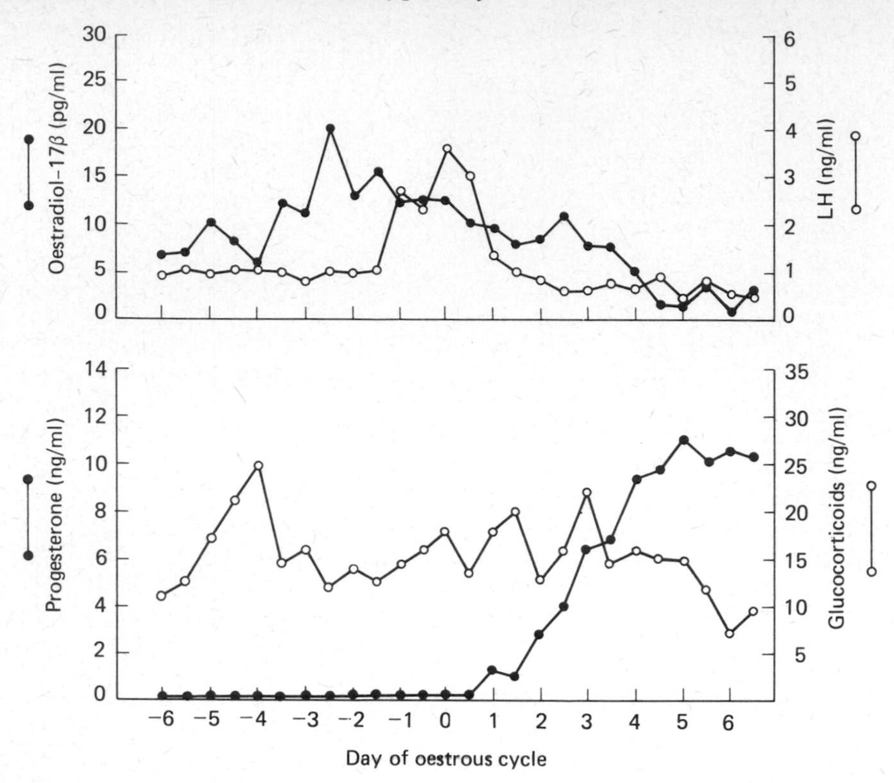

Figure 7.1 Mean plasma hormone concentrations at the onset of puberty in relocated gilts (n = 9). From Esbenshade *et al.* (1982)

11 ng/ml by day 5 due to the formation of corpora lutea. Cortisol levels fluctuated widely, ranging from 7.4–24.5 ng/ml and they did not appear to be related to oestrus or ovulation.

In a second experiment blood was collected from four gilts every six hours from boar exposure until after puberty. The hormonal patterns were similar in each gilt, although they were out of phase with each other as the gilts reached puberty at different times after exposure to the boars. *Figure 7.2* shows the mean LH and E_2 levels in these gilts arranged with respect to the onset of oestrus (0 hours). Plasma E_2 rose from basal levels and remained elevated for a mean time of 83 hours before the onset of oestrus. Peak levels were measured at —18 hours and E_2 had returned to basal levels by +12 hours. Plasma LH was basal throughout with the exception of the pre-ovulatory surge.

The hormonal pattern of oestradiol and LH observed during the follicular phase leading to first oestrus in these prepubertal gilts was similar to the pattern observed during the follicular phase of the oestrous cycle (Henricks, Guthrie and Handlin, 1972; Guthrie, Henricks and Handlin, 1972; Shearer *et al.*, 1972) and during the follicular phase of gilts synchronized with a synthetic progestin (Redmer and Day, 1981). These data indicate that the increase in plasma oestrogen prior to first oestrus is due to the initiation of follicular growth.

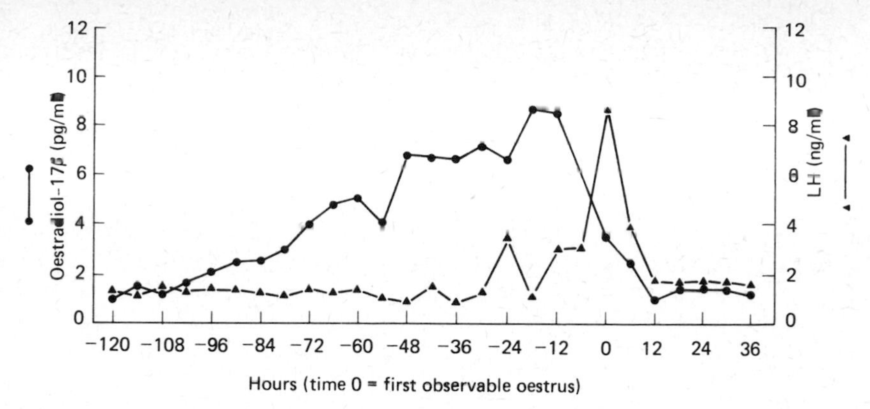

Figure 7.2 Mean plasma hormone concentrations in gilts reaching puberty after exposure to boars (n = 4). From Esbenshade *et al.* (1982)

In all the gilts reaching puberty in these studies the only changes in LH seen were those associated with the pre-ovulatory surge; that is, no changes in LH in response to surgery, relocation or boar exposure which could be considered causative of the initiation of follicle growth were detected. Such changes have been detected in the mouse (Bronson and Maruniak, 1976) and the ewe (Martin, Oldham and Lindsay, 1980). It is possible that the extremely low levels of circulating LH in the gilt compared with those in the mouse precluded detection of a subtle shift in the basal levels of this hormone. Alternatively, it may be that the change in LH is not in mean levels but in the frequency and/or amplitude of episodic pulses of LH as has been recently described in the ewe by Foster (1980). Such episodes are known to occur in the immature gilt (Elsaesser and Foxcroft, 1978) and the sampling regime used in our studies would make such changes impossible to detect.

We have also measured the hormonal profiles in two other gilts which responded to the introduction of boars by showing elevated levels of oestradiol but did not show oestrus or ovulation. Some follicle growth must have taken place in these gilts but it did not continue to ovulation and there was no positive feedback of oestrogen or LH. The elevated oestrogen levels did not induce oestrus. The data suggest that the ability to respond to a stimulus by initiating follicle growth and the ability to respond to the resulting elevated oestradiol levels do not develop simultaneously. This finding may explain the phenomenon of vulval development without oestrus or ovulation which is often observed in young gilts, particularly after mixing or relocation.

It appears that boars stimulate puberty in gilts by affecting a rise in oestradiol levels. Whether the sequence of events leading to this rise, especially the change in basal LH secretion, is the same as has been described in other species such as the mouse still remains to be determined.

Induction of puberty with exogenous gonadotrophins

Casida (1935) first demonstrated that exogenous gonadotrophins can cause ovulation in immature gilts. Since then many workers have used pregnant

Table 7.1 THE RESPONSE OF PREPUBERTAL GILTS TO HIGH DOSES OF PMSG FOLLOWED BY HCG

Reference	*Age* (days)	*Weight* (kg)	*PMSG* (iu)	*HCG* (iu)	*Oestrus* (%)	*Ovulation* (%)	*Ovulation rate*	*Farowing* (%)	*Litter size*
Baker and Coggins (1966)	100–180		1000	500 (48)[a]	0	0			
			2000	500 (48)	0	64			
			250	500 (48)	0	100	7.2		
			500	500 (48)	0	100	12.5		
			1000	500 (48)	0	100	19.6		
			2000	500 (48)	0	100	45.8		
Dziuk and Gehlbach (1966)	110		500	500 (96)	40	100	7.6		
	130		500	500 (96)	20	80	7.7		
	100		1000	500 (96)	41	91	8.1		
	105		1250	500 (96)	63	100	5.8		
	125		1250	500 (96)	60	100	10.2		
Dziuk and Dhindsa (1969)	250–340	136–170	500	–	75	80	13.0	67	
Shaw *et al.* (1971)		45–55	1000	500 (48)		100	18.0		
Phillipo (1968)	160–190	55–65	750		9	100	11.2		
			1000		9	100	25.2		
			1500		9	100	31.9		
Rampacek *et al.* (1976)	155		1000	500 (72)		100	27.8	20	1.0
	175		1000	500 (72)		100	27.3	40	5.5
	165		750	500 (72)		100	15.0	20	5.0
	185		750	500 (72)		100	17.3	80	4.0
Segal and Baker (1973)		55	1000	500 (48/72)	0	100	24.6		
Guthrie (1977)	160	64–86	1000	–	60	100	6.8		
	160	64–86	1000	GnRH	33	100	17.2		
	155	65–100	1000	500 (72)	69	100	13.5		
	155	65–100	1000	GnRH	69	100	15.2		

[a]Numbers in parentheses indicate the number of hours between administration of PMSG and HCG.

mare's serum gonadotrophin (PMSG) and human chorionic gonadotrophin (HCG) in attempts to stimulate precocious puberty in gilts.

Data from a number of experiments using PMSG followed by HCG to induce ovulation in prepubertal gilts are summarized in *Table 7.1*. Doses of 500–2000 iu PMSG followed by 500 iu HCG 48–96 hours later caused ovulation in up to 100% of gilts over a range of ages and weights. Some exceptions have been reported, including the failure of any gilts to ovulate after 1000 iu PMSG in the work of Baker and Coggins (1966). However in general ovulation rate is related to the dose of PMSG given. Both Baker and Coggins (1966) and Phillipo (1968) established positive linear relationships between dose of PMSG and ovulation rate and Rampacek *et al.* (1976) found higher ovulation rates after 1000 iu than after 750 iu PMSG.

In the range considered, age did not appear to affect markedly the response to PMSG. Gilts differing in age by 20 days had similar ovulation rates in the work of Rampacek *et al.* (1976). Dziuk and Gehlbach (1966) found similar ovulation rates in gilts treated with 500 iu PMSG at 110 or 130 days of age. A difference with age was found in their work for gilts given 1250 iu but it should be noted that the younger gilts had a lower ovulation rate than any other group, including those of the same age given 500 iu, suggesting that their response was abnormal. The gilts treated by Dziuk and Dhindsa (1969) had ovulation rates comparable with much younger gilts given the same dose by Baker and Coggins (1966). Ovulation rate varies among experiments but this is probably due to factors other than chronological age, such as environment, hormone preparations and physiological differences among and within groups of gilts.

The ova produced in response to PMSG/HCG are capable of being fertilized (Dziuk and Polge, 1965) but reproductive performance after induction of puberty by this method has been poor. The data in *Table 7.1* show that only a small percentage of gilts which ovulated exhibited oestrus. The highest incidence of oestrus was 75% (Dziuk and Dhindsa, 1969) but this was recorded in gilts which were much older than those treated by other workers, having failed to reach puberty naturally by 250 days of age or 136 kg. In gilts around 160 days of age, the highest incidence of oestrus was 69% (Guthrie, 1977) but only 9% of those studied by Phillipo (1968) displayed oestrus. Segal and Baker (1973) showed that standing oestrus after PMSG treatment can be enhanced by injections of diethylstilboestrol given at the same time as HCG (0 versus 73%). However, this compound can influence sperm transport (Dziuk and Polge, 1965) which may adversely affect reproductive performance by causing asynchrony between ova and sperm at the time of fertilization.

Even when oestrus is displayed and breeding takes place, great variation in farrowing rate and litter size has been reported although acceptable farrowing rates have been obtained, e.g. 67% (Dziuk and Dhindsa, 1969) and 80% (Holtz *et al.*, 1977), but there are reports of farrowing rates as low as 20% (Rampacek *et al.*, 1976; Hühn, Heidler and Ressin, 1977) and a litter size as low as one live piglet has been recorded (Rampacek *et al.*, 1976).

There have been reports that pregnancy in gilts induced to ovulate with PMSG may fail due to premature regression of the corpora lutea (Shaw, McDonald and Baker, 1971; Ellicott, Dziuk and Polge, 1973). In addition,

Table 7.2 THE RESPONSE OF PREPUBERTAL GILTS TO LOW DOSES OF PMSG GIVEN IN COMBINATION WITH HCG

Reference	*Age* (days)	*Weight* (kg)	*PMS/HCG* (iu)	*Oestrus* (%)	*Ovulation* (%)	*Ovulation rate*	*Ovulation maintained* (%)	*Fertilization rate* (%)	*Litter size* (No. alive)
Schilling and Cerne (1972)	140–154		400/200	100			95		
	168–182		400/200	97			93	82	9.8
	154–168		400/200	–	80	11.6		66	8.5[a]
Baker and Rajamahendran (1973)	165–180	85–90	400/200	75	88	4–19			
	165–180	85–90	400 PMS only	38	13	8.0			
Hühn *et al.* (1977)	168		400/200	94				17	6.8
	196		500/250	83				62	6.2
Miskovic *et al.* (1977)	131–134	50–60	300/200	100	100	12.2			
			400/200	100	100	23.8			
	151–157	60–64	300/200	100	100	17.7			
			400/200	100	100	25.1			
Guthrie (1977)	160		400/200	100	100	23.2			
Holtz *et al.* (1977)		70–100	400/200		90	14.3		80	9.0[a]
Breeuwsma (1974)			400/200	43	81	17.0			
			800/200+ODB[b]	55	100	25.0			
Bielánski (1978)	180–200	50–80	400/200					28	8.1
Schlegel *et al.* (1978)	211–253		500/250		99		51		
Paterson and Lindsay (1981)	157	74	400/200	84	100	12.8	39[c]		
	157	72	400/200	78	89	11.7	81[d]		
	162	73	400/200	83	100	–	52[c]		
	162	73	400/200	88	100	–	87[d]		

[a]At 30 days post coitum.
[b]ODB = oestradiol benzoate
[c]Isolated from boars
[d]Housed with boars

maintenance of cyclic activity to a second ovulation and oestrus in non-pregnant gilts may be poor (Dziuk and Gehlbach, 1966; Segal and Baker, 1973). However, this facet of the use of gonadotrophins to induce early puberty has not been systematically studied.

The use of PMSG to induce ovulation in prepubertal gilts, and mating at that ovulation, is not a satisfactory means of improving reproductive performance at this time. Mating at the second cycle after treatment offers an alternative but cyclic activity is not always maintained to a second oestrus and ovulation, which therefore precludes this practice.

Recently there has been interest in using PMSG and HCG in combination as a single injection. The combination most commonly used has been 400 iu PMSG and 200 iu HCG and these doses are considerably lower than those used in earlier work. Attention has centred on gilts around 150–180 days of age, which would be expected to be approaching the age of natural puberty. The results of some of the work in this area are summarized in *Table 7.2.*

In each experiment considered, 80% or more of treated gilts ovulated. Ovulation rates were generally lower than those found when high doses of PMSG followed by HCG were given but wide variation between experiments is still apparent. More gilts showed oestrus with several reports of up to 100% of the animals responding, although reports as low as 43% do exist (Breeuwsma, 1974). Data on the reproductive performance of these gilts is limited. Schilling and Cerne (1972) and Holtz *et al.* (1977) reported acceptable farrowing rates and litter sizes. However, other workers have found much lower fertilization rates and there is clearly a need for further research before the breeding of gilts after induction of ovulation with combinations of PMSG and HCG could be considered on a commercial basis.

Maintenance of cyclic activity in PMSG/HCG treated gilts has not been widely studied. Schilling and Cerne (1972) reported that over 90% of their gilts maintained cyclic activity but Schlegel, Wahner and Stenzel (1978) found a much lower incidence of second ovulation. In studies at the University of Western Australia (Paterson and Lindsay, 1981) it has also been found that maintenance of cyclic activity in gilts induced to ovulate with 400 iu PMSG and 200 iu HCG is poor. Of 41 gilts which ovulated in response to this treatment, 24 (58%) ovulated a second time and only 19 of them (46%) showed oestrus. Failure to maintain cyclic activity would severely limit the usefulness of this technique for the routine induction of puberty. We are able to show that this problem could be substantially reduced by using mature boars as an additional stimulatory factor. When gilts were housed in the presence of boars and exposed to them daily following the induction of ovulation, significantly more gilts displayed a second oestrus (33/39 cf. 19/41, $\chi^2 = 11.2$, $P<0.001$) and had a second ovulation (32/39 cf. 24/41, $\chi^2 = 4.2$, $P<0.05$) than when they were isolated from boars.

Based on these findings a management regime using induced ovulation, the presence of boars to facilitate the maintenance of cyclic activity and breeding of the gilts at their second oestrus may prove to be a viable commercial practice. Further study is warranted in this area, particularly with respect to maintenance of cyclic activity and reproductive performance.

According to Gates and Bozarth (1978) the ovulatory process in response to PMSG in rodents is initiated by the FSH-like activity of the gonadotrophin which stimulates the maturation of the ovarian follicles. The growing follicles produce oestrogens, the circulating levels of which rise until, after a critical period, the hypothalamus is stimulated to release LH releasing hormone. This releasing hormone acts on the anterior pituitary gland to release LH and ovulation takes place several hours later. Similar hormonal data have recently been obtained in the pig. Esbenshade *et al.* (1982) gave prepubertal gilts 1000 iu PMSG and collected blood samples twice daily for eight days. All the gilts showed oestrus five days after treatment and *Figure 7.3* shows the mean hormone levels arranged

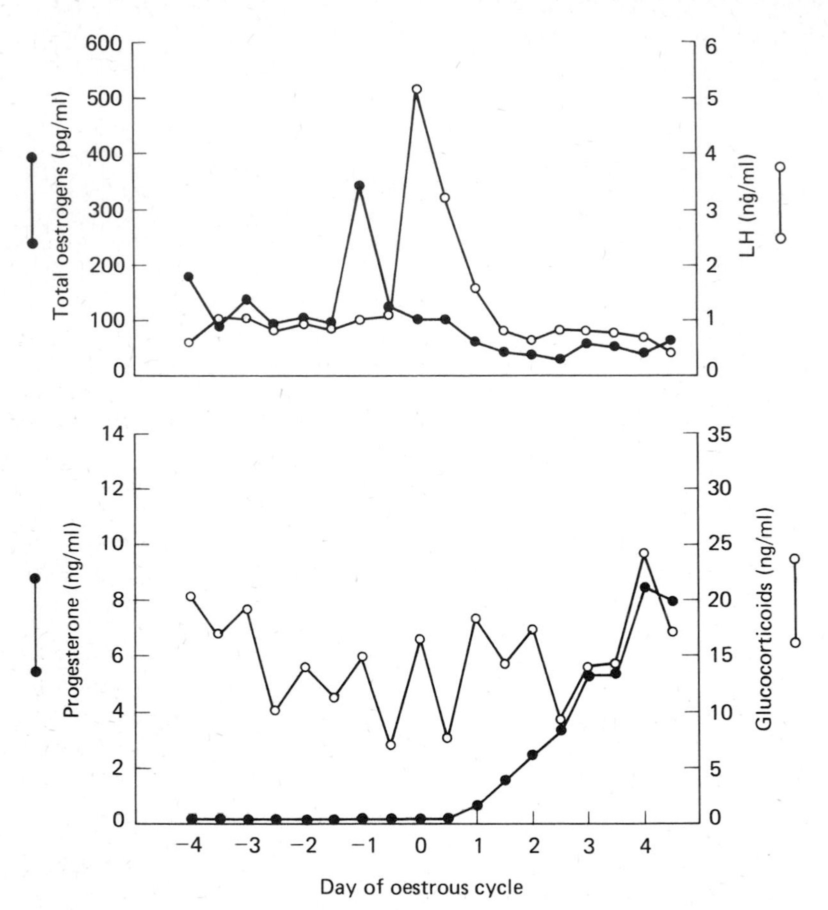

Figure 7.3 Mean plasma hormone concentrations in gilts induced to ovulate with 1000 iu PMSG (n = 5). From Esbenshade *et al.* (1982)

with respect to the onset of oestrus (day 0). Plasma total oestrogens reached peak levels on day —1 and the pre-ovulatory surge of LH took place on the first day of oestrus. Paterson and Martin (1981) measured the levels of plasma oestradiol in gilts treated with 400 iu PMSG and 200 iu HCG in combination. The data in *Table 7.3* show that oestradiol was

Table 7.3 CHARACTERISTICS OF THE PLASMA OESTRADIOL-17β PATTERN OF GILTS INDUCED TO OVULATE WITH 400 iu PMSG AND 200 iu HCG

Gilt no. (± Boars)	*Basal concentration*[a] (pg/ml)	*Day of 1st rise*	*Day of peak*	*Peak concentration* (pg/ml)	*Day back to basal*
296+	3.0	2	3	20.0	5
297+	U.D.[b]	2	4	14.2	5
307+	18.5	1	3	56.2	5
309+	31.6	1	2	51.0	4
295−	17.2	2	2	33.6	3
280+	27.7	2	3	77.2	6
300+	4.4	2	3	15.6	5
308+	9.6	1	3	47.6	4
288−	39.7	1	2	90.8	5
298−	22.3	1	1	77.8	5
304−	U.D.	1	3	27.0	5
306−	13.6	3	6	203.1	10
287+	21.0	1	2	34.4	4
290+	26.7	–	3	54.6	4
299+	4.5	1	2	41.4	5
305−	8.6	1	1	21.0	5

[a]Basal concentration is that of −4, −2, 0.
[b]Below limit of detection of the assay.
From Paterson (1979)

elevated two days after injection in 15/16 gilts and peak levels were recorded on day 2 or 3. The mean height of the peak was 52.9±12.51 pg/ml and the interval from the first rise until oestradiol concentration returned to basal levels was 3–4 days.

There are some similarities in the oestrogen response to PMSG and male stimuli. In both cases oestradiol rises quickly and remains elevated for some time before the pre-ovulatory surge of LH takes place. The hypothesis that a long period of elevated oestrogen titres is necessary for maturation of the hypothalamic positive feedback systems (Bronson, 1975; Bronson and Maruniak, 1976) fits the data for PMSG-induced ovulation as well as male-induced ovulation.

In an attempt to determine an endocrine basis for the failure of gonadotrophin-treated gilts to maintain cyclic activity, Paterson and Martin (1981) continued blood sampling through the luteal phase until after the time of the expected second oestrus. The hormonal patterns in the first follicular phase and during the luteal phase were similar to those reported in cyclic sows (Guthrie, Henricks and Handlin, 1972; Henricks, Guthrie and Handlin, 1972), and show that the ovaries of the prepubertal gilt around 160 days of age produce normal patterns and levels of oestradiol-17β (E_2) and progesterone in response to this combination of PMSG and HCG. Following the luteal phase three different types of cyclic activity were observed and each had a typical hormonal profile associated with it. A representative profile of each type is shown in *Figure 7.4*. Five gilts (type A) were able to produce a surge of E_2 which was associated with oestrus and ovulation while gilts (type B) were not able to do so, thus precluding any chance of oestrus or ovulation. Four other gilts (type C) produced a surge of E_2 and exhibited vulval development but ovulation did not take place. The E_2 produced may have failed to induce a pre-ovulatory surge of

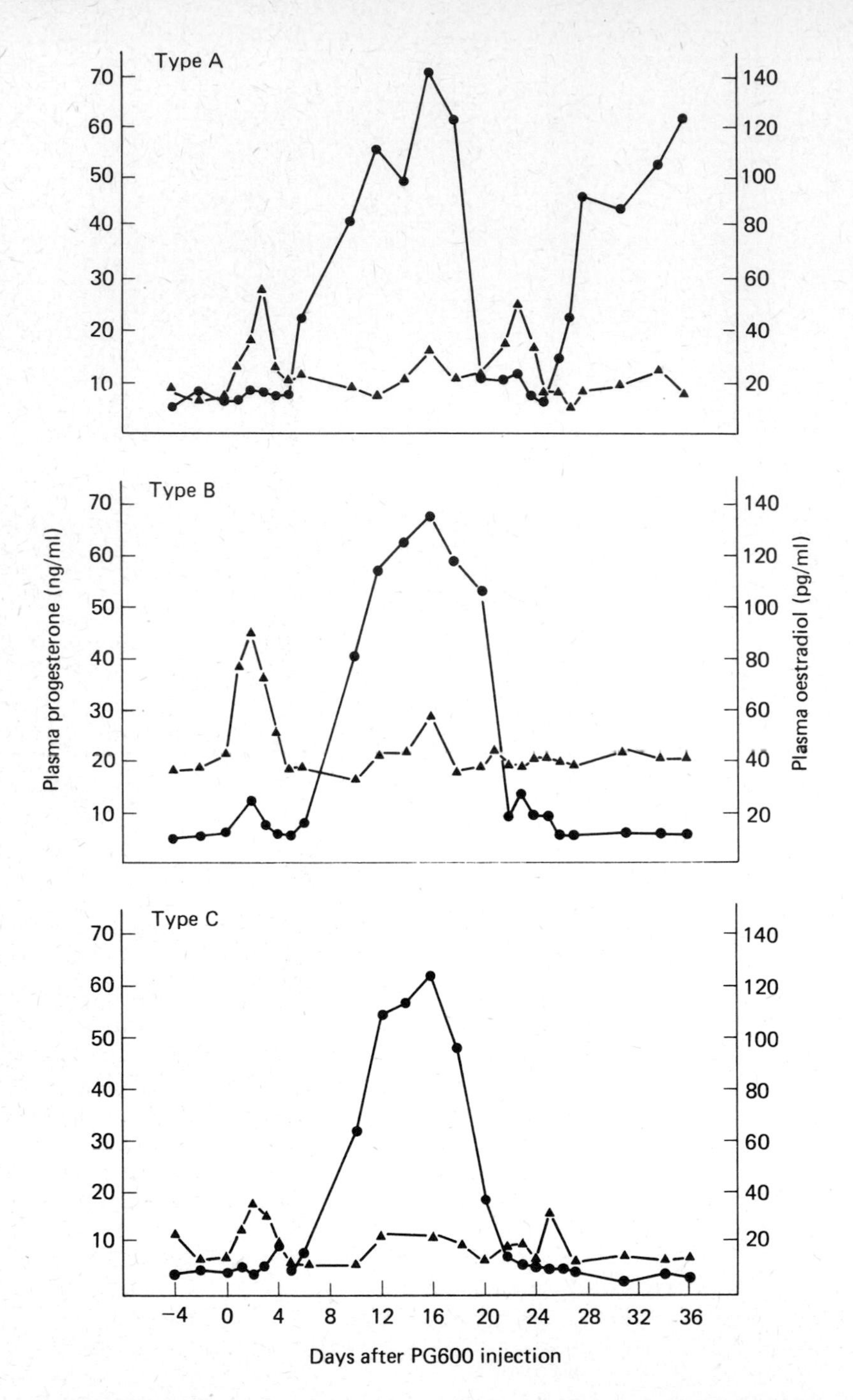

Figure 7.4 Plasma progesterone (•) and plasma oestradiol-17β (▲) in a gilt representing each type of cyclic activity observed after induction of ovulation with 400 iu PMSG and 200 iu HCG. From Paterson and Martin (1981).

LH or too little LH may have been released to cause ovulation. Alternatively, the ovary may have become insensitive to the pre-ovualtory surge of LH. It may be that the basal LH secretory pattern in those animals which continue to cycle differ from those that do not, and this may be a possible explanation for the effect of the male on maintenance of cyclic activity. However, LH was not measured in this study and experimental confirmation of the role of LH in this process is still lacking.

Induction of puberty with oestrogens

As we have seen, the first changes in plasma hormones which have been observed at the onset of puberty in the gilt are changes in oestradiol. Bronson (1975) has shown that, in mice, the puberty-stimulating effect of the male can be replaced by exogenous oestrogen therapy. As long as plasma oestradiol remains elevated for two days, whether this is achieved by the presence of the male, the use of oestrogen injections or by a combination of the two, precocious puberty will result. This suggests that it should be possible to initiate cyclic activity in the gilt with oestradiol treatments. However, the few studies in which this has been attempted have produced equivocal results.

Dziuk (1965) reported that 15/20 gilts showed oestrus after oral administration of 20 mg ethinyl oestradiol/gilt/day for five days. Of the 10 gilts examined, eight had ovulated but the mean number of corpora lutea was only four/gilt and there were numerous unovulated follicles. This treatment regime failed to initiate normal recurring cycles. Baker and Downey (1975) reported similar results; of eight gilts injected with 40 mg ethinyl oestradiol one showed oestrus and five ovulated. When 200 mg ethinyl oestradiol was administered in a subcutaneous implant 7/8 gilts showed oestrus and six ovulated. Treatment with the same doses of oestrone had no effect. In contrast, Hughes and Cole (1978) were successful in initiating cyclic activity in 6/10 gilts around 140 days of age with injections of 0.4 mg of oestradiol benzoate (ODB) on three successive days.

At the University of Missouri we have recently carried out a series of experiments examining the effect of ODB on oestrus and ovulation in prepubertal gilts around 180 days of age (Paterson and Day, 1980). In the first experiment 2/9 gilts given 0.4 mg ODB and 5/9 given 0.8 mg ODB showed oestrus, but 1/9 and 0/9 at each dose respectively had corpora lutea when examined on day 10.

Since Hughes and Cole (1978) achieved their success using repeated injections it was decided to compare the effect on oestrus and ovulation of a range of dose regimes of ODB. Forty eight gilts received a total of 1.2 mg of ODB spread over either 1, 2 or 3 days and given as either one or two injections per day, generating six treatment groups ($n = 8$). The results of this experiment are shown in *Table 7.4*. Most treated gilts showed vulval development but only six displayed oestrus. At laparoscopy 10–12 days post treatment 34 had quiescent ovaries, eight had ovulated and six had large unovulated follicles. Neither oestrus nor the distribution of these classes of ovarian activity were associated with ODB treatment.

Table 7.4 OESTRUS AND OVULATION IN PREPUBERTAL GILTS TREATED WITH OESTRADIOL BENZOATE (ODB) IN THE ABSENCE OF BOARS

Treatment (mg ODB × injections)	*Number of gilts*					
	n	*Oestrus*	*Ovaries immature*	*Ovulating normally*	*Ovulation with cysts*	*Cysts/ follicles*
A : 0.4 mg × 3	8	1	4	2	0	2
B : 0.6 mg × 2	8	0	7	0	0	1
C : 1.2 mg × 1	8	2	5	2	1	0
D : 0.2 mg × 6	8	1	4	1	1	2
E : 0.3 mg × 4	8	1	8	0	0	0
F : 0.6 mg × 2	8	1	6	0	1	1
Control	8	0	8	0	0	0

Table 7.5 OESTRUS AND OVULATION IN PREPUBERTAL GILTS TREATED WITH OESTRADIOL BENZOATE (ODB) IN THE PRESENCE OF BOARS

Treatment (mg ODB × injections)	*Number of gilts*				
			Ovulating normally	*Ovulation with cysts*	*Ovaries immature*
0.4 mg × 1	Oestrus	1	1	0	0
(*n* = 8)	No oestrus	7	0	0	7
0.4 mg × 3	Oestrus	3	0	1	2
(*n* = 8)	No oestrus	5	2	0	3
1.2 mg × 1	Oestrus	4	1	1	2
(*n* = 8)	No oestrus	4	1	1	2
Control	Oestrus	2	2	0	0
(*n* = 8)	No oestrus	6	1	0	5

Both these experiments were conducted in the absence of boars to avoid the possibility of confounding their stimulatory effects with those of the ODB. The poor oestrous response obtained suggested that failure to detect oestrus may have been a factor in these experiments. To test this hypothesis groups of eight gilts were relocated, exposed to boars and given a range of ODB treatments. The data from this experiment (*Table 7.5*) clearly shows that the poor oestrous response observed in the first two experiments was not due to failure to detect oestrus. When similar ODB treatments were compared among the three experiments, the presence of boars (B+) had no significant effect on the proportion of gilts displaying oestrus (B+ 8/24 cf. B— 5/25, $\chi^2 = 0.36$) or the proportion ovulating (B+ 5/24 cf. B— 6/25, $\chi^2 = 0.002$).

Plasma hormone levels were not measured in these gilts but it may be suggested from data from ovariectomized gilts (Paterson *et al.*, 1982) that treatments in which all the ODB was given on one day did not elevate plasma E_2 levels for long enough to produce an effect. Similarly, treatments using 0.2 or 0.3 mg of ODB may not have caused a large enough rise in E_2 to stimulate oestrus and ovulation. However, the higher doses (0.4 or 0.6 mg) spread over 2 or 3 days should have been sufficient to keep plasma E_2 at concentrations above those seen in the normal prepubertal elevation (Esbenshade *et al.*, 1982) for at least 80 hours. In addition Elsaesser and Foxcroft (1978) showed that similar doses of ODB to 160 days old gilts would induce an LH surge. The failure of these treatments to elicit oestrus and ovulation was surprising and the exact explanation for it is not apparent. Perhaps the natural rise in oestrogens prior to puberty is associated with changes in other hormones (for example LH, FSH or prolactin) which did not take place when oestrogens were artificially elevated with ODB but are necessary for oestrus and ovulation to take place.

In these experiments a wide range of ODB treatments failed to initiate cyclic activity. Overall 21/90 gilts showed oestrus and 13/90 ovulated. These results are in conflict with those of Hughes and Cole (1978) and no other published data are available for comparison. However, some recent work at the University of Nottingham initially reported by Foxcroft (1980) may provide an explanation. Their results suggest that the response of prepubertal gilts to ODB may be affected by season. In one experiment a total of 252 prepubertal gilts (65–70 kg liveweight) were treated with 1.2 mg ODB/gilt over the period May to December. The results in *Figure 7.5* show that the percentage of animals showing follicular development

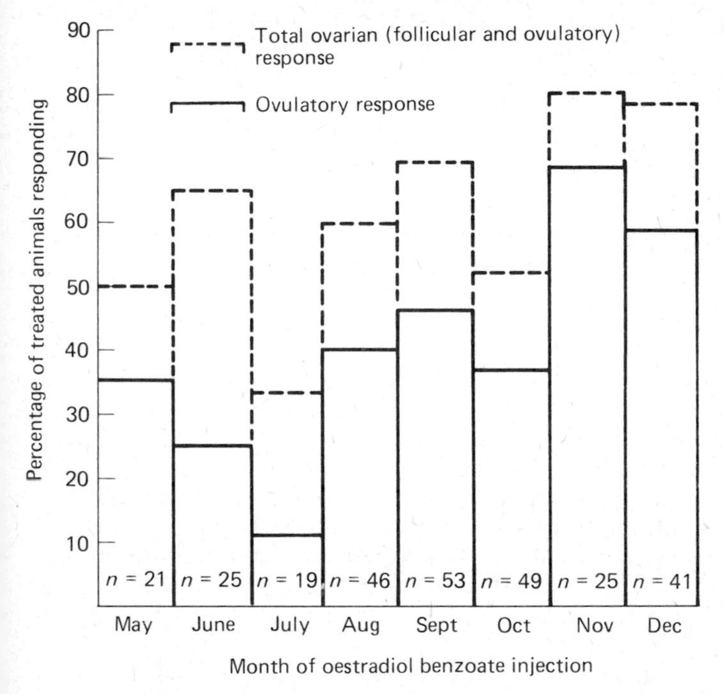

Figure 7.5 Variation with the month of treatment on the ovarian response of prepubertal gilts to 1.2 mg oestradiol benzoate. From Stickney and Foxcroft (unpublished data)

and the percentage ovulating followed a marked seasonal trend. Much greater responses were obtained in the winter months, when up to 70% of treated gilts ovulated, compared with about 10% in July.

In a second experiment conducted between February and December gilts given 1.2 mg ODB again exhibited a marked decline in ovulation response in the summer months (*Figure 7.6*). In contrast, gilts given

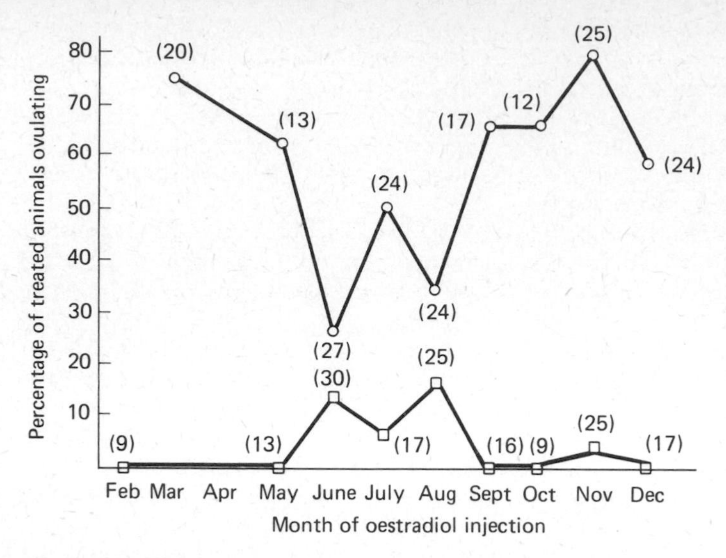

Figure 7.6 Variation with the month of treatment on the ovarian response of prepubertal gilts to 1.2 (○——○) or 0.12 (□——□) mg oestradiol benzoate. Figures in parentheses are number of gilts treated. Stickney and Foxcroft (unpublished data)

0.12 mg ODB demonstrated an ovulation response only in the summer months. Although the response to 0.12 mg was much lower than that to 1.2 mg ODB at the same time of year, it is consistent with the hypothesis that the seasonal effect may be mediated via variation in plasma oestrogen levels. Data from endocrine studies (Stickney and Foxcroft, unpublished data) indicate that plasma E_2 in response to the same dose of ODB is higher in summer than winter. Thus, the use of a high dose of ODB in winter may supply the correct steroid environment for inducing ovulation whereas such a dose in summer may inhibit ovulation. Conversely, a dose of 0.12 mg ODB in winter is not sufficient to elicit a response, but in summer this dose does have some effect. From a practical point of view it seems likely that the optimum ODB dosage for maximum responses throughout the year is somewhere within the range tested and that the dose used should be varied according to the month of administration. Clearly, this area requires further investigation, but it does hold promise as a means of reliably inducing synchronized puberty in the gilt.

Conclusions

As the gilt approaches puberty the frequency of episodic LH secretion increases and this is believed to provide the trigger for final follicular maturation and ovulation. When puberty is induced by exposure to a boar the first endocrine change so far observed is an increase in the levels of circulating oestradiol. Whether this increase in oestrogens is mediated via an increase in episodic LH secretion has yet to be determined, but evidence from the sheep and the mouse, in which LH levels have been shown to change in animals exposed to males, suggests that this is likely.

Treatment with exogenous hormones has not yet provided a useful method for the controlled induction of puberty in the gilt. The use of large doses of PMSG followed by HCG has many problems associated with it which precludes its adoption as a routine procedure. These include variable oestrous responses, poor reproductive performance, failure of the corpora lutea to persist in pregnant animals and failure to maintain cyclic activity in non-pregnant gilts. The use of low doses of PMSG and HCG in combination shows some promise as a method for the controlled induction of puberty, but further studies particularly on the maintenance of cyclic activity are required before this technique could be applied on a commercial basis.

Limited work with oestrogens has produced conflicting results and these compounds cannot be considered for the routine induction of puberty in practice at this time. However, the recent finding that the response to oestrogens follows a seasonal trend has increased our knowledge of this technique and promises to lead to a commercially viable system for controlling the onset of puberty. Finally, it also emphasizes the need to consider carefully the conditions under which induction experiments are carried out because many factors may interact to affect the response of gilts to such treatments. For this reason particular care in interpretation should be taken when making recommendations for the practical application of such techniques.

References

BAKER, R.D. and COGGINS, E.G. (1966). Induced ovulation and fertility in immature gilts. *J. Anim. Sci.* **25**, 918 (Abstract)

BAKER, R.D. and DOWNEY, B.R. (1975). Induction of estrus, ovulation and fertility in prepubertal gilts. *Annl Biol. anim. Biochim. Biophys.* **15**, 375–382

BAKER, R.D. and RAJAMAHENDRAN, R. (1973). Induction of estrus, ovulation and fertilization in prepubertal gilts by a single injection of PMSG, HCG and PMSG:HCG combination. *Can. J. Anim. Sci.* **53**, 693–694

BIELANSKI, A. (1978). The possibilities of a reduction in the age at first farrowing. *Anim. Breed. Abstr.* **46**, 581

BOOTH, W.D. (1975). Changes with age in the occurrence of C_{19} steroids in the testes and submaxillary gland of the boar. *J. Reprod. Fert.* **42**, 459–472

BOOTH, W.D. (1977). Metabolism of androgens *in vitro* by the submaxillary salivary gland of the mature domestic boar. *J. Endocr.* **75**, 145–154

BREEUWSMA, A.J. (1974). Induction of oestrus and ovulation in prepubertal and anoestrous pigs. *3rd Int. Congr. Int. Pig Vet. Soc., Lyon*, pp. G14, 1–5

BRONSON, F.H. (1975). Male-induced precocial puberty in female mice: confirmation of the role of estrogen. *Endocrinology* **96**, 511–514

BRONSON, F.H. and DESJARDINS, C. (1974). Circulating concentrations of FSH, LH, estradiol and progesterone associated with acute, male-induced puberty in female mice. *Endocrinology* **94**, 1658–1668

BRONSON, F.H. and MARSDEN, H.M. (1964). Male-induced synchrony of estrus in deermice. *Gen. Comp. Endocr.* **4**, 634–637

BRONSON, F.H. and MARUNIAK, J.A. 1975). Male-induced puberty in female mice: evidence for a synergistic action of social cues. *Biol. Reprod.* **13**, 94–98

BRONSON, F.H. and MARUNIAK, J.A. (1976). Differential effects of male stimuli on follicle-stimulating hormone, luteinizing hormone, and prolactin secretion in pubertal female mice. *Endocrinology* **98**, 1101–1108

BRONSON, F.H. and WHITTEN, W.K. (1968). Oestrus-accelerating pheromone of mice: assay, androgen-dependency and presence in bladder urine. *J. Reprod. Fert.* **15**, 131–134

BROOKS, P.H. and COLE, D.J.A. (1969). The effect of boar presence on the age at puberty of gilts. *Rep. Sch. Agric. Univ. Nott.* (1968–69) pp. 74–77

BROOKS, P.H. and COLE, D.J.A. (1970). The effect of the presence of a boar on attainment of puberty in gilts. *J. Reprod. Fert.* **23**, 435–440

CASIDA, L.E. (1935). Prepubertal development of the pig ovary and its relation to stimulation with gonadotrophic hormones. *Anat. Rec.* **61**, 389–396

COOPER, K.J. and HAYNES, N.B. (1967). Modification of the oestrous cycle of the underfed rat associated with the presence of the male. *J. Reprod. Fert.* **14**, 317–320

DAVIDSON, J.M. (1974). Hypothalamic–pituitary regulation of puberty: evidence from animal experimentation. In *Control of the Onset of Puberty*, (M.M. Grumbach, G.D. Grave and F.E. Mayer, Eds.), pp. 79–103. New York, John Wiley and Sons

DEBELJUK, L., ARIMURA, A. and SCHALLY, A.V. (1972). Studies on the pituitary responsiveness to luteinizing hormone-releasing hormone (LH-RH) in intact male rats of different ages. *Endocrinology* **90**, 585–588

DOMINIC, C.J. (1965). The origin of the pheromones causing pregnancy block in mice. *J. Reprod. Fert.* **10**, 469–472

DU MESNIL DU BUISSON, F. and SIGNORET, J.P. (1962). Influences de facteurs externes sur le déclenchement de la puberté chez la truie. *Ann. Zootech.* **11**, 53–59

DUNCAN, D.L. and LODGE, G.A. (1960). Diet in relation to reproduction and viability of the young. Part III, Pigs. *Commonwealth Bureau of Animal Nutrition Tech. Comm. No. 21*. Rowett Research Institute, Bucksburn, Aberdeen.

DYRMUNDSSON, O.R. and LEES, J.L. (1972). Effect of rams on the onset of breeding activity in Clun Forest ewe lambs. *J. agric. Sci., Camb.* **79**, 269–271

DZIUK, P.J. (1965). Response of sheep and swine to treatments for the control of ovulation. *USDA Mis. Pub.* **1005**, 50–53

DZIUK, P.J. and DHINDSA, D.S. (1969). Induction of heat, ovulation and fertility in gilts with delayed puberty. *J. Anim. Sci.* **29**, 39–41

DZIUK, P.J. and GEHLBACH, G.D. (1966). Induction of ovulation and fertilization in the immature gilt. *J. Anim. Sci.* **25**, 410–413

DZIUK, P.J. and POLGE, C. (1965). Fertility in gilts following induced ovulation. *Vet. Rec.* **77**, 236–238

ELLICOTT, A.R., DZIUK, P.J. and POLGE, C. (1973). Maintenance of pregnancy in prepubertal gilts. *J. Anim. Sci.* **37**, 971–973

ELSAESSER, F. and FOXCROFT, G.R. (1978). Maturational changes in the characteristics of oestrogen-induced surges of luteinizing hormone in immature domestic gilts. *J. Endocr.* **78**, 455–456

ESBENSHADE, K.L., PATERSON, A.M., CANTLEY, T.C. and DAY, B.N. (1982). Changes in plasma hormone concentrations associated with the onset of puberty in the gilt. *J. Anim. Sci.* **54**, 320–324

FOSTER, D.L. (1980). Comparative development of mammalian females: proposed analogies among patterns of LH secretion in various species. In *Proceedings of the Serono Symposia 32*, (C. La Cauza and A.W. Root, Eds.), pp. 193–209. London, Academic Press.

FOSTER, D.L. COOK, B. and NALBANDOV, A.V. (1972). Regulation of luteinizing hormone (LH) in the fetal and neonatal lamb: effect of castration during the early postnatal period on levels of LH in sera and pituitaries of neonatal lambs. *Biol. Reprod.* **6**, 253–257

FOXCROFT, G.R. (1978). The development of pituitary gland function. In *The Control of Ovulation* (D.B. Crighton, G.R. Foxcroft, N.B. Haynes and G.E. Lamming, Eds.), pp. 117–138. London, Butterworths

FOXCROFT, G.R. (1980). Growth and breeding performance in animals and birds. In *Growth in Animals* (T.L.J. Lawrence, Ed.), pp. 229–247. London, Butterworths

GATES, A.H. and BOZARTH, J.L. (1978). Ovulation in the PMSG-treated immature mouse: effect of dose, age, weight, puberty, season and strain (BALBc 129 and C129F, Hybrid). *Biol. Reprod.* **18**, 497–505

GORSKI, R.A. (1974). Extrahypothalamic influences on gonadotrophin regulation. In *Control of the Onset of Puberty* (M.M. Grumbach, G.D. Grave and F.E. Mayer, Eds.), pp. 182–207. New York, John Wiley and Sons

GOWER, D.B. (1972). 16-unsaturated C_{19} steroids. A review of their chemistry, biochemistry and possible physiological role. *J. Steroid Biochem.* **3**, 45–103

GRUMBACH, M.M., ROTH, J.C., KAPLAN, S.L. and KELCH, R.P. (1974). Hypothalamic–pituitary regulation of puberty in man: evidence and concepts derived from clinical research. In *Control of the Onset of Puberty* (M.M. Grumbach, G.D. Grave and F.E. Mayer, Eds.), pp. 115–181. New York, John Wiley and Sons

GUTHRIE, H.D. (1977). Induction of ovulation and fertility in prepubertal gilts. *J. Anim. Sci.* **45**, 1360–1367

GUTHRIE, H.D., HENRICKS, D.M. and HANDLIN, D.L. (1972). Plasma estrogen, progesterone and luteinizing hormone prior to estrus and during early pregnancy in pigs. *Endocrinology* **91**, 675–679

HENRICKS, D.M., GUTHRIE, H.D. and HANDLIN, D.L. (1972). Plasma estrogen, progesterone and luteinizing hormone levels during the estrous cycle in pigs. *Biol. Reprod.* **6**, 210–218

HOLTZ, W., POLANCO, A., von KAUFMANN, F. and HERRMANN, H.H. (1977). Induction of precocious ovulation in gilts. *Proceedings: Society for the Study of Fertility, Nottingham,* Abstract 10

HUGHES, P.E. and COLE, D.J.A. (1978). Reproduction in the gilt. 3. The effect of exogenous oestrogen on the attainment of puberty and subsequent reproductive performance. *Anim. Prod.* **27**, 11–20

HÜHN, U., HEIDLER, W. and RESSIN, E. (1977). Induction of sexual maturity in prepubertal gilts by gonadotrophins. 2. Reproductive performance of puberty-induced gilts following oestrous synchronization and artificial insemination. *Anim. Breed. Abstr.* **45**, 242

KINSEY, R.E., CARLSON, R., PROUD, C. and ZIMMERMAN, D.R. (1976). Influence of boar stimuli on age at puberty in gilts. *J. Anim. Sci.* **42**, 1362 (Abstract)

KIRKWOOD, R.N. and HUGHES, P.E. (1980). A note on the influence of 'boar effect' component stimuli on puberty attainment in the gilt. *Anim. Prod.* **31**, 209–211

KIRKWOOD, R.N., FORBES, J.M. and HUGHES, P.E. (1981). Influence of boar contact on attainment of puberty in gilts after the removal of the olfactory bulbs. *J. Reprod. Fert.* **61**, 193–196

LEVASSEUR, M.C. (1977). Thoughts on puberty. Initiation of gonadotrophic function. *Annls Biol. anim. Biochim. Biophys.* **17**, 345–361

MARTIN, G.B., OLDHAM, C.M. and LINDSAY, D.R. (1980). Increased plasma LH levels in seasonally anovular Merino ewes following the introduction of rams. *Anim. Reprod. Sci.* **3**, 125–132

MISKOVIC, M., SIMIC, M., JOJKIC, M. and STANCIC, B. (1977). Ovulation induction in prepubertal gilts using PMS and HCG. *Anim. Breed. Abstr.* **45**, 118

PARKES, A.S. and BRUCE, H.M. (1962). Pregnancy-block in female mice placed in boxes soiled by males. *J. Reprod. Fert.* **4**, 303–308

PATERSON, A.M. (1979). The reproductive performance of sows and gilts under intensive conditions. PhD Thesis. University of Western Australia

PATERSON, A.M. and DAY, B.N. (1980). The effect of estradiol benzoate on estrus and ovulation in prepubertal gilts. *J. Anim. Sci.* **51**, (*Suppl.* **1**) 88

PATERSON, A.M. and LINDSAY, D.R. (1981). Induction of puberty in gilts. 2. The effect of boars on maintenance of cyclic activity in gilts induced to ovulate with PMSG and HCG. *Anim. Prod.* **32**, 51–54

PATERSON, A.M. and MARTIN, G.B. (1981). Induction of puberty in gilts. 3. Ovulation, plasma oestradiol and progesterone in gilts injected with PMSG and HCG. *Anim. Prod.* **32**, 55–59

PATERSON, A.M., CANTLEY, T.C., ESBENSHADE, K.L. and DAY, B.N. (1982). Glucocorticoids and estrus in swine. II. Plasma levels of estradiol-17β glucocorticoids and LH in ovariectomized gilts given estradiol benzoate and triamcinolone acetonide. *J. Anim. Sci.* (submitted for publication)

PHILLIPO, M. (1968). Superovulation in the pig. *Adv. Reprod. Physiol.* **3**, 147–166

RAMIREZ, V.D. and McCANN, S.M. (1963). Comparison of the regulation of luteinizing hormone (LH) secretion in immature and adult rats. *Endocrinology* **72**, 452–464

RAMPACEK, G.B., SCHWARTZ, F.L., FELLOWS, R.E., ROBISON, O.W. and ULBERG, L.C. (1976). Initiation of reproductive function and subsequent activity of the corpora lutea in prepubertal gilts. *J. Anim. Sci.* **46**, 881–887

REDMER, D.A. and DAY, B.N. (1981). Ovarian activity and hormonal patterns in gilts fed allyl trenbolone. *J. Anim. Sci.* **53**, 1088–1094

REED, H.C.B., MELROSE, D.R. and PATTERSON, R.L.S. (1974). Androgen steroids as an aid to the detection of oestrus in pig artificial insemination. *Br. vet. J.* **130**, 61–67

RYAN, K.D. and FOSTER, D.L. (1980). Neuroendocrine mechanisms involved in the onset of puberty in the female: concepts derived from the lamb. *Fed. Proc.* **9**, 2372–2377

SCHILLING, E. and CERNE, F. (1972). Induction and synchronization of oestrus in prepubertal gilts and anoestrous sows by a PMS/HCG compound. *Vet. Rec.* **91**, 471–474

SCHLEGEL, W., WAHNER, M. and STENZEL, S. (1978). Studies into the further course of the cycle following biotechnical induction of puberty in gilts. *Anim. Breed. Abstr.* **46**, 585

SEGAL, D.H. and BAKER, R.D. (1973). Maintenance of corpora lutea in prepubertal gilts. *J. Anim. Sci.* **37**, 762–767

SHAW, G.A., McDONALD, B.E. and BAKER, R.D. (1971). Fetal mortality in the prepubertal gilt. *Can. J. Anim. Sci.* **51**, 233–236

SHEARER, I.J., PURVIS, K., JENKINS, G. and HAYNES, N.B. (1972). Peripheral plasma progesterone and oestradiol-17β levels before and after puberty in gilts. *J. Reprod. Fert.* **30**, 347–360

VANDENBERGH, J.G. (1969). Male odor accelerates female sexual maturation in mice. *Endocrinology* **84**, 658–660

VANDENBERGH, J.G. (1975). Pheromonal stimulation of puberty in female mice. *J. Endocr.* **64**, 38p (Abstract)

WHITTEN, W.K. (1956a). Modification of the oestrous cycle of the mouse by external stimuli associated with the male. *J. Endocr.* **13**, 399–404

WHITTEN, W.K. (1956b). The effect of removal of the olfactory bulbs on the gonads of mice. *J. Endocr.* **14**, 160–163

WILSON, E.O. and BOSSERT, W.H. (1963). Chemical communication among animals. *Recent Progr. Horm. Res.* **19**, 673–710

8

ENDOCRINE CONTROL OF THE OESTROUS CYCLE

G.R. FOXCROFT
Physiology and Environmental Studies, University of Nottingham School of Agriculture, Sutton Bonington, Loughborough, UK

D.F.M. VAN DE WIEL
Research Institute for Animal Husbandry "Schoonoord", Driebergseweg 10 D, Zeist, The Netherlands

The study of the mechanisms controlling the oestrous cycle of the sow has already illustrated some important comparative differences from other species, although even the direct measurement of those circulating hormones involved in controlling ovarian activity is still incomplete. While some technical difficulties exist in using the pig for *in vivo* experimentation, the relatively high levels of circulating ovarian steroids make it an excellent animal in which to study steroid–gonadotrophin interactions. Furthermore the ready availability of porcine ovarian tissue has resulted in the accumulation of a considerable body of evidence from *in vitro* experimentation related to the intra-ovarian mechanisms involved in the integrated control of the hypothalamic–pituitary–ovarian axis.

The major objective of this Chapter is to establish a possible model for the control of the oestrous cycle in the pig with reference to pertinent data from both *in vivo* and *in vitro* studies. To some extent this precludes an exhaustive coverage of the available literature, but where possible, reference is made to papers in which more detailed information is available. In attempting to achieve these objectives the oestrous cycle will be divided into five phases which will be considered in a chronological sequence as follows:

(a) the established luteal phase of the cycle;
(b) luteal regression;
(c) the early follicular phase;
(d) the late follicular phase and the pre-ovulatory surge of gonadotrophins;
(e) the periovulatory period (including the early luteal phase)

The established luteal phase of the cycle

Several aspects of hypothalamo–hypophysial–ovarian function peculiar to those mammals that exhibit long luteal phase cycles have been discussed in the general context of cyclic control by Greenwald (1979). As the maintenance of luteal function appears to dictate the length of the cycle,

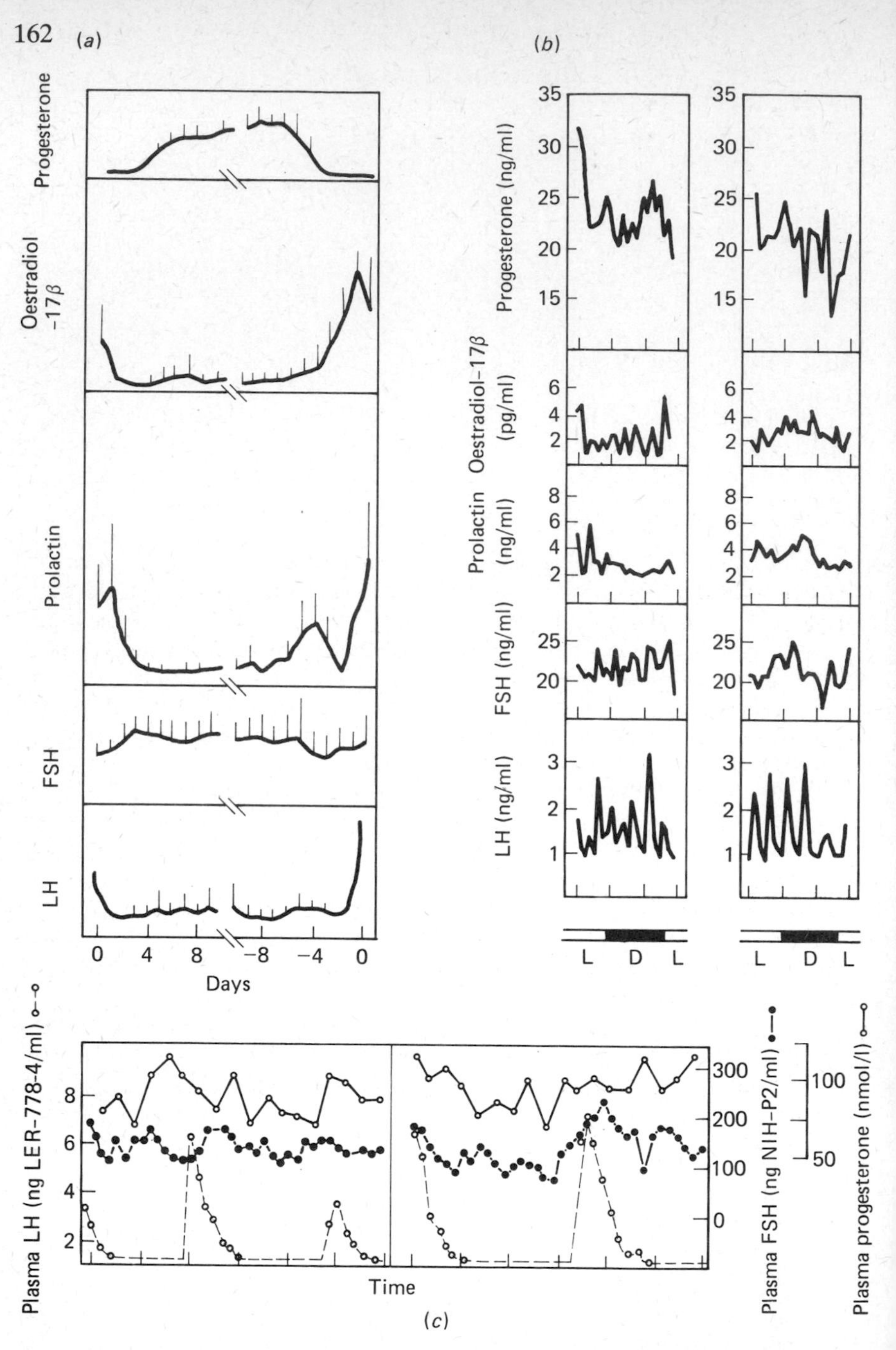

Figure 8.1 (a) Mean relative changes in circulating hormones (+SDM) throughout the oestrous cycle of the pig (day 0 = day of LH surge), as determined by daily sampling. (b) Changes in circulating hormones on days 11–12 of the oestrous cycle in two intact pigs sampled at hourly intervals for 24 hours. L = light, D = dark periods. (c) Changes in circulating LH (○---○), FSH (●——●) and progesterone (○——○) during the mid-luteal phase of the oestrous cycle in two pigs sampled at 10 minute intervals for 6 hours. (a) and (b) after Van de Wiel *et al.* (1981); (c) Foxcroft and Edwards, unpublished observations

the choice of the 'established' luteal phase as a starting point for a consideration of cyclic control seems logical.

The changes in peripheral progesterone concentrations are the best documented endocrine events of the pig oestrous cycle (Tillson and Erb, 1967; Stabenfeldt *et al.*, 1969; Tillson, Erb and Niswender, 1970; Edquist and Lamm, 1971; Shearer *et al.*, 1972; Henricks, Guthrie and Handlin, 1972; Parvizi *et al.*, 1976; Van de Wiel *et al.*, 1981), although some inconsistencies still exist in these data. As shown in *Figure 8.1(a)*, a rise in progesterone secretion is generally observed on days 3–4 of the oestrous cycle (the first day of standing heat being designated as day 0), but considerable variability exists as to the time of maximum circulating concentrations. Some data suggest that peripheral levels may begin to decline as early as day 10 of the cycle as opposed to day 15 in other work; variations in sampling frequency may explain some of these differences but Perotti *et al.* (1979) have also observed differences in the pattern of progesterone secretion that appear to be related to season.

As it is important to obtain a clearer indication of the timing of the changes in progesterone production in developing a model for the onset of luteolysis, estimates of progesterone secretion based on ovarian vein sampling are of particular interest. Gomes, Herschler and Erb (1965) reported that the ovarian vein concentration of progesterone was maximal on days 10–12, declined slowly to days 13–15 and then fell rapidly. These findings are consistent with the report of Masuda *et al.* (1967) that the production rate (involving measurements of both ovarian vein concentration and blood flow) of progesterone was maximal on day 8, declined slowly to day 12 and then again fell more rapdidly.

These data suggest that a decline in progesterone secretion as early as day 10 should be reflected in an earlier decline in progesterone in the peripheral circulation than is frequently observed. However, recent data (Hillbrand, F.W. and Elsaesser, F., personal communication) demonstrate that a very substantial amount of progesterone (up to 100 times that in the circulation) may be sequestered in the body fat of the sow. Following a fall in ovarian progesterone output the return of this fat 'depot' into the circulation will result in a latent fall in circulating progesterone levels. These observations have important implications for the control of the cycle as the initial decrease in progesterone secretion by the corpora lutea (the onset of luteolysis?) may occur as early as day 10 of the cycle, and yet any withdrawal of an effect of progesterone acting at the hypothalamic–pituitary level will be seen considerably later.

The necessity for luteotrophic support during the oestrous cycle has been the subject of considerable debate and much of the early evidence (Nalbandov, 1970; Hansel, Concannon and Lukaszewska, 1973) suggested that once the signal for luteinization has occurred in the form of the pre-ovulatory surge of luteinizing hormone (LH), the corpora lutea are essentially autonomous for the duration of the luteal phase of the cycle. Thus, neither hypophysectomy early in the cycle (du Mesnil du Buisson and Leglise, 1963) nor treatment with LH antisera (Spies, Slyter and Quadri, 1967) appear to affect luteal function.

Nevertheless a number of more recent observations suggest that the corpora lutea of the cycle may be responsive to luteotrophic stimuli.

Watson and Leask (1975) used an *in vitro* superfusion system to study the steroidogenic activity of porcine luteal tissue obtained during the cycle and reported an increase in progesterone secretion in response to both pulses and infusions of LH. Furthermore, an increase in oestrogen was also observed, suggesting that in agreement with the results of Weiss, Brinkley and Young (1976), the corpora lutea may also secrete oestrogens during the luteal phase of the cycle. From their study of LH–progesterone interactions during the oestrous cycle of the miniature pig Parvizi *et al.* (1976) also suggested that episodes of LH tended to precede and be associated with, elevations in peripheral progesterone, although no such relationship was clearly established in the domestic sow in the data of Van de Wiel *et al.* (1981), nor in unpublished observations at Nottingham University (see *Figure 8.1(c)*). However, during the mid-luteal phase Van de Wiel *et al.* (1981) did suggest a possible relationship between LH episodes and oestradiol secretion (see *Figure 8.1(b)*). Thus, although a distinct pattern of high amplitude, low frequency LH episodes has been reported by a number of authors during the luteal phase of the cycle (Rayford, Brinkley and Young, 1971; Parvizi *et al.*, 1976; Foxcroft, 1978; Van de Wiel *et al.*, 1981), as shown in *Figure 8.1*, conclusive evidence is still needed for a physiological role for LH at this time. Specific luteal binding of both prolactin and LH have been reported (Rolland, Gunsalus and Hammond, 1976; Ziecik, Shaw and Flint, 1980) again suggesting that such hormones may be involved in controlling luteal activity.

Luteal regression

As in other species prostaglandin $F_{2\alpha}$ (PGF) has been postulated to be the major signal for luteolysis in the pig and the evidence to support this hypothesis is extensively reviewed by Bazer (Chapter 12). In contrast to other species, exogenous PGF will not cause luteolysis before day 12 in the pig, suggesting that the PGF-sensitive mechanism is inoperative in the early stages of the cycle. The means by which PGF of uterine origin is transferred to the ovarian artery has also been extensively investigated and in addition to an accepted vascular route of transfer, Kotwica (1980) has recently reported a possible involvement of the lymphatic circulation.

The mechanism of action of prostaglandins in the pig is unresolved. Exogenous gonadotrophin is ineffective in maintaining the corpora lutea of the cycle in the presence of an intact uterus (Anderson, 1966), suggesting that PGF inactivates some component of the LH stimulatory mechanism that is effective in maintaining luteal function in hysterectomized females. A dramatic decrease in unoccupied LH receptors was observed by Ziecik, Shaw and Flint (1980) after day 12 of the cycle and an increase in an LH receptor binding inhibitor of luteal origin has been associated with increasing age of the corpus luteum (Tucker, Kumari and Channing, 1979). These data would therefore be consistent with the theory that at the time of luteal regression the corpora lutea may be LH-dependent and that loss of the LH receptor mechanism within luteal tissue results in the decline in progesterone synthesis that occurs as early as day 12 of the cycle. In addition there is evidence that the initial luteolytic stimulus from the uterus may also activate the local production of PGF within the corpus luteum itself, thus reinforcing the luteolytic effect of prostaglandins (Guthrie, Rexroad and Bolt, 1979).

Two other phenomena exist in the late luteal phase of the cycle of the sow that contrast with the situation in other large domestic species. The first is the ability of exogenous oestrogen to block the luteolytic effects of prostaglandin and considerably extend luteal function (see Bazer, Chapter 12). The mechanism by which oestrogen exerts this luteotrophic effect may involve the redirection of prostaglandin as is proposed for the early stages of pregnancy, but preliminary evidence discussed during this meeting (Garverick, Flint and Polge, unpublished observations) indicates that oestrogen may also act to increase LH receptors in the corpora lutea. Although the luteotrophic effect of oestrogen appears to have little physiological significance for the control of the normal cycle, these data again suggest that the presence of adequate LH receptors may be critical to continued luteal function at the end of the cycle. Secondly, there are consistent reports of markedly elevated levels of prolactin at the time of luteal regression (Wilfinger, 1974; Van Landeghem and Van de Wiel, 1977; Van de Wiel *et al.*, 1981). The close relationship between this rise in prolactin and the decline in peripheral progesterone concentrations invites speculation concerning a luteolytic role for prolactin and Rolland, Gunsalus and Hammond (1976) have suggested that evidence exists for an inhibitory effect of prolactin on progesterone secretion, at least in porcine granulosa cells. However, Van de Wiel *et al.* (1981) also emphasize the close relationship between the same prolactin rise and the initial increase in oestradiol secretion and therefore a possible role in the stimulation of follicular growth and steroidogenesis as discussed in the next section.

The early follicular phase

Although the adopted sequence in this review suggests that the follicular phase of the cycle can be discretely separated from the luteal phase of the cycle, this may be misleading. The data currently available indicate that important changes occur as a continuum in developing follicles between the time of antral formation and the time that they achieve a size of 6–12 mm and become pre-ovulatory Graafian follicles at the end of the oestrous cycle. Thus, the number of granulosa LH receptors increase approximately 100-fold, LH-stimulated increases in cyclic AMP increase in magnitude, the sensitivity of the cyclic AMP response to prostaglandin E decreases, the ability of the granulosa to secrete progesterone and to convert androgen to oestrogen increases and the ability of follicle stimulating hormone (FSH) to stimulate the aromatization of androgen in the granulosa only becomes apparent in large follicles (Anderson, Schaerf and Channing, 1979; Leung, Tsang and Armstrong, 1979; Schwartz-Kripner and Channing, 1979). Taken together these observations suggest that the control of follicular growth and steroidogenesis in the pig may follow a similar pattern to that proposed for the rat (Armstrong and Dorrington, 1977; Richards, Rao and Ireland, 1978) in which the initiation of responses by the follicle involve an LH-induced increase in androgen production by the thecal tissue followed by an increase in the aromatization of this androgen to oestrogen by the granulosa cells. Such changes are mediated

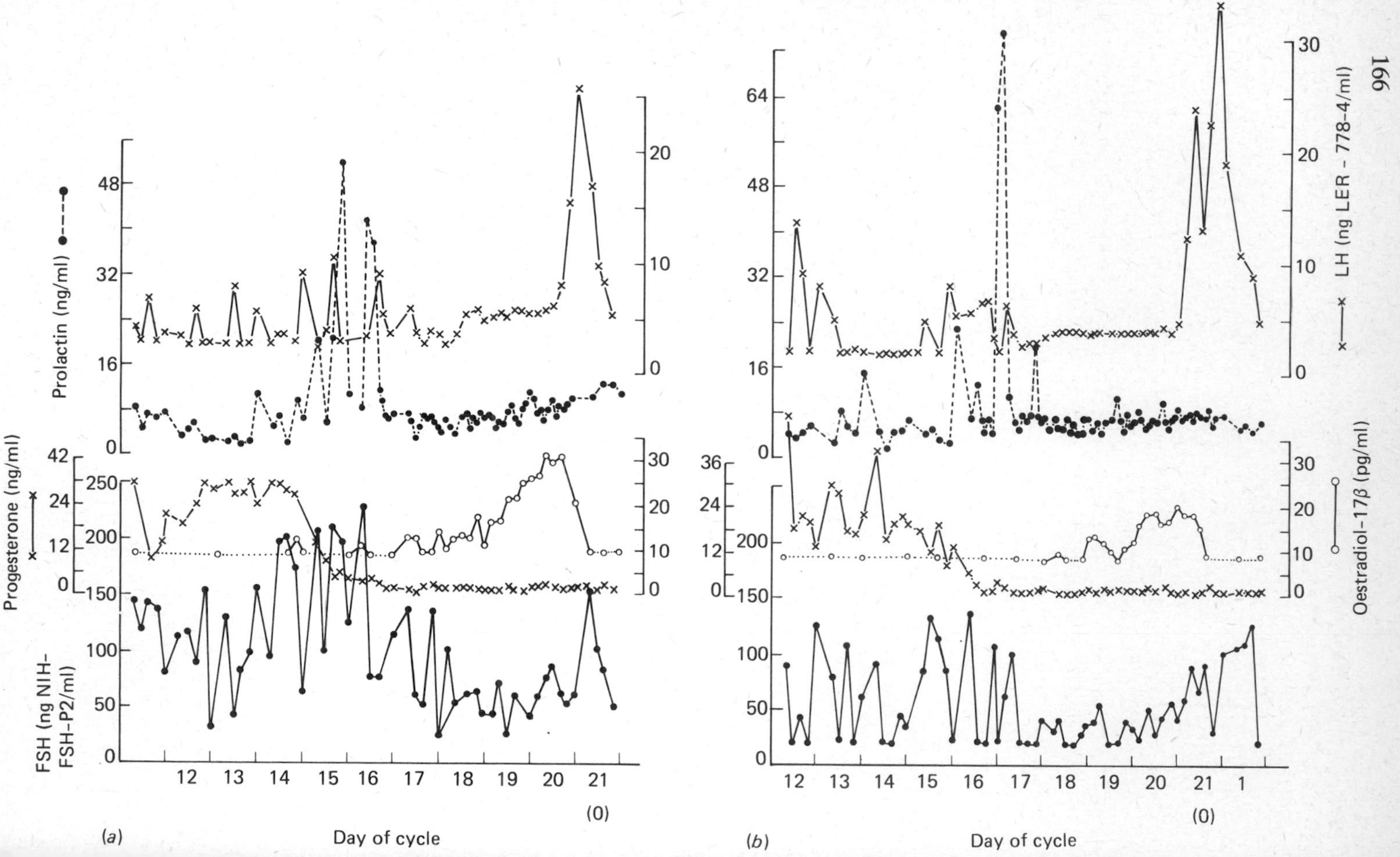

Prolactin (ng/ml)
Progesterone (ng/ml)
FSH (ng NIH-FSH-P2/ml)
LH (ng LER - 778-4/ml)
Oestradiol-17β (pg/ml)

(a) Day of cycle

(b) Day of cycle

by oestrogen-dependent changes in receptors for both LH and FSH in the granulosa and through the induction of the aromatase enzymes at an intracellular level.

In view of the timing of many of these initial changes in the follicle it is difficult to determine those changes in gonadotrophin secretion, if any, that act as the stimuli for the onset of follicular development. As discussed later, some evidence suggests that the pre-ovulatory gonadotrophin surge of the previous cycle may be responsible for the recruitment of a new crop of follicles and, in the absence of any further specific gonadotrophin signal, ordered sequences of maturational changes may occur that are entirely dependent on intra-ovarian control mechanisms. However, as is well known, such changes only result in limited growth of the follicles and the absence of a major increase in oestradiol production, if progesterone levels are maintained beyond day 14–17 of the cycle. This suggests that high levels of progesterone block further maturation of the follicle and it would be of considerable interest to know whether this is a direct, or indirect, effect. Initial observations (Foxcroft, 1978) demonstrate that, as in other species, the decline in progesterone secretion leads to a gradual increase in the frequency, and a decrease in the amplitude, of LH episodes as well as a gradual decline in circulating FSH levels. However as discussed on p. 165 (Henricks, Guthrie and Handlin, 1972; Van de Wiel *et al.*, 1981), the initial fall in progesterone is also concurrent with both the initial rise in peripheral oestradiol and an increase in prolactin secretion and it is not possible to determine cause or effect from these data.

In the sow, therefore, although important intra-ovarian changes have been described that are undoubtedly associated with normal follicular development during the cycle, the trigger for the onset of increased oestrogen secretion in the late luteal/early follicular phase of the cycle is unknown. If the decline in progesterone is related to the appearance of a trigger of pituitary origin, then this can only be associated with a qualitative change in the pattern of LH/FSH secretion rather than a simple quantitative increase in gonadotrophin release. The possibility exists however that progesterone may exert its block to follicular development entirely at the ovarian level and that the changing patterns of LH/FSH secretion are merely indirect consequences of the changes in circulating steroids and have no physiological significance as ovarian stimuli other than in maintaining a minimum level of gonadotrophin in the circulation. In discussing such indirect effects of progesterone it is relevant to note that in the luteal phase of the cycle no positive feedback response to oestrogen can be elicited (Foxcroft and Edwards, unpublished data); thus even in the presence of oestrogen secretion a central block to ovulation exists. The overall changes in circulating hormones at the time of luteal regression are shown in *Figure 8.2* to give an indication of the complexity of the relationships that exist at this time. Further detailed studies relating such changes to intra-ovarian developments at this critical stage of the cycle are clearly required.

The late follicular phase and the pre-ovulatory surge of gonadotrophins

The net effect of the changes described in the previous section is a major rise in circulating oestrogen between days 18 and 20 of the cycle. This

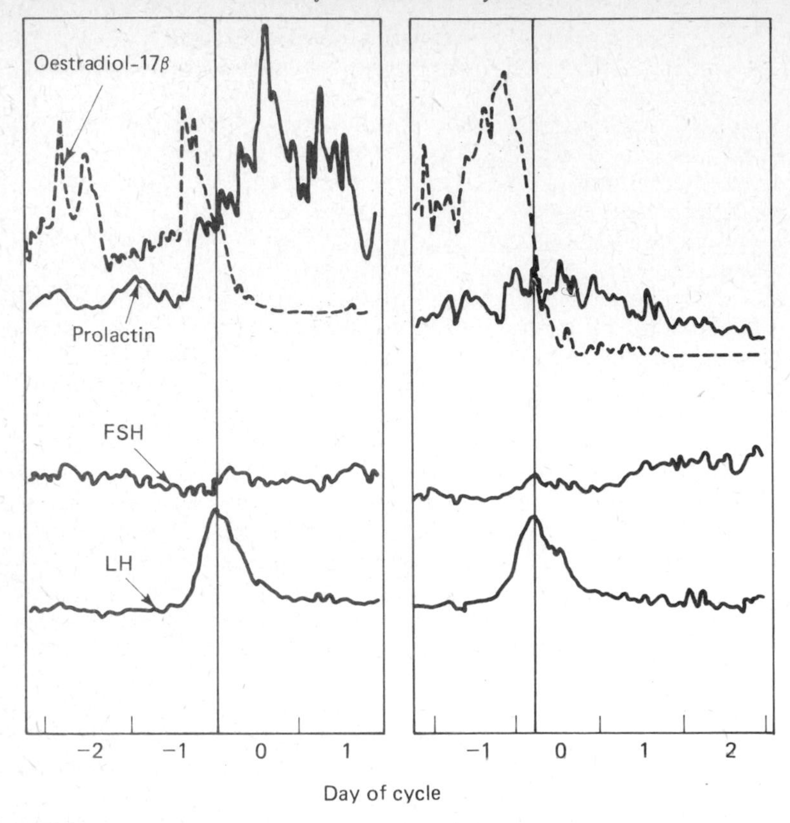

Figure 8.3 Relative changes in plasma levels of circulating hormones in the late follicular phase of the oestrous cycle in two intact pigs sampled at hourly intervals. Day 0 = first day of oestrous; vertical line = time of maximum LH levels. From Van de Wiel *et al.* (1981)

pattern of oestrogen release was initially determined indirectly by the measurement of urinary oestrone clearance (Lunaas, 1962; Raeside, 1963; Liptrap and Raeside, 1966) and subsequently confirmed by the direct estimation of oestradiol-17β in peripheral plasma by radioimmunoassay (Henricks, Guthrie and Handlin, 1972; Shearer *et al.*, 1972; Van de Wiel *et al.*, 1981). The temporal relationship between this rise in oestrogen and the pre-ovulatory surge of LH and FSH (shown in *Figure 8.3*) suggested that as in other species the rise in oestrogen triggered a positive feedback mechanism within the hypothalamus. Confirmation that oestrogen could exert this effect in the mature domestic sow has come from the studies of Edwards (1980) and the latency of the response appears to be approximately 50–55 hours in this species. The increase in circulating oestradiol results initially in complete and immediate suppression of episodic LH release, whilst its precise effect on the basal levels of LH needs clarification (Foxcroft, 1978). In contrast the inhibitory effect on FSH is much more latent, although FSH levels typically reach minimal levels within individual animals immediately preceding the pre-ovulatory surge of gonadotrophins (Vandalem *et al.*, 1979; Van de Wiel *et al.*, 1981). A detailed assessment of FSH levels during the late follicular phase demonstrates, however, that there are periods of active FSH secretion in the presence of high concentrations of oestradiol, at a time when marked suppression of LH release exists (see *Figures 8.2* and *8.3*). This differential feedback effect on LH/FSH release seems to merit further study and as this pattern of gonadotrophin secretion provides a considerable contrast to that in other large domestic species with a low litter size, it may be of considerable comparative significance.

As shown by Van de Wiel *et al.* (1981) in the cyclic gilt and by Edwards (1980) in the weaned domestic sow, the initial rise in pre-ovulatory LH precedes the rapid decline in oestradiol secretion, resulting in the occurrence of peak levels of oestradiol in the circulation approximately 8–15 hours before peak LH levels are observed (see *Figure 8.3*). The characteristics of the pre-ovulatory LH surge at the time of oestrus were initially determined by bioassay and in retrospect the observation of Liptrap and Raeside (1966) that a rise in plasma LH occurred some 40–48 hours before ovulation suggests that the ovarian cholesterol depletion assay used by these authors provided a reliable estimate of LH changes. The first data obtained by radioimmunoassay were those of Niswender, Reichert and Zimmerman (1970) who reported an LH surge with a duration of approximately 20 hours coincident with the onset of oestrus. These data have been confirmed in subsequent studies of cyclic gilts and sows (Henricks, Guthrie and Handlin, 1972; Rayford, Brinkley and Young, 1971; Parvizi *et al.*, 1976; Vandalem *et al.*, 1979; Van de Wiel *et al.*, 1981) and in the weaned sow (Edwards, 1980).

Two points are perhaps worthy of comment with respect to the LH surge. Firstly, on the basis of published data, the precise relationship between the onset of behavioural oestrus and the time of the LH surge appears to be variable and these events may be displaced by as much as 12 hours in either direction (Foxcroft, Tilton, Ziecik and Coombs, unpublished data). Therefore, assuming a fixed interval between the LH surge and ovulation (36–40 hours), the possibility exists that the time of ovulation within the heat period may vary considerably and this may be of consequence if the number of matings per sow is restricted and related entirely to the onset of heat.

Secondly, a comparison of maximum LH levels associated with tonic episodic secretion during the luteal phase of the cycle (approximately 3 ng/ml) and the maximum levels observed during the LH surge of 6 ng/ml (Van de Wiel *et al.*, 1981) demonstrates that compared with other species the LH surge of the pig is characterized by a relatively moderate percentage increase in LH, but that this elevation is of considerable duration. Again in comparative terms it will be very interesting to determine whether this pattern of LH secretion is of physiological significance.

Compared with the pre-ovulatory LH surge, the FSH response to oestradiol positive feedback is variable and Van de Wiel *et al.* (1981) suggest that it might be explained solely on the basis of the withdrawal of the inhibitory effect of oestrogens at this time. However, consideration of the data of other authors (Rayford *et al.*, 1974; Vandalem *et al.*, 1979) suggests that an FSH 'surge' is generally a consistent event and is frequently followed by a temporary fall in FSH levels prior to a subsequent rise as discussed on p. 171. Furthermore in studies of both gilts (Elsaesser and Foxcroft, 1978) and sows (Edwards, 1980) in which a pre-ovulatory gonadotrophin surge was induced by exogenous oestrogen, an FSH rise occurred at a time when oestradiol levels were still elevated and able to block the rise in FSH observed subsequently in control animals (see *Figure 8.4*). These data are considered to provide conclusive evidence for a direct oestrogen-induced surge of FSH release coincident with the LH surge, though of reduced magnitude.

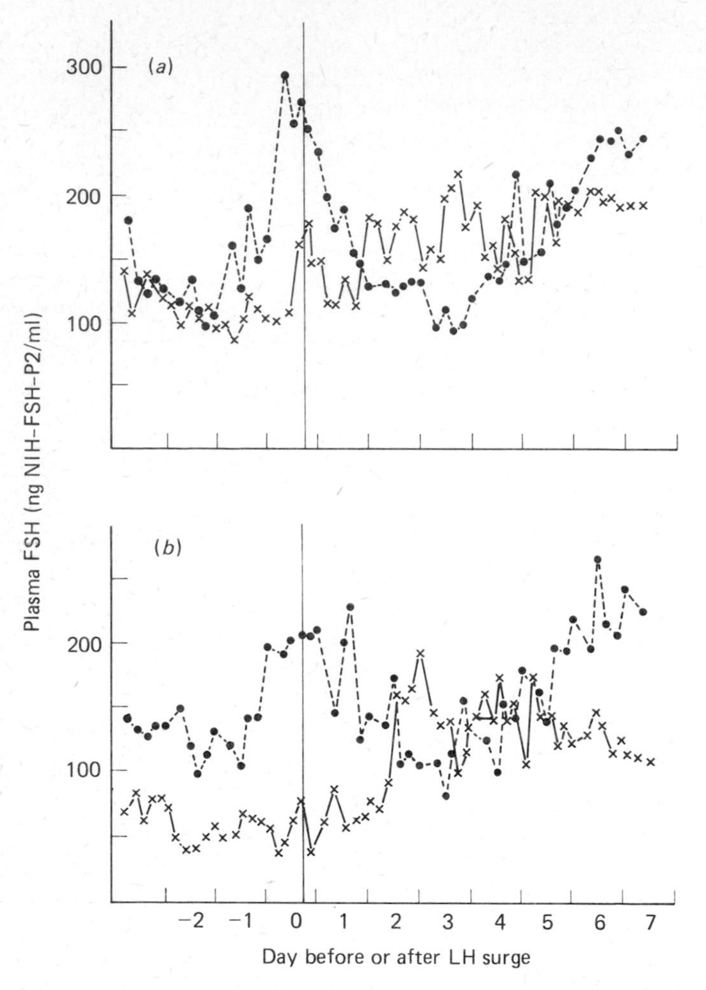

Figure 8.4 Mean plasma levels of FSH measured at 4-hourly intervals in groups of multiparous sows following weaning. Data are standardized to the time that maximal LH levels were observed in the same animals (shown by vertical line on day 0). Standard errors are omitted for clarity of presentation. (a) Sows weaned after a 5-week lactation and (b) after a 3-week lactation. Continuous lines represent data from sows returning to oestrus naturally; broken lines represent data from sows treated with 30 μg/kg oestradiol benzoate (OB) i.m. on the second day after weaning. Following 3-week weaning (b), a pre-ovulatory surge of FSH was observed in only 1/6 (OB treated) and 0/6 (untreated) sows compared with 8/12 (untreated) and 5/6 (OB treated) sows weaned at 5 weeks. OB treatment consistently suppressed a rise in FSH seen in the periovulatory period (day+1 to day+4) in untreated sows. From Edwards (1981)

Some of the immediate consequences of the rise in LH within the ovary have been well documented from *in vitro* studies of granulosa tissue from the pig and other species (see reviews of Hunzicker-Dunn, Bockaert and Birnbaumer, 1978; Hunzicker-Dunn *et al.*, 1979; Catt *et al.*, 1979), and result in the luteinization of the theca and granulosa tissue, a rapid decline in oestradiol production followed by a gradual increase in progesterone synthesis and ultimately in ovulation.

The surge of LH itself is also thought to directly act by overriding an inhibition of oocyte maturation exerted by a non-steroidal Oocyte Maturation Inhibitor (OMI) found in follicular fluid (Channing, 1979) which is present at higher concentrations in small as compared with large follicles (Van de Wiel *et al.*, 1982). In addition the surge may also result in an efflux of follicular fluid steroids that accumulate at almost pharmacological levels within the developing follicle and may also act as important intra-ovarian regulators (Eiler and Nalbandov, 1977).

An absolute requirement for a concomitant rise in FSH in stimulating such changes appears to be questionable in the pig, as a normal sequence of periovulatory events is observed in early weaned sows in which an FSH 'surge' is frequently absent (Edwards, 1980; see *Figure 8.4*).

The periovulatory period

Although dependent on the exact relationship between the onset of oestrus and the time of the LH surge (see above), ovulation generally occurs during the latter part of day 2 of the oestrous cycle. At this time LH levels in both cyclic gilts and weaned sows are consistently low, in contrast to a marked increase in the release of FSH during days 2–3 of the cycle (Rayford *et al.*, 1974; Vandalem *et al.*, 1979; Edwards, 1980; Van de Wiel *et al.*, 1981). This pattern of FSH release has also been observed in the rat (Chappel and Barraclough, 1976) and is considered by Van de Wiel (1981) to relate possibly to the low levels of oestradiol at this stage of the cycle; the data of Edwards (1980) shown in *Figure 8.4* which indicate that in the pig this post-ovulatory rise in FSH secretion can be inhibited by oestradiol are consistent with this hypothesis. However, the presence of an ovarian 'inhibin' within porcine follicular fluid has also been extensively documented (Channing, 1979) and the removal of this source of inhibin at the time of ovulation has also been postulated to be the trigger for a rise in FSH in the immediate post-ovulatory period; in view of the specific feedback of inhibin on FSH secretion this would provide a more satisfactory explanation for the lack of an LH response at this time. Evidence for a pituitary site of action for inhibin from porcine follicular fluid has also been reviewed (Channing, 1979).

The biological significance of this post-ovulatory rise in FSH has been the subject of considerable speculation and at least in short-cycle species has been associated with the recruitment of the crop of follicles destined to ovulate at the subsequent oestrus (Schwartz, 1979). Whether this applies to long-cycle species such as the pig needs to be examined, although recent data from a small number of sows treated with oestradiol benzoate after weaning showed that even if these animals failed to conceive at the first post-treatment oestrus, they still returned to heat 21 days later. Thus, the presence of the post-ovulatory FSH rise is not an absolute requirement for subsequent follicular development. At the intra-ovarian level ovulation may also be associated with the removal of a non-steroidal luteinization inhibitor also shown to be present in porcine follicular fluid (Channing, 1979).

Following the immediate post-ovulatory period the rising levels of progesterone are associated with a stabilization of, and generally a slight decline in, FSH secretion (Vandalem *et al.*, 1979; Edwards, 1980; Van de Wiel *et al.*, 1981; see *Figure 8.4*). Over the same period there is the gradual development of episodic LH release of increasing amplitude and decreasing frequency, an effect that is considered to be a consequence of, rather than a stimulus for, the rising levels of peripheral progesterone (Foxcroft, 1978).

The rise in pre-ovulatory gonadotrophins has also been reported to be coincident with an increase in prolactin secretion by both Wilfinger (1974) and Van de Wiel *et al.* (1981). Although the role of prolactin at this time is uncertain the latter authors suggest a possible involvement in the endocrine trigger for the development of behavioural oestrus.

Conclusions

A consideration of even the limited amount of literature reviewed in this Chapter serves to illustrate the complexity of changes at all levels of the reproductive axis that may contribute to the endocrine control of the oestrous cycle. In many cases, however, critical experimental evidence is required to confirm what initially seem to be plausible functional interrelationships between the different hormones. Even if specific hormones are shown to be interdependent, the question still remains as to whether such changes are of direct consequence to the control of ovarian function.

An attempt to identify the possible critical regulators of cyclic ovarian function will, however, be made with particular reference to the recent data of Van de Wiel *et al.* (1979) and Van de Wiel and Pierantoni (personal communication), which describe the response of the pituitary to luteinizing hormone releasing hormone (LHRH) stimulation at different stages of the oestrous cycle (see *Figure 8.5*).

1. The dominant regulators of the pig oestrous cycle appear to be those factors that maintain the corpus luteum and in consequence block the final stages of follicular growth and steroidogenesis.
2. A requirement for luteotrophic support for the corpora lutea early in the oestrous cycle remains unresolved, but as a decline in luteal activity as early as day 10–12 of the cycle has been associated with changes in LH receptor levels, receptor inhibitory factors and increases in PGF production, LH support may be critical at this time.
3. No characteristic change in the pattern of LH secretion is obvious at the onset of luteal regression and LH secretion may be directly modulated by the feedback effects of progesterone, which only shows a latent decline on days 15–18 of the cycle. However both FSH basal levels and LHRH-induced LH and FSH responses are reported to decline before the fall in circulating progesterone (see *Figure 8.5*). This suggests that the pituitary may gradually become unresponsive as a result of the continuous secretion of progesterone during the luteal

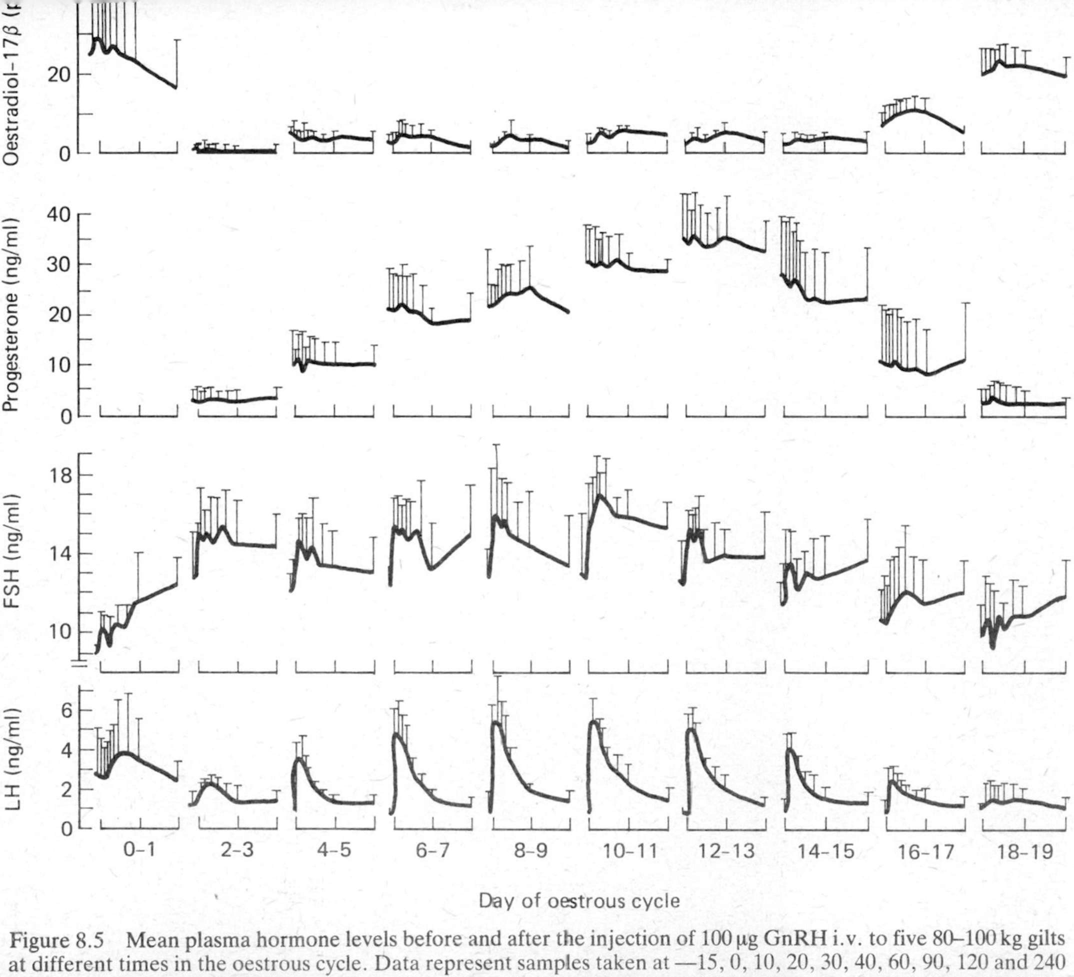

Figure 8.5 Mean plasma hormone levels before and after the injection of 100 µg GnRH i.v. to five 80–100 kg gilts at different times in the oestrous cycle. Data represent samples taken at —15, 0, 10, 20, 30, 40, 60, 90, 120 and 240 minutes with respect to the times of injection. From Van de Wiel *et al.* (1979), with unpublished data of D.F.M. Van de Wiel and R. Pierantoni

phase of the cycle (similar to the long-term effects of progesterone during pregnancy) or that another ovarian factor, possibly oestrogen, controls such changes at the hypothalamic–pituitary level. The trigger for luteal regression appears, however, to originate at the utero-ovarian level with little evidence for the involvement of a critical change in the gonadotrophic stimulus received from the pituitary.

4. The block to continued follicular development and oestradiol secretion can be directly related to the presence of high progesterone levels in the circulation. The removal of this block could have effects at two levels of the reproductive axis. At the hypothalamic–pituitary level characteristic changes in the pattern of LH and FSH secretion occur, coincident with a period of enhanced prolactin release. It should be emphasized, however, that the nature of the change in gonadotrophin secretion involves a substantial decline in total secretion which may not appear to be consistent with the observation of an increase in gonadotrophin secretion as the usual trigger for gonadal development. Nevertheless the changes in LH/FSH release could result either in a particular pattern of episodic LH secretion and/or the achievement of a particular LH:FSH ratio that is of functional significance. Alternatively the withdrawal of progesterone at the ovarian level may remove an inhibition of one or more intra-ovarian factors which are known to be critical for the initiation of late follicular growth and steroidogenesis. Whatever the mechanisms involved, either one or both of these changes probably initially results in enhanced follicular secretion of oestradiol; this is then responsible for initiating a sequence of changes mainly regulated at the intra-ovarian level culminating in follicular growth, a rise in circulating oestrogen and oocyte maturation.
5. The oestrogen-induced surge of gonadotrophins usually involves both LH and FSH with a coincident rise in prolactin, followed in the immediate post-ovulatory period by an increase in FSH secretion alone. The LH surge directly induces intrafollicular changes leading to the luteinization of the granulosa cells and indirectly by causing ovulation, removes inhibitory effects at both the ovarian and pituitary level of a number of steroidal and non-steroidal regulators of follicular fluid origin. The biological significance of both the pre-ovulatory and post-ovulatory rises in FSH is uncertain.
6. During the follicular phase of the cycle the initial inhibition and subsequent enhancement of both LH and FSH responses to LHRH is consistent with the hypothesis that changes in pituitary sensitivity play a major role in the control of the surge secretion of gonadotrophins. Furthermore the marked suppression of LH and LHRH-induced LH responses during the initial inhibitory period of oestrogen feedback, and the overall characterisitcs of the LH surge, are in considerable contrast to the pattern of endocrine changes observed in other large domestic species and may be of physiological significance.

References

ANDERSON, L.D., SCHAERF, F.W. and CHANNING, C.P. (1979). Effects of follicular development on the ability of cultured porcine granulosa cells to convert androgens to estrogens. *Adv. exp. Med. Biol.* **112**, 187–195

ANDERSON, L.L. (1966). Pituitary–ovarian–uterine relationships. In *Ovarian Regulatory Mechanisms. J. Reprod. Fert., Suppl.* **1**, 21–32

ARMSTRONG, D.T. and DORRINGTON, J.H. (1977). In *Regulatory mechanisms affecting gonadal hormone action, Vol. 3, Advances in Sex Hormone Research,* (J.A Thomas and R.L. Singhal, Eds.), p. 217. Baltimore, University Press

CATT, K.J., HARWOOD, J.P., RICHERT, N.D., CONN, P.M., CONTI, M. and DUFAU, F. (1979). Luteal desensitization: hormonal regulation of LH receptors, adenylate cyclase and steroidogenic responses in the luteal cell. *Adv. exp. Med. Biol.* **112**, 647–662

CHANNING, C.P. (1979). Follicular non-steroidal regulators. *Adv. exp. Med. Biol.* **112**, 327–343

CHAPPEL, S.C. and BARRACLOUGH, C.A. (1976). Hypothalamic regulation of pituitary FSH secretion. *Endocrinology* **98**, 927–935

COLENBRANDER, B. (1978). Aspects of sexual differentiation in the pig. PhD Thesis. State University of Utrecht

DU MESNIL DU BUISSON and LEGLISE, P.C. (1963). Effect de l'hypophysectomie sur les corps jaunes de la truis. Resultats préliminaires. *C.r. hebd. Séanc. Acad. Sci., Paris* **257**, 261–263

EDQUIST, L.E. and LAMM, A.M. (1971). Progesterone levels in plasma during the oestrous cycle of the sow measured by a rapid competitive protein binding assay. *J. Reprod. Fert.* **25**, 447–449

EDWARDS, S. (1980). Reproductive physiology of the postparturient sow. PhD Thesis. University of Nottingham

EILER, H. and NALBANDOV, A.V. (1977). Sex steroids in follicular fluid and blood plasma during the estrous cycle of pigs. *Endocrinology* **100**, 331–338

ELSAESSER, F. and FOXCROFT, G.R. (1978). Maturational changes in the characteristics of oestrogen-induced surges of luteinizing hormone in immature gilts. *J. Endocr.* **78**, 455–456

FOXCROFT, G.R. (1978). The development of pituitary gland function. In *Control of Ovulation*, (D.B. Crighton, G.R. Foxcroft, N.B. Haynes and G.E. Lamming, Eds.), p. 129. London, Butterworths

GOMES, W.R., HERSCHLER, R.C. and ERB, R.E. (1965). Progesterone levels in ovarian venous effluent of the non-pregnant sow. *J. Anim. Sci.* **24**, 722–727

GREENWALD, G.S. (1979). Introductory remark; Ruminations on ovarian function. *Adv. exp. Med. Biol.* **112**, 3–8

GUTHRIE, H.D., REXROAD, C.E. and BOLT, D.J. (1979). *In vitro* release of progesterone and prostaglandins F and E by porcine luteal and endometrial tissue during induced luteolysis. *Adv. exp. Med. Biol.* **112**, 627–632

HANSEL, W., CONCANNON, P.W. and LUKASZEWSKA, J.H. (1973). Corpora lutea of the large domestic animals. *Biol. Reprod.* **8**, 222–245

HENRICKS, D.M., GUTHRIE, H.D. and HANDLIN, D.L. (1972). Plasma oestrogen, progesterone and LH levels during the oestrous cycle in pigs. *Biol. Reprod.* **6**, 210–218

HUNZICKER-DUNN, M., BOCKAERT, J. and BIRNBAUMER, L. (1978). In *Receptors and Hormone Action, Vol. 3*, (B.W. O'Malley and L. Birnbaumer, Eds.), p. 393. New York, Academic Press

HUNZICKER-DUNN, M., JUNGMANN, R., DERDA, D. and BIRNBAUMER, L.

(1979). LH-induced desensitization of the adenylyl cyclase system in ovarian follicles. *Adv. exp. Med. Biol.* **112**, 27–44

KOTWICA, J. (1980). Mechanism of prostaglandin $F_{2\alpha}$ penetration from the horn of the uterus to the ovaries in pigs. *J. Reprod. Fert.* **59**, 237–241

LEUNG, P.C.K., TSANG, B.K. and ARMSTRONG, D.T. (1979). Estrogen inhibits porcine thecal androgen production *in vitro*. *Adv. exp. Med. Biol.* **112**, 241–243

LIPTRAP, R.M. and RAESIDE, J.I. (1966). LH activity in blood and urinary oestrogen excretion by the sow at oestrus and ovulation. *J. Reprod. Fert.* **11**, 439–446

LUNAAS, T. (1962). Urinary oestrogen levels in the sow during oestrous cycle and early pregnancy. *J. Reprod. Fert.* **4**, 13–20

MASUDA, H., ANDERSON, L.L., HENRICKS, D.M. and MELAMPY, R.M. (1967). Progesterone in ovarian venous plasma and corpora lutea of the pig. *Endocrinology* **80**, 240–246

NALBANDOV, A.V. (1970). Comparative aspects of corpus luteum function. *Biol. Reprod.* **2**, 7–13

NISWENDER, G.D., REICHERT, L.E. and ZIMMERMAN, D.R. (1970). Radioimmunoassay of serum levels of luteinizing hormone throughout the estrous cycle in pigs. *Endocrinology* **87**, 576–580

PARVIZI, N., ELSAESSER, F., SMIDT, D. and ELLENDORFF, F. (1976). Plasma luteinizing hormone and progesterone in the adult female pig during the oestrous cycle, late pregnancy and lactation, and after ovariectomy and pentobarbitone treatment. *J. Endocr.* **69**, 193–203

PEROTTI, L., ENNE, G., MEGGIOLARO, D. and DELRIO, G. (1979). Concentrazione plasmatica de progesterone in scrofe durante cicli estrali estivi ed invernali. *Revta Zootech. Vet.* **1**, 10–12

RAESIDE, J.I. (1963). Urinary oestrogen excretion in the pig at oestrus and during the oestrous cycle. *J. Reprod. Fert.* **6**, 421–426

RAYFORD, P.L., BRINKLEY, H.J. and YOUNG, E.P. (1971). Radioimmunoassay determination of LH concentration in the serum of female pigs. *Endocrinology* **88**, 707–713

RAYFORD, P.L., BRINKLEY, H.J., YOUNG, E.P. and REICHERT, L.E. (1974). Radioimmunoassay of porcine FSH. *J. Anim. Sci.* **39**, 348–354

RICHARDS, J.S., RAO, M.C. and IRELAND, J.J. (1978). In *Control of Ovulation,* (D.B. Crighton, G.R. Foxcroft, N.B. Haynes and G.E. Lamming, Eds.), pp. 197–216. London, Butterworths

ROLLAND, R., GUNSALUS, G.L. and HAMMOND, J.M. (1976). Demonstration of specific binding of prolactin by porcine corpora lutea. *Endocrinology* **98**, 1083–1091

SCHWARTZ, N.B. (1979). In discussion. *Adv. exp. Med. Biol.* **112**, 399–400

SCHWARTZ-KRIPNER, A. and CHANNING, C.P. (1979). Changes in responsiveness of porcine granulosa cells to prostaglandins and luteinizing hormone in terms of cyclic-AMP accumulation during follicular maturation. *Adv. exp. Med. Biol.* **112**, 137–143

SHEARER, I.J., PURVIS, K., JENKIN, G. and HAYNES, N.B. (1972). Peripheral plasma progesterone and oestradiol-17β before and after puberty in gilts. *J. Reprod. Fert.* **30**, 347–360

SPIES, H.G., SLYTER, A.L. and QUADRI, S.K. (1967). Regression of corpora lutea in pregnant gilts administered antiovine LH rabbit serum. *J. Anim. Sci.* **26**, 768–771

STABENFELDT, G.H. AKINS, E.L., EWING, L.L. and MORRISSETTE, M.C. (1969). Peripheral plasma progesterone levels in pigs during the oestrous cycle. *J. Reprod. Fert.* **20**, 443–449

TILLSON, S.A. and ERB, R.E. (1967). Progesterone concentration in peripheral blood plasma of the domestic sow prior to and during early pregnancy. *J. Anim. Sci.* **26**, 1366–1368

TILLSON, S.A., ERB, R.E. and NISWENDER, G.D. (1970). Comparison of LH and progesterone in urine of domestic sows during the oestrous cycle and early pregnancy. *J. Anim. Sci.* **30**, 795–805

TUCKER, S., KUMARI, L., and CHANNING, C.P. (1979). Evidence of a greater activity of LH/HCG binding inhibitor present in aqueous extracts from old compared to young porcine corpus luteum. *Adv. exp. Med. Biol.* **112**, 723–728

VANDALEM, J.L., BODART, Ch., PIRENS, G., CLOSSET, J. and HENNEN, G. (1979). Development and application of homologous radiommunoassays for porcine gonadotrophins. *J. Endocr.* **81**, 1–10

VAN DE WIEL, D.F.M., BAR-AMI, S., TSAFRIRI, A. and DE JONG, F.H. (1982). Oocyte maturation inhibitor, inhibin and steroid concentrations in porcine follicular fluid of various stages of the oestrous cycle. Submitted for publication

VAN DE WIEL, D.F.M., VAN DE BOEZEM, E.A., VERGROESEN, L.P.B.M. and DE VRIES, F.P.W. (1979). Feedback action of progesterone and oestrogen on LH responses after administration of GnRH in cycling gilts. In *Research on Steroids, Vol. VIII,* (A. Klopper, L. Lerner, H.J. Van der Molen and F. Sciarra, Eds.), pp. 221–224. London, Academic Press

VAN DE WIEL, D.F.M., ERKENS, J., KOOPS, W., VOS, E. and VAN LANDEGHEM, A.A.J. (1981). Perioestrous and midluteal time courses of circulating LH, FSH, prolactin, estradiol-17β and progesterone in the domestic pig. *Biol. Reprod.* **24**, 223–233

VAN LANDEGHEM, A.A.J. and VAN DE WIEL, D.F.M. (1977). Plasma prolactin levels in gilts during the oestrous cycle and at hourly intervals around the time of oestrus. *Acta endocr., Copenh., Suppl.* **212**, 143

WATSON, J. and LEASK, J.T.S. (1975). Superfusion *in vitro* in the study of ovarian steroidogenesis. *J. Endocr.* **64**, 163–173

WEISS, J.R., BRINKLEY, H.J. and YOUNG, E.P. (1976). *In vitro* steroidogenesis in porcine corpora lutea. *J. Anim. Sci.* **42**, 121–130

WILFINGER, W.W. (1974). Plasma concentrations of luteinizing hormone, follicle stimulating hormone and prolactin in ovariectomized, hysterectomized and intact swine. PhD Thesis. University of Maryland

ZIECIK, A., SHAW, H.J. and FLINT, A.P.F. (1980). Luteal LH receptors during the oestrous cycle and early pregnancy in the pig. *J. Reprod. Fert.* **60**, 129–137

9

THE CENTRAL NERVOUS SYSTEM AND THE CONTROL OF PITUITARY HORMONE RELEASE IN THE PIG

F. ELLENDORFF and N. PARVIZI
Institut für Tierzucht und Tierverhalten Mariensee (FAL), 3057 Neustadt 1, Federal Republic of Germany

Domestic animals have been recognized as Nobel prize winning sources of biological material. For example, the ox served to identify oxytocin (du Vingneaud, Barlett and Johl, 1954), the sheep (Burgus *et al.*, 1971) and the pig (Schally *et al.*, 1971) to characterize gonadotrophin releasing hormones. As yet physiologists, in particular neuro-endocrinologists, have only taken limited advantage of them experimentally. Since this symposium is devoted to pig reproduction, we shall attempt to present what is

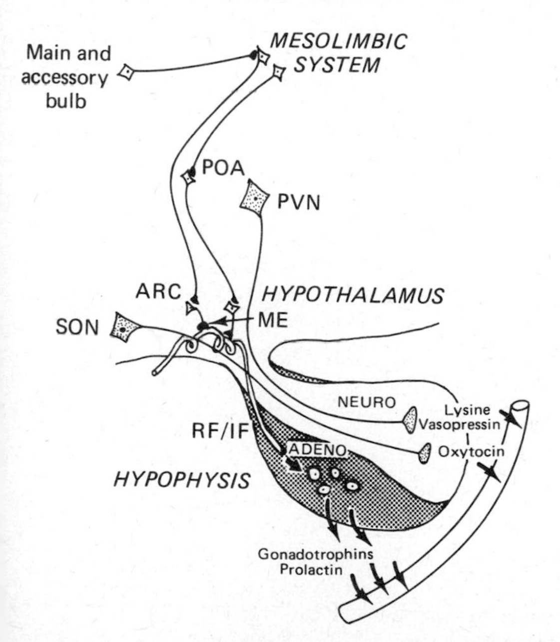

Figure 9.1 Schematic diagram of neuroendocrine systems known to be involved in pituitary hormone secretion and reproductive functions in the pig. ARC: arcuate nucleus; Main and accessory bulb: comprises the main and accessory olfactory system. ME: median eminence; forms together with ARC, ME and ventromedial nucleus of the hypothalamus the medio-basal hypothalamus; Mesolimbic System: comprising the limbic system including amygdala and hippocampus and the mesencephalon; POA: preoptic area; PVN: paraventricular nucleus; RF/IF: releasing/inhibiting factors; SON: supraoptic nucleus

known on the control of pituitary hormone secretion by the central nervous system (CNS) in this species. Two systems will be of concern both pre- and postnatally: CNS control over luteinizing hormone (LH) secretion by the anterior pituitary and mechanisms leading to the release of the neurohormones oxytocin and vasopressin.

The systems controlling pituitary hormone secretion are beyond any doubt situated within the brain but not exclusively within the hypothalamus (*Figure 9.1*). Extrahypothalamic areas, especially the limbic system and the mesencephalon, play a definite role and the portal vessels serve to channel releasing and inhibiting hormones from the region of the median eminence into the pituitary. Paraventricular and supraoptic neurohormones are sent via axons of the pituitary stalk into the posterior pituitary for storage and release. Although the electrical and neurochemical events of the brain constitute the common underlying mechanisms for the stimulation or inhibition of synthesis and release of neurosecretory materials, such as the releasing or inhibiting hormones, or oxytocin and vasopressin, factors other than those inherent to the CNS participate in their release. Thus peripherally-produced steroids, or nervous reflexes originating peripherally, are able to modulate neurosecretory activity and the pheromones may also be part of the control system. In reviewing current knowledge of the neuroendrocinology of the pig we shall limit our presentation to direct, rather than circumstantial, evidence for CNS control over hypophysial function and thus to an involvement of the brain in the control of reproduction in the pig.

Foetal and adult brain control of LH secretion

Brain control over pituitary hormone secretion probably commences well before the foetus turns to independent life. For the pig our laboratory investigated a number of elements of the secretion of LH both in chronically catheterized, unanaesthetized foetuses and in acute experiments. Measurable levels of LH were present during foetal life (Elsaesser *et al.*, 1976; Colenbrander *et al.*, 1977); furthermore the foetal pituitary could be stimulated by synthetic LH releasing hormone under *in vivo* and *in vitro* conditions (for details and references see Elsaesser, Chapter 5). However, this does not necessarily imply that the hypothalamus is already capable of releasing sufficient LH releasing hormone to trigger the release of LH into the circulation and even in the adult a single pulse of LH releasing hormone does not seem to reflect the natural mechanisms of release. However, essential differences in the electrical activity (*Figure 9.2*) of cortical brain structures between the late prenatal and the pubertal pig are not obvious (Konda *et al.*, 1979) and this reduces the likelihood that the late prenatal brain might not be capable of functioning in a postnatal fashion.

We have therefore stimulated the foetal hypothalamus electrically and electrochemically at various foetal ages (Bruhn, Parvizi and Ellendorff, 1981 and unpublished). An LH response was virtually absent on day 60; however detectable stimulation of foetal plasma LH levels could be evoked at day 80 and were clearly observed at 105 days of age. It is thus evident

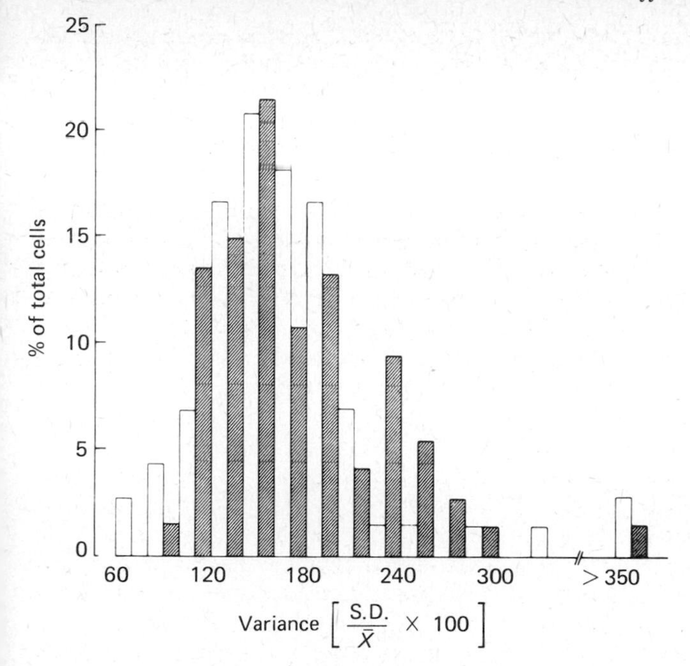

Figure 9.2 Comparison of distribution of variance in firing rate of single units recorded from cerebral cortex of prepubertal (hatched bars, n = 77 units) and foetal (open bars, n = 72 units) pigs. No significant difference existed. By courtesy of Konda *et al.* (1979)

that somewhere between days 60 and 80 the brain–pituitary axis advances sufficiently in its maturation to be capable of exerting control over the secretion of LH releasing hormone from hypothalamic nerve terminals. The pituitary portal vessels must be able to pass this information on to the anterior pituitary which also becomes responsive prior to and around day 80. These observations have provided us with a time period over which to study the ontogeny of the other mechanisms involved in brain control over pituitary hormone secretion.

The only evidence that manipulation of the postnatal but prepubertal brain subsequently affects reproduction in the female pig indicates that mediobasal and anterior hypothalamic lesions, as well as constant illumination, result in anovulation (Döcke and Busch, 1974). This is probably due to an absence of the LH surge since polyfollicular ovaries were present.

In the adult pig more extensive investigations have been carried out. Undoubtedly, a number of hypothalamic and extrahypothalamic brain areas can be stimulated in the pig with consequent changes in the blood levels of pituitary secretions. However, such responses are not always as uniform as might be expected when compared with a multitude of experiments in the rat and other smaller species, although there are possible experimental reasons for such differences, the most important being the much higher structural resolution that can be achieved in the pig's brain when compared with smaller species. Although changes in plasma LH in response to electrical stimulation of discrete brain areas are indicative of general brain control over pituitary LH release, it is known from many studies in small laboratory species that apart from the mediobasal hypothalamus and the preoptic area, the limbic and mesencephalic structures must be taken into account when considering the regulation of gonadotrophin secretion and this topic has been discussed elsewhere (Ellendorff, 1978; Ellendorff and Parvizi, 1980). Moreover, at any level (be it the amygdala, hippocampus, mesencephalon or the hypothalamus), the steroids, the classical neurotransmitters and the neuropeptides are essential in contributing to any changes in LH releasing hormone and finally LH release, and there is no reason to believe that the pig should differ in basic principles of design.

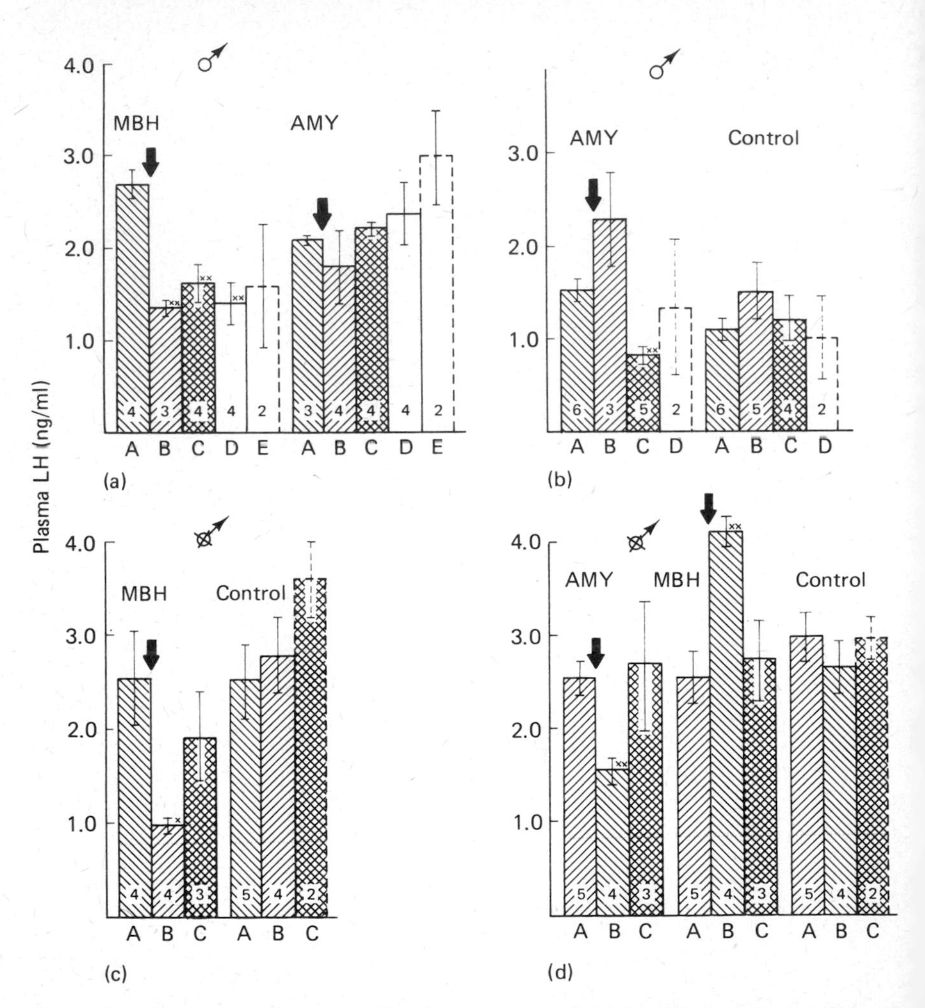

Figure 9.3 Examples of plasma LH response to electrical stimulation of the mediobasal hypothalamus (MBH) and the amygdala (AMY) in (a) and (b) intact male, (c) and (d) castrated male. All animals were chronically implanted and unanaesthetized. Stimulus parameters: 100–200 μA, 100 Hz, 0.5 ms 30 s on/off, 60 min, each stimulation at least two days apart. A: prior to stimulation; B: up to 60 min; C: 60–120 min; D: 140–200 min; E: 220–240 min post stimulation. Numbers in columns indicate numbers of samples taken at 10–20 min intervals. Arrows indicate onset of stimulation. From Ellendorff and Parvizi (unpublished)

In testing such interactions experimentally the intact and castrate male pig has been the model of preference. Electrical stimulation of various brain structures in the unanaesthetized male pig resulted in altered plasma levels of LH (*Figure 9.3*) (Ellendorff *et al.*, 1973). Five of six orchidectomized boars had lowered plasma LH concentrations following amygdala stimulation and one did not change relative to controls. The response of intact males was not as clearly defined; of five boars stimulated two displayed increased, and one decreased, plasma LH levels and in one there was no change. Stimulation of the mediobasal hypothalamus (MBH), however, produced more equivocal results. We hypothesize therefore that gonadal steroids and/or differences in the levels of circulating gonadotrophins which have been reported previously in castrate and intact boars (Pomerantz *et al.*, 1974) were responsible for the differences observed.

It is now well established in all species investigated that steroid receptors are present in at least the amygdala and hypothalamus. However, if functional significance is to be attributed to such binding or localization studies, peripheral as well as local application of steroids should result in measurable responses, e.g. changes in plasma LH levels. In the orchidectomized pig intramuscular injections (*Figure 9.4 (c)* and *(d)*) of testosterone (T), as well as oestradiol (E_2), lowered plasma levels of LH within 24 hours, irrespective of the two dose levels given (15mg and 6.0 mg for T and 1.5mg and 0.6 mg for E_2 per kg body weight). 5α-Dihydrotestosterone (5α-DHT) evoked an increase in plasma LH when given at a lower dose (6 mg/kg); however at considerably higher concentrations (15 mg/kg) LH levels were significantly depressed when compared with levels prior to treatment and in untreated controls (Parvizi *et al.*, 1977). Although the responses to testosterone and oestradiol were consistent with results from other species, the stimulatory effect of 5α-DHT on LH secretion was unexpected and appears to be unique to the pig.

At least some of these effects should be due to the action of steroids on the hypothalamus and/or the amygdala. Microinjections of testosterone, oestradiol or 5α-DHT were therefore first placed into the amygdala of castrated males (*Figure 9.4(e)*). The outcome for testosterone and oestradiol was initially rather disappointing: 60 ng testosterone or 6 ng 17β-oestradiol did not alter plasma LH levels in 6/7 and in 5/6 animals respectively 2, 4, 24 or 48 hours after application. In contrast, when 5α-DHT (60 ng) was given to six animals, significantly elevated plasma LH was recorded in two animals within 4 hours, in four animals within 24 hours and in five animals within 48 hours after application to the amygdala. Initially we concluded that in the castrated male pig 5α-DHT participates in the regulation of LH secretion by a stimulatory role that is localized in the amygdala and becomes effective within 48 hours of exposure. It has been shown in other Chapters that rather striking effects of gonadal steroids (oestrogens) on LH secretion take much longer to develop. It is possible, however, that single microinjections of testosterone and of oestradiol in the amygdala are ineffective due to their rapid metabolism, or that the amygdala is not a location in which it is possible to provoke testosterone-mediated changes in LH. In order to test these possibilities crystalline testosterone or 5α-DHT was implanted into the mediobasal hypothalamus or the amygdala (*Figure 9.4(a)* and *(b)*) to assure a longer lasting exposure

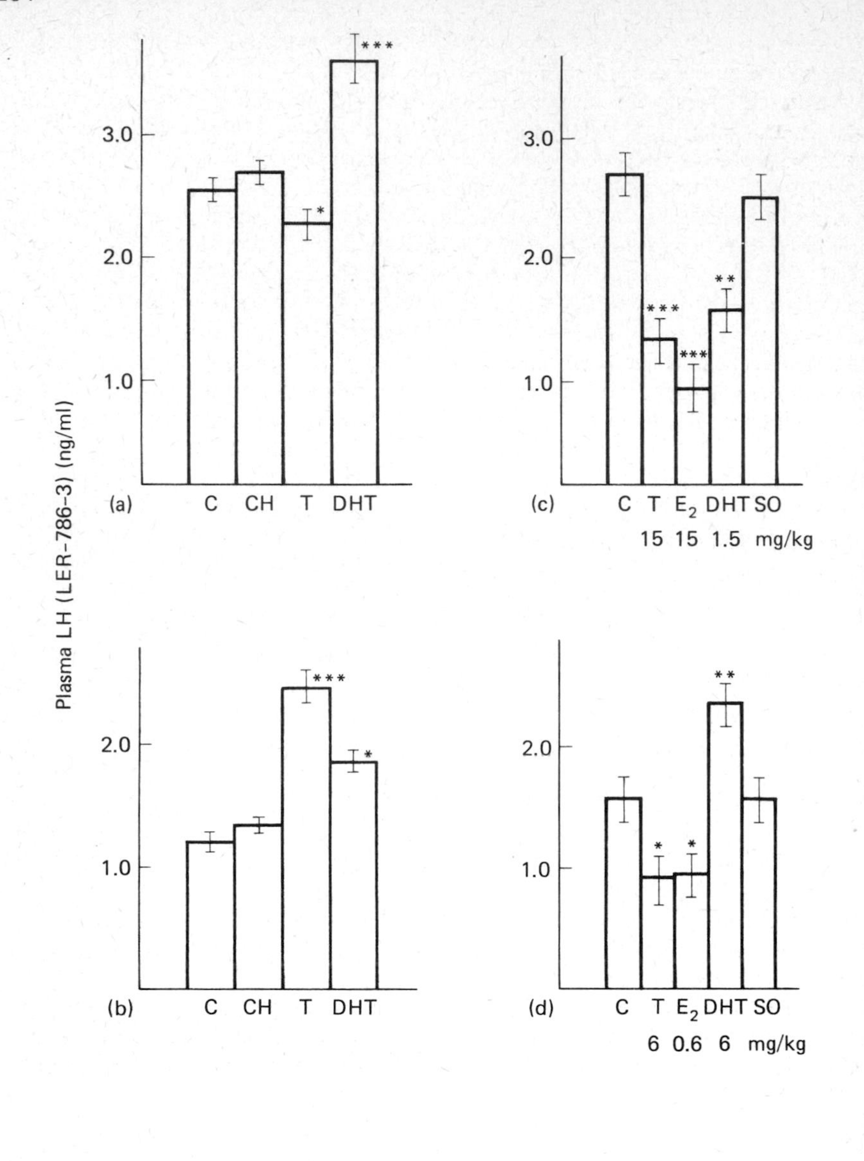
Plasma LH (LER-786-3) (ng/ml)
3.0
2.0
1.0
(a)
C CH T DHT
(c)
C T E2 DHT SO
15 15 1.5 mg/kg
(b)
C CH T DHT
(d)
C T E2 DHT SO
6 0.6 6 mg/kg

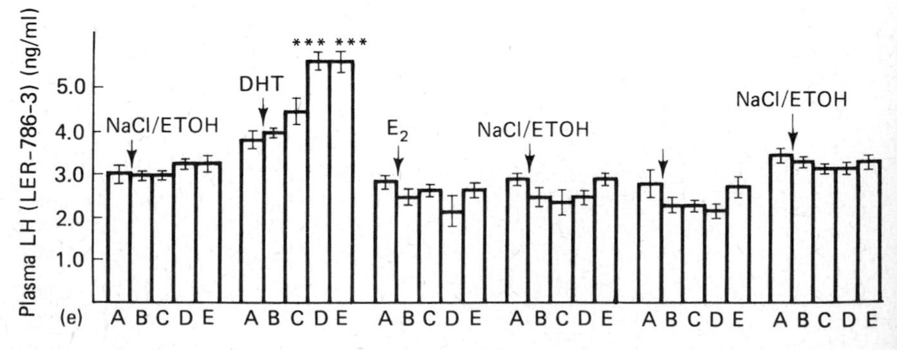
Plasma LH (LER-786-3) (ng/ml)
5.0
4.0
3.0
2.0
1.0
(e)
NaCl/ETOH
DHT
E2
NaCl/ETOH
NaCl/ETOH
A B C D E

of each structure (Parvizi *et al.*, 1977). This time, 5/7 animals with testosterone implants in the amygdala reacted with slightly, but significantly, lowered plasma LH levels for a period of 10 days after the implant had been introduced (two animals showed no response). On the other hand, a clear increase in LH was observed when testosterone was placed into the mediobasal hypothalamus. 5α-DHT caused a stimulation of LH levels from both locations. We further concluded, therefore, that in the pig 5α-DHT exerts its positive effects on plasma LH both at the amygdala and MBH in short and long-term exposure situations.

Testosterone, on the other hand, is more difficult to characterize. It can be inhibitory or stimulatory to LH secretion depending on the site of accumulation and the duration of action. Alternatively, testosterone or its metabolites as well as other steroids, may induce immediate changes in the neurons to which they are attached, but these changes do not necessarily find an immediate expression in changing LH levels under normal conditions.

One hypothesis is that steroids either alter the sensitivity of neurons to incoming electrical impulses or alter the threshold for outgoing signals (e.g. action potentials). If this is true, then alterations in LH levels induced by electrical stimulation should be modifiable by prior exposure to steroids of the area to be stimulated. We have already mentioned that electrical stimulation of the amygdala in the orchidectomized pig *per se* usually decreases plasma LH levels and that microinjections of testosterone alone have no effect on LH. If, however, testosterone was microinjected prior to electrical stimulation, the expected decreasing effects of electrical stimulation on LH were not only abolished, but plasma LH was significantly elevated in 5/6 animals within 210 minutes after application of an electrical current to the amygdala (*Figure 9.5*). Oestradiol and 5α-DHT only abolished the LH decline (Parvizi and Ellendorff, 1980a) evoked by electrical stimulation alone. Thus a modulatory role of testosterone and its metabolites 5α-DHT and oestradiol can be postulated for the amygdala. Under what circumstances the fast or slow components become effective is not known.

Very little is known about mechanisms involved in these modulating effects. A link to neurotransmitter metabolism is suggested since some oestrogen metabolites (hydroxylated oestrogens, catecholoestrogens)

Figure 9.4 (*opposite*) (a), (b): Plasma LH after successive implantation of 5α-dihydrotestosterone (DHT: cholesterol, 1:2); testosterone (T: cholesterol, 1:2), cholesterol (CH) or no treatment (C) into (a) the amygdala, (b) the mediobasal hypothalamus of two individual miniature pigs.
(c), (d): Different effects of 5α-dihydrotestosterone (DHT) and the effects of testosterone (T), 17β-oestradiol (E_2), sesame oil (SO) or no treatment (C) on plasma LH secretion in the castrated male miniature pig. Samples were taken 24 hours after injection.
(e): Responses of an individual animal to microinjection into the amygdala of 2 μl NaCl–EtOH (7 ml 0.9% NaCl + 3 ml 20% ethanol), 60 ng 5α-dihydrotestosterone (DHT), 6 ng 17β-oestradiol (E_2) and 60 ng testosterone (T) each dissolved in 2 μl NaCl–EtOH. Arrows indicate time of injection. Seven to nine samples per block, obtained at 15 minute intervals. Each column (B–E) was compared with the corresponding control period (A). A = 0–2 hours before microinjection; B = 0–2 hours after microinjection; C = 2–4 hours after microinjection; D = 24–25 hours after microinjection; E = 48–49.5 hours after microinjection. Asterisks denote degree of significance: * $P<0.05$; ** $P<0.01$; *** $P<0.001$. By courtesy of Parvizi *et al.* (1977)

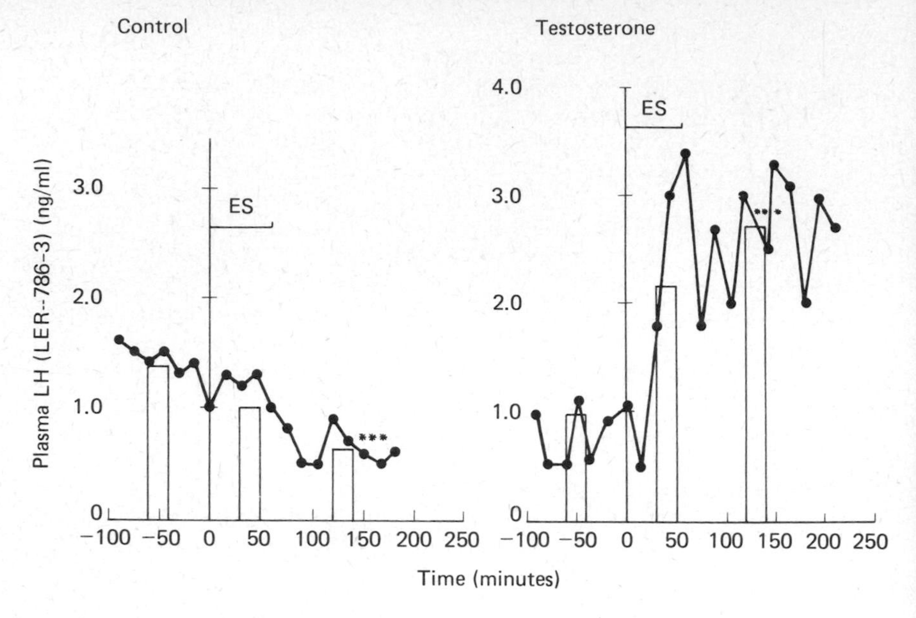

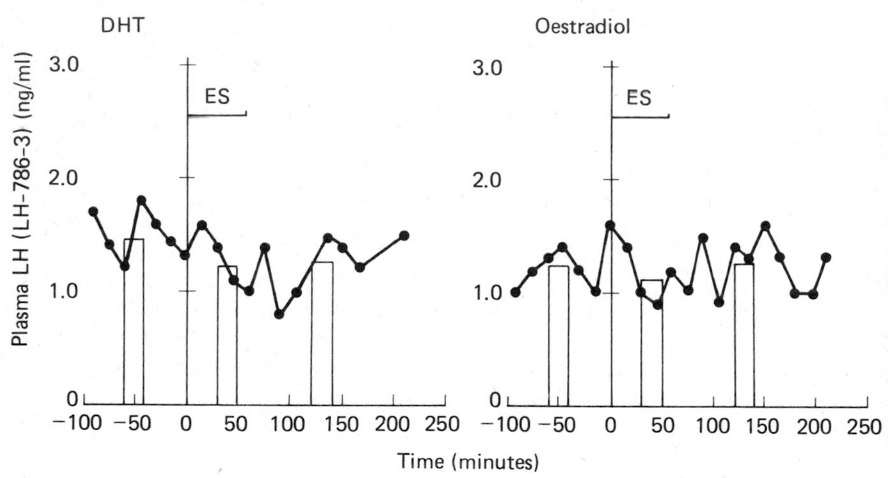

Figure 9.5 Plasma LH levels in single animals which had been microinjected with steroids into the amygdala 210 minutes prior to the onset of electrical stimulation of the amygdala. Control: No prior microinjection; Testosterone, DHT: Animals received 60 ng of either substance in 1 μl of solvent; Oestradiol: Animal received 6 ng in 1 μl. ES: Electrical stimulation 10 Hz, 100 μA, 0.1 ms, 30 s on/off, 60 min ***$P<0.001$. From Parvizi and Ellendorff (1980a)

compete with catecholamines for catechol-O-methyl-transferase (COMT) (Breuer, Vogel and Knuppen, 1962). Higher affinity of COMT for 2-OH-oestrogens should, for instance, result in the accumulation of norepinephrine and therefore induce similar effects to norepinephrine. The pig was the first species in which a direct application of catecholoestrogens into the brain was attempted (Parvizi and Ellendorff, 1975). Microinjections of 60 ng 2-OH-oestradiol (2-OHE_2) into the amygdala

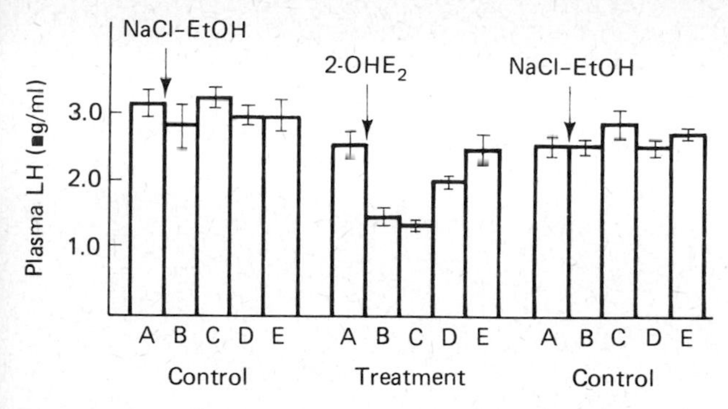

Figure 9.6 Plasma LH (mean ±S.E.M.; LER-786-3) after microinjection of 2-OH-oestradiol (2-OHE_2) into the amygdala of an individual orchidectomized adult miniature pig. NaCl–EtOH: 2 μl of a stock made from 7 ml 0.9% NaCl and 3 ml 20% EtOH; 2-OHE_2: 2 μl NaCl–EtOH containing 60 ng OHE_2 (seven to nine samples per block obtained at 15 minute intervals). Each column (B–E) was compared with the corresponding control period (A) by Student's *t*-test. 2-OHE_2 treatment resulted in a significant decrease in columns (B) and (C), ($P \leqslant 0.001$) and in column (D), ($P \leqslant 0.01$). A = 0–2 hours before microinjection; B = 0–2 hours after microinjection; C = 2–4 hours after microinjection; D = 24–25.5 hours after microinjection, E = 48–49.5 hours after microinjection. By courtesy of Parvizi and Ellendorff (1975)

were followed by a decrease in plasma LH within 90 minutes (*Figure 9.6*). This relatively rapid response fitted well with the effects observed after testosterone, but not after oestradiol microinjections into the amygdala. Electrical stimulation following 2-OHE_2 microinjections affected the inhibition induced by electrical stimulation alone but in equivocal fashion (*Figure 9.7*). Microinjection of 2-OHE_2 into the hypothalamus also inhibited plasma LH levels in the intact male. If the above hypothesis is

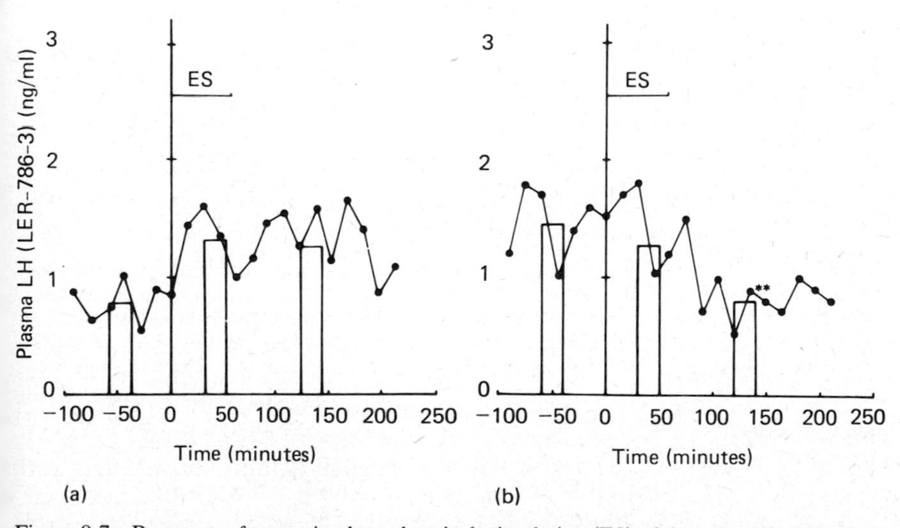

Figure 9.7 Response of two animals to electrical stimulation (ES) of the amygdala following microinjection of 2-OHE_2 210 minutes prior to ES. (a) 2-OHE_2 had lowered LH prior to the onset of ES; ES did not alter LH. (b) 2-OHE_2 had not altered LH levels prior to ES; ES lowered LH. ** $P<0.01$. By courtesy of Parvizi and Ellendorff (1980)

true, we would also expect inhibition of LH secretion when the hypothalamus is exposed to norepinephrine and, indeed, when 60 ng of norepinephrine was microinjected into the hypothalamus, 7/9 males responded with reduced plasma LH levels (*Figure 9.8*). Microinjections of 60 ng norepinephrine into the third ventricle, on the other hand, induced a surge of LH in 4/4 males similar to data reported for the rat (Schneider and McCann, 1970).

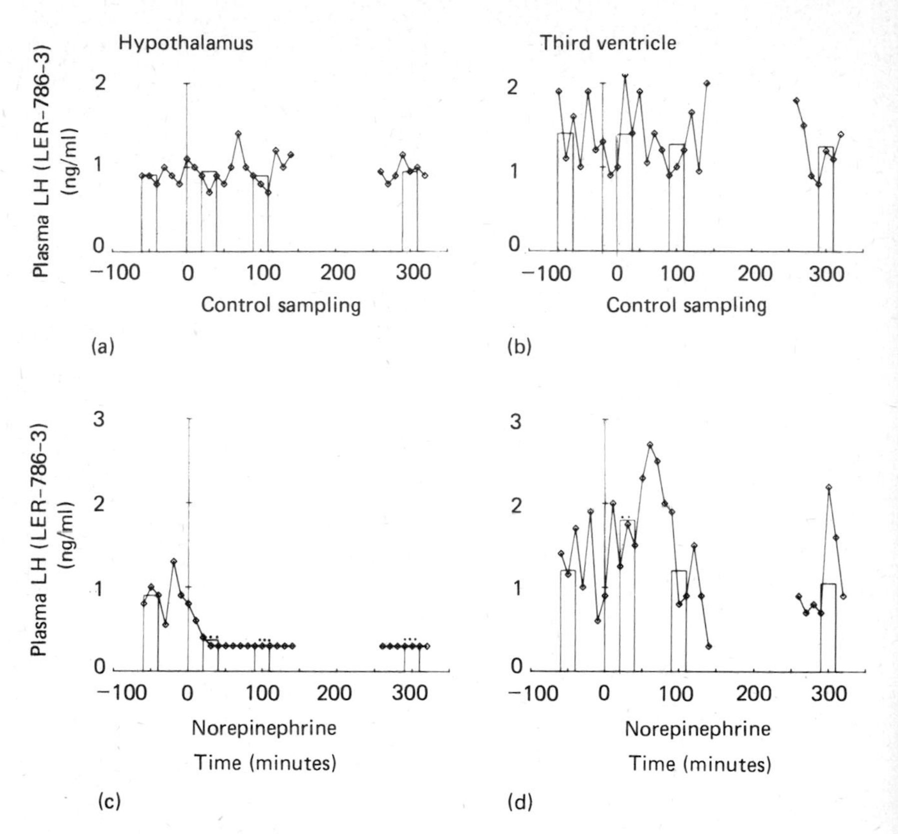

Figure 9.8 Effect of norepinephrine on LH secretion when microinjected into the mediobasal hypothalamus (a) versus (c) and the third ventricle (b) versus (d). By courtesy of Parvizi and Ellendorff (1978)

Apart from steroid–neurotransmitter interaction, the interference of neurobiologically-active peptides with the regulation of LH has become a focus of discussion. Peripheral injections of endorphins into small laboratory species affect the secretion of pituitary hormones (Bruni *et al.*, 1977) and the hypothalamus has been suggested as the site of action. The microinjection of β-endorphin into the hypothalamus or amygdala of the pig (Parvizi and Ellendorff, 1980b) suggests that it is not the hypothalamus but the amygdala in which β-endorphin becomes effective (*Figure 9.9*). If β-endorphin was microinjected simultaneously into the amygdala and hypothalamus, a small enhancement of the LH decline was observed when compared with a microinjection into the amygdala alone. This may give a

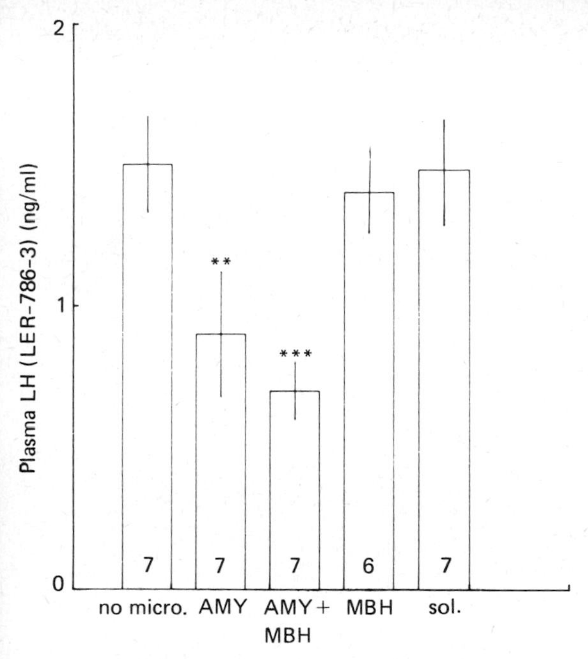

Figure 9.9 Decrease of plasma LH levels 100–180 minutes after microinjection of 30 ng β-endorphin into the amygdala (AMY) or mediobasal hypothalamus (MBH) or both (AMY+MBH) of ovariectomized miniature pigs. 'No micro' = control blood sampling without any treatment; 'Sol.' = microinjection of 1 μl of 9.7% saline. Numbers in the bars represent the number of animals. For statistical analysis the period of sampling was divided into three blocks each of nine successive samples taken at 10 minute intervals. The mean LH value for each animal within each block was calculated. A one-way analysis of variance was then carried out to detect differences among mean LH values of treatments and time blocks. **P<0.01; ***P<0.001; $\bar{x}$±S.E.M. By courtesy of Parvizi and Ellendorff (1980b)

first insight into a most complex system of regulation of gonadotrophin secretion by neuroactive peptides in the pig as well as in other species.

OLFACTION

A discussion of the central nervous system and the control of reproduction in the pig would be incomplete without mentioning the olfactory system. It is unique because the steroidal pheromones produced by the boar have been identified and are available synthetically. The interest of the neuroendocrinologist centres around its uptake and the possible transmission of pheromonal information within the CNS. The olfactory system of the pig is essentially the same as that of other mammals (Reinhardt *et al.*, 1981). The mitral cell layer of the main olfactory bulb receives input from the olfactory nerves and passes the information on via the lateral olfactory tract (LOT) to higher centres, including the amygdala and, to a lesser extent, the hypothalamus. These connections are often reciprocal. If 5α-androst-16-ene-3-one (Δ-16) and 3α-hydroxy-5α-androst-16-ene are pheromones, they could become effective via the olfactory system, though other forms of uptake, e.g. the nasal mucosa, should not be discarded.

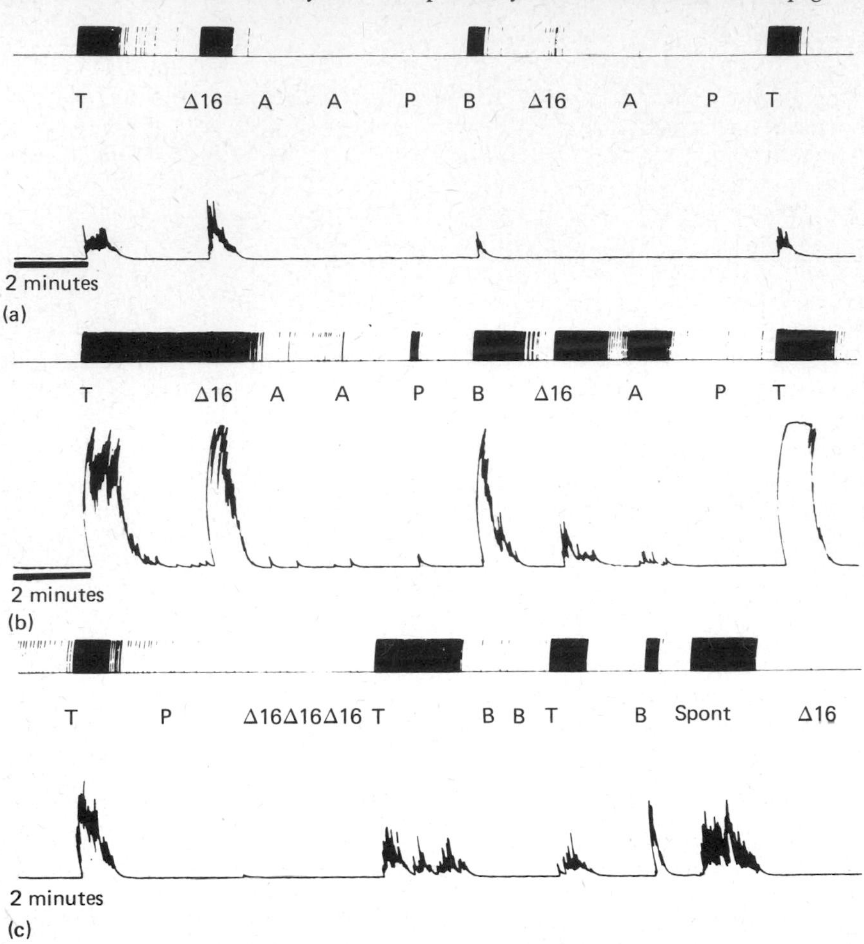

Figure 9.10 Polygraph recording of neural activity. Each pen deflection on the upper traces represents four action potentials; the lower trace is an integrated record of this activity derived from cell action potentials. (a) and (b) two neurons responded to both 5α-androst-16-ene-3-one (Δ-16) and testosterone (T). In (c) neurons responded to T but only 2/10 stimuli responded to Δ-16. Other organic substances tested: A = Amylacetate; P = Pyridine; B = Benzene. Spont = a spontaneous burst of firing. By courtesy of MacLeod, Reinhardt and Ellendorff (1979)

In the pig both Δ-16 and testosterone alter the electrical activity of mitral neurons after exposure of the olfactory system to aerosols of these substances (*Figure 9.10*), and the neurons may be activated by both or either one of the steroids indicating discriminatory abilities of the olfactory system for steroids. The connections to the amygdala could be one pathway by which behavioural or endocrine changes may be brought about. In addition to the main olfactory system, the accessory olfactory system may be involved in reproductive functions (Ladewig, Price and Hart, 1980). This is well developed in the pig with a nasopalatine duct allowing access to the vomeronasal organ from where fibres project over the surface of the main bulb into the accessory bulb.

Oxytocin and vasopressin secretion

The neurohypophysis, like the anterior pituitary, is also able to secrete its hormones prenatally, so the magnocellular neurosecretory system of the paraventricular and supraoptic nuclei must be functional in the foetal pig. Circulating levels of oxytocin and lysine vasopressin (LVP) are detectable at the foetal age of 75 and 109 days respectively. In fact, concentrations of both hormones exceed those found in simultaneous maternal samples (MacDonald *et al.*, 1979). In addition, the foetus responds with elevated LVP levels when exposed to haemorrhage (Forsling, Macdonald, and Ellendorff, 1979).

In the adult sow both parturition and nursing are associated with massive oxytocin release (Forsling *et al.*, 1979). In late pregnancy oxytocin concentrations of sow plasma are close to the lower level of detection and only a few hours (<7 hours) before foetal expulsion does the range of oxytocin release increase to about 24 μU oxytocin/ml. Immediately after expulsion this value could be exceeded almost threefold and during delivery of the placentas a similar high surge of oxytocin occurs (*Figure 9.11*). It is likely that the extremely high concentrations of oxytocin are due

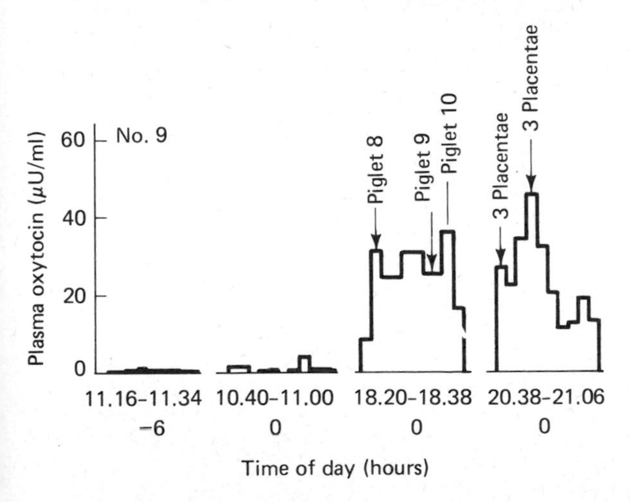

Figure 9.11 Oxytocin secretion before, during and after parturition in the miniature pig. By courtesy of Forsling *et al.* (1979)

to mechanical stimulation of the cervix thus producing a Ferguson reflex release of oxytocin. It is not easy to explain the mechanisms leading to the initial release of oxytocin prior to expulsion of the first foetus, although several endocrine changes take place at this time. Plasma prostaglandin levels display a surge of release (Silver *et al.*, 1979) and prostaglandins stimulate oxytocin release in the sow (Ellendorff *et al.*, 1979). Oestrogens clearly reach their maximal values immediately prior to parturition (Shearer *et al.*, 1972; Ash and Heap, 1975; Taverne *et al.*, 1979) and progesterone declines rapidly just prior to and during parturition, when it is closely related to changing oxytocin levels (Forsling *et al.*, 1979). It is also likely that the release pattern of oxytocin follows a cascade effect that has been proposed for the steep rise of a number of hormones prior to parturition (Thorburn, Challis and Currie, 1977).

The first evidence that oxytocin is responsible for milk ejection in the sow as in other species came from experiments in which injections of oxytocin (Whittlestone, 1954) produced a subsequent increase in intramammary pressure. Later some oxytocin measurements by bioassay (Folley and Knaggs, 1966) indicated a rise of oxytocin during the act of suckling. We have also observed that oxytocin levels were highest during suckling and lowest during periods of no suckling and higher peak levels of oxytocin occurred in early lactation when compared with later stages (Forsling *et al.*, 1979).

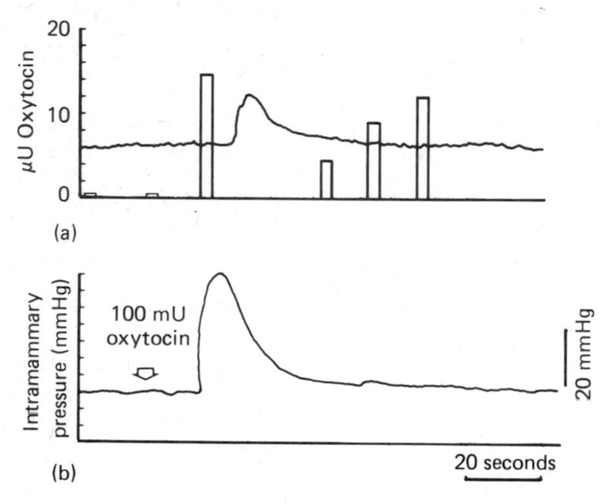

Figure 9.12 (a) Maternal milk ejection reflex in the sow displaying intramammary pressure changes and oxytocin secretion (open bars). (b) Induction of intramammary pressure changes and milk ejection with a single rapid injection of oxytocin. The relatively high dose evoked a much stronger response by the mammary gland. From Ellendorff, Forsling and Poulain (1981)

More recently a detailed analysis of oxytocin and time-linked events was undertaken (Ellendorff, Forsling and Poulain, 1981) and it was found that the suckling event that occurs every 40–50 minutes is associated with a surge in oxytocin (*Figure 9.12*) in the absence of any appreciable increase in vasopressin secretion. The characteristic grunting pattern described previously in detail (Fraser, 1973) reaches its crescendo around the time of oxytocin release but is not always indicative of oxytocin release. The amount of oxytocin secreted during a successful milk ejection corresponds to about 10 mU of oxytocin given as a single rapid injection. The signal initiating the secretion of oxytocin from the neurohypophysis is probably very similar to that described in the rat (for review see Cross *et al.*, 1975). Oxytocin release and intramammary pressure changes are almost identical to the natural milk ejection when the posterior pituitary of the rat is stimulated at frequencies between 25 and 40 Hz under such experimental conditions; however, there is also significant release of vasopressin. A distinct difference to the rat exists in the pig (as probably in many other mammals) with respect to sleep patterns related to nursing. In contrast to the rat which invariably displays a slow wave sleep EEG (Lincoln *et al.*, 1980; Voloschin and Tramezzani, 1979), a similar sleep pattern is not predictably associated with the onset of each suckling period in the pig (Poulain, Rodriguez and Ellendorff, 1981).

References

ASH, R.W. and HEAP, R.B. (1975). Oestrogen, progesterone and corticosteroid concentration in peripheral plasma of sows during pregnancy, parturition, lactation and after weaning. *J. Endocr.* **64**, 141–154

BREUER, H., VOGEL, W. and KNUPPEN, R. (1962). Enzymatische Methylierung von 2-Hydroxy-Östradiol-(17β) durch eine S-Adenosyl-Methionine-Acceptor-O-Methyltransferase der Rattenleber. *Hoppe-Seyler's Z. physiol. Chem.* **327**, 217–222

BRUHN, T., PARVIZI, N. and ELLENDORFF, F. (1981). Ability of the fetal hypothalamus to alter LH-secretion in response to electrical and electrochemical stimulation. *Acta endocr.* **96**, *Suppl.* **240**, 46

BRUNI, J.F., van VUGT, D., MARSHALL, S. and MEITES, J. (1977). Effects of naloxone, morphine and methionine-enkephalin on serum prolactin, luteinizing hormone, follicle stimulating hormone, thyroid stimulating hormone and growth hormone. *Life Sci.* **21**, 461–466

BURGUS, R., BUTCHER, M., LING, N., MONAHAN, M., RIVIER, J., FELLOWS, R., AMOSS, M., BLACKWELL, R., VALE, W. and GUILLEMIN, R. (1971). Structure moleculaire du facteur hypothalamique (LRF) d'origine ovine controlant la secretion de l'hormone de gonadotrope hypo-physaire du luteinisation. *C. R. hebd. Seanc. Sci., Paris* **273**, 1611–1613

COLENBRANDER, B., DRUIP, TH.A.M., DIELEMAN, S.J. and WENSING, C.J.G. (1977). Changes in serum LH concentrations during normal and abnormal sexual development in the pig *Biol. Reprod.* **17**, 506–513

CROSS, B.A., DYBALL, R.E.J., DYER, R.G., JONES, C.W., LINCOLN, D.W. MORRIS, J.F. and PICKERING, B.T. (1975). Endocrine neurons. *Recent Prog. Horm. Res.* **31**, 243–294

DÖCKE, F. and BUSCH, W. (1974). Evidence for anterior hypothalamic control of cyclic gonadotrophin secretion in female pigs. *Endocrinology* **63**, 415–421

ELLENDORFF, F. (1978). Extra-hypothalamic centres involved in the control of ovulation. In *Control of Ovulation*, (D.B. Crighton, N.B. Haynes, G.R. Foxcroft and G.E. Lamming, Eds.), pp. 7–19. London, Butterworths

ELLENDORFF, F. and PARVIZI, N. (1980). Role of extrahypothalamic centres in neuroendocrine integration. In *The Endocrine Functions of the Brain*, (M. Motta, Ed.), pp. 297–325. New York, Raven Press

ELLENDORFF, F., FORSLING, M. and POULAIN, D. (1981). Plasma oxytocin levels associated with the milk ejection reflex in the sow. *Acta endocr.* **96**, *Suppl.* **240**, 49

ELLENDORFF, F., KREIKENBAUM, K., PARVIZI, N. and SMIDT, D. (1973). LH-secretion in response to electrical stimulation of the chronically implanted miniature pigs brain. *Acta endocr. Suppl.* **177**, 73

ELLENDORFF, F., FORSLING, M., PARVIZI, N. and WILLIAMS, H., TAVERNE, M. and SMIDT, D. (1979). Plasma oxytocin and vasopressin concentrations in response to prostaglandin injection into the pig. *J. Reprod. Fert.* **56**, 573–577

ELSAESSER, F., ELLENDORFF, F., POMERANTZ, D.K., PARVIZI, N. and SMIDT, D. (1976). Plasma levels of luteinizing hormone, progesterone, testosterone and 5α-dihydrotestosterone in male and female pigs during sexual maturation. *J. Endocr.* **68**, 147–148

FOLLEY, S.J. and KNAGGS, G.S. (1966). Milk-ejection activity (oxytocin) in the external jugular vein blood of the cow, goat and sow, in relation to the stimulus of milking or suckling. *J. Endocr.* **34**, 197–214

FORSLING, M.L., MACDONALD, A.A. and ELLENDORFF, F. (1979). The neurohypophysial hormones. *Anim. Reprod. Sci.* **2**, 43–56

FORSLING, M.L., TAVERNE, M.A.M., PARVIZI, N., ELSAESSER, F., SMIDT, D. and ELLENDORFF, F. (1979). Plasma oxytocin and steroid concentrations during later pregnancy, parturition and lactation in the miniature pig. *J. Endocr.* **82**, 61–69

FRASER, D. (1973). The nursing and suckling behaviour of pigs. I. The importance of stimulation of the anterior teats. *Br. vet. J.* **129**, 324–336

KONDA, N., DYER, R.G., BRUHN, T., MACDONALD, A.A. and ELLENDORFF, F. (1979). A method for recording single unit activity from the brains of foetal pigs *in utero*. *J. Neurosci. Methods* **1**, 289–300

LADEWIG, J., PRICE, E.O. and HART, B.L. (1980). Flehmen in male goats: role in sexual behaviour. *Behavl Neur. Biol.* **30**, 312–322

LINCOLN, D.W. HENTZEN, K., HIN, T., VAN DER SCHOOT, P., CLARKE, G. and SUMMERLEE, A.J.S. (1980). Sleep a prerequisite for reflex milk ejection in the rat. *Exp. Brain Res.* **38**, 151–162

MACDONALD, A.A., FORSLING, M.L., WILLIAMS, H. and ELLENDORFF, F. (1979). Plasma vasopressin and oxytocin concentrations in the conscious pig foetus: response to haemorrhage. *J. Endocr.* **81**, 124P–125P

MACLEOD, N., REINHARDT, W. and ELLENDORFF, F. (1979). Olfactory bulb neurons of the pig respond to an identified steroidal pheromone and testosterone. *Brain Res.* **164**, 323–327

PARVIZI, N. and ELLENDORFF, F. (1975). 2-Hydroxy-oestradiol-17β as a possible link in steroid brain interaction. *Nature, Lond.* **256**, 59–60

PARVIZI, N. and ELLENDORFF, F. (1978). Norepinephrine and luteinizing hormone secretion: intrahypothalamic and intraventricular microinjection of norepinephrine. *Brain Res.* **148**, 521–525

PARVIZI, N. and ELLENDORFF, F. (1980a). Gonadal steroids in the amygdala – differential effects on LH. *Brain Res.* **195**, 363–372

PARVIZI, N. and ELLENDORFF, F. (1980b). β-Endorphin alters luteinizing hormone secretion via the amygdala but not the hypothalamus. *Nature, Lond.* **286**, 812–813

PARVIZI, N., ELSAESSER, F., SMIDT, D. and ELLENDORFF, F. (1977). Effects of intracerebral implantation, microinjection, and peripheral application of sexual steroids on plasma luteinizing hormone levels in the male miniature pig. *Endocrinology* **101**, 1078–1087

POMERANTZ, D.K., ELLENDORFF, F., ELSAESSER, F., KONIG, A. and SMIDT, D. (1974). Plasma LH changes in intact adult, castrated adult and pubertal male pigs following various doses of synthetic luteinizing hormone-releasing hormone (LH-RH). *Endocrinology* **94**, 330–335

POULAIN, D.A., RODRIGUEZ, F. and ELLENDORFF, F. (1981). Sleep is not a prerequisite for the milk ejection reflex in the pig. *Exp. Brain Res.* **43**, 107–110

REINHARDT, W., KONDA, N., MACLEOD, N. and ELLENDORFF, F. (1981). Electrophysiology of olfacto–limbic–hypothalamic connections in the pig. *Exp. Brain Res.* **43**, 1–10

SCHALLY, A.V., ARIMURA, A., BABA, Y., MAIR, R.M.G., MATSUO, H., REDDING, T.W., DEBELJUK, L. and WHITE, W.F. (1971). Isolation and properties of the FSH and LH-releasing hormone. *Biochem. Biophys. Res. Commun.* **43**, 393–399

SCHNEIDER, H.P.G. and McCANN, S.M. (1970). Mono- and indolamines and control of LH-secretion. *Endocrinology* **86**, 1127–1133

SHEARER, I.G., PURVIS, K., JENKINS, G. and HAYNES, N.B. (1972). Peripheral plasma progesterone and oestradiol-17β levels before and after puberty in gilts. *J. Reprod. Fert.* **30**, 347–360

SILVER, M., BARNES, R.J., COMLINE, R.S., FOWDEN, A.L., CLOVER, L. and MITCHELL, M.D. (1979). Prostaglandins in the foetal pig and prepartum endocrine changes in mother and foetus. *Anim. Reprod. Sci.* **2**, 305–322

TAVERNE, M.A.M., NAAKTGEBOREN, C., ELSAESSER, F., FORSLING, M.L., VAN DER WEYDEN, G.C., ELLENDORFF, F. and SMIDT, D. (1979). Myometrial electrical activity and plasma concentrations of progesterone, estrogens and oxytocin during late pregnancy and parturition in the miniature pig. *Biol. Reprod.* **21**, 1125–1134

THORBURN, G.D., CHALLIS, J.R.G. and CURRIE, W.B. (1977). Control of parturition in domestic animals. *Biol. Reprod.* **16**, 18–27

VINGNEAUD DU, V., BARLETT, M.F. and JOHL, A. (1954–1955). Hormones of the posterior pituitary gland: Oxytocin and vasopressin. *Harvey Lect.* **50**, 1–26

VOLOSCHIN, L.M. and TRAMEZZANI, J.H. (1979). Milk ejection reflex linked to slow wave sleep in nursing rats. *Endocrinology* **105**, 1202–1207

WHITTLESTONE, W.B. (1954). Intramammary pressure changes in the lactating sow. I. Effect of oxytocin. *J. Dairy Res.* **21**, 19–30

10

THE CONTROL OF OVULATION

S.K. WEBEL
Reproductive Consultant Services, Illinois, USA

B.N. DAY
University of Missouri, USA

Extensive studies have been directed towards the effective control of oestrus and ovulation in the pig. Management procedures, including the simultaneous exposure of gilts to the boar or the weaning of pigs from a group of sows, have been developed for the synchronization of oestrus and ovulation in the pig herd. Other investigators have concentrated on the use of exogenous hormones or other compounds to interrupt the normal oestrous cycle as an additional approach toward oestrus and ovulation control.

The oestrous cycle may be altered by either suppressing ovarian activity to delay oestrus or by inducing premature regression of corpora lutea to hasten the onset of oestrus. Spontaneous follicular development usually occurs following these treatments in sexually mature animals. Regulation of luteal function or suppression of follicular growth will be referred to as *control of the luteal phase.* Precise control of the initiation of follicular development in anoestrous animals and of the occurrence of ovulation can be obtained by treatment with gonadotrophins. The use of exogenous hormones for this purpose will be discussed as *control of the follicular phase.* A review will be presented of reported studies that have used hormones and other compounds to control ovulation in the pig followed by results from recent investigations on the use of a synthetic progestagen, altrenogest, for the control of ovulation in pigs on commercial farms.

Control of the luteal phase

Progesterone and synthetic progestagens have been used to suppress the oestrous cycle. Daily injections of progesterone inhibits oestrus and, if adequate doses are given, results in normal fertility (Ulberg, Grummer and Casida, 1951; Baker *et al.*, 1954; Gerrits *et al.*, 1963). Likewise, several synthetic progestagens administered either orally or by injection have inhibited follicular growth and oestrus.

High doses of 6-methyl-17-acetoxyprogesterone (MAP) have been reported to inhibit oestrus without producing cystic follicles, but oestrus was usually not well synchronized and litter size was often reduced (Baker *et al.*, 1954; Dziuk, 1960, 1964; Nellor, 1960; Nellor *et al.*, 1961; Dziuk and Baker, 1962; First *et al.*, 1963; Dziuk and Polge, 1965). Oral administration of other progestational compounds including 6-chloro-Δ^6-17-acetoxyprogesterone (Wagner and Seerley, 1961; Veenhuizen *et al.*, 1965; Ray and Seerley, 1966), 17α-acetoxy-6-methylpregna-4,6-dien-3,20-dione (Pond *et al.*, 1965), or injections of norethandrolone (Martinat-Botte, 1975a) have produced results similar to those obtained with MAP. In general, the administration of progesterone or progestational compounds has not been a satisfactory treatment for controlling oestrus and ovulation because of the increased incidence of cystic follicles, decreased fertility at the first post-treatment oestrus and a lack of precise synchronization.

Two new progestagen compounds, SA-45249 (Mayer and Schutze, 1977a) and altrenogest (Webel, 1976, 1978; Davis *et al.*, 1980) offer a new opportunity for regulating the oestrous cycle. These compounds, when administered orally in sufficient doses, suppress oestrus and result in a synchronized return to oestrus following withdrawal, without a reduction in fertility or litter size. The two compounds are similar in structure and dosage requirements.

Low doses of 3 mg of SA-45249 (Mayer and Schutze, 1977a) or 2.5 mg of altrenogest (Webel, 1978; Redmer and Day, 1981) results in a high incidence of cystic follicles. Levels of 6 mg of SA-45249 or 10 mg of altrenogest are effective for oestrus synchronization with a low incidence of cystic follicles. Recommended doses for general use of altrenogest are 15–20 mg which effectively synchronizes oestrus with minimal numbers of cystic follicles (Webel, 1978; Redmer *et al.*, 1979; Davis *et al.*, 1980; Kraeling *et al.*, 1981). Treatment durations of 10–18 days were investigated with the longer durations producing shorter intervals to oestrus and closer synchronization of oestrus after withdrawal (Schutze and Mayer, 1977; Webel, 1978). A minimal 14-day feeding period appears to be necessary to control oestrus in gilts at all phases of the oestrous cycle when treatment is initiated (Stevenson and Davis, 1981). As expected, individual feeding where each animal receives the prescribed dose produces closer synchronization of oestrus than group feeding (Zerobin, 1977; Martinat-Botte *et al.*, 1980).

The study of endogenous hormones associated with SA-45249 (Mayer and Schutze, 1977b) or altrenogest (Redmer and Day, 1981) suggest that both compounds work in a similar manner by inhibiting follicular growth. There is no influence on the lifespan of the corpus luteum.

Effective control of the oestrous cycle has been obtained by inhibiting ovarian function with a non-steroidal compound, ICI 33828 (a dithiocarbamoylhydrazine derivative). A high proportion of sows or gilts exhibited oestrus 5–8 days after withdrawal of the compound following an 18–20-day feeding period and fertility was not reduced following treatment with this compound (Polge, 1965, 1966; Gerrits and Johnson, 1965; Stratman and First, 1965; Groves, 1967; Polge, Day and Groves, 1968). Use of ICI 33828 was curtailed and regulatory approvals withdrawn in many countries following reports of teratogenic effects in pregnant gilts.

Another method for regulating the oestrous cycle is to induce accessory corpora lutea in cycling animals and then allow them to regress normally. Injections of pregnant mare's serum gonadotrophin (PMSG) followed by human chorionic gonadotrophin (HCG) induces ovulation at any stage of the oestrous cycle. The accessory corpora lutea then regress after an approximately normal life span with oestrus occurring 18–24 days after the HCG injection (Neill and Day, 1964; Day *et al.*, 1965; Caldwell *et al.*, 1969). Although this treatment offers some degree of oestrus synchronization, it is not very precise because of the variability in the duration of luteal function and the early regression of accessory corpora lutea induced during the first six days of the cycle.

Interruption of the oestrous cycle by shortening the life span of the corpus luteum is another method of oestrous cycle control. The injection of oestrogen in the pig often has luteotropic effects (Gardner, First and Casida, 1963; Dziuk, 1964) as contrasted to the luteolytic influence in the cow. It is not, therefore, an effective treatment for reducing the length of the oestrous cycle. Prostaglandins are not luteolytic in the pig until about day 11 or 12 of the cycle, so they also do not offer a practical means of synchronizing oestrus (Diehl and Day, 1974; Hallford *et al.*, 1975; Guthrie and Polge, 1976a; Lindloff *et al.*, 1976). However, effective synchronization has been obtained when prostaglandins were used to regress corpora lutea during pregnancy (Guthrie, 1975; Guthrie and Polge, 1976b), or following oestrogen treatments to prolong luteal function (Guthrie, 1975; Guthrie and Polge, 1976b; Kraeling and Rampacek, 1977). Regression of accessory corpora induced by gonadotrophins has also been reported by Guthrie (1979) to provide a means for the control of oestrus.

Control of follicular phase

Pituitary gonadotrophic preparations, pregnant mare serum gonadotrophin (PMSG), human chorionic gonadotrophin (HCG), hypothalamic releasing hormones (GnRH), or combinations of these hormones have been widely used to induce follicular growth or ovulation. These treatments have been used in prepubertal gilts, during the luteal and follicular phase of the oestrous cycle, in anoestrous gilts or sows, in lactating or early weaned sows and following suppression of the oestrous cycle with other exogenous hormones.

As early as 1935, Casida demonstrated that ovulation could be induced in prepubertal gilts by giving multiple injections of PMSG or purified pituitary preparations. These observations were confirmed by du Mesnil du Buisson (1954), Dziuk and Gehlbach (1966) and Baker and Coggins (1968). Although injection of PMSG followed by HCG 48–96 hours later induced a fertile ovulation, pregnancy was not usually associated with such a treatment in gilts 4–5 months of age unless exogenous progestagens or gonadotrophins were given after breeding, as the corpora lutea normally regressed by day 20–25 of pregnancy (Shaw, McDonald and Baker, 1971; Segal and Baker, 1973; Ellicott, Dziuk and Polge, 1973; Rampacek *et al.*, 1976). Other workers using a combination of PMSG and HCG given as a single injection have reported synchronized oestrus and ovulation in prepubertal gilts (Schilling and Cerne, 1972; Baker and Rajamahendran, 1973; Guthrie, 1977). The sometimes conflicting results on pregnancy rates in the prepubertal pig need further clarification regarding such variables as breed, age, dose and combinations of hormones.

It has also been well established that superovulation can be induced in pigs by the injection of appropriate gonadotrophins (Tanabe *et al.*, 1949; Gibson *et al.*, 1963; Hunter, 1964, 1966; Longenecker, Lasley and Day, 1965; Day *et al.*, 1967; Longenecker and Day, 1968; Phillipo, 1968; Christenson *et al.*, 1973). However, litter size at farrowing has not consistently been increased and may only be significantly increased by PMSG in sows which have lower than average litter sizes when untreated (Schilling and Cerne, 1972).

The time of ovulation can be precisely controlled by the injection of HCG (Dziuk and Baker, 1962; Dziuk, Polge and Rowson, 1964; Hunter, 1964, 1966; Buttle and Hancock, 1967), or GnRH (Baker, Downey and Brinkley, 1973; Webel and Rippel, 1975; Guthrie, 1977; Guthrie, Pursel and Frobish, 1978) 48–96 hours after PMSG. Attempts to induce follicular stimulation, superovulation or oestrus with gonadotrophin-releasing hormone have apparently been unsuccessful (Baker and Downey, 1975; Guthrie, 1977; Webel, 1978). However, recent unpublished work in the author's laboratories suggest that whereas a single injection of large doses of GnRH is ineffective, frequent repeated injections of small doses promote follicular growth and development. J.H. Britt (personal communication) reported induction of oestrus and ovulation in lactating sows injected with 2.5 μg GnRH every two hours for seven days. More data will be needed, however, before practical application can be recommended.

The ability of PMSG to stimulate follicular development and of HCG to control precisely the time of ovulation has been utilized to synchronize ovulation and to allow insemination at a fixed time. This combined treatment (PMSG + HCG) has been used following inhibition or suppression of the oestrous cycle with compounds such as ICI 33828 (Polge, Day and Groves, 1968; Webel, Peters and Anderson, 1970; Christenson *et al.*, 1973; Baker, Shaw and Dodds, 1970) or following oral administration of progestagens (Dziuk and Baker, 1962; Dziuk and Polge, 1965). The sequence of treatments (ICI 33828/progestagen + PMSG + HCG) is effective for either controlling the time of ovulation to allow a single insemination or to induce superovulation. Other uses of PMSG and HCG have been to synchronize oestrus in sows by injecting PMSG on the day of weaning to shorten the interval to the first oestrus, or the injection of PMSG followed by HCG 80–96 hours later in order to synchronize ovulation and permit a single insemination (Longenecker and Day, 1968; Christenson and Teague, 1975; Soma and Speer, 1975).

In recent years, an increasing number of farms have placed the breeding herd in controlled-environment, limited-space housing. Several reproductive problems have been associated with these confined conditions including delayed puberty in gilts, and the failure of sows to return to oestrus following weaning, particularly in the late summer. Injection of PMSG or a combination of PMSG/HCG as a single injection to sows at weaning or to non-cycling gilts has overcome these anoestrous problems and shortened the time of oestrus (Dziuk and Dhindsa, 1969; Schilling and Cerne, 1972; Hurtgen, 1976; Hurtgen and Leman, 1979).

The use of gonadotrophins to induce ovulation during lactation has recently attracted renewed attention. Pregnancy has been induced in lactating animals but with quite variable results, especially when administered soon after parturition. The ovarian response to PMSG was increased as the interval from parturition to injection increased, but consistently good results were not obtained (Heitman and Cole, 1956; Epstein and Kadmon, 1969; Crighton, 1970a, 1970b; Martinat-Botte, 1975b; Hausler *et al.*, 1980). Follicular development may occur without ovulation when sows are given only PMSG (Martinat-Botte, 1975b; Guthrie, Pursel and Frobish, 1976, 1978; Webel, unpublished). Normal fertility has been obtained in lactating sows with an injection of PMSG at 25 days post-partum followed 96 hours later by HCG (Kuo, Hodson and Hausler, 1976; Hausler *et al.*, 1980). The injection of a follicle stimulant such as PMSG or PMSG/HCG combination followed by HCG or GnRH to induce ovulation seems to offer a possible means for inducing pregnancy in the lactating sow. However, questions regarding the breed of sow, the stage post-partum and the doses of the hormones must be answered before a management system can be widely recommended.

Use of altrenogest for reproductive management in gilts

Altrenogest may be a useful tool in the management of the breeding herd because of its effectiveness for regulating and synchronizing the oestrous cycle. Altrenogest (17α-allyl-estratiene-4-9-11, 17-β-ol-3-one) is a synthetic steroid with progestagenic activity. This compound was previously identified as allyl trenbolone and is also identified by the numbers A-35957 and RU-2267. The proprietary name is REGU-MATE®. Altrenogest is a product of Roussel-Uclaf and is being developed for use in synchronizing oestrus in gilts and sows. This synchronization permits planned breeding within a preplanned period of 3–5 days and facilitates introduction of gilts into the breeding herd. Oestrus in weaned sows and gilts can be scheduled to occur simultaneously by coordinating the dates for weaning sows and feeding altrenogest to gilts. Oestrus synchronization is obtained by feeding the compound for 18 days at levels of 15 or 20 mg/gilt/day.

The effectiveness of altrenogest for synchronizing oestrus in gilts which have exhibited at least one previous oestrus is illustrated by two histograms in *Figure 10.1*. Up to 100% of treated gilts have expressed oestrus within a 3-day period (Mauleon, Martinat-Botte and Scheid, 1979) and in some individual trials with small numbers of gilts, all gilts were in oestrus on a single day (Webel, unpublished). However, these histograms depict more typical results observed with larger numbers of gilts. Pursel *et al.* (1981) observed 80% of the treated gilts in oestrus within a two-day period and 97% within four days. In trials on French farms 65% of the gilts were in oestrus on days 6 and 7 and 93% in oestrus on days 5–8 (A. Jobard, J.M. Boisson and J.P. Scheid, personal communication). In the study reported by Pursel *et al.* (1981), all gilts were fed 15 mg daily on a single farm, whereas the French trial was conducted on different farms with a daily dose of 20 mg. These variations in the trials may explain the difference in precision of synchronization since a more consistent response would be expected in a single herd. Also, the increase in interval and oestrus

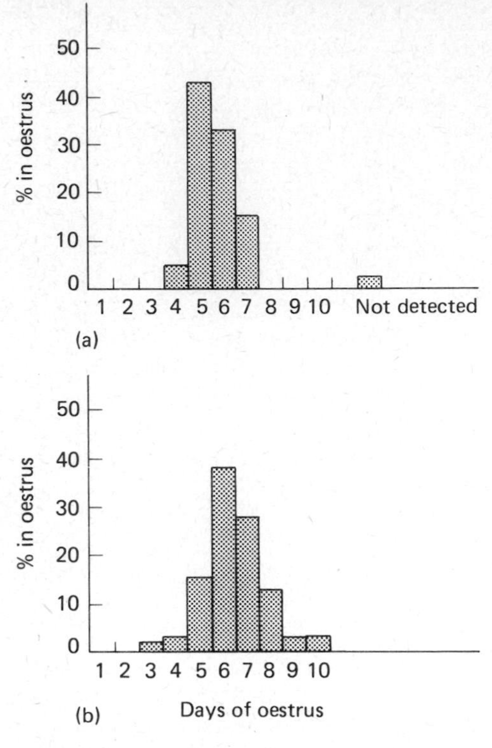

Figure 10.1 Oestrus synchronization following altrenogest in gilts with a previous oestrous cycle recorded. (a) Dosage: 15 mg/gilt/day; $n = 60$. By courtesy of Pursel *et al.* (1981). (b) Dosage: 20 mg/gilt/day; $n = 175$. By courtesy of Jobard *et al.* (unpublished)

following the higher dose is consistent with earlier studies (Webel, 1978, 1980; Kraeling *et al.*, 1981).

Fertility at the first oestrus following synchronization with altrenogest has not been different from controls; however, there has been a tendency for an increase in ovulation rate and litter size in treated gilts (Webel, 1978; Davis *et al.*, 1980). A summary of reported farrowing rates and litter sizes is shown in *Table 10.1*. Increased litter sizes have not been consistently

Table 10.1 FERTILITY FOLLOWING SYNCHRONIZATION OF OESTRUS WITH ALTRENOGEST IN SEXUALLY MATURE GILTS

Treatment	*Number of animals bred*	*Percent farrowed*	*Litter size*	*Reference*
Control	145	83	10.0	Jobard *et al.*, unpublished
Altrenogest (20 mg)	175	83	10.7	
Control	70	60	10.0	Webel, 1978
Altrenogest (15 mg)	68	75	11.3[a]	
Control	68	74	9.1	Pursel *et al.*, 1981
Altrenogest (15 mg)	58	71	10.5	
Control	29	86	9.8	Britt, 1980
Altrenogest (15 mg)	48	85	9.9	

[a]Significantly ($P<0.05$) different from control group.

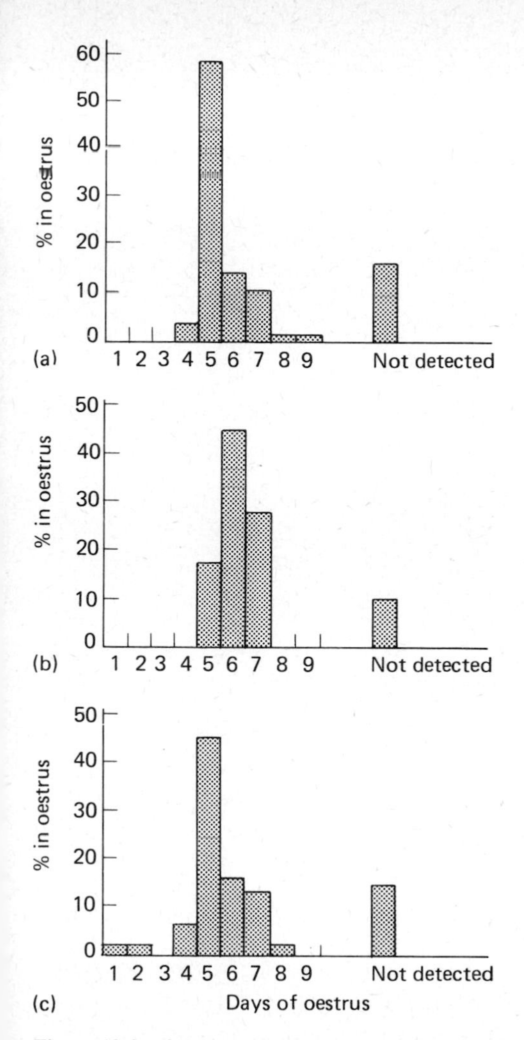

Figure 10.2 Oestrus synchronization following altrenogest in gilts on commercial pig farms without previous records of oestrous cycles. (a) Dosage: 15 mg/gilt/day; n = 175. (b) Dosage: 15 mg/gilt/day; n = 29. (c) Dosage: 15 mg/gilt/day; n = 62

observed in all trials, but in some cases the difference has been statistically significant.

In contrast to controlled research studies, previous oestrous records are seldom available on commercial pig breeding farms, and in many cases the gilts have not reached puberty or are in a state of anoestrus. If only gilts with a previous record of oestrus are used, the synchronization results on commercial farms following altrenogest are similar to those shown in *Figure 10.1*. However, most trials have been conducted on farms where the previous history was not available. A summary of results from these farms is shown in *Figure 10.2*. Although the precision of synchronization for those gilts that exhibited oestrus was similar to previous observations, 10–15% of the gilts were not observed in oestrus following treatment. These gilts were similar in age (7–9 months) to those used in the studies shown in *Figure 10.1*, but may not have exhibited a previous oestrus. These results agree with other reports (O'Reilly *et al.*, 1979; Webel, Scheid and Bouffault, 1980) where 65–75% of those gilts in which previous records of oestrus behaviour were not known exhibited a synchronized oestrus.

In a trial at the University of Missouri, 12 of 49 gilts allotted to altrenogest had not been observed in oestrus. Seven of these (58%) were synchronized following treatment compared with 78% of those which had shown oestrus previously. These results provide evidence that a lower proportion of non-cycling animals are synchronized compared with gilts that had been detected in oestrus prior to treatment. In other studies the ovaries of gilts which did not exhibit oestrus following withdrawal of altrenogest were examined. In many cases these gilts had inactive ovaries indicating that they were not cycling or had not yet attained puberty (Webel, 1978; Britt, 1980; Martinat-Botte *et al.*, 1980). These reports are consistent with the observations that altrenogest effectively synchronizes oestrus in cycling gilts with active ovaries but is less effective in inactive gilts. On commercial farms a similar proportion of animals treated with altrenogest have been observed in oestrus within a 4-day period as compared with a 25-day period for controls. When the proportion of controls which expressed oestrus within 25 days was low the proportion of gilts synchronized following altrenogest was also low. In contrast, if a high proportion of controls were in oestrus a high proportion of treated gilts were effectively synchronized. This proportion varies from about 50% up to 100%, averaging around 85%. In general, approximately 95% of those gilts which had exhibited a previous oestrus were synchronized to a 4-day period whereas only about 75–85% of similar gilts whose reproductive history was not known were synchronized.

The farrowing rate and litter size for three farms are shown in *Table 10.2*. The lower farrowing rate for controls differs from previous observations in controlled laboratory trials. Although there is no clear explanation for these differences the close synchronization of oestrus in treated gilts may facilitate management. Perhaps the breeder took greater care when

Table 10.2 FERTILITY FOLLOWING ALTRENOGEST SYNCHRONIZATION ON COMMERCIAL PIG BREEDING FARMS

Treatment	*Number of animals bred*	*Percent farrowed*	*Mean litter size*
Trial 1			
Control	43	30[(a)]	9.3
Altrenogest (15 mg)	53	58[(b)]	10.9
Altrenogest (20 mg)	54	52[(b)]	11.0
Trial 2			
Control	26	54[(a)]	9.8
Altrenogest (15 mg)	26	78[(b)]	9.2
Altrenogest (20 mg)	26	85[(b)]	10.7
Trial 3			
Control	25	40	10.7
Altrenogest (15 mg)	25	68	9.4
Altrenogest (20 mg)	23	57	10.3

[(a),(b)]Means within trials with different superscripts are significantly different ($P<0.05$).

breeding synchronized gilts for 3–5 days than when breeding controls during a 3-week period.

The effectiveness of altrenogest in young gilts that had not been observed in oestrus was evaluated on a farm where gilts were bred at approximately 6 months of age rather than at 8–9 months. Gilts were selected at approximately 100 kg body weight and moved from the growing/finishing building to the breeding building on a Friday or Saturday and altrenogest feeding began on the following Monday. The probability that these gilts had exhibited a previous oestrus was low. Controls were handled in a similar manner. The cumulative percentages in oestrus by day 10, 20, 30 and 40 are shown in *Table 10.3*. These calculations are based on

Table 10.3 OESTROUS RESPONSE FOLLOWING ALTRENOGEST IN THE YOUNG GILT

Observation	*Control*	*Altrenogest*	
		7 days	*14 days*
Number of gilts	26	28	36
Cumulative percent in oestrus by day:			
10	15[a]	57[b]	61[b]
20	23[a]	64[b]	67[b]
30	27[a]	64[b]	67[b]
40	54[c]	68[d]	72[d]

Days to oestrus were calculated from the last day fed for altrenogest or from the date moved to the breeding area for controls.
[a],[b]Numbers in rows with different superscripts are different ($P<0.05$).
[c],[d]Numbers in rows with different superscripts are different ($P<0.10$).

the number of days following the last altrenogest administration for treated gilts and from the date moved to the breeding area for controls. A higher proportion ($P<0.10$) of treated animals were in oestrus at each time period. To compare the total number of days from selection to oestrus, one must add 7 or 14 days to the treated animals to compensate for the feeding period. However, even when the proportion of controls in oestrus by 30 days (27%) is compared with the percentage of treated gilts in oestrus by 10 days (57% and 61%), an advantage still exists for gilts fed altrenogest. In fact, there was a similar proportion in oestrus within 10 days post-treatment (57% and 61%) as within 40 days (54%) in controls. Altrenogest appeared to promote an earlier and more synchronized attainment of puberty in these gilts.

Even though the proportion of young gilts exhibiting a synchronized oestrus (57% or 61%) was lower than observed in older gilts in farm trials (*Figure 10.2*; approximately 85%) or in cycling gilts (*Figure 10.1*; 95% to 100%), treatment of young gilts with altrenogest would appear to provide a useful management tool to the pig producer.

Christenson and Ford (1979) have found that depending on the breed and confinement conditions, between 52% and 94% of gilts tested with boars from 145 days of age had exhibited regular oestrous cycles by 9 months of age. Therefore, altrenogest may be a tool for selecting gilts that will show puberty at an early age and eliminate the need to maintain large numbers of gilts for several months before introduction into the breeding herd.

References

BAKER, R.D. and COGGINS, E.G. (1968). Control of ovulation rate and fertilization in prepubertal gilts. *J. Anim. Sci.* **27**, 1607–1610

BAKER, R.D. and DOWNEY, B.R. (1975). Induction of estrus, ovulation and fertility in prepubertal gilts. *Annls Biol. anim. biochim. biophys.* **15**, 375–382

BAKER, R.D. and RAJAMAHENDRAN, R. (1973). Induction of estrus, ovulation, and fertilization in prepubertal gilts by a single injection of PMSG, HCG and PMSG:HCG combination. *Can. J. Anim. Sci.* **53**, 693–694

BAKER, R.D., DOWNEY, B.R. and BRINKLEY, H.J. (1973). Induction of ovulation in pigs with gonadotrophin releasing hormone. *J. Anim. Sci.* **37**, 1376–1379

BAKER, R.D., SHAW, G.A. and DODDS, J.S. (1970). Control of estrus and litter size in gilts with aimax (ICI 33,828) and pregnant mare's serum. *Can. J. Anim. Sci.* **50**, 25–29

BAKER, L.N., ULBERG, L.C., GRUMMER, R.H. and CASIDA, L.E. (1954). Inhibition of heat by progesterone and its effect on subsequent fertility in gilts. *J. Anim. Sci.* **13**, 648–657

BRITT, J.H. (1980). Synchronization of estrus in swine. *1980 Report, Department of Animal Science, North Carolina State University*

BUTTLE, H.L. and HANCOCK, J.L. (1967). The control of ovulation in the sow. *J. Reprod. Fert.* **14**, 485–487

CALDWELL, B.V., MOOR, R.M., WILMUT, I., POLGE, C. and ROWSON, L.E.A. (1969). The relationship between day of formation and functional life span of induced corpora lutea in the pig. *J. Reprod. Fert.* **18**, 107–113

CASIDA, L.E. (1935). Prepubertal development of the pig ovary and its relation to stimulation with gonadotrophic hormones. *Anat. Rec.* **61**, 389–396

CHRISTENSON, R.K. and FORD, J.J. (1979). Puberty and estrus in confinement-reared gilts. *J. Anim. Sci.* **49**, 743–751

CHRISTENSON, R.K. and TEAGUE, H.S. (1975). Synchronization of ovulation and artificial insemination of sows after lactation. *J. Anim. Sci.* **41**, 560–563

CHRISTENSON, R.K., POPE, C.E., ZIMMERMAN-POPE, V.A. and DAY, B.N. (1973). Synchronization of estrus and ovulation in superovulated gilts. *J. Anim. Sci.* **36**, 914–918

CRIGHTON, D.B. (1970a). Induction of pregnancy during lactation in the sow. *J. Reprod. Fert.* **22**, 223–231

CRIGHTON, D.B. (1970b). The induction of pregnancy during lactation in the sow: The effects of a treatment imposed at 21 days of lactation. *Anim. Prod.* **12**, 611–617

DAVIS, D.L., KNIGHT, J.W., KILLIAN, D.B. and DAY, B.N. (1980). Control of estrus in gilts with a progestagen. *J. Anim. Sci.* **49**, 1506–1509

DAY, B.N., LONGENECKER, D.E., JAFFE, S.C., GIBSON, E.W. and LASLEY, J.F. (1967). Fertility of swine following superovulation. *J. Anim. Sci.* **26**, 777–780

DAY, B.N., NEILL, J.D., OXENREIDER, S.L., WAITE, A.B. and LASLEY, J.F. (1965). Use of gonadotrophins to synchronize estrous cycles in swine. *J. Anim. Sci.* **24**, 1075–1079

DIEHL, J.R. and DAY, B.N. (1974). Effect of prostaglandin $F_{2\alpha}$ on luteal function in swine. *J. Anim. Sci.* **39**, 392–396

DU MESNIL DU BUISSON, F. (1954). Possibilté d'ovulation et de fécondation chez la truie avant la puberté. *Annls Endocr.* **15**, 333–340

DZIUK, P.J. (1960). Influence of orally administered progestin on estrus and ovulation in swine. *J. Anim. Sci.* **19**, 1319–1320

DZIUK, P.J. (1964). Response of sheep and swine to treatments for control of ovulation. *Proc. Conf. Estrus Cycle Control in Domestic Animals,* USDA Miscellaneous Publication 1005, 50–57

DZIUK, P.J. and BAKER, R.D. (1962). Induction and control of ovulation in swine. *J. Anim. Sci.* **21**, 697–699

DZIUK, P.J. and DHINDSA, D.A. (1969). Induction of heat, ovulation and fertility in gilts with delayed puberty. *J. Anim. Sci.* **29**, 39–40

DZIUK, P.J. and GEHLBACH, G.D. (1966). Induction of ovulation and fertilization in the immature gilt. *J. Anim. Sci.* **25**, 410–413

DZIUK, P.J. and POLGE, C. (1965). Fertility in gilts following induced ovulation. *Vet. Rec.* **77**, 236–239

DZIUK, P.J., POLGE, C. and ROWSON, L.E.A. (1964). Intra-uterine migration and mixing of embryos in swine following egg transfer. *J. Anim. Sci.* **23**, 37–42

ELLICOTT, A.R., DZIUK, P.J. and POLGE, C. (1973). Maintenance of pregnancy in prepubertal gilts. *J. Anim. Sci.* **37**, 971–973

EPSTEIN, H. and KADMON, S. (1969). Physiological and economic feasibility of hormonally induced ovulation in lactating large white sows. *J. agric. Sci., Camb.* **72**, 365–370

FIRST, N.L., STRATMAN, F.W., RIGOR, E.M. and CASIDA, L.E. (1963). Factors affecting ovulation and follicular cyst formation in sows and gilts fed 6-methyl-17-acetoxyprogesterone. *J. Anim. Sci.* **22**, 66–71

GARDNER, M.L., FIRST, N.L. and CASIDA, L.E. (1963). Effect of exogenous estrogens on corpus luteum maintenance in gilts. *J. Anim. Sci.* **22**, 132–134

GERRITS, R.J. and JOHNSON, L.A. (1965). Synchronization of estrus in gilts fed two levels of I.C.I. 33,828 and the effect of fertility, embryo survival and litter size. *J. Anim. Sci.* **24**, 917–918

GERRITS, R.J., FAHNING, M.L., MEADE, R.J. and GRAHAM, E.F. (1963). Effect of synchronization of estrus on fertility in gilts. *J. Anim. Sci.* **21**, 1022

GIBSON, E.W., JAFFE, S.C., LASLEY, J.F. and DAY, B.N. (1963). Reproductive performance in swine following superovulation. *J. Anim. Sci.* **22**, 858

GROVES, T.W. (1967). Methallibure in the synchronization of oestrus in gilts. *Vet. Rec.* **80**, 470–475

GUTHRIE, H.D. (1975). Estrous synchronization and fertility in gilts treated with estradiol-benzoate and prostaglandin $F_{2\alpha}$. *Theriogenology* **4**, 69–75

GUTHRIE, H.D. (1977). Induction of ovulation and fertility in prepubertal gilts. *J. Anim. Sci.* **45**, 1360–1367

GUTHRIE, H.D. (1979). Fertility after estrous cycle control using gonadotrophin and $PGF_{2\alpha}$ treatment of sows. *J. Anim. Sci.* **49**, 158–162

GUTHRIE, H.D. and POLGE, C. (1976a). Luteal function and oestrus in gilts treated with a synthetic analogue of prostaglandin F-2α (ICI 79,939) at various times during the oestrous cycle. *J. Reprod. Fert.* **48**, 423–425

GUTHRIE, H.D. and POLGE, C. (1976b). Control of oestrus and fertility in gilts with accessory corpora lutea by prostaglandin analogues ICI 79,939 and ICI 80,996. *J. Reprod. Fert.* **48**, 427–430

GUTHRIE, H.D., PURSEL, V.G. and FROBISH, L.T. (1976). Attempts to initiate conception in lactating sows. *J. Anim. Sci.* **43**, 287

GUTHRIE, H.D., PURSEL, V.G. and FROBISH, L.T. (1978). Attempts to induce conception in lactating sows. *J. Anim. Sci.* **47**, 1145

HALLFORD, D.M., WETTEMAN, R.P., TURMAN, E.J. and OMTVEDT, I.T. (1975). Luteal function in gilts after prostaglandin $F_{2\alpha}$. *J. Anim. Sci.* **41**, 1706–1710

HAUSLER, C.L., HODSON, H.H., Jr., KUO, D.C., KINNEY, T.J., RAUWOLF, C.A. and STRACK, L.E. (1980). Induced ovulation and conception in lactating sows. *J. Anim. Sci.* **50**, 773–778

HEITMAN, H. and COLE, H.H. (1956). Further studies in the induction of estrus in lactating sows with equine gonadotrophin. *J. Anim. Sci.* **15**, 970–977

HUNTER, R.H.F. (1964). Superovulation and fertility in the pig. *Anim. Prod.* **6**, 189–194

HUNTER, R.H.F. (1966). The effect of superovulation on fertilization and embryonic survival in the pig. *Anim. Prod.* **8**, 457–465

HURTGEN, J.P. (1976). Seasonal anestrus in a Minnesota swine breeding herd. *4th Int. Congr. Int. Pig Vet. Soc., Section D*, Abstract 22

HURTGEN, J.P. and LEMAN, A.D. (1979). Use of PMSG in the prevention of seasonal post-weaning anestrus in sows. *Theriogenology* **12**, 207–214

KRAELING, R.R. and RAMPACEK, G.B. (1977). Synchronization of estrus and ovulation in gilts with estradiol and prostaglandin $F_{2\alpha}$. *Theriogenology* **8**, 103

KRAELING, R.R., DZIUK, P.J., PURSEL, V.G., RAMPACEK, G.B. and WEBEL, S.K. (1981). Synchronization of estrus in swine with allyl trenbolone (RU-2267). *J. Anim. Sci.* **52**, 831–835

KUO, D., HODSON, H.H. and HAUSLER, C.L. (1976). Induction of ovulation, artificial insemination and conception in lactating sows. *4th Int. Congr. Int. Pig Vet. Soc. Section D*, Abstract 23

LINDLOFF, G., HOLTZ, W. ELSAESSER, F., KREIKENBAUM, K. and SMIDT, D. (1976). The effect of prostaglandin $F_{2\alpha}$ on corpus luteum function in the Göttingen miniature pig. *Biol. Reprod.* **15**, 303–310

LONGENECKER, D.E. and DAY, B.N. (1968). Fertility level of sows superovulated on post weaning estrus. *J. Anim. Sci.* **27**, 709–711

LONGENECKER, D.E., LASLEY, J.F. and DAY, B.N. (1965). Fecundity in gilts and sows administered PMS. *J. Anim. Sci.* **24**, 924

MARTINAT-BOTTE, F. (1975a). Estrus control in gilts with norethandrolone injections and an analogue of prostaglandins (ICI 80996). *Annls Biol. anim. biochim. biophys.* **15**, 383–384

MARTINAT-BOTTE, F. (1975b). Induction of gestation during lactation in the sow. *Annls Biol. anim. biochim. biophys.* **15**, 369–374

MARTINAT-BOTTE, F., BARITEAU, F., SCHEID, J.P. and MAULEON, P. (1980). Use of progestagen (RU-2267) for oestrus control in gilts. *Int. Pig Vet. Soc., Copenhagen, Denmark*, p. 47.

MAULEON, P., MARTINAT-BOTTE, F. and SCHEID, J.P. (1979). Oestrus control in nulliparous gilts with a progestagen treatment (RU-2267). *J. Anim. Sci.* **49**, 317

MAYER, P. and SCHUTZE, E. (1977a). A new progestin (SA-45249) for cycle control in pigs. Communication I: Hormone activities and dosage. *Theriogenology* **8**, 357–366

MAYER, P. and SCHUTZE, E. (1977b). A new progestin (SA-45249) for cycle control in pigs. Communication IV: Analysis of hormone blood levels. *Theriogenology* **8**, 389–401

NEILL, J.D. and DAY, B.N. (1964). Relationship of developmental stage to regression of the corpus luteum in swine. *Endocrinology* **74**, 355–360

NELLOR, J.E. (1960). Control of estrus and ovulation in gilts by orally effective progestational compounds. *J. Anim. Sci.* **19**, 412–420

NELLOR, J.E., AHRENHOLD, J.E., FIRST, N.L. and HOEFER, J.A. (1961). Control of estrus and ovulation in gilts by orally effective progestational compounds. *J. Anim. Sci.* **20**, 22–30

O'REILLY, P.J., McCORMACK, R. and O'MAHONEY, K. (1979). Oestrus synchronization and fertility in gilts using a synthetic progestagen (allyl trenbolone) and inseminated with fresh stored or frozen semen. *Theriogenology* **12**, 131–137

PHILLIPO, M. (1968). Superovulation in the pig. *Adv. Reprod. Physiol.* **3**, 147–166

POLGE, C. (1965). Effective synchronization of oestrus in pigs after treatment with ICI compound 33828. *Vet. Rec.* **77**, 232–236

POLGE, C. (1966). Recent advances in controlled breeding of pigs. *Outlook on Agriculture* **5**, 44–48

POLGE, C., DAY, B.N. and GROVES, T.W. (1968). Synchronization of ovulation and artificial insemination in pigs. *Vet. Rec.* **83**, 136–142

POND, W.G., HANSEL, W., DUNN, J.A., BRATTON, R.W. and FOOTE, R.H. (1965). Estrous cycle synchronization and fertility in gilts fed progestational and estrogen compounds. *J. Anim. Sci.* **24**, 536–540

PURSEL, V.G., ELLIOTT, D.O., NEWMAN, C.W. and STAIGMILLER, R.B. (1981). Synchronization of estrus in gilts with allyl trenbolone: Fecundity after natural service and insemination with frozen semen. *J. Anim. Sci.* **52**, 130

RAMPACEK, G.B., SCHWARTZ, F.L., FELLOWS, R.E., ROBINSON, O.W. and ULBERG, L.C. (1976). Initiation of reproductive function and subsequent activity of the corpora lutea in prepubertal gilts. *J. Anim. Sci.* **42**, 881–887

RAY, D.E. and SEERLEY, R.W. (1966). Oestrus and ovarian morphology in gilts following treatment with orally effective steroids. *Nature, Lond.* **211**, 1102–1103

REDMER, D.A. and DAY, B.N. (1981). Ovarian activity and hormonal patterns in gilts fed allyl trenbolone. *J. Anim. Sci.* **53**, 1088–1094

REDMER, D.A., MEREDITH, S., BALL, G.D., YANKOWSKY, A. and DAY, B.N. (1979). Estrus and ovulation in gilts fed a synthetic progestogen (RU-2267). *J. Anim. Sci.* **49**, 116

SCHILLING, E. and CERNE, F. (1972). Induction and synchronization of oestrus in prepubertal gilts and anoestrus sows by PMS/HCG-compound. *Vet. Rec.* **91**, 471–474

SCHUTZE, E. and MAYER, P. (1977). A new progestin (SA-45249) for cycle control in pigs. Communication II: Duration of treatment and effectiveness as an estrous cycle synchronizer. *Theriogenology* **8**, 367–377

SEGAL, D.H. and BAKER, R.D. (1973). Maintenance of corpora lutea in prepubertal gilts. *J. Anim. Sci.* **37**, 762–767

SHAW, G.A., McDONALD, B.E. and BAKER, R.D. (1971). Fetal mortality in the prepubertal gilt. *Can. J. Anim. Sci.* **51**, 233–236

SOMA, J.A. and SPEER, V.C. (1975). Effects of pregnant mare serum and chlortetracycline on the reproductive efficiency of sows. *J. Anim. Sci.* **41**, 100–105

STEVENSON, J.S. and DAVIS, D.L. (1981). Estrous synchronization in gilts fed allyl trenbolone 14 or 18 days beginning 0 to 21 days post-estrus. *Am. Soc. Anim. Sci., 1981 Ann. Meeting* **53**(*Suppl.* **1**), 368

STRATMAN, F.W. and FIRST, N.L. (1965). Estrus inhibition gilts fed a dithiocarbamoylhydrazine (ICI 33828). *J. Anim. Sci.* **24**, 930

TANABE, T.Y., WARNICK, A.C., CASIDA, L.E. and GRUMMER, R.H. (1949). The effects of gonadotrophins administered to sows and gilts during different stages of the estrual cycle. *J. Anim. Sci.* **8**, 550

ULBERT, L.C., GRUMMER, R.H. and CASIDA, L.E. (1951). The effects of progesterone upon ovarian function in gilts. *J. Anim. Sci.* **10**, 665–671

VEENHUIZEN, E.L., WAGNER, J.F., WAITE, W.P. and TONKINSON, L. (1965). Estrous control in gilts treated sequentially with DES and CAP. *J. Anim. Sci.* **24**, 931

WAGNER, J.F. and SEERLEY, R.W. (1961). Synchronization of estrus in gilts with an orally active progestin. *J. Anim. Sci.* **20**, 980–981

WEBEL, S.K. (1976). Estrous control in swine with a progestagen. *J. Anim. Sci.* **42**, 1358

WEBEL, S.K. (1978). Control of ovulation in the pig. In *Control of Ovulation*, (D.G. Crighton, N.B. Haynes, G.R. Foxcroft and G.E. Lamming, Eds.), pp. 421–434. London, Butterworths

WEBEL, S.K. (1980). Control of the estrous cycle in the pig with allyl trenbolone. *Ninth Int. Congr. Anim. Reprod. A.I.* **1**, 114

WEBEL, S.K. and RIPPEL, R.H. (1975). Ovulation in the pig with releasing hormones. *J. Anim. Sci.* **41**, 385

WEBEL, S.K., PETERS, J.B. and ANDERSON, L.L. (1970). Control of estrus and ovulation in the pig by ICI 33828 and gonadotrophins. *J. Anim. Sci.* **30**, 791–794

WEBEL, S.K., SCHEID, J.P. and BOUFFAULT, J.C. (1980). Estrus control in pigs. *Int. Pig Vet. Soc., Copenhagen, Denmark,* p.49

ZEROBIN, K. (1977). A new progestin (SA-45429) for cycle control in pigs. Communication III; Confirmation of effectiveness in field trials. *Theriogenology* **8**, 379–388

11

THE GILT FOR BREEDING AND FOR MEAT

P.H. BROOKS
Seale-Hayne College, Newton Abbot, Devon, U.K.

The management of the replacement gilt and her successful integration into the breeding herd continue to present severe problems to both individual producers and the industry as a whole. Every year vast numbers of sows are culled, many before they might have been expected to reach optimum performance, to be replaced by gilts whose performance is generally so indifferent that at best they will only maintain herd performance and at worst may reduce it considerably. This chapter considers the impact of the gilt within the herd and attempts to indicate how some of the worst features of gilt performance may be ameliorated by modifying the management of this capricious animal.

The influence of gilts on herd productivity

The problem presented by gilts can be summed up as 'too many gilts producing too few piglets with too little predictability'. The high numbers of gilts in herds is a reflection of the culling rate. Recent surveys in Britain, the Netherlands and France estimated culling rates of 34.6%, 43% and 50% respectively (MLC, 1980a; 1980b; Kroes and Van Male, 1979; Dagorn and Aumaitre, 1979). Culling rate tends to increase as herd size increases (*Table 11.1*) and as lactation length is reduced (*Table 11.2*).

Table 11.1 RELATIONSHIP BETWEEN CULLING OF SOWS AND HERD SIZE

Herd size (sows)	*Culled sows* (%/annum)
49	29.4
50–99	33.3
100–149	36.5
150–249	38.1
250+	37.9
Average all herds	36.9

From Meat and Livestock Commission (1980b)

The increased annual culling rate with earlier weaning does not appear to result from sows having fewer litters but reflects the reduction in time spent lactating. For every sow that is culled a gilt must enter the herd and with culling rates of 30–50% this means that 15–25% of all litters are born

Table 11.2 RELATIONSHIP BETWEEN CULLING OF SOWS AND AGE AT WEANING

Age at weaning (days)	*Culled sows* (%/annum)	*Litters/sow/year*	*Average herd life* (years)	*Average no. litters/sow*
Below 19	42.2	2.3	2.37	5.45
19–25	36.5	2.2	2.74	6.03
26–32	39.5	2.1	2.53	5.32
33–39	35.1	2.1	2.85	5.98
39+	33.7	2.0	2.96	5.93
Average	36.9	2.2	2.71	5.96

From Meat and Livestock Commission (1980b)

Table 11.3 THE INFLUENCE OF PARITY ON TOTAL LITTER SIZE

Litter No.	*Total litter size*		*No. born alive*	
	A	*B*	*A*	*B*
1	9.9	9.7	9.3	9.2
2	10.7	10.7	10.2	10.2
3	11.6	11.2	10.8	10.6
4	11.5	11.3	10.6	10.8
5	12.0	11.4	11.1	10.7
6	11.5	11.4	10.3	10.8
7	11.7	11.6	10.6	10.8
8	11.4	11.4	10.4	10.7
9	11.3	11.9	10.2	10.9
10+	12.1	12.5	10.2	11.0
Unweighted sow mean	11.5	11.5	10.5	10.7
Mean % superiority of sows	16.2	18.6	12.9	16.3

A—MLC (1980)
B—Kroes and Van Male (1979).

Table 11.4 RELATIONSHIP BETWEEN NUMBER OF LITTERS/CULLED SOW AND REPRODUCTIVE PERFORMANCE

Mean litters/ culled sow	*Weaning to conception interval* (days)	*Mean pigs born/litter*	*Pigs weaned/ sow/year*
<3	19.9	10.17	15.5
3–3.99	18.3	10.58	16.5
4–4.99	17.4	10.58	16.7
5–5.99	17.4	10.69	17.1
≥6	17.1	10.79	17.6

From Dagorn and Aumaitre (1979)

to gilts. Unfortunately the litter productivity of the gilt is generally inferior to that of the sow (*Table 11.3*). Furthermore, the interval to service is generally longer following the first litter than following subsequent litters (Rasbech, 1969). As a consequence of these two factors average herd productivity tends to decline as the culling rate increases (or average herd age decreases). This was clearly demonstrated in the survey of Dagorn and Aumaitre (1979) who found that in herds where the average number of litters produced per culled sow was less than three, annual production was

Table 11.5 EFFECT OF CULLING RATE ON HERD PRODUCTIVITY

	Culling rate		
	Low	*Average*	*High*
Culling rate (%)	31.3	43.4	55.4
Litters per sow	6.56	4.55	3.42
Litters/sow/year	2.06	1.97	1.89
Weaners/sow/year	17.9	17.1	16.4
Cost/weaner (% of average group)	96.6	100.0	103.8
Labour income/sow[a] (% of average group)	114.6	100.0	85.2

[a]Labour income = all income minus all costs excluding labour.
From Kroes and Van Male (1979)

Table 11.6 THE INFLUENCE OF PARITY ON CULLING RATE

Litter No.	*Culling rate (%)*		
	Kroes and Van Male (1979)	*Dagorn and Aumaitre (1979)*	*MLC (1980)*
1	19.6	21.2	14.4
2	16.3	15.4	14.1
3	13.8	12.8	8.6
4	11.6	11.2	10.9
5	9.9	10.0	8.3
6	8.6	8.9	8.6
7	7.0	7.4	8.3
8	5.3	5.4	9.6
9	3.7	3.6	9.3
10+	4.2	4.1	7.8

15.5 pigs/sow, whereas in herds where culled sows had averaged six or more litters productivity was 17.6 pigs/sow/year (*Table 11.4*). As Kroes and Van Male (1979) have demonstrated, changes in culling rate can have an appreciable influence on both the productivity and profitability of a herd (*Table 11.5*).

Another feature of gilt performance which should give cause for concern is the high culling rate of gilts following weaning (*Table 11.6*). In the survey of Dagorn and Aumaitre (1979) the two most important reasons for culling gilts were reproductive failure (38% of cullings) and lameness (15%). Survey data gives little evidence of the extent to which producers discard primiparous animals for poor litter performance. Comments of producers suggest that this practice is still widespread despite evidence showing the low repeatability of litter size (Strang and King, 1970; Eikje, 1974; Bognor *et al.*, 1974). This being the case, there would seem to be very little justification for removing a primiparous female from the herd on the basis of a poor first litter when the size of her second litter has every chance of exceeding that of a gilt introduced as a replacement for her.

Mating age and productivity

The appropriate age at which to mate a gilt depends upon the criteria we choose to evaluate her performance and her contribution to overall herd productivity. Unfortunately inappropriate conclusions are often drawn because of the singlemindedness with which most producers adopt first litter performance as the sole measure of gilt productivity. There is ample evidence in the literature to demonstrate that the number of piglets born in the gilt litter increases with age at farrowing (e.g. Squiers, Dickerson and Mayer, 1952; Omtvedt, Stanislaw and Whatley, 1965; Milojić and Simović, 1968; Strang, 1970; Legault and Dagorn, 1973; Stanković *et al.*, 1973; Beremski and Germanova, 1974; MacPherson, Hovell and Jones, 1977). Unfortunately in all these studies chronological age has been confounded with sexual age (i.e. the number of heat periods experienced) so the relative importance of these two components has been obscured. Nevertheless experiments in which gilts have been subjected to comparable management but mated at different heat periods (and hence different ages) are valuable as they indicate the responses which may be anticipated if producers make conscious decisions to delay mating until later heat periods (*Table 11.7*).

Table 11.7 EFFECT OF NO. OF HEAT PERIODS AT MATING ON LITTER SIZE OF GILTS

Author	*No. of heat periods at mating*			*Increase in litter size per day delay in mating*[a]
	1	*2*	*3*	
Brooks and Cole (1973)	8.8	–	9.9	0.026
Pay and Davies (1973)	7.9	–	9.3	0.033
MacPherson *et al.* (1977)	8.4	9.8	10.4	0.062

[a] Assuming two 21-day oestrus cycles

Recently Bichard and Coates (1981, personal communication) studied the relationship between mating age and litter performance in large populations of purebred gilts and derived the following equations:

For Large White gilts, $y = 0.16x + 6.38$ (1)

For Landrace gilts, $y = 0.019x + 5.31$ (2)

where y = number born and x = age at effective service.

The increase in litter size for each day's delay in mating implied by these equations (0.016–0.019 piglets) is considerably lower than the rate of increase implied by the data in *Table 11.7* (0.026–0.062).

On the basis of this evidence there can be little doubt that a delay in mating will result in an increase in the size of the first litter. However, it must be questioned whether such an increase represents a real improvement in productivity of the gilt and more importantly whether it will improve overall herd productivity. The results of some theoretical calculations presented in *Table 11.8* suggest that the increase in litter size likely to be achieved by delaying mating is insufficient to make up for the time lost by keeping the gilt out of production for an extra 21 or 42 days. Indeed these calculations indicate that to produce a similar output per gilt housed per day, the average performance of gilts mated at the third oestrus would have to be comparable with that found for third litter sows (*Table 11.3*).

So far the effect of mating age has only been considered in terms of first parity performance. It is also important to consider whether there are any

Table 11.8 LITTER SIZE NEEDED FOR GILTS MATED AT SECOND AND THIRD HEAT TO MAINTAIN EQUIVALENCE WITH GILTS MATED AT PUBERTY

	Heat period at mating		
	First	*Second*	*Third*
Mating age (days)	190	211	232
Days from entry to herd to farrowing[a]	129	150	171
A: Predicted litter size using Bichard and Coates equation[b]	9.4	9.8	10.1
A: Litter size required to maintain daily production equivalent to gilts mated at puberty	–	11.0	12.5
B: Litter size after MacPherson *et al.* (1977)	8.4	9.8	10.4
B: Litter size required to maintain daily production equivalent to gilts mated at puberty	–	9.8	11.1

[a]Entry to herd assumed to be at 175 days of age; average interval to puberty 15 days.
[b]Equation for Large White gilts (see p.214).

Table 11.9 THE EFFECT OF MATING AT PUBERTY, SECOND OR THIRD HEAT ON THE PERFORMANCE OF SOWS OVER THREE PARITIES

	Heat mated		
	1	*2*	*3*
Pigs weaned (1st litter)	7.8	8.3	8.6
Pigs weaned (litters 1–3)	26.5	26.4	26.9
Total weight weaner produced litters 1–3 (kg)	280.7	282.4	284.8
Weight at mating (kg)			
Parity 1	88.1	98.2	115.1
Parity 4	165.5	168.9	165.8

From MacPherson, Hovell and Jones (1977)

Table 11.10 EFFECT OF MATING AGE ON THE PERFORMANCE OF SOWS OVER FIVE PARITIES

	Early mated	*Conventionally mated*
Mean age at mating (days)	198	237
Pigs born (1st litter)	8.6	9.5
Pigs born (litters 1–5)	53.7	53.8
Pigs born alive (litters 1–5)	51.6	50.4
Pigs weaned	42.6	43.8
Mean piglet birth weight (kg)	1.20	1.13
Mean piglet weaning weight (kg)	9.16	9.13
Sow food/kg weaner (kg)	6.1	6.5

From Brooks and Smith (1977)

long-term effects of mating at different ages. Although MacPherson, Hovell and Jones (1977) found considerable differences in first litter performance for gilts mated at different heat periods their performance over three litters was almost identical (*Table 11.9*).

Brooks and Smith (1977) induced puberty at different ages by the use of boar stimuli then mated gilts at their second heat period. Gilts mated at an average age of 198 days produced smaller first litters than gilts mated at 237 days but over five litters the number of piglets born differed by only 0.2% (*Table 11.10*). However, the gilts mated at a younger age consumed 6.2% less food/kg weaner produced. These data are consistent with the survey data of Legault and Dagorn (1973) who found that neither the number of litters produced nor the herd life of the sow were affected by age at first farrowing and that for each day that mating age was increased, annual sow productivity was reduced by 0.02–0.03 pigs. They also noted a slight increase in farrowing interval as mating age increased.

It must be concluded from these results that little is to be gained by delaying mating once the gilt has reached puberty. However in practice it may be prudent to delay mating of certain gilts in order to maintain continuity of throughput in the unit. In such cases the 'gilt pool' approach suggested by Brooks (1978) has much to commend it.

Nutrition of the breeding gilt

Over the last decade both the management of the gilt and her genetic constitution has changed appreciably. The combined effects of genetic change and earlier mating mean that gilts now start their breeding lives at lighter weights and with smaller fat deposits than they did a decade ago. It is important to consider whether this should influence the nutritional management of the animal. At present there seems to be little reason to revise recommendations for the nutrition of the gilt around the time of mating and in early pregnancy. The relationships between nutrition and reproduction probably do not differ significantly from those outlined by Anderson and Melampy (1972), Brooks and Cooper (1972) and Brooks and Cole (1974). However there would appear to be a need to review the nutrition of the gilt during the growing period and throughout her reproductive life.

One of the consequences of mating gilts at younger ages is that they generally weigh less. Initially it was thought that this might inhibit their growth and reduce their ultimate liveweight. Such fears (or hopes) have been shown to be groundless. MacPherson, Hovell and Jones (1977) found that initial differences in liveweight had disappeared by the end of the third parity (*Table 11.9*) and Brooks and Smith (1980) found that early mated gilts caught up with initially heavier control gilts by the middle of the second pregnancy (*Figure 11.1*) and had a similar pattern of weight change thereafter. Of rather more significance may be the nature of gains and losses in modern gilts. Current feeding recommendations are based on nutritional studies conducted mainly in the late sixties and have as their underlying premise the depletion of fat reserves during the first two or three parities. Whittemore, Franklin and Pearce (1980) have rightly pointed out that such regimes may not be appropriate to modern gilts starting their breeding life with limited fat reserves. In their studies of gilts on the MLC Commercial Product Evaluation Scheme, they found that although gilts made a net liveweight gain of 22 kg over their first five parities they actually lost 7.4 mm of backfat (which they estimated to be equivalent to 8 kg fatty tissue), so that at the end of the second parity they contained only 5–12 kg of body fat. Clearly if fat losses of a similar magnitude occurred in succeeding parities most of the sows would have dissipated all their fat reserves by the end of their fourth parity.

There are two ways in which this problem could be overcome; either the fat reserves of the gilt should be increased prior to first farrowing to

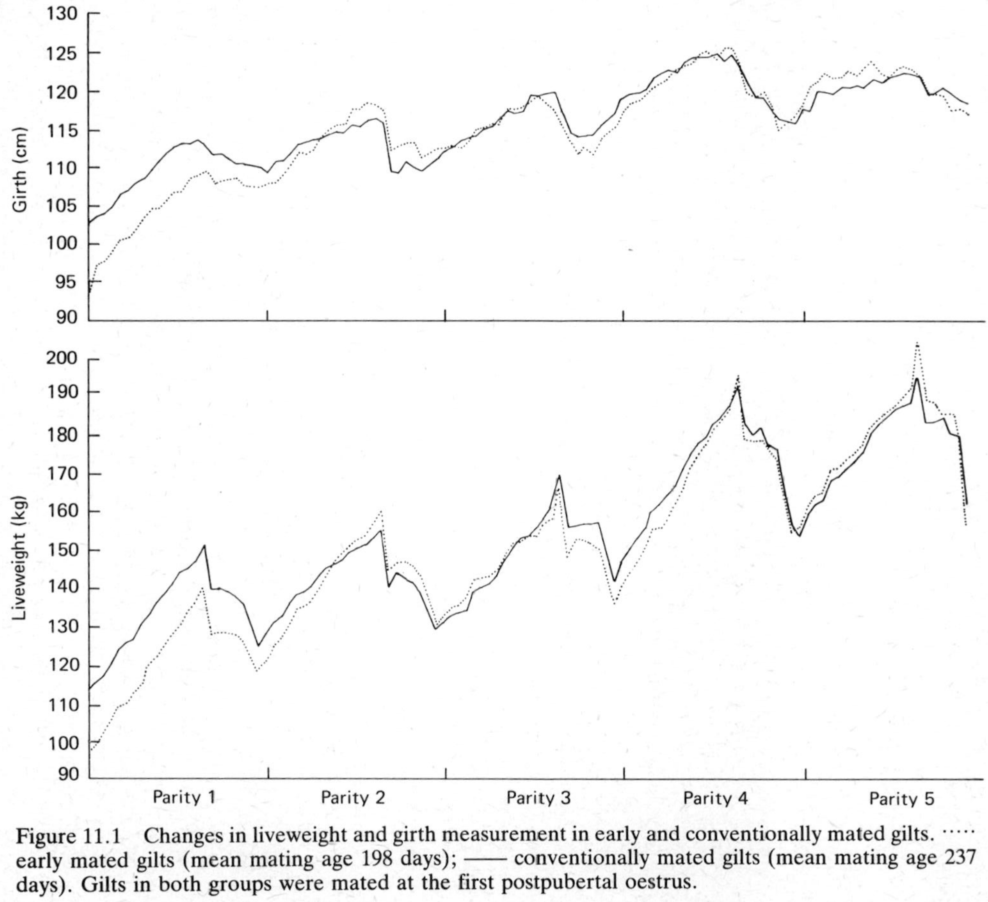

Figure 11.1 Changes in liveweight and girth measurement in early and conventionally mated gilts. ····· early mated gilts (mean mating age 198 days); —— conventionally mated gilts (mean mating age 237 days). Gilts in both groups were mated at the first postpubertal oestrus.

provide fat for later depletion, or the feeding regime of sows should be revised in order to avoid fat depletion. It is doubtful whether increasing fat reserves prior to first farrowing is a practicable solution, for two reasons. First, if the increase in fat intake is to be achieved by a higher feed intake in pregnancy, this is likely to reduce lactation feed intake as it has been clearly demonstrated that increased food intake in pregnancy leads to reduced voluntary feed intake in lactation (Dean and Tribble, 1961; Salmon-Legagneur and Rerat, 1962; Baker *et al.*, 1969). In the experiment of Baker *et al.* (1969) this resulted in a linear decrease in lactation weight gain with increase in gestation feed consumption (*Table 11.11*). Even when

Table 11.11 EFFECT OF GESTATION FEED LEVEL ON LACTATION FEED INTAKE AND WEIGHT CHANGE

Daily feed intake in gestation	*Lactation diet intake* (kg)	*Gestation weight gain* (kg)	*Lactation weight gain* (kg)
0.9	89.4	5.9	6.1
1.4	90.3	30.3	0.9
1.9	90.5	51.2	–4.4
2.4	81.1	62.8	–7.6
3.0	71.7	74.4	–8.5

From Baker *et al.* (1969)

pigs receive the same gestation allowances there is a tendency for gilts which farrow at heavier weights to lose more weight in the following lactation as shown in *Figure 11.1* (Brooks and Cole, 1973; Brooks and Smith, 1980). The effect in these two trials was not an effect on appetite as the gilts were fed to scale; nor could it be attributed totally to a higher maintenance requirement in the heavier gilts. In the trial of Brooks and Smith (1980) one result of this phenomenon was that heavier (and fatter) conventionally mated gilts lost more fat during the first lactation and between weaning and remating so that they started their second parity with similar fat deposits to the early mated animals (*Table 11.12*). It is interesting that although these animals continued to gain weight in subsequent parities (*Figure 11.1*), fat thickness appeared to stabilize after the second lactation.

From these results it would appear that a more appropriate approach to the problem of maintaining fat reserves may be to prevent their depletion by differential feeding of the sow from first farrowing onwards. Problems are frequently encountered when attempting to rebreed gilts following

Table 11.12 CHANGES IN MIDBACK FAT DEPTH (mm)[a] FOR CONVENTIONAL AND EARLY MATED GILTS

	Early mated	*Conventionally mated*[b]
Post partum (1st litters)	18.8	25.0
Weaning (1st litter)	16.8	18.2
Remating (2nd parity)	18.2	18.0
3rd parity	14.6	15.3
4th parity	14.0	15.1

From Brooks and Smith, unpublished data (1980)
[a]Minimum fat depth over the spine at the last rib
[b]For details of animals and management see Brooks and Smith (1980)

weaning. This problem has been considered in earlier papers (Brooks and Cole, 1974; Brooks, 1978). It has been suggested that these difficulties might be induced by the large weight losses which gilts often exhibit during their first lactation and that the problem might be ameliorated by the provision of generous feed allowances after weaning. To the rebreeding problem has now been added the difficulty that some modern hybrid strains tend to produce smaller litters in their second litters than in their first. A re-examination of the data presented by Whittemore, Franklin and Pearce (1980) demonstrates this point. Of the nine breed groups examined six showed an increase in litter size from first to second litter, one showed no change and two showed a decrease. If the percentage change in litter size is plotted against the change in backfat thickness between weaning and

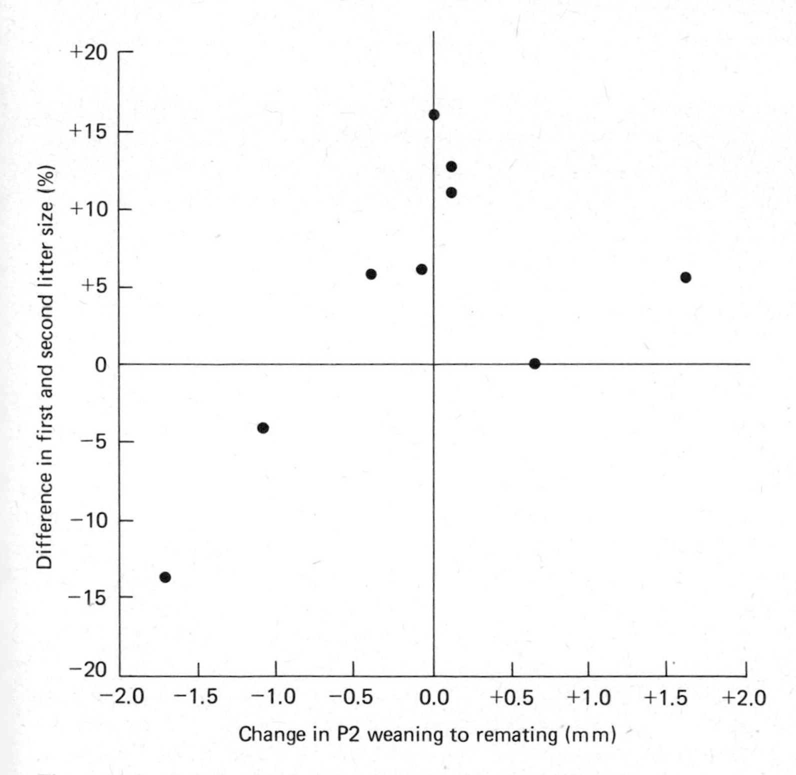

Figure 11.2 Relationship between backfat change and litter size

remating (*Figure 11.2*), there is an indication that the dynamic changes in body condition may be influencing subsequent litter size. This suggests that some gilts having become catabolic during lactation do not immediately revert to an anabolic state after weaning and as a consequence have reduced ovulation rates. This may well explain the responses to post-weaning nutrition reported in earlier papers (Brooks and Cole, 1974; Brooks, 1978). Work in progress at the moment suggests that it is not uncommon for fat depletion to continue after weaning and that on some nutritional regimes repletion may not be apparent until 30–40 days after weaning (Hardy, 1981, personal communication). These dynamic changes may well influence ovulation rate at the post-weaning oestrus. Love (1979) has shown that ovulation rate increases from 11.7 at the first post-weaning oestrus to 12.3 at the second. These differences may provide a partial explanation for the increase in litter size with longer weaning to mating intervals reported by Love (1979) and confirmed in our own animals (Brooks, 1980, unpublished data) although by no means all the animals mated 12+ days after weaning were experiencing their second oestrus (*Table 11.13*).

Table 11.13 EFFECT OF WEANING TO REMATING INTERVAL ON SUBSEQUENT LITTER SIZE

	Mated within 12 days of weaning	*Mated 12+ days after weaning*
Love (1979)	9.0±2.5	10.4±2.1
Brooks (1980)	9.34±2.26	10.54±1.56

When taken together all these elements tend to indicate that feed regimes for gilts should pay particular attention to the short term and that problems arising from limited fat reserves should be solved by the development of more appropriate feeding regimes for gilts from first farrowing onwards.

The gilt as a meat animal (once-bred gilt)

The possibility of producing a litter of piglets, which could be fattened for slaughter, from gilts that were themselves destined for slaughter has been considered by a number of workers in recent years (Brooks and Cole, 1973; Kotarbinska and Kielanowski, 1973; Pay and Davies, 1973; Brooks, Cole and Jennings, 1975; MacPherson, Hovell and Jones, 1977; Hovell *et al.*, 1977a,b; Brooks and Smith, 1977; Friend *et al.*, 1979). The original impetus for these recent studies was the realization that gilts could be stimulated using the 'boar effect' to achieve puberty at younger ages and lower liveweights (see Chapter 6). It was reasoned that if acceptable litters could be produced from such gilts they might produce a litter and still yield a carcass in the weight range normally associated with heavy manufacturing pigs (77.5–100 kg deadweight). It was further considered possible that pregnancy anabolism might result in the more efficient conversion of food into carcass gains.

Mean litter size of gilts mated at puberty ranged from 7.1–10.5 in the papers listed above, indicating that piglet production from gilts mated at low liveweights is satisfactory. It also proved possible to produce carcasses within the required weight range from a variety of different nutritional and management regimes. It is interesting that in the studies made by Hovell *et al.* (1977a,b) there was no lasting effect of pregnancy on maternal growth. Although there were apparent increases in gain during gestation they did not persist beyond the first nine days after farrowing. Despite this Brooks and Cole (1973), MacPherson, Hovell and Jones (1977) and Hovell *et al.* (1977a,b) all found the efficiency of the once-bred gilt to be higher than that of unmated animals when allowance was made for litter production. MacPherson, Hovell and Jones (1977), after allowing for piglet production calculated that bred gilts had a food conversion ratio of 2.9:1 compared with 3.6:1 for unmated gilts while Hovell *et al.* (1977) calculated effective food conversion ratios (after allowance for the litter) ranging from 1.8–3.4:1 for bred gilts compared with 4.7–5.3:1 for unmated animals. Brooks and Cole (1973) used a somewhat different calculation and estimated the food required to produce a weaner pig additional to the requirement for fattening an unmated female to a similar weight. The value was 22.1 kg/weaner with early weaning and 37.5 kg/weaner with conventionally weaned pigs, implying a considerable reduction in the food required for piglet production.

Different authors have used different ways of assessing the carcasses from bred gilts. Hovell *et al.* (1977a) found little difference between the carcass composition of bred gilts slaughtered at nine days post-partum and unmated gilts of the same weight. They also concluded that although there was little difference between the groups in terms of total protein deposited, mated gilts deposited less protein than unmated gilts when corrected to a constant level of fat deposition (Hovell *et al.*, 1977b). Brooks, Cole and Jennings (1975) found that carcasses from bred gilts were significantly less fat than unmated controls and that yield of primal joints was only 0.62% less. The extent to which trimming of mammary tissue is required varies according to the views of the wholesaler and the interval between weaning and slaughter. Brooks and Smith (1977) found that removal of unregressed or damaged mammary tissue resulted in a loss of 4.9% of the total weight of the middle.

The value of the carcass depends upon its classification by the purchaser and this in turn is dependent upon the purpose to which the carcass will be put. The results obtained by Brooks and Smith (1977) suggested that the bred gilt is unlikely to be considered as a direct substitute for traditional heavy manufacturing pigs. There are two reasons for this. First the streak block has to be trimmed reducing its yield and altering its shape. In addition the streak block does not produce bacon of acceptable quality, having poor colour and texture. The back block cures satisfactorily but the rashers produced tend to have fat separation and are subject to moisture loss which make them unattractive when packaged. Despite this unsuitability for curing there seem to be few discernible or significant differences in acceptability of cooked fresh meat from bred gilts (Friend *et al.*, 1979).

These results suggest that unless novel manufacturing approaches are developed, for which the once-bred gilt is particularly suited and hence able to command a price premium, the gilt will continue to be classified as a 'sow' for payment purposes.

It appears that in terms of biological efficiency a production system based on once-bred gilts could prove more efficient than conventional systems of production. However, to achieve this improved biological efficiency gilt management has to be of a very high standard. Failure to meet output targets can quickly erode any biological advantage and will have an adverse effect upon the productivity and profitability of the dependent fattening enterprise. Management problems would increase still further on a unit in which all piglets were derived from once-bred gilts as some form of criss-cross breeding system would be needed to maintain the genetic stability of the herd.

Finally, a question which cannot yet be answered is whether the rapid turnover of breeding stock would reduce herd immunity levels and hence increase health problems. If experience in breeding units is any guide this could be a factor of some significance.

References

ANDERSON, L.L. and MELAMPY, R.M. (1972). Factors affecting ovulation rate in the pig. In *Pig Production*, (D.J.A. Cole, Ed.), pp. 329–366. London, Butterworths

BAKER, D.H., BECKER, D.E., NORTON, H.W., SASSE, C.E., JENSEN, A.H. and HARMON, B.G. (1969). Reproductive performance and progeny development in swine as influenced by feed intake during pregnancy. *J. Nutr.* **97**, 489–495

BEREMSKI, S. and GERMANOVA, L. (1974). The optimum age and body weight at first mating and reproductive performance of sows. I. The effect of age. *Zhivotnov''dni Nauki* **11**, 57–63

BOGNOR, H., IVANOUSKI, R., LEDERER, J. and AVERDUNK, G. (1974). Relationship of litter size in successive litters of breeding sows. *Zuchthygiene* **9**, 77

BROOKS, P.H. (1978). Early sexual maturity and mating of gilts. *ADAS Quart. Rev.* **30**, 139–152

BROOKS, P.H. (1979). Rearing gilts. *Proc. MLC/Farmers Weekly Conf., Cambridge*, February 1979

BROOKS, P.H. and COLE, D.J.A. (1973). Meat production from pigs which have farrowed. 1. Reproductive performance and food conversion efficiency. *Anim. Prod.* **17**, 305–315

BROOKS, P.H. and COLE, D.J.A. (1974). The effect of nutrition during the growing period and the oestrous cycle on the reproductive performance of the pig. *Livest. Prod. Sci.* **1**, 7–20

BROOKS, P.H. and COOPER, K.J. (1972). Short term nutrition and litter size. In *Pig Production*, (D.J.A. Cole, Ed.), pp. 385–398. London, Butterworths

BROOKS, P.H. and SMITH, D.A. (1977). Meat production from pigs which have farrowed. 3. The effect of weaning to slaughter interval on food utilization and carcass quality. *Anim. Prod.* **25**, 247–254

BROOKS, P.H. and SMITH, D.A. (1980). The effect of mating age on the reproductive performance, food utilization and liveweight change of the female pig. *Livest. Prod. Sci.* **7**, 67–78

BROOKS, P.H., COLE, D.J.A. and JENNINGS, W.J.N. (1975). Meat production from gilts which have farrowed. 2. Carcass characteristics. *Anim. Prod.* **20**, 123–131

DAGORN, J. and AUMAITRE, A. (1979). Sow culling: reasons for and effect on productivity. *Livest. Prod. Sci.* **6**, 167–177

DEAN, B.T. and TRIBBLE, L.F. (1961). Reproductive performance of swine fed different planes of energy during gestation. *Res. Bull. Mo. agric. Exp. Stn* No. 774

EIKJE, E.D. (1974). Phenotypic and genetic parameters of litter size in pigs. *Meld. Norg. LandbrHøisk.* **53**, 23pp.

FRIEND, D.W., LARMOND, E., WOLYNETZ, M.S. and PRICE, K.R. (1979). Piglet and pork production from gilts bred at puberty: Chemical composition of the carcass and assessment of meat quality. *J. Anim. Sci.* **49**, 330–341

HOVELL, F.D.DeB., MACPHERSON, R.M., CROFTS, R.M.J. and PENNIE, K. (1977). The effect of energy intake and mating weight on growth, carcass yield and litter size of female pigs. *Anim. Prod.* **25**, 233–245

HOVELL, F.D.DeB., MACPHERSON, R.M., CROFTS, R.M.J. and SMART, R.I. (1977). The effect of pregnancy, energy intake and mating weight on protein deposition and energy retention of female pigs. *Anim. Prod.* **25**, 281–290

KOTARBINSKA, M. and KIELANOWSKI, J. (1973). A note on meat production from pigs slaughtered after first weaning a litter. *Anim. Prod.* **17**, 317–320

KROES, Y. and VAN MALE, J.P. (1979). Reproductive lifetime of sows in relation to economy of production. *Livest. Prod. Sci* **6**, 179–183

LEGAULT, C. and DAGORN, J. (1973). Incidence de l'âge à la première mise-bas sur la productivité de la truie. *Journées Rech. Porcine en France* 1973, pp. 227–237. Paris, L'Institut Technique du Porc

LOVE, R.J. (1900). Reproductive performance of first parity sows. *Vet. Rec.* **104**, 238–240

MACPHERSON, R.M., HOVELL, F.D.DeB. and JONES, A.S. (1977). Performance of sows mated at puberty or second or third oestrus and carcass assessment of once-bred gilts. *Anim. Prod.* **24**, 333–342

MEAT AND LIVESTOCK COMMISSION (1980a). Newsletter No. 14, March 1980

MEAT AND LIVESTOCK COMMISSION (1980b). Commercial Pig Production Yearbook 1979

MILOJIĆ, M. and SIMOVIĆ, B. (1968). The age of gilts at first farrowing and their fertility. *VI Congr. Reprod. A.I., Paris,* Resumés, 336

OMTVEDT, I.T., STANISLAW, C.M. and WHATLEY, J.A. (1965). Relationship of gestation length, age and weight at breeding and gestation gain to sow productivity at farrowing. *J. Anim. Sci.* **24**, 531–535

PAY, M.G. and DAVIES, T.E. (1973). Growth, food consumption and litter production of female pigs mated at puberty and at low body weights. *Anim. Prod.* **17**, 85–91

RASBECH, N.O. (1969). A review of causes of reproductive failure in swine. *Br. vet. J.* **125**, 599–616

SALMON-LEGAGNEUR, E. and RERAT, A. (1962). Nutrition of the sow during pregnancy. In *Nutrition of Pigs and Poultry,* (J.T. Morgan and D. Lewis, Eds.), pp. 207–237. London, Butterworths

SQUIERS, C.D., DICKERSON, G.E. and MAYER, D.T. (1952). Influence of inbreeding age and growth rate of sows on sexual maturity, rate of ovulation, fertilization and embryo survival. *Res. Bull. Mo. agric. Exp. Stn* No. 494, 40pp.

STANKOVIĆ, M., ZALETEL, I., STANKOUVIC, J. and GROZDANIC, G. (1973). Effect of age at first mating on longevity and lifetime performance of Swedish Large White sows. *Arh. poljopr. Nauke Teh.* **26**, 33–44

STRANG, G.S. (1970). Litter productivity in Large White pigs. 1. The relative importance of some sources of variation. *Anim. Prod.* **12**, 285–333

STRANG, G.S. and KING, J.W.B. (1970). Litter productivity in Large White pigs. 2. Heritability and repeatability estimates. *Anim. Prod.* **12**, 235–243

WHITTEMORE, C.T., FRANKLIN, M.F. and PEARCE, B.S. (1980). Fat changes in breeding sows. *Anim. Prod.* **31**, 183–190

III

PREGNANCY

12

THE ESTABLISHMENT AND MAINTENANCE OF PREGNANCY

F.W. BAZER, R.D. GEISERT, W.W. THATCHER† and R.M. ROBERTS‡
Departments of Animal Science, Dairy Science†, and Biochemistry and Molecular Biology‡, University of Florida, USA

In many submammalian species considerable energy is expended by the female to produce liver and/or reproductive tract secretions which are incorporated into ova prior to ovulation. Fertilization of these ova may or may not occur. Females of these submammalian species, therefore, expend energy for the reproductive process regardless of whether offspring result.

In mammalian species, such as the pig, restrictions are placed upon the extent to which energy is expended for support of endometrial secretory activity. In the absence of fertilized ova and blastocysts developing in synchrony with the uterine endometrium, the period of endometrial secretory activity is limited primarily to the mid to late luteal phase of the oestrous cycle. In the pig, endometrial secretory activity is dependent upon maintenance of functional corpora lutea (CL). The lifespan of the CL, in turn, is limited during the oestrous cycle by endometrial production of a uterine luteolytic factor which is assumed to be prostaglandin $F_{2\alpha}$ ($PGF_{2\alpha}$). Release of $PGF_{2\alpha}$ from non-pregnant uterine endometrium results in morphological regression of CL, progesterone secretion ceases and, therefore, progesterone-dependent endometrial secretory activity is terminated in the late luteal phase of the oestrous cycle.

In pigs having normally developing blastocysts, endometrial–blastocyst interactions begin to occur by day 11 of pregnancy. Blastocysts signal their presence, presumably through oestrogen production, which results in CL maintenance and continued endometrial development and secretory activity. Species, such as the pig, that have central or fusion implantation (Schlafke and Enders, 1975) appear to depend upon endometrial histotroph (Amoroso, 1952) for a major portion of pregnancy.

Endometrial secretions of the pig are assumed to contain both a luteolytic factor ($PGF_{2\alpha}$) and embryotrophic (histotroph) factors. Blastocysts produce oestrogen which is believed to initiate events which result in CL maintenance, i.e. oestrogen acts as a luteostatic factor. This chapter will discuss evidence for the theory that oestrogens, of blastocyst origin, allow for continued secretion of $PGF_{2\alpha}$ and other components of endometrial histotroph into the uterine lumen, i.e. in an exocrine direction, which prevents their release toward the uterine vascular bed, i.e. in an endocrine direction. This mechanism appears to be essential for the establishment and maintenance of pregnancy in pigs.

The uterus and corpus luteum maintenance

Sexually mature female pigs (*Sus scrofa domesticus*) have recurring oestrous cycles of 18 to 21 days throughout the year. Behavioural oestrus lasts for 24–72 hours and ovulation occurs 36–42 hours after onset of oestrus. Corpora lutea are well formed by day 4 or 5 of the oestrous cycle and progesterone secretions increase from that time to maximum production between days 12 and 14 of the oestrous cycle (Guthrie, Henricks and Handlin, 1972). Luteal regression begins on about day 15 in non-pregnant females and plasma progesterone concentrations decline rapidly to basal levels (1 ng/ml or less) by days 17 to 18, leading to recurrent oestrous cycles unless interrupted by pregnancy or events leading to endocrine dysfunction.

The pig uterine endometrium has been identified as the source of a substance (uterine luteolysin) which causes morphological regression of CL and cessation of progesterone secretion. Loeb (1923) first noted that hysterectomy of guinea pigs during the luteal phase of the oestrous cycle allowed prolonged CL maintenance. This was later demonstrated in pigs by Spies *et al.* (1958) and du Mesnil du Buisson and Dauzier (1959). Anderson, Butcher and Melampy (1961) found that bilateral hysterectomy in the early to mid luteal phase of the oestrous cycle resulted in CL maintenance for periods equal to or longer than the 114 days of a normal pregnancy. Introduction of corrosive and toxic agents into the uterine lumen resulting in destruction of the endometrial epithelium or congenital absence of the uterine endometrium leads to prolonged CL maintenance (see review by Anderson, Bland and Melampy, 1969).

The nature of the pig uterine endometrial luteolysin has been and is subject to some controversy. Schomberg (1967) initially reported that a high molecular weight component of pig uterine flushings had a luteolytic effect on cultured granulosa cells, but this effect was later shown to be due to a cytolytic factor (Schomberg, 1969).

Pharriss and Wyngarden (1969) reported that $PGF_{2\alpha}$ was luteolytic in rats and, subsequently, it was evaluated as a luteolytic agent in swine. Diehl and Day (1973) found 2–5 mg $PGF_{2\alpha}$ to be ineffective in shortening the oestrous cycle when administered intraluminally on day 10 or intramuscularly on day 12 of the cycle. Hallford *et al.* (1974) injected gilts intramuscularly with 20 mg $PGF_{2\alpha}$ at 12 hour intervals on days 4 and 5 of the cycle (total dose of 80 mg $PGF_{2\alpha}$) with no effect on oestrous cycle length. However, when the same $PGF_{2\alpha}$ treatment protocol was used on days 12 and 13, oestrous cycle length was reduced in treated (17.2 days) versus control (18.6 days) gilts. Douglas and Ginther (1975) reported that intramuscular injections of 10 or 20 mg $PGF_{2\alpha}$ on day 12 of the oestrous cycle resulted in reduced CL weight, but neither oestrus nor ovulation had occurred by day 16 of the cycle. In contrast to these results, Diehl and Day (1974) determined that an intramuscular injection of 5 mg $PGF_{2\alpha}$ in early pregnant (days 25–30) gilts resulted in a marked decline in plasma progesterone concentrations within 12 hours, followed by abortion and expression of behavioural oestrus within 72 hours.

At this point, it seems necessary to digress from the discussion of $PGF_{2\alpha}$

so that results presented in previous and subsequent paragraphs can be considered relative to previous work by du Mesnil du Buisson and Leglise (1963). They reported that CL development in pigs hypophysectomized on the first day of oestrus was apparently normal up to day 12 of the oestrous cycle, but CL were undergoing regression by day 16. It was concluded that pig CL require only initial pituitary support, at or about the time of onset of oestrus, for formation and function for the first 12 days of the oestrous cycle. Later studies indicated that CL maintenance could be extended beyond day 12 in hysterectomized pigs which were later hypophysectomized and given either bovine luteinizing hormone (LH, 5 mg/day), human chorionic gonadotrophin (1000 iu/day) or 250 mg/day of an acid–acetone extract of ovine pituitary (du Mesnil du Buisson and Leglise, 1963; du Mesnil du Buisson, 1966; Anderson *et al.*, 1965). The gonadotrophin preparations were not effective in gilts which had intact uteri which indicated that the uterine luteolytic factor could 'override' pituitary gonadotrophins after day 12 of the oestrous cycle.

Since pig CL appear to be 'autonomous' for the first 12 days of the oestrous cycle, failure of exogenous $PGF_{2\alpha}$ to exert a marked effect on oestrous cycle length may be due to refractoriness of CL until day 12. If so, pig CL only become susceptible to exogenous $PGF_{2\alpha}$ at about the time they would normally regress and the only effect of exogenous $PGF_{2\alpha}$ would be a slight reduction in length of the oestrous cycle. This conclusion is supported by results of Diehl and Day (1973), Hallford *et al.* (1974) and Douglas and Ginther (1975), which were discussed previously. Several other studies, as discussed later (p.231), also support this conclusion.

Caldwell *et al.* (1969) reported that CL induced with human chorionic gonadotrophin on days 6, 8, 10 and 16 of the oestrous cycle remained functional for an average of 12.5, 14.5, 15.0 and 19.3 days, respectively, from the day they were induced. Therefore, only CL induced on day 6 were old enough to undergo regression at about the same time as the natural CL. Moeljono, Bazer and Thatcher (1976) treated hysterectomized gilts with intramuscular $PGF_{2\alpha}$ (10 mg at 08.00 hours and 10 mg at 20.00 hours) on either day 8, 11, 14 or 17 after onset of oestrus. None of the six gilts treated on either day 8 or day 11 exhibited oestrus for at least 60 days thereafter. Two of three gilts treated on day 14 and all three of the gilts treated on day 17 exhibited oestrus at 116.0±9.8 hours ($\bar{x}$±S.E.M.) post-treatment. Krzymowski *et al.* (1976) infused a total of 2 mg of $PGF_{2\alpha}$ into the anterior uterine vein of sows over a 10 hour period on either day 6, 8, 10, 12, 14 or 15 of the oestrous cycle. The $PGF_{2\alpha}$ had no effect on CL weight or plasma progesterone concentrations when given on days 6, 8 or 10, but luteolysis was initiated in sows treated on either day 12, 14 or 15. These data clearly support the concept that pig CL are refractory to the luteolytic effect of $PGF_{2\alpha}$ up to day 12 of the oestrous cycle. Henderson and McNatty (1975) suggested that pig CL may remain refractory to $PGF_{2\alpha}$ until LH begins to dissociate from luteal cell membrane receptors. At this time, presumably on day 12, conformational changes within the luteal cell membrane may facilitate $PGF_{2\alpha}$ binding. The $PGF_{2\alpha}$, in turn, is then proposed to alter the adenyl cyclase system to inhibit progesterone secretion and activate lysosomal enzymes to cause morphological regression of the luteal cells.

IS PROSTAGLANDIN F_2 THE UTERINE LUTEOLYSIN IN PIGS?

Moeljono, Bazer and Thatcher (1976) administered $PGF_{2\alpha}$ intramuscularly to hysterectomized gilts on either day 17 or 38 post oestrus and reported that all gilts expressed oestrus at an average of 88.0±13.5 hours ($\bar{x}$±S.E.M.) after treatment. Saline treated controls did not exhibit oestrus for at least 60 days post treatment. In a subsequent experiment, two gilts were bilaterally hysterectomized and the saphenous artery catheterized on day 7 after the onset of oestrus. On day 17 they received $PGF_{2\alpha}$ (10 mg intramuscularly at 08.00 hours and 10 mg at 20.00 hours). Plasma progesterone concentrations decreased precipitously after the first injection and gilts exhibited oestrus between 90 and 110 hours after the first $PGF_{2\alpha}$ injection. An additional gilt was bilaterally hysterectomized and the CL marked with Indian ink. Treatment with $PGF_{2\alpha}$, as before, resulted in oestrus at about 92 hours after the first $PGF_{2\alpha}$ injection. On day 4 of the treatment-induced oestrous cycle, regression of CL to corpora albicantia and presence of newly formed CL were confirmed at laparotomy. These data indicated that $PGF_{2\alpha}$ is, by definition, luteolytic in swine, i.e. it causes morphological regression of CL and cessation of progesterone secretion followed by onset of behavioural oestrus. The luteolytic effect of exogenous $PGF_{2\alpha}$ has been demonstrated in hysterectomized gilts (Moeljono, Bazer and Thatcher, 1976), gilts in which CL maintenance has been induced with oestrogen treatment (Kraeling, Barb and Davis, 1975) and pregnant gilts (Diehl and Day, 1974).

Endometrial extracts from days 13 to 17 of the oestrous cycle and day 19 of pregnancy exert a luteolytic effect in unilaterally pregnant gilts having induced CL (Christenson and Day, 1972). Patek and Watson (1976) reported that PGF production *in vitro* was greater for endometrial tissues from the mid and late luteal phase, compared with tissue obtained earlier in the oestrous cycle, and that this production could be blocked by indomethacin. Watson and Patek (1979) later reported similar conclusions in that *in vitro* $PGF_{2\alpha}$ production was higher for uterine endometrium taken from the late luteal phase of the oestrous cycle than for endometrium obtained during the mid luteal stage of the cycle and early pregnancy (days 16–22). However, $PGF_{2\alpha}$ production was similar for endometrium from the mid luteal stage of the oestrous cycle and early pregnancy. Guthrie, Rexroad and Bolt (1978) have also demonstrated *in vitro* PGF production by pig endometrium representing days 7–8 and 15–16 of the oestrous cycle, but they did not detect a difference in PGF production between the two stages. Prostaglandin F is produced *in vitro* by pig endometrium taken from either day 8, 12, 14, 16 or 18 of the oestrous cycle (Guthrie and Rexroad, 1980); its production increased from days 8 to 16 and then decreased slightly on day 18, but only production values for days 16 and 18 were significantly greater than those for days 8, 12 and 14.

In support of these *in vitro* studies, other research has been directed toward determining concentrations of immunoreactive PGF in utero-ovarian vein plasma. Gleeson, Thorburn and Cox (1974), Moeljono *et al.* (1977), Frank *et al.* (1977) and Killian, Davis and Day (1976) all concluded that utero-ovarian vein plasma PGF concentrations were significantly greater during the period of expected luteolysis in non-pregnant gilts.

Halton and First (unpublished data) reported that intrauterine infusion of indomethacin (an inhibitor of $PGF_{2\alpha}$ synthesis) resulted in a prolonged luteal phase of the oestrous cycle. This effect of indomethacin could be overcome by intramuscular injection of $PGF_{2\alpha}$, which indicated that treatment had no effect on CL sensitivity to the luteolytic effect of exogenous $PGF_{2\alpha}$.

Collectively, available data suggest that endometrial production of PGF is greater during the period of expected luteolysis and that exogenous $PGF_{2\alpha}$ is luteolytic in gilts by day 12 after onset of oestrus. Controversy on the question of whether or not $PGF_{2\alpha}$ is *the* pig endometrial luteolytic factor continues. Watson and Walker (1977), for example, reported studies of superfusion of pig CL *in vitro* with either $PGF_{2\alpha}$, endometrial extracts or uterine flushings. Because endometrial superfusates from late luteal phase tissue were more effective than $PFG_{2\alpha}$ alone, or uterine flushings, in permanently suppressing CL progesterone production, they suggested that some factor other than $PGF_{2\alpha}$ or acting in concert with $PGF_{2\alpha}$ must be required for luteolysis. Their conclusion is not supported by the fact that $PGF_{2\alpha}$ is luteolytic in bilaterally hysterectomized gilts (Moeljono, Bazer and Thatcher, 1976) and that $PGF_{2\alpha}$ infusion into the anterior uterine vein of gilts is luteolytic from days 12 to 15 of the oestrous cycle (Krzymowski *et al.*, 1976). Because of the substantial amount of evidence suggesting that $PGF_{2\alpha}$ is the uterine luteolysin in pigs, that assumption will be made relative to subsequent discussion this chapter.

THE LUTEOSTATIC EFFECT OF OESTROGENS IN THE PIG

In pregnant gilts, embryos move from the oviducts and into the uterine horns at about the 4-cell stage, i.e. about 60–72 hours after onset of oestrus. Embryos reach the blastocyst stage by day 5 and emerge from the zona pellucida (hatching) between days 6 and 7. Blastocysts expand from about 0.5–1 mm diameter at hatching, to 2–6 mm diameter on day 10 and then elongate rapidly to a threadlike organism 700–1000 mm in length by days 14–16 of pregnancy (Perry and Rowlands, 1962; Anderson, 1978). These long filamentous blastocysts lie end to end after initial elongation and follow the contour of the endometrial folds along the entire length of the uterine horns. In order for maintenance of apposition between blastocysts and uterine endometrium, it seems likely that the uterine endometrium and blastocysts grow in a parallel manner between days 10.5 to 11 and 14 to 16 of pregnancy. Perry and Rowlands (1962) reported that the average length of each uterine horn was 360 cm for gilts on days 13–18 of pregnancy as compared to 190 cm on day 3.

Rapid expansion and development of the embryo, and its associated membranes and fluids occurs between days 18 and 30 of gestation and by day 60 placental development is complete. This period of pregnancy will not be discussed.

Several investigators have suggested that the porcine conceptus is the source of oestrogens during gestation (Velle, 1960; Raeside, 1963; Molokwu and Wagner, 1973; Choong and Raeside, 1974; Robertson and King, 1974; Wettemann *et al.*, 1977; Knight *et al.*, 1977). However, Perry, Heap and Amoroso (1973) first demonstrated conversion of tritium (^{3}H)-labelled dehydroepiandrosterone (DHEA), androstenedione, progesterone and oestrone sulphate to unconjugated oestradiol and oestrone by pig blastocysts. These findings have been extended and confirmed by Heap and Perry (1974), Perry *et al.* (1976), Flint *et al.* (1979), Heap *et al.* (1979) and Gadsby, Heap and Burton (1980).

Evidence for conversion of acetate, cholesterol, progesterone, DHEA, androstenedione and conjugated oestrogens by the conceptus to unconjugated oestrone and oestradiol is also available (Heap *et al.*, 1979). These precursors may be derived, in part, from the maternal circulation; however, conversion of progesterone to androgens and conjugated oestrogens by the uterine endometrium may also serve to provide pig blastocysts with precursors for conversion to free oestrogens. Data from our laboratory indicate that the endometrium from pigs can convert tritium-labelled progesterone to androstenedione, testosterone, oestrone, oestradiol and conjugated oestrone and oestradiol (Dueben *et al.*, 1977; Dueben *et al.*, 1979; Fischer and Bazer, unpublished data). In addition, Henricks and Tindall (1971) have reported metabolism of ^{3}H-progesterone by pig endometrium to at least ten metabolites, of which two have been identified as 5α-dihydroprogesterone and its 3β-hydroxylated counterpart. Henricks and Tindall (1971) and Dueben *et al.* (1979) found about one third of the radioactivity in the aqueous fraction of homogenized endometrial tissue incubated after three extractions with diethyl ether, which suggests that steroid sulphates and glucuronides are also major products of the endometrial metabolism of ^{3}H-progesterone.

The possibility that the uterine endometrium of pregnant gilts metabolizes progesterone is suggested by two lines of evidence. First, Knight *et al.* (1977) observed that uterine artery–uterine vein (A–V) differences in plasma progesterone concentrations were positive at all stages of gestation studied between days 20 and 100. Conversely, A–V differences in oestrone and oestradiol were negative at all stages of gestation studied. These data suggest that progesterone is being taken up and/or metabolized by the pregnant uterus and/or its contents. The second line of evidence is from a study in our laboratory in which gilts were bilaterally ovariectomized on day 4 of pregnancy and treated with 25, 50, 100 or 200 mg progesterone/day to maintain pregnancy until day 60 of gestation. Gilts, at day 60, which were pregnant (viable conceptuses were present) had plasma progestin concentrations ranging from 7–26.5 ng/ml. However, gilts for which there was no evidence that pregnancy had been established had plasma progestin concentrations of 163–428 ng/ml. Thus, events associated with establishment of pregnancy have a marked effect on progesterone metabolism by the uterus and/or conceptuses.

In normal pregnancy in pigs there is a 30–70% decrease in maternal plasma progestin concentrations between days 14 and 30 of gestation (Guthrie, Henricks and Handlin, 1972; Robertson and King, 1974; Moeljono *et al.*, 1977). This decrease in plasma progestin concentrations has been attributed to partial CL regression followed by CL 'rescue' during the period of establishment of pregnancy. However, the decrease in plasma progestins may reflect initiation of progesterone metabolism by the pregnant uterine endometrium and/or conceptuses.

In considering products of endometrial progesterone metabolism, those products which can be most efficiently converted to free oestrogens may be most important. The percentage conversion of conjugated oestrogens, e.g. oestrone sulphate to oestrone (82%) and oestradiol (16%) by pig blastocysts is very efficient compared with their conversion of androstenedione to oestrone (14%) and oestradiol (2%) or their conversion of progesterone to oestrone (2%) and oestradiol (0.4%) (Perry, Heap and Amoroso, 1973; Perry *et al.*, 1976). The pig endometrium is known to have 5α-reductase enzymatic activity (Henricks and Tindall, 1971) which would result in 5α reduction of progestins and androgens. The 5α reduced steroids cannot be aromatized to oestrogens (Wilson, 1972). Metabolism of progesterone or other potential oestrogen precursors to conjugated oestrogens by the uterine endometrium would circumvent the need for oestrogen precursor pools of androgens which could be reduced and rendered non-aromatizable and also provide the most efficient form of precursor for 'free' oestrogen production by pig blastocysts. Heap and Perry (1974) have demonstrated aromatase and sulphatase enzymatic activity in blastocyst tissues.

Heap and Perry (1974) have suggested that oestrone and oestradiol are produced by pig blastocysts and that these 'free' oestrogens are acted upon by 17β-hydroxysteroid dehydrogenase and sulphotransferase in the uterine endometrium as these oestrogens move towards the maternal circulation. Therefore, oestrone sulphate is the primary oestrogen in maternal plasma (Robertson and King, 1974) and allantoic fluid (Dueben *et al.*, 1979). This concept explains the high unconjugated oestrogen concentration acting locally at the interface between blastocyst and endometrium with no systemic effects due to high concentrations of 'free' oestrogens. The 'free' oestrogens may act on the uterus to bring about increased uterine blood flow (Ford and Christenson, 1979), water and electrolyte movement (Goldstein, Bazer and Barron, 1980), maternal recognition of pregnancy phenomena (Bazer and Thatcher, 1977), uterine secretory activity (Roberts and Bazer, 1980) and other events associated with placentation. After exerting their local effect, but before leaving the uterus, the unconjugated oestrogens are conjugated and enter the maternal circulation in a biologically inactive form.

Robertson and King (1974) have reported that oestrone sulphate is detectable in peripheral maternal plasma by day 16 of pregnancy (60 pg/ml) and concentrations increase up to days 23–30 (3 ng/ml at peak). However, they could not detect unconjugated oestrogens in peripheral plasma until days 70–80 of gestation. In our laboratory, Moeljono *et al.* (1977) detected greater concentrations of oestradiol (20 pg/ml) in utero-ovarian vein plasma of pregnant gilts between days 12 and 17 than in utero-ovarian vein plasma of non-pregnant gilts (5 pg/ml) for the same days of the oestrous cycle. Considerable variability in oestradiol and oestrone utero-ovarian vein plasma concentrations was detected in pregnant pigs during this period. Oestradiol concentrations of 240 pg/ml were detected, but duration and day of occurrence varied considerably. Oestrone concentrations were of lower magnitude.

Total recoverable oestrone and oestradiol in uterine flushings obtained on days 6, 8, 10, 12, 14, 15, 16 and 18 of the oestrous cycle and pregnancy have been reported by Zavy *et al.* (1980). Both total recoverable oestrone and oestradiol were greater in uterine flushings from pregnant gilts from days 12–18, which supports the concept that pig blastocysts begin to produce oestrogen by day 12 (Perry *et al.*, 1976).

At this point, it is necessary to discuss oestrogen production by pig blastocysts during critical stages of pig blastocyst elongation. Anderson (1978) reported that pig blastocysts were spherical on days 9 (0.8 mm diameter) and 10 (2.9 mm diameter), achieved a tubular form on day 11 (58 mm long) and then elongated to the filamentous form by day 12 (100 mm long). Pig blastocysts in this study reached a maximum length of 700–900 mm between days 14 and 16 of pregnancy. The rate of change in blastocyst development during this period is highly variable and extremely rapid during the transition from tubular to filamentous form. Geisert and Bazer (unpublished data) estimated that the increase in diameter of spherical blastocysts was at a rate of 0.3–0.4 mm/hour between the 2.5 mm diameter and 10 mm diameter stage. However, once blastocysts reached the tubular form of 10–50 mm length, they elongated rapidly at rates of up to 45 mm/ hour. This estimate was based on recovery of blastocysts from one uterine horn of a gilt on day 11 that had an average length of 51 mm. Two hours later the other uterine horn was found to contain blastocysts with an average length of 141 mm. In another gilt, blastocysts changed from 11 mm in diameter to greater than 150 mm in length within 6 hours, for a rate of change of at least 20–25 mm/hour. The transition period from large spherical blastocysts to the filamentous form represents the time that close apposition is achieved between blastocysts and endometrium (Crombie, 1972) and oestrogen production begins (Geisert, Bazer and Thatcher, unpublished data; Heap *et al.*, 1979).

In order to closely examine the relationship between stage of blastocyst development and oestrogen production, uterine flushings were obtained from pregnant and non-pregnant gilts between days 10.5 and 12 after onset of oestrus. Uterine flushings were obtained from one uterine horn from all females on day 10.5 and the second uterine horn was flushed later on either day 11, 11.5 or 12 of the oestrous cycle. Variability in blastocyst development among gilts over this 36 hour period was great, as has been noted by Anderson (1978). For example, filamentous blastocysts were recovered from one gilt on day 10.5 while all other gilts had blastocysts of less than 10 mm diameter. On day 12, most gilts had filamentous blastocysts, but some gilts had blastocysts which were still in the early tubular stage. In examining the data, it was obvious that total recoverable oestrogens in uterine flushings were closely associated with stage of blastocyst development rather than the actual day on which the uterine flushing was obtained. The data in *Table 12.1* are, therefore, organized and summarized relative to the stage of blastocyst development.

Total recoverable oestrone (E_1), oestradiol (E_2), oestriol (E_3), oestrone sulphate (E_1S), oestradiol sulphate (E_2S) and oestriol sulphate (E_3S) were measured in uterine flushings from pregnant and non-pregnant pigs. Total recoverable E_1 and E_2 were greater in flushings having tubular and filamentous blastocysts, but an increase in E_3 was not apparent except in uterine flushings containing filamentous blastocysts. Total recoverable E_1S and E_2S also were greater in uterine flushings containing tubular and

Table 12.1 TOTAL RECOVERABLE OESTRONE (E_1), OESTRADIOL (E_2), OESTRIOL (E_3), OESTRONE SULPHATE (E_1S), OESTRADIOL SULPHATE (E_2S) AND OESTRIOL SULPHATE (E_3S) PER UTERINE HORN ($\bar{x}\pm$S.E.M.) IN UTERINE FLUSHINGS FROM PREGNANT (P) AND NON-PREGNANT (NP) GILTS BETWEEN DAYS 10.5 AND 12.0 AFTER DAY OF ONSET OF OESTRUS (DAY 0)

		Stage of blastocyst development				
Oestrogen	*Pregnancy*[a]	*Spherical*		*Tubular*	*Filamentous*	*Filamentous*
(ng)	*status*	*5 mm*	*5–8 mm*	*9–50 mm*	*50 mm*	*Day 14*
Total E_1	P	0.5±0.1 (8)[b]	0.5±0.1 (7)	1.6±0.7 (8)	2.6±1.0 (10)	0.6±0.1 (2)
	NP	0.4±0.1 (9)	0.4±0.1 (3)	0.3±0.1 (3)	0.4±0.1 (7)	0.2±0.04 (3)
Total E_2	P	0.7±0.3	0.5±0.1	2.0±0.3	4.2±1.5	0.4±0.1
	NP	0.4±0.1	0.5±0.3	0.2±0.1	0.3±0.1	0.2±0.1
Total E_3	P	2.2 (1)	2.1 (1)	1.1±0.5 (4)	8.3±6.0 (6)	ND
	NP	1.5 (1)	1.5 (1)	ND[c]	ND	ND
Total E_1S	P	0.8±0.2	1.3±0.4	3.4±0.7	2.2±0.3	0.8±0.2
	NP	0.9±0.5	0.9±0.5	0.2±0.1	0.6±0.1	1.6±0.8
Total E_2S	P	0.5±0.2	0.5±0.2	1.7±0.4	1.6±0.3	0.2±0.1
	NP	1.0±0.4	0.3±0.1	0.6±0.2	0.3±0.1	0.08±0.07
Total E_3S	P	1.9±0.6 (2)	1.4±0.2 (2)	1.6±0.3 (6)	1.5±0.5 (6)	ND
	NP	0.6±0.2 (3)	ND	1.1 (1)	ND	ND

[a]Non-pregnant uterine flushings represent days 10.5, 11, 11.5 and 12 of the oestrous cycle which were considered equivalent periods to those from which the four stages of blastocyst development were recorded. Overall values greater ($P<0.05$) for P than NP gilts.

[b]The numbers in parentheses represent the total number of uterine flushings analyzed for each of the oestrogens. E_3 and E_3S the numbers in parentheses are the number of flushings of that total containing E_3 and E_3S.

[c]Samples in which no E_3 or E_3S was detected; ND = not detected.

filamentous blastocysts. However, total recoverable E_3S was greater for uterine flushings containing tubular (6 of 8) and filamentous (6 of 10) blastocysts compared with flushings containing spherical (4 of 15) blastocysts.

There was no discernible trend for total recoverable E_1, E_2, E_3, E_2S or E_3S in uterine flushings from days 10.5, 11, 11.5 and 12 of the oestrous cycle. The E_1S did, however, tend to be present in less quantity in uterine flushings from days 11.5 and 12 of the oestrous cycle. That E_1S, E_2S and, in a few cases, E_3S are present in uterine flushings from non-pregnant animals supports the possibility of their serving as readily available precursors for conversion to 'free' E_1, E_2 and E_3 by potential blastocysts.

It should be noted that total recoverable E_2 is greater than E_1, and E_3 is greater than E_1 and E_2 in uterine flushings containing filamentous blastocysts. Zavy *et al.* (1980) also found total recoverable E_2 to be greater than E_1 in uterine flushings recovered on days 6, 10 and 12 of pregnancy, but total E_1 was greater than E_2 in uterine flushings collected on days 14, 15, 16 and 18 of gestation.

Observations on total recoverable E_1, E_2, E_3, E_1S, E_2S and E_3S in uterine flushings from day 14 of pregnancy indicate a decrease for all of these oestrogens relative to values found on about day 12 (early filamentous blastocysts). This implies that a transient increase in oestrogen production may accompany the transition from the spherical to early filamentous stages of pig blastocyst development and decline thereafter. This implication is supported by preliminary results from a study of E_1S concentrations in utero-ovarian vein plasma of three pregnant gilts. The E_1S concentrations ($\bar{x}\pm$S.E.M.) were 95.0±22.8, 73.0±2.2, 153.4±21.5, 104.0±9.5, 75.3±16.9 and 55.8±12.4 pg/ml on days 9, 10, 11, 12, 13 and 14 of pregnancy, respectively. For non-pregnant pigs (two gilts) utero-ovarian vein plasma E_1S concentrations ($\bar{x}\pm$S.E.M.) were 54.2±11.5, 71.8±10.3, 34.4±6.9, 68.6±21.7, 76.9±12.9 and 40.0±9.1 pg/ml on days 9, 10, 11, 12, 13 and 14, respectively. For pregnant gilts, therefore, greater ($P<0.05$) plasma E_1S concentrations were detected on days 11 and 12. This is consistent with data on total recoverable unconjugated and conjugated oestrogens previously described for pregnant uterine flushings.

Assuming that free oestrogens provide the signal for maternal recognition of pregnancy in pigs, the collective data of Heap *et al.* (1979), Zavy *et al.* (1980) and Geisert, Bazer, Thatcher and Roberts (unpublished data) suggest that this signal is provided on days 11 and 12 of pregnancy. If unconjugated oestrogens are the porcine conceptus' signal for the maternal recognition of pregnancy, injection of exogenous oestrogens should allow for continued CL maintenance (luteostasis) and uterine function. Kidder, Casida and Grummer (1955), Gardner, First and Casida (1963), Chakraborty, England and Stormshak (1972), Frank *et al.* (1977) and Frank *et al.* (1978) have demonstrated that injection of oestrogen into gilts on or before day 11 of the oestrous cycle results in luteal maintenance. Anderson *et al.* (1965) have reported that daily injections of LH (5 mg/day) results in luteal maintenance in hypophysectomized sows only if the uterus is removed or if LH and oestrogen are administered concurrently. Du Mesnil du Buisson (1966) examined the effect of oestrogen in non-pregnant, hypophysectomized gilts in which the uterus was intact. Injection of 5 mg oestradiol valerate (E_2V) and 5 mg LH daily from days 12–20 allowed maintenance of CL of normal size. He concluded that, since the pituitary was absent, oestrogen was not acting by modifying a gonadotrophic factor, but was inhibiting the luteolytic action of the uterus by blocking secretion or excretion of the uterine luteolysin.

In our laboratory, the term pseudopregnant gilt is used to refer to females that have received 5 mg E_2V daily from days 11–15 of the oestrous cycle. These gilts have interoestrous intervals ranging from 50 to over 300 days, but mean ($\bar{x}\pm$S.E.M.) interoestrous intervals were 146.5±74.8 days (Frank *et al.*, 1977) and 92±11.2 days (Frank *et al.*, 1978) in two studies. Termination of pseudopregnancy is often preceded by swelling of the vulva and engorgement of mammary glands from which milk can be expressed (Frank and Bazer, unpublished data).

Progesterone (P_4) and E_1S concentrations in jugular vein plasma have been determined for five gilts at selected intervals between days 10 and 110 of pseudopregnancy and compared with values from contemporary pregnant gilts having either 4 to 7 (three gilts) or 8 to 11 (three gilts) foetuses at day 110 of pregnancy (Kensinger, Bazer and Collier, unpublished data). Plasma P_4 concentrations were not significantly different between the pseudopregnant and pregnant gilts. Plasma total conjugated oestrogen concentrations ($\bar{x}\pm$S.E.M.) increased from 476±54 pg/ml on day 10 to 1727±721 pg/ml on day 30 of pseudopregnancy and then declined to concentrations of 420±92 to 1249±492 pg/ml on days 40 and 110, respectively. Data for plasma progesterone and total conjugated oestrogen concentrations are depicted in *Figure 12.1*. The pseudopregnant and pregnant gilts had similar patterns of change in plasma E_1S between days 10 and 70 after onset of oestrus, but E_1S did not increase in pseudopregnant gilts between days 70 and 110 as was the case for pregnant gilts.

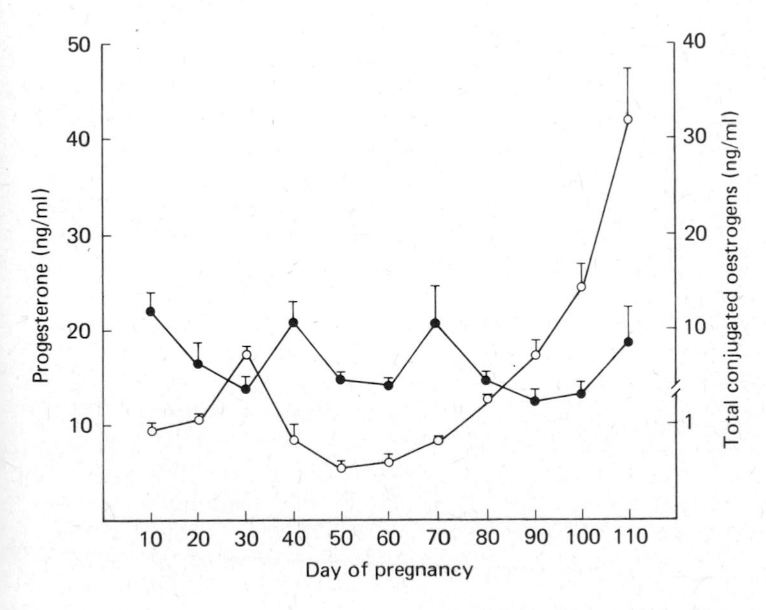

Figure 12.1 Concentrations of progesterone (●——●) and total conjugated oestrogens (○——○) in peripheral plasma of gilts between days 10 and 110 of pregnancy.

Furthermore, E_1S concentrations for pseudopregnant gilts were much lower ($P<0.01$) than those for pregnant gilts.

Uterine endometrial function, based on quantitative and qualitative aspects of proteins secreted into the uterine lumen (Basha *et al.*, 1980) or by endometrial explants *in vitro* (Basha, Bazer and Roberts, 1980), appears to be identical for pregnant and pseudopregnant pigs on days 60–75 after onset of oestrus. Inhibition of uteroferrin secretion by the pregnant uterine endometrium occurs as placental oestrogen production increases between day 70 of gestation and term (Basha, Bazer and Roberts, 1979). This oestrogen-associated inhibition does not occur in pseudopregnant gilts which have an accumulation of uteroferrin within the uterine lumen from days 45–110 (Bazer, Kensinger, Collier and Roberts, unpublished data).

PROSTAGLANDIN F IN NON-PREGNANT VERSUS PREGNANT AND PSEUDOPREGNANT GILTS

Prostaglandin $F_{2\alpha}$ has been shown to exert a luteolytic effect on pig CL by about day 12 after onset of oestrus, as previously discussed (p.229). The studies described in this section were designed to examine the relationship between utero-ovarian vein concentrations of immunoreactive prostaglandin F(PGF) and CL function during the oestrous cycle and the first three to four weeks of either pregnancy or pseudopregnancy.

Gleeson, Thorburn and Cox (1974) first reported a temporal relationship between elevated PGF concentrations in utero-ovarian vein plasma of pigs and declining plasma progesterone concentrations associated with CL regression. Moeljono *et al.* (1977) later compared utero-ovarian vein plasma PGF and peripheral plasma progestins in non-pregnant and pregnant pigs. Utero-ovarian vein samples were collected at 15 minute intervals between 07.00 and 10.00 hours, and 19.00 and 22.00 hours each day from day 12 until either onset of oestrus or day 25 of pregnancy. Utero-ovarian vein PGF concentrations were significantly elevated ($P<0.01$) between days 13 and 17 of the oestrous cycle when plasma progestin concentrations were declining rapidly (*Figure 12.2*). For pregnant gilts however, there were no significant changes in utero-ovarian vein PGF concentrations between days 12 and 25 of pregnancy (*Figure 12.3*). When compared to data from pregnant gilts, utero-ovarian vein mean PGF concentrations, the number of PGF peaks (concentrations greater than $\bar{x}\pm2$S.D. for each gilt) and the frequency of PGF peaks were greater for non-pregnant gilts.

Killian, Davis and Day (1976) reported comparison of utero-ovarian vein plasma concentrations of PGF and progesterone between days 10 and 18 of the oestrous cycle and pregnancy. Blood samples were collected three times daily between 08.00 and 24.00 hours from six non-pregnant and six pregnant gilts. For non-pregnant gilts, mean plasma PGF concentrations varied significantly ($P<0.02$) among days and increased from 0.26 ng/ml on day 10 to 1.48 ng/ml on day 16 and then declined to 0.61 ng/ml on day 18. A correlation coefficient of —0.96 ($P<0.01$) was reported for the temporal relationship between plasma progesterone and PGF concentrations in

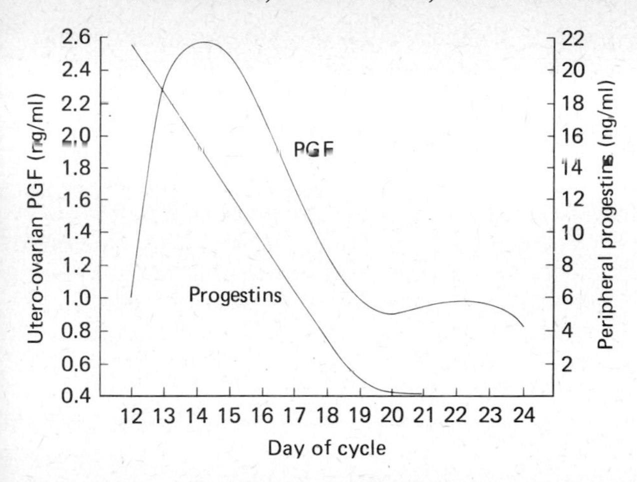

Figure 12.2 The temporal relationship between increasing concentrations of PGF in utero-ovarian vein plasma and decreasing concentrations of progesterone in peripheral plasma in non-pregnant gilts suggest that PGF, of uterine endometrial origin, is involved in luteolysis

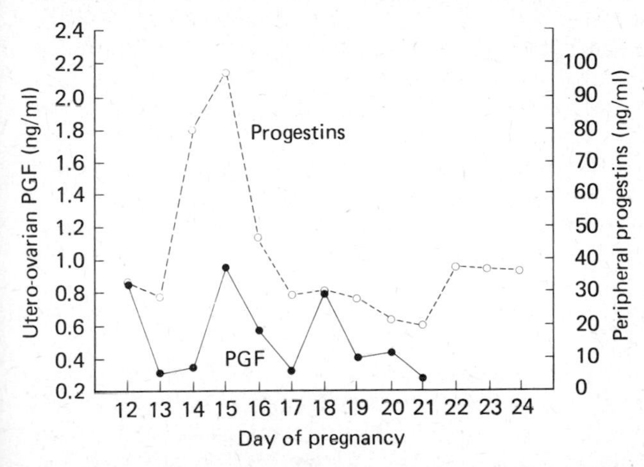

Figure 12.3 The concentrations of PGF in utero-ovarian vein plasma of pregnant gilts do not change significantly between days 12 and 24 of gestation and corpora lutea are maintained as indicated by progesterone concentrations in peripheral plasma.

non-pregnant gilts. For pregnant gilts, plasma PGF concentrations were not significantly affected by day of pregnancy and varied only between 0.42 and 0.78 ng/ml. A non-significant correlation coefficient of only 0.02 was found between plasma progesterone and PGF concentrations in pregnant gilts.

Collectively, the data of Gleeson, Thorburn and Cox (1974), Killian, Davis and Day (1976), Moeljono *et al.* (1977) and Frank *et al.* (1977) clearly indicate a strong temporal relationship between elevated utero-ovarian vein plasma PGF concentrations and decreasing plasma progesterone concentrations associated with luteolysis. However, plasma progesterone and PGF were not related in pregnant gilts (Killian, Davis and Day, 1976; Moeljono *et al.*, 1977).

Subsequent to the reports of Killian, Davis and Day (1976) and Moeljono *et al.* (1977), Ford and Christenson (1979) observed that uterine arterial blood flow increases 2- to 4-fold from day 11 (50 ml/min) to days 12 and 13 (150–200 ml/min) of pregnancy. By day 14 of pregnancy, uterine blood flow decreases slightly to 100–150 ml/min but remains higher than for comparable days of the oestrous cycle. These data raise the question of whether or not utero-ovarian vein concentrations are lower between days 12 and 16 of pregnancy because of the dilution effect of increased uterine blood flow. Data are not available which compare rate of release of PGF from the pregnant and non-pregnant uterus. However, results of studies of the PGF metabolite 13,14-dihydro-15-keto-PGF_2 (PGFM) in peripheral plasma of non-pregnant and pregnant gilts have been reported (Shille *et al.*, 1979; Terqui, Martinat-Botte and Thatcher, 1979). In both studies, samples were obtained from peripheral veins and PGFM concentrations reflect sampling from a plasma pool that should not be affected by changes in uterine blood flow. Data from both studies support the conclusions of Killian, Davis and Day (1976) and Moeljono *et al.* (1977) that there is greater PGF release from the uterus during the period of luteal regression. PGFM concentrations were elevated between days 12 or 13 and days 15 to 17 of the oestrous cycle. However, only basal PGFM concentrations (0.1–0.2 ng/ml) were detected during comparable periods of pregnancy, except for occasional transient increases.

Collectively, these data indicate that there is greater release of PGF into the utero-ovarian vein during the period of luteolysis in non-pregnant pigs. However, release of PGF into the utero-ovarian vein of pregnant pigs remains at a 'basal' level and CL maintenance results.

The results of the previous experiments raise the question of whether or not PGF production by the uterine endometrium is altered during pregnancy. An experiment was designed, therefore, to examine total recoverable PGF in uterine flushings representing days 8, 10, 12, 14, 15, 16 and 18 of the oestrous cycle and pregnancy. Total recoverable PGF was greater ($P<0.01$) in uterine flushings from pregnant gilts than from non-pregnant gilts (Zavy *et al.*, 1980). For non-pregnant gilts, total recoverable PGF ($\bar{x}\pm$S.E.M.) of 464.5 ± 37.6 ng on day 18 was maximum, compared with $22\,688.1\pm1772.4$ ng on day 18 of pregnancy. These data indicate that PGF synthesis and secretion is not being inhibited during pregnancy and suggest that PGF of endometrial and/or conceptus origin is sequestered within the uterine lumen of pregnant gilts. Utero-ovarian vein plasma PGF and peripheral vein plasma PGFM concentrations support this interpretation. The PGF produced by the uterus appears to enter the uterine venous drainage during the mid to late luteal phase of the oestrous cycle, but not during early pregnancy. Walker *et al.* (1977) have suggested that the conceptus (amnion in particular) may be involved in metabolizing PGF_2 and PGE_2 to their respective PGFM; however, data comparing PGF and PGFM in uterine flushings and utero-ovarian vein plasma of non-pregnant and pregnant pigs are not available.

Since prolonged CL maintenance (pseudopregnancy) can be induced in gilts by administration of 5 mg E_2V daily on days 11–15 of the oestrous cycle, Frank *et al.* (1977) have studied utero-ovarian vein plasma PGF and peripheral plasma progestin concentrations. The experimental protocol was essentially identical to that used by Moeljono *et al.* (1977) for comparisons between non-pregnant and pregnant gilts. Similar to the results of Moeljono *et al.* (1977), utero-ovarian vein plasma PGF concentrations were significantly greater between days 12 and 17 than on either days 10 and 11 or 18 of the oestrous cycle for control pigs. Plasma progesterone concentrations decreased rapidly between days 13 and 15 and the average interoestrous interval for the control gilts was 19±0.6 ($\bar{x}$±S.E.M.) days. Mean utero-ovarian vein plasma PGF concentrations, number of PGF peaks (concentrations greater than $\bar{x}$±2 S.D. for each gilt) and frequency of PGF peaks were all significantly lower for E_2V-treated gilts than for control gilts. The mean interoestrous interval was 146.5±74.8 ($\bar{x}$±S.E.M.) days for E_2V-treated gilts.

In a subsequent experiment, Frank *et al.* (1978) have examined total recoverable PGF in uterine flushings recovered on days 11, 13, 15, 17 and 19 of the oestrous cycle and E_2V-induced pseudopregnancy. Total recoverable PGF per uterine horn ($\bar{x}$±S.E.M.) was maximal (210.2±58.7 ng) on day 17 of the oestrous cycle. For pseudopregnant gilts, however, there was a linear increase in total recoverable PGF from 1.9±0.8 ng on day 11 to 5113.3±1735.1 ng on day 19. These data indicate that PGF secretion is not inhibited in pseudopregnant pigs. Rather, instead of PGF entering the utero-ovarian venous drainage, as appears to occur in non-pregnant gilts, PGF is sequestered within the uterine lumen as previously discussed for pregnant gilts. It should be noted that total recoverable PGF was only 5113 ng per uterine horn or about 10 226 ng for both horns, as compared to 22 688 ng from both uterine horns of pregnant gilts. This may reflect some contribution of PGF by pig blastocysts as has been reported for cow blastocysts (Lewis *et al.*, 1979).

UTEROFERRIN

Uteroferrin is the purple coloured, progesterone-induced, intrauterine glycoprotein of pig uterine secretions (histotroph) which appears to be involved in transport of iron from the uterine endometrium to the porcine conceptus (see review by Roberts and Bazer, 1980). Uteroferrin is synthesized and secreted by the surface and glandular epithelium of the pig uterine endometrium (Chen *et al.*, 1975). In non-pregnant gilts, uteroferrin is secreted into the lumen of the uterine glands between days 9 and 13 of the oestrous cycle. Beginning on day 14, however, uteroferrin begins to become localized within the uterine endometrial stroma surrounding the basement membrane of uterine glands. It is possible that uteroferrin enters the interstitial fluid and/or vascular system within the endometrial stroma for transport to the spleen or liver for degradation and conservation of the iron. In contrast to the changing pattern of localization of uteroferrin in non-pregnant gilts, uteroferrin has not been found in the endometrial stroma at any of the stages of pregnancy studied, i.e. days 6, 8, 12, 14, 16, 18, 30, 50, 70 and 90 of gestation. Uteroferrin is always localized within the epithelial cells and lumen of uterine glands and within the placental areolae after day 30 of pregnancy. In the pregnant animal, uteroferrin, one component of uterine histotroph, continues to be secreted into the uterine lumen where it is available to provide a nutrient source to the developing pig conceptus for a major portion of gestation. These observations on uteroferrin localization by Chen *et al.* (1975) have provided the basis for the question of whether or not oestrogens might prevent CL regression in the pig by affecting the direction of secretion of PGF by the uterine endometrium.

Theory of maternal recognition of pregnancy in the pig

The theory of maternal recognition of pregnancy in the pig to be discussed is the one proposed by Bazer and Thatcher (1977). In non-pregnant pigs it is proposed that progesterone enhances and/or induces PGF synthesis by the uterine endometrium and secretion is associated with elevated plasma oestradiol concentrations between days 12 and 18 of the oestrous cycle (Zavy *et al.*, 1980). This oestrogen is likely to be of ovarian origin and is not present in adequate concentration in the endometrium to elicit a pseudopregnant response. Secretion of PGF during the mid to late luteal phase of the oestrous cycle is primarily in an endocrine direction, i.e. toward the uterine venous drainage. This is based on the observations that utero-ovarian vein PGF concentrations are greatest between days 12 and 18 of the oestrous cycle when events leading to luteolysis are initiated. It has also been pointed out that there is little PGF accumulation within the uterine lumen during the oestrous cycle.

In pregnant pigs, it is also assumed that progesterone enhances and/or induces PGF synthesis by the uterine endometrium and that secretion is enhanced by markedly elevated oestrogen concentrations due to blastocyst production of oestrogens. However, oestrogens produced by pig blastocysts as early as day 11 (tubular and filamentous blastocysts) of pregnancy, alter the direction of movement of PGF so that its secretion remains in an exocrine direction, i.e. toward the uterine lumen. Maintenance of PGF secretion in an exocrine direction would prevent PGF from entering the uterine venous drainage and exerting a luteolytic effect on the CL. That is, the endocrine role of the endometrium would be negated by the effect(s) of oestrogens produced by the blastocysts. Of comparable significance is the fact that maintenance of histotroph secretion in an exocrine direction is essential for nourishment of the pig conceptus.

The proposed theory requires that mechanisms exist for altering the direction of movement of PGF from the epithelial cells of the uterine endometrium where it is assumed to be produced, and that PGF can be compartmentalized and sequestered when the uterus is under appropriate hormonal stimulation (see *Figure 12.4*).

Ogra *et al.* (1974) used immunofluorescent antibody procedures to localize PGF in the oviductal epithelium of women. PGF was localized in the mucosa during the pre-ovulatory phase of the menstrual cycle, but during the post-ovulatory phase it was localized in the lamina propria. Bito, Wallenstein and Baroody (1976) have suggested that the movement of prostaglandins is carrier-mediated and that they do not necessarily freely penetrate cell membranes. Bito (1972) and Bito and Spellane (1974) have provided evidence for the accumulation of prostaglandins in uterine tissues

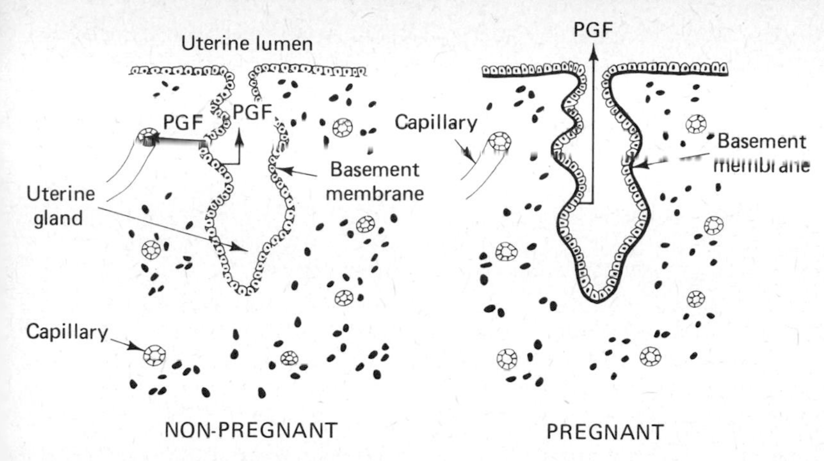

Figure 12.4 In non-pregnant pigs PGF and at least some components of histotroph, e.g. uteroferrin, move toward the endometrial stroma (endocrine direction) during the period of corpora lutea regression. For pregnant pigs, secretion of PGF and histotroph into the uterine lumen (exocrine direction) is maintained. This allows sequestering of the uterine luteolysin as well as histotroph for support of conceptus development

and for the facilitated transport of $PGF_{2\alpha}$ from the lumen of the rabbit vagina.

There is abundant evidence that PGF can be compartmentalized and sequestered within reproductive tissues during pregnancy. Data indicating PGF accumulation within the uterine lumen of pregnant (Zavy *et al.*, 1980) and pseudopregnant gilts (Frank *et al.*, 1978) have been discussed. PGF also appears to be sequestered in allantoic fluid of pig conceptuses (Moeljono, Bazer and Thatcher, unpublished data). Based on the assay of PGF in six to eight samples from various days of pregnancy, allantoic PGF content/conceptus (PGF concentration × allantoic fluid volume/conceptus, $\bar{x}$±S.E.M.) was 0.4±0.04 µg, 98.4±4.5 µg, 46.1±1.9 µg, 76.0±4.5 µg and 8.1±0.2 µg on days 20, 35, 40, 60 and 100 of gestation, respectively.

Harrison *et al.* (1972) recovered from 150 ng to 7.6 mg of authentic $PGF_{2\alpha}$ (analysis by gas chromatography–mass spectrometry) from fluid within the uterine horn of sheep exhibiting prolonged CL maintenance after autotransplantation of the ovary to the neck. This finding was later confirmed in pregnant ewes having a surgically prepared uterine pouch. Fluid from the pouches contained 0.1–333 µg total PGF_2 (Harrison, Heap and Poyser, 1976). Bazer *et al.* (1979) detected 222.8±68 µg PGF in uterine fluid of unilaterally pregnant ewes on day 140 of gestation.

Total recoverable PGF in bovine uterine flushings increased from 13.9 ng on day 8 to 111.0 ng on day 19 of the oestrous cycle and from 17 ng to 1188 ng between days 8 and 19 of pregnancy (Thatcher *et al.*, 1979). Bovine allantoic fluid total PGF was reported to increase from 48.9±12 ng on day 27 to 4228.9±177.2 ng on day 111 of pregnancy while total PGF in amniotic fluid increased from 206.9±109.9 ng on day 60 to 8111.2±1918.9 ng on day 111 (Eley, Thatcher and Bazer, 1979).

Data on PGF accumulation have been presented to emphasize the fact that PGF can be sequestered and compartmentalized with the pregnant and pseudopregnant uterus. However, data are not available concerning either the mechanism(s) whereby this is accomplished or the hormonal control of such mechanisms.

Bazer and Thatcher (1977) have suggested that the luteostatic effect of oestrogens of blastocyst origin or of exogenous oestrogen is on the uterine endometrium. This concept is based on the following evidence:

(a) it is well established that simple removal of the uterus from otherwise intact gilts, without hormonal therapy, allows for prolonged CL maintenance;
(b) CL of pregnant gilts and pseudopregnant gilts are susceptible to the luteolytic effect of exogenous $PGF_{2\alpha}$;
(c) production of $PGF_{2\alpha}$ by endometrium from pregnant and pseudopregnant gilts is not inhibited and appears to be equal to that for mid-luteal phase endometrium of non-pregnant gilts;
(d) LaMotte (1977) has concluded that $PGF_{2\alpha}$ is secreted into the uterine lumen and transferred from there to the uterine venous circulation in non-pregnant gilts;
(e) the production of a systemic luteostatic factor seems unlikely since establishment of pregnancy in only one uterine horn does not have a protective effect on the CL of both ovaries.

POTENTIAL ROLES FOR PGF IN THE PREGNANT UTERUS

Prostaglandins elicit an array of biochemical, physiological and endocrinological effects on various cellular processes and hormonal secretions (Thatcher and Chenault, 1976). Of particular significance in the establishment of pregnancy are the effects of prostaglandins on:

(a) blood flow mediated through changes in the vasodilation of arterioles and the capillary bed (Janson, Albrecht and Ahren, 1975);
(b) changes in the permeability of the endometrial vascular bed (Kennedy, 1980);
(c) fluid and electrolyte transport across epithelia (Heintze *et al.*, 1975; Barry, Hall and Martin, 1975; Biggers *et al.*, 1978);
(d) cellular proliferation as mediated through cyclic-AMP (MacManus and Whitfield, 1974) and steroid biosynthesis (Batta, 1975).

With respect to the effect of PGF on steroid biosynthesis, data from our laboratory have indicated that $PGF_{2\alpha}$ (75 ng/ml medium) tends to increase ($P<0.05$) secretion of oestradiol by day 20 pig conceptuses incubated for 4 hours *in vitro* (Valdivia, Bazer and Thatcher, unpublished data).

Although evidence for these various effects of prostaglandins have been documented in other species, data are not available to indicate their specific role(s) within the pregnant uterus of pigs.

CHANGES IN SELECTED COMPONENTS OF UTERINE SECRETIONS DURING ESTABLISHMENT OF PREGNANCY

Uterine flushings obtained, as previously described, to obtain data on unconjugated and conjugated oestrogens (*Table 12.1*) during the period of

Table 12.2 TOTAL RECOVERABLE PROTEIN, ACID PHOSPHATASE ENZYMATIC ACTIVITY (Pi), CALCIUM AND PROSTAGLANDIN F PER UTERINE HORN ($\bar{x}\pm$S.E.M.) IN UTERINE FLUSHINGS FROM PREGNANT (P) AND NON-PREGNANT (NP) GILTS BETWEEN DAYS 10.5 AND 12.0 AFTER ONSET OF OESTRUS (DAY 0)

Item	*Pregnancy*[a] *status*	*Stage of blastocyst development*				
		Spherical		*Tubular*	*Filamentous*	*Filamentous*
		5 mm	*5–8 mm*	*9–50 mm*	*50 mm*	*Day 14*
Total protein	P	12.7±2.6 (8)[b]	13.7±1.9 (7)	31.4±5.3 (8)	34.7± 3.5 (10)	69.1±11.8 (2)
(mg)	NP	12.6±2.8 (9)	7.7±2.3 (3)	15.5±5.6 (3)	20.5± 7.3 (7)	55.3±11.0 (3)
Total acid	P	2.5±1.1	3.8±1.5	22.9±7.1	17.3± 2.8	16.3± 2.5
phosphatase[c]	NP	2.7±0.7	8.5±3.0	5.0±2.1	25.9±16.9	33.9± 4.8
Total calcium	P	0.14±0.07	0.63±0.17	1.68±0.15	0.95±0.21	0.06±0.02
(mg)	NP	0.20±0.08	0.46±0.11	0.16±0.06	0.36±0.11	0.47±0.22
Total PGF (ng)	P	2.6±0.4	3.3±0.3	5.7±0.5	13.9±3.0	106.7±22.4
	NP	1.7±0.2	1.7±0.2	3.0±0.9	4.3±1.9	4.2± 2.2

[a]Non-pregnant uterine flushings represent days 10.5, 11, 11.5 and 12 of the oestrous cycle which were considered equivalent periods to those from which the four stages of blastocyst development were recovered. Overall values greater ($P<0.05$) for P than NP gilts.
[b]Numbers in parentheses represent the total number of uterine flushings analyzed for each of the items.
[c]μmole Pi released/minute using P-nitrophenylphosphate as substrate at 30°C ($\times 10^{-3}$).

blastocyst elongation have also been analyzed for total recoverable calcium, protein, acid phosphatase (uteroferrin marker) and PGF (Geisert, Bazer, Thatcher and Roberts, unpublished data). Data are summarized for these components in *Table 12.2*, using the same format as for *Table 12.1*, since they were also more closely related to stage of blastocyst development than day of pregnancy. Total recoverable calcium, protein and PGF were greater ($P<0.05$) in uterine flushings from pregnant gilts, but differences in total recoverable acid phosphatase (index for uteroferrin) activity were not significant. The increase in calcium was marked in flushings containing tubular and day 12 filamentous blastocysts as compared with those containing either spherical or day 14 filamentous blastocysts. Total recoverable protein and PGF increased in a similar pattern except that both of these components continued to increase to day 14 of pregnancy. In comparing *Tables 12.1* and *12.2*, temporal events suggest that oestrogens of blastocyst origin may stimulate release of free calcium which would, in turn, induce events which lead to the release of products in secretory vesicles of the uterine epithelium and activate synthesis and secretion of PGF into the uterine lumen. This would result in support for continued blastocyst development and the establishment of pregnancy in the pig.

Events associated with establishment of pregnancy in the pig are similar in many ways to those occurring during the period of termination of embryonic diapause in the roe deer (Aitken, 1979). He reported negligible amounts of protein, carbohydrate, calcium and α-amino nitrogen in uterine flushings from roe deer having embryos in diapause, but these increased in association with blastocyst elongation.

Aitken (1974) suggested that secretions accumulate in the apical region of the epithelial cells of uterine glands of roe deer during the period of embryonic diapause which are released to induce and support blastocyst development. The possibility of oestrogens, of blastocyst origin, stimulating this secretory process was also noted by Aitken (1974). Gadsby, Heap and Burton (1980) later demonstrated that elongated roe deer blastocysts can convert androstenedione to oestrone and oestradiol.

Acknowledgements

Research summarized in this paper has been supported by U.S.D.A. Cooperative Agreement 12–14–7001–1119 and National Institute of Health Grants HD08564 and HD10436.

References

AITKEN, R.J. (1974). Delayed implantation in roe deer (*Capreolus capreolus*). *J. Reprod. Fert.* **39**, 225–233

AITKEN, R.J. (1979). The hormonal control of implantation. In *Maternal Recognition of Pregnancy*, (J. Whelan, Ed.), pp. 53–84. Amsterdam, Excerpta Medica

AMOROSO, E.C. (1952). Placentation. In *Marshall's Physiology of Reproduction*, (A.S. Parkes, Ed.), pp. 127–311. Boston, Little, Brown and Co.

ANDERSON, L.L. (1978). Growth, protein content and distribution of early pig embryos. *Anat. Rec.* **190**, 143–154

ANDERSON, L.L. and MELAMPY, R.M. (1967). Hypophysial and uterine influence on pig luteal function. In *Reproduction in the Female Mammal*, (G.E. Lamming and E.C. Amoroso, Eds), pp. 285–316. New York, Plenum Press

ANDERSON, L.L., BLAND, K.P. and MELAMPY, R.M. (1969). Comparative aspects of uterine-luteal relationships. *Recent Prog. Horm. Res.* **25**, 57–104

ANDERSON, L.L., BUTCHER, R.L. and MELAMPY, R.M. (1961). Subtotal hysterectomy and ovarian function in gilts. *Endocrinology,* **69**, 571–580

ANDERSON, L.L., LEGLISE, P.C., DU MESNIL DU BUISSON, F. and ROMBAUTS, P. (1965). Interaction des hormones gonadotropes et de l'uterus dans le maintien du tissu luteal ovarien chez la truie. *C. r. hebd. Seanc. Acad. Sci., Paris* **261**, 3675–3678

BARRY, E., HALL, W.J. and MARTIN, J.D.G. (1975). Prostaglandin E_1 and movement of salt and water in frog skin (*Rana temporaria*). *Gen. Pharmacol.* **6**, 141–150

BASHA, S.M.M., BAZER, F.W. and ROBERTS, R.M. (1979). The secretion of a uterine-specific, purple phosphatase by cultured explants of porcine endometrium. Dependency upon the state of pregnancy of the donor animal. *Biol. Reprod.* **20**, 431–441

BASHA, S.M.M., BAZER, F.W. and ROBERTS, R.M. (1980). Effect of the conceptus on quantitative and qualitative aspects of uterine secretion in pigs. *J. Reprod. Fert.* **60**, 41–48

BASHA, S.M.M., BAZER, F.W., GEISERT, R.D. and ROBERTS, R.M. (1980). Progesterone-induced uterine secretions in pigs. Recovery from pseudopregnant and unilaterally pregnant gilts. *J. Anim. Sci.* **50**, 113–123

BATTA, S.K. (1975). Effect of prostaglandins on steroid biosynthesis. *J. Ster. Biochem.* **6**, 1075–1080

BAZER, F.W. and THATCHER, W.W. (1977). Theory of maternal recognition of pregnancy in swine based on estrogen controlled endocrine versus exocrine secretion of prostaglandin $F_{2\alpha}$ by the uterine endometrium. *Prostaglandins* **14**, 397–401

BAZER, F.W., ROBERTS, R.M., BASHA, S.M.M., ZAVY, M.T., CATON, D. and BARRON, D.H. (1979). Method for obtaining ovine uterine secretions from unilaterally pregnant ewes. *J. Anim. Sci.* **49**, 1522–1527

BIGGERS, J.D., LEONOV, B.V., BASKAR, J.F. and FRIED, J. (1978). Inhibition of hatching of mouse blastocysts *in vitro* by prostaglandin antagonists. *Biol. Reprod.* **19**, 519–533

BITO, L.Z. (1972). Comparative study of concentrative prostaglandin accumulation by various tissues of mammals and marine vertebrates and invertebrates. *Comp. Biochem. Physiol.* **43A**, 65–82

BITO, L.Z. and SPELLANE, P.J. (1974). Saturable, 'carrier-mediated', absorption of prostaglandin F_2 from the *in vivo* rabbit vagina and its inhibition by prostaglandin F_2. *Prostaglandins* **8**, 345–352

BITO, L.Z., WALLENSTEIN, M. and BAROODY, R. (1976). The role of transport processes in the distribution and disposition of prostaglandins. *Adv. Prostaglandin Thromboxane Res.* **1**, 297–303

CALDWELL, B.V., MOOR, R.M., WILMUT, I., POLGE, C. and ROWSON, L.E.A. (1969). The relationship between day of formation and functional lifespan of induced corpora lutea in the pig. *J. Reprod. Fert.* **18**, 107–113

CHAKRABORTY, P.K., ENGLAND, D.C. and STORMSHAK, R. (1972). Effect of 17β-estradiol on pituitary gonadotrophins and luteal function in gilts. *J. Anim. Sci.* **34**, 427–429

CHEN, T.T., BAZER, F.W., GEBHARDT, B.M. and ROBERTS, R.M. (1975). Uterine secretion in mammals: Synthesis and placental transport of a purple acid phosphatase in pigs. *Biol. Reprod.* **13**, 304–313

CHOONG, C.H. and RAESIDE, J.I. (1974). Chemical determination of estrogen distribution in the fetus and placenta of the domestic pig. *Acta Endocr.* **77**, 171–185

CHRISTENSON, R.K. and DAY, B.N. (1972). Luteolytic effects of endometrial extracts in the pig. *J. Anim. Sci.* **34**, 620–626

CROMBIE, P.R. (1972). The morphology and ultrastructure of the pig's placenta throughout pregnancy. PhD Dissertation. University of Cambridge

DIEHL, J.R. and DAY, B.N. (1973). Effect of prostaglandin $F_{2\alpha}$ on luteal function in swine. *J. Anim. Sci.* **37**, 307 (Abstract)

DIEHL, J.R. and DAY, B.N. (1974). Effect of prostaglandin $F_{2\alpha}$ on luteal function in swine. *J. Anim. Sci.* **39**, 392–396

DOUGLAS, R.H. and GINTHER, O.J. (1975). Effect of prostaglandin $F_{2\alpha}$ on estrous cycles or corpus luteum in mares and gilts. *J. Anim. Sci.* **40**, 518–522

DUEBEN, B.D., WISE, T.H., BAZER, F.W. and FIELDS, M.J. (1977). Metabolism of H^3-progesterone to androgens by pregnant gilt endometrium. *Proc. Am. Soc. Anim. Sci., Madison, Wisconsin,* p. 153

DUEBEN, B.D., WISE, T.H., BAZER, F.W., FIELDS, M.J. and KALRA, P.S. (1979). Metabolism of H^3-progesterone to estrogens by pregnant gilt endometrium and conceptus. *Proc. Am. Soc. Anim. Sci., Tuscon, Arizona*, p. 263

DU MESNIL DU BUISSON, F. (1966). Contribution a l'etude du maintien du corps jaune de la truie. PhD Dissertation. Institut National de la Recherche Agronomique, Nouzilly, France

DU MESNIL DU BUISSON, F. and DAUZIER, L. (1959). Controle mutuel de l'uterus et de l'ovarie chez la truie. *Annls Zootech., Serie D, Suppl.* 149–151

DU MESNIL DU BUISSON, F. and LEGLISE, P.C. (1963). Effet de l'hypophysectomie sur les corps jaunes de la truie. Résultats préliminaires. *C.r. hebd. Seanc. Acad. Sci., Paris* **257**, 261–263

ELEY, R.M., THATCHER, W.W. and BAZER, F.W. (1979). Hormonal and physical changes associated with bovine conceptus development. *J. Reprod. Fert.* **55**, 181–190

FLINT, A.P.F., BURTON, R.D., GADSBY, J.E., SAUNDERS, P.T.K. and HEAP, R.B. (1979). Blastocyst oestrogen synthesis and the maternal recognition of pregnancy. In *Maternal Recognition of Pregnancy*, (J. Whelan, Ed.), pp. 209–238. Amsterdam, Excerpta Medica

FORD, S.P., and CHRISTENSON, R.K. (1979). Blood flow to uteri of sows during the estrous cycle and early pregnancy: Local effect of the conceptus on the uterine blood supply. *Biol. Reprod.* **21**, 617–624

FRANK, M., BAZER, F.W., THATCHER, W.W. and WILCOX, C.J. (1977). A study of prostaglandin $F_{2\alpha}$ as the luteolysin in swine: III. Effects of estradiol valerate on prostaglandin F, progestins, estrone, and estradiol concentrations in the utero-ovarian vein of nonpregnant gilts. *Prostaglandins* **14**, 1183–1196

FRANK, M., BAZER, F.W., THATCHER, W.W. and WILCOX, C.J. (1978). A study of prostaglandin $F_{2\alpha}$ as the luteolysin in swine: IV. An explanation for the luteotrophic effect of estradiol. *Prostaglandins* **15**, 151–160

GADSBY, J.E., HEAP, R.B. and BURTON, R.D. (1980). Oestrogen production by blastocyst and early embryonic tissue of various species. *J. Reprod. Fert.* **60**, 409–417

GARDNER, M.L., FIRST, N.L. and CASIDA, L.E. (1963). Effect of exogenous estrogens on corpus luteum maintenance in gilts. *J. Anim. Sci.* **22**, 132–134

GLEESON, A.R., THORBURN, G.D. and COX, R.I. (1974). Prostaglandin F concentrations in the utero-ovarian vein plasma of the sow during the late luteal phase of the estrous cycle. *Prostaglandins* **5**, 521–530

GOLDSTEIN, M.H., BAZER, F.W. and BARRON, D.H. (1980). Characterization of changes in volume, osmolarity and electrolyte composition of porcine fetal fluids during gestation. *Biol. Reprod.* **22**, 1168–1180

GUTHRIE, H.D. and REXROAD, C.E., Jr. (1980). Progesterone secretion and prostaglandin F release *in vitro* by endometrial and luteal tissue of cyclic pigs. *J. Reprod. Fert.* **60**, 157–163

GUTHRIE, H.D., HENRICKS, D.M. and HANDLIN, D.L. (1972). Plasma estrogen, progesterone and luteinizing hormone prior to estrus and during early pregnancy in pigs. *Endocrinology* **91**, 675–679

GUTHRIE, H.D., REXROAD, C.E., Jr and BOLT, D.J. (1978). *In vitro* synthesis of progesterone and prostaglandin F by luteal tissue and prostaglandin F by endometrial tissue from the pig. *Prostaglandins* **16**, 433–440

HALLFORD, D.M. WETTEMANN, R.P., TURMAN, E.J. and OMTVEDT, I.T. (1974). Luteal function in gilts after prostaglandin $F_{2\alpha}$. *J. Anim. Sci.* **28**, 213 (Abstract)

HARRISON, F.A., HEAP, R.B. and POYSER, N.L. (1976). Production, chemical composition and prostaglandin $F_{2\alpha}$ content of uterine fluid in pregnant sheep. *J. Reprod. Fert.* **48**, 61–67

HARRISON, F.A., HEAP, R.B., HORTON, E.W. and POYSER, N.L. (1972). Identification of prostaglandin $F_{2\alpha}$ in uterine fluid from the non-pregnant sheep with an autotransplanted ovary. *J. Endocr.* **53**, 215–222

HEAP, R.B. and PERRY, J.S. (1974). The maternal recognition of pregnancy. *Br. J. Hosp. Med.* **12**, 8–14

HEAP, R.B., FLINT, A.P.F., GADSBY, J.E. and RICE, C. (1979). Hormones, the early embryo and the uterine environment. *J. Reprod. Fert.* **55**, 267–275

HEINTZE, K., LEINESSER, W., PETERSEN, K.U. and HEIDENREICH, O. (1975). Triphasic effect of prostaglandins E_1, E_2 and F_2 on the fluid transport of isolated gall-bladder of guinea-pigs. *Prostaglandins* **9**, 309–322

HENDERSON, K.M. and McNATTY, K.P. (1975). A biochemical hypothesis to explain the mechanism of luteal regression. *Prostaglandins* **9**, 779–797

HENRICKS, D.M. and TINDALL, D.J. (1971). Metabolism of progesterone-4-C^{14} in porcine uterine endometrium. *Endocrinology* **89**, 920–924

JANSON, P.O., ALBRECHT, I. and AHREN, K. (1975). Effects of prostaglandin $F_{2\alpha}$ on ovarian blood flow and vascular resistance in the pseudopregnant rabbit. *Acta Endocr.* **79**, 337–350

KENNEDY, T.G. (1980). Estrogen and uterine sensitization for the decidual cell reaction: Role of prostaglandins. *Biol. Reprod.* **23**, 955–962

KIDDER, H.E., CASIDA, L.E. and GRUMMER, R.H. (1955). Some effects of estrogen injections on estrual cycle of gilts. *J. Anim. Sci.* **14**, 470–474

KILLIAN, D.B., DAVIS, D.L. and DAY, B.N. (1976). Plasma PGF and hormonal changes during the estrous cycle and early pregnancy in the gilt. *Proc. Int. Pig Vet. Soc., Ames, Iowa*, pp. D.1

KNIGHT, J.W., BAZER, F.W., THATCHER, W.W., FRANKE, D.E. and WALLACE, H.D. (1977). Conceptus development in intact and unilaterally hysterectomized-ovariecomized gilts: Interrelations between hormonal status, placental development, fetal fluids and fetal growth. *J. Anim. Sci.* **44**, 620–637

KRAELING, R.R., BARB, C.R. and DAVIS, B.J. (1975). Prostaglandin induced regression of porcine corpora lutea maintained by estrogen. *Prostaglandins* **9**, 459–462

KRZYMOWSKI, T., KOTWICA, J., OKRASA, S. and DOBOSZYNSKA, T. (1976). The function and regression of corpora lutea during the sow's estrous cycle after 10 hours of prostaglandin $F_{2\alpha}$ infusion into the anterior uterine vein. *Proc. VIII Int. Congr. Anim. Reprod. A. I., Krakow*, p. 143

LaMOTTE, J.O. (1977). Prostaglandins: I. Presence of a prostaglandin-like substance in porcine uterine flushings; II. A new chemical assay for PGF_2 alpha. *Anim. Breed. Abstr.* **45**, 572

LEWIS, G.S., THATCHER, W.W., BAZER, F.W., ROBERTS, R.M. and WILLIAMS, W.F. (1979). Metabolism of arachidonic acid by bovine blastocysts and endometrium. *Proc. Am. Soc. Anim. Sci. Tuscon, Arizona*, pp. 313–334

LOEB. L. (1923). The effect of extirpation of the uterus on the life and function of the corpus luteum in the guinea pig. *Proc. Soc. exp. Biol. Med.* **20**, 441–464

MacMANUS, J.P. and WHITFIELD, J.F. (1974). Cyclic AMP, prostaglandins and the control of cell proliferation. *Prostaglandins* **6**, 475–487

MOELJONO, M.P.E., BAZER, F.W. and THATCHER, W.W. (1976). A study of prostaglandin $F_{2\alpha}$ as the luteolysin in swine: I. Effect of prostaglandin $F_{2\alpha}$ in hysterectomized gilts. *Prostaglandins* **11**, 737–743

MOELJONO, M.P.E. THATCHER, W.W., BAZER, F.W., FRANK, M., OWENS, L.J. and WILCOX, C.J. (1977). A study of prostaglandin $F_{2\alpha}$ as the luteolysin in swine: II. Characterization and comparison of prostaglandin F, estrogen and progestin concentrations in utero-ovarian vein plasma of nonpregnant gilts. *Prostaglandins* **14**, 543–555

MOLOKWU, E.C. and WAGNER, W.C. (1973). Endocrine physiology of the puerperal sow. *J. Anim. Sci.* **36**, 1158–1163

OGRA, S.S., KIRTON, K.T., TOMASI, T.B. and LIPPES, J. (1974). Prostaglandins in the human fallopian tube. *Fert. Steril.* **25**, 250–255

PATEK, C.E. and WATSON, J. (1976). Prostaglandin F and progesterone secretion by porcine endometrium and corpus luteum *in vitro*. *Prostaglandins* **12**, 97–111

PERRY, J.S. and ROWLANDS, I.W. (1962). Early pregnancy in the pig. *J. Reprod. Fert.* **4**, 175–188

PERRY, J.S., HEAP, R.B. and AMOROSO, E.C. (1973). Steroid hormone production by pig blastocysts. *Nature, Lond.* **245**, 45–47

PERRY, J.S., HEAP, R.B., BURTON, R.D. and GADSBY, J.E. (1976). Endocrinology of the blastocyst and its role in the establishment of pregnancy. *J. Reprod. Fert. Suppl.* **25**, 85–104

PHARRIS, B.B. and WYNGARDEN, I.J. (1969). The effect of prostaglandin $F_{2\alpha}$ on the progesterone content of ovaries from pseudopregnant rats. *Proc. Soc. exp. Biol. Med.* **130**, 92–94

RAESIDE, J.I. (1963). Urinary estrogen excretion in the pig during pregnancy and parturition. *J. Reprod. Fert.* **6**, 427–431

ROBERTS, R.M. and BAZER, F.W. (1980). The properties, function and hormonal control of synthesis of uteroferrin, the purple protein of the pig uterus. In *Steroid Induced Uterine Proteins*, (M. Beato, Ed.), pp. 133–149. Amsterdam, Elsevier/North Holland Biomedical Press

ROBERTSON, H.A. and KING, G.J. (1974). Plasma concentrations of progesterone, estrone, estradiol-17β and estrone sulphate in the pig at implantation, during pregnancy and at parturition. *J. Reprod. Fert.* **40**, 133–141

SCHLAFKE, S. and ENDERS, A.C. (1975). Cellular basis of interaction between trophoblast and uterus at implantation. *Biol. Reprod.* **12**, 41–65

SCHOMBERG, D.W. (1967). A demonstration *in vitro* of luteolytic activity in pig uterine flushings. *J. Endocr.* **38**, 359–360

SCHOMBERG, D.W. (1969). The concept of a uterine luteolytic hormone. In *The Gonads*, (K.W. McKerns, Ed.), pp. 282–397. New York, Appleton-Century-Croft

SHILLE, V.M., KARLBOM, I., EINARSSON, S., LARSSON, K. KINDAHL, H. and EDQVIST, L.E. (1979). Concentrations of progesterone and 15-keto-13,14-dihydroprostaglandin $F_{2\alpha}$ in peripheral plasma during the estrous cycle and early pregnancy in gilts. *Zentbl. VetMed.* **26**, 169–181

SPIES, H.G., ZIMMERMAN, D.R., SELF, H.L. and CASIDA, L.E. (1958). Influence of hysterectomy and exogenous progesterone on size and progesterone content of the corpora lutea in gilts. *J. Anim. Sci.* **17**, 1234 (Abstract)

TERQUI, M., MARTINAT-BOTTE, F. and THATCHER, W.W. (1979). Early diagnosis of pregnancy in sows: Determination of $PGF_{2\alpha}$ levels on the 14th and 15th days after insemination. *Journées de la recherche porcine en France*, pp. 365–371. Paris, L'Institut Technique du Porc

THATCHER, W.W. and CHENAULT, J.R. (1976). Reproductive physiological responses of cattle to exogenous prostaglandin $F_{2\alpha}$. *J. Dairy Sci.* **59**, 1366–1375

THATCHER, W.W., WILCOX, C.J., BAZER, F.W., COLLIER, R.J., ELEY, R.M. STOVER, D.G. and BARTOL, F.F. (1979). Bovine conceptus effects prepartum and potential carryover effects postpartum. In *Animal Reproduction 3, Beltsville Symposia in Agriculture Research*, (H.H. Hawk, Ed.), pp. 259–275. New Jersey, Osmum and Company

VELLE, W. (1960). Early pregnancy diagnosis in the sow. *Vet. Rec.* **72**, 116–118

WALKER, F.M.M. PATEK, C.E., LEAF, C.F. and WATSON, J. (1977). The metabolism of prostaglandins $F_{2\alpha}$ and E_2 by nonpregnant porcine endometrial tissue, luteal tissue and conceptuses *in vitro*. *Prostaglandins* **14**, 557–562

WATSON, J. and PATEK, C.E. (1979). Steroid and prostaglandin secretion by the corpus luteum, endometrium and embryos of cyclic and pregnant gilts. *J. Endocr.* **82**, 425–428

WATSON, J. and WALKER, F.M.M. (1977). Effect of prostaglandin $F_{2\alpha}$ and uterine extracts on progesterone secretion *in vitro* by superfused pig corpora lutea. *J. Reprod. Fert.* **51**, 393–398

WETTEMANN, R.P., HALLFORD, D.M., KREIDER, D.L. and TURMAN, E.J. (1977). Influence of prostaglandin $F_{2\alpha}$ on endocrine changes at parturition in gilts. *J. Anim. Sci.* **44**, 107–111

WILSON, J.D. (1972). Recent studies on the mechanism of action of testosterone. *New Engl. J. Med.* **287**, 1284–1291

ZAVY, M.T. (1979). A comparison of the nonpregnant and pregnant uterine luminal environments in the porcine and equine. PhD Dissertation. University of Florida

ZAVY, M.T., BAZER, F.W., THATCHER, W.W. and WILCOX, C.J. (1980). A study of prostaglandin $F_{2\alpha}$ as the luteolysin in swine: V. Comparison of prostaglandin F, progestins, estrone and estradiol in uterine flushings from pregnant and nonpregnant gilts. *Prostaglandins* **20**, 837–851

13

BLASTOCYST–ENDOMETRIUM INTERACTIONS AND THEIR SIGNIFICANCE IN EMBRYONIC MORTALITY

A.P.F. FLINT, P.T.K. SAUNDERS and A.J. ZIECIK†
A.R.C. Institute of Animal Physiology, Babraham, Cambridge, UK and Institute of Animal Physiology and Biochemistry, 10-718 Olsztyn, Poland†

One of the most important economic considerations for the commercial pig breeder is the number of pigs marketed per sow per year. With adequate nutritional and environmental regimes, the realization of maximal productivity is limited by the sow's reproductive potential, which in theory can be increased by reducing the farrowing interval (for instance through early weaning) and by achieving a high litter size (by raising ovulation rate and reducing embryonic, perinatal and pre-weaning losses). For a given ovulation rate, litter size is determined by losses at fertilization, during gestation and in the perinatal period. Of these, loss during gestation is quantitatively the most relevant; the majority of this loss occurs before or during embryogenesis, and is termed *embryonic mortality*.

It has been known for some time that embryonic mortality is relatively high in the pig (Hammond, 1914; Corner, 1923; Perry, 1954; Hanly, 1961;

Table 13.1 ESTIMATES OF PRENATAL LOSS IN PIGS: TIME OF EMBRYONIC OR FOETAL DEATH

Type of animal	*Stage of pregnancy at slaughter* (days)[a]	*Mortality* (%)[b]	*References*
Sows	F	27	Hammond (1914)
Sows	14–60	33	Hammond (1921)
Sows	F	44	Casida (1956)
Sows+gilts	25/F	33/40	Perry (1954)
Gilts	25/70	43/50	Baker *et al.* (1956)
Sows	F	41	Lasley (1957)
Sows	28	39	King and Young (1957)
Gilts	17/25	25/34	Lerner, Mayer and Lasley (1957)
Gilts	55	23	Reddy, Mayer and Lasley (1958)
Gilts	25/70	30/48	Baker *et al.* (1958)
Gilts	25/40	33/38	Day *et al.* (1959)
Sows	F	39	Pomeroy (1960)
Gilts+sows	13–18/26–40	28/35	Perry and Rowlands (1962)
Sows	36–109	46	Marrable and Ashdown (1967)
Sows	9/13	21/52	Scofield, Clegg and Lamming (1974)
Sows	25	17	Dyck (1974)
Gilts	9–18	17	Anderson (1978)

[a]Where animals were slaughtered at two different times during gestation, both are given, together with corresponding mortality rates. F = data obtained at farrowing.
[b]Mortality rates determined from numbers of ova ovulated (taken as equal to number of corpora lutea).

Scofield, 1972; see *Table 13.1*). There is little evidence to suggest it is decreasing, and no generally accepted explanation has been put forward for it (Wrathall, 1971). Rates of embryonic mortality, calculated by counting embryos and using numbers of corpora lutea as a measure of the number of eggs ovulated, usually range from 20–45%; and although this figure may under some circumstances be raised (for instance by mating late in oestrus or by exposure to high ambient temperatures), attempts to reduce embryonic losses have almost consistently failed. The purpose of this chapter is to describe some of the physiological and endocrine events occurring at the time when most embryonic loss takes place, and thereby to attempt to identify processes whose failure may cause embryos to die. It will be seen that although the study of early embryonic development has yet to lead to the identification of a generally applicable cause of embryonic mortality, it provides information which is likely to be of great value to those involved in reducing this form of loss.

To some degree, embryonic loss can be associated with known abnormalities in the fertilized ovum and with conditions of animal husbandry. Chromosomal aberrations in the conceptus are probably a major cause of embryonic death, which Bishop (1964) has described as 'unavoidable and should be regarded as a normal way of eliminating unfit genotypes in each generation'. This is the only form of embryonic mortality associated with a demonstrable abnormality in the conceptus, and it can be induced by procedures allowing polyspermic fertilization, such as mating late during oestrus (Hunter, 1967) or progesterone treatment before mating (Polge and Dziuk, 1965; Day and Polge, 1968). Such structural abnormalities in the conceptus are, of course, distinct from genetic traits which may lead to increased embryonic death through inheritance of homozygous recessive genes coding for lethal characteristics expressed in early pregnancy (Bishop, 1964). The existence of genetic factors of this kind is suggested by the observations that high loss rates are sometimes found in offspring of the same boar (Perry and Rowlands, 1962), and that losses in inbred strains are reduced by outcrossing (Squires, Dickerson and Mayer, 1952; Rampacek, Robison and Ulberg, 1975). Conditions such as heat stress (Omtvedt *et al.*, 1971; Wildt, Riegle and Dukelow, 1975; Cameron, 1980), plane of nutrition (Robertson *et al.*, 1951; Tassell, 1967) and intrauterine infections (Schofield, Clegg and Lamming, 1974), are also associated with increased embryonic mortality, though it is not certain how these conditions cause death of the conceptuses. In addition there are seasonal variations in fecundity (Stork, 1979), such as those thought to occur in European wild pigs (Mauget, 1978), which may be due in part to increased embryonic loss. However, it seems unlikely that conditions such as these account for all the embryonic deaths observed, and because of this it is concluded that some loss may occur which is of unknown aetiology, hereafter referred to as 'basal loss'. Embryonic death due to chromosomal aberrations and environmental conditions may be presumed to be either unavoidable, or susceptible to reduction by changes in animal husbandry; therefore it would appear that basal loss is the category in which improvements may be made through the study of the physiology of early pregnancy.

Given that it is required to reduce basal embryonic loss, then it is desirable to know what proportion of loss is represented by this category.

Unfortunately this is difficult to estimate. From the examination of karyotypes of blastocysts flushed from uteri of pigs at day 10 post coitum, McFeely (1967) and Day and Polge (1968) suggested that 8–12% of fertilized eggs bear abnormalities probably reflecting polyspermy, and which are likely to be lethal. The occurrence of identified chromosomal abnormalities leading to embryonic death before day 9 generally appears to be low; recoveries of blastocysts indicate most loss occurs after day 10 (McFeely, 1967; but see Bouters, Bonte and Vandeplassche, 1974). If loss due to environmental factors is minimized, and the embryonic death rate is 30–40%, subtracting a 10% loss due to chromosomal abnormalities leaves a basal loss rate of 20–30%. However, a major flaw in this argument is the assumption that loss due to environmental factors was low in the herds in which total loss was shown to be 30–40%; much of this may be represented by animals suffering subclinical intrauterine infections, which could occur in as many as 30–45% of sows and double the rate of embryonic death in them (Scofield, 1972; Scofield, Clegg and Lamming, 1974). Thus it seems difficult at present to assign an upper limit to the proportion of embryonic death likely to be caused by chromosomal aberrations, and this leads to the possibility that the majority of loss may be accounted for by chromosomal *plus* identified environmental factors, with no contribution due to basal loss. However, there is independent evidence in the pig, as in many other polytocous animals, that endometrial factors influence blastocyst growth and survival. Transplantation of pre-elongation porcine blastocysts to the ureter or to the outer wall of the uterus shows that in such sites trophoblast growth is invasive and abnormal (Samuel, 1971; Samuel and Perry, 1972). In fact the ectopic trophoblast appears to undergo a syncytial transformation reminiscent of the invasive trophoblast of ruminants, a situation not known to exist *in utero*. This suggests there are influences acting on developing embryos *in utero* to reduce or control their growth, and this conclusion is supported by the lack of success of embryo transfer experiments in increasing litter size by the number of embryos transferred (Webel, Peters and Anderson, 1970), and by *in vitro* evidence which suggests the existence of endometrial factors required for morula development to proceed past the 4- to 8-cell stage (see Polge, Chapter 14).

As shown in *Table 13.1*, experiments in which embryos have been counted in the uterus at varying stages of pregnancy reveal that the majority of embryonic loss occurs before day 25 of gestation. This is perhaps not surprising, in view of the complexity of the physiological events occurring at this stage of gestation; this period includes the time of the maternal recognition of pregnancy (when on day 11 post coitum, the corpora lutea of the cycle are prevented from regressing, and so become the corpora lutea of pregnancy; Dhindsa and Dziuk, 1968), the formation of mesoderm and the elongation of the blastocyst, the attachment of the blastocyst to the endometrial epithelium, and increased secretion of several specific endometrial proteins. Since increases in the number of ova by superovulation and/or superinduction have failed to result in a consistent increase in litter size it has been proposed that the uterus imposes a limit on the number of developing embryos. Although physical overcrowding may result in foetal death during the last trimester, results obtained when embryos were restricted to parts of the uterus (Dziuk, 1968; Ulberg and Rampacek, 1974) indicate that physical overcrowding is unlikely to account for a significant proportion of the loss observed prior to day 25. Consequently emphasis has been placed on biochemical interactions between the embryo and the endometrium and it has been postulated that litter size is limited by the availability of an essential biochemical factor (Bazer, 1975). In subsequent sections we will deal with substances known to be produced by the endometrium and blastocyst which may be involved in such interactions.

Components of blastocyst–endometrial interactions

ENDOMETRIAL PRODUCTS

The non-invasive nature of placentation in the pig might be expected to be associated with the secretion, by the endometrial glands and epithelium, of a high proportion of the nutritional requirements of the trophoblast (Amoroso, 1952; Dantzer, Björkman and Hasselager, 1981). Therefore it would not be surprising if endometrial secretions (histiotrophe) were complex in this species.

Proteins

A number of characteristic proteins appear in uterine flushings during the luteal phase of the oestrous cycle, when progesterone levels are high (Murray *et al.*, 1972; Squire, Bazer and Murray, 1972) and the same endometrial secretory proteins can be induced in ovariectomized pigs by treatment with progesterone (Knight, Bazer and Wallace, 1973; Roberts *et al.*, 1976). They are also produced in early pregnancy (Zavy *et al.*, 1977). Three of these proteins (an iron-containing purple acid phosphatase, uteroferrin, which is also known as purple protein, lysozyme and leucine aminopeptidase) are enzymes, and have been shown to accumulate in allantoic fluid after day 30 of pregnancy (Bazer *et al.*, 1975; Roberts *et al.*, 1976). Results of immunofluorescence studies support the concept that components of histiotrophe, such as uteroferrin, are secreted by the endometrial glands and pass into the foetus via the chorio-allantoic areolae (Chan *et al.*, 1975).

Prostaglandins

In common with other species the pig endometrium produces prostaglandin $F_{2\alpha}$ *in vitro* (Patek and Watson, 1976; Guthrie and Rexroad, 1980; 1981), and the raised levels of prostaglandin $F_{2\alpha}$ reported in the uterine venous effluent towards the end of the oestrous cycle (Gleeson and Thorburn, 1973; Moeljono *et al.*, 1977) are likely to be endometrial in origin. Prostaglandin $F_{2\alpha}$ and its analogues are luteolytic on administration to pigs after day 12 post coitum and it is postulated that prostaglandin $F_{2\alpha}$ is the uterine luteolysin in this species (Gleeson, 1974; Hallford *et al.*, 1975; Guthrie and Polge, 1976; Moeljono, Bazer and Thatcher, 1976; Guthrie and Polge, 1978).

Steroids

Porcine endometrium contains enzymes catalyzing the reductive metabolism of progesterone to pregnanolones and pregnanediols (Henricks and Tindall, 1971). In addition it has recently been shown that the endometrium may synthesize C_{21} and C_{19} steroids, and that this may be physiologically significant. In particular 3β-hydroxysteroid dehydrogenase is present in the endometrium on day 17 post coitum in pregnant animals, and at every stage of pregnancy examined thereafter (V.A. Craig, unpublished observations). Small amounts of androgens and oestrogen have also been shown to be synthesized in pregnant endometrium (Dueben *et al.*, 1979), and this may provide aromatizable steroid to the blastocyst, thereby contributing towards oestrogen synthesis in early pregnancy.

Oestrogens circulate in early pregnancy predominantly in the form of oestrone sulphate (Robertson and King, 1974) which is thought to be formed as a result of sulphation in the endometrium of unconjugated oestrogens produced by the blastocyst (and also, possibly, the endometrium itself). The sulphokinase responsible for this process is a progesterone-dependent enzyme, the activity of which alters during the oestrous cycle in parallel with circulating progesterone concentrations (Pack and Brooks, 1974). Measurement of oestrone sulphate in peripheral plasma between days 26–29 post coitum has been proposed as a pregnancy test (Saba and Hattersley, 1981); furthermore, recent studies of Stoner *et al.* (1980) have revealed that circulating oestrone sulphate levels at day 30 of pregnancy may be correlated with litter size and total litter weight at birth, suggesting that measurement of oestrone sulphate in blood in early pregnancy may provide a useful indication of placental functions affecting foetal development and/or survival. However it is not established whether high oestrone sulphate levels simply reflect increased placental weight in view of the uncertain contribution from the endometrium. Furthermore, other authors have failed to confirm the relationship between oestrone sulphate levels and litter size (Hattersley *et al.*, 1980).

Unidentified endometrial products

In an attempt to evaluate effects of endometrial products on blastocyst metabolism, co-culture techniques have been applied, utilizing blastocyst and endometrium explants and determining blastocyst protein or DNA synthesis. In the technique used by Wyatt (1976) and by Rice, Ackland and Heap (1981) blastocysts dissected from the endometrium on day 16 post coitum were cultured (5 mm lengths on lens tissue at a fluid/gas interface) with or without explants (2 mm^3) of maternal tissues. Protein synthesis was monitored by incorporation of [^{3}H]leucine. Leucine incorporation into blastocyst tissue proteins and proteins recovered from the medium was significantly raised if embryonic tissue was cultured together with endometrium, when blastocyst protein synthesis was low; however this effect was absent when the basal rate of protein synthesis was higher. Other maternal tissues were without effect. Disc gel electrophoresis of blastocyst extracts after culture revealed that stimulated leucine incorporation reflected increased synthesis of specific pre-albumins (with molecular weights 25 000–30 000). It is not certain what product of the endometrium is responsible for increasing leucine incorporation into blastocyst prealbumins; experiments with a variety of purified proteins and serum have failed to identify stimulants, with the possible exception of uteroferrin (Rice, Ackland and Heap, 1981).

Similar explant culture techniques have been used to investigate effects of endometrial products on blastocyst DNA synthesis, using [^{3}H]thymidine as precursor (Flint, 1981). A component of uterine flushings obtained from ovariectomized gilts treated chronically with progestagen has been shown to stimulate incorporation of [^{3}H]thymidine into DNA; the active component is present in a small molecular weight, non-basic fraction, and is unstable during storage lyophilized at 4 °C. Preparations in which the stimulatory activity has decomposed are inhibitory when tested in the same culture system, suggesting either that the stimulatory materials break down to yield inhibitors of thymidine incorporation, or that stimulators mask the effects of inhibitors, the latter being present throughout. Further work will be required to decide between these two possibilities.

An attempt has also been made to isolate proteins which may play a role in the maintenance of the preimplantation blastocyst by employing a number of immunological techniques (P.T.K. Saunders, unpublished observations). Antisera were raised in rabbits against endometrial tissue cytosols and maternal plasmas taken from pregnant and non-pregnant sows 12 and 16 days after the onset of oestrus. Proteins not unique to the endometrial tissue were selectively absorbed by chromatography on Affigel blue (Bio-Rad Laboratory Ltd) followed by absorption with an immobilized antiserum immunoadsorbent (Arrameas and Ternynck, 1969). Unabsorbed proteins were visualized by crossed immunoelectrophoresis into gels containing immunoglobulins raised against tissue and fluids from pregnant and non-pregnant sows. Whilst preliminary studies resulted in the isolation of protein(s) with α electrophoretic mobility at pH 8.6 unique to gravid endometrial tissue (day 16), contradictory results were obtained from a more detailed examination of the proteins present in both gravid and non-gravid uteri. Therefore this work and that of Zavy *et al.* (1977) and Basha, Bazer and Roberts (1979) provides no clear evidence that the endometrium synthesizes any unique proteins during pregnancy.

Other components of endometrial secretion

In addition to the components described above, endometrial secretions may be assumed to contain other compounds required for the growth of the conceptus. Those isolated from uterine flushings to date include the vitamins riboflavin (Murray, Moffatt and Grifo, 1980; Moffatt *et al.*, 1980) and retinol (vitamin A) (Adams, Bazer and Roberts, 1981).

BLASTOCYST PRODUCTS

Steroids

One of the best known properties of the pig blastocyst is its ability to synthesize oestrogens. Since Perry, Heap and Amoroso (1973) first demonstrated aromatase activity in preimplantation pig blastocysts, the initial findings have been both confirmed and extended (Perry *et al.*, 1976; Flint *et al.*, 1979; Gadsby, Heap and Burton, 1980). Aromatase activity becomes measurable between days 12 and 14 after mating and has been found as early as day 10 (i.e. before blastocyst elongation) in ovariectomized gilts treated with medroxyprogesterone acetate (Heap *et al.*, 1981). Other enzymes of oestrogen synthesis (3β-hydroxysteroid dehydrogenase, 17α-hydroxylase and C-17,20-lyase) are also present in the blastocyst on days 16 and 20, and blastocysts in culture produce progesterone, androstenedione and oestrogens (Heap, Flint and Gadsby, 1981). They also contain high concentrations of these steroids (Gadsby and Heap, 1978).

Since all the enzymes of oestrogen synthesis from pregnenolone are present in the elongated blastocyst, it seems likely that the conceptus may be capable of sufficient *de novo* steroid synthesis to account for the rise in oestrogens in the peripheral plasma on days 25 to 30 post coitum. This has been tested in ovariectomized gilts in which pregnancy was maintained by administering a synthetic progestin (medroxyprogesterone acetate); the results show that peripheral circulating oestrone sulphate concentrations are normal in ovariectomized animals, rising to levels identical to those in intact controls (Heap *et al.*, 1981). However in view of recent evidence that the endometrium has the potential to contribute to oestrogen synthesis, it is not certain to what extent this represents steroidogenesis confined to the blastocyst.

Proteins

Among the trophoblast-specific proteins identified in other domestic animals are hormones such as chorionic gonadotrophins (found in the horse) and placental lactogens (which are present in the blastocysts of the sheep and cow). Although no placental lactogen has been demonstrated in the Suidae (the pig and the hippopotamus have been investigated; Kelly *et al.*, 1976; A.P.F. Flint, unpublished observations; W.B. Currie, personal communication), there is some evidence for a chorionic gonadotrophin. An LH-like material, which cross reacts in a radioreceptor assay but not in a radioimmunoassay has been demonstrated in the pig blastocyst (Saunders, Ziecik and Flint, 1980): this activity is neither due to the presence of a protease, nor to a protein binding, the [^{125}I]-labelled porcine LH used as tracer, and is present in blastocyst extracts (but not in liver or skeletal muscle) as early as day 10 post coitum. A similar compound has been partially purified by ion exchange chromatography from porcine placentas; however it has not proved consistently possible to obtain a preparation of sufficiently high potency to demonstrate biological activity (one of five preparations were active by rat ovarian ascorbic acid depletion test). At present, therefore, it is not certain whether this material represents a biologically active gonadotrophin.

Blastocyst–endometrium interactions

MAINTENANCE OF LUTEAL FUNCTION

One of the better known relationships between the blastocyst and the endometrium is that leading to maintenance of luteal function; in the pig the blastocyst must exert an antiluteolytic, or luteotrophic, effect before day 11 in order for this to occur, and since the endometrium is the source of the uterine luteolysin, this signal is most probably directed at the endometrium. The maintenance of luteal function ensures a continued supply of progesterone, which in turn is necessary for the secretion of specific endometrial products such as uteroferrin and the constituents of histiotrophe required for blastocyst growth.

In order for luteal function to be maintained in early pregnancy it is important that the effects of the uterine luteolysin are prevented, and there has been much discussion as to how this is brought about. The utero-ovarian venous concentration of prostaglandin F is reduced in early pregnancy relative to the levels found at luteal regression (Moeljono *et al.*, 1977; Zavy *et al.*, 1980) and this reduction is reflected in peripheral levels of the prostaglandin F metabolite, 13,14-dihydro-15-ketoprostaglandin $F_{2\alpha}$ (Zavy *et al.*, 1980; Guthrie and Rexroad, 1981). Therefore it seems likely that uterine secretion of prostaglandin $F_{2\alpha}$ is reduced in the presence of embryos on days 16–22 post oestrus, and this is consistent with the reduction in endometrial prostaglandin F production observed *in vitro*, in pregnant animals (Watson and Patek, 1979; Guthrie and Rexroad, 1981). However a reduction in endometrial synthesis of prostaglandin $F_{2\alpha}$ at this time may not be the only explanation for the decline in utero-ovarian venous concentrations in early gestation, since Bazer and Thatcher (1977) and Zavy *et al.* (1980) have provided compelling evidence for a redirection of release of prostaglandin $F_{2\alpha}$ away from the vasculature and into the lumen of the uterus in pregnant animals (see Bazer, Chapter 12). Whatever the mechanism underlying the reduction in blood prostaglandin levels, it should be noted that this effect is unlikely simply to reflect increased utero-ovarian blood flow during pregnancy (which Ford and Christensen, 1979, have demonstrated on days 11–13 post coitum), since peripheral prostaglandin metabolite levels are also reduced (Zavy *et al.*, 1980).

Decreased endometrial synthesis of prostaglandin $F_{2\alpha}$, or a redirection of its secretion, with a resulting reduction in utero-ovarian venous prostaglandin concentrations, provides a satisfactory explanation for the lack of a uterine luteolytic effect during pregnancy. However, it is difficult to rule out the possibility that the decline in prostaglandin F production is a result, rather than a cause, of luteal maintenance, as has been suggested previously in the sheep (Heap, Flint and Jenkin, 1978). As in the sheep, uterine prostaglandin release is stimulated by inducing premature luteal regression with a synthetic prostaglandin, and can be reduced by treatment with chorionic gonadotrophin on day 12 of the cycle, which causes luteal maintenance (Guthrie and Rexroad, 1981). Therefore it seems likely that a large proportion of the prostaglandin F released at luteal regression may result from stimulation by declining levels of progesterone, and it is not certain whether release of the rest is inhibited in pregnancy. If the early luteolytic surges are relatively small and transient during the cycle, it will be difficult to show they are inhibited during early gestation.

Whatever the nature of the antiluteolytic signal from the embryo (see below), it appears to act locally on the endometrium underlying the expanded blastocyst. Pregnancy failure (i.e. 100% embryonic loss) occurs in pigs bearing fewer than five embryos (Polge, Rowson and Chang, 1966) presumably because when conceptus numbers are low, blastocyst tissue fails to influence an adequate proportion of the endometrium to prevent the luteolytic signal. Fewer than five embryos is consistent with pregnancy only when that part of the uterus not bearing blastocysts is removed by subtotal hysterectomy (du Mesnil du Buisson and Rombauts, 1963); the presence of a sterile uterine horn always leads to pregnancy failure (du Mesnil du Buisson, 1961; Anderson, Rathmacher and Melampy, 1966; Niswender *et al.*, 1970). In sheep the embryo produces a proteinaceous antiluteolysin (trophoblastin) which is responsible for prolonging luteal function when embryonic extracts are infused into the uterus, and similar findings have been made in unilaterally pregnant pigs in which fertilized ova were flushed from one uterine horn two days after mating (Longenecker and Day, 1972). In the pig, the active material has been shown to be heat-stable and absorbed by charcoal (Ball and Day, 1979) and may therefore be a steroid (such as an oestrogen) rather than a protein. These experiments suggest the signal for luteal maintenance arises in the embryo, and that its action depends on a close association between the embryo and the uterus.

In the pig the uterine luteolysin is thought to act systemically as well as unilaterally; the unilateral effect, which occurs in the pig as in other species, precedes the systemic one. Luteal function is prolonged after either complete or partial hysterectomy (du Mesnil du Buisson and Dauzier, 1959; Spies *et al.*, 1960; Anderson, Butcher and Melampy, 1961); however if embryos are confined to one horn the unilateral pregnancy so established is not maintained unless the non-pregnant (contralateral) horn is removed before the 14th day after mating (see above). If left *in situ* the non-pregnant horn exerts a luteolytic influence firstly over the ipsilateral ovary, and shortly thereafter causes regression of corpora lutea in the contralateral ovary. Evidence for a systemic action of the uterus has been found in cyclic sows bearing an autotransplanted ovary (Harrison, 1979), and in sows with autotransplanted ovaries at parturition (Martin, Bevier and Dziuk, 1978). It was at first thought this systemic action reflected production of a luteolysin other than prostaglandin $F_{2\alpha}$ since in many animals prostaglandin $F_{2\alpha}$ is unable to act systemically by virtue of its rapid metabolism in the lung; in fact evidence has been obtained in superfusion studies for a luteolysin distinct from prostaglandin $F_{2\alpha}$ (Watson and Maule Walker, 1977). However more recent work has explained this discrepancy in terms of the rate of metabolism of prostaglandin $F_{2\alpha}$; pulmonary clearance of prostaglandin $F_{2\alpha}$ is much less rapid in the pig than the sheep (Davis *et al.*, 1979).

The most important, and earliest, secretion of the embryo yet shown to be involved in the maintenance of luteal function is an oestrogen. Oestrogens are luteotrophic in the pig (Kidder, Casida and Grummer, 1955; Gardner, First and Casida, 1963) and are produced sufficiently early by the blastocyst (by day 12, see above) to be candidates for the antiluteolytic signal. Administration of oestradiol-17β to cyclic animals reduces utero-ovarian venous concentrations of prostaglandin $F_{2\alpha}$ and raises prostaglandin $F_{2\alpha}$ levels in the uterine lumen, changes which are suggested to reflect the redirection of prostaglandin $F_{2\alpha}$ secretion (Frank *et al.*, 1977; 1978; Zavy *et al.*, 1980). If oestrogens produced by the blastocyst act directly on the endometrium, there would be no requirement to postulate a systemic effect; therefore the fact that peripheral plasma oestrogen levels have not been shown to be raised before day 15 post coitum is not an important objection to this hypothesis (Robertson, King and Dyck, 1978). An alternative mechanism, which would likewise obviate the need for a systemic effect, might arise from local counter current transfer of oestradiol from the utero-ovarian vein to the ovarian artery. Such a transfer has recently been reported for testosterone (Krymowski, Kotwica and Stefanćzyk, 1981), and would be consistent with the relatively high levels of oestradiol present in the utero-ovarian vein from day 12 post coitum (Moeljono *et al.*, 1977). Oestradiol acts synergistically with human chorionic gonadotrophin in stimulating progesterone synthesis by incubated porcine granulosa cells (Goldenberg, Bridson and Kohler, 1972) and stilboestrol has been reported to increase binding of LH to these cells (Nakano *et al.*, 1977); if similar effects are exerted on luteal cells, then raised levels of oestrogens in ovarian arterial blood might be expected to have a direct effect on luteal function. However there is evidence against a blood-borne signal being involved in luteal maintenance, since Robertson, Dwyer and King (1980) were unable to prevent luteal maintenance by passive immunization of sows against both oestrone and oestradiol between days 10 and 21 after mating; the best explanation for this lack of effect is that the antisera administered were ineffective in interfering with events inside the uterus.

In addition to the changes brought about by oestrogen in the endometrium, important changes are occurring between days 12–14 in the factors controlling luteal activity. The corpus luteum functions autonomously before day 14, so that neither hypophysectomy nor lowering LH levels by progesterone administration cause immediate luteal regression (Anderson, 1966; du Mesnil du Buisson, 1961; Woody, First and Pope, 1967). Furthermore, corpora lutea are refractory to the luteolytic effect of prostaglandin $F_{2\alpha}$ before day 12, whereas prostaglandin administration after day 8 does cause a transient decline in progesterone secretion (Guthrie and Polge, 1976); an irreversible effect is not found until some four days later. However, after day 14 of pregnancy luteal maintenance becomes dependent on LH, as indicated by the effects of hypophysectomy and administration of progesterone or antisera to LH (Sammelwitz, Aldred and Nalbandov, 1961; Brinkley, Norton and Nalbandov, 1964; Spies, Slyter and Quadri, 1967; Short *et al.*, 1968). LH dependency then continues from 14 to approximately 70 days gestation, during which time unilateral ovariectomy results in hypertrophy of the contralateral corpora lutea (Staigmiller, First and Casida, 1972); later in pregnancy, luteal function can be maintained after hypophysectomy with prolactin alone (du Mesnil du Buisson and Denamur, 1968). It is doubtful, however, whether the influence of LH is ever lost during gestation, since temporal interrelationships have been demonstrated between surges in circulating LH level and progesterone secretion as late as day 90 (Parvizi *et al.*, 1976). The onset of prolactin dependency appears to be associated with increased concentrations of prolactin receptors in the corpora lutea (Rolland, Gunsalus and Hammond, 1976). The appearance of LH dependency in the corpora lutea after day 14 implies that plasma LH concentrations must be maintained after this time and, in fact, circulating LH levels are raised in pregnant compared with non-pregnant pigs (Guthrie, Henricks and Handlin, 1972; Ziecik, Tilton and Williams, 1981). In contrast there appears to be no influence of pregnancy on prolactin concentrations until shortly before term, when there is a dramatic surge, presumably involved in lactogenesis (Dusza and Krzymowska, 1981).

Measurements of the luteal LH receptor in the pig have revealed dramatic changes in its concentration during the cycle and in early pregnancy (Ziecik, Shaw and Flint, 1980). Although the corpus luteum is independent of control by LH before day 14 of the cycle, the LH receptor is present as early as day 8; its concentration peaks on days 10 and 12 before declining as the corpora lutea regress. LH receptor concentrations in early pregnancy are lower than those during the cycle, possibly as a result of down regulation by the raised LH levels in pregnancy relative to the cycle. Subsequently luteal LH receptor concentrations rise dramatically between days 20 and 30 of gestation, at which time there is a decline in receptor occupancy by LH; this may reflect blastocyst production of a chorionic gonadotrophin which displaces LH from the receptor but does not interfere in the radioimmunoassay used to determine receptor occupancy. The demonstration, reviewed above, that the early blastocyst contains material which cross reacts in an LH radioreceptor assay is consistent with this view. Alternatively, the LH receptor may be controlled by oestrogen, or by a combination of oestrogen and prolactin. The rise in LH receptor between days 20 and 30 coincides with the rise in circulating oestrogen concentrations, and unpublished observations (H.A. Garverick, C. Polge and A.P.F. Flint, 1981) suggest that administration of oestrogen may raise luteal LH receptor concentrations in non-pregnant animals.

ENDOMETRIAL STEROID RECEPTORS

Another potentially important interaction between the blastocyst and endometrium is that involving endometrial enzymes catalysing the deactivation of progesterone and oestrogens, the synthesis of unconjugated oestrogen by the blastocyst, and the occupancy and nuclear translocation of endometrial progesterone and oestrogen receptors. Measurements of endometrial oestrogen receptors show that cytoplasmic concentrations are similar in non-pregnant and pregnant pigs, with levels reaching maximum values during the early or mid-luteal phase (Pack *et al.*, 1978; Deaver and Guthrie, 1980). Experiments with ovariectomized pigs have shown that oestradiol increases both the synthesis of its receptor and receptor translocation to the nucleus (Jungblut *et al.*, 1976). These effects are similar to those found in the rat and sheep, in which progesterone has also been shown to modulate oestrogen receptor concentrations, by reducing receptor synthesis; therefore it seems probable (though it has not been tested) that progesterone has a comparable action in the pig. However it is not certain that receptor concentrations are controlled by these hormones alone, since endometrial levels of progesterone and oestradiol have not been found to alter during either the cycle or in early pregnancy in a manner consistent with the observed changes in receptor levels. Concentrations of oestradiol in endometrial tissue are low during the cycle except for a rise on day 20; levels in pregnancy are also low until day 15 or 16 (Deaver and Guthrie, 1980). Endometrial progesterone levels, as expected on the basis of concentrations in blood, are similar in non-pregnant and pregnant pigs with the exception of a fall in non-pregnant animals towards the end of the cycle (Deaver and Guthrie, 1980). Endometrial oestradiol and progesterone concentrations are presumably influenced by activities of oestrogen sulphokinase (Pack and Brooks, 1974) and enzymes metabolizing progesterone to less active pregnanediols (Henricks and Tindall, 1971); the sulphokinase is a progesterone-dependent enzyme (Pack and Brooks, 1974). Thus rising titres of progesterone after ovulation are associated with increased levels of endometrial sulphokinase which might be expected to result in a decreased intracellular oestrogen concentration; this would tend to reduce translocation of cytoplasmic receptors to the nucleus, and may contribute to the rise in endometrial cytoplasmic oestrogen receptors at this time. It has been suggested that these cytoplasmic receptors may interact with blastocyst oestrogen on days 10–12, if the animal is pregnant, and thereby mediate the antiluteolytic effect (Deaver and Guthrie, 1980). On the other hand such a hypothesis does not appear to be consistent with the lack of rise in endometrial oestrogen concentrations until day 15 or 16, although raised levels of both oestradiol and oestrone have been found in uterine flushings between days 10 and 12 post coitum (Zavy *et al.*, 1980).

GROWTH FACTORS

Although the existence of endometrial products that influence blastocyst growth may be postulated on the basis of culture experiments, these have not been identified or measured, and there is no conclusive immunoelectrophoretic (see above) or chromatographic (Basha, Bazer and Roberts, 1979) evidence for pregnancy-specific endometrial proteins which might be involved in stimulating blastocyst growth. The proteins produced by the endometrium in early pregnancy appear to be identical to those secreted late in the cycle by the non-pregnant endometrium, though the time course of secretion may be altered by the presence of conceptuses (Basha, Bazer and Roberts, 1980). Nonetheless, a possible involvement of blastocyst oestrogen in controlling the secretion of endometrial embryotrophic factors should not be overlooked, particularly in view of their apparent importance in some other species (Flint, 1981). One possible hypothesis is that blastocyst oestrogen acts on the endometrium to increase secretion of substances that stimulate blastocyst growth (or to inhibit production of blastocyst growth inhibitors), as well as exerting an antiluteolytic effect. Evidence that some such interaction may exist is provided by the findings of Vandeplassche (1969), who postulated that the occurrence of superfetation in pigs reflects a temporary cessation of growth on the part of some of the blastocysts in a large litter. The growth promoting (and possibly inhibitory) substances demonstrated in culture experiments represent components of endometrial secretions that may be involved in this process, and competition for growth factors of this kind may be involved in deciding which embryos are lost. Such a mechanism might be envisaged if the conceptuses which elongate earliest obtain a disproportionately high share of available growth-supporting substances, and therefore deprive those elongating later; this is consistent with the variation between blastocyst sizes which occurs during the time of elongation (Anderson, 1978). If this kind of competition for growth factors were to operate, it might lead to considerable advantages to the species if the last blastocysts to elongate arose from the last ova fertilized, as fertilization late during oestrus leads to an increased incidence of polyploidy. It should be noted however that no evidence for such a mechanism has been found (Anderson, 1978).

Attempts to reduce embryonic mortality

Because of the evident requirement for them in early pregnancy, many attempts have been made to reduce or prevent embryonic death by administration of steroid hormones, particularly progesterone. Glasgow, Mayer and Dickerson (1951) reported a significant correlation between production of excretory metabolites of progesterone and embryonic survival, but attempts to correlate luteal function with embryonic survival have not been so promising (Mayer, Glasgow and Gawienowski, 1961; Erb *et al.*, 1962; Phillippo, 1968). Furthermore, efforts to reduce embryonic loss by progesterone administration have generally been unsuccessful or inconsistent (Sammelwitz, Dziuk and Nalbandov, 1956; Haines, Warnick and Wallace, 1958; Spies *et al.*, 1959; Day, Romack and Lasley, 1963), although there have been reports of success, particularly using potent synthetic progestagens (Schultz *et al.*, 1966) or progesterone administered with oestrogens (Reddy, Mayer and Lasley, 1958; Day *et al.*, 1959; Gentry, Anderson and Melampy, 1973; Wildt *et al.*, 1976). As pointed out by Scofield (1975), there is a poor correlation between ovulation rate and embryonic mortality within the normal range of ovulation rates (Perry, 1954), and since progesterone concentrations are closely associated with numbers of corpora lutea (Brinkley and Young, 1970) progesterone deficiency as a cause of embryonic death in the majority of normal animals would seem unlikely. This is supported by the survival to day 25 of large numbers of embryos in embryo transfer experiments in animals bearing more embryos than corpora lutea (Pope *et al.*, 1972). High doses of progesterone are in fact likely to be detrimental to pregnancy by reducing circulating LH levels at a time (from the start of the third week of gestation) when the corpora lutea are becoming dependent on this hormone. Despite such negative results, it seems possible that severe progesterone deficiency may occur in some cases, possibly as a result of incomplete inhibition of secretion of uterine luteolysin, and that these may benefit from administered progesterone (see Day, Romack and Lasley, 1963).

In contrast to the lack of effect of progesterone before attachment, subsequent placental development may be enhanced by progesterone administration. Bazer (1975) showed that treatment with progesterone increased placental length and weight and allantoic fluid volume, and this has been confirmed recently by McGovern *et al.* (1981). These authors obtained a 15% increase in mean chorionic area and allantoic fluid volume, as determined between days 30 and 35, when 25 mg progesterone *plus*

12.5 μg oestrone were administered daily between days 14 and 23. This effect is presumably mediated through increased secretion of proteins, including uteroferrin, by the endometrium, since passive immunization of gilts against uteroferrin reduces placental development (Bazer, 1975). Placental growth in early pregnancy may be important in determining embryo survival rates later in gestation, since Fenton *et al.* (1970) and Knight *et al.* (1977) have shown in unilaterally ovariectomized gilts in which one uterine horn was also removed, that artificial overcrowding of embryos (which reduced placental size) leads to increased embryonic loss after day 25 of gestation. It should be appreciated therefore that even if early embryonic death were reduced as a result of some treatment administered during the first two weeks of pregnancy, it is by no means certain that this would lead to increases in litter size unless it also raised placental size.

Comment

It may one day be possible to reduce embryonic mortality by administering, during early pregnancy, a specific substance, perhaps an 'embryo growth factor', developed from a knowledge of the chemical dialogue between the blastocyst and the endometrium. As will be recognized from the foregoing discussion, however, our understanding of the complex, two-way, interaction between these tissues is still too rudimentary at present to allow such treatment.

Nevertheless, whilst we await a more physiological solution, considerable improvements in fecundity, including reduced embryonic mortality, can be achieved by attending to relevant aspects of husbandry (such as breeding programmes involving out-crossing, time of mating, flush feeding, etc.).

References

ADAMS, K.L., BAZER, F.W. and ROBERTS, R.M. (1981). Progesterone-induced secretion of a retinol-binding protein in the pig uterus. *J. Reprod. Fert.* **62**, 39–47

AMOROSO, E.C. (1952). Placentation. In *Marshall's Physiology of Reproduction*, Third Edition, Volume 2 (A.S. Parkes, Ed.) pp. 127–311. London, Longman, Green and Co.

ANDERSON, L.L. (1966). Pituitary–ovarian–uterine relationships in pigs. *J. Reprod. Fert. Suppl.* **1**, 21–32

ANDERSON, L.L. (1978). Growth, protein content and distribution of early pig embryos. *Anat. Rec.* **190**, 143–154

ANDERSON, L.L., BUTCHER, R.L. and MELAMPY, R.M. (1961). Subtotal hysterectomy and ovarian function in gilts. *Endocrinology* **69**, 571–580

ANDERSON, L.L., RATHMACHER, R.P. and MELAMPY, R.M. (1966). The uterus and unilateral regression of corpora lutea in the pig. *Am. J. Physiol.* **210**, 611–614

ARRAMEAS, S. and TERNYNCK, T. (1969). The cross-linking of proteins with glutaraldehyde and its use for the preparation of immunoadsorbents. *Immunochemistry* **6**, 53–66

BAKER, L.N., CHAPMAN, A.B., GRUMMER, R.H. and CASIDA, L.E. (1958). Some factors affecting litter size and fetal weight in pure-bred and reciprocal cross matings of Chester White and Poland-China swine. *J. Anim. Sci.* **17**, 612–621

BAKER, L.N., SELF, H.L., CHAPMAN, A.B., GRUMMER, R.H. and CASIDA, L.E. (1956). Interrelationships of factors affecting litter size and fetus weight and development in swine. *J. Anim. Sci.* **15**, 1227 (Abstract)

BALL, G.D. and DAY, B.N. (1979). Effects of embryonic extracts on luteal function in the pig. *J. Anim. Sci.* **49**, (Suppl. **1**), 279 (Abstract)

BASHA, S.M.M., BAZER, F.W. and ROBERTS, R.M. (1979). The secretion of a uterine-specific purple phosphatase by cultured explants of porcine endometrium. Dependency upon the state of pregnancy of the donor animal. *Biol. Reprod.* **20**, 431–441

BASHA, S.M.M., BAZER, F.W. and ROBERTS, R.M. (1980). Effect of the conceptus on quantitative and qualitative aspects of uterine secretions in pigs. *J. Reprod. Fert.* **60**, 41–48

BAZER, F.W. (1975). Uterine protein secretions: relationship to development of the conceptus. *J. Anim. Sci.* **41**, 1376–1382

BAZER, F.W. and THATCHER, W.W. (1977). Theory of maternal recognition in swine based on oestrogen controlled endocrine versus exocrine secretion of prostaglandin F_2 alpha by the uterine epithelium. *Prostaglandins* **14**, 397–401

BAZER, F.W., CHEN, T.T., KNIGHT, J.W., SCHLOSNAGLE, D., BALDWIN, N.J. and ROBERTS, R.M. (1975). Presence of a progesterone-induced, uterine specific, acid phosphatase in allantoic fluid of gilts. *J. Anim. Sci.* **41**, 1112–1119

BISHOP, M.W.H. (1964). Paternal contribution to embryonic death. *J. Reprod. Fert.* **7**, 383–396

BOUTERS, R., BONTE, P. and VANDEPLASSCHE, M. (1974). Chromosomal abnormalities and embryonic death in pigs. In *Proceedings of the 1st World Congress on Genetics Applied to Livestock Production*, pp. 169–171. Madrid, Garsi

BRINKLEY, H.J., NORTON, H.W. and NALBANDOV, A.V. (1964). Role of a hypophysial luteotrophic substance in the function of porcine corpora lutea. *Endocrinology* **74**, 9–13

BRINKLEY, H.J. and YOUNG, E.P. (1970). Structural determinants of porcine luteal function. *J. Anim. Sci.* **31**, 218 (Abstract)

CAMERON, R.D.A. (1980). The effect of heat stress on reproductive efficiency in breeding pigs. In *The Veterinary Annual*, 12th issue (C.S.G. Grunsell and F.W.G. Hill, Eds.), pp. 265–274. Bristol, Scientechnica

CASIDA, L.E. (1956). Variables in the maternal environment affecting prenatal survival and development. *Proceedings of the Third International Congress in Animal Reproduction, Cambridge,* Plenary Papers, pp. 19–25

CHEN, T.T., BAZER, F.W., GEBHARDT, B.M. and ROBERTS, R.M. (1975). Synthesis and movement of porcine purple uterine protein. *J. Anim. Sci.* **40**, 170 (Abstract)

CORNER, G.W. (1923). The problem of embryonic pathology in mammals with observations upon intra-uterine mortality in the pig. *Am. J. Anat.* **31**, 523–545

DANTZER, V., BJÖRKMAN, N. AND HASSELAGER, E. (1981). An electron microscopic study of histiotrophe in the interareolar part of the porcine placenta. *Placenta* **2**, 19–28

DAVIS, A.J., FLEET, I.R., HARRISON, F.A. and MAULE WALKER, F.M. (1979). Pulmonary metabolism of prostaglandin $F_{2\alpha}$ in the conscious non-pregnant ewe and sow. *J. Physiol.* **301**, 86P (Abstract)

DAY, B.N. and POLGE, C. (1968). Effects of progesterone on fertilization and egg transport in the pig. *J. Reprod. Fert.* **17**, 227–230

DAY, B.N., ROMACK, F.E. and LASLEY, J.F. (1963). Influence of progesterone–estrogen implants on early embryonic mortality in swine. *J. Anim. Sci.* **22**, 637–639

DAY, B.N., ANDERSON, L.L., EMMERSON, M.A., HAZEL, L.N. and MELAMPY, R.M. (1959). Effect of estrogen and progesterone on early embryonic mortality in ovariectomized gilts. *J. Anim. Sci.* **18**, 607–613

DEAVER, D.R. and GUTHRIE, H.D. (1980). Cytoplasmic estrogen receptor, estradiol and progesterone concentrations in endometrium of non-pregnant and pregnant gilts. *Biol. Reprod.* **23**, 72–77

DHINDSA, D.S. and DZIUK, P.J. (1968). Effects of pregnancy in the pig after killing embryos or fetuses in one uterine horn in early pregnancy. *J. Anim. Sci.* **27**, 122–126

DUEBEN, B.D., WISE, T.H., BAZER, F.W., FIELDS, M.J. and KALRA, P.S. (1979). Metabolism of ^{3}H-progesterone to estrogens by pregnant gilt endometrium and conceptus. *J. Anim. Sci.* **49**, (Suppl. **1**), 293 (Abstract)

DU MESNIL DU BUISSON, F. (1961). Possibilité d'un functionnement dissemblable des ovaries pendant le gestation chez la truie. *C. r. hebd. Séanc. Acad. Sci., Paris* **253**, 724–729

DU MESNIL DU BUISSON, F. and DAUZIER, L. (1959). Controle mutuel de l'uterus et de l'ovaire chez la truie. *Annls Zootech.* (Suppl.) 147–159

DU MESNIL DU BUISSON, F. and DENAMUR, R. (1968). Méchanismes du contrôle de la fonction lutéale chez la truie, la brebis et la vache. In *Proceedings of 3rd International Congress of Endocrinology, Mexico* (International Congress Series, No. 184) (L. Gual and F.J. Ebling, Eds.) pp. 927–934. Amsterdam, Excerpta Medica

DU MESNIL DU BUISSON, F. and ROMBAUTS, P. (1963). Réduction expérimentale du nombre de foetus au cours de la gestation de la truie et maintien des corps jaune. *Annls Biol. anim. Biochim. Biophys.* **3**, 445–448

DUSZA, L. and KRZYMOWSKA, H. (1981). Plasma prolactin levels in sows during pregnancy, parturition and early lactation. *J. Reprod. Fert.* **61**, 131–134

DYCK, G.W. (1974). The effects of stage of pregnancy, mating at the first and second estrus after weaning and level of feeding on fetal survival in sows. *Can. J. Anim. Sci.* **54**, 277–285

DZIUK, P.J. (1968). Effect of number of embryos and uterine space on embryo survival in the pig. *J. Anim. Sci.* **27**, 673–676

ERB, R.E., NOFZIGER, J.C., STORMSHAK, F. and JOHNSON, J.B. (1962). Progesterone in corpora lutea, ovaries and adrenals of pregnant sows and its relationship to number of implants. *J. Anim. Sci.* **21**, 562–567

FENTON, F.R., BAZER, F.W., ROBISON, O.W. and ULBERG, L.C. (1970). Effect of quantity of uterus on uterine capacity in gilts. *J. Anim. Sci.* **31**, 104–106

FLINT, A.P.F. (1981). A unifying hypothesis for the control of blastocyst growth based on observations of the pig. *J. Reprod. Fert., Suppl.* **29**, 215–227

FLINT, A.P.F., BURTON, R.D,, GADSBY, J.E., SAUNDERS, P.T.K. and HEAP, R.B. (1979). Blastocyst oestrogen synthesis and the maternal recognition of pregnancy. In *Maternal Recognition of Pregnancy,* Ciba Foundation Symposium No. 64 (J. Whelan, Ed.), pp. 209–288. Amsterdam, Excerpta Medica

FORD, S.P. and CHRISTENSEN, R.K. (1979). Blood flow to uteri of sows during the estrous cycle and early pregnancy: local effect of the conceptus on the uterine blood supply. *Biol. Reprod.* **21**, 617–624

FRANK, M., BAZER, F.W., THATCHER, W.W. and WILCOX, C.J. (1977). A study of prostaglandin $F_{2\alpha}$ as the luteolysin in swine. III. Effects of estradiol valerate on prostaglandin F, progestins, estrone and estradiol concentrations in the utero-ovarian vein of non-pregnant gilts. *Prostaglandins* **14**, 1183–1196

FRANK, M., BAZER, F.W., THATCHER, W.W. and WILCOX, C.J. (1978). A study of prostaglandin $F_{2\alpha}$ as the luteolysin in swine. IV. An explanation for the luteotrophic effect of estradiol. *Prostaglandins* **15**, 151–160

GADSBY, J.E. and HEAP, R.B. (1978). Steroid hormones and their synthesis in the early embryo. In *Novel Aspects of Reproductive Physiology* (C.H. Spilman and J.W. Wilks, Eds.), pp. 263–283. New York, SP Medical and Scientific Books

GADSBY, J.E., HEAP, R.B. and BURTON, R.D. (1980). Oestrogen synthesis by blastocyst and early embryonic tissue of various species. *J. Reprod. Fert.* **60**, 409–417

GARDNER, M.L., FIRST, N.L. and CASIDA, L.E. (1963). Effect of exogenous estrogens on corpus luteum maintenance in gilts. *J. Anim. Sci.* **22**, 132–134

GENTRY, B.E., ANDERSON, L.L. and MELAMPY, R.M. (1973). Exogenous progesterone and estradiol benzoate on early embryonic survival in the pig. *J. Anim. Sci.* **37**, 722–727

GLASGOW, B.R., MAYER, D.T. and DICKERSON, G.E. (1951). The colorimetric determination of the excretion products of progesterone and their correlation with certain phenomena of pregnancy in the sow. *J. Anim. Sci.* **10**, 1076 (Abstract)

GLEESON, A.R. (1974). Luteal function in the cyclic sow after infusion of prostaglandin $F_{2\alpha}$ through a uterine vein. *J. Reprod. Fert.* **36**, 487–488

GLEESON, A.R. and THORBURN, G.D. (1973). Plasma progesterone and prostaglandin F concentrations in the cyclic sow. *J. Reprod. Fert.* **32**, 343–344

GOLDENBERG, R.L., BRIDSON, W.E. and KOHLER, P.O. (1972). Estrogen stimulation of progesterone synthesis by porcine granulosa cells in culture. *Biochem. biophys. Res. Commun.* **48**, 101–107

GUTHRIE, H.D. and POLGE, C. (1976). Luteal function and oestrus in gilts treated with a synthetic analogue of prostaglandin $F_{2\alpha}$ (ICI 79,939) at various times during the oestrous cycle. *J. Reprod. Fert.* **48**, 423–425

GUTHRIE, H.D. and POLGE, C. (1978). Treatment of pregnant gilts with a prostaglandin analogue, cloprostenol, to control oestrus and fertility. *J. Reprod. Fert.* **52**, 271–273

GUTHRIE, H.D. and REXROAD, C.E., Jr. (1980). Progesterone secretion and prostaglandin F release *in vitro* by endometrial and luteal tissue of cyclic pigs. *J. Reprod. Fert.* **60**, 157–163

GUTHRIE, H.D. AND REXROAD, C.E., Jr. (1981). Endometrial prostaglandin F release *in vitro* and plasma 13,14-dihydro-15-keto-prostaglandin $F_{2\alpha}$ in pigs with luteolysis blocked by pregnancy, estradiol benzoate or human chorionic gonadotrophin. *J. Anim. Sci.* **52**, 330–339

GUTHRIE, H.D., HENRICKS, D.M. and HANDLIN, D.L. (1972). Plasma estrogen, progesterone and LH prior to estrus and during early pregnancy in pigs. *Endocrinology* **91**, 675–679

HAINES, C.E., WARNICK, A.C. and WALLACE, H.D. (1958). The effect of exogenous progesterone and level of feeding on prenatal survival in gilts. *J. Anim. Sci.* **17**, 879–885

HALLFORD, D.M., WETTEMANN, R.P., TURMAN, E.J. and OMTVEDT, I.T. (1975). Luteal function in gilts after prostaglandin $F_{2\alpha}$. *J. Anim. Sci.* **41**, 1706–1716

HAMMOND, J. (1914). On some factors controlling fertility in domestic animals. *J. agric. Sci.* **6**, 263

HAMMOND, J. (1921). Further observations on the factors controlling fertility and fetal atrophy. *J. agric. Sci.* **11**, 337–366

HANLY, S. (1961). Prenatal mortality in farm animals. *J. Reprod. Fert.* **2**, 182–194

HARRISON, F.A. (1979). Luteolysin in the pig. *J. Physiol.* **290**, 36P.

HATTERSLEY, J.P., DRANE, H.M., MATHEWS, J.G., WRATHALL, A.E. and SABA, N. (1980). Estimation of oestrone sulphate in the serum of pregnant sows. *J. Reprod. Fert.* **58**, 7–12

HEAP, R.B., FLINT, A.P.F. and GADSBY, J.E. (1981). Embryonic signals and maternal recognition. In *Cellular and Molecular Aspects of Implantation*, (S. Glasser and D.W. Bullock, Eds.), pp. 311–322. New York, Plenum

HEAP, R.B., FLINT, A.P.F. and JENKIN, G. (1978). Control of ovarian function during the establishment of gestation. In *Control of Ovulation*, (D.B. Crighton, N.B. Haynes, G.R. Foxcroft and G.E. Lamming, Eds.), pp. 295–318. London, Butterworths

HEAP, R.B., FLINT, A.P.F., HARTMAN, P.E., GADSBY, J.E., STAPLES, L.D., ACKLAND, N. and HAMON, M. (1981). Oestrogen production in early pregnancy. *J. Endocr., Suppl.* **89**, 77P–94P

HENDERSON, K.M., SCARAMUZZI, R.J. and BAIRD, D.T. (1977). Simultaneous infusion of prostaglandin E_2 antagonizes the luteolytic effect of prostaglandin $F_{2\alpha}$ *in vivo*. *J. Endocr.* **72**, 379–383

HENRICKS, D.M. and TINDALL, D.J. (1971). Metabolism of progesterone-4-^{14}C in porcine endometrium. *Endocrinology* **89**, 920–924

HUNTER, R.H.F. (1967). The effects of delayed insemination on fertilization and early cleavage in the pig. *J. Reprod. Fert.* **13**, 133–147

JUNGBLUT, P.W., GAUES, J., HUGHES, A., KALLWEIT, E., SIEERALTA, W., SZENDRO, P. and WAGNER, R.K. (1976). Activation of transcription-regulating protein by steroids. *J. Steroid Biochem.* **7**, 1109–1116

KELLY, P.A., TSUSHIMA, T., SHIU, R.P.C. and FRIESEN, H.G. (1976). Lactogenic and growth hormone-like activities in pregnancy determined by radioreceptor assays. *Endocrinology* **99**, 765–774

KIDDER, H.E., CASIDA, L.E. and GRUMMER, R.H. (1955). Some effects of estrogen injection on the estrual cycle of gilts. *J. Anim. Sci.* **14**, 470–474

KING, J.W.B. and YOUNG, G.B. (1957). Maternal influences on litter size in pigs. *J. agric. Sci.* **48**, 457–463

KNIGHT, J.W., BAZER, F.W. and WALLACE, H.D. (1973). Hormonal regulation of porcine uterine protein secretion. *J. Anim. Sci.* **36**, 546–553

KNIGHT, J.W., BAZER, F.W., THATCHER, W.W., FRANKE, D.E. and WALLACE, H.D. (1977). Conceptus development in intact and unilaterally hysterectomized-ovariectomized gilts: interrelations among hormonal status, placental development, fetal fluids and fetal growth. *J. Anim. Sci.* **44**, 620–637

KRZYMOWSKI, T., KOTWICA, J. and STEFAŃCZYK, S. (1981). Venous-arterial countercurrent transfer of [^{3}H]testosterone in the vascular pedicle of the sow ovary. *J. Reprod. Fert.* **61**, 317–323

LASLEY, E.L. (1957). Ovulation, prenatal mortality and litter size in swine. *J. Anim. Sci.* **16**, 335–340

LERNER, E.H., MAYER, D.T. and LASLEY, J.F. (1957). Early embryonic mortality in strain crossbred gilts. *Res. Bull. Mo. agric. Exp. Stn* No.629

LONGENECKER, D.E. and DAY, B.N. (1972). Maintenance of corpora lutea and pregnancy in unilaterally pregnant gilts by intrauterine infusion of embryonic tissue. *J. Reprod. Fert.* **31**, 171–177

McFEELY, R.A. (1967). Chromosomal abnormalities in early embryos of the pig. *J. Reprod. Fert.* **13**, 579–581

McGOVERN, P.T., MORCOM, C.B., DE SA, W.F. and DUKELOW, W.R. (1981). Chorionic surface area in conceptuses from sows treated with progesterone and oestrogen during early pregnancy. *J. Reprod. Fert.* **61**, 439–442

MARRABLE, A.W. and ASHDOWN, R.R. (1967). Quantitative observations on pig embryos of known ages. *J. agric. Sci.* **69**, 443–447

MARTIN, P.A., BEVIER, G.W. and DZIUK. P.J. (1978). The effect of disconnecting the uterus and ovary on the length of gestation in the pig. *Biol. Reprod.* **18**, 428–433

MAUGET, R. (1978). Seasonal reproductive activity in the European wild boar. Comparison with the domestic sow. In *Environmental Endocrinology*, (I. Assenmacher and D.S. Farner, Eds.), pp. 79–80. Berlin, Springer Verlag

MAYER, D.T., GLASGOW, B.R. and GAWIENOWSKI, A.M. (1961). Metabolites of progesterone and their physiological significance in the urine of pregnant and non-pregnant sows and gilts. *J. Anim. Sci.* **20**, 66–70

MOELJONO, M.P.E., BAZER, F.W. and THATCHER, W.W. (1976). A study of prostaglandin $F_{2\alpha}$ as the luteolysin in swine: I. Effect of prostaglandin $F_{2\alpha}$ in hysterectomized gilts. *Prostaglandins* **11**, 737–743

MOELJONO, M.P.E., THATCHER, W.W., BAZER, F.W., FRANK, M., OWENS, M.J. and WILCOX, C.J. (1977). A study of prostaglandin $F_{2\alpha}$ as the luteolysin in swine: II. Characterization and comparison of prostaglandin F, oestrogen and progestin concentrations in utero-ovarian vein plasma of non-pregnant and pregnant gilts. *Prostaglandins* **14**, 543–555

MOFFATT, R.J., MURRAY, F.A., GRIFO, A.P., Jr., HAYNES, L.W., KINDER, J.E. and WILSON, G.R. (1980). Identification of riboflavin in porcine uterine secretions. *Biol. Reprod.* **23**, 331–335

MURRAY, F.A., MOFFAT, R.J. and GRIFO, A.P. (1980). Secretion of riboflavin by the porcine uterus. *J. Anim. Sci.* **50**, 926–929

MURRAY, F.A., Jr., BAZER, F.W., WALLACE, H.D. and WARNICK, A.C. (1972). Quantitative and qualitative variation in the secretion of protein by the porcine uterus during the estrous cycle. *Biol. Reprod.* **7**, 314–320

NAKANO, R., AKAHORI, T., KATAYAMA, K. and TOJO, S. (1977). Binding of LH and FSH to porcine granulosa cells during follicular maturation. *J. Reprod. Fert.* **51**, 23–27

NISWENDER, G.D., DZIUK, P.J., KALTENBACH, C.C. and NORTON, H.W. (1970). Local effects of embryos and the uterus on corpora lutea in gilts. *J. Anim. Sci.* **30**, 225–228

OMTVEDT, I.T., NELSON, R.E., EDWARDS, R.L., STEPHENS, D.F. and TURMAN, E.J. (1971). Influence of heat stress during early, mid and late pregnancy of gilts. *J. Anim. Sci.* **32**, 312–317

PACK, B.A. and BROOKS, S.C. (1974). Cyclic activity of estrogen sulfotransferase in the gilt uterus. *Endocrinology* **95**, 1680–1690

PACK, B.A, CHRISTENSON, C., DOURAGHY, M. and BROOKS, S.C. (1978). Nuclear and cytosolic estrogen receptor in gilt endometrium throughout the estrous cycle. *Endocrinology* **103**, 2129–2136

PARVIZI, N., ELSAESSER, F., SMIDT, D. and ELLENDORFF, F. (1976). Plasma luteinizing hormone and progesterone in the adult female pig during the oestrous cycle, late pregnancy and lactation, and after ovariectomy and pentobarbitone treatment. *J. Endocr.* **69**, 193–203

PATEK, C.E. and WATSON, J. (1976). Prostaglandin F and progesterone secretion by porcine endometrium and corpus luteum *in vitro*. *Prostaglandins* **12**, 97–111

PERRY, J.S. (1954). Fecundity and embryonic mortality in pigs. *J. Embryol. exp. Morph.* **2**, 308–322

PERRY, J.S. and ROWLANDS, I.W. (1962). Early pregnancy in the pig. *J. Reprod. Fert.* **4**, 175–188

PERRY, J.S., HEAP, R.B. and AMOROSO, E.C. (1973). Steroid hormone production by pig blastocysts. *Nature, Lond.* **245**, 45–47

PERRY, J.S., HEAP, R.B., BURTON, R.D. and GADSBY, J.E. (1976). Endocrinology of the blastocyst and its role in the establishment of pregnancy. *J. Reprod. Fert., Suppl.* **25**, 85–104

PHILLIPPO, M. (1968). Superovulation in the pig. *Adv. Reprod. Physiol.* **3**, 147–166

POLGE, C. and DZIUK, P.J. (1965). Recovery of immature eggs penetrated by spermatozoa following induced ovulation in the pig. *J. Reprod. Fert.* **9**, 357–358

POLGE, C., ROWSON, L.E.A. and CHANG, M.C. (1966). The effect of reducing the number of embryos during early stages of gestation on the maintenance of pregnancy in the pig. *J. Reprod. Fert.* **12**, 395–397

POMEROY, R.W. (1960). Infertility and neonatal mortality in the sow. IV. Further observations and conclusions. *J. agric. Sci.* **54**, 57–66

POPE, C.E., CHRISTENSON, R.K., ZIMMERMAN-POPE, V.A. and DAY, B.N. (1972). Effect of number of embryos on embryonic survival in recipient gilts. *J. Anim. Sci.* **35**, 805–808

RAMPACEK, G.B., ROBISON, D.W. and ULBERG, L.C. (1975). Uterine capacity and progestin levels in superovulated gilts. *J. Anim. Sci.* **41**, 564–567

REDDY, V.B., MAYER, D.T. and LASLEY, J.F. (1958). Hormonal modification of the intra-uterine environment in swine and its effect on embryonic viability. *Res. Bull. Mo. agric. Exp. Stn*, No. 667

RICE, C., ACKLAND, N. and HEAP, R.B. (1981). Blastocyst–endometrial interactions and protein synthesis during pre-implantation development in the pig studied *in vitro*. *Placenta* **2**, 129–142

ROBERTS, R.M., BAZER, F.W., BALDWIN, N. and POLLARD, W.E. (1976). Progesterone induction of lysozyme and peptidase activities in the porcine uterus. *Archs Biochem. Biophys.* **177**, 499–507

ROBERTSON, G.L., CASIDA, L.E., GRUMMER, R.H. and CHAPMAN, A.B. (1951). Some feeding and management factors affecting age at puberty and related phenomena in Chester White and Poland China gilts. *J. Anim. Sci.* **10**, 841–866

ROBERTSON, H.A. and KING, G.J. (1974). Plasma concentrations of progesterone, oestrone, oestradiol-17β and of oestrone sulphate in the pig at implantation, during pregnancy and at parturition. *J. Reprod. Fert.* **40**, 133–141

ROBERTSON, H.A., DWYER, R.J. and KING, G.J. (1980). Effect of oestrogen antisera early in gestation on pregnancy maintenance in the pig. *J. Reprod. Fert.* **58**, 115–120

ROBERTSON, H.A., KING, G.J. and DYCK, G.W. (1978). The appearance of oestrone sulphate in the peripheral plasma of the pig early in pregnancy. *J. Reprod. Fert.* **52**, 337–338

ROLLAND, R., GUNSALUS, G.L. and HAMMOND, J.M. (1976). Demonstration of specific binding of prolactin by porcine corpora lutea. *Endocrinology* **98**, 1083–1091

SABA, N. and HATTERSLEY, J.P. (1981). Direct estimation of oestrone sulphate in sow serum for a rapid pregnancy diagnosis test. *J. Reprod. Fert.* **62**, 87–92

SAMMELWITZ, P.H., ALDRED, J.P. and NALBANDOV, A.V. (1961). Mechanisms of maintenance of corpora lutea in pigs and rats. *J. Reprod. Fert.* **2**, 387–393

SAMMELWITZ, P.H., DZIUK, P.J. and NALBANDOV, A.V. (1956). Effects of progesterone on embryonal mortality of rats and swine. *J. Anim. Sci.* **15**, 1211–1212 (Abstract)

SAMUEL, C.A. (1971). The development of the pig trophoblast in ectopic sites. *J. Reprod. Fert.* **27**, 494–495

SAMUEL, C.A. and PERRY, J.S. (1972). The ultrastructure of pig trophoblast transplanted to an ectopic site in the uterine wall. *J. Anat.* **113**, 139–149

SAUNDERS, P.T.K., ZIECIK, A.J. and FLINT, A.P.F. (1980). Gonadotrophin-like substance in pig placenta and embryonic membranes. *J. Endocr.* **85**, 25P

SCHULTZ, J.R., SPEER,.V.C., HAYS, V.W. and MELAMPY, R.M. (1966). Influence of feed intake and progesterone on reproductive performance in swine. *J. Anim. Sci.* **25**, 157–160

SCOFIELD, A.M. (1972). Embryonic mortality. In *Pig Production*, (D.J.A. Cole, Ed.), pp. 367–383. London, Butterworths

SCOFIELD, A.M. (1975). Embryonic mortality in the pig. In *Veterinary*

Annual Vol. 15, (C.S.G. Grunsell and F.W.G. Hill, Eds.), pp. 91–94. Bristol, Wright

SCOFIELD, A.M., CLEGG, F.G. and LAMMING, G.E. (1974). Embryonic mortality and uterine infection in the pig. *J. Reprod. Fert.* **36**, 353–361

SHORT, R.E., PETERS, J.B., FIRST, N.L. and CASIDA, L.E. (1968). Effect of exogenous progesterone and unilateral ovariectomy on ovarian and pituitary gland activity. *J. Anim. Sci.* **27**, 705–708

SPIES, H.G., SLYTER, A.L. and QUADRI, S.K. (1967). Regression of corpora lutea in pregnant gilts administered antiovine LH rabbit serum. *J. Anim. Sci.* **26**, 768–771

SPIES, H.G., ZIMMERMAN, D.R., SELF, H.L. and CASIDA, L.E. (1959). The effect of exogenous progesterone on formation and maintenance of the corpora lutea and on early embryo survival in pregnant swine. *J. Anim. Sci.* **18**, 163–172

SPIES, H.G., ZIMMERMAN, D.R., SELF, H.L. and CASIDA, L.E. (1960). Maintenance of early pregnancy in ovariectomized gilts treated with gonadal hormones. *J. Anim. Sci.* **19**, 114–118

SQUIRE, G.D., BAZER, F.W. and MURRAY, F.A., Jr. (1972). Electrophoretic patterns of porcine uterine protein secretions during the estrous cycle. *Biol. Reprod.* **7**, 321–325

SQUIRES, C.D., DICKERSON, G.E. and MAYER, D.T. (1952). Influence of inbreeding, age and growth rate on sow's sexual maturity, rate of ovulation, fertilization and embryonic survival. *Res. Bull. Mo. agric. Exp. Stn*, No. 494

STAIGMILLER, R.B., FIRST, N.L. and CASIDA, L.E. (1972). Ovarian compensatory hypertrophy following unilateral ovariectomy in hysterectomized and early pregnant gilts. *J. Anim. Sci.* **35**, 809–813

STONER, C.S., BAZER, F.W., THATCHER, W.W., WILCOX, C.J., COMBS, G.E., KNIGHT, J.W., WETTEMAN, R.P. and WHITE, C.E. (1980). Relationship between plasma estrone sulfate and reproductive performance in the sow. *J. Anim. Sci.* **51**, Suppl. **1**, 280 (Abstract)

STORK, M.G. (1979). Seasonal reproductive inefficiency in large pig breeding units in Britain. *Vet. Rec.* **104**, 49–52

TASSELL, R. (1967). The effects of diet on reproduction in pigs, sheep and cattle. I. Plane of nutrition in pigs. *Br. vet. J.* **123**, 76–83

ULBERG, L.C. and RAMPACEK, G.B. (1974). Embryonic and fetal development: uterine components and influences. *J. Anim. Sci.* **38**, 1013–1017

VANDEPLASSCHE, M. (1969). The physiological explanation of split parturition in the pig and other mammalian species. *Ann. Endocrinol.* **3**, 328–341

WATSON, J. and MAULE WALKER, F.M. (1977). Effect of prostaglandin $F_{2\alpha}$ and uterine extracts on progestrone secretion *in vitro* by superfused pig corpora lutea. *J. Reprod. Fert.* **51**, 393–398

WATSON, J. and PATEK, C.E. (1979). Steroid and prostaglandin secretion by the corpus luteum, endometrium and embryos of cyclic and pregnant pigs. *J. Endocr.* **82**, 425–428

WEBEL, S.K., PETERS, J.B. and ANDERSON, L.L. (1970). Synchronous and asynchronous transfer of embryos in the pig. *J. Anim. Sci.* **30**, 565–568

WILDT, D.E., RIEGLE, G.D. and DUKELOW, W.R. (1975). Physiological temperature response and embryonic mortality in stressed swine. *Am. J. Physiol.* **229**, 1471–1475

WILDT, D.E., CULVER, A.A., MORCOMB, C.B. and DUKELOW, W.R. (1976). Effect of administration of progesterone and oestrogen on litter size in pigs. *J. Reprod. Fert.* **48**, 209–211

WOODY, C.O., FIRST, N.L. and POPE, A.L. (1967). Effect of exogenous progesterone on estrous cycle length. *J. Anim. Sci.* **26**, 139–141

WRATHALL, A.E. (1971). *Prenatal survival in pigs. Part I. Ovulation rate and its influence on prenatal survival and litter size in pigs.* Slough, England, Commonwealth Agricultural Bureau

WYATT, C. (1976). Endometrial components involved in protein synthesis by 16-day blastocyst tissue in culture. *J. Physiol.* **260**, 73–74P

ZAVY, M.T., BAZER, F.W., THATCHER, W.W. and WILCOX, C.J. (1980). A study of prostaglandin $F_{2\alpha}$ as the luteolysin in swine: V. Comparison of prostaglandin F, progestins, estrone and estradiol in uterine flushings from pregnant and non-pregnant gilts. *Prostaglandins* **20**, 837–851

ZAVY, M.T., BAZER, F.W., CLARK, W.R., ROBERTS, R.M. and WILCOX, C.J. (1977). Comparison of constituents of uterine flushings from non-pregnant and pregnant gilts. *Proc. Am. Soc. Anim. Sci. Madison, Wisconsin,* p. 220

ZIECIK, A.J., SHAW, H.J. and FLINT, A.P.F. (1980). Luteal luteinizing hormone receptors during the oestrous cycle and early pregnancy in the pig. *J. Reprod. Fert.* **60**, 129–137

ZIECIK, A., TILTON, J.E. and WILLIAMS, G.L. (1981). Effect of mating on the LH surge in the pig. *J. Anim. Sci.* **53**, 434–438

14

EMBRYO TRANSPLANTATION AND PRESERVATION

C. POLGE
A.R.C. Institute of Animal Physiology, Animal Research Station, Huntingdon Road, Cambridge, UK

Effective techniques for embryo transplantation in some laboratory and farm animals are now well established. Methods for collection and transfer of embryos in the pig were first developed in the early 1960s (Hancock and Hovell, 1962; Dziuk, Polge and Rowson, 1964; Vincent, Robison and Ulberg, 1964) and since then they have been applied mainly in research. Embryo transplantation has proved to be a valuable experimental tool in a number of studies concerned with early embryonic development, the survival of embryos *in vivo* or *in vitro*, migration and spacing of embryos within the uterus and factors affecting the maintenance of pregnancy. Future research is also likely to be concerned increasingly with cellular and genetic manipulation of eggs and embryos *in vitro* and the application of these techniques depends to a large extent on having reliable methods for the culture of embryos and their subsequent transfer.

Practical applications, particularly in farm animals, are also important and the best example is in cattle where methods developed for research have now been extended very successfully into the practice of animal breeding. Applications in pig husbandry have so far been on a relatively small scale. The high fecundity and reproductive rate in pigs compared with cattle will never provide the same economic incentive to apply such methods for the purpose of genetic improvement or to get more offspring from a few superior animals. On the other hand, strict control of disease, especially in large intensive units, is a most important aspect of modern pig husbandry and embryo transplantation should provide the safest method of introducting new genetic material into closed herds. It is mainly for this reason, and perhaps also for the possibility of transporting embryos between countries, that embryo transplantation in pigs is likely to be applied as a practical measure.

Methods of embryo collection and transfer

Techniques for collection and transfer of embryos used at different laboratories are basically very similar to those described by Hancock and Hovell (1962). The methods used routinely in experiments at the Animal Research Station are described here.

COLLECTION OF EMBRYOS

The tortuous nature of the cervix and uterus in the pig virtually precludes the collection of embryos from the uterus by non-surgical means. The surgical approach, however, is relatively simple and operations on donor animals can usually be completed in less than 30 minutes. Following mid-ventral laparotomy under general anaesthesia the reproductive tract is exposed and a region appropriate to the developmental stage of the embryos to be collected (time after ovulation) is then flushed. One-, 2- and early 4-cell embryos can be collected from the oviducts up to about 40 hours after ovulation. A fine glass cannula is inserted into the isthmus through a small hole made in the tip of the uterine horn. Flushing fluid passed down the oviduct from the fimbriated end is collected via the cannula into a glass cup or petri dish. Pig embryos normally enter the uterus much sooner and at an earlier stage of development than in many other species examined. This means that flushing the oviducts alone should be attempted only when the time of ovulation is known quite precisely. Estimates of ovulation time based on the assumption that the endogenous luteinizing hormone (LH) surge coincides with the onset of oestrus can sometimes be misleading. More precise timing of ovulation can be obtained by injection of human chorionic gonadotrophin (HCG) during late pro-oestrus (Dziuk and Baker, 1962; Hunter, 1972). When embryos enter the uterus at the 4-cell stage, it is then virtually impossible to flush them back through the oviducts due to the valve-like nature of the utero-tubal junction. The most efficient technique for obtaining uterine embryos, or indeed those that may be tubal or uterine, is to flush fluid down the oviduct and wash all the embryos away from the tip of the horn. The uterus is clamped with bowel forceps at an appropriate length from the tip of the horn and the flushing medium entering from the oviduct is massaged towards the clamp. A cannula with a larger bore than that used for insertion into the isthmus is introduced into the uterine lumen through a small incision made near the tip of the horn and the fluid is 'milked' back out. Up to 5–6 days after oestrus the embryos remain close to the tips of the horns and it is only necessary to flush a small section. Later, when the embryos have started to migrate throughout the uterus, the whole of each individual horn should be flushed. Several different media have been used successfully for embryo collection, but the medium now used routinely at the Animal Research Station is Dulbecco's phosphate buffered saline enriched with lactate, pyruvate and bovine serum albumin (Whittingham, 1971). An advantage of a medium buffered with phosphate over some others that may be buffered with bicarbonate is that pH is quite well maintained when the medium is exposed to air.

The methods of embryo recovery described are suitable for animals up to about 12 days after the onset of oestrus. If the uterus is flushed at later stages when the embryos have become extremely elongated, the embryonic membranes become entangled together. The efficiency of the technique is very high and the majority of the embryos within the reproductive tract are generally recovered. For example, in a recent experiment 205 gilts were used as embryo donors and the operations were performed 3–9 days after the onset of oestrus. Eggs or embryos were recovered from all animals except one. In 57% of animals the recovery rate, estimated by counting corpora lutea and embryos, was 100% and in the remaining animals only a few embryos were missing. The average recovery rate for all animals (3234 corpora lutea) was 95%.

Repeated operations on donors inevitably tends to build up scar tissue and adhesions, and in our experience 3–4 operations are about the maximum that can be performed successfully on one individual.

Superovulation in donor animals can be achieved by administration of gonadotrophic hormones at an appropriate time in the cycle. A dose of 1000–1500 iu pregnant mare's serum gonadotrophin (PMSG) given early in the follicular phase (day 15 or 16 of the oestrous cycle) usually produces 25–30 ovulations in mature gilts (Hunter, 1964) although the response is quite variable. A mixture of 600 iu PMSG + 200 iu HCG given as a single dose will produce a similar result. Gonadotrophins can also be given after some treatments used for the synchronization of oestrus (Polge, Day and Groves, 1968). Heat normally occurs 3.5–4 days after PMSG treatment, but 500 iu HCG given during the third day after PMSG will synchronize the time of ovulation in a group of animals. Fertilization in superovulated animals is generally normal except when the ovulation rate is excessively high and a number of immature oocytes may be ovulated. Similar treatments can be used to induce ovulation and obtain embryos from prepubertal gilts (Dziuk and Gehlbach, 1966; Baker and Coggins, 1968).

EMBRYO TRANSFER

A similar surgical approach to that applied in donors is used for recipient animals. After exposure of the reproductive tract, embryos can be transferred either to the oviduct or uterus depending on the stage of embryonic development. When transferring to the oviduct the embryos are picked up in about 0.2 ml of fluid in a fine Pasteur pipette which is threaded down the lumen via the fimbria to a depth of about 5 cm. When embryos are transferred to the uterus, a small puncture is made in the isthmus region of the oviduct about 2 cm from the tip of the horn and the tip of the pipette is slid down into the uterine lumen through the utero-tubal junction. This method avoids making a puncture wound in the uterus itself and the consequent possibility of causing endometrial haemorrhage.

In early experiments (Polge, 1966) care was taken to transfer embryos at very early stages of development (2-cell and early 4-cell) only to the oviducts and all later stages to the uterus. In some species the site of transplantation in relation to the stage of embryonic development appears to be important. In cattle, for example, embryonic survival is reduced if embryos collected from the oviducts of donors are transplanted to the uterus of synchronized recipients one or two days earlier than the time they would normally enter the uterus (Newcomb and Rowson, 1975). In pigs, however, the length of time that embryos normally remain within the oviducts is quite short and in recent experiments it has been found that embryonic survival is not reduced when 2-cell embryos collected from the oviducts are transplanted to the uterus of recipients at the same stage of the reproductive cycle. An alternative method of transfer for all embryos can therefore be used and this is simply to flush them down the oviduct and into the uterine lumen in a larger volume of fluid. It is generally only necessary to transfer the embryos to one side of the uterus since they will later migrate and become evenly spaced throughout both horns (Dziuk, Polge and Rowson, 1964).

Attempts at non-surgical transfer of embryos via the cervix in the pig have not been very successful (Polge and Day, 1968). It is possible, however, that embryo transfer could be achieved by means of laparoscopy (B.N. Day, personal communication). This approach could be most useful on farms where suitable facilities for surgery might be lacking.

Synchronization of oestrous cycles

An important factor affecting success in embryo transfer is the degree of synchrony between recipient and donor animals. In most species examined it has been found that pregnancy rate and embryonic survival are reduced if the stage of the reproductive cycle of the recipient is more than about one day out of phase with that of the donor.

In many of the early experiments on embryo transfer in the pig, methallibure was used to control the time of onset of oestrus in donor and recipient gilts. This drug was a most effective synchronizing agent and, in addition, gonadotrophins could be used following treatment either to stimulate superovulation or induce ovulation at a predetermined time (Polge, Day and Groves, 1968). Since the use of methallibure was withheld in many countries in the early 1970s on the grounds that it was a teratogenetic agent, alternative methods for oestrus synchronization have been sought. In pigs, analogues of prostaglandin $F_{2\alpha}$ do not induce luteolysis effectively when given earlier than day 12 or 13 of the oestrous cycle (Guthrie and Polge, 1976a) and there is no way therefore that these compounds alone can be used to synchronize oestrus in randomly cycling animals. Various approaches have been attempted in order to create groups of gilts in which the ovaries of all animals contained corpora lutea which were old enough to respond to exogenous prostaglandin. The methods included the induction of accessory corpora lutea (Guthrie and Polge, 1976b) and the extension of luteal function by oestrogen. Perhaps the most successful method was simply to prolong luteal function by means of pregnancy. In animals injected with cloprostenol (ESTRU-MATE, ICI Ltd) 12–40 days after mating, abortion was induced and a very high proportion returned to oestrus 4–7 days later. Insemination at the synchronized oestrus resulted in normal levels of fertilization and embryonic survival (Guthrie and Polge, 1978).

A simpler and more effective method of oestrus synchronization is now available however. In the past, several orally active progestational agents had been examined in pigs, but there were always major drawbacks to their uses, including reduced fertility after treatment and the induction of follicular cysts. Encouraging preliminary results with a new orally active progestin, allyl trenbolone or altrenogest (REGU-MATE, Roussel Ltd) were first described at a recent Nottingham Easter School (Webel, 1978) and since then trials have been carried out at the Animal Research Station.

In these trials two hundred and twenty eight mature, crossbred gilts that had had at least one oestrus were divided into groups of six and fed a normal diet containing altrenogest. Treatment was started on any day of the oestrous cycle and a dose of 15 or 20 mg altrenogest/pig/day was given for a period of 18 days. Twice daily testing for oestrus with a boar was carried out during and after treatment. Oestrus was effectively suppressed in all animals during treatment, but following withdrawal of the compound there was a spontaneous rebound to follicular activity. Oestrus was detected in 98% of animals within eight days after treatment and about 80% of the heats occurred on the fifth or sixth day (*Table 14.1*). The maximum oestrus

Table 14.1 OESTRUS SYNCHRONIZATION IN GILTS

		Altrenogest treatment for 18 days	
		15 (mg/day)	*20* (mg/day)
Number treated		144	84
Number showing oestrus after treatment		141	83
Day of onset of oestrus after treatment (% of animals)	4	8.5	–
	5	58.9	25.3
	6	27.0	53.0
	7	5.7	16.9
	8	–	4.8
Average number of corpora lutea/animal		15.5	15.6
% eggs fertilized		89.8	92.3

response was on the fifth day after treatment in animals that had received 15 mg/day and on the sixth day after treatment in animals that had received 20 mg/day. Most of the gilts in this trial were used in embryo transfer experiments. About half of them, which were used as donors, were therefore artificially inseminated (two inseminations with 50 ml fresh undiluted semen on the second day of oestrus). The fertilization rate of eggs collected 3–9 days later was over 90%. At laparotomy in donor and recipient animals it was also possible to examine the state of the ovaries and the incidence of follicular cysts was found to be negligible. Altrenogest, therefore, appears to be an exceedingly effective drug for oestrus synchronization in pigs. The only problem relating to its use in embryo transfer experiments, however, may be that the induction of superovulation in donor animals by means of PMSG and HCG has so far produced rather variable results (Polge, 1981).

Early embryonic development

Early embryonic mortality is a common feature in most animals and in the pig it has been estimated that about 30% of the potential embryos are lost by the 25th day of gestation (Hanly, 1961). In a study on early pregnancy in the pig, Perry and Rowlands (1962) made observations on gilts and sows killed between 2 and 40 days after mating and determined the incidence of embryonic loss and the time and stage of development when the loss occurred. An important observation was that 22% of the embryos recovered from the uterus between the sixth and ninth days after mating appeared to be degenerating.

During the course of experiments on embryo transplantation at the Animal Research Station an opportunity has been afforded to examine a large number of embryos collected during the first 10 days after oestrus. These studies are interesting from the point of view of the efficiency of fertilization and the extent of early embryonic degeneration in pigs. In the experiment referred to in an earlier section, the embryos collected from 205 donors 3–9 days after the onset of oestrus were examined. Since the average recovery rate was 95%, very few embryos were lost through technical reasons and in the majority of cases the whole 'litter' representing all the eggs that had been ovulated was available. A total of 3085 embryos was examined. Animals had been treated with altrenogest and inseminated on the second day of oestrus as described above. Out of the 204 animals from which eggs or embryos were recovered, none of the eggs was fertilized in 11 (5%) pigs. In 82% of the remaining animals all the eggs were fertilized whereas in 18% there was a mixture of fertilized and unfertilized eggs. Eggs were classified as unfertilized if they were single cell or fragmenting and contained no sperm in the zona pellucida. The proportion of unfertilized eggs in individual animals with partial fertilization varied quite considerably; in some it was just one or two eggs out of the total, but in others the number was much higher. The main cause of fertilization failure related to unilateral sperm transport within the uterus and fertilized eggs were recovered from one uterine horn only. Overall, 34% of the eggs were unfertilized in the animals in which fertilization was not complete. Apart from fertilization failure there was very little evidence of embryonic loss up to nine days after the onset of oestrus. In all embryos classified as fertilized, less than 1% showed any obvious signs of degeneration. These observations are clearly not in agreement with those of Perry and Rowlands (1962). It should be noted, however, that these authors treated their data with some caution and stated that it was possible that some of the ova collected 6–9 days after mating and classified as degenerating were, in fact, viable. They agree that if this were the case, then mortality among fertilized eggs up to the ninth day must be very small.

Our data certainly suggest that embryonic mortality up to the ninth day is very small, but these observations should also be treated cautiously. When embryos are used in transfer experiments it is possible only to examine them in a relatively superficial manner under a stereomicroscope. It may be that a more detailed examination of fixed and stained embryos could reveal abnormalities which are not immediately obvious. For example it was often noted that some of these embryos at the late morula and early blastocyst stage appeared to have some cells developing outside the main body of the embryos themselves. These were not classified as degenerate. Also, after the blastocysts had hatched, there were frequently considerable discrepancies in the size of embryos within a 'litter', although all appeared to be viable. Whether such features reflect the ability of embryos to survive at a later stage is not known. Genetic abnormalities resulting from errors arising around the time of fertilization have been suggested as a cause for early embryonic mortality. However, in a detailed cytogenetic study of pig embryos collected from animals up to 12 days after ovulation, very few chromosomal abnormalities could be detected (Lupse, 1973).

Other observations, also recorded in the study of Perry and Rowlands (1962), suggest that a large amount of embryonic loss, at least during the first 2–2.5 weeks of pregnancy, is not characteristic of most animals. Embryonic loss in thirteen pigs killed between the 13th and 18th days was 28.4%, but the greater part of this loss occurred in two of the animals. The average loss in the remaining 11 animals was only 12%. Undoubtedly, however, embryonic losses are higher in the majority of animals by the 30th day of pregnancy.

Results achieved in embryo transfer

When embryos have been collected from donors 2–5 days after the onset of oestrus and transferred to synchronous unmated recipients, the pregnancy rate achieved has been 60–70% with embryonic survival in pregnant animals (usually determined about the 30th day of pregnancy) also around 60–70% (Dziuk, Polge and Rowson, 1964; Vincent, Robison and Ulberg, 1964). Transfer of embryos at later stages, seven or eight days after the onset of oestrus, has resulted either in a complete failure to establish pregnancy (Webel, Peters and Anderson, 1970) or in much reduced pregnancy rates (Hunter, Polge and Rowson, 1967). Relatively little is known about the importance of synchrony between donors and recipients in the pig, although in one experiment there was evidence that transfers in which oestrus in the donor was one or two days earlier or one day later than the recipient were as successful as synchronous transfers (Webel, Peters and Anderson, 1970).

Quite a large scale experiment has recently been undertaken at the Animal Research Station in order to determine more precisely the effects relating to degree of synchrony between donors and recipients and stage of embryonic development at which transfers are made. Although this experiment has not yet been completed, enough has been achieved to provide some interesting results. Embryos have been collected from donors on days 3, 4, 5, 6, 7, 8 and 9 (onset of oestrus = day 0). Transfers have been made to recipients in which the onset of oestrus was either synchronous with that of the donors or one or two days earlier or later. Five recipients have been allocated to each group and thus the experiment involves 175 transfers of which results are now available for 140. The time of onset of oestrus in groups of animals was controlled by feeding altrenogest and the methods for collection and transfer of embryos were those described earlier. The average number of embryos transferred to recipients was 14 (range 11–18) and embryonic survival in pregnant animals was determined at slaughter on the 30th day of pregnancy.

The results are shown in *Figures 14.1* and *14.2*. Pregnancy rate was over 70% when transfers were made to recipients in which the onset of oestrus was either synchronous with that of the donors or one or two days later. In fact, the highest pregnancy rate (86%) was achieved in recipients which came on heat two days later than the donors *Figure 14.1(a)*. By contrast,

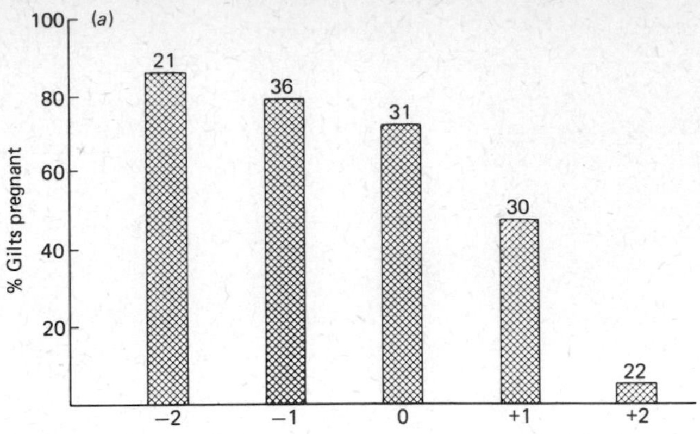

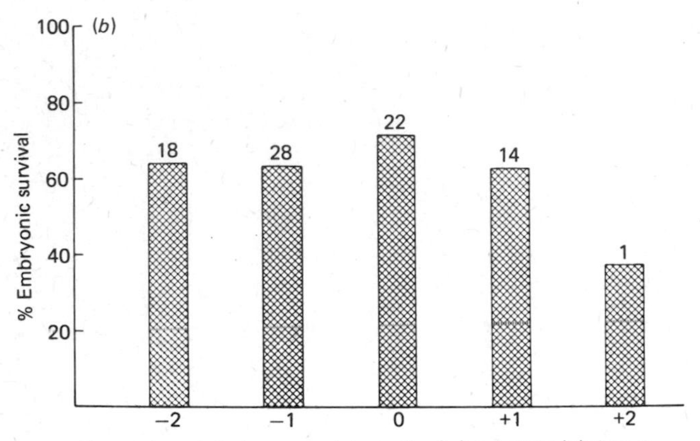

Figure 14.1 Embryo transplantation in gilts: differences in synchrony between recipients and donors. (a) Pregnancy rate following transfer of embryos collected from donors 3–9 days after the onset of oestrus. (b) Embryonic survival rate in pregnant gilts on the 30th day of pregnancy. Numbers above columns refer to the numbers of recipients in (a) and the numbers of pregnant animals in (b)

pregnancy rate fell dramatically in recipients which were ahead of the donors and only one animal out of 22 became pregnant in the groups that came on heat two days before the donors. Results were similar when embryos were collected from donors at any stage of the cycle from days 3–9. The average embryonic survival rate in pregnant animals was 65% and was lowest in the one animal that was two days ahead of the donor (*Figure 14.1(b)*. In *Figure 14.2*, the results have been presented according to the day of the cycle of the donor on which embryos were collected. The figures include results from all recipients except those which were two days ahead of the donors since virtually no pregnancies were achieved in these groups. The main point of interest is that, although the results were somewhat lower in groups receiving embryos collected from donors on days 8 and 9, there was by no means a dramatic fall in pregnancy rate or embryonic survival following transfer of these older embryos.

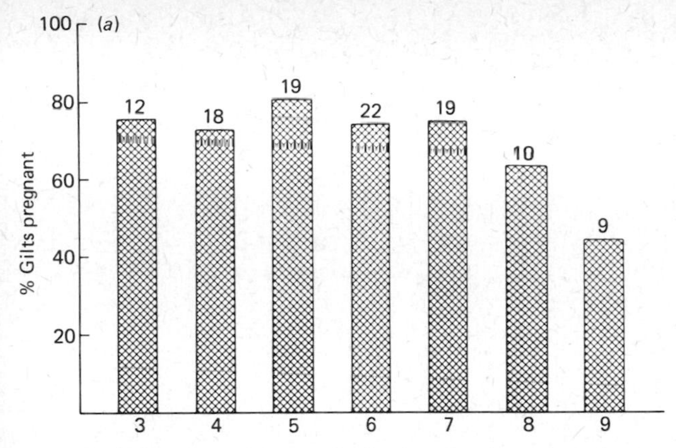

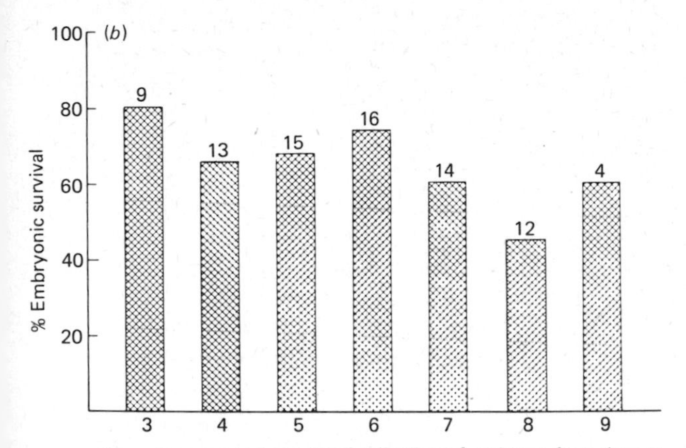

Figure 14.2 Embryo transplantation in gilts: transfer of embryos collected from donors 3–9 days after the onset of oestrus. (a) Pregnancy rate in recipients in which onset of oestrus varied from one day before to two days after that of donors. (b) Embryonic survival in pregnant gilts on the 30th day of pregnancy. Numbers above columns refer to the numbers of recipients in (a) and the numbers of pregnant animals in (b)

Embryonic survival following asynchronous transfer in pigs now confirms observations made in some other species which show that deviation on the side of transferring 'older' eggs to 'younger' uteri is often better tolerated than the reverse situation. How much 'younger' than two days the uteri could be without jeopardizing embryonic survival remains to be determined. In the rabbit it has been shown that exposure of the early embryo to a more advanced progestational uterus for less than 24 hours is incompatible with survival (Adams, 1971). Pig embryos also fail to survive in uteri which are 48 hours in advance even when transfers are done relatively early in the cycle. It would be interesting to know whether the embryos die soon after transfer or whether, being so much 'younger' than the uterus, they are unable to prevent luteolysis.

Earlier experiments (Webel, Peters and Anderson, 1970) in which no embryos survived when transferred to recipients on days 7 and 8, led to the suggestion that perhaps embryos need to be present in the uterus of the pig before day 7 for pregnancy to be established and maintained. Our recent results do not confirm this suggestion. A possible explanation of the poor results achieved following transfer of older embryos in some earlier experiments could relate to the type of medium used for collection. We have noted that when embryos have hatched from the zona pellucida they are very easily damaged by adverse environmental conditions. Media that will support development of early embryos *in vitro* may be unsuitable for the storage of embryos at a later stage (Robl and Davis, 1981).

Embryo transplantation in research

Embryo transplantation has been used as an experimental tool in a number of studies on relationships between the uterus and embryos during early stages of gestation. In most pregnancies some embryos probably migrate from one uterine horn to the other before implantation in order to establish even spacing throughout the uterus. The incidence of transuterine migration and the mixing of embryos within the uterus has been studied experimentally by transferring genetically marked embryos to each uterine horn (Dziuk, Polge and Rowson, 1964). When embryos from black donors were transferred to the tip of one uterine horn and embryos from white donors were transferred to the tip of the opposite horn, it was found that not only did they migrate from the horn of origin, but in most cases they became interspersed throughout the uterus.

Migration of embryos from the tips of the horns usually occurs soon after the sixth day of gestation. In experiments in which embryos were permitted to enter the uterus from one side only it was found that they started to enter the opposite horn around day 8 or 9 and the uterus was occupied completely by day 15 (Dhindsa, Dziuk and Norton, 1967). The time of cessation of intrauterine migration was studied by restricting the embryos to anterior sections of the uterine horns by means of ligatures applied soon after mating. Removal of the ligatures between days 8 and 13 showed that embryos could still migrate beyond the ligature sites up to day 11 and pregnancy was then maintained. Gilts in which the ligatures were removed on day 13 did not remain pregnant (Polge and Dziuk, 1970).

There is now much evidence to show that the presence of embryos throughout a large part of the uterus by days 12 or 13 is essential for the maintenance of pregnancy. Non-gravid sections of the uterus have been established experimentally, in some cases by transferring embryos to isolated segments (du Mesnil du Buisson, 1966; Anderson, Rathmacher and Melampy, 1966; Day *et al.*, 1967; Dhindsa and Dziuk, 1968). It has been shown that a relationship exists between the length of the non-pregnant uterine segment and maintenance of pregnancy. Pregnancy is usually interrupted when over one half of a uterine horn does not contain embryos. However, pregnancy continues when a smaller sterile uterine segment is isolated adjacent to one ovary even though corpora lutea in the ipsilateral ovary may regress. Effects of non-pregnant uterine segments are similar whether they are located at the tip or base of the uterine horns.

Restriction of embryos into isolated segments of the uterus has shown that pregnancy fails regardless of the nubmer of embryos present within the isolated segment. Maintenance of pregnancy does not therefore depend upon having a sufficient number of embryos to provide a luteotrophic stimulus. It is more important that the embryos are in contact with the greater part of the endometrium in order to prevent luteolysis (see F.W. Bazer, Chapter 12). In pregnancies with a small number of embryos, the embryonic membranes of individual embryos are longer than those of embryos which are more crowded within the uterus. However, when the number of viable embryos is very small around days 12–15, there is still likely to be a considerable area of the endometrium which is not in contact with the embryonic membranes and the chances of maintaining pregnancy are reduced. When the number of embryos present within the uterus has been reduced experimentally by means of embryo transfer, it has been shown that more than four embryos are necessary in gilts for the consistent maintenance of luteal function during early pregnancy (Polge, Rowson and Chang, 1966; Polge *et al.*, 1967).

Embryo transplantation has also been used to increase above normal the number of embryos present within the uterus and to determine whether 'uterine capacity' is an important factor affecting embryonic survival and litter size (Dziuk, 1968; Bazer *et al.*, 1969; Pope *et al.*, 1972). The best evidence supports a conclusion that uterine crowding is not a major factor limiting litter size in pigs during the first 25–30 days of pregnancy. Embryonic survival rates were similar at about day 25 in gilts to which either 12 or 24 embryos had been transferred (Pope *et al.*, 1972). A similar result was obtained in an experiment at the Animal Research Station (*Table 14.2*). Thus, early embryonic mortality appears to be more closely

Table 14.2 EMBRYONIC SURVIVAL IN RECIPIENT GILTS TO WHICH 12 OR 24 EMBRYOS WERE TRANSPLANTED (SYNCHRONOUS TRANSFER ON DAY 5)

Number of embryos/ recipient	*Number of recipients*	*Number pregnant on day 30*	*% pregnant*	*Embryo survival*			
				Total transferred	*Total live embryos on day 30*	*% survived*	*% survived in pregnant gilts*
12	21	15	71.4	244	102	41.8	58.6
24	20	18	90.0	408	247	60.5	61.9
Total	41	33	80.5	652	349	53.5	60.9

associated with intrinsic embryonic factors interacting with the uterine environment than with the ability of the uterus to support only a limited number of embryos. Nevertheless, uterine capacity is an important factor affecting foetal survival during later stages of gestation since litter size at parturition has not generally been increased above normal either by transferring additional embryos or by superovulation (Fenton *et al.*, 1970; Longenecker and Day, 1968).

Culture and preservation of embryos

In experiments on embryo transplantation, the time elapsing between collection of embryos from donors and their transfer to recipients is generally not more than about one hour. Since the survival rate of embryos has been normal, at least with embryos collected from donors up to seven days after the onset of oestrus, the phosphate-buffered saline medium (Whittingham, 1971) which is used routinely, appears to be quite satisfactory for temporary storage. However, this medium may be less suitable for the storage of hatched blastocysts collected at later stages.

For longer-term preservation of embryos, media and conditions are required which will support normal development *in vitro*. Various media have been examined for the culture of pig embryos; for a review, see Wright and Bondioli (1981). A characteristic finding in the majority of experiments has been that development *in vitro* of 1- and 2-cell embryos is generally very limited and few progress beyond the 4-cell stage (Polge and Frederick, 1968; Rundell and Vincent, 1969). By contrast, a variety of media has been found to support development *in vitro* of embryos collected at later stages. Late 4-cell embryos and morulae have been cultured to hatched blastocysts (Wright, 1977; Lindner and Wright, 1978). Indeed, very simple media appear to support cleavage and development. Using a modified Krebs Ringer bicarbonate medium with various additives, Davis and Day (1978) found that more eggs cultured in medium without lactate and pyruvate formed blastocysts than when lactate and pyruvate were included. Pyruvate alone inhibited development and 4-cell eggs formed blastocysts when only bovine serum albumin was added to the inorganic salt solution.

The ability of embryos to continue to cleave *in vitro*, however, may be a poor indication of their ability to survive when transplanted back to recipients. Transplantation of embryos cultured *in vitro* for 24 hours has resulted in normal embryonic survival. By contrast, embryo survival has been negligible in gilts to which embryos have been transferred after 48 or 72 hours culture (Pope and Day, 1977; Davis and Day, 1978).

The lack of development beyond the 4-cell stage of 1- and 2-cell embryos cultured *in vitro* is an interesting phenomenon. We have noted that cleavage *in vivo* is also arrested at the 4-cell stage. Following fertilization, the first cleavage *in vivo* normally occurs at 16–18 hours and by 30 hours after fertilization the majority of embryos have cleaved to 4-cells. In many pigs, however, embryos recovered at 60–70 hours after ovulation are still at the 4-cell stage. Soon after this, mixtures of 4- to 8-cell embryos are recovered (Lupse, 1973). Studies on protein synthesis of pig embryos during early stages of development *in vivo* show that a number of new proteins are synthesized during the 'lag' phase at 4-cells. In some 1- and 2-cell embryos cultured *in vitro*, however, the new pattern of protein synthesis is not observed (Osborn and Polge, unpublished observations). If this new pattern of protein synthesis is essential for further normal development of the embryos, it would provide an explanation for the lack of development *in vitro* beyond the 4-cell stage. Similar problems could be involved in longer-term culture of embryos *in vitro* which could account for their poor survival following transplantation. Changing patterns of metabolism and protein synthesis during early development might also be related to early embryonic mortality. Some embryos, for example, may not 'switch on' as soon as others. It has already been shown that embryonic survival is reduced when they are transplanted to a uterine environment which is more advanced than that from which they have been recovered. It seems unlikely, however, that there is any very specific uterine factor in the pig which regulates early embryonic development since pig embryos survive and develop normally when transferred to the oviduct of a rabbit (Polge, Adams and Baker, 1972).

Preservation of embryos by freezing and storage in liquid nitrogen has now been achieved in a number of species. So far, however, pig embryos have not been frozen and thawed successfully. A major problem is their sensitivity to cooling *per se* and few have survived after exposure to temperatures below +15 °C (Polge, Wilmut and Rowson, 1974). During cooling and rewarming it has been observed that there is a loss of intracellular lipids, which are very abundant in the pig embryo. It would be interesting to speculate that it is the lipid composition of the embryonic membranes which affects their sensitivity to cooling (Polge, 1977).

Despite the inability to freeze pig embryos, some successful long-distance shipments of embryos have been achieved (Baker and Dziuk, 1969; James *et al.*, 1980). In the latter experiment, embryos were transported in culture medium from the USA to England and transplanted to recipients within 20–27 hours after collection from donors. Seven of the 12 recipients farrowed producing 58 piglets from 227 transferred embryos.

References

ADAMS, C.E. (1971). The fate of fertilized eggs transferred to the uterus or oviduct during advancing pseudopregnancy in the rabbit. *J. Reprod. Fert.* **26**, 99–111

ANDERSON, L.L., RATHMACHER, R.P. and MELAMPY, R.M. (1966). The uterus and unilateral regression of corpora lutea in the pig. *Am. J. Physiol.* **210**, 611–614

BAKER, R.D. and COGGINS, E.G. (1968). Control of ovulation rate and fertilization in prepubertal gilts. *J. Anim. Sci.* **27**, 1607–1610

BAKER, R.D. and DZIUK, P.J. (1970). Aerial transport of fertilized pig ova. *Can. J. Anim. Sci.* **50**, 215–216

BAZER, F.W., ROBISON, O.W., CLAWSON, A.J. and ULBERG, L.C. (1969). Uterine capacity at two stages of gestation in gilts following embryo superinduction. *J. Anim. Sci.* **29**, 30–34

DAVIS, D.L. and DAY, B.N. (1978). Cleavage and blastocyst formation by pig eggs *in vitro*. *J. Anim. Sci.* **46**, 1043–1053

DAY, B.N., POLGE, C., MOOR, R.M., ROWSON, L.E.A. and BOOTH, D. (1967). Local effect of the uterus on the corpora lutea of early pregnancy in swine. *J. Anim. Sci.* **26**, 499 (Abstract)

DHINDSA, D.S. and DZIUK, P.J. (1968). Influence of varying the proportion of uterus occupied by embryos on maintenance of pregnancy in the pig. *J. Anim. Sci.* **27**, 668–672

DHINDSA, D.S., DZIUK, P.J. and NORTON, H.W. (1967). Time of transuterine migration and distribution of embryos in the pig. *Anat. Rec.* **159**, 325–330

DU MESNIL DU BUISSON, F. (1966). Contribution à l'étude du maintien du corps jaune de la truie. Thesis. University of Paris

DZIUK, P.J. (1968). Effect of number of embryos and uterine space on embryo survival in the pig. *J. Anim. Sci.* **27**, 673–676

DZIUK, P.J. and BAKER, R.D. (1962). Induction and control of ovulation in swine. *J. Anim. Sci.* **21**, 697–699

DZIUK, P.J. and GEHLBACH, G.D. (1966). Induction of ovulation and fertilization in the immature gilt. *J. Anim. Sci.* **25**, 410–413

DZIUK, P.J., POLGE, C. and ROWSON, L.E.A. (1964). Intra-uterine migration and mixing of embryos in swine following egg transfer. *J. Anim. Sci.* **23**, 37–42

FENTON, F.R., BAZER, F.W., ROBISON, O.W. and ULBERG, L.C. (1970). Effect of quantity of uterus on uterine capacity in gilts. *J. Anim. Sci.* **31**, 104–106

GUTHRIE, H.D. and POLGE, C. (1976a). Luteal function and oestrus in gilts treated with a synthetic analogue of prostaglandin $F_{2\alpha}$ (ICI 79,939) at various times during the oestrous cycle. *J. Reprod. Fert.* **48**, 423–425

GUTHRIE, H.D. and POLGE, C. (1976b). Control of oestrus and fertility in gilts with accessory corpora lutea by prostaglandin analogues ICI 79,939 and ICI 80,996. *J. Reprod. Fert.* **48**, 427–430

GUTHRIE, H.D. and POLGE, C. (1978). Treatment of pregnant gilts with a prostaglandin analogue, cloprostenol, to control oestrus and fertility. *J. Reprod. Fert.* **52**, 271–273

HANCOCK, J.L. and HOVELL, G.J.R. (1962). Egg transfer in the sow. *J. Reprod. Fert.* **4**, 195–201

HANLY, S. (1961). Prenatal mortality in farm animals. *J. Reprod. Fert.* **2**, 182–194

HUNTER, R.H.F. (1964). Superovulation and fertility in the pig. *Anim. Prod.* **6**, 189–194

HUNTER, R.H.F. (1972). Ovulation in the pig: timing of the response to injection of human chorionic gonadotrophin. *Res. vet. Sci.* **13**, 356–361

HUNTER, R.H.F., POLGE, C. and ROWSON, L.E.A. (1967). The recovery, transfer and survival of blastocysts in pigs. *J. Reprod. Fert.* **14**, 501–502

JAMES, J.E., REESER, P.D., DAVIS, D.L., STRAITON, E.C., TALBOT, A.C. and POLGE, C. (1980). Culture and long-distance shipment of swine embryos. *Theriogenology* **14**, 463–469

LONGENECKER, D.E. and DAY, B.N. (1968). Fertility level of sows superovulated at post weaning estrus. *J. Anim. Sci.* **27**, 709–711

LINDNER, G.M. and WRIGHT, R.W. (1978). Morphological and quantitative aspects of the development of swine embryos *in vitro*. *J. Anim. Sci.* **46**, 711–718

LUPSE, R.M. (1973). Early embryonic development in the pig: A cleavage timing and cytogenetic study. Thesis. The Graduate School of Arts and Sciences, George Washington University, Washington, D.C.

NEWCOMB, R. and ROWSON, L.E.A. (1975). Conception rate after uterine transfer of cow eggs in relation to synchronization of oestrus and age of eggs. *J. Reprod. Fert.* **43**, 539–541

PERRY, J.S. and ROWLANDS, I.W. (1962). Early pregnancy in the pig. *J. Reprod. Fert.* **4**, 175–188

POLGE, C. (1966). Egg transplantation in the pig. *Wld Rev. Anim. Prod.* **4**, 79–86

POLGE, C. (1977). The freezing of mammalian embryos: perspectives and possibilities. In *The Freezing of Mammalian Embryos* (K. Elliott and J. Whelan, Eds.), pp. 3–13. Amsterdam, Elsevier/Exerpta Medica, North-Holland

POLGE, C. (1981). An assessment of techniques for the control of oestrus and ovulation in pigs. In *Steroids in Animal Reproduction* (H. Jasiorowski, Ed.), pp. 73–84. Warsaw Agricultural University: SGGW-AR, Roussel-Uclaf, Warsaw

POLGE, C. and DAY, B.N. (1968). Pregnancy following non-surgical egg transfer in pigs. *Vet. Rec.* **82**, 712

POLGE, C. and DZIUK, P.J. (1970). Time of cessation of intrauterine migration of pig embryos. *J. Anim. Sci.* **31**, 565–566

POLGE, C. and FREDERICK, C.L. (1968). Culture and storage of fertilized pig eggs. *VIth Int. Congr. Anim. Reprod. A.I., Paris,* **1**, 211 (Abstract)

POLGE, C., ADAMS, C.E. and BAKER, R.D. (1972). Development and survival of pig embryos in the rabbit oviduct. *VIIth Int. Congr. Anim. Reprod. A.I., Munich,* **1**, 513–517

POLGE, C., DAY, B.N. and GROVES, T.W. (1968). Synchronization of ovulation and artificial insemination in pigs. *Vet. Rec.* **83**, 136–142

POLGE, C., ROWSON, L.E.A. and CHANG, M.C. (1966). The effect of reducing the number of embryos during early stages of gestation on the maintenance of pregnancy in the pig. *J. Reprod. Fert.* **12**, 395–397

POLGE, C., WILMUT, I. and ROWSON, L.E.A. (1974). Low temperature preservation of cow, sheep and pig embryos. *Cryobiology,* **11**, 560

POLGE, C., MOOR, R.M., DAY, B.N., BOOTH, W.D. and ROWSON, L.E.A. (1967). Embryo numbers and luteal maintenance during early pregnancy in swine. *J. Anim. Sci.,* **26**, 1499 (Abstract)

POPE, C.E. and DAY, B.N. (1977). Transfer of preimplantation pig embryos following *in vitro* culture for 24 or 48 hours. *J. Anim. Sci.* **44**, 1036–1040

POPE, C.E., CHRISTENSEN, R.K., ZIMMERMAN-POPE, V.A. and DAY, B.N. (1972). Effect of number of embryos on embryonic survival in recipient gilts. *J. Anim. Sci.* **35**, 805–808

ROBL, J.M. and DAVIS, D.L. (1981). Effects of serum on swine morulae and blastocysts *in vitro*. *J. Anim. Sci.* **52**, 1450–1456

RUNDELL, J.W. and VINCENT, C.K. (1969). *In vitro* culture of swine ova. *J. Anim. Sci.* **27**, 1196 (Abstract)

VINCENT, C.K., ROBISON, O.W. and ULBERG, L.C. (1964). A technique for reciprocal embryo transfer in swine. *J. Anim. Sci.* **23**, 1084–1088

WEBEL, S.K. (1978). Ovulation control in the pig. In *Control of Ovulation* (D.B. Crighton, N.B. Haynes, G.R. Foxcroft and G.E. Lamming, Eds.), pp. 421–434. London, Butterworths

WEBEL, S.K., PETERS, J.B. and ANDERSON, L.L. (1970). Synchronous and asynchronous transfer of embryos in the pig. *J. Anim. Sci.* **30**, 565–568

WHITTINGHAM, D.G. (1971). Survival of mouse embryos after freezing and thawing. *Nature, Lond.* **233**, 125–126

WRIGHT, R.W. (1977). Successful culture *in vitro* of swine embryos to the blastocyst stage. *J. Anim. Sci.* **44**, 854–858

WRIGHT, R.W. and BONDIOLI, K.R. (1981). Aspects of *in vitro* fertilization and embryo culture in domestic animals. *J. Anim. Sci.* **53**, 702–729

15

PREGNANCY DIAGNOSIS

G.W. DYCK
Animal Science Section, Agriculture Canada Research Station, Brandon, Manitoba, Canada

The diagnosis of pregnancy in the pig has taken several diverse forms involving tests for the presence of the products of conception or tests for the absence of oestrous activity. Thus, some tests determine specifically that the animal is pregnant while others indicate that she is not pregnant. The methods currently available to confirm pregnancy or non-pregnancy are effective only after the beginning of the expected time of implantation, with accuracy of the diagnosis increasing with time after mating. In some animals it may be necessary to employ two or three different tests to obtain an accurate diagnosis. Excluding disease problems which may result in death of the entire litter, the accuracy of pregnancy confirmation approaches 100% by the end of the second trimester. In contrast, the accuracy of detection of the non-pregnant female is extremely variable and dependent on the nature of the reason(s) for her failure to become pregnant.

The problems of pregnancy diagnosis may be divided into the three areas of:

(1) Early confirmation of pregnancy,
(2) Detection of the non-pregnant females,
(3) Rapid 'on farm' confirmation of pregnancy and detection of non-pregnant animals.

Thus, the desired system of pregnancy diagnosis is one that can be operated by the farmer and will confirm pregnancy or non-pregnancy by day 18 after mating. While such a system is not available currently, extensive progress has been made in the last ten years.

Pregnancy diagnosis during the pre-implantation stage

Starting with fertilization of the ova, the developing conceptuses secrete specific biochemical messengers that indicate their presence to the host female. A relatively complete review of the subject has been made by Dickmann, Dey and Sen Gupta (1976). These chemicals are probably closely related to the suppression of immunorejection, to blastocyst growth and later to the maintenance of pregnancy. This assumption is based on observations that blastocysts may be transferred into cycling females only up to eight days after oestrus (Hunter, Polge and Rowson, 1967) and that embryos must be present in both uterine horns between days 10 and 12 (Dhindsa and Dziuk, 1968) for the maintenance of pregnancy. Therefore any diagnostic parameters that could be used before day 8 would identify the presence of the conceptuses while after day 8 detection of the implanted embryos is possible (pregnancy).

At the present time, there are no readily identifiable parameters in the pig for the diagnosis of pregnancy during the pre-implantation stage of development. Changes have been observed in the metabolism of progesterone with an increase in urinary levels of 5β-pregnan-3αol-20-one and 5β-pregnan-3α6α-diol-20-one (Tillson, Erb and Niswender, 1970). Further research is required to determine if these changes in progesterone metabolism could constitute an effective method of pregnancy diagnosis or if there are other hormonal changes that could be used. In the mouse (Morton, Hegh and Clunie, 1976) and sheep (Evinson *et al.*, 1977; Nancarrow *et al.*, 1980) specific pregnancy-associated immunosuppressive activity has been observed in plasma by four hours after fertilization. In the pig similar immunosuppressive activity has been observed (H. Morton, personal communication). Further research is required to confirm its efficacy as a very early pregnancy diagnosis.

Pregnancy diagnosis during embryonic and foetal development

Blastocysts elongate and begin the process of implantation about twelve days after mating. The hormonal interrelationships involved and the nature of their action are discussed in Chapters 12, 13 and 14. The resulting growth of the uterus and embryos, the failure to exhibit oestrus, and the changes in hormone levels in the blood all provide accurate means for the detection of pregnancy.

BLOOD PROSTAGLANDINS

Recently, the determination of blood concentration of prostaglandins by radioimmunoassay has been suggested as an accurate means of pregnancy diagnosis (Martinat-Botte *et al.*, 1980). This procedure is based on the observation of Bazer and Thatcher (1977) that viable embryos prevent the secretion of prostaglandins into the uterine vein between the eleventh and sixteenth day of the cycle. Thus, low blood levels of prostaglandins at this time should be indicative of pregnancy. In an experiment using 389 sows with blood samples collected on day 13, 14 or 15, Martinat-Botte *et al.* (1980) reported an accuracy of 90% in diagnosing 292 of the sows pregnant. They also reported an accuracy of 68% for the detection of the remaining 97 non-pregnant sows.

The failure to detect 10% of the pregnant sows and the low accuracy of detecting non-pregnant sows suggests that further research is required to determine the efficacy of this procedure as a means of early pregnancy diagnosis.

LAPAROSCOPY

For the detection of pregnancy, direct observation of the reproductive tract by laparoscopy provides the earliest accurate diagnosis. In the technique as described by Wildt *et al.* (1973) the gilt is placed head down and supine on a 30° sloped table with the hind legs loosely tied in a vertical position. After surgical preparation, a trocar and cannula are inserted along the mid-line just anterior to the position of the ovaries. The trocar is then withdrawn and a 5 mm laparoscope inserted. A tactile probe for manipulation of the internal organs is inserted about 10 mm lateral to the laparoscope and the abdominal cavity insufflated with 5% CO_2 in air. The entire procedure takes 30–45 minutes. Wildt, Morcom and Dukelow (1975) reported a 100% accuracy in pregnancy diagnosis by the end of the second week after mating. Differences between pregnant and non-pregnant gilts were observed in uterine colouration as a result of increased blood flow to the uterus and in uterine contractions as a result of hormonal stimulation. Changes in luteal regression were observed as early as day 13, and by day 16 differences in follicular development were apparent. Laparoscopy had no effect on foetal survival. While no observations have been reported on the effect of an extended luteal phase on pregnancy diagnosis, the utilization of both uterine colouration differences and ovarian morphology suggests that these animals would be correctly diagnosed as non-pregnant. Other abnormalities such as cystic ovaries, ova-testes and anoestrous sows would be detected and thus culled from the breeding herd. Non-pregnant animals could be mated again at the next oestrus. The only animals that would not be accurately diagnosed are those with late embryonic death.

Laparoscopy, while accurate for the diagnosis of pregnancy, is probably only applicable to research situations. The time involved for each animal, the cost of equipment and the need for personnel trained in surgical procedures all combine to make the technique uneconomical.

BLOOD PROGESTERONE

The analysis of blood, collected between days 17 and 24 after mating, for progesterone concentration has been found to provide a relatively accurate means of diagnosing pregnancy (Robertson and Sarda, 1971; Meyer, Elsaesser and Ellendorff, 1975). A small sample of blood (1ml) is collected from the pig and the plasma analysed for its progesterone concentration using either competitive protein binding or radioimmunoassay procedures. A sow is deemed to be pregnant if the plasma concentration of progesterone is greater than 5 ng/ml. Stabenfeldt *et al.* (1969) have observed plasma progesterone concentrations of less than 5 ng/ml from four days before to three days after the first day of oestrus. In ovariectomized gilts, Ellicot and Dziuk (1973) reported that the minimum daily progesterone requirement for the maintenance of pregnancy was 28.6 mg which corresponded to a plasma concentration of 6 ng/ml. Pregnancy was not maintained at a plasma concentration below 4 ng/ml. Thus, progesterone as a means of pregnancy diagnosis relies on the maintenance of corpus luteum function rather than pregnancy although the two are usually analogous.

In their initial observations Robertson and Sarda (1971) reported an accuracy of 88% for diagnosing pregnancy in 25 sows. The three sows wrongly diagnosed as pregnant had oestrous cycles of 28, 37 and 40 days. In a larger population of animals Williamson, Hennessy and Cutler (1980) reported an accuracy of 97% for confirming pregnancy in 217 sows and 60% for detecting 136 non-pregnant sows. This was a herd with reproductive problems and a further evaluation of the results revealed that the 17 sows with a 'normal' oestrous cycle length were detected. Only 54% of the sows with a delayed return to oestrus were detected. The remaining 46% all had progesterone concentrations similar to those of pregnant sows. In large scale commercial application to 9920 sows the accuracy of pregnancy confirmation was 92% while the detection of non-pregnant sows was 99% (MacNeil, 1979).

Thus, blood progesterone analysis has a definite place as a method of pregnancy diagnosis. In the current commercial application blood samples are collected on day 17, the samples analysed on day 18 or 19 and the results sent back to the producers on day 21 when the open sows could be expected to be in oestrus (MacNeil, 1979). The technique is limited by the need to collect blood samples and the proximity of a laboratory to conduct the analysis. The accuracy of the results for detecting non-pregnant animals is limited by the occurrence of longer or shorter than normal oestrous cycles and late embryonic death. The failure to detect all pregnant sows can be attributed to the occasional sow that has a low blood progesterone concentration.

OESTRUS DETECTION

The failure of the sow to return to oestrus after mating is the most widely used method of diagnosing pregnancy. The value of this technique is based on the knowledge that essentially all pregnant sows will not exhibit oestrus during pregnancy while non-pregnant sows may be expected to return to oestrus between 18 and 25 days after the previous oestrus. Oestrus is confirmed when the sow will assume the 'standing reflex' posture. However, this posture does not always occur in all sows in the absence of a boar regardless of other stimuli such as back pressure, boar odour, boar sounds, or a boar in an adjacent pen (Signoret, 1970). Thus, the physical presence of a boar is the most accurate method of oestrus detection and the confirmation of non-pregnancy. The use of oestrogenic preparations (Nishikawa, 1953) or androgen–oestrogen preparations (Jöchle and Schilling, 1965) have been recommended as an aid to stimulating oestrus in non-pregnant sows. Bosc, Martinat-Botte and Nicolle (1975) found no improvement in the detection of non-pregnant sows following the injection of androgen–oestrogen preparations.

Although widely used for pregnancy diagnosis there are very few reports on the value of oestrus detection as a diagnostic tool. Bosc, Martinat-Botte and Nicolle (1975) reported an accuracy of 39% for the detection of oestrus in 154 non-pregnant sows between 19 and 25 days after insemination. The 94 non-pregnant sows not detected in oestrus comprised 9.6% of the total animal population. The reason for the low rate of oestrus detection is that the sows were not in oestrus at the times of checking. There are many reasons for the failure of the animals to be in oestrus. Several reports (Pásztor and Tóth, 1964; Signoret, 1967; Dyck, 1971) have indicated that between 7 and 25% of the oestrous cycles are outside the 18–25 day range. In addition, sows with silent oestrus or cystic ovaries, anoestrous sows, and sows with late embryonic death, etc. would not be detected. Continued daily checking for oestrus would eventually detect all of these sows but it is extremely time-consuming.

VAGINAL BIOPSY

The initial observations that the histological structure of the vaginal epithelium of the sow showed cyclical changes were made by McKenzie (1924) and Wilson (1926). The utilization of these changes as a tool in pregnancy diagnosis was first recommended by Ciurea *et al.* (1955). The tissue sample for analysis should be collected from the dorsal or lateral wall of the vagina approximately 8 cm back from the cervix and then immediately frozen or fixed in formalin for transport to the laboratory for analysis. Standard techniques for the histological examination of frozen or fixed tissues all provide results of equal accuracy (Diehl and Day, 1973). During proestrus, mitosis and cellular proliferation commence with growth continuing during oestrus when a depth of 5–20 epithelial cells may be observed. During metoestrus a sloughing of the outer cell layers occurs and continues until a 3–5 cell depth is reached. During dioestrus the cell depth remains at 3–5 cells. In the pregnant sow the same changes occur as are found from oestrus through to dioestrus. However, as pregnancy progresses there is a further sloughing of cells to a depth of 2–3 cells by day 22. The differences in the vaginal epithelium of pregnant sows and cycling sows during proestrus, oestrus or metoestrus (days 18–25 after mating) provide an accurate means of pregnancy diagnosis (Morton and Rankin, 1969). Thus, this method of pregnancy diagnosis is based on the absence of oestrous cycle activity.

In a review of the results published between 1963 and 1970, Walker (1972) reported the accuracy of pregnancy confirmation at between 94% and 100% and the accuracy of detecting non-pregnant animals at between 50% and 100%. Examples of more recent studies are shown in *Table 15.1*. Bosc, Martinat-Botte and Nicolle (1975) found no difference between days in the ability to correctly diagnose pregnancy. However, in the non-pregnant sows the accuracy of the diagnosis increased with time after mating. Among sows with a normal return to oestrus (days 18–25), 81% were correctly diagnosed compared with 69% of the sows with a longer than normal oestrous cycle. In this latter group the accuracy of the diagnosis increased from 0% to 69% to 81% and to 91% on days 18–21, 22–25, 26–27 and 28–29 respectively.

Table 15.1 ACCURACY OF THE DIAGNOSIS OF PREGNANCY AND NON-PREGNANCY IN PIGS BY THE VAGINAL BIOPSY METHOD

Authors	*Time after mating* (days)	*Accuracy of diagnosis (%)*	
		Pregnant	*Non-pregnant*
Diehl and Day (1973)	20–25	95.5 (156)	94.4 (18)
Bosc, Martinat-Botte and Nicolle (1975)	18–29	96.7 (573)	71.5 (144)
McCaughey (1979)	17–22	99.4 (168)	77.3 (22)

Numbers in parentheses indicate number of animals examined

The results show that vaginal biopsy can be used effectively as a means of pregnancy diagnosis. The tissue sample requires approximately two minutes to collect and identify. In practice the results are reported back to the producer within 48 hours of receipt at the laboratory. The technique is limited to the collection of samples close to the expected time of oestrus and as such the results are not generally available until after the sow would be expected in oestrus (McCaughey, 1979). The failure to detect all pregnant animals suggests an error in tissue sampling or that in some animals the sloughing of the epithelial cell layers does not proceed as fast as in others. The failure to detect all non-pregnant sows may be attributed to an extended dioestrous condition which could easily be mistaken for pregnancy. This could occur with a longer than normal luteal phase and with anoestrus. Although the earlier work reviewed by Walker (1972) indicated that an accurate diagnosis can be made after day 18 and up to at least day 90, the potential for confusion of dioestrus and pregnancy suggests that the most accurate results can be obtained around the expected time of oestrus.

BLOOD OESTROGENS

The observation that the pregnant sow excretes large quantities of oestrogens in the urine was first demonstrated by Küst (1931). Since that time numerous methods have been developed to measure urinary oestrogen concentration as a means of confirming pregnancy. This research has been reviewed by Kawata and Fukui (1977). These earlier techniques of analysis were found to be of limited value because of the time involved in urine collection and analysis, the complexity and cost of the analysis, and the variability in the results that were obtained. The development of radioimmunoassays for oestrogens and the observations of Perry, Heap and Amoroso (1973) that the increase in oestrogen concentration was the result of production by the embryos has renewed interest in blood oestrogen analysis as a pregnancy diagnosis technique. Robertson and King (1974) reported that the initial increase in blood oestrogen concentration is apparent by day 16 of pregnancy and reaches an initial maximum (>3 ng/ml) between days 23 and 30, followed by a decline to 35 pg/ml on day 46 and an increase to the maximum on the day before parturition. The initial peak in oestrogen concentration is almost entirely oestrone sulphate while the increase observed after day 46 is composed of similar quantities of oestrone and oestrone sulphate. The maximum concentration of oestrone on the day before parturition was sixteen-fold greater than that of oestradiol-17β.

In view of these recent findings the analysis of blood for oestrone sulphate between days 23 and 30 of pregnancy and the analysis for oestrone and oestrone sulphate after day 70 should provide an accurate means of pregnancy diagnosis. The radioimmunoassay system used by Robertson and King (1974) was too complex for routine analysis. Recently Wright *et al.* (1978), Guthrie and Deaver (1979) and Saba and Hattersley (1981) have all developed radioimmunoassay procedures that are suitable for routine analysis. The assay procedure of Saba and Hattersley (1981) appears to have advantages over the others in that the assay is conducted directly on the blood serum sample and thus measures both oestrone and oestrone sulphate. They reported a reduction in accuracy of approximately 12% compared with oestrone sulphate alone. In 87 serum samples collected 26–28 days after mating the mean oestrone plus oestrone sulphate concentration was 2.7 ng/ml. Eleven of these serum samples had less than 1.0 ng/ml of oestrone plus oestrone sulphate but none were less than 0.5 ng/ml. In a comparable group of 28 non-pregnant sows, 21 of the sows had oestrone plus oestrone sulphate concentrations of less than 0.3 ng/ml while none of the sows had a concentration greater than 0.5 ng/ml. Three of these latter blood samples were collected during oestrus.

The analysis of blood for oestrone sulphate appears to have sufficient accuracy to be an effective means of pregnancy diagnosis. However, more extensive field trials are required before an accurate assessment can be made. The observation by Guthrie and Deaver (1979) of an initial rise in oestrone concentration on day 22 for three animals with delayed luteolysis suggests late embryonic death which would have been detected and the sows diagnosed as non-pregnant. Similarly, sows with silent oestrus and anoestrous sows would be identified as non-pregnant. The major disadvantage of the assay is that the blood samples must be obtained after the expected time of oestrus. Potentially the only animals that would not be properly diagnosed are those with late embryonic death. In addition, the oestrone/oestrone sulphate analysis may be of value in determining litter size. Guthrie and Deaver (1979), Chew *et al.* (1979) and Horne and Dziuk (1979) all reported significant correlations between litter size and oestrone sulphate concentration. Further research is still needed to determine the accuracy of this estimate of litter size as approximately one third of the intrauterine death losses occur after day 25 (Dyck, 1974).

RECTAL PALPATION

The use of rectal examination to diagnose pregnancy was first described by Huchzermeyer and Plonait (1960). A detailed description of the procedure and the changes observed by palpation of the cervix and uterus throughout the oestrous cycle and pregnancy have been reported by Cameron (1977). The hand is passed through the anal ring at least 30 cm into the rectum and the cervix found below the rectum. After examination of the cervix and uterus, the middle uterine artery is detected where it crosses the external iliac artery. The external iliac artery appears to be about 1 cm in diameter while the middle uterine artery varies from 2–4 mm in diameter during the oestrous cycle. During pregnancy the middle uterine artery increases in size from about 5 mm at the end of the third week of pregnancy to more than 1 cm as pregnancy progresses beyond 60 days. During pregnancy a distinct fremitus is detected as a continuous pulse when compared with the normal pulse in the external iliac artery. The size of the artery and the condition of the pulse are both positive diagnostic criteria for pregnancy. Palpation of the uterus for the presence of embryos is not possible until late pregnancy.

For the confirmation of pregnancy, rectal palpation has been found to have an accuracy approaching 100% after 28 days (Meredith, 1976). Accuracy of pregnancy diagnosis during the third and fourth weeks was 32% and 75%, respectively. Other examples of the accuracy of diagnosing pregnancy and non-pregnancy are shown in *Table 15.2*.

Table 15.2 ACCURACY OF THE DIAGNOSIS OF PREGNANCY AND NON-PREGNANCY IN PIGS BY THE RECTAL PALPATION METHOD

Authors	*Time after mating* (days)	*Accuracy of diagnosis (%)*	
		Pregnant	*Non-pregnant*
Fritzsch and Hühn (1976)	30–80 (gilts)	80.6 (636)	29.7 (60)
	30–80 (sows)	89.0 (1023)	15.8 (55)
Cameron (1977)	21–30	84.6 (26)	20.0 (5)
	31–60	93.2 (336)	55.8 (52)
	>60	99.0 (98)	66.7 (3)
Benjaminsen and Karlberg (1980)	30–60	100.0 (86)	100.0 (34)

Numbers in parentheses indicate number of animals examined.

Rectal palpation is most limited by the size of the pubic symphysis and thus the size of the palpator's hand. Huchzermeyer and Plonait (1960) recommended a minimum animal weight of 150 kg for ease of palpation. While most reports suggest that pregnancy diagnosis is not accurate before 30 days of pregnancy, Yamanouchi (1973) was able to detect the fremitus two weeks after service. Thus, it may be possible in some animals to confirm pregnancy before the next expected oestrus. The low accuracy of detecting non-pregnant animals in some of the reports suggests that a certain minimum experience is required which would not be possible with occasional palpations. The major advantages are an immediate diagnosis within a few minutes and a minimum of restraint of the animal.

ULTRASONIC ANALYSES

Ultrasonic techniques of pregnancy diagnosis are based either on foetal heart rate and the foetal pulse (Doppler ultrasound), or the detection of foetal fluids in the uterus (A-mode or amplitude-depth ultrasound). Thus, both tests identify the products of conception and are potentially more accurate for the confirmation of pregnancy than techniques based on secondary effects of conception or oestrous activity.

Doppler techniques

The Doppler ultrasound technique, as described by Fraser and Robertson (1967), is based on the alteration of the reflected signal as a result of movement within the body and detected as sounds heard through ear-phones, loudspeaker or visible signal on an oscilloscope. While various sounds may be heard it is easy to detect differences in pulse rate and those of the foetus and mother can be differentiated. The foetal heart rate was found by Too *et al.* (1974) to be 2–3 times the maternal heart rate with the difference decreasing as pregnancy progressed. Using a transducer applied on the skin of the lower abdomen or flank Too *et al.* (1974) determined the accuracy of pregnancy diagnosis from 22 to 114 days of gestation. In the period between days 22 to 39 of pregnancy confirmation was 68% correct while after 40 days the accuracy was 100%. Pierce, Middleton and Phillips (1976) used an intrarectal transducer and were able to detect the foetal heart beat 19 days after mating although not all animals were correctly diagnosed as pregnant until 34 days after mating. Recently, Benjaminsen and Karlberg (1980) reported a 94% accuracy in confirming pregnancy and a 96% accuracy in detecting non-pregnant sows 30 to 60 days after mating.

Amplitude-depth techniques

The A-mode (amplitude-depth) ultrasonic technique as described by Lindahl *et al.* (1975) is similar to the Doppler technique except that it detects an echo from the fluid-filled uterus. The external transducer is placed against the abdominal wall in the area of the flank and in the direction of the uterus. Using an oscilloscope to detect the reflected signal they found that non-pregnant sows produced signals only to a depth of 5 cm. Pregnant animals produced signals at a depth of 15–20 cm. O'Reilly (1976) using 686 pregnant animals found a 55% accuracy in confirming pregnancy between 0 and 30 days after mating, 98% from 31 to 90 days and 84% from 91 days to term. In a group of 69 non-pregnant animals 81% were correctly diagnosed. Lindahl *et al.* (1975) reported a 99% accuracy in diagnosing pregnancy in a group of 801 sows and an accuracy of 98% in diagnosing 189 of these sows as non-pregnant. In a field survey on 21 000 animals the accuracy of diagnosing non-pregnant sows was greater than 95%. A similarly high degree of accuracy was reported by Hansen and Christiansen (1976) with 96 observations on 70 sows. Sixty-two observations were correctly diagnosed as pregnant and 31 observations correctly diagnosed as non-pregnant. The remaining three observations on two sows were incorrectly diagnosed as pregnant due to an abnormal accumulation of fluid in the uterus. In contrast to these results, Benjaminsen and Karlberg (1980) reported an 89% accuracy in confirming pregnancy in 116 sows and a 56% accuracy in detecting 63 non-pregnant sows.

The accuracy of ultrasonic pregnancy diagnosis appears to be most limited by the ability of the operator to properly place and direct the transducer towards the uterus. For the Doppler technique experience in differentiating the specific sounds received also is required. Errors in placement of the A-mode transducer may result in the detection of a fluid-filled bladder which would give false positive results or conversely miss the uterus entirely and thus produce false negative results. The technique however provides immediate results without requiring the assistance of specially trained personnel although accurate results for the diagnosis of non-pregnant animals are often not possible before 40–60 days of pregnancy. The large number of commercially available A-mode units may make the selection of the most desirable unit difficult.

Table 15.3 TECHNIQUES AVAILABLE FOR PREGNANCY DIAGNOSIS

Method	*Stage of pregnancy (days)*	*Basis of diagnosis*	*Accuracy (%)*		*Advantages*	*Limitations*		
			Pregnant	*Non-pregnant*		*Cost*	*Time/sow*	*Other*
Laparoscopy	16–term	Uterine size and condition. Ovarian morphology.	100	100	Immediate results. Done on the farm.	High	30 min	Surgical equipment and trained personnel are required to evaluate the animals.
Blood progesterone	17–24	High concentration (>5 ng/ml plasma). Absence of cyclic oestrous activity.	92–97	55–99 (95)	Results within 2–3 days. Checking for oestrus is not required.	Low 1–2 £/sow	<5 min	Chemical laboratory analysis of the blood is required. Results are not available until after the expected time of oestrus. The occasional pregnant sow with a low progesterone concentration may be called non-pregnant. Non-pregnant sows with extended or short oestrous cycles may be diagnosed as pregnant.
Oestrus detection	18–25 (or 18–term)	Return to oestrus after mating in non-pregnant sows.	100	40–99 (85)	Immediate results. Done on the farm by the operator. Sows may be mated again at the same time.	Nil	<5 min per day	Daily checking for oestrus may take considerable time if the checking is done to term. A boar is required for checking. Animals that are anoestrous or with long cycles, or late embryonic death are not detected when oestrus checking is limited to days 18–25.
Vaginal biopsy	18–25 (or 18–term)	2–4 layers of cells on the vaginal epithelium. Absence of cyclic oestrous activity.	90–100	50–100 (75)	Tissue samples can be collected at any time. Results within 2–3 days. Checking for oestrus is not required.	Low >1 £/sow	<5 min	Histological laboratory analysis of the tissue is required. Results are not available until after the expected time of oestrus. The most accurate results are obtained when tissue samples are collected during proestrus, oestrus and metoestrus. Dioestrous and anoestrous sows may be incorrectly diagnosed as

Blood oestrogens	20–34 (or 20–term) (best at 20–30)	High concentration of oestrone and oestrone sulphate produced by the embryos or foetuses (>0.5 ng/ml serum).	>95	>95	Results within 2–3 days and may also indicate probable litter size.	Low 1–2 £/sow	<5 min	Chemical laboratory analysis of the blood is required. Blood samples are not collected until after the expected time of oestrus. The occasional animal may be pregnant with a low oestrogen concentration. Animals in oestrus may have sufficiently high oestrogen concentrations to be incorrectly diagnosed as pregnant.
Rectal palpation	30–term	Size of the middle uterine artery and the pulse of the blood flow through the artery.	80–100 (95)	20–100 (50)	Immediate results. Done on the farm by the operator.	Nil	<5 min	Palpation is limited by the size of the palpator's hand and the size of the sow (minimum sow weight 150 kg). Moderate restraint of the sow is required. Experience is required for the detection of the increased blood flow during pregnancy.
Doppler ultrasonic analysis	30–term	Detection of the foetal pulse by the reflected sound waves.	95–99	95–99	Immediate results. Done on the farm by the operator.	Moderate initial	<5 min	Errors in the alignment of the transducer may produce errors in pregnancy detection. Experience is required to recognize the foetal pulse. A second analysis at mid-pregnancy (on negative results) is desirable.
A-mode ultrasonic analysis	30–80	Detection of the foetal fluids by reflected sound waves.	95–99	80–96 (92)	Immediate results. Done on the farm by the operator.	Moderate initial	<5 min	Errors in the alignment of the transducer may produce errors in pregnancy detection. A full bladder, with the wrong alignment, would give a false positive result. Low foetal fluid volumes may not be detected. A second analysis at mid-pregnancy (on negative results) is desirable.

Conclusion

This chapter has been directed towards an evaluation of the current status of our ability to accurately detect pregnant and non-pregnant sows from the time of mating to parturition. It has not been the intention of this review to recommend specific techniques in preference to others, but rather to highlight the advantages and limitations of the various methods. At the present time there is no accepted accurate method of determining the reproductive status of the pig from the time of mating to the expected time of embryo implantation. The measurement of immunosuppressive activity may provide an acceptable means of very early pregnancy diagnosis. During the early stages of embryonic growth (days 13–15) the determination of blood prostaglandin levels may also provide an accurate means of early pregnancy diagnosis. The major emphasis on pregnancy diagnosis has been directed at the embryonic and early foetal stage of development. The problem has been to accurately identify the small proportion of sows that fail to become pregnant after mating. This group usually comprises about 5% of the sows mated and may occasionally be in excess of 20% in some herds. The difficulties arise because of the variability of the reasons for the failure of conception. Thus, while any one diagnostic procedure will detect sows that are not pregnant for one or more reasons, it will not provide accurate results for all causes of conception failure. These observations are summarized in *Table 15.3*.

The variety of techniques developed for the confirmation of pregnancy and the detection of the non-pregnant sow after the beginning of implantation emphasizes the extent of the research that has been undertaken. The techniques of laparoscopy, oestrous detection, vaginal biopsy and rectal palpation appear to have been fully developed. However, in the use of ultrasonic techniques, further refinement of the electronics to discriminate among the signals received and the sensitivity of the receiver may be anticipated. These advances should shorten the time interval from mating to detection of pregnancy. The determination of blood concentrations of progesterone and oestrone sulphate may be developed to allow on-farm evaluation utilizing 'test-tube' analysis procedures.

References

BAZER, F.W. and THATCHER, W.W. (1977). Theory of maternal recognition of pregnancy in swine based on estrogen controlled endocrine versus exocrine secretion of prostaglandin $F_{2\alpha}$ by the uterine endometrium. *Prostaglandins* **14**, 397–401

BENJAMINSEN, E. and KARLBERG, K. (1980). Pregnancy examination in the sow: A comparison of two types of ultrasound equipment and rectal examination. *Nord. VetMed.* **32**, 417–422

BOSC, M.J., MARTINAT-BOTTE, F. and NICOLLE, A. (1975). Étude de deux technique de diagnostic de gestation chez la Truie. *Annls. Zootech.* **24**, 651–660

CAMERON, R.D.A. (1977). Pregnancy diagnosis in the sow by rectal examination. *Aust. vet. J.* **53**, 432–435

CHEW, B.P., DZIUK, P.J., THOMFORD, P.J. and KESLER, D.J. (1979). Relationships between blood estrone sulfate and fetal number in gilts between days 22 and 80 of pregnancy. *Proc. 71st Ann. Meeting, Am. Soc. Anim. Sci.*, p.285 (Abstract No. 356)

CIUREA, V., NEUMANN, F., PASTEA, Z. and OLARIAN, E. (1955). Diagnosticul histologic al gestatiei la scroafa – *Acad. R.P.R. Baza Timisoara, Ann. II (Seria 2)* 113. Quoted by Kawata and Fukui (1977)

DHINDSA, D.S. and DZIUK, P.J. (1968). Effect on pregnancy in the pig after killing embryos or fetuses in one uterine horn in early gestation. *J. Anim. Sci.* **27**, 122–126

DICKMANN, Z., DEY, S.K. and SEN GUPTA, J. (1976). A new concept: Control of early pregnancy by steroid hormones originating in the preimplantation embryo. *Vitams. Horm.* **34**, 215–242

DIEHL, J.R. and DAY, B.N. (1973). Utilization of frozen sections with the vaginal biopsy technique for early pregnancy diagnosis in swine. *J. Anim. Sci.* **37**, 114–117

DYCK, G.W. (1971). Puberty, post weaning estrus and estrous cycle length in Yorkshire and Lacombe swine. *Can. J. Anim. Sci.* **51**, 135–140

DYCK, G.W. (1974). The effects of stage of pregnancy, mating at the first and second estrus after weaning and level of feeding on fetal survival in sows. *Can. J. Anim. Sci.* **54**, 277–285

EVINSON, B., NANCARROW, C.D., MORTON, H., SCARAMUZZI, R.J. and CLUNIE, J.A. (1977). Detection of early pregnancy and embryo mortality in sheep by the rosette inhibition test. *Theriogenology* **8**, 157 (Abstract)

ELLICOT, A.R. and DZIUK, P.J. (1973). Minimum daily dose of progesterone and plasma concentration for maintenance of pregnancy in ovariectomized gilts. *Biol. Reprod.* **9**, 300–304

FRASER, A.F. and ROBERTSON, J.G. (1967). The detection of foetal life in ewes and sows. *Vet. Rec.* **80**, 528–529

FRITZSCH, Von M. and HÜHN, U. (1976). Untersuchungen zur rektalen trächtigkeitsdiagnostik bei jung-und Altsauen vom 30. bis 80. tag post inseminationem. *Mh. Vet. Med.* **31**, 569–571

GUTHRIE, H.D. and DEAVER, D.R. (1979). Estrone concentration in the peripheral plasma of pregnant and non-pregnant gilts. *Theriogenology* **11**, 321–329

HANSEN, L.H. and CHRISTIANSEN, Ib. J. (1976). Frühe trächtigkeitsdiagnose beim schwein mit eimen neuentwickelten ultraschall-A-skan-gerät. *Zuchthygiene* **11**, 19–21

HORNE, C. and DZIUK, P.J. (1979). Relationship between level of estrone sulfate and number of fetuses in gilts from day 10 to 32. *Proc. 71st Ann. Meeting, Am. Soc. Anim. Sci.*, pp. 304–305 (Abstract No. 403)

HUCHZERMEYER, F. and PLONAIT, H. (1960). Trächtigkeitsdiagnose und rectaluntersuchung beim schwein. *Tierärztl. Umsch.* **15**, 399–401

HUNTER, R.H.F., POLGE, C. and ROWSON, L.E.A. (1967). The recovery, transfer and survival of blastocysts in pigs. *J. Reprod. Fert.* **14**, 501–502

JÖCHLE, W. and SCHILLING, E. (1965). Improvement of conception rate and diagnosis of pregnancy in sows by an androgen-oestrogen-depot preparation. *J. Reprod. Fert.* **10**, 439–440

KAWATA, K. and FUKUI, Y. (1977). Pregnancy diagnosis in the pig: A review. *Folia Vet. Lat.* **7**, 91–110

KÜST, D. (1931). Die trächtigkeitsfeststellung bei unseren haustieren durch den nachweis dis sexualhormons in harn. *Dt. tierarztl. Wschr.* **39**, 738–741

LINDAHL. I.L., TOTSCH, J.P., MARTIN, P.A. and DZIUK, P.J. (1975). Early diagnosis of pregnancy in sows by ultrasonic amplitude-depth analysis. *J. Anim. Sci.* **40**, 220–222

McCAUGHEY, W.J. (1979). Pregnancy diagnosis in sows: A comparison of the vaginal biopsy and doppler ultrasound techniques. *Vet. Rec.* **104**, 255–258

McKENZIE, F.F. (1924). Correlations of external signs and vaginal changes with the ovarian cycle in swine. *Anat. Rec.* **27**, 185–186

MACNEIL, F. (1979). Blood testing for pregnancy. *Pig Fmg.* September, 40–43

MARTINAT-BOTTE, F., GAUTIER, J., DEPRES, P. and TERQUI, M. (1980). Application d'un diagnostic très précoce de gestation en élevage porcin. *Journées de la Recherche Porcine en France*, pp. 167–170. Paris, L'Institut Technique du Porc

MEREDITH, M.J. (1976). Pregnancy diagnosis in the sow by examination of the uterine arteries: *Proc. 4th Int. Congr. Pig Veterinary Society, Ames, Iowa, U.S.A.* p.D.5

MEYER, J.N., ELSASSER, F. and ELLENDORFF, F. (1975). Trächtigkeits und fertilitätstest beim schwein mit hilfe der plasma progesteronbestimmung. *Dt. tierärztl. Wschr.* **12**, 473–475

MORTON, D.B. and RANKIN, J.E.F. (1969). The histology of the vaginal epithelium of the sow in oestrus and its use in pregnancy diagnosis. *Vet. Rec.* **84**, 658–662

MORTON, H., HEGH, V. and CLUNIE, C.J.A. (1976). Studies of the rosette inhibition test in pregnant mice: evidence of immunosuppression? *Proc. Roy. Soc., Lond.* **B193**, 413–419

NANCARROW, C.D., QUINN, P.J., RIGBY, N.W., WALLACE, A.L.C. and GREWAL, A.S. (1980). Involvement of the zygote, oviduct and ovary in the production of an early pregnancy factor in sheep. *Biol. Reprod.* **22 (Suppl. 1)**, 28A (Abstract)

NISHIKAWA, Y. (1953). A method of early pregnancy diagnosis and detection of luteal phase in domestic animals by injection of synthetic estrogen "euvestin". *Suikai (The Veterinary World)* **30**, 8–11

O'REILLY, P.J. (1976). Pregnancy diagnosis in pigs by ultrasonic amplitude depth analysis – A field evaluation. *Ir. vet. J.* **24**, 165–167

PÁSZTOR, L. and TÓTH, S. (1964). Data on the oestrous cycle of the sow. *Kisérl. Közl. Allatten.* **57 B**, 105–109. (In *Anim. Breed. Abstr.* **35**, 122, 1967.)

PERRY, J.S., HEAP, R.B. and AMOROSO, E.C. (1973). Steroid hormone production by pig blastocysts. *Nature, Lond.* **245**, 45–47

PIERCE, J.E., MIDDLETON, C.C. and PHILLIPS, J.M. (1976). Early pregnancy diagnosis in swine using doppler ultrasound. *Proc. 4th Int. Congr. Pig Veterinary Society, Ames, Iowa, U.S.A.*, p.D.3

ROBERTSON, H.A. and KING, G.J. (1974). Plasma concentrations of progesterone, oestrone, oestradiol-17β and of oestrone sulphate in the pig at implanation, during pregnancy and at parturition. *J. Reprod. Fert.* **40**, 133–141

ROBERTSON, H.A. and SARDA, J.R. (1971). A very early pregnancy test for mammals: Its application to the cow, ewe and sow. *J. Endocr.* **49**, 407–419

SABA, N. and HATTERSLEY, J.P. (1981). Direct estimation of oestrone sulphate in sow serum for a rapid pregnancy diagnosis test. *J. Reprod. Fert.* **62**, 87–92

SIGNORET, J.P. (1967). Durée du cycle oestrien et de l'oestrus chez la truie. Action du benzoate d'oestradiol chez la femelle ovariectomizée. *Annls. Biol. anim. Biochim. Biophys.* **7**, 407–421

SIGNORET, J.P. (1970). Reproductive behavior of pigs. *J. Reprod. Fert. Suppl.* **11**, 105–117

STABENFELDT, G.H., AKINS, E.L., EWING, L.L. and MORRISSETTE, M.C. (1969). Peripheral plasma progesterone levels in pigs during the oestrous cycle. *J. Reprod. Fert.* **20**, 443–449

TILLSON, S.A., ERB, R.E. and NISWENDER, G.D. (1970). Comparison of luteinizing hormone and progesterone in blood and metabolites of progesterone in urine of domestic sows during the estrus cycle and early pregnancy. *J. Anim. Sci.* **30**, 795–805

TOO, K., KAWATA, K., FUKUI, Y., SATO, K., KAGOTO, K. and KAWABE, K. (1974). Studies on pregnancy diagnosis in domestic animals by an ultrasonic doppler method: I. Pregnancy diagnosis in the pig and fetal heart rate changes during pregnancy. *Jap. J. vet. Res.* **22**, 61–71

WALKER, D. (1972). Pregnancy diagnosis in pigs. *Vet. Rec.* **90**, 139–144

WILDT, D.E., FUJIMOTO, S., SPENCER, J.L. and DUKELOW, W.R. (1973). Direct ovarian observation in the pig by means of laparoscopy. *J. Reprod. Fert.* **35**, 541–543

WILDT, D.E., MORCOM, C.B. and DUKELOW, W.R. (1975). Laparoscopic pregnancy diagnosis and uterine fluid recovery in swine. *J. Reprod. Fert.* **44**, 301–304

WILLIAMSON, P., HENNESSEY, D.P. and CUTLER, R. (1980). The use of progesterone and oestrogen concentrations in the diagnosis of pregnancy, and in the study of seasonal infertility in sows. *Aust. J. agric. Res.* **31**, 233–238

WILSON, K.M. (1926). Histological changes in the vaginal mucosa of the sow in relation to the oestrous cycle. *Am. J. Anat.* **37**, 417–430

WRIGHT, K., COLLINS, D.C. MUSEY, P.I. and PREEDY, J.R.K. (1978). A specific radioimmunoassay for estrone sulfate in plasma and urine without hydrolysis. *J. clin. Endocr. Metab.* **47**, 1092–1098

YAMANOUCHI, G. (1973). Rectal examination in the pig: Pregnancy diagnosis and palpation of the ovary. *Jui-chikusam Shimpo (J. Vet. Med.)* **585**, 203–207

IV

PARTURITION

16

THE ENDOCRINE CONTROL OF PARTURITION

N.L. FIRST, J.K. LOHSE and B.S. NARA
Department of Meat and Animal Science and Endocrinology–Reproductive Physiology Program, University of Wisconsin, USA

The pig evolved as a litter bearing species with the maintenance of pregnancy dependent on the continued presence of the litter *in utero* and on continued production of ovarian progesterone. The length of gestation is reasonably precise, being approximately 112–116 days depending on the breed, size of litter and season (Cox, 1964; Bichard *et al.*, 1976; Aumaitre, Deglaire and LeBost, 1979). Parturition occurs slightly more frequently in the late afternoon and at night (Bichard *et al.*, 1976; Boning, 1979) but no differences are seen in the frequency of day and night delivery when parturition is artificially induced (Hammond and Matty, 1980). For an individual herd a knowledge of the average length of gestation and exact breeding dates are essential for optimal piglet survival and growth when parturition is to be induced.

The entire parturition process requires 2–5 hours with piglets being delivered at approximately 12–16 minute intervals (Sprecher *et al.*, 1974; see *Table 16.1*). Piglets are delivered randomly from the two uterine horns (Dziuk and Harmon, 1969; Taverne *et al.*, 1977). They sometimes pass each other in birth order (Taverne *et al.*, 1977) and the placentas are delivered either in part after the emptying of one uterine horn or within approximately four hours after the last piglet is delivered (Jones, 1966). This is at a time when plasma levels of oxytocin are elevated (Taverne, 1979). Parturition is normally preceded by udder oedema, attempted maternal nest building and a milk ejection response (First and Bosc, 1979).

Parturition is not without complications. Body temperature increases 13±4.1 hours before delivery of the first piglet, reaches a peak of 0.6 °C–1.2 °C above the normal of 38.3±0.3 °C and in healthy sows returns to near normal within 24 hours (Elmore *et al.*, 1979). Parturition is often complicated by a disease called Mastitis–Metritis–Agalactia in which the sow's temperature remains greatly elevated for a prolonged period and she refuses to provide milk for the piglets.

Not all piglets survive farrowing; at least 6% are born dead (Randall, 1972; Sprecher *et al.*, 1974; Leman, Hurtgen and Hilley, 1979) and the last piglet in each uterine horn has less than a 50% chance of survival (Bevier and Dziuk, 1976). The factors influencing piglet survival were reviewed recently by Leman, Hurtgen and Hilley (1979). It is apparent from their

Table 16.1 A SUMMARY OF FIELD TRIALS WHERE PARTURITION HAS BEEN INDUCED IN PIGS BY PROSTAGLANDIN $F_{2\alpha}$ OR ITS ANALOGUES[a]

Prostaglandin used	*No. sows*	*Farrowing within 48 hours after treatment* (%)	*Interval treatment to first birth* (hours)	*Farrowing duration* (hours)	*Stillborn piglets* (%)	*Piglet weight*		*MMA* (%)
						birth (kg)	*weaning* (kg)	
$PGF_{2\alpha}$[b]	607	87	28±5.5	4.8±4.0	9.4	1.24±0.4	5.1±0.8	18
Control	404	12	108±39	5.1±1.2	9.4	1.33±0.2	5.7±0.7	27
Cloprostenol[c]	2473	95.3	26.4±5.7	4.2±2.8	5.3	1.30	5.6	36.7
Control	612	19	81.5±18	4.4±4.2	7.3	1.33	5.1	40.7
PGF Ay24655[d]	54	93	27.2±3		7.3	1.3		
Control	45				3.5	1.4		
Prostalene, 4 mg[e]	19	100	25.2±3.7	5.1±0.9	8.2	1.18		
5 mg	20	100	23.8±2.4	4.9±0.9		1.26		
Control	5		122±44	2.2±0.9	1.7	1.5±0.2		

[a]The data summarized are only from experiments in which $PGF_{2\alpha}$ compounds were administered at effective doses on or after day 10 of gestation.
[b]The dose of $PGF_{2\alpha}$ ranged from 7.5 to 12.5 mg i.m./sow. These data were obtained from the following publications: Ehnvall *et al.*, 1976; Backstrom *et al.*, 1976; B.N. Day, unpublished; King, Robertson and Elliott, 1979; Hagner, Elze and Michel, 1979; Hühn, Lutter and Hühn, 1980.
[c]The dose of cloprostenol ranged from 150 to 200 μg i.m./sow. The data were obtained from the following publications: Bosc and Martinat-Botte, 1976; Hammond and Carlyle, 1977; Walker, 1977; Hammond and Matty, 1980; Jainudeen and Brandenburg, 1980; Lynch and Langley, 1977; Willemse *et al.*, 1979; Pool, Copeland and Godke, 1979; Holtz, Welp and Spangenberg, 1979; Elze *et al.*, 1979.
[d]The dose of Ay24655 was 50 mg/kg/sow i.m. on day 111–113. The data are from Downey, Conlon and Baker, 1976.
[e]Prostalene was administered i.m. at doses of either 4 mg or 5 mg/sow. The data are from Holtz *et al.*, 1979.

review, and that of Dziuk (1979), that pre- and post-partum losses of piglets are sizable and may be as high as 20–25%.

Understanding the birth process and the mechanisms controlling parturition should result in ways of reducing this loss and of controlling the moment of delivery so that attendants might be present. The presence of an attendant and resuscitation of piglets has been shown to save as many as one additional piglet per litter (Milosavljevic *et al.*, 1972; Hammond and Matty, 1980). Synchronization of the moment of farrowing also results in increased efficiency of production by allowing management and handling of groups rather than individual litters.

Normal parturition in the pig requires termination of the forces maintaining pregnancy, active preparation of the foetuses for birth and of the mother for opening of the birth canal, uterine contraction, nest building and lactation. The mechanisms controlling maintenance of late pregnancy and the initiation of parturition, as well as methods for the induction of parturition in pigs, have been subjects of several recent reviews (First, 1979; First and Bosc, 1979; Dziuk, 1979; Taverne, 1979; Ellendorff *et al.*, 1979b; Silver *et al.*, 1979; Ellendorff, 1980) and will be discussed in the following sections. Evidence will be provided to show the way the foetuses initiate the birth process.

Maintenance of late pregnancy

The establishment of pregnancy and the mechanisms maintaining early pregnancy have been discussed by Bazer (Chapter 12). The maintenance of late pregnancy is graphically described in *Figure 16.1*. The reader is encouraged to consult this figure while reading the following section.

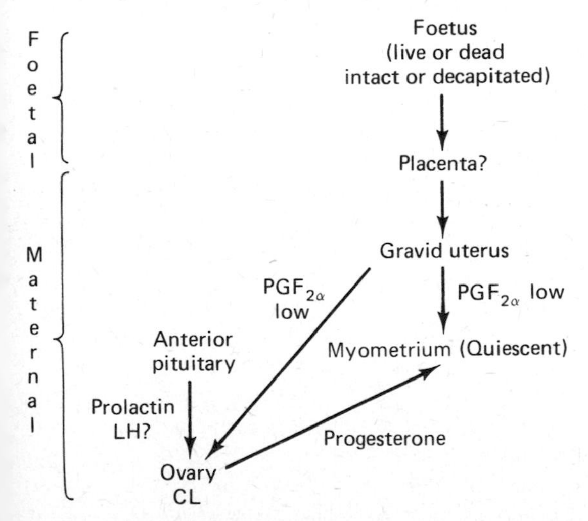

Figure 16.1 Graphic representation of the maintenance of late pregnancy in pigs. Pregnancy maintenance is dependent on a progesterone block of uterine contraction. Progesterone is from the corpora lutea. Maintenance of the corpora lutea and continued production of progesterone requires either LH or prolactin from the sow's pituitary gland and prevention by the foetuses, *in utero*, of prostaglandin $F_{2\alpha}$ secretion into the blood of the mother. If prostaglandin $F_{2\alpha}$ reached the ovaries, regression of the corpora lutea (CL) would occur.

In the pig, progesterone is the hormone responsible for preventing uterine contraction and maintaining gestation to term. When progesterone levels in peripheral plasma are maintained by the administration of progesterone, parturition at term (Curtis, Roger and Martin, 1969; Minar and Schilling, 1970; Nellor *et al.*, 1975; First and Staigmiller, 1973) and parturition induced by dexamethasone (a synthetic glucocorticoid) or prostaglandin $F_{2\alpha}$ (Coggins, Van Horn and First, 1977) are prevented. The principal factor in the maintenance of pregnancy in pigs seems to be the level of progesterone. The main source of progesterone throughout gestation is the corpora lutea, since ovariectomy (du Mesnil du Buisson and Dauzier, 1957; Ellicott and Dziuk, 1973; First and Staigmiller, 1973; Nara, Darmadja and First, 1975; 1981) or removal of corpora lutea (Nara, Darmadja and First, 1975; 1981) decrease progesterone levels and terminate pregnancy. Approximately 4–6 corpora lutea are needed for the production of sufficient progesterone to maintain pregnancy (Martin, Norton and Dziuk, 1977).

The identity and specificity of the luteotrophic hormone responsible for maintaining the corpora lutea and late pregnancy is not well established for pigs. Both prolactin and luteinizing hormone (LH) have been implicated. Hypophysectomy during late gestation terminates pregnancy in sows (du Mesnil du Buisson and Denamur, 1969; Kraeling and Davis, 1974) and prolactin, but not LH, maintains pregnancy after hypophysectomy at day 70 (du Mesnil du Buisson and Denamur, 1969). However, in intact pigs the corpora lutea remain responsive to LH stimulation until immediately before parturition, i.e. 41–17 hours before parturition (Parvizi *et al.*, 1976). During the two days preceding parturition, progesterone levels in plasma are independent of episodic increases in LH, whereas at three weeks pre-partum each episodic peak of LH is followed by an episodic peak of progesterone (Parvizi *et al.*, 1976).

Live or dead foetal tissue must be present in the uterus for maintenance of the corpora lutea of pregnancy. The amount of non-pregnant uterus present determines the requirement of a foetal contribution to luteal maintenance. At least four foetuses are required to establish and maintain early pregnancy (Polge, Rowson and Chang, 1966), whereas when one uterine horn is removed, pregnancy is maintained by two foetuses (Dhindsa and Dziuk, 1968) or by only one foetus when all the uterus except that portion occupied by the foetus is removed before the 14th day of pregnancy (du Mesnil du Buisson and Rombauts, 1963). Killing all foetuses after day 30 does not prevent maintenance of corpora lutea or lower plasma concentrations of progesterone up to day 60 (Webel, Reimers and Dziuk, 1975) or from day 100 to day 120 of gestation (Coggins and First, 1977) if the foetuses have not been resorbed.

However, removal of all foetuses at day 102 causes delivery of the placentas in less than 48 hours (Chiboka, Casida and First, 1976) and removal of 2–4 foetuses terminates the pregnancy in 42–72 hours (K.A. Martin and R.M. Liptrap, personal communication). These facts indicate that foetal tissue or a product of the intact conceptus, such as oestrogen (Bazer and Thatcher, 1977) must be present *in utero* to prevent the uterus from initiating luteolysis.

Hormonal changes associated with parturition and uterine contractability

Parturition, i.e. expulsion of the foetus or foetuses, is brought about by uterine contractions. Activation of the myometrium at parturition (*Figure 16.2*) is preceded by a multitude of hormonal changes in the blood of the mother and increased plasma cortisol in the foetuses (*Figures 16.3* and *16.4*). Parturition is preceded by increased foetal plasma cortisol (Fevre, Terqui and Bosc, 1975; Silver *et al.*, 1979) and by changes in maternal plasma that include: increased concentrations of oestrone and oestradiol

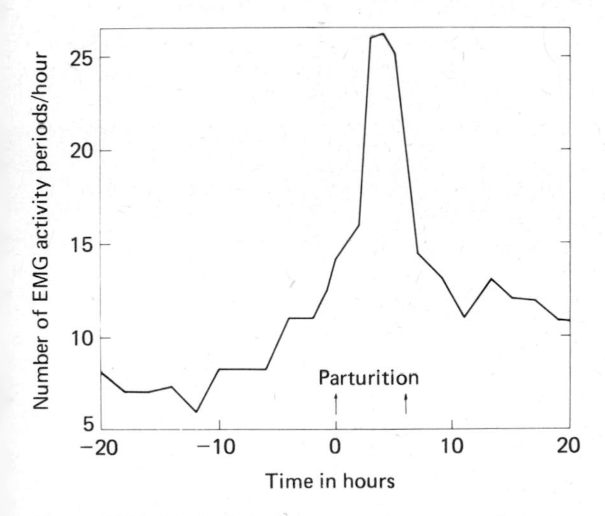

Figure 16.2 Typical changes in the frequency of uterine contractions of a sow around the time of parturition (delivery occurred at day 116). By courtesy of Dr Marie Jeanne Prud'homme, unpublished

(Ash and Heap, 1975; Fevre, Terqui and Bosc, 1975; Forsling *et al.*, 1979; Gustafsson *et al.*, 1976; Molukwu and Wagner, 1973; Robertson and King, 1974; Taverne *et al.*, 1979a,b; Wettemann *et al.*, 1977), relaxin (Sherwood *et al.*, 1975), corticosteroid (Ash and Heap, 1975; Molukwu and Wagner, 1973; Silver *et al.*, 1979), prolactin (Taverne *et al.*, 1979a; Van Landeghem and Van de Wiel, 1978), prostaglandin $F_{2\alpha}$ metabolite (Nara, 1979; Nara and First, 1977; 1981b) and oxytocin (Forsling *et al.*, 1979) at the time of delivery.

Maternal concentrations of progesterone decline within 1–2 days before parturition (Ash and Heap, 1975; Baldwin and Stabenfeldt, 1975; Coggins, Van Horn and First, 1977; Gustafsson *et al.*, 1976; Molukwu and Wagner, 1973; Nara, 1979; Nara and First, 1977; 1981b; Robertson and King, 1974; Taverne *et al.*, 1979a,b; Wettemann *et al.*, 1977; Silver *et al.*, 1979) and this rapid decline is dependent on the synthesis of prostaglandin $F_{2\alpha}$ (*Figure 16.5a*, Nara and First, 1977; 1981b). Prostaglandin $F_{2\alpha}$ increases rapidly to high concentrations during delivery (Nara, 1979; Nara and First, 1977; 1981b; Silver *et al.*, 1979) and has been shown to cause release of prolactin (Taverne *et al.*, 1979) and oxytocin in the pig (Ellendorff *et al.*, 1979a).

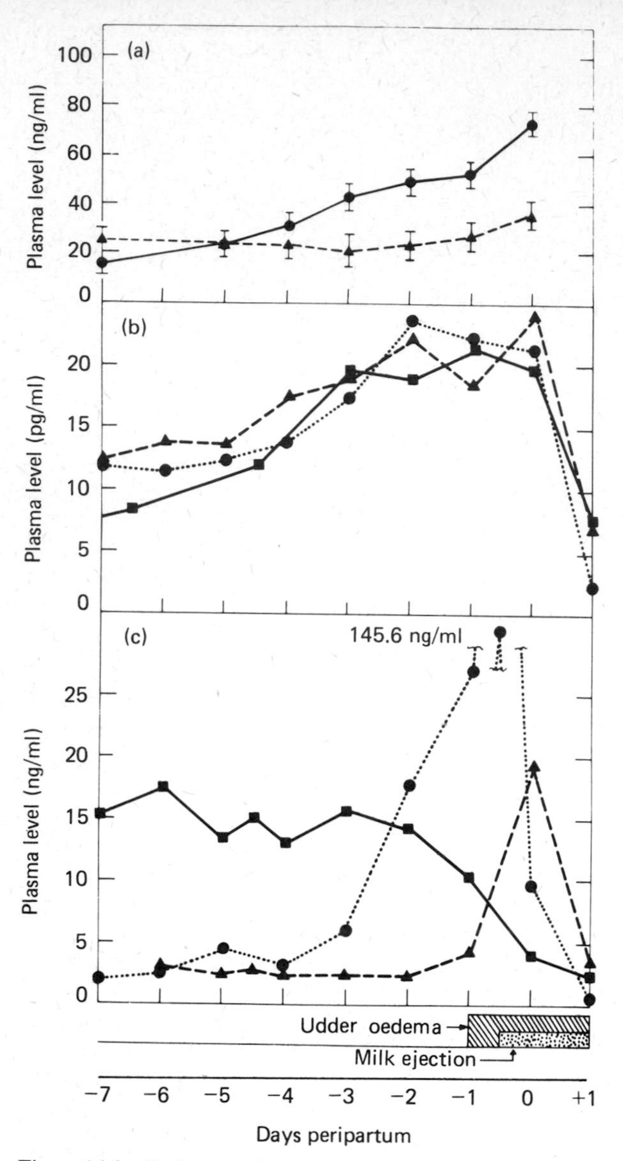

Figure 16.3 Endocrine changes preceding parturition in pigs. (a) Corticosteroids: –●–●–●– foetal cortisol, --▲--▲--▲-- maternal corticoids (Silver *et al.*, 1979); (b) –■–■–■– total oestrogens (pg × 333/ml) (Baldwin and Stabenfeldt, 1975), --▲--▲--▲-- oestradiol-17β (pg × 0.33/ml), ●...●...● oestrone (pg × 100/ml) (Molukwu and Wagner, 1973); (c) ..●...●...● relaxin (×2) (Sherwood *et al.*, 1975; 1979), –■—■—■– progesterone and --▲--▲--▲-- $PGF_{2\alpha}$ metabolite. From Nara and First (1981a,b)

Oxytocin concentrations in peripheral plasma were elevated between 9 and 4 hours before birth of the first piglet and reached their highest values during delivery of the piglets (Forsling *et al.*, 1979). The initial increase in prostaglandin $F_{2\alpha}$ concentration also causes release of relaxin (*Figures 16.5* and *16.6*; Nara, 1979; Nara and First, 1981b; Sherwood *et al.*, 1976;

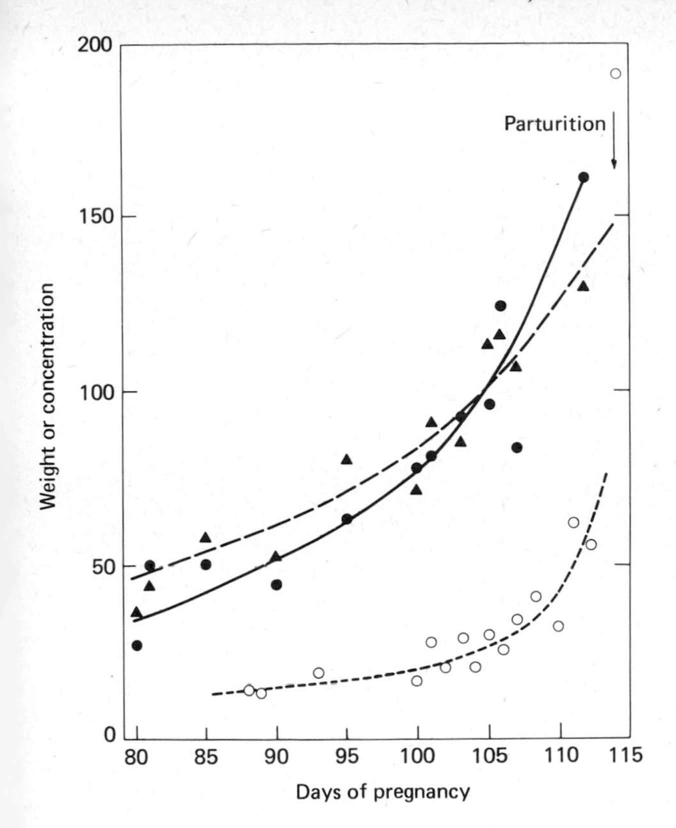

Figure 16.4 Foetal weight, ▲-- (10 g), foetal adrenal weight, ●— (mg), and foetal plasma cortisol level, ○-- (ng/ml), during late pregnancy in pigs. From Bosc (1973); Fevre *et al.* (1975)

Sherwood *et al.*, 1979) and this release is prevented by inhibition of prostaglandin synthesis (Sherwood *et al.*, 1979; Nara and First, 1981b).

It has been proposed that relaxin softens or loosens the cervix (Zarrow *et al.*, 1956; Kertiles and Anderson, 1979) and thus may facilitate delivery of the foetuses. Results of a recent experiment (Nara *et al.*, 1981) show that (1) absence of relaxin in ovariectomized gilts with pregnancies maintained by exogenous progesterone causes prolonged deliveries with high incidences of stillbirths after withdrawal of the progesterone, and (2) supplementation of the progesterone treatment with highly purified porcine relaxin returns duration of delivery and frequency of live births to values similar to controls.

Udder oedema, which is indicative of milk formation, begins about 24 hours before delivery (Coggins, 1975; Coggins, Van Horn and First, 1977; Diehl *et al.*, 1974; Zerobin and Sporri, 1972) and milk ejection, an oxytocin-induced response, within 12 hours pre-partum (Nara and First, 1981a,b; see *Figure 16.3*). Exogenous oxytocin will induce labour, but only after the time that milk can be ejected (Muhrer, Shippen and Lasley, 1955).

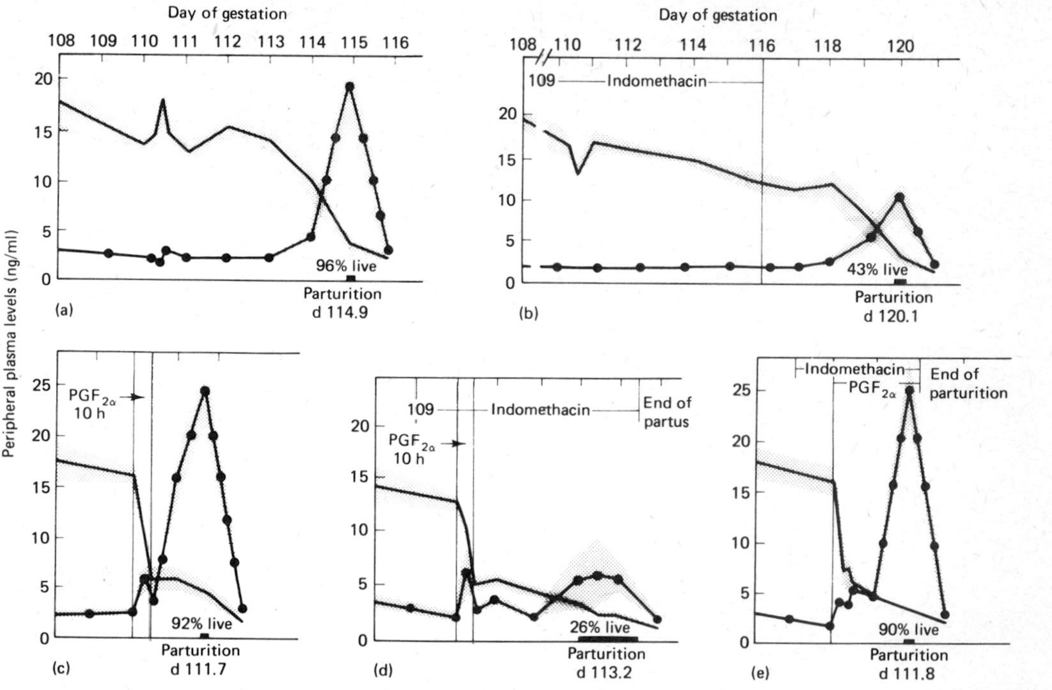

Figure 16.5 Peripheral plasma concentrations ±S.E.M. (shaded areas) of progesterone, ——, and $PGF_{2\alpha}$-metabolite ($PGF_{2\alpha}$-M) ●—●, percentage of live births (% live), duration of parturition (■) and length of gestation in sows receiving the following treatments (each, n = 5): (a) Control, vehicles of indomethacin and $PGF_{2\alpha}$; (b) Indomethacin (I), 4 mg I/kg 2×/day from day 109 to 116; (c) $PGF_{2\alpha}$-low (P), 0.5 P/hour infused for 10 hours on day 110; Indo + $PGF_{2\alpha}$-low (IP), treatments I and P; (d) Indo + $PGF_{2\alpha}$-low + high (IP+), treatment I + $PGF_{2\alpha}$ continuously infused from day 110 until the end of parturition starting at a rate of 0.5 mg/hour for 24 hours, followed by 3 mg/hour for 3 hours, 6 mg/hour for 3 hours and 9 mg/hour until the end of parturition. Indomethacin prevented a rise in plasma concentrations of prostaglandin $F_{2\alpha}$ metabolite and decline in progesterone. This inhibition was overcome by infusion of a luteolytic dose of $PGF_{2\alpha}$. However, normal delivery and live birth did not occur in sows receiving indomethacin until a dose of $PGF_{2\alpha}$, sufficient to cause a rise in $PGF_{2\alpha}$ metabolite comparable to that of the controls, was infused.

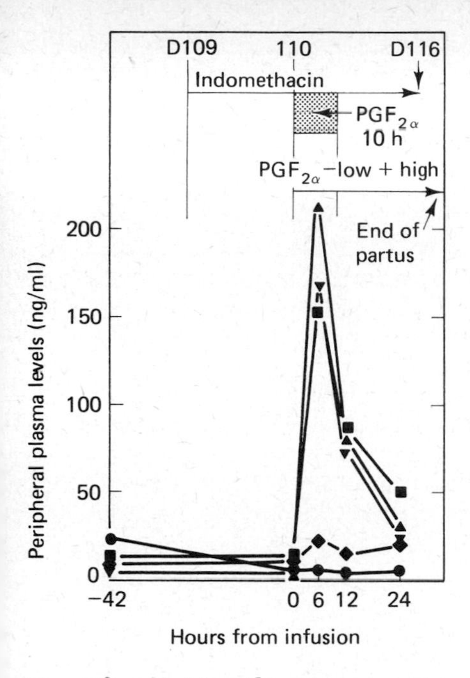

Figure 16.6 Peripheral plasma concentrations of relaxin in the sows of Figure 16.5 at –42, 0, 6, 12 and 24 hours relative to the start of infusion with $PGF_{2\alpha}$ (treatments: $PGF_{2\alpha}$-low ■, Indo + $PGF_{2\alpha}$-low ▲, and Indo + $PGF_{2\alpha}$-low + high ▼) or saline (treatments: Control ♦ and Indomethacin ●). Relaxin levels surged to peak levels ($P<0.01$) within 6 hours after infusion with $PGF_{2\alpha}$ was begun. Adapted from Nara and First (1981b)

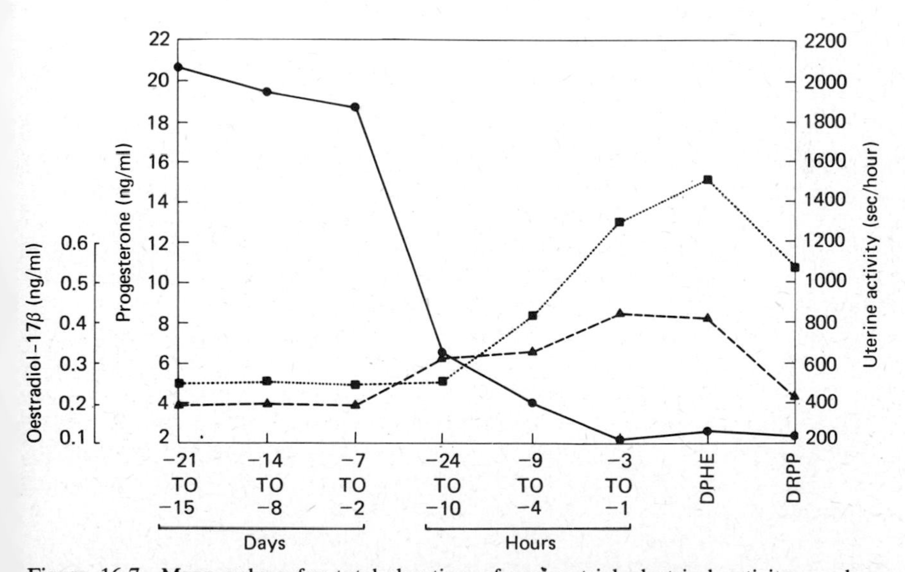

Figure 16.7 Mean values for total duration of myometrial electrical activity per hour (·■···■··■·) and plasma concentrations of progesterone (–●–●–●–), and oestradiol-17β (--▲--▲--▲--) in seven sows from 21 days before, until the end of, parturition. DPHE: hours of *D*elivery until all *P*iglets from the *H*orn with *E*lectrodes are born; DRPP: hours of *D*elivery of the *R*emaining *P*iglets and *P*lacentae. Adapted from Taverne *et al.* (1979b)

Myometrial activity during late pregnancy has been shown to consist of irregular episodes of prolonged activity in those uterine segments containing a foetus, while empty parts of the uterus remain relatively inactive (Taverne *et al.*, 1979b). At this time, plasma concentrations of oxytocin remain below 1.3 μU/ml and plasma concentrations of progesterone and oestrogen remain unchanged (Forsling *et al.*, 1979; Taverne *et al.*, 1979b). Between 24 to 10 hours before expulsion of the first piglet, when concentrations of progesterone have significantly decreased and oestrogens have increased, myometrial activity is still similar to that recorded on the previous days. Only between 9 and 4 hours before the birth of the first piglet does myometrial activity increase in all parts of the uterus (*Figure 16.7*). This increase in myometrial activity coincides with elevated concentrations of oxytocin in peripheral plasma. Uterine contractions are most frequent during delivery at the time that oxytocin concentrations are at their highest levels (Taverne *et al.*, 1979b). Release of oxytocin seems to be related to decreased concentrations of progesterone and not to increased levels of oestrogens, since oestrogen levels are already declining when oxytocin reaches its highest levels during delivery (Forsling *et al.*, 1979).

An increase in myometrial sensitivity to oxytocin near parturition has been shown to exist in rats and guinea pigs and the increased sensitivity corresponds with increased concentrations of myometrial oxytocin receptors in rats (Soloff, Alexandrova and Fernstrom, 1979) and guinea pigs (Alexandrova and Soloff, 1980c). The increase in myometrial oxytocin receptors in rats and guinea pigs appears to be correlated with a decrease in the ratio of progesterone to oestradiol concentration (P/E ratio) in the blood (Soloff, Alexandrova and Fernstrom, 1979; Alexandrova and Soloff, 1980a,b). In rats, a species requiring a decline in progesterone for the initiation of parturition, the decrease in the P/E ratio is primarily due to a decline in progesterone before spontaneous (Soloff, Alexandrova and Fernstrom, 1979; Alexandrova and Soloff, 1980a) or $PGF_{2\alpha}$-induced deliveries (Alexandrova and Soloff, 1980b). In guinea pigs, a species not experiencing a decrease in progesterone before parturition, the decrease in the P/E ratio is primarily caused by an increase in oestrogens near term (Alexandrova and Soloff, 1980b). Soloff and Swartz (1974) demonstrated the presence of oxytocin receptors in the porcine myometrium at one week before expected term. Like rats, pigs require a decrease in progesterone before parturition can be initiated and like rats, the increase in concentration of oxytocin receptors in the myometrium may be dependent on a decrease in progesterone. Lack of oxytocin receptors in the myometrium before the decline of progesterone in the blood may be the reason that oxytocin is effective in inducing deliveries in the pig only after milk can be ejected from the teats and within 24 hours before expected parturition (Muhrer, Shippen and Lasley, 1975; Welk and First, 1979; see section on *Induction of Parturition* beginning on p.330).

Maternal plasma concentrations of $PGF_{2\alpha}$ metabolite (13,14-dihydro-15-keto-$PGF_{2\alpha}$) were shown to be elevated 10–20-fold once labour had started and uterine contractions could be detected by allantoic fluid pressure changes (Silver *et al.*, 1979). It has been shown (Nara, 1979; Nara and First, 1977; 1981b) that high levels of $PGF_{2\alpha}$ are needed at parturition for normal rapid delivery of live piglets. When $PGF_{2\alpha}$ synthesis is inhibited by adminstration of indomethacin to sows during parturition, delivery is prolonged and most of the piglets are stillborn. Infusion of high doses of $PGF_{2\alpha}$ during parturition in indomethacin-treated sows returns normal rapid delivery and live birth (*Figure 16.5*). Since $PGF_{2\alpha}$ stimulates release of oxytocin in pigs (Ellendorff *et al.*, 1979) and oxytocin has been shown to stimulate uterine production of prostaglandins (Chan, 1977; Mitchell and Flint, 1978; Mitchell, Flint and Turnbull, 1975), their interrelationship may have a cascading effect on the development of uterine contractions.

Relaxin has been shown to have an inhibitory effect on uterine activity (Porter, 1979) and the increasing levels of relaxin near parturition (Sherwood *et al.*, 1975) may thus serve as a supplementary regulatory mechanism preventing the uterus from contractions until delivery is due.

Termination of progesterone production

Since in the pig progesterone is the principal factor maintaining pregnancy (Coggins, Van Horn and First, 1977; First and Staigmiller, 1973) and corpora lutea are the main source of progesterone (Nara, 1979; Nara, Darmadja and First, 1981), termination of luteal production of progesterone (functional luteolysis) is necessary before parturition can occur. Prostaglandin $F_{2\alpha}$ has been implicated as the uterine luteolytic agent during the oestrous cycle and exogenous $PGF_{2\alpha}$ has induced parturition. Parturition occurred approximately 30 hours after initiation of a 9–10 hour infusion of 2–5 mg of $PGF_{2\alpha}$ (Nara and First, 1981a,b; Nara, 1979), and after an intramuscular injection of 10–12 mg of $PGF_{2\alpha}$ (see *Table 16.1*). The induced parturition resulting from $PGF_{2\alpha}$ treatment is preceded by a rapid decline in plasma progesterone from 10–19 ng/ml to 3–4 ng/ml at the time of parturition (Coggins, 1975; Coggins, Van Horn and First, 1977; Diehl *et al.*, 1974; Nara, 1979; Nara and First, 1981b; Wettemann *et al.*, 1977).

Proof that $PGF_{2\alpha}$ is a natural luteolytic agent at term comes from an experiment (Nara and First, 1977; 1981b) in which the increase of $PGF_{2\alpha}$ metabolite, together with luteolysis (decrease in progesterone) and parturition, were prevented by inhibiting prostaglandin synthesis with indomethacin (*Figure 16.6*). Exogenous $PGF_{2\alpha}$ replaced the inhibited endogenous production and caused a progesterone decline and parturition. Additionally, $PGF_{2\alpha}$ metabolite was shown to be increased in the sow's blood at the same time as the decline in progesterone. A dual role of $PGF_{2\alpha}$ in causing parturition was shown in indomethacin-treated sows by the fact that low doses of $PGF_{2\alpha}$ caused luteolysis but very high doses of $PGF_{2\alpha}$ were required following infusion of the initial low luteolytic dose for normal rapid delivery of the piglets.

The key role of endogenous $PGF_{2\alpha}$ in the mechanism causing prepartum luteolysis is emphasized by the fact that dexamethasone induces luteolysis and parturition in swine through stimulation of $PGF_{2\alpha}$ synthesis (Nara, 1979; Nara and First, 1978; 1981a). In this study (*Figure 16.8*), sows treated with dexamethasone had a premature increase in $PGF_{2\alpha}$ metabolite, with a premature decrease of progesterone and the induction of

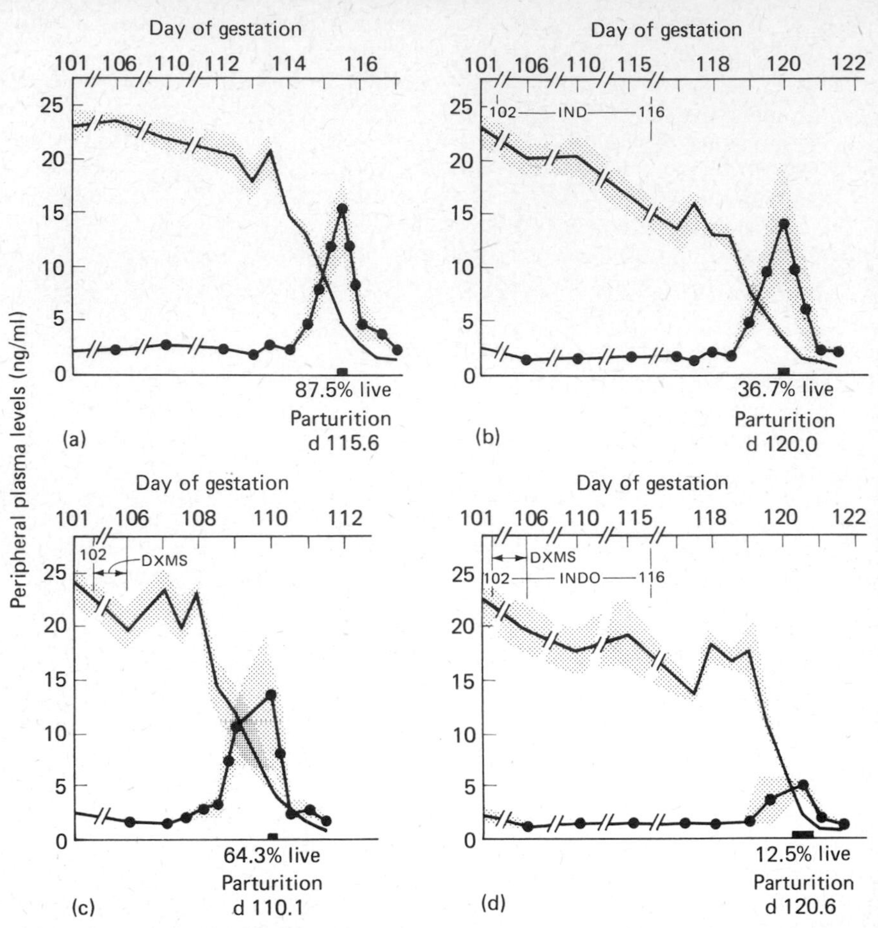

Figure 16.8 Peripheral plasma concentrations ±S.E.M. (shaded areas) of progesterone (——) and $PGF_{2\alpha}$ metabolite (—●—), percentage of live births (% live), duration of parturition (■) and length of gestation in sows receiving the following treatments (each, n = 5): (a) Control: vehicles of indomethacin and dexamethasone; (b) Indomethacin: 3 mg/kg twice daily from days 102–116 + vehicle of dexamethasone; (c) Dexamethasone: 75 mg/sow twice daily from days 102–106 + vehicle of indomethacin; (d) Indomethacin + dexamethasone: combination of treatments (b) and (c). Pre-term administration of dexamethasone caused a premature rise in plasma concentration of $PGF_{2\alpha}$ metabolite and a decline in progesterone as well as pre-term delivery. Indomethacin prevented a rise in plasma concentration of $PGF_{2\alpha}$ metabolite, a decline in progesterone and early delivery. It also prevented dexamethasone from inducing these parturient changes. From Nara and First (1981a)

parturition. The dexamethasone-induced increase in $PGF_{2\alpha}$ metabolite, progesterone decline and premature parturition were prevented when the sows were also treated with indomethacin.

This luteolytic agent at term comes from the uterus, because hysterectomy at day 112 prolongs the life of old corpora lutea and new corpora lutea that have been induced about one week before the expected term (Bosc, du Mesnil du Buisson and Locatelli, 1974). This agent travels from the uterus to the ovaries at least in part by a systemic route, since parturition occurred after the ovaries were transplanted to the body wall in late pregnancy (Torres, 1975; Martin, Bevier and Dziuk, 1978).

These are the maternal endocrine changes known to precede and accompany parturition. The initiation of these pre-partum maternal endocrine changes is controlled by the foetuses through a sequence of events controlled by the developing foetal endocrine system.

Role of the foetuses in initiation of parturition

The way in which foetal pigs control the initiation of parturition is not completely understood, but evidence that they do is most convincing.

The source of the impetus for parturition in the pig is the foetal brain, since decapitated (Stryker and Dziuk, 1975; Coggins and First, 1977), or hypophysectomized (Bosc, du Mesnil du Buisson and Locatelli, 1974) pig foetuses will not initiate parturition.

This is similar to the situation in sheep, in which foetal pituitary ablation prevents initiation of parturition (Liggins, Kennedy and Holm, 1967), but unlike the situation in primates. Although anencephaly in human foetuses (Honnebier and Swaab, 1973) or decapitation of rhesus monkey foetuses (Novy, Walsh and Kittinger, 1977) increases variation in gestation lengths, the mean gestation length remains normal. Only one intact pig foetus in an otherwise empty uterus can initiate parturition, but if the ratio of decapitated to intact foetuses reaches 4:1, parturition is delayed (Stryker and Dziuk, 1975).

Several lines of evidence point to increased glucocorticoid production by the foetal adrenal as a crucial step in the process by which the foetus initiates parturition. The adrenalectomized sheep foetus does not initiate parturition (Drost and Holm, 1968; Liggins, 1969). Foetal adrenalectomy has not been done in pigs but foetal adrenal atrophy brought about by hypophysectomy (Bosc, du Mesnil du Buisson and Locatelli, 1974) or decapitation (Stryker and Dziuk, 1975) of pig foetuses *in utero* does prevent parturition at term. Parturition, accompanied by live birth and milk ejection, can be induced in the pig by administration of dexamethasone to the foetuses (North, Hauser and First, 1973) or sow (North, Hauser and First, 1973; First and Staigmiller, 1973; Hühn, Hühn and Konig, 1976; Coggins and First, 1977; Hühn, Konig and Hühn, 1978; Hühn and Kiupel, 1979). In addition, exogenous adrenocorticotrophic hormone (ACTH) given to the pig foetuses in the last 10 days of gestation causes hypertrophy of the foetal adrenal cortex, followed by premature parturition (Bosc, 1973).

The classic pituitary ablation/ACTH replacement experiment has not been done in the pig. Jones *et al.* (1978) hypophysectomized four foetal sheep at days 110–118 of gestation, and then infused them with Synacthen (synthetic ACTH) starting 7–16 days later. Delivery occurred within 77–112 hours of the start of infusion, almost three weeks before normal term. Here again, the mechanisms operating in the pig appear similar to those in the sheep, in which ACTH or glucocorticoids will induce pre-term parturition (Liggins, 1968), but dissimilar to those in humans, in which glucocorticoids will not induce pre-term labour (Liggins and Howie, 1972).

The porcine foetal adrenal greatly increases its capacity for glucocorticoid production late in gestation. Foetal adrenal weight increases along with increased foetal plasma cortisol levels (see *Figure 16.4*; Bosc, 1973; Fevre, Terqui and Bosc, 1975). In the sheep, most of the increase in adrenal weight is due to enlargement of the adrenal cortex, particularly the zona fasciculata (Durand, Bosc and Niedle, 1978). Recent studies in our laboratory have shown that this is also the case in the pig. Although some increase in cortex cell size occurs, most of the increase in adrenal cortex size is due to an increase in cell number, and many mitotic figures appear in the adrenal cortex at day 110 and later (*Figure 16.9*). The porcine foetal

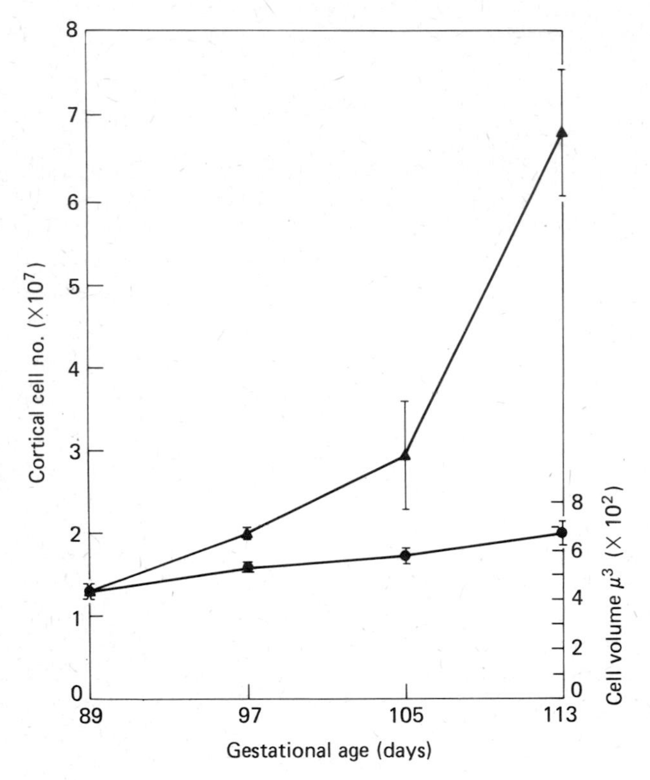

Figure 16.9 The effect of gestational age on the number and size of porcine foetal adrenal cortical cells. Each data point represents the mean ±S.E.M. of adrenal cortical cell number or volume from the paired adrenals of two foetuses from each of four litters. From Lohse and First (1981)

adrenal cortex also acquires adult type zonation and vascularity around day 110 (Lohse and First, 1981). In the light of these histological changes, ultrastructural changes were expected similar to those reported in the sheep by Robinson, Rowe and Wintour (1979). Surprisingly, in the porcine foetal adrenal, at least as early as day 89, mitochondria, smooth endoplasmic reticulum, and lipid droplets characteristic of the mature, active adrenal cortex were consistently present (Lohse and First, 1981; *Figure 16.10*).

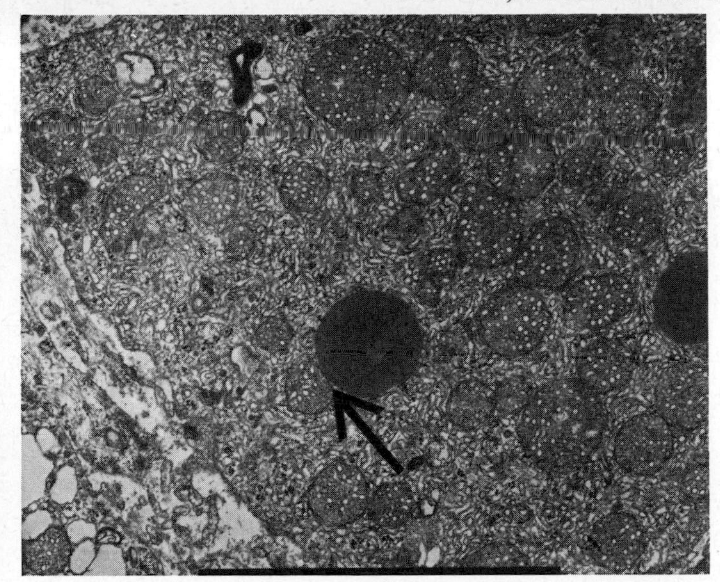

Figure 16.9 Electron micrograph of the adrenal cortex of a foetus at day 89 of gestation × 12000 (reduced to two-thirds in reproduction). Note the abundant smooth endoplasmic reticulum, typical adrenal cortex-type mitochondria and lipid droplets. Note also area of contact between lipid droplet and mitochondrion (arrow). From Lohse and First (1981)

The porcine foetal adrenal also increases its capacity for cortisol production *in vitro* as term approaches. Dvorak (1972) measured fluorogenic steroid production by adrenal tissue from late term foetal, neonatal and young pigs *in vitro* in the presence of 0.5 units ACTH/ml. This steroid production *in vitro* reached a peak at term along with foetal plasma cortisol and the ratio of adrenal weight to foetal weight. We have confirmed and expanded on the report of Dvorak (1972). Fluorogenic steroid production by porcine foetal adrenal tissue from days 89, 97, 105 and 113 of gestation was measured *in vitro* with or without 0.5 units ACTH/ml. Although the main effects of age and ACTH were both significant ($P<0.0001$), their interaction was not; the response ratio (+ACTH/—ACTH) was the same at all ages examined (*Figure 16.11*; Lohse and First, 1981). Since this result disagreed with analogous studies in the sheep (Wintour *et al.*, 1975), the dose-response to ACTH of porcine foetal adrenal tissue *in vitro* was further investigated. The same four ages were used, and following a 40 minute pre-incubation period, the tissue was incubated with 0, 1, 10 or 100 mu ACTH/ml. Although the effect of treatment on fluorogenic steroid production was again significant ($P<0.0001$), with significant increases between 0 to 1, and 1 to 10 mu/ml, the dose-response curves were the same for all ages examined. Foetal pituitary homogenate was also tested in this study, and its effects at all ages on fluorogenic steroid production was the same as that of a maximally effective dose of ACTH (Lohse and First, 1981).

This evidence that the foetal pituitary adrenal axis is functional well before the pre-partum cortisol surge is supported by earlier work from our

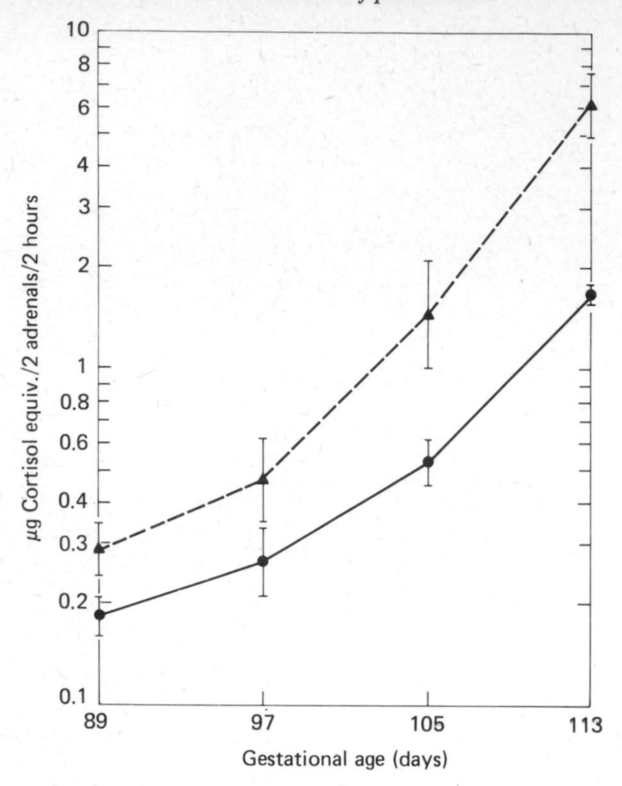

Figure 16.11 The effect of ACTH and foetal age on the *in vitro* production of adrenal fluorogenic steroids; -▲--▲- 500 mu ACTH/ml, -●—●- buffer only. The data represent the mean ±S.E.M. of the logs of μg cortisol equivalents of fluorogenic steroids produced by the adrenals of a foetus during 2 hours of incubation. Each data point is derived from two foetuses of each of four litters. From Lohse and First (1981)

laboratory. North, Hauser and First (1973) found that dexamethasone injected into foetal pigs on day 102 of gestation lowered foetal adrenal weights, probably through a negative feedback effect on pituitary corticotrophin. Hühn and Kiupel (1979) reported a suppression of foetal adrenal weight at parturition induced by administration of dexamethasone to the sow. Unlike North, Hauser and First (1973), however, these investigators did not adjust for the reduced birthweight of those piglets whose premature birth was induced by dexamethasone.

Because the fluorogenic steroid assay does not distinguish between cortisol and corticosterone, and since earlier studies (Madill and Bassett, 1973; Wintour *et al.*, 1975) reported a change with gestational age in the proportions of steroid produced by the foetal sheep adrenal, samples from each age that had been incubated with 0 or 100 mu ACTH/ml or with foetal pituitary homogenate were subjected to steroid separation and analysis by high pressure liquid chromatography (HPLC). No change in steroid proportions was observed until day 113, when there was a large increase in the relative proportion of cortisol (*Figure 16.12*; Lohse and First, 1981).

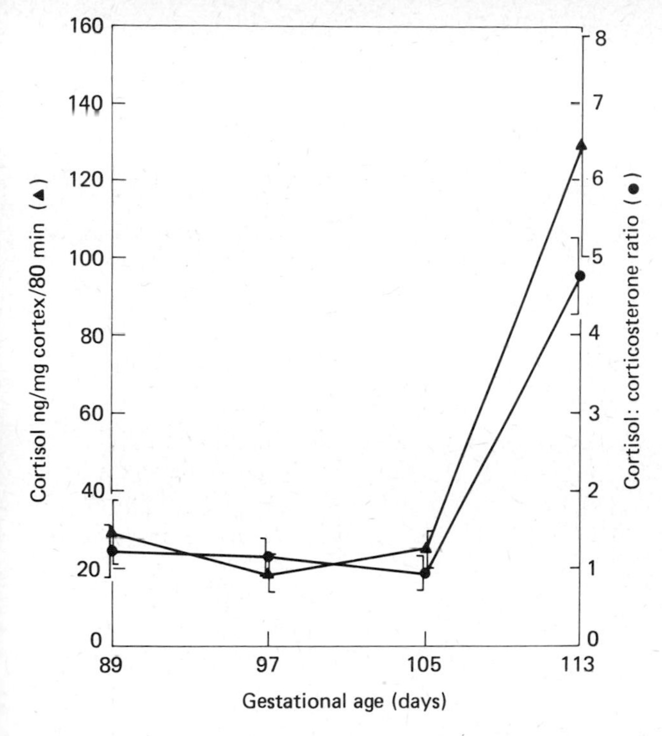

Figure 16.12 The effect of gestational age on the production *in vitro* of cortisol and the cortisol:corticosterone ratio of porcine foetal adrenals. Each data point is derived from the mean ±S.E.M. of two litters in which both adrenals of all piglets in the litter were incubated and the data pooled to provide a value for a litter. From Lohse and First (1981)

Consideration together of results from the various parts of this study suggests that increased adrenal cortex weight, due primarily to increased mitosis, accounts for the increase in cortisol production from days 89–105, but not from days 105–113. It was between days 105 and 113 that the change in steroid ratios occurred (*Figure 16.12*).

The means by which the foetal brain controls the foetal adrenal is not yet completely known. Although ACTH can bring about adrenal hypertrophy and parturition in the last 10 days of gestation, it is not usually associated with the hyperplasia that was observed in the late term foetal adrenal. Some other hormonal and/or neural factors may be involved as well. In addition, even if ACTH is the sole means by which the foetal brain controls the foetal adrenal, what controls foetal ACTH? If foetal ACTH secretion is regulated by foetal hypothalamic corticotrophin releasing factor (CRF), what controls foetal CRF secretion? To further complicate matters, control of adrenal cortisol production by the foetal pituitary may not be the only way the foetal brain regulates steps in the induction of parturition. Parturition in pigs was not induced by maternal administration of dexamethasone until after day 100 of gestation (Coggins, 1975; *Table 16.2*) and dexamethasone injected after day 100 was incapable of inducing parturition when foetuses were decapitated (Coggins and First, 1977). Adrenocorticotrophic hormone given to the foetuses also failed to induce

Table 16.2 EFFECT OF TREATMENT WITH DEXAMETHASONE AT DIFFERENT STAGES OF GESTATION ON LENGTH OF GESTATION

	Day of gestation at start of treatment[a]				
	81	*91*	*101*	*Control*	*S.D.*
No. of sows	4	4	4	4	
Gestation length (days)[b]	116.3	115.3	109.7	115.8	1.75

[a]100 mg of dexamethasone/day injected intramuscularly on four consecutive days (adapted from Coggins, 1975)
[b]Means not significantly different from each other are underlined; day 101 group different ($P<0.05$)

parturition until day 100 (Bosc, 1973). These facts imply that a parallel or later step, perhaps a target for the glucocorticoids, must be developed before dexamethasone can induce parturition. This step, like increased foetal adrenal glucocorticoid secretion, depends on the foetal brain.

The concept of the placenta as the target organ for the pre-partum glucocorticoid action is supported by studies in sheep and rabbits. In sheep, the source of pregnancy-maintaining progesterone is the placenta (Linzell and Heap, 1968). Before either spontaneous or glucocorticoid-induced parturition, progesterone levels decline, 17,20α-dihydroxy-4-pregn-3-one (17,20αP) levels increase, and oestrogen levels increase (Flint *et al.*, 1975). Placental minces taken at these times increase their metabolism of ^{3}H-pregnenolone to 17,20αP at the expense of progesterone (Anderson, Flint and Turnbull, 1975).

There also appears to be an increase in C-17,20-lyase activity *in vitro* in ovine placental tissue collected following endogenous or exogenous glucocorticoid elevation (Steele, Flint and Turnbull, 1976a). This is confirmed by a reported rise in uterine androstenedione production at parturition (Steele, Flint and Turnbull, 1976b). It has been suggested (Flint, Ricketts and Craig, 1979) that this increased conversion of C-21 to C-19 steroids may be a result of decreased placental progesterone concentration, since progesterone inhibits C-17,20 lyase in some tissues. The possibility that the foetal or maternal adrenals are the source for pre-partum aromatizable androgens has been ruled out by adrenalectomy experiments (Flint *et al.*, 1976; Flint and Ricketts, 1979) and by the lack of effect of hypoxic stress on maternal or foetal androgen levels (Jones *et al.*, 1977).

The importance of placental participation in parturition has been demonstrated by a study in the rabbit, in which the close connection between the placenta and endometrium permits the placenta to be maintained in the absence of a foetus. When foetuses were removed on or before day 25 of gestation, parturition at term did not occur and could not be induced by dexamethasone. When foetuses were removed on or after day 26, the placentas were delivered at normal term on day 32 (Chiboka, Casida and First, 1977). Unfortunately, complete foetectomy in the pig results in prompt abortion (Chiboka, Casida and First, 1976).

It seems apparent from the studies of Nara and First (1978; 1981a) that in the pig glucocorticoids induce parturition by causing an increase in uterine $PGF_{2\alpha}$ secretion, which in turn brings about luteolysis. The mechanism by which glucocorticoids elevate $PGF_{2\alpha}$ secretion has not yet been worked out. In the sow, plasma oestrogens (Robertson and King,

1974) and urinary oestrogens (Rombauts, 1962) increase considerably in late gestation. The source of these oestrogens appears to be intrauterine, since neither ovariectomy, hypophysectomy (Fevre, Leglise and Rombauts, 1968) nor adrenalectomy (Fevre, Leglise and Reynaud, 1972) of pregnant sows alters the pattern of oestrogen excretion during pregnancy. Also, the amount of urinary oestrogens is proportional to litter size and drops off sharply following parturition (Fevre, Leglise and Rombauts, 1968).

Oestrogens stimulate the release of $PGF_{2\alpha}$ by the gravid uterus and can be used to induce parturition in the sheep (Currie, 1977; Liggins *et al.*, 1973; 1977) and goat (Currie and Thorburn, 1973). However, in three different studies, large doses of oestrogen to the sow (B.S. Nara and N.L. First, unpublished; Nellor *et al.*, 1975; Flint, Ricketts and Craig, 1979) or the foetuses (*Table 16.3*) failed to induce parturition. Exogenous oestrogen

Table 16.3 EFFECT OF INTRAMUSCULAR INJECTIONS OF OESTRADIOL CYPIONATE (OCP) IN COTTONSEED OIL (CSO) ON PREGNANT PIGS[(a)]

OCP dose (mg/day)	*Days of gestation injected*	*No. sows*	*Gestation length* (days)	*Live births* (%)	*Litter size*	*Onset of lactation*
Maternal application:						
50	103–105	4	115.5±0.6	92.1±5.9	9.3±1.9	4/4
25	110–pt.[(b)]	4	115.8±0.5	96.1±3.8	8.8±1.4	4/4
0 (CSO 15 ml)	80–97	5	115.8±0.4	94.4±3.4	7.6±1.7	5/5
15	80–97	4	115.0±0.4	88.6±11.4	9.0±1.2	4/4
0 (CSO 15 ml)	90–107	5	115.2±0.9	87.7±6.4	7.6±1.7	5/5
15	90–107	5	113.8±0.6	95.6±2.9	11.2±1.4	5/5
Fetal application:						
6 foetuses/litter						
2.5 mg/foetus						
(15 mg total)	103	4	112.5±1.7	78.6±21.4	6.8±0.9	4/4
2.5 ml CSO/foetus						
(0 mg total)	103	4	111.5±2.1	77.9±15.0	8.5±1.6	4/4

[(a)]Adapted from Nara, 1979: No significant effects were detected among treatments with either maternal or foetal application of oestradiol cypionate.
[(b)]pt = parturition.

administration may be ineffective in the pig in the last 35 days of gestation, since circulating oestrogens in the sow are quickly inactivated by conjugation at this time (Rombauts, 1962). The effect of infusion of oestrogen into the placental circulation has not been examined.

Another surprising change in porcine placental steroid metabolism is an apparent increase in progesterone production at delivery, as indicated by foetal and umbilical venous progesterone levels at birth (Silver *et al.*, 1979).

Although the exact mechanisms by which increased foetal cortisol elevates uterine $PGF_{2\alpha}$ production and by which the foetal brain controls the foetal adrenal are not yet completely understood, the available information is sufficient to formulate a model describing parturition in the pig. This proposed model is presented in *Figure 16.13*.

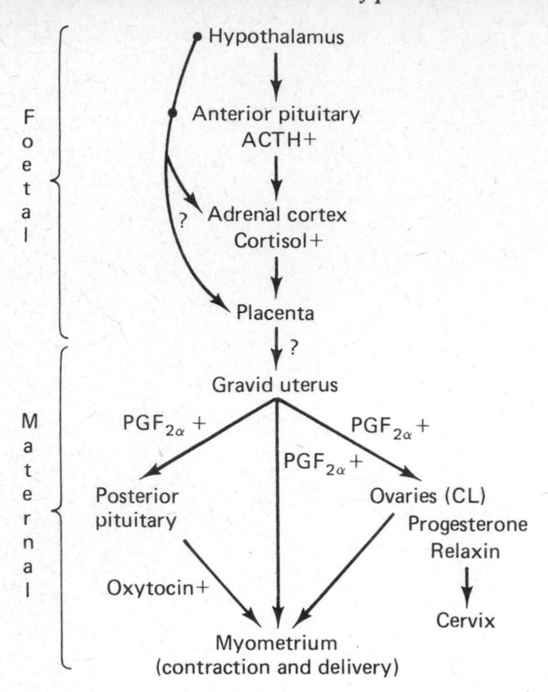

Figure 16.13 A suggested sequence of events leading to and associated with parturition in the pig. Hormones having a stimulatory effect on a target are designated with a (+); unknown steps or compounds are indicated by a (?). From First and Bosc (1979)

The induction of parturition

If the proposed model (*Figure 16.12*) for the steps and mechanisms initiating parturition is valid, it should be possible to induce parturition before term by exogenous administration of the hormonal messengers eliciting the sequence of events leading to parturition. This has been shown possible for four of the messengers: ACTH to the foetus (Bosc, 1973); cortisol analogues to foetuses (North, Hauser and First, 1973) or the mother (North, Hauser and First, 1973; First and Staigmiller, 1973; Coggins and First, 1977; Coggins, Van Horn and First, 1977; Hühn, Konig and Hühn, 1978; Hühn and Kiupel, 1979; Brenner *et al.*, 1979; Nara and First, 1981a); $PGF_{2\alpha}$ and its analogues to the mother (*Table 16.1*); and within the last 12 hours of gestation, oxytocin to the mother (Muhrer, Shippen and Lasley, 1955; Welk and First, 1979).

This knowledge has provided the basis for the development of methods for the induction of parturition. The potential benefits derived from the ability to induce parturition at a precise time are:

(a) more efficient use of farrowing facilities and labour,
(b) avoidance of parturition on weekends, holidays or at late hours of the night,
(c) more efficient cross-fostering of litters,

(d) more uniformity in the time of oestrus for lactating sows after weaning and in the age and weight of piglets in feedlots, and
(e) reduction in the length of gestation.

Unfortunately, the latter is of limited feasibility. Attempts to cause delivery before day 109 have resulted in death of the piglets by day 1 post-partum (Wierzchos and Pejsak, 1976; N.L. First, unpublished). Piglets born as late as day 110 (Jainudeen and Brandenburg, 1980) or in some cases day 111 (Gilchrist-Shirlaw, Hillyer and Miller, 1978; Aumaitre, Deglaire and LeBost, 1979; Bosc and Martinat-Botte, 1976; Hammond and Carlyle, 1976) have reduced survival and reduced birth and weaning weight. However, piglets born two, and in some experiments, three days before term have normal survival and normal birth and weaning weights (Aumaitre, Deglaire and LeBost, 1979; *Table 16.1*).

Of the potential substances which might be used to induce parturition in commercial pig herds, ACTH or Synacthen, seems not to be practical for commercial use because it is effective only by foetal administration (Bosc, 1973; Brenner *et al.*, 1979).

Oxytocin is effective at a time when the blood plasma is nearly depleted of progesterone and after milk ejection can be elicited (Muhrer, Shippen and Lasley, 1955). At this stage, injection of 50 iu oxytocin will cause delivery within 0.6±0.6 hours (Welk and First, 1979).

The glucocorticoid dexamethasone has been used in several experiments to successfully induce parturition. While a single small dose of 6 mg to each of seven foetuses was effective (North, Hauser and First, 1973), a large dose of 75–100 mg/day for 3–4 days was required for parturition to be induced by maternal intramuscular injection (First and Staigmiller, 1973; Coggins and First, 1977; Hühn, Konig and Hühn, 1978; Hühn and Kiupel, 1979; Brenner *et al.*, 1979; Nara and First, 1981a). Besides being expensive and requiring repeated injection, there is more variation in the interval from treatment to parturition than after the injection of $PGF_{2\alpha}$.

The most effective, efficient and widely accepted method to date for inducing parturition in pigs is an intramuscular injection of $PGF_{2\alpha}$ or one of its analogues. The number of studies in which $PGF_{2\alpha}$ or its analogues has been used is far greater than can be presented in one review. *Table 16.1* represents an attempt to summarize several such studies in which similar data were collected and where untreated sows provided contemporary control data. Only data derived from doses of drugs proved to be effective and from sows treated after day 110, have been considered. There seems to be little difference in the response of sows and litters to $PGF_{2\alpha}$ or its analogues except for the magnitude of the effective dose and for the studies summarized here, a tendency for the analogues to cause more sows to farrow within 48 hours after treatment. This difference may relate more to the dose of drug used than to differences between the compounds. When both $PGF_{2\alpha}$ and cloprostenol were compared in the same experiment, there was no difference (Boland, Craig and Kellcher, 1979).

Approximately 87% of sows injected after day 110 with $PGF_{2\alpha}$ or 93–100% with its analogues farrowed within 48 hours. The average time from injection to delivery of the first piglet was approximately 28±5.5 hours for $PGF_{2\alpha}$ and 26.4±5.7 hours for cloprostenol, the most commonly used analogue of $PGF_{2\alpha}$. Why parturition is not induced in some sows is unknown. It may be pertinent that all are induced when $PGF_{2\alpha}$ is infused intravenously for 10 hours (Nara and First, 1981a,b). Additionally, some investigators found that sows not responding to the injected $PGF_{2\alpha}$ were not pregnant (Jainudeen and Brandenburg, 1980) while others observed those not responding had an excessively long gestation period (Hansen, 1979).

When $PGF_{2\alpha}$ compounds are administered after day 110, most experiments show no significant difference between treated and control sows in duration of labour, the frequency of piglets born dead, birth weight, survival to weaning or weaning weight although the means suggest slightly greater birth weight for the controls. This seems to be the case for the data summarized in *Table 16.1*.

The interval from injection of $PGF_{2\alpha}$ compounds to the initiation of parturition is shortened when the drug is injected very close to expected parturition (Bosc and Martinat-Botte, 1976; Willemse *et al.*, 1979). This is likely to be due to the endogenous initiation of the events leading to parturition before the exogenous administration of $PGF_{2\alpha}$.

The sows return to oestrus and have normal post-weaning reproductive performance after the induction of parturition (Bosc and Martinat-Botte, 1976; Walker, 1979; Lynch and Langley, 1977; Robertson, King and Elliott, 1978).

Prostaglandin $F_{2\alpha}$ may not be the only prostaglandin capable of inducing parturition. Of considerable interest is the recent study of Vaje *et al.* (1980) in which parturition was induced on day 110 by intravenous infusion for 10 hours on day 109 of a prostaglandin E analogue, sulproston, or 47±11 hours after two intramuscular injections of sulproston. How this is accomplished is unknown.

The distribution of sows farrowing after injection of $PGF_{2\alpha}$ compounds is compared with control sows in *Figure 16.14*. While most sows farrow on one day (~70%), those which do not complicate the management of a farrowing unit and reduce the benefits of induced parturition.

There have been two attempts to develop methods for making the time of prostaglandin-induced parturition more precise. When oxytocin was injected on the expected day of cloprostenol-induced parturition and at a time when milk could be ejected, the interval from injection of prostaglandin to delivery of the first piglet was 27.7±2 hours and the variance in time of delivery was significantly reduced from that due to cloprostenol alone (Welk and First, 1979). This combination of drugs allows synchronization and supervision of farrowings during a specified half day.

In a second attempt, the treatment involved the daily administration of 100 mg of progesterone on days 112, 113 and 114, accompanied by 200 μg of cloprostenol on day 115 (Gooneratne *et al.*, 1979). 80% of the sows farrowed between 08.00 and 17.00 on day 116 and a precise time of parturition, 25.4±1 hour, after injection of cloprostenol was achieved. Although this study was without a control group in which cloprostenol alone was used, the initiation of lactation, piglet survival and weight, and post-partum reproductive performance of the sow were not different from untreated controls.

The frequency of sows showing prolonged elevation of body temperature or clinical symptoms of the Metritis–Mastitis–Agalactia (MMA)

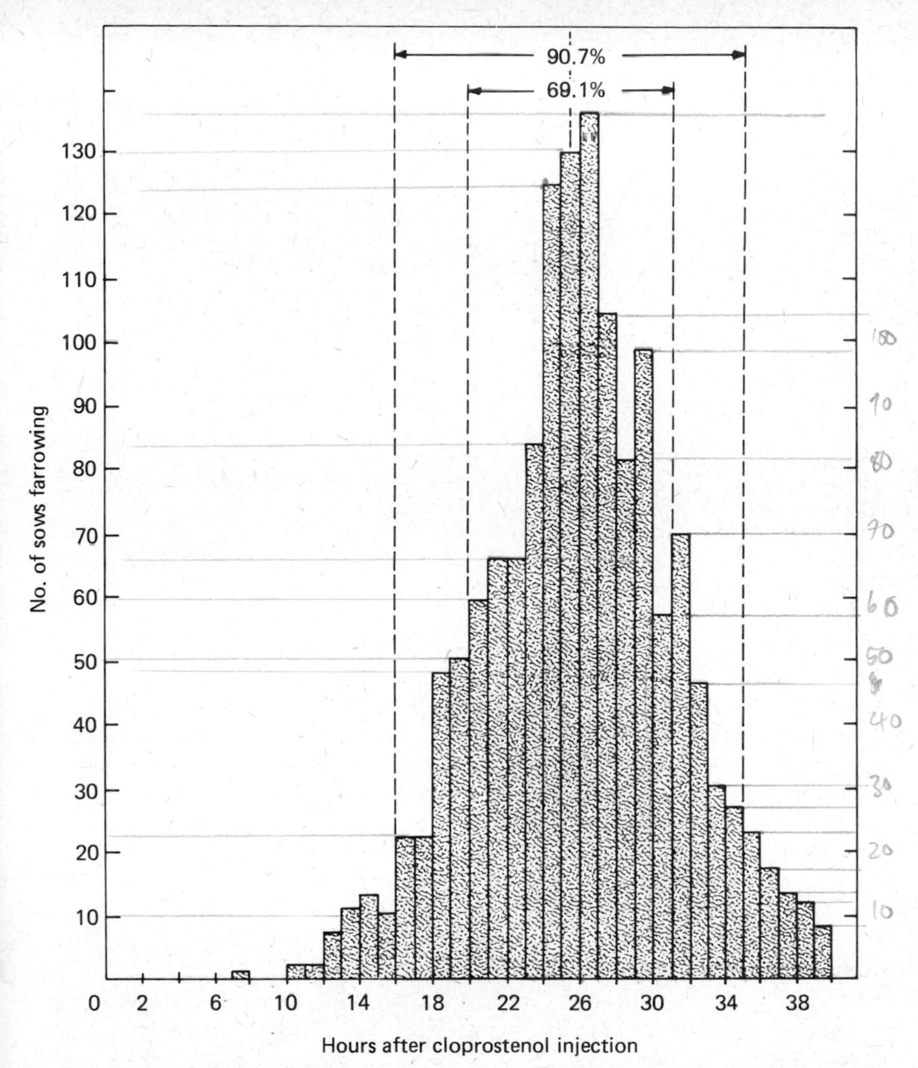

Figure 16.14 Distribution of 1459 sows farrowing at specific times after a single intramuscular injection of 175 μg cloprostenol. Approximately 69% commenced farrowing between 20 and 30 hours and 90.7% between 16 and 34 hours after treatment. From Hammond and Matty (1980)

syndrome has been reported in several studies to be less after the induction of parturition than for control sows (Einarsson, Gustafsson and Larsson, 1975; Backstrom *et al.*, 1976; Ehnvall *et al.*, 1976; Hansen and Jacobsen, 1976; Bogataj, 1979; Humke, Seidel and Scharp, 1979; Hühn, Lutter and Hühn, 1980). A slight difference in the same direction is evident from the studies summarized in *Table 16.1.* Whether this is a consistent benefit from the induction of parturition or mainly reported by investigators finding a difference remains to be determined. Some have found no reduction in the frequency of MMA after induction of parturition (Hansen, 1979; Samol, 1980).

The ultimate value of $PGF_{2\alpha}$ or its analogues as agents for induction of parturition is enhanced by the fact that these compounds induce a normal parturition as well as a series of parturition-related events including initiation of lactation and expulsion of the placenta (First and Bosc, 1979). Whether the induction of parturition will become a widely accepted pig management tool remains to be determined.

Acknowledgements

Research from the authors' laboratory has been supported by USDA CSRS Grant No. 616-15-152, USDA SEA Grant No. 2529, the National Pork Producers Council, Public Health Service Training Grant No. 5-T01-HD-00104-10 from the National Institute of Child Health and Human Development Grant No. 630-0505B from the Ford Foundation, and by the College of Agricultural and Life Sciences, University of Wisconsin-Madison.

This is Department of Meat and Animal Science Paper No. 787.

References

ALEXANDROVA, M. and SOLOFF, M.S. (1980a). Oxytocin receptors and parturition. I. Control of oxytocin receptor concentration in the rat myometrium at term. *Endocrinology* **106**, 730–735

ALEXANDROVA, M. and SOLOFF, M.S. (1980b). Oxytocin receptors and parturition. III. Increases in estrogen receptor and oxytocin receptor concentrations in the rat myometrium during $PGF_{2\alpha}$-induced abortion. *Endocrinology* **106**, 739–743

ALEXANDROVA, M. and SOLOFF, M.S. (1980c). Oxytocin receptors and parturition in the guinea pig. *Biol. Reprod.* **22**, 1106–1111

ANDERSON, A.B.M., FLINT, A.P.F. and TURNBULL, A.C. (1975). Mechanism of action of glucocorticoids in induction of ovine parturition: Effect on placental steroid metabolism. *J. Endocr.* **66**, 61–70

ASH, R.W. and HEAP, R.B. (1975). Oestrogen, progesterone and corticosteroid concentration in peripubertal plasma of sows during pregnancy, parturition, lactation and after weaning. *J. Endocr.* **64**, 141–154

AUMAITRE, A., DEGLAIRE, B. and LeBOST, J. (1979). Premature parturition in sows and the significance of birth weight of piglets. (French) *Annls Biol. anim. Biochim. Biophys.* **19**, 1B, 267–275

BACKSTROM, L., EINARSSON, S., GUSTAFSSON, B. and LARSSON, K. (1976). Prostaglandin $F_{2\alpha}$-induced parturition for prevention of the agalactia syndrome in the sow. *Proc. Fourth Int. Pig Vet. Soc. Congr., Ames, Iowa*, p. 5

BALDWIN, D.M. and STABENFELDT, G.H. (1975). Endocrine changes in the pig during late pregnancy, parturition and lactation. *Biol. Reprod.* **12**, 508–515

BAZER, F.W. and THATCHER, W.W. (1977). Theory of maternal recognition of pregnancy in swine based on estrogen controlled endocrine versus exocrine secretion of prostaglandin $F_{2\alpha}$ by uterine endometrium. *Prostaglandins* **14**, 397–401

BEVIER, G.W. and DZIUK, P.J. (1976). The effect of number of fetuses and their location in the uterus on the incidence of stillbirth. *Proc. IVth Int. Pig Vet. Soc. Congr., Ames, Iowa*, p.21

BICHARD, M., STORK, M.G., RICKATSON, S. and PEASE, A.H.R. (1976). The use of synchronized farrowing in large pig units. *Anim. Prod.* **22**, 138–139. (Abstract)

BOGATAJ, M. (1979). Effect of prostaglandin $F_{2\alpha}$ treatment on parturition, puerperium and fertility of breeding sows. *Inaugural Dissertation Tierartzliche Hochschule, Hannover*

BOLAND, M.P., CRAIG, J. and KELLCHER, D.L. (1979). Induction of farrowing: Comparison of the effects of prostaglandin $F_{2\alpha}$ (Lutalyse) and an analogue (Cloprostenol). *Ir. vet. J.* **33**, 45–47

BONING, J. (1979). Investigations on onset of parturition in pigs. (German) *Tierzucht* **33**, 561–563

BOSC, M.J. (1973). Modification de la durée de gestation de la Truie après administration d'ACTH aux foetus. *C. r. hebd. Séanc. Acad. Sci., Paris* **276**, 3183–3186

BOSC, M.J. and MARTINAT-BOTTE, F. (1976). Induction de la parturition chez la Truie au moyen de prostaglandines. *Econ. Med. Anim.* **17**, 235–244

BOSC, M.J., du MESNIL du BUISSON, F. and LOCATELLI, A. (1974). Mise en evidence d'un controle foetal de la parturition chez la Truie. Interactions avec la fonction luteale. *C.r. hebd. Séanc. Acad. Sci., Paris* **278**, 1507–1510

BRENNER, K.V., NITZSCHE, K., GURTLER, H. and MULLER, N. (1979). Effect of administration of ACTH or corticoids to sows at the end of pregnancy on duration of pregnancy and parturition. (German) *Mh. VetMed.* **34**, 91–95

CHAN, W.Y. (1977). Relationship between the uterotonic action of oxytocin and prostaglandin: Oxytocin action and release of PG activity in isolated nonpregnant and pregnant rat uteri. *Biol. Reprod.* **7**, 541–548

CHIBOKA, O., CASIDA, L.E. and FIRST, N.L. (1976). Effect of fetectomy on pregnancy maintenance in swine. *J. Anim. Sci.* **42**, 1363 (Abstract)

CHIBOKA, O., CASIDA, L.E. and FIRST, N.L. (1977). Role of rabbit fetuses and placentas in the maintenance of gestation and parturition. *J. Anim. Sci.* **46**, 776–783

COGGINS, E.G. (1975). Mechanisms controlling parturition in swine. PhD Thesis. University of Wisconsin-Madison

COGGINS, E.G. and FIRST, N.L. (1977). Effect of dexamethason, methallibure and fetal decapitation on porcine gestation. *J. Anim. Sci.* **44**, 1041–1049

COGGINS, E.G., Van HORN, D. and FIRST, N.L. (1977). Influence of prostaglandin $F_{2\alpha}$, dexamethasone, progesterone and induced CL on porcine parturition. *J. Anim. Sci.* **46**, 754–762

COX, D.F. (1964). Genetic variation in the gestation period of swine. *J. Anim. Sci.* **23**, 746–751

CURRIE, W.B. (1977). Endocrinology of pregnancy and parturition in sheep and goats. In *Management of Reproduction in Sheep and Goats Symposium*, pp. 72–78. Madison, Wisconsin,

CURRIE, W.B. and THORBURN, G.D. (1973). Release of prostaglandin F, regression of corpora lutea and induction of premature parturition in goats treated with estradiol-17β. *Prostaglandins* **12**, 1093–1103

CURTIS, S.E., ROGER, J.C. and MARTIN, T.G. (1969). Neonatal thermostability and body composition of piglets from experimentally prolonged gestation. *J. Anim. Sci.* **29**, 335–340

DHINDSA, D.S. and DZIUK, P.J. (1968). Effect on pregnancy in the pig after killing embryos or fetuses in one uterine horn in early gestation. *J. Anim. Sci.* **27**, 122–126

DIEHL, J.R., GODKE, R.A., KILLIAN, D.B. and DAY, B.N. (1974). Induction of parturition in swine with prostaglandin $F_{2\alpha}$. *J. Anim. Sci.* **38**, 1229–1234

DOWNEY, B.R., CONLON, P.D. and BAKER, R.D. (1976). Controlled farrowing program using a prostaglandin analogue Ay24655. *Can. J. Anim. Sci.* **56**, 655

DROST, J. and HOLM, L.W. (1968). Prolonged gestation in ewes after fetal adrenalectomy. *J. Endocr.* **40**, 293–295

DU MESNIL DU BUISSON, F. and DAUZIER, L. (1957). Influence de l'ovariectomie chez la Truie pendant la gestation. *C.r. Séanc. Soc. Biol.* **151**, 311–313

DU MESNIL DU BUISSON, F. and DENAMUR, R. (1969). Mechanismes des controle de la fonction luteale chez la Truie, la Brebis et al Vache. *3rd Int. Congr. Endo. Mexico, Excerpta Med. Int. Congr. Ser.* **184**, 927–934

DU MESNIL DU BUISSON, F. and ROMBAUTS, P. (1963). Reduction experimentale du nombre des foetus au course de la gestation. *Annls Biol. anim. Biochem. Biophys.* **3**, 445–449

DURAND, P., BOSC, M. and NIEDLE, A. (1978). Croissance de surrenales de foetus ovin en fin de gestation: evolution de l'ADN et des proteenes membrainares. *C.r. hebd. Séanc. Acad. Sci., Paris* **287**, Serie D, 297–300

DVORAK, K. (1972). Adrenocortical function in fetal, neonatal and young pigs. *J. Endocr.* **54**, 473–481

DZIUK, P.J. (1979). Control and mechanics of parturition in the pig. *Anim. Reprod. Sci.* **2**, 335–342

DZIUK, P.J. and HARMON, B.G. (1969). Succession of fetuses at parturition in the pig. *Am. J. vet. Res.* **30**, 419–421

EHNVALL, R., EINARSSON, S., GUSTAFSSON, B. and LARSSON, K. (1976). A field study of prostaglandin-induced parturition in the sow. *Proc. Fourth Int. Pig Vet. Soc. Cong., Ames, Iowa,* p. D.6

EINARSSON, S., GUSTAFSSON, B. and LARSSON, K. (1975). Prostaglandin induced parturition in swine with some aspects on prevention of the MMA (Metritis–Mastitis–Agalactia) Syndrome. *Nord. VetMed.* **27**, 429–436

ELLENDORFF, F. (1980). Recent findings on physiology of parturition in the sow – importance for the completion of parturition. *Tierzüchter* **32**, 138–140

ELLENDORFF, F., FORSLING, M., PARVIZI, N., WILLIAMS, H., TAVERNE, M. and SMIDT, D. (1979a). Plasma oxytocin and vasopressin concentrations in response to prostaglandin injection into the pig. *J. Reprod. Fert.* **56**, 573–577

ELLENDORFF, F., TAVERNE, M.A.M., ELSAESSER, F., FORSLING, M.L., PARVIZI, N., NAAKTGEBOREN, C. and SMIDT, D. (1979b) Endocrinology of parturition in the pig. *Anim. Reprod. Sci.* **2**, 323–334

ELLICOTT, A.R. and DZIUK, P.J. (1973). Minimum daily dose of progesterone and plasma concentration for maintenance of pregnancy in ovariectomized gilts. *Biol. Reprod.* **9**, 300–304

ELMORE, R.G., MARTIN, C.E., RILEY, J.L. and LITTLEDIKE, T. (1979). Body temperatures of farrowing swine. *J. Am. vet. med. Ass.* **174**, 620–622

ELZE, K., HAGNER, H.J., SCHNEIDER, F., KOHLER, R., EULENBERGER, K., HÜHN, R., MICHEL, G. and RANWOLF, A. (1979). Control of parturition in swine. 2. Study results of parturition synchronization in swine with the prostaglandin analog cloprostenol (ICI 80996, Estrumate). *Arch. exp. VetMed.* **33**, 791–806

FEVRE, J., LEGLISE, P.C. and REYNAUD, O. (1972). Role des surrenales maternelles dans la production d'oestrogenes par la truie gravide. *Annls Biol. anim. Biochem. Biophys.* **12**, 559–569

FEVRE, J., LEGLISE, P.C. and ROMBAUTS, P. (1968). Du role de l'hypophyse et des ovaries dans la biosynthese des oestrogenes aux cours de la gestation chez la Truie. *Annls Biol. anim. Biochem. Biophys.* **8**, 225–233

FEVRE, J., TERQUI, M. and BOSC, M.J. (1975). Mechanismes de la naissance chez la Truie. Equilibres hormonaux avant et pendant de parturition. *Journees Rech. Porcine en France*, pp. 393–398. Paris, L'Institut Technique du Porc

FIRST, N.L. (1979). Mechanisms controlling parturition in farm animals. In *Animal Reproduction*, (H. Hawk, Ed.), pp. 215–257. Montclair, New Jersey, Allanheld Osmun

FIRST, N.L. and BOSC, M.J. (1979). Proposed mechanisms controlling parturition and the induction of parturition in swine. *J. Anim. Sci.* **48**, 1407–1421

FIRST, N.L. and STAIGMILLER, R.B. (1973). Effects of ovariectomy, dexamethasone and progesterone on the maintenance of pregnancy in swine. *J. Anim. Sci.* **37**, 1191–1194

FLINT, A.P.F. and RICKETTS, A.P. (1979). Control of placental endocrine function: Role of enzyme activation in the onset of labour. *J. Ster. Biochem.* **11**, 493–500

FLINT, A.P.F., RICKETTS, A.P. and CRAIG, V.A. (1979). The control of placental steroid synthesis at parturition in domestic animals. *Anim. Reprod. Sci.* **2**, 239–251

FLINT, A.P.F., ANDERSON, A.B.M., STEELE, P.A. and TURNBULL, A.C. (1975). The mechanism by which fetal cortisol controls the onset of parturition in sheep. *Biochem. Soc. Trans.* **3**, 1189–1194

FLINT, A.P.F., ANDERSON, A.B.M., GOODSON, J.D., STEELE, P.A. and TURNBULL, A.C. (1976). Bilateral adrenalectomy of lambs *in utero*: Effect on maternal hormone levels at induced parturition. *J. Endocr.* **69**, 433–444

FORSLING, M.L., TAVERNE, M.A.M., PARVIZI, N. ELSAESSER, F., SMIDT, D. and ELLENDORFF, F. (1979). Plasma oxytocin and steroid concentrations during late pregnancy, parturition and lactation in the miniature pig. *J. Endocr.* **82**, 61–69

GILCHRIST-SHIRLAW, D.W., HILLYER, G.M. and MILLER, J.K. (1978). Induction of farrowing in gilts and sows. *Annual Report*, Edinburgh School of Agriculture, pp. 56–57

GOONERATNE, A., HARTMANN, P.E., McCAULEY, I. and MARTIN, C.E. (1979). Control of parturition in the sow using progesterone and prostaglandin. *Aust. J. biol. Sci.* **32**, 587–595

GUSTAFSSON, B., EINARSSON, S., LARSSON, K. and EDQVIST, L.E. (1976). Sequential changes of estrogens and progesterone and prostaglandin-induced parturition in the sow. *Am. J. vet. Res.* **37**, 1017–1020

HAGNER, H.J., ELZE, K. and MICHEL, G. (1979). Control of parturition in pigs. Part 1. Results of clinical, histological and haematological studies of induction of parturition in pigs by means of prostaglandin $F_{2\alpha}$. (German) *Arch. exp. VetMed.* **33**, 475–488

HAMMOND, D. and CARLYLE, W.W. (1976). Controlled farrowing on commercial pig breeding units using cloprostenol, a synthetic analogue of prostaglandin $F_{2\alpha}$. *8th Int. Congr. Anim. Reprod. A.I., Krakow* **3**, 365–368

HAMMOND, D. and MATTY, G. (1980). A farrowing management system using cloprostenol to control the time of parturition. *Vet. Rec.* **106**, 72–75

HANSEN, L.M. (1979). Reproductive efficiency and incidence of MMA after controlled farrowing using a prostaglandin analogue cloprostenol. *Nord. VetMed.* **31**, 122–128

HANSEN, L.M. and JACOBSEN, M.I. (1976). The course of the puerperium in an MMA herd after induction of parturition with prostaglandins. *Nord. VetMed.* **28**, 357–360

HOLTZ, W., WELP, C. and SPANGENBERG, W. (1979). Induction of parturition in pigs with a prostaglandin $F_{2\alpha}$ analog, cloprostenol (ICI 80996). (German) *Zentbl. VetMed.* **26A**, 815–825

HOLTZ, W., DIALLO, T., SPANGENBERG, B., ROCKEL, P., BOGNER, H., SMIDT, D. and ZEIDLE, W. (1979) Induction of parturition in sows with a prostaglandin $F_{2\alpha}$-analog. *J. Anim. Sci.* **49**, 367–373

HONNEBIER, A.J. and SWAAB, J.F. (1973). The influence of anencephaly upon intrauterine growth of the fetus and placenta and upon gestation length. *J. Obstet. Gynaec. Br. Commonw.* **80**, 577–588

HÜHN, R. and KIUPEL, H. (1979). Experimental use of dexamethasone to control parturition in sows. II. Influence of dexamethasone on adrenal weight and cell nucleus volume of the zona fasciculata in the piglet. (German) *Arch. exp. VetMed.* **33**, 247–251

HÜHN, U., HÜHN, R. and KONIG, I. (1976). Synchronization of parturition in pigs with dexamethasone and prostaglandin $F_{2\alpha}$. *8th Int. Congr. Anim. Reprod. A.I., Krakow* **3**, 369–372

HÜHN, R., HÜHN, U. and KONIG, I. (1980). Effectiveness of various treatment regimes for inducing parturition in sows with prostaglandin $F_{2\alpha}$. (German) *Arch. exp. VetMed.* **34**, 167–173

HÜHN, R., KONIG, I. and HÜHN, U. (1978). Experiment on synchronization of parturition in swine. *Arch. Tierz.* **21**, 409–415

HÜHN, R., LUTTER, K. and HÜHN, U. (1980). Influence of induction of parturition on selected performance values of sows and piglets. (German) *Arch. exp. VetMed.* **34**, 175–180

HUMKE, R., SEIDEL, L. and SCHARP, H. (1979). Trials on induction of parturition in pigs with the luteolytic prostaglandin $F_{2\alpha}$-analogue HR 837. (German) *Dt. tierarztl. Wschr.* **86**, 221–225

JAINUDEEN, M.R. and BRANDENBURG, A.C. (1980). Induction of parturition in crossbred sows with cloprostenol, an analogue of prostaglandin $F_{2\alpha}$. *Anim. Reprod. Sci.* **3**, 161–166

JONES, C.T., BODDY, K., ROBINSON, J.S. and RATCLIFFE, J.G. (1977). Developmental changes in the responses of the adrenal glands of the fetal sheep to endogenous adrenocorticotropin as indicated by the hormone response to hypoxaemia. *J. Endocr.* **72**, 279–293

JONES, C.T., KENDALL, J.Z., RITCHIE, J.W.K., ROBINSON, J.S. and THORBURN, G.D. (1978). Adrenocorticotrophin and corticosteroid changes during dexamethasone infusion to intact and synacthen infusion to hypophysectomized fetuses. *Acta endocr.* **87**, 203–211

JONES, J.E.T. (1966). Observations on parturition in the sow. Part II. The parturient and post-parturient phases. *Br. vet. J.* **122**, 471–478

KERTILES, L.P. AND ANDERSON, L.L. (1979). Effect of relaxin on cervical dilation, parturition and lactation in the pig. *Biol. Reprod.* **21**, 57–68

KING, G.J., ROBERTSON, H.A. and ELLIOTT, J.I. (1979). Induced parturition in swine herds. *Can. vet. J.* **20**, 157–160

KRAELING, R.R. and DAVIS, B.J. (1974). Termination of pregnancy by hypophysectomy in the pig. *J. Reprod. Fert.* **36**, 215–217

LEMAN, A.D., HURTGEN, J.P. and HILLEY, H.D. (1979). Influence of intrauterine events on postnatal survival in the pig. *J. Anim. Sci.* **49**, 221–224

LIGGINS, G.C. (1968). Premature parturition after infusion of corticotrophin or cortisol into foetal lambs. *J. Endocr.* **42**, 323–329

LIGGINS, G.C. (1969). The foetal role in the initiation of parturition in the ewe. In *Foetal Autonomy*, (G.E.W. Wolstenholme and M. O'Connor, Eds.), pp. 218–244. London, Churchill

LIGGINS, G.C. and HOWIE, R.N. (1972). A controlled trial of antepartum glucocorticoid treatment for prevention of the respiratory distress syndrome in premature infants. *Pediatrics* **50**, 515–525

LIGGINS, G.C., KENNEDY, P.C. and HOLM, L.W. (1967). Failure of initiation of parturition after electrocoagulation of the pituitary of the fetal lamb. *Am. J. Obstet. Gynec.* **98**, 1080–1086

LIGGINS, G.C., FAIRCLOUGH, R.J., GRIEVES, S.A., KENDALL, J.Z. and KNOX, B.S. (1973). The mechanism of initiation of parturition in the ewe. *Recent Prog. Horm. Res.* **29**, 111–150

LIGGINS, G.C., FAIRCLOUGH, R.J., GRIEVES, S.A., FORSTER, C.S. and KNOX, B.S. (1977). Parturition in the sheep. In *The Fetus and Birth*, pp. 5–30. Amsterdam, Elsevier–Excerpta Medica–North Holland

LINZELL, J.L. and HEAP, R.B. (1968). A comparison of progesterone metabolism in the pregnant sheep and goat: Sources of production and an estimation of uptake by some target organs. *J. Endocr.* **41**, 433–438

LOHSE, J.K. and FIRST, N.L. (1981). Development of the porcine fetal adrenal in late gestation. *Biol. Reprod.* **25**, 181–190

LYNCH, B.P. and LANGLEY, O.H. (1977). Induced parturition in sows using prostaglandin analogue (ICI 80996). *Ir. J. agric. Res.* **16**, 259–265

MADILL, D. and BASSETT, J.M. (1973). Corticosteroid release by adrenal tissue from fetal and newborn lambs in response to corticotrophin in a perfusion system *in vitro*. *J. Endocr.* **38**, 75–87

MARTIN, P.A., BEVIER, G.W. and DZIUK, P.J. (1978). The effect of disconnecting the uterus and ovary on the length of gestation in the pig. *Biol. Reprod.* **18**, 428–433

MARTIN, P.A., NORTON, H.W. and DZIUK, P.J. (1977). The effect of corpora lutea induced during pregnancy on the length of gestation in the pig. *Biol. Reprod.* **17**, 712–717

MILOSAVLJEVIC, S., MILJKOVIC, V., SAVLJANSKI, B., RODOVIC, B., TRBOJEVIC, G. and STANKOV, M. (1972). The revival of apparently stillborn piglets. *Acta vet., Beogr.* **22**, 71–76

MINAR, M. and SCHILLING, E. (1970). Die beeinflussung des geburts-termins beim schwein durch gestagene hormone. *Dt. tierarztl. Wschr.* **77**, 421–444

MITCHELL, M.D. and FLINT, A.P.F. (1978). Prostaglandin production by intrauterine tissues from periparturient sheep: Use of a superfusion technique. *J. Endocr.* **76**, 111–121

MITCHELL, M.D., FLINT, A.P.F. and TURNBULL, A.C. (1975). Stimulation by oxytocin of prostaglandin F in uterine venous effluent in pregnant and puerperal sheep. *Prostaglandins* **9**, 47–56

MOLUKWU, E.C.I. and WAGNER, W.C. (1973). Endocrine physiology of the puerperial sow. *J. Anim. Sci.* **36**, 1158–1163

MUHRER, M.E., SHIPPEN, O.F. and LASLEY, J.F. (1955). The use of oxytocin for initiating parturition and reducing farrowing time in sows. *J. Anim. Sci.* **14**, 1250 (Abstract)

NARA, B.S. (1979). Mechanisms controlling prepartum luteolysis in swine. PhD Thesis. University of Wisconsin-Madison

NARA, B.S. and FIRST, N.L. (1977). Effect of indomethacin and prostaglandin $F_{2\alpha}$ on porcine parturition. *J. Anim. Sci.* **45** (Suppl. 1), 191 (Abstract)

NARA, B.S. and FIRST, N.L. (1978). Effect of indomethacin on dexamethasone-induced parturition in swine. *J. Anim. Sci.* **47** (Suppl. 1), 3 (Abstract)

NARA, B.S. and FIRST, N.L. (1981a). Effect of indomethacin on dexamethasone-induced parturition in swine. *J. Anim. Sci.* **52**, 788–793

NARA, B.S. and FIRST, N.L. (1981b). Effect of indomethacin and prostaglandin $F_{2\alpha}$ on parturition in swine. *J. Anim. Sci.* **52**, 1360–1370

NARA, B.S., DARMADJA, D. and FIRST, N.L. (1975). Ovary, follicle or CL removal in prepartum sows. *J. Anim. Sci.* **41**, 371 (Abstract)

NARA, B.S., DARMADJA, D. and FIRST, N.L. (1981). Effect of removal of follicles, corpora lutea or ovaries on maintenance of pregnancy in swine. *J. Anim. Sci.* **52**, 794–801

NARA, B.S., WELK, F.A., RUTHERFORD, J.E., SHERWOOD, O.D. and FIRST, N.L. (1982). Effect of relaxin on parturition and frequency of live birth in swine. *J. Reprod. Fert.* (in press)

NELLOR, J.E., DANIELS, R.W., HOEFER, J.A., WILDT, D.E. and DUKELOW, W.R. (1975). Influence of induced delayed parturition on foetal survival in pigs. *Theriogenology* **4**, 23–31

NORTH, S.A., HAUSER, E.R. and FIRST, N.L. (1973). Induction of parturition in swine and rabbits with the corticosteroid dexamethasone. *J. Anim. Sci.* **36**, 1170–1174

NOVY, M.J., WALSH, S.W. and KITTINGER, G.W. (1977). Experimental foetal anencephaly in the Rhesus monkey: Effect on gestational length and maternal and foetal plasma steroids. *J. clin. Endocr. Metab.* **45**, 1031–1038

PARVIZI, N., ELSAESSER, F., SMIDT, D. and ELLENDORFF, F. (1976). Plasma luteinizing hormone and progesterone in the adult female pig during the oestrous cycle, late pregnancy and lactation, and after ovariectomy and pentobarbitone treatment. *J. Endocr.* **69**, 193–203

POLGE, C., ROWSON, L.E.A. and CHANG, M.C. (1966). The effect of reducing the number of embryos during early stages of gestation on the maintenance of pregnancy in the pig. *J. Reprod. Fert.* **12**, 395–397

POOL, S.H., COPELAND, D.D. and GODKE, R.A. (1979). Induced farrowing in commercial gilts with three dose levels of cloprostenol (ICI 80996). *J. Anim. Sci.* **49** (Suppl. 1), 327 (Abstract)

PORTER, D.G. (1979). The myometrium and the relaxin enigma. *Anim. Reprod. Sci.* **2**, 77–96

RANDALL, G.C.B. (1972). Observations on parturition in the sow. II. Factors influencing stillbirth and perinatal mortality. *Vet. Rec.* **90**, 183–186

ROBERTSON, H.A. and KING, G.J. (1974). Plasma concentrations of progesterone, oestrone, oestradiol-17β and of oestrone sulphate in the pig at implantation during pregnancy and at parturition. *J. Reprod. Fert.* **40**, 133–141

ROBERTSON, H.A., KING, G.J. and ELLIOT, J.I. (1978). Control of the time of parturition in sows with prostaglandin $F_{2\alpha}$. *Can. J. comp. Med.* **42**, 32–34

ROBINSON, P.M., ROWE, E.J. and WINTOUR, E.M. (1979). The histogenesis of the adrenal cortex in the foetal sheep. *Acta endocr.* **91**, 134–149

ROMBAUTS, P. (1962). Excretion urinaire d'oestrogenes chez la Truie pendant la gestation. *Annls Biol. anim. Biochim. Biophys.* **2**, 151–156

SAMOL, S. (1980). Prostaglandin as a stimulator of parturition in sows for the prevention of the Metritis, Mastitis, Agalactia (MMA) Syndrome. *Medycyna wet.* **36**, 171–173

SHERWOOD, O.D., CHANG, C.C., BEVIER, G.W. and DZIUK, P.J. (1975). Radioimmunoassay of plasma relaxin levels throughout pregnancy and at parturition in the pig. *Endocrinology* **97**, 834–837

SHERWOOD, O.D, NARA, B.S., CRNEKOVIC, V.E. and FIRST, N.L. (1979). Relaxin concentrations in pig plasma following administration of indomethacin and prostaglandin $F_{2\alpha}$ during late pregnancy. *Endocrinology* **104**, 1716–1721

SHERWOOD, O.D., CHANG, C.C., BEVIER, G.W., DIEHL, J.R. and DZIUK, P.J. (1976). Relaxin concentration in pig plasma following the administration of prostaglandin $F_{2\alpha}$ during late pregnancy. *Endocrinology* **98**, 875–879

SILVER, M., BARNES, R.J., COMLINE, R.S., FORDEN, A.L., CLOVER, L. and MITCHELL, M.D. (1979). Prostaglandins in the foetal pig and prepartum endocrine changes in mother and foetus. *Anim. Reprod. Sci.* **2**, 305–322

SOLOFF, M.S. and SWARTZ, T.L. (1974). Characterization of a proposed oxytocin receptor in the uterus of the rat and sow. *J. biol. Chem.* **249**, 1376–1381

SOLOFF, M.S., ALEXANDROVA, M. and FERNSTROM, M.J. (1979). Oxytocin response: triggers for parturition and lactation. *Science* **204**, 1313–1315

SPRECHER, D.J., LEMAN, A.D., DZIUK, P.J., CROPPER, M. and DEDECKER, M. (1974). Causes and control of swine stillbirths. *J. Am. vet.med. Ass.* **165**, 698–701

STEELE, P.A., FLINT, A.P.F. and TURNBULL, A.C. (1976a). Activity of steroid C-17,20 lyase in the ovine placenta: Effect of exposure to foetal glucocorticoid. *J. Endocr.* **69**, 239–246

STEELE, P.A., FLINT, A.P.F. and TURNBULL, A.C. (1976b). Increased utero-ovarian androstenedione production before parturition in sheep. *J. Reprod. Fert.* **46**, 443–445

STRYKER, J. and DZIUK, P.J. (1975). Effects of foetal decapitation on foetal development, parturition and lactation in pigs. *J. Anim. Sci.* **40**, 282–287

TAVERNE, M. (1979). Physiological aspects of parturition in the pig. Doctoral Thesis. Ryksuniversiteit Utrecht, The Netherlands

TAVERNE, M.A.M., VAN DER WEYDEN, G.C., FONTIJNE, P., ELLENDORFF, F., NAAKTGEBOREN C. and SMIDT, D. (1977). Uterine position and presentation of mini-pig foetuses and their order and presentation at birth. *Am. J. vet. Res.* **38**, 1761–1774

TAVERNE, M., WILLEMSE, A.H., SIELEMAN, S.J. AND BEVERS, M. (1979a). Plasma prolactin, progesterone and oestradiol-17β concentrations around parturition in the pig. *Anim. Reprod. Sci.* **1**, 257–263

TAVERNE, M.A.M. NAAKTGEBOREN, C., ELSAESSER, F., FORSLING, M.L., VAN DER WEYDEN, G.C., ELLENDORFF, F. and SMIDT, D. (1979b). Myometrial electrical activity and plasma concentrations of progesterone, estrogens and oxytocin during late pregnancy and parturition in the miniature pig. *Biol. Reprod.* **21**, 1125–1134

TORRES, C.A.A. (1975). Effects of utero-ovarian separation on estrus, pregnancy maintenance and parturition in swine and the effects of blocked pregnenolone synthesis on gestation length in rabbits. PhD Thesis. University of Wisconsin-Madison

VAJE, S., ELSAESSER, F., ELGER, W. AND ELLENDORFF, F. (1980). Induction of parturition in the sow with prostaglandin E (sulprostone). *Dt. tierarztl. Wschr.* **87**, 77–79

VAN LANDEGHEM, A.A.M. and VAN DE WIEL, D.F.M. (1978). Radioimmunoassay for porcine prolactin: plasma levels during lactation, suckling and weaning and after TRH administration. *Acta endocr.* **88**, 653–667

WALKER, N. (1979). The induction of parturition in sows on a twice weekly basis using cloprostenol. *Rec. Agric. Res. (N.Ir.)* **27**, 5–10

WEBEL, S.K., REIMERS, T.J. and DZIUK, P.J. (1975). The lack of relationship between plasma progesterone levels and number of embryos and their survival in the pig. *Biol. Reprod.* **13**, 177–186

WELK, F. and FIRST, N.L. (1979). Effect of oxytocin on the synchrony of parturition induced by $PGF_{2\alpha}$ (ICI 80996) cloprostenol in sows. *J. Anim. Sci.* **49** (Suppl. 1), 347–348 (Abstract)

WETTEMANN, R.P., HALLFORD, D.M., KREIDER, D.L. and TURMAN, E.J. (1977). Influence of prostaglandin $F_{2\alpha}$ on endocrine changes at parturition in gilts. *J. Anim. Sci.* **44**, 106–111

WIERZCHOS, E. and PEJSAK, Z. (1976). Induction of parturition in sows with Prostin $F_{2\alpha}$ in commercial swine farms. *8th Int. Congr. Anim. Reprod. A.I., Krakow* **3**, 418–420

WILLEMSE, A.H., TAVERNE, M.A.M., ROPPE, L.J.J.A. and ADAMS, W.M. (1979). Induction of parturition in the sow with prostaglandin analogue (ICI 80996). *Vet. Quart.* **1**, 145–149

WINTOUR, E.M., BROWN, E.H., DENTON, D.A., HARVEY, K.J., McDOUGALL, J.G. ODDIE, C.J. and WHIPP, G.T. (1975). The ontogeny and regulation of corticosteroid secretion by the ovine foetal adrenal. *Acta endocr.* **79**, 301–316

ZARROW, M.L., NEHER, G.M., SIKES, D., BRENNAN, D.M. and BULLARD, J.F. (1956). Dilation of the uterine cervix of the sow following treatment with relaxin. *Am. J. Obstet. Gynec.* **72**, 260–263

ZEROBIN, K. and SPORRI, H. (1972). Motility of the bovine and porcine uterus and Fallopian tubes. *Adv. vet. Sci. comp. Med.* **16**, 303–354

17

RELAXIN AT PARTURITION IN THE PIG

O.D. SHERWOOD
School of Basic Medical Sciences and Department of Physiology and Biophysics, University of Illinois, USA

Relaxin is a polypeptide hormone which is found in the blood and reproductive organs during pregnancy in many species (Schwabe *et al.*, 1978; Porter, 1979b). The hormone was discovered in 1926 when Hisaw found that injection of serum obtained from pregnant rabbits or pregnant guinea pigs into adult female guinea pigs at oestrus caused a relaxation of the pubic symphysis which was similar to that which occurs during late pregnancy (Hisaw, 1926). Within three years of the discovery of these hormonal effects, Hisaw (1929) reported that the relaxative hormone was also present in the blood of pregnant sows and that sow corpora lutea were a 'very excellent source for the substance'. A crude aqueous extract of the hormone was first obtained from pregnant sow ovaries by Fevold, Hisaw and Meyer (1930), and they named the hormone 'relaxin'. From 1930 until the present, most of our knowledge about the chemistry and physiology of relaxin has been learned from studies which employed porcine relaxin.

This chapter will be largely confined to the physiology of relaxin at parturition in the pig. For more comprehensive descriptions of the chemistry of relaxin and physiological studies of relaxin in species other than the pig, the reader is referred to the following recent reviews: Schwabe *et al.* (1978), Porter (1979b), Steinetz, O'Byrne and Kroc (1980), Niall *et al.* (1982) and Sherwood (1982a,b).

Isolation and chemistry of porcine relaxin

Fevold, Hisaw and Meyer (1930) first showed that an aqueous extract of sow corpora lutea contained relaxin, and they concluded that relaxin might be a peptide-like molecule since it was amphoteric and vulnerable to trypsin digestion. Nearly all subsequent efforts to isolate relaxin employed ovaries obtained from pigs during late pregnancy, since this source of the hormone was known to have a high content of relaxin activity (Hisaw and Zarrow, 1948) and was relatively easy to acquire in large quantities. Progress toward the isolation of porcine relaxin was slow. Well documented progress toward the purification of relaxin was first made in the 1950s and 1960s, and this progress was largely attributable to the availability of new techniques for the isolation and characterization of proteins, as well as new or improved methods for the bioassay of relaxin. The relaxin bioassays most commonly used to monitor relaxin isolation efforts involved either *in vivo* stimulation of interpubic ligament formation in oestrogen-treated mice or guinea pigs, or *in vitro* inhibition of spontaneous contractions of uteri obtained from oestrogen-treated rats (Steinetz, Beach and Kroc, 1969). During the 1960s three laboratories reported procedures whereby they obtained preparations of porcine relaxin which contained high biological activity (Cohen, 1963; Frieden, Stone and Layman, 1960; Griss *et al.*, 1967). Although the homogeneity of these relaxin preparations was not rigorously established and the physicochemical properties of porcine relaxin were not precisely described, it is important to appreciate that these workers correctly indicated that porcine relaxin is a protein with a molecular weight between 4000 and 10 000 (Frieden, Stone and Layman, 1960; Cohen, 1963; Griss *et al.*, 1967), that it has a basic isoelectric point (Cohen, 1963; Griss *et al.*, 1967), and that it contains disulphide bonds which are essential for biological activity (Frieden and Hisaw, 1953; Cohen, 1963).

In 1974 we reported (Sherwood and O'Byrne, 1974) a simple isolation procedure whereby three essentially homogeneous preparations of porcine relaxin, designated CMB, CMa, and CMa′, are obtained in quantities sufficient for detailed chemical and physiological studies (*Figure 17.1*). Physicochemical studies indicated that the three porcine relaxin preparations are nearly identical to one another. They have molecular weights of approximately 6000, nearly identical amino acid compositions, contain no histidine, proline, or tyrosine, and consist of two chains of similar size, designated A and B, which are linked by disulphide bonds (Sherwood and

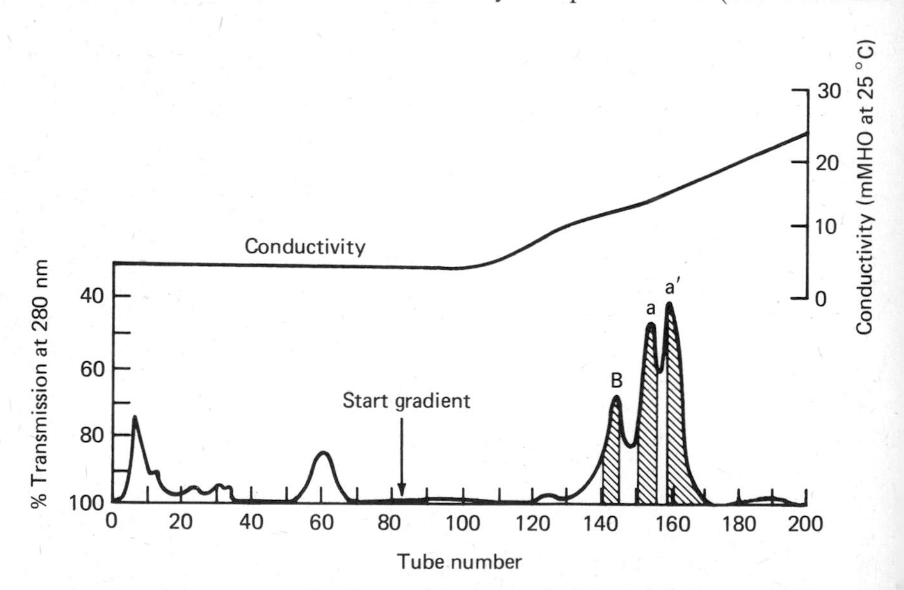

Figure 17.1 The final step in the isolation of porcine relaxin. Partially purified porcine relaxin was adsorbed to a column of carboxymethylcellulose. A linear gradient, which consisted of the column equilibrating buffer plus NaCl, brought about the elution of three contiguous peaks which contained relaxin activity. The contents of the tubes denoted by hatching were pooled to form the three highly purified relaxin preparations CMB, CMa, and CMa′. From Sherwood and O'Byrne (1974)

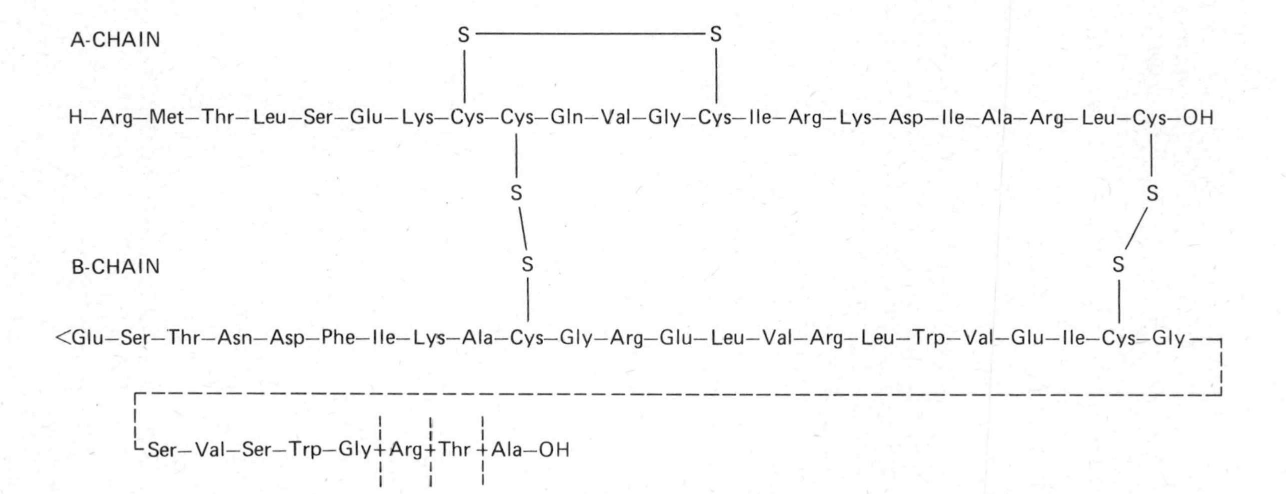

Figure 17.2 The primary structure of porcine relaxin reported by Niall *et al.* (1981). These workers attribute the multiple forms of porcine relaxin to differences among their B chains which range from 28–31 amino acids as shown in the figure above. The N-terminal amino acid in the B chain is pyroglutamic acid. The position of the disulphide linkages are as reported by Schwabe and McDonald (1977). Small differences remain in the amino acid sequences which have been reported for porcine relaxin. Schwabe, McDonald and Steinetz (1976, 1977) have reported that the A chain contains glutamic acid rather than glutamine in position 10, and that the B chain contains only 26 amino acids with C-terminal sequence Ile-Cys-Gly-Val-Trp-Ser

O'Byrne, 1974). We also reported that porcine relaxin preparation CMa has 22 amino acids in the A chain and 28–31 amino acids in the B chain. Two groups (Schwabe, McDonald and Steinetz, 1976, 1977; James *et al.*, 1977; Niall *et al.*, 1981) have determined the amino acid sequences of the A and B chains of porcine relaxin (*Figure 17.2*). Niall and his coworkers recently reported that the multiple forms of porcine relaxin are attributable to slight differences in the lengths of B chains which may result from varying degrees of proteolysis of the C-terminus during the isolation procedure (Walsh and Niall, 1980; Niall *et al.*, 1982).

Relaxin has structural features which are similar to those of insulin, and it has been suggested that relaxin and insulin may have evolved from a common ancestral gene (Schwabe and McDonald, 1977; James *et al.*, 1977). Schwabe and McDonald (1977) showed that the locations of the disulphide bonds within porcine relaxin are homologous to those of insulin (*Figure 17.2*). When the disulphide bonds (half-cysteines) of porcine relaxin and porcine insulin are aligned in register, only five residues in addition to the half-cysteines are identical in the two hormones. However, in many cases the amino acid residues which are in comparable positions in the two hormones have similar structures, i.e. are 'conservative substitutions' (Schwabe and McDonald, 1977; James *et al.*, 1977); and it has been suggested on the basis of studies which employed model building (Bedarkar *et al.*, 1977) and computer graphics (Isaacs *et al.*, 1978) that the aminoacid sequence of porcine relaxin permits the hormone to adopt a three-dimensional conformation similar to that of insulin. However, if relaxin and insulin evolved from a common ancestral gene, the following observations seem to indicate that considerable evolutionary divergence has occurred between the two hormones. There is evidence that the biosynthetic precursors for relaxin are larger than those for insulin. Whereas pre-proinsulin and proinsulin have molecular weights of 12 000 and 9000 daltons respectively, putative relaxin precursors with molecular weights of 10 000, 13 000, 19 000 (Kwok, Chamley and Bryant-Greenwood, 1978) and 42 000 daltons (Frieden and Yeh, 1977), which could be converted into about 6000 dalton relaxin, have been reported. Recently, Gast *et al.* (1980) demonstrated in a cell-free system that porcine luteal mRNA directs the synthesis of a protein which is immunologically related to relaxin. This 'relaxin-containing protein' has an apparent molecular weight of approximately 23 000.

Additionally, studies which employed radioimmunoassays for porcine relaxin (Sherwood, Rosentreter and Birkhimer, 1975a) and porcine insulin (Rawitch, Moore and Frieden, 1980) showed no apparent homology between the antigenic determinants of relaxin and insulin. Finally, there is evidence that relaxin does not bind to insulin receptors since highly purified porcine relaxin failed to compete with radioiodinated porcine insulin for insulin receptors on mononuclear leukocytes (Rawitch, Moore and Frieden, 1980).

Relaxin levels in the corpora lutea throughout pregnancy and at parturition

During the luteal phase of the oestrous cycle, relaxin biological activity within pig ovaries is detectable but extremely low compared with the

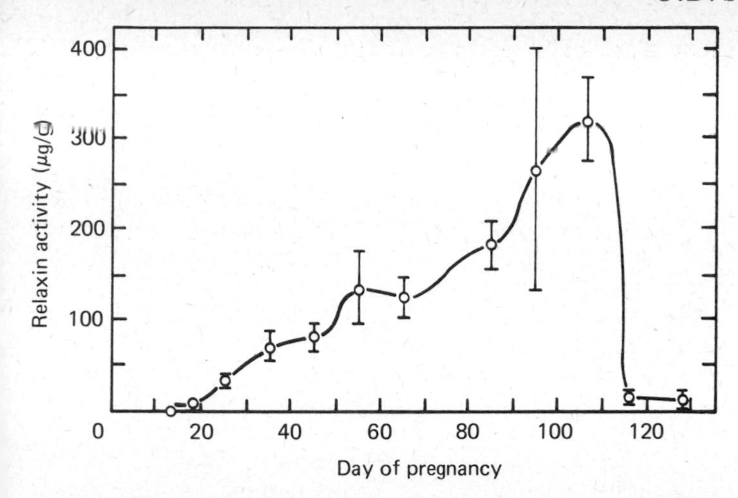

Figure 17.3 Relaxin activity in extracts of porcine corpora lutea obtained throughout pregnancy. Relaxin biological activity was determined by the mouse interpubic ligament bioassay (Steinetz, Beach and Kroc, 1969). The biological activity is expressed in μg of a partially purified NIH porcine relaxin preparation per gram of luteal tissue. From Anderson *et al.* (1973)

relaxin levels present within the ovaries during most of pregnancy (Hisaw and Zarrow, 1948; Anderson *et al.*, 1973; Sherwood and Rutherford, 1981). Hisaw and Zarrow (1948) reported that the amount of relaxin biological activity in pig ovaries was low during early pregnancy and increased approximately 20-fold to maximal concentrations by the time the foetuses attained a length of 5 or 6 inches. More recently, Anderson *et al.* (1973) reported that the levels of relaxin biological activity within corpora lutea rose steadily from day 20 of pregnancy to maximal levels between days 110 to 115 and then declined rapidly within 16 hours of parturition (*Figure 17.3*). Ultrastructural studies of the corpora lutea throughout pregnancy by Belt *et al.* (1971) have also contributed much to an understanding of the synthesis, storage and release of relaxin during pregnancy in the pig. These workers indicated that the most evident change of fine structure of the granulosa lutein cells during pregnancy was in the population of dense membrane-limited granules. These granules, which were not apparent during the oestrous cycle (Cavazos *et al.*, 1969), became a conspicuous constituent of the cytoplasm by day 28, were maximal at days 105 to 110 (*Figure 17.4*), and declined markedly during the 24 hours before parturition (*Figure 17.5*). The appearance, accumulation, and disappearance of these granules during pregnancy closely paralleled the concentrations of relaxin bioactivity in the corpora lutea (*Figure 17.3*) and it was suggested that the cytoplasmic granules are storage sites for relaxin (Belt *et al.*, 1971; Anderson *et al.*, 1973). This view has been strengthened by recent immunocytochemical studies at the ultrastructural level which showed that relaxin immunoactivity is associated with the dense granules which occur in the corpora lutea of pregnant pigs (Corteel, Lemon and Dubois, 1977; Kendall, Plopper and Bryant-Greenwood, 1978).

The regulation of relaxin synthesis and storage in pig corpora lutea is not well understood. Anderson *et al.* (1973) reported that pig corpora lutea

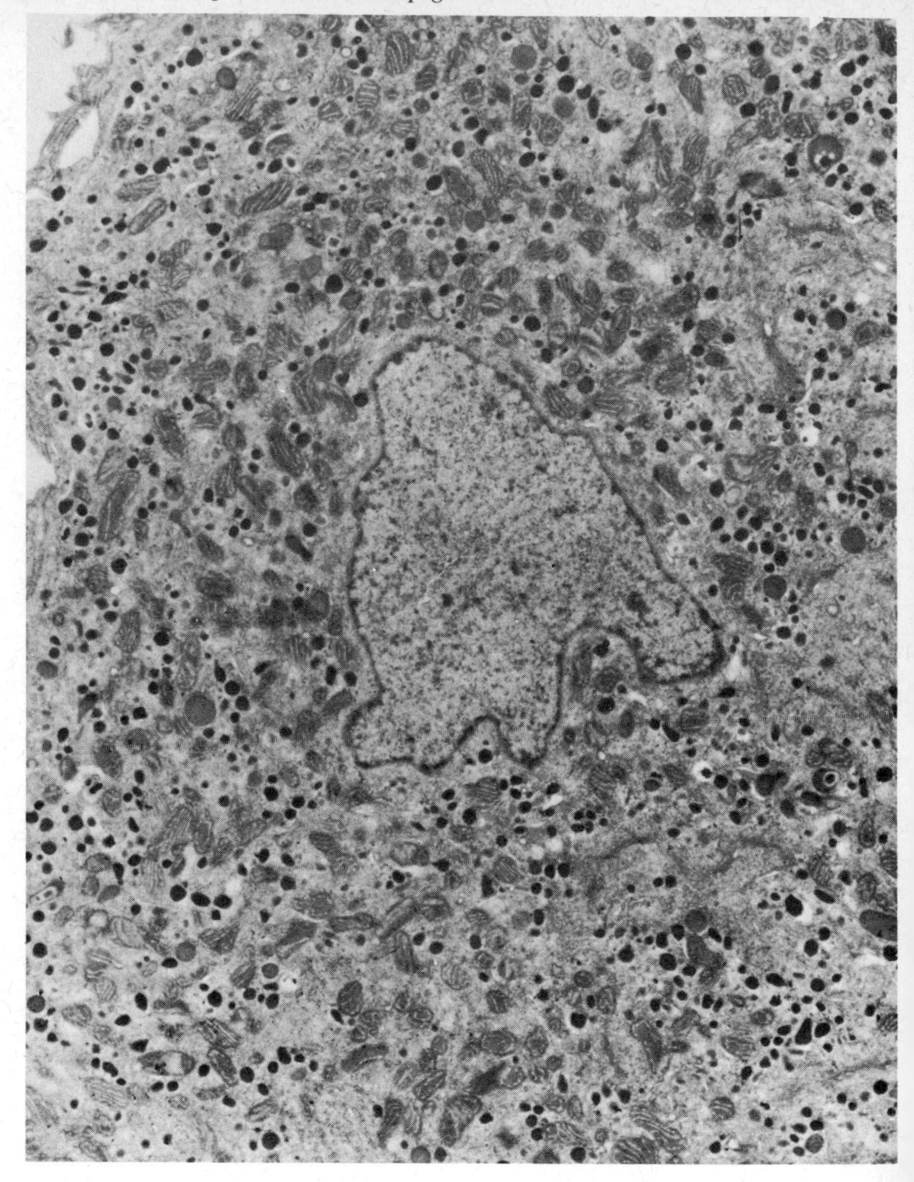

Figure 17.4 A portion of a porcine granulosa lutein cell at day 110. (×6800, reduced to two-thirds in reproduction). The granule content is maximal and the Golgi apparatus is well developed. From Belt *et al.* (1971)

maintained beyond 100 days by hysterectomy or administration of oestrogen during the oestrous cycle contained relaxin bioactivity levels which approached those observed in pregnant pigs. The observation that relaxin levels increased as the age of the corpora lutea increased led these workers to conclude that relaxin levels in pig corpora lutea may be an indication of an ageing process (Anderson *et al.*, 1973).

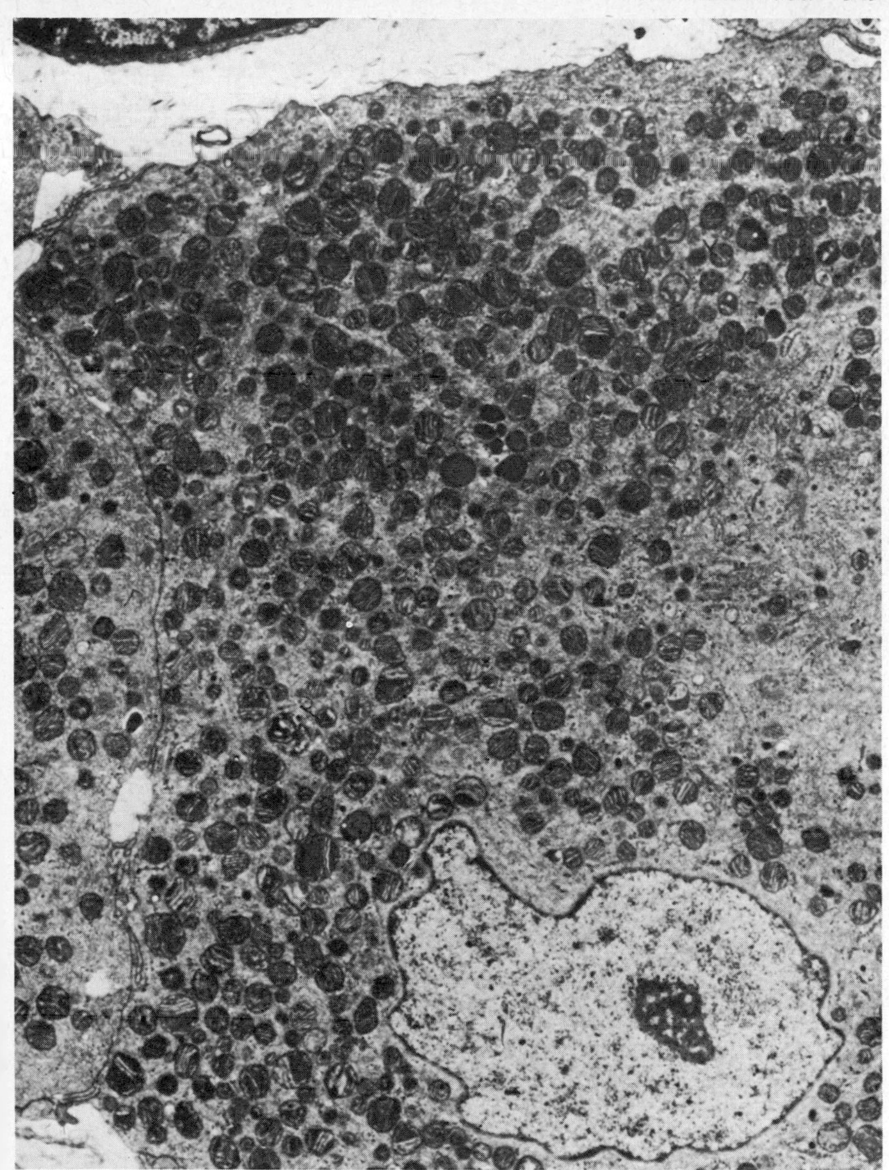

Figure 17.5 A portion of a porcine granulosa lutein cell at day 116, 6 hours prior to parturition (×8500, reduced to two-thirds in reproduction). The granule depletion of the cell is nearly total. From Belt *et al.* (1971)

Relaxin may be produced in ovarian sites other than the corpora lutea in the pig. Anderson *et al.* (1973) reported that extremely low levels of relaxin bioactivity were found in follicular and interstitial tissue of ovaries removed from pregnant pigs and suggested that the relaxin in these ovarian components may result from the transport of the hormone from adjacent corpora lutea. In contrast, an immunofluorescent study of the ovaries of pregnant pigs which employed an antiserum to highly purified porcine relaxin demonstrated that relaxin immunoactivity was confined to the corpora lutea (Larkin, Fields and Oliver, 1977). Recently, small amounts of relaxin immunoactivity were reported to be present in ovarian follicular fluid obtained from non-pregnant and pregnant sows and also in ovaries of immature sows (Bryant-Greenwood *et al.*, 1980; 1981; Matsumoto and Chamley, 1980).

Relaxin levels in the blood throughout pregnancy and at parturition

BIOASSAY

Efforts to measure the levels of relaxin in peripheral blood were long hindered by the insensitivity and imprecision of relaxin bioassays. Nevertheless, by concentrating ovarian venous blood, Belt *et al.* (1971) demonstrated a sharp increase in relaxin bioactivity in pig plasma between 44 and 26 hours before parturition, i.e. concomitant with the rapid degranulation of granulosa lutein cells in the corpora lutea.

RADIOIMMUNOASSAY

The advent of the radioimmunoassay in the 1960s and the isolation of highly purified porcine relaxin in the mid 1970s (Sherwood and O'Byrne, 1974) made possible the development of assays which are specific for relaxin and sufficiently sensitive to measure the levels of relaxin found in the peripheral blood. In 1975 Sherwood, Rosentreter and Birkhimer (1975) reported the development of a homologous radioimmunoassay for porcine relaxin which routinely detects as little as 32 pg of porcine relaxin and is more than 1000 times more sensitive than the commonly employed mouse interpubic ligament bioassay.

The levels of relaxin in the peripheral blood of pigs throughout gestation and at parturition have been determined with this radioimmunoassay. Relaxin immunoactivity concentrations in peripheral plasma remain below 2 ng/ml until about day 100 and then rise gradually to approximately 5 ng/ml on day 110 (Sherwood *et al.*, 1975). From day 110 to day 112, mean relaxin concentrations increase to approximately 15 ng/ml, and in most pigs small surges in relaxin levels occur during these three days (*Figure 17.6*). During the two days which precede parturition, relaxin levels increase markedly and attain maximal concentrations which generally range from 60 to 250 ng/ml (*Figure 17.6*; Sherwood *et al.*, 1975, 1976, 1978, 1979, 1981). In nearly all cases, this pre-partum elevation in relaxin consists of two or three sustained surges which last for 10–20 hours. The peak which immediately precedes parturition has maximal relaxin concentrations and generally occurs from 14–22 hours before delivery (Sherwood *et al.*, 1975, 1976, 1978, 1979, 1981). This maximum is followed by a decline in relaxin concentrations which does not appear to be interrupted during parturition

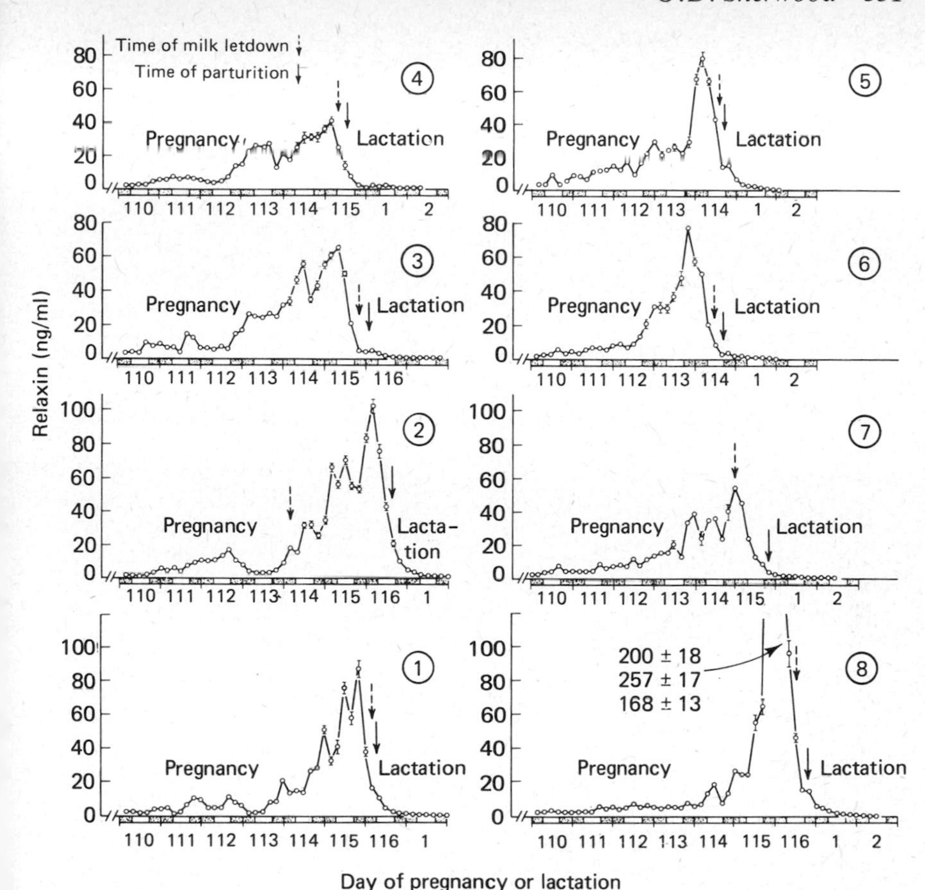

Figure 17.6 Relaxin concentrations (±S.E.M.) in peripheral plasma obtained from eight pigs at 4-hour intervals from 04.00 hours on day 110 until approximately 37 hours following parturition. The alternating dark and light bars along the abscissas indicate the periods of darkness and artificial lighting, respectively. From Sherwood *et al.* (1981)

(*Figure 17.7*) and continues following birth of the piglets (*Figure 17.8*). By 37 hours after parturition, relaxin levels in the plasma are less than 0.5 ng/ml. The range of relaxin concentrations obtained in the plasma during delivery (*Figure 17.7*) is in general agreement with that reported by Afele *et al.* (1979); however, unlike those workers this work (Sherwood *et al.*, 1981) which included blood samplings taken as frequently as 2-minute intervals, failed to show marked fluctuations in relaxin levels during parturition or during suckling. Since the clearance $t_{1/2}$ for relaxin in the pig is approximately 60 minutes (Sherwood, 1982b), it seems likely that marked fluctuations in serum relaxin levels during parturition or suckling would have been detected in the study of Sherwood *et al.* (1982) had they occurred.

One must be careful in the interpretation of polypeptide hormone levels obtained by radioimmunoassay since the radioimmunoassay does not necessarily measure biologically active molecules. It seems likely, however, that the immunoactive 'substance' measured with our porcine relaxin

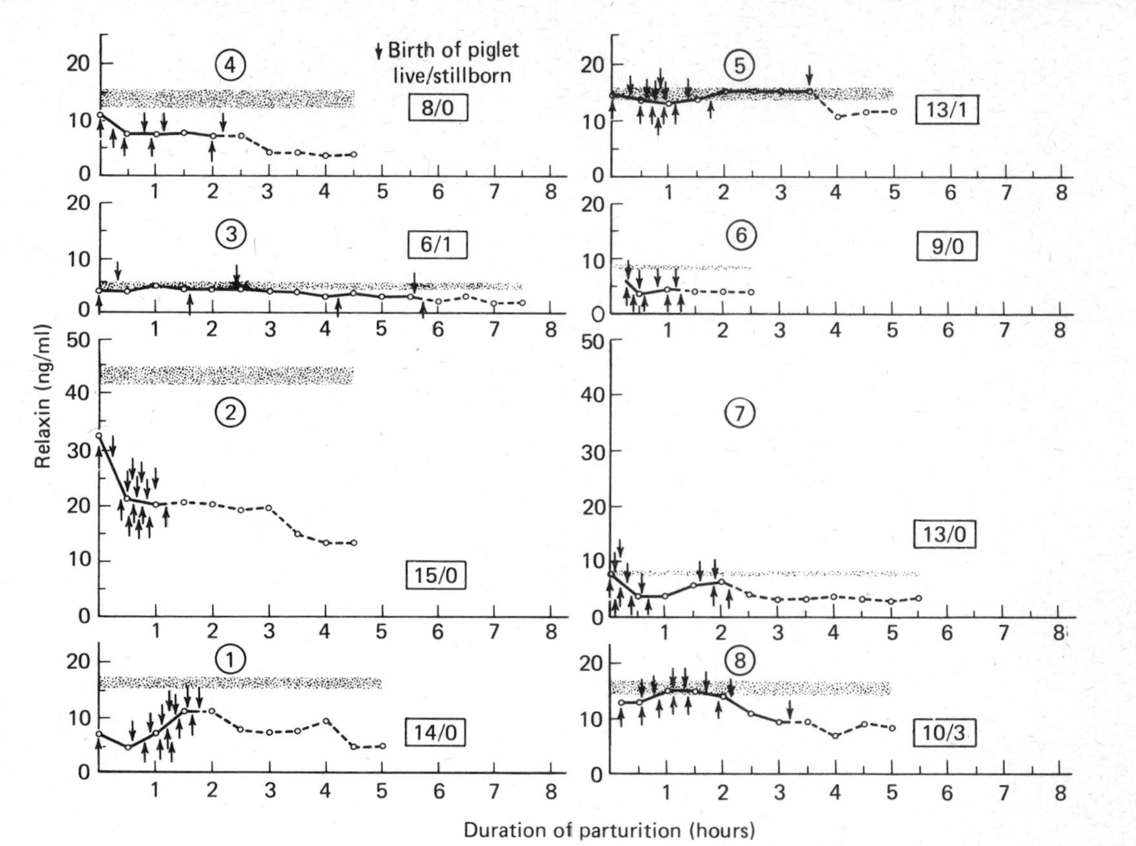

Figure 17.7 Relaxin concentrations in peripheral plasma obtained from eight pigs at 30 minute intervals throughout parturition. The horizontal shaded bars indicate the relaxin level (±S.E.M.) in the last plasma sample obtained before parturition. These bars have been extended to facilitate comparison with relaxin levels obtained during parturition. The solid lines indicate the duration of delivery of the piglets and the intermittent extension of

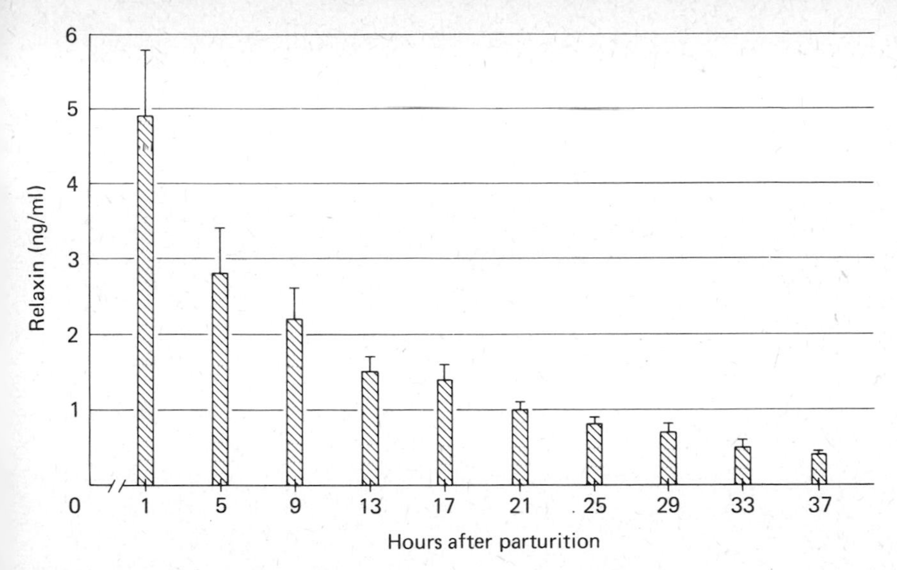

Figure 17.8 Relaxin concentrations in peripheral plasma obtained at 4 hour intervals from 1 hour to 37 hours following parturition. This figure contains the means (±S.E.M.) of the eight animals shown in *Figures 17.6* and *17.7.* From Sherwood *et al.* (1981)

radioimmunoassay is, at least in part, biologically active relaxin since the pre-partum occurrence of high blood levels of relaxin immunoactivity (*Figure 17.6*) and relaxin bioactivity (Belt *et al.*, 1971) coincide.

Association of the pre-partum release of relaxin from the corpora lutea with luteal regression

In the pregnant pig the corpora lutea are the primary source of both progesterone and relaxin (Belt *et al.*, 1971; Anderson *et al.*, 1973; Corteel, Lemon and Dubois, 1977; Larkin, Fields and Oliver, 1977; Sherwood *et al.*, 1977a; Dubois and Dacheux, 1978; Kendall, Plopper and Bryant-Greenwood, 1978; Arakaki, Kleinfeld and Bryant-Greenwood, 1980). Since the corpora lutea are undergoing functional regression, as judged by a rapid fall in progesterone levels during the two days which precede parturition (Molokwu and Wagner, 1973; Killian, Garverick and Day, 1973; Baldwin and Stabenfeldt, 1975; Ash and Heap, 1975), it seemed evident that the simultaneous pre-partum elevation in peripheral plasma relaxin levels might also be associated with luteal regression. This possibility has been explored. Pregnant pigs were given surgical or pharmacological treatments which might influence luteal function during late pregnancy in order to determine whether these treatments also influenced pre-partum levels of relaxin in the peripheral blood.

INFLUENCE OF ALTERED UTERO-OVARIAN RELATIONSHIP

There is evidence that the factor(s) which bring about luteolysis during late pregnancy can be carried in the systemic circulation. Martin, BeVier and Dziuk (1978) demonstrated that luteolysis and parturition occurred at the

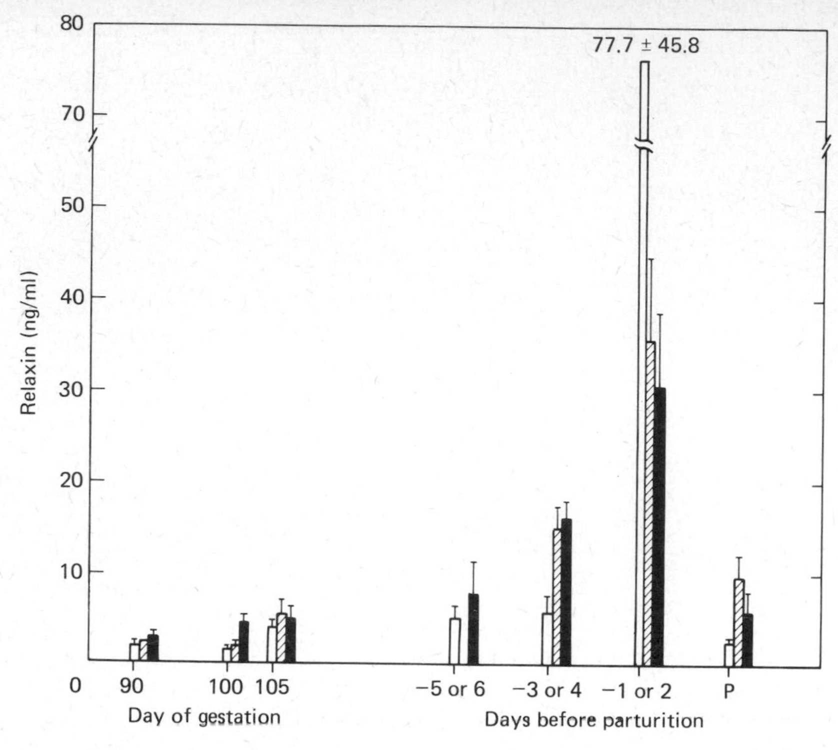

Figure 17.9 Mean relaxin concentrations (±S.E.M.) in peripheral plasma obtained during late pregnancy in pigs with altered utero-ovarian connections. □—Group I comprising seven gilts in which the ovaries were transplanted to the adjacent uterine wall. Blood was collected at two day intervals. ▨—Group II comprising three gilts in which the ovaries were transplanted to the exterior abdominal wall. ■—Group III comprising four gilts in which one uterine horn and its contralateral ovary were removed. P = day of parturition. From Sherwood *et al.* (1977b)

expected time in pigs whose normal utero-ovarian relationship was altered so that there were no direct utero-ovarian connections. Likewise, the pre-partum elevation of relaxin levels in the blood of these pigs occurred at the expected time and concomitant with luteolysis (*Figure 17.9*). Therefore, it appears that the factor(s) which brings about the pre-partum elevation of relaxin levels, as well as that which brings about luteolysis, can be carried in the systemic circulation.

INFLUENCE OF PROSTAGLANDINS

The nature of the factors which bring about luteolysis and the pre-partum elevation in relaxin levels are not known with certainty. However, there is evidence that prostaglandins may be involved with both phenomena (Diehl *et al.*, 1974; Sherwood *et al.*, 1976; 1979; Nara and First, 1981). *Figure 17.10* shows that the infusion of sufficient $PGF_{2\alpha}$ on day 110 to induce parturition on day 111 brought about a drop in progesterone concentrations and a surge in relaxin levels in peripheral plasma. Additionally, daily

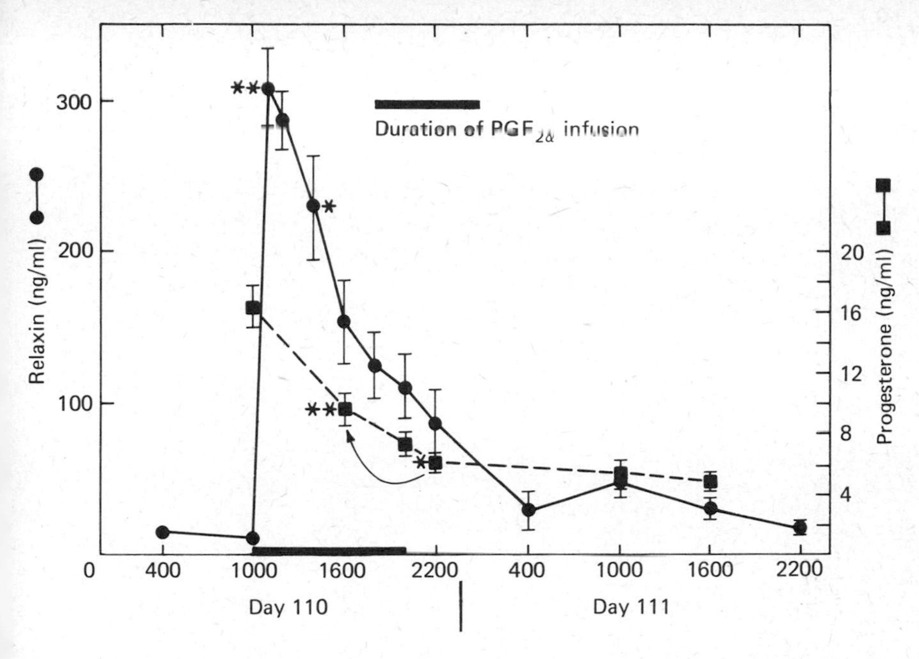

Figure 17.10 Mean relaxin and progesterone levels (± S.E.M.) in peripheral plasma obtained from five pigs which were infused with prostaglandin $F_{2\alpha}$ at a rate of 0.5 mg/hr over a 10 hour period on day 110. Asterisks denote those mean hormone concentrations which differ significantly (* $P<0.05$; ** $P<0.01$) from those which immediately precede them or those indicated with an arrow. From Sherwood *et al.* (1979)

injection of the prostaglandin synthesis inhibitor indomethacin from day 109 to day 116 delayed both the elevation of relaxin levels and parturition, which normally occur between day 112 and 116, until 2–4 days after the termination of indomethacin administration (*Figure 17.11*). Likewise, the decline in progesterone concentrations in indomethacin-treated pigs was delayed and was concurrent with the surge in relaxin levels which occurred on approximately day 119 (*Figure 17.12*). It is not known whether luteolysis and the pre-partum release of relaxin into the blood are initiated by common or separate prostaglandin-mediated mechanisms.

INFLUENCE OF THE FOETUS

There is evidence that the foetal pituitary–adrenal–placental system controls the initiation of parturition in the pig (First and Bosc, 1979). Destruction of the foetal pituitary (Bosc, du Mesnil du Buisson and Locatelli, 1974) or foetal decapitation *in utero* (Stryker and Dziuk, 1975; Coggins and First, 1977) prolongs pregnancy. The ablation of the foetal pituitary apparently prolongs pregnancy by preventing the rapid and sustained luteolysis which normally occurs on approximately days 113–115 (Fevre, Terqui and Bosc, 1975; Coggins and First, 1977). Ablation of the foetal pituitary by foetal decapitation (Sherwood, Hagen, Dial and Dziuk,

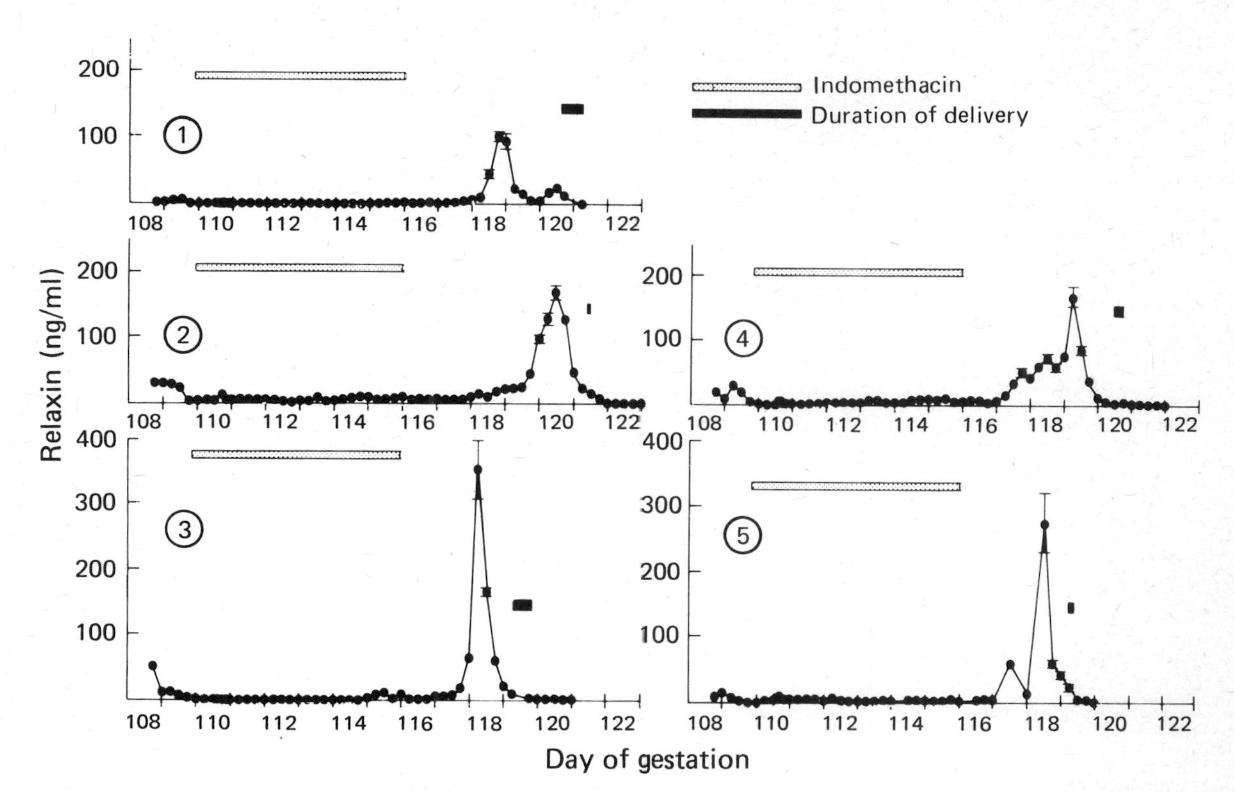

Figure 17.11 Relaxin concentrations (± S.E.M.) in peripheral plasma obtained at 6 hour intervals from 16.00 hours on day 108 until 24 hours after parturition in five pigs injected i.m. twice each day with indomethacin at a dose of 4 mg/kg from day 109 to day 116. From Sherwood *et al.* (1979)

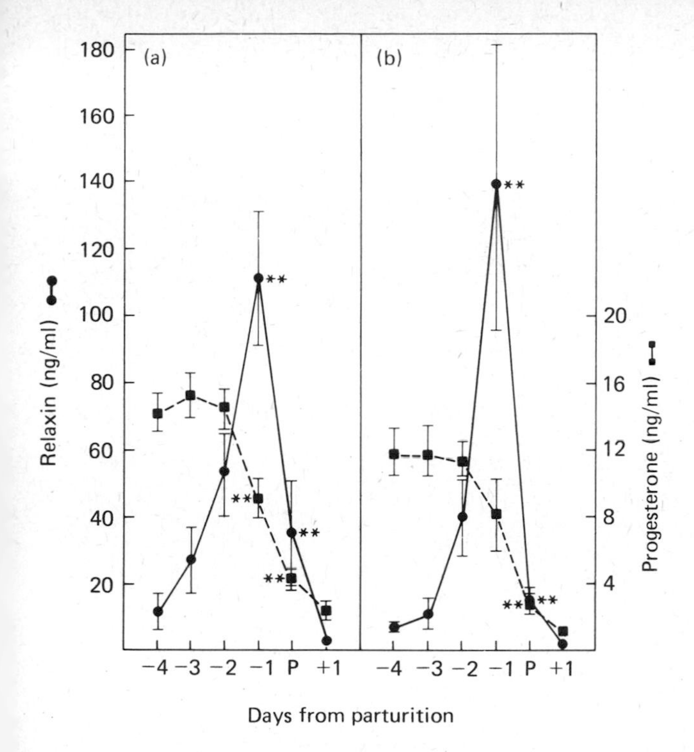

Figure 17.12 Mean relaxin and progesterone concentrations (±S.E.M.) in peripheral plasma obtained from control (a) and indomethacin-treated (b) pregnant pigs. The times of the initiation of delivery for control and the indomethacin groups were 114.9±0.5 (S.E.M.) and 120.1±0.4 (S.E.M.) days, respectively. Asterisks denote those mean hormone concentrations which differ significantly (**$P<0.01$) from those which immediately precede them. P = day of parturition. From Sherwood *et al.* (1979)

unpublished) or foetal hypophysectomy (Kendall *et al.*, 1980) also influences the pre-partum release of relaxin. *Figure 17.13(b)* shows that gilts which contained decapitated foetuses and failed to deliver by day 117 had elevated levels of relaxin in the peripheral plasma on one or more days near the time of normal parturition. However, unlike the gilts which contained intact foetuses (*Figure 17.13(a)*), the peaks of relaxin occurred randomly and were not related to any particular day of gestation or interval before parturition. Therefore, it appears that the foetus may play a role in the control of both luteolysis and the release of relaxin which normally occur during the two days which precede parturition.

The mechanisms which bring about luteolysis in the pregnant pig are not well understood. The apparent association of the pre-partum elevation in relaxin levels with luteolysis may provide a specific indicator for the study of luteal regression. It may be that the multiple and generally progressively greater surges in relaxin levels detected during the last 2–4 days of pregnancy (*Figure 17.6*) are a manifestation of an underlying intermittent and increasingly effective luteolytic mechanism which ultimately brings about the rapid decline in plasma progesterone levels which occurs during

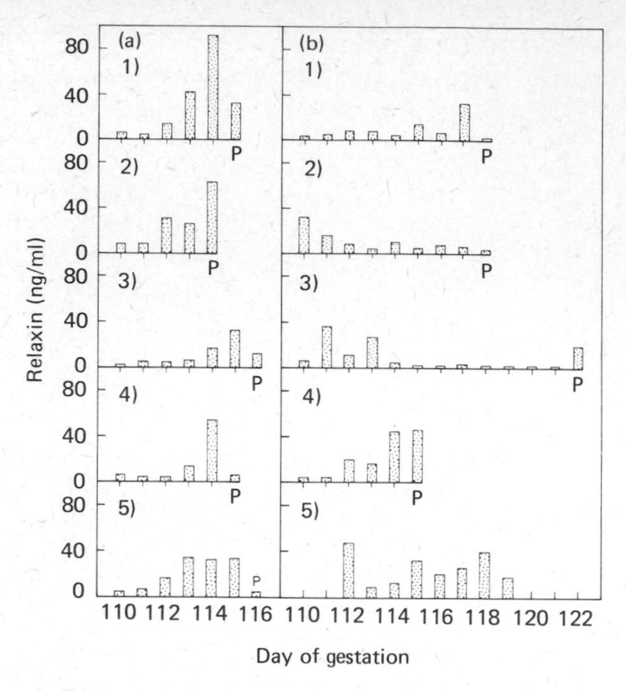

Figure 17.13 Relaxin concentrations in peripheral plasma obtained at daily intervals during late pregnancy from (a) control pigs and (b) pigs containing decapitated foetuses. P = day of parturition. With animal number 5 in treatment group (b) the foetuses were removed surgically by hysterotomy on day 119. From Sherwood, Hagen, Dial and Dziuk (unpublished)

this time. The author is not aware that an intermittent luteolytic mechanism has been previously identified or suggested for the pig during late pregnancy.

Physiological effects of relaxin

The observations that relaxin levels increase in the corpora lutea during pregnancy (Hisaw and Zarrow, 1948; Belt *et al.*, 1971; Anderson *et al.*, 1973) and that relaxin levels increase in the blood during the last 10–14 days of pregnancy (Sherwood, Rosentreter and Birkhimer, 1975) indicate that relaxin may have important physiological functions during pregnancy and parturition in the pig. However, the physiological effects of relaxin in the pregnant pig have not been extensively studied and are not well understood. Accordingly, to a large degree, this section contains conclusions and hypotheses which are based on limited studies, which in most cases employed impure porcine relaxin preparations and species other than the pig, or inferences drawn from indirect observations. Nevertheless, consideration of those studies which have been done seems important since additional research concerning the physiological effects of relaxin in the pig are needed, and the insights and inferences drawn from past studies may contribute toward well-designed and productive future experiments.

EFFECT OF RELAXIN ON PARTURITION

There is evidence that relaxin promotes a high rate of livebirths and may do so by contributing to a short duration of delivery of the piglets. Kertiles and Anderson (1979) reported that daily intramuscular injections of partially purified porcine relaxin beginning on days 105 or 107 and before surgical enucleation of corpora lutea (lutectomy) on day 110 significantly reduced the duration of delivery of all neonates in the litter compared with lutectomized controls. In a recent experiment we abolished the elevated levels of relaxin normally experienced during late pregnancy by bilaterally ovariectomizing pigs on day 105 and then maintained pregnancy by progesterone administration (Nara *et al.*, submitted for publication). When

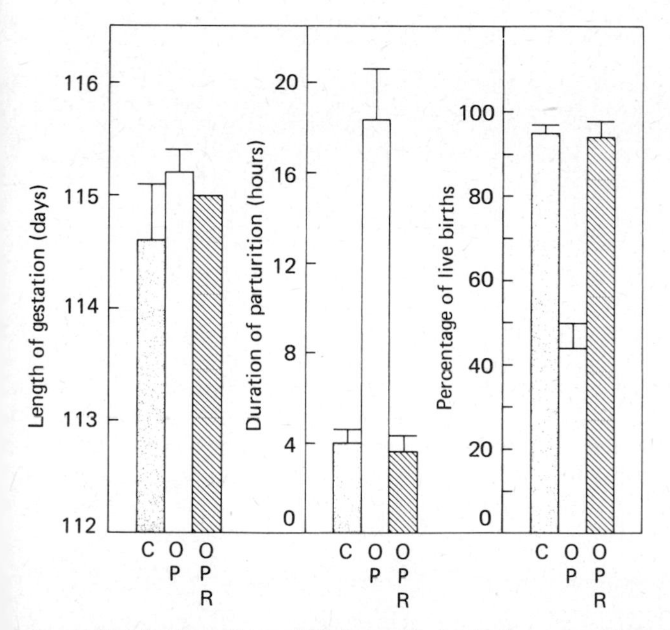

Figure 17.14 The effects of highly purified relaxin on parturition. The five gilts in control group C were sham ovariectomized on day 105 and given intramuscular injections of (1) 4 ml of corn oil at 12 hour intervals from 18.00 hours on day 105 to 18.00 hours on day 112, and (2) 1 ml of 0.9% saline at 6 hour intervals from 12.00 hours on day 105 until parturition was completed. The five gilts in treatment group OP were bilaterally ovariectomized and were injected with (1) 100 mg progesterone in 4 ml of corn oil and with (2) 1 ml of 0.9% saline according to the injection regimen used with treatment C for corn oil and saline alone. The five gilts in treatment group OPR were ovariectomized and treated with progesterone as those in treatment group OP, but were given intramuscular injections of 1 mg of highly purified porcine relaxin in 1 ml of 0.9% saline instead of the injections of saline alone. From Nara *et al.* (submitted for publication)

parturition was induced by progesterone withdrawal on day 112, the duration of parturition was prolonged, and the incidence of piglets born alive was much lower than controls (*Figure 17.14*). Replacement therapy with progesterone plus physiological levels of highly purified porcine relaxin restored the duration of parturition and the incidence of livebirths to values similar to those of controls. It seems likely that the capacity of relaxin to promote livebirths is associated with its effect on the duration of parturition, since a high incidence of stillbirths has been reported to be associated with a prolonged duration of farrowing (Friend, Cunningham and Nicholson, 1962; Randall, 1972).

EFFECT OF RELAXIN ON THE UTERINE CERVIX

The mechanism(s) whereby relaxin promotes a short duration of delivery and high incidence of livebirths in the pig is not well understood. There is evidence that relaxin may do so, at least in part, through direct effects on the uterine cervix which will hereafter be referred to as the cervix. McMurtry, Kwok and Bryant-Greenwood (1978) reported that ^{125}I-labelled porcine relaxin bound with the characteristics of a hormone–receptor interaction to a homogenate of guinea pig cervix. Additionally, it has recently been demonstrated that the levels of cAMP increased in rat (Cheah and Sherwood, 1980; Sanborn *et al.*, 1980) and pig (Judson, Pay and Bhoola, 1980) cervical tissue following *in vitro* incubation with porcine relaxin.

Several lines of evidence indicate that relaxin may be associated with changes in the physical properties of the cervix which occur during pregnancy. The literature describes these cervical changes in a variety of ways including 'ripening', 'softening', 'increased compliance', 'increased dilatation', 'increased extensibility', and 'increased distensibility'. In the interest of brevity, the cervical changes characteristic of pregnancy will generally be referred to as increased distensibility. Several ultrastructural and biochemical changes have been reported which may influence the changes in the physical properties of the cervix which occur during pregnancy. These include increased cervical water content, increased secretion of proteolytic enzymes by cervical fibroblasts, loosening of the collagenous fibre network, and alterations in the extracellular glycosaminoglycans composition (Steinetz, O'Byrne and Kroc, 1980; Golichowski, 1980; Veis, 1980). The biochemical mechanisms associated with the cervical changes which occur during pregnancy and the influence of hormones on these mechanisms remain poorly understood and will not be further discussed in this chapter.

Relaxin appears to be associated with the increased distensibility of the cervix which occurs during pregnancy in rodents. In the rat increased cervical distensibility occurs after day 11 or 12 (Uyldert and DeVaal, 1947; Harkness and Harkness, 1959; Kroc, Steinetz and Beach, 1959; Zarrow and Yochim, 1961; Hollingsworth, Isherwood and Foster, 1979) and coincides with the occurrence of elevated levels of relaxin in the peripheral blood (O'Byrne and Steinetz, 1976; Sherwood *et al.*, 1980). Similar close correlations between increased cervical distensibility and serum levels of relaxin immunoactivity have been described for the pregnant mouse (Steinetz, O'Byrne and Kroc, 1980) and hamster (O'Byrne *et al.*, 1976). When rats were ovariectomized on days 12, 15 or 16 and pregnancy was maintained by injections of progesterone and oestradiol, the marked increase in cervical distensibility normally observed during late pregnancy did not occur (Kroc, Steinetz and Beach, 1959; Steinetz, Beach and Kroc,

1959; Zarrow and Yochim, 1961; Hollingsworth, Isherwood and Foster, 1979). However, when porcine relaxin was given to similarly treated rats, there was a marked increase in cervical distensibility (Kroc, Steinetz and Beach, 1959; Steinetz, Beach and Kroc, 1959; Zarrow and Yochim, 1961; Hollingsworth, Isherwood and Foster, 1979).

Relaxin may also influence cervical changes in the pregnant human being. MacLennan *et al.* (1980) recently reported that the placement of a viscous gel containing 2 mg of highly purified porcine relaxin into the posterior vaginal fornix of women the evening before surgical induction of labour resulted in improved cervical scores. Similar cervical changes have also been observed in non-pregnant ovariectomized rats, mice, monkeys and heifers following treatment with relaxin and oestradiol or relaxin,

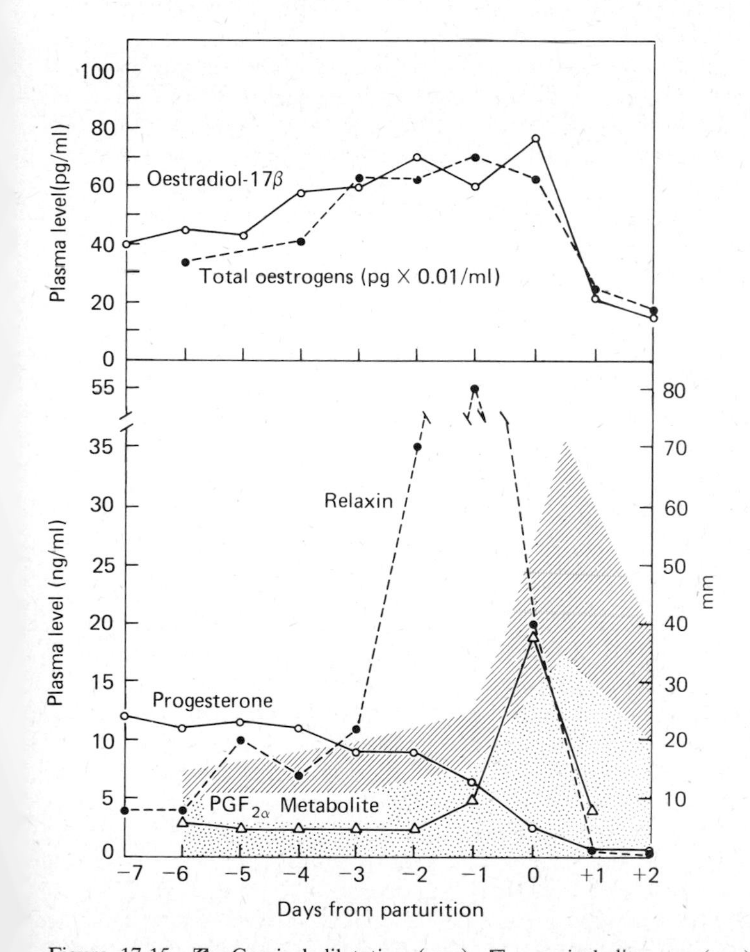

Figure 17.15 ▨—Cervical dilatation (mm), ▩—cervical diameter (mm), and peripheral plasma levels of relaxin, progesterone, $PGF_{2\alpha}$ metabolite, oestradiol-17β, and total oestrogens. Dilatation of the cervix of four gilts was determined by inserting aluminum rods of different diameters into the cervix (from Kertiles and Anderson, 1979). Data from the following reports were used to prepare this figure: Oestradiol-17β (Molokwu and Wagner, 1973), total oestrogens (Baldwin and Stabenfeldt, 1975), progesterone (Molokwu and Wagner, 1973), $PGF_{2\alpha}$ metabolite (Nara and First, 1981), relaxin (Sherwood *et al.*, 1975b)

oestradiol and progesterone, but not following treatment with these steroids alone (Graham and Dracy, 1953; Zarrow, Sikes and Neher, 1954; Kroc, Steinetz and Beach, 1959; Steinetz, Beach and Kroc, 1959; Cullen and Harkness, 1960; Hall, 1960; Zarrow and Yochim, 1961; Leppi, 1964; Hisaw and Hisaw, 1964; Kennedy, 1974; Fields and Larkin, 1980).

There have been few studies concerning the effects of relaxin on the cervix in the pig. Zarrow *et al.* (1956) reported that three daily intramuscular injections of partially purified porcine relaxin (approximately 5 mg relaxin/day) for four days to non-pregnant ovariectomized sows which had been pretreated for seven days with 5 mg of diethylstilboestrol caused a dilatation of the cervix which was accompanied by an increase in cervical water content and depolymerization of cervical glycoproteins. A recent study supports the possibility that relaxin may promote increased distensibility of the cervix during late pregnancy in the pig. Kertiles and Anderson (1979) demonstrated that the cervical diameter in pregnant pigs did not increase until late pregnancy when relaxin levels in the blood are elevated (*Figure 17.15*). Additionally, these workers reported that daily intramuscular injections of a small amount of partially purified porcine relaxin (approximately 250 μg relaxin) beginning on days 105 or 107 induced premature cervical dilatation in pigs which were lutectomized on day 110.

EFFECT OF OTHER HORMONES ON THE UTERINE CERVIX

Studies which have largely been conducted with species other than the pig indicate that the cervical modifications which occur near parturition may be influenced not only by relaxin but also directly or indirectly by other hormones. *Figure 17.15* shows pre-partum blood levels of relaxin, prostaglandin $F_{2\alpha}$ metabolite, progesterone, and oestrogen—three hormones and the metabolite of a hormone which may influence cervical function at parturition in the pig.

Prostaglandins

Evidence acquired from studies with several species indicates that the prostaglandins and perhaps prostacyclin may influence cervical distensibility during late pregnancy and at parturition. (For reviews see McInnes, Naftolin and van der Rest, 1980; Fitzpatrick and Liggins, 1980.) Hollingsworth, Gallimore and Isherwood (1980) reported that $PGF_{2\alpha}$ and PGE_2 caused increased cervical distensibility in pregnant rats. These workers suggested that $PGF_{2\alpha}$ may act on the ovaries to decrease progesterone secretion and perhaps release relaxin, whereas PGE_2 may act directly on cervical tissue. It has also been suggexted that prostaglandins may mediate the effect of relaxin on the rat cervix. Kennedy (1976) administered oestrogen in order to constrict the cervix of ovariectomized non-pregnant rats and then found that the effectiveness with which relaxin released accumulated intraluminal secretions was blocked by the administration of indomethacin. There is evidence that prostaglandins may have direct effects on the cervix in human beings. Intravaginal administration of PGE_2

or $PGF_{2\alpha}$ (Dingfelder *et al.*, 1975; Mackenzie and Embrey, 1977) or *in vitro* administration of PGE_2 (Najak, Hillier and Karim, 1970; Conrad and Ueland, 1976) enhanced ripening or elasticity of human cervical tissue. Likewise, continuous infusion or gel suspension of PGE_2 into the extra-amniotic space has been used for the induction of cervical ripening in women (Calder, 1980). Prostaglandins and/or prostacyclin may promote cervical distensibility in sheep. Infusion of PGE_2 or $PGF_{2\alpha}$ directly into the cervical lumen of pregnant sheep during late pregnancy was reported to produce cervical softening (Fitzpatrick, 1977). More recently, however, Fitzpatrick and Liggins (1980) reported that it may be prostacyclin and not prostaglandins which normally promotes cervical distensibility in pregnant sheep; when parturition was induced by means of a continuous five-day infusion of ACTH into the foetus, the levels of the stable prostacyclin breakdown product 6-keto-$PGF_{1\alpha}$ increased in uterine veins approximately three days before elevated levels of PGF were detected. There are also data indicating that $PGF_{2\alpha}$ may not be associated with cervical distensibility which occurs during late pregnancy in goats. Increased compliance of the cervix occurred several hours before increased levels of PGF were detectable in the blood of goats in which parturition was induced with a synthetic prostaglandin (Fitzpatrick and Liggins, 1980). Neither the effects of prostaglandins nor prostacyclin on the cervix of the pregnant pig have been reported.

Progesterone

The effects of progesterone on the cervix are not clearly understood. Studies with ovariectomized rats and mice indicate that the administration of progesterone alone has little effect on the distensibility of the cervix (Kroc, Steinetz and Beach, 1959; Cullen and Harkness, 1960; Zarrow and Yochim, 1964; Leppi, 1964). However, progesterone has been reported to bring about some augmentation of the stimulating effects of relaxin on cervical distensibility in ovariectomized rats and mice which were primed with oestrogen and treated with relaxin and progesterone (Kroc, Steinetz and Beach, 1959; Cullen and Harkness, 1960; Leppi, 1964). Progesterone does not appear to inhibit the cervical distensibility which occurs at parturition in sheep. Stys *et al.* (1980) demonstrated that the daily administration of 200 mg of progesterone to ewes in which parturition was induced by the infusion of dexamethasone into the foetus failed to inhibit an increase in cervical compliance. The effects of progesterone on the cervix during late pregnancy in the pig have not been extensively studied. The limited data available indicate that progesterone does not promote cervical distensibility. The marked increase in cervical distensibility which occurs during late pregnancy coincides with a decline in progesterone levels in the peripheral plasma (*Figure 17.15*). In pregnant pigs in which parturition was delayed for 3–7 days by the administration of exogenous progestin, there was a marked decline in livebirths (Nellor *et al.*, 1975; Sherwood *et al.*, 1978) which may have been, at least in part, attributable to incomplete cervical dilatation (Nellor *et al.*, 1975). The apparent failure of normal cervical dilatation at parturition in progestin-treated pigs (Nellor *et al.*, 1975) may not be attributable to a relaxin deficiency, since elevated levels of relaxin occurred at the expected time (approximately day 113) in similarly treated animals (Sherwood *et al.*, 1978). Results obtained with non-pregnant pigs are consistent with the view that progesterone does not increase the distensibility of the cervix. The daily injection of progesterone did not promote an increase in the distensibility of the cervix in sows in oestrus (Smith and Nalbandov, 1958) or in ovariectomized sows regardless of whether the animals were treated with progesterone alone (Zarrow *et al.*, 1956) or with progesterone and oestrogen (Smith and Nalbandov, 1958; Zarrow *et al.*, 1956).

Oestrogens

Oestrogens also influence cervical distensibility. In rodents, oestrogens stimulate an increase in cervical weight (Kroc, Steinetz and Beach, 1959; Cullen and Harkness, 1960; Zarrow and Yochim, 1961; Leppi, 1964) and apparently bring about changes which enable relaxin to increase cervical distensibility. When non-pregnant ovariectomized mice (Leppi, 1964) or rats (Kroc, Steinetz and Beach, 1959; Cullen and Harkness, 1960) were injected with either oestrogen or porcine relaxin alone, there was little effect on cervical distensibility, whereas when relaxin was administered to similar animals which had been pretreated with oestrogen, there was a marked increase in cervical distensibility (Leppi, 1964; Kroc, Steinetz and Beach, 1959; Cullen and Harkness, 1960; Kennedy, 1974). In sheep the subcutaneous injection of 20 mg of diethylstilboestrol on approximately day 130 of pregnancy promoted cervical compliance (Stys *et al.*, 1980), and a single subcutaneous injection of 5 mg of oestradiol benzoate between days 127 and 128 of pregnancy increased cervical weight and dilatation (Fitzpatrick and Liggins, 1980). Oestrogens also promote cervical ripening in pregnant women. Systemic administration of oestradiol-17β (Pinto *et al.*, 1964) or dehydroepiandrosterone sulphate (Mochizuki and Tojo, 1980) during late pregnancy increased ripening of the cervix. Oestrogen may act directly on the cervix, since the extra-amniotic placement of oestradiol in a viscous gel promoted increased cervical ripening (Gordon and Calder, 1977).

The above observations raise fundamental and important questions concerning the role of relaxin on the pig cervix. Does relaxin play an important role in promoting the increased distensibility of the cervix which occurs during late pregnancy? If so, what are the effects of relaxin on the morphology and biochemistry of the pig cervix? Do these effects involve interactions with other hormones such as prostaglandins, progesterone and oestrogen and, if so, what is the nature of these interactions? Intensified efforts to answer these questions seem worthwhile.

EFFECT OF RELAXIN ON UTERINE CONTRACTILE ACTIVITY

It seems likely that the uterine horns are also a target tissue for relaxin. Intact uterine horns (Cheah and Sherwood, 1980) or slices of uterine horns

(Mercado-Simmen, Bryant-Greenwood and Greenwood, 1980) obtained from rats bound ^{125}I-labelled porcine relaxin with characteristics of a hormone–receptor interaction. Relaxin may exert its effects on the uterus through cAMP, since levels of cAMP increased in pieces of rat (Cheah and Sherwood, 1980; Sanborn *et al.*, 1980; Judson, Pay and Bhoola, 1980) and pig (Judson, Pay and Bhoola, 1980) uterine horn following incubation with porcine relaxin.

The effects of relaxin on the uterine horns are not well understood. Relaxin may influence uterine contractile activity during pregnancy. Both *in vivo* and *in vitro* studies have demonstrated that partially purified preparations of porcine relaxin reduce the frequency of uterine contractile activity in the guinea pig (Krantz, Bryant and Carr, 1950; Porter, 1971; 1972), rat (Sawyer, Frieden and Martin, 1953; Porter, Downing and

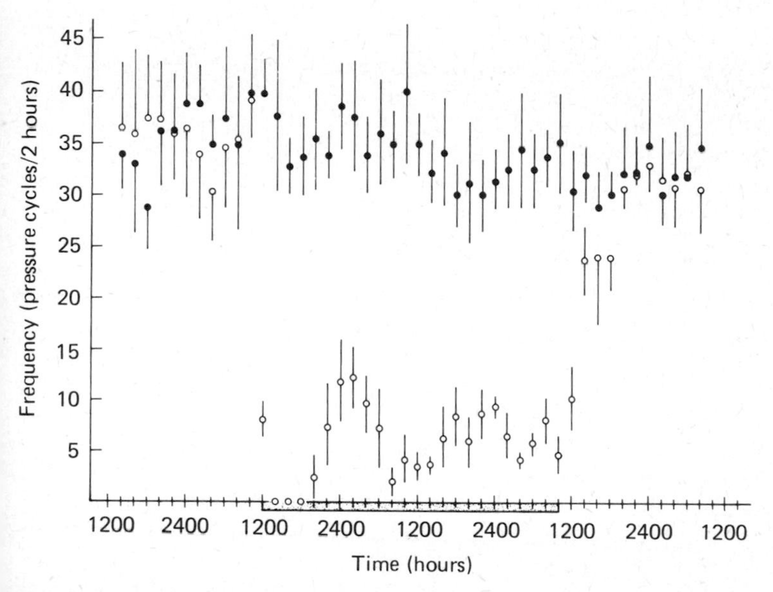

Figure 17.16 The effect of infusion of highly purified porcine relaxin (20 μg/hour) on the frequency of contractions of the uterus in conscious and unrestrained oestrogen-treated ovariectomized rats. The frequency of contractions (±S.E.M.) in four relaxin-treated rats (○) and three control rats (●) are shown. The period of infusion of the relaxin is shown by the stippled bar. From Cheah and Sherwood (1981)

Bradshaw, 1979; Chamley, Bagoyo and Bryant-Greenwood, 1977), mouse (Wiqvist, 1959) and hamster (Khaligh, 1968). The inhibitory effect of relaxin on rat uterine myometrial activity has recently been confirmed by studies which employed highly purified porcine relaxin (*Figure 17.16*; Sanborn *et al.*, 1980). Porter and his colleagues provided compelling evidence that relaxin may restrain uterine contractile activity during late pregnancy. They demonstrated with cross-circulation techniques that a myometrial inhibitor(s) with a more rapid onset of activity than oestrogen or progesterone is present in the blood of the guinea pig (Porter, 1972), rabbit (Porter, 1974), and rat (Porter and Downing, 1978) during late pregnancy when the levels of relaxin in the blood of these species are elevated (Zarrow and Rosenberg, 1953; O'Byrne and Steinetz, 1976; Sherwood *et al.*, 1980). Additionally, it has been reported that the frequency of uterine contractions diminishes during the second half of pregnancy in the rat (Fuchs, 1978) when relaxin immunoactivity levels in the blood are elevated. Based largely on the results of the studies described above, it has been hypothesized that relaxin may provide a mechanism which protects the foetuses and placentas during the period of progesterone withdrawal just before parturition in rabbits and rats by restraining spontaneous uterine contractile activity (Porter, 1974; 1979a). Such a protective mechanism may be important during late pregnancy. It is known that in the rat as the uterus progressively comes under greater oestrogen domination during late pregnancy (Yoshinaga, Hawkins and Stocker, 1969), its capacity for strong highly coordinated contractions increases (Fuchs, 1978). In the presence of elevated blood levels of oestrogen, there is an increase in electrical conductivity (Melton and Saldivar, 1964), uterine prostaglandin production (Harney, Sneddon and Williams, 1974; Carminati, Luzzani and Lerner, 1976), uterine oxytocin receptor concentrations (Alexandrova and Soloff, 1980a; 1980b) and responsiveness of the uterus to oxytocin (Fuchs and Poblete, 1970; Fuchs, 1978). Relaxin may inhibit uterine contractions until overriden by stimulating agents such as oxytocin and prostaglandins or other mechanisms which bring about the onset of strong, highly coordinated uterine contractions and the delivery of the foetuses.

At this time there is no direct evidence that relaxin inhibits uterine contractions in the pig. However, the results of a recent study of the myometrial electrical activity of the miniature pig during the last 21 days of pregnancy are consistent with the hypothesis that relaxin may suppress uterine contractions during late pregnancy in the pig. Taverne *et al.* (1979) reported that as late as the period between 24 hours and 10 hours before expulsion of the first piglet, the electrical activity of the myometrium had not changed even though the peripheral blood levels of progesterone had fallen and those of oestrogen had increased. The evolution of regular and frequent phases of myometrial activity which occurred during the 9 hour period before parturition in these pigs coincided with increasing maternal plasma oxytocin concentrations. In agreement with Porter (1979a), these workers suggested that relaxin may act as a secondary myometrial inhibitor during the period of progesterone withdrawal.

The effects of relaxin on uterine contractile activity may be influenced by several hormones including progesterone, oestrogens, oxytocin, catecholamines, and prostaglandins whose concentrations change during late pregnancy. In order to gain a good understanding of the effect(s) of relaxin on uterine contractile activity in the pregnant pig, experiments are needed which recognize that these fluctuations in hormone concentrations are probably accompanied by changes in the responsiveness of the uterus to these hormones.

OTHER EFFECTS OF RELAXIN

There is evidence which has been acquired almost entirely from studies on species other than the pig which indicates that relaxin may have effects in addition to those on cervical distensibility and uterine contractile activity. It has been reported that the administration of porcine relaxin to ovariectomized oestrogen-primed rats brings about a rapid increase in water content, dry weight, nitrogen content and glycogen content in the uterus (Steinetz *et al.*, 1957; Vasilenko, Frieden and Adams, 1980). Hall (1960) reported that most of the glycogen was found in the myometrium in mice. It is not known whether the effects of relaxin on glycogen levels in the rat uterus reflect a direct stimulation of glycogen synthesis or an indirect reduction in glycogen breakdown which occurs in response to an inhibition of uterine contractile activity.

There are considerable data which indicate that in some species, including the guinea pig and mouse, relaxin plays a role in bringing about modifications in the pubic symphysis which enable the pelvic girdle to be distended during parturition (for reviews see Schwabe *et al.*, 1978; Porter, 1979b).

Relaxin may also influence mammary gland growth and function. Treatment of ovariectomized rats with partially purified porcine relaxin together with oestrogen and progesterone reportedly increased growth, lobulation and DNA content of the mammary glands (Hamolsky and Sparrow, 1945; Smith, 1954; Harness and Anderson, 1975). The growth-promoting effects of relaxin on rat mammary glands has recently been confirmed with highly purified porcine relaxin (Wright and Anderson, 1981). Partially purified porcine relaxin has also been reported to depress milk yield in rats (Knox and Griffith, 1970) and goats (Cowie *et al.*, 1965). Recently, Kertiles and Anderson (1979) reported that lactation was reduced severely in intact and lutectomized pigs which were given partially purified porcine relaxin for several days during late pregnancy.

References

AFELE, S., BRYANT-GREENWOOD, G.D., CHAMLEY, W.A. and DAX, E.M. (1979). Plasma relaxin immunoactivity in the pig at parturition and during nuzzling and suckling. *J. Reprod. Fert.* **56**, 451–457

ALEXANDROVA, M. and SOLOFF, M.S. (1980a). Oxytocin receptors and parturition. I. Control of oxytocin receptor concentration in the rat myometrium at term. *Endocrinology* **106**, 730–735

ALEXANDROVA, M. and SOLOFF, M.S. (1980b). Oxytocin receptors and parturition. II. Concentrations of receptors for oxytocin and estrogen in the gravid and nongravid uterus at term. *Endocrinology* **106**, 736–738

ANDERSON, L.L., FORD, J.J., MELAMPY, R.M. and COX, D.F. (1973). Relaxin in porcine corpora lutea during pregnancy and after hysterectomy. *Am. J. Physiol.* **225**, 1215–1219

ARAKAKI, R.F., KLEINFELD, R.G. and BRYANT-GREENWOOD, G.D. (1980). Immunofluorescence studies using antisera to crude and to purified porcine relaxin. *Biol. Reprod.* **23**, 153–159

ASH, R.W. and HEAP, R.B. (1975). Oestrogen, progesterone and corticosteroid concentrations in peripheral plasma of sows during pregnancy, parturition, lactation and after weaning. *J. Endocr.* **64**, 141–154

BALDWIN, D.M. and STABENFELDT, G.H. (1975). Endocrine changes in the pig during late pregnancy, parturition and lactation. *Biol. Reprod.* **12**, 508–515

BEDARKAR, S., TURNELL, W.G., BLUNDELL, T.L. and SCHWABE, C. (1977). Relaxin has conformational homology with insulin. *Nature, Lond.* **270**, 449–451

BELT, W.D., ANDERSON, L.L., CAVAZOS, L.F. and MELAMPY, R.M. (1971). Cytoplasmic granules and relaxin levels in porcine corpora lutea. *Endocrinology* **89**, 1–10

BOSC, M.J., du MESNIL du BUISSON, F. and LOCATELLI, A. (1974). Mise en évidence d'un contrôle foetal de la parturition chez la Truie. Interactions avec le fonction lutéale. *C. r. Acad. Sci., Paris* **278**, 1507–1510

BRYANT-GREENWOOD, G.D., JEFFREY, R., RALPH, M.M. and SEAMARK, R.F. (1980). Relaxin production by the porcine ovarian graafian follicle *in vitro*. *Biol. Reprod.* **23**, 792–800

BRYANT-GREENWOOD, G.D., MERCADO-SIMMEN, R., YAMAMOTO, S., ARAKAKI, R.F., UCHIMA, F.D.A. and GREENWOOD, F.C. (1982). Relaxin receptors and a study of the physiological roles of relaxin. In *Proceedings of the 15th Midwest Conference on Endocrinology and Metabolism—Relaxin*, (R.R. Anderson, Ed.). New York, Plenum Press

CALDER, A.A. (1980). Pharmacological management of the unripe cervix in the human. In *Dilatation of the Uterine Cervix*, (F. Naftolin and P.G. Stubblefield, Eds.), pp. 317–333. New York, Raven Press

CARMINATI, P., LUZZANI, F., and LERNER, L.J. (1976). Synthesis and metabolism of prostaglandins in rat placenta, uterus and ovary during various stages of pregnancy. In *Advances in Prostaglandin and Thromboxane Research*, Vol. 2, (B. Sammuelsson and R. Paoletti, Eds.), pp. 627–632. New York, Raven Press

CAVAZOS, L.F., ANDERSON, L.L., BELT, W.D., HENRICKS, D.M., KRAELING, R.R. and MELAMPY, R.M. (1969). Fine structure and progesterone levels in the corpus luteum of the pig during the estrous cycle. *Biol. Reprod.* **1**, 83–106

CHAMLEY, W.A., BAGOYO, M.M. and BRYANT-GREENWOOD, G.D. (1977). *In vitro* response of relaxin treated rat uterus to prostaglandins and oxytocin. *Prostaglandins* **14**, 763–769

CHEAH, S.H. and SHERWOOD, O.D. (1980). Target tissues for relaxin in the rat: Tissue distribution of injected ^{125}I-labeled relaxin and tissue changes in adenosine 3′,5′-monophosphate levels after *in vitro* relaxin incubation. *Endocrinology* **106**, 1203–1209

CHEAH, S.H. and SHERWOOD, O.D. (1981). Effects of relaxin on uterine contractions *in vivo* in conscious and unrestrained estrogen-treated and steroid-untreated ovariectomized rats. *Endocrinology* **109**, 2076–2083

COGGINS, E.G. and FIRST, N.L. (1977). Effect of dexamethasone, methallibure and fetal decapitation on porcine gestation. *J. Anim. Sci.* **44**, 1041–1049

COHEN, H. (1963). Relaxin: Studies dealing with isolation, purification, and characterization. *Trans. N.Y. Acad. Sci.* **25**, 313–330

CONRAD, J.T. and UELAND, K. (1976). Reduction of the stretch modulus of human cervical tissue by prostaglandin E_2. *Am. J. Obstet. Gynec.* **126**, 218–223

CORTEEL, M., LEMON, M. and DUBOIS, M. (1977). Évolution de la réaction immunocytologique du corps jaune de truie au cours de la gestation. *J. Physiol, Paris* **73**, 63A–64A

COWIE, A.T., COX, C.P., FOLLEY, S.J., HOSKING, Z.D. and TINDAL, J.S. (1965). Relative efficiency of crystalline suspensions of hexoestrol and of oestradiol monobenzoate inducing mammary development and lactation in the goat; and effects of relaxin on mammogenesis and lactation. *J. Endocr.* **31**, 165–172

CULLEN, B.M. and HARKNESS, R.D. (1960). The effect of hormones on the physical properties and collagen content of the rat's uterine cervix. *J. Physiol.* **152**, 419–436

DIEHL, J.R., GODKE, R.A., KILLIAN, D.B. and DAY, B.N. (1974). Induction of parturition in swine with prostaglandin $F_{2\alpha}$. *J. Anim. Sci.* **38**, 1229–1234

DINGFELDER, J.R., BRENNER, W.E., HENDRICKS, C.H. and STAUROVSKY, L.G. (1975). Reduction of cervical resistance by prostaglandin suppositories prior to dilatation for induced abortion. *Am. J. Obstet. Gynec.* **122**, 25–30

DUBOIS, M.P. and DACHEUX, J.L. (1978). Relaxin, a male hormone? Immuno-cytological localization of a related antigen in the boar testis. *Cell Tiss. Res.* **187**, 201–214

FEVOLD, H.L., HISAW, F.L. and MEYER, R.K. (1930). The relaxative hormone of the corpus luteum. Its purification and concentration. *J. Am. Chem. Soc.* **52**, 3340–3348

FEVRE, J., TERQUI, M. and BOSC, M.J. (1975). Mechanismes de la naissance chez la Truie. Equilibres hormonaux avant et pendant de parturition. *Journées de la Recherche Porcine en France*, pp. 393–398. Paris, L'Institut Technique du Porc

FIELDS, P.A. and LARKIN, L.H. (1980). Enhancement of uterine cervix extensibility in oestrogen-primed mice following administration of relaxin. *J. Endocr.* **87**, 147–152

FIRST, N.L. and BOSC, M.J. (1979). Proposed mechanisms controlling parturition and the induction of parturition in swine. *J. Anim. Sci.* **48**, 1407–1421

FITZPATRICK, R.J. (1977). Dilatation of the uterine cervix. In *The Fetus and Birth*, pp. 31–39. Amsterdam, Elsevier/Excerpta Medica/North Holland

FITZPATRICK, R.J. and LIGGINS, G.C. (1980). Effects of prostaglandins on the cervix of pregnant women and sheep. In *Dilatation of the Uterine Cervix*, (F. Naftolin and P.G. Stubblefield, Eds.), pp. 287–300. New York, Raven Press

FRIEDEN, E.H. and HISAW, F.L. (1953). The biochemistry of relaxin. *Recent Prog. Horm. Res.* **8**, 333–372

FRIEDEN, E.H. and YEH, L. (1977). Evidence for a 'prorelaxin' in porcine relaxin concentrates. *Proc. Soc. exp. Biol. Med.* **154**, 407–411

FRIEDEN, E.H., STONE, N.R. and LAYMAN, N.W. (1960). Nonsteroid ovarian hormones. III. The properties of relaxin preparations purified by counter-current distribution. *J. Biol. Chem.* **235**, 2267–2271

FRIEND, D.W., CUNNINGHAM, H.M. and NICHOLSON, J.W.G. (1962). The duration of farrowing in relation to the reproductive performance of Yorkshire sows. *Can. J. comp. Med. Vet. Sci.* **26**, 127–130

FUCHS, A.R. (1978). Hormonal control of myometrial function during pregnancy and parturition. *Acta endocr.* **89**, Suppl. **221**, 1–70

FUCHS, A.R. and POBLETE, V.R. JR. (1970). Oxytocin and uterine function in pregnant and parturient rats. *Biol. Reprod.* **2**, 387–400

GAST, M.J., MERCADO-SIMMEN, R., NIALL, H. and BOIME, I. (1980). Cell-free synthesis of a high molecular weight relaxin-related protein. *Ann. N.Y. Acad. Sci.* **343**, 148–154

GOLICHOWSKI, A. (1980). Cervical stromal interstitial polysaccharide metabolism in pregnancy. In *Dilatation of the Uterine Cervix*, (F. Naftolin and P.G. Stubblefield, Eds.), pp. 99–112. New York, Raven Press

GORDON, A.J. and CALDER, A.A. (1977). Oestradiol applied locally to ripen the unfavourable cervix. *Lancet* **2**, 1319–1321

GRAHAM, E.F. and DRACY, A.E. (1953). The effect of relaxin and mechanical dilatation on the bovine cervix. *J. Dairy Sci.* **36**, 772–777

GRISS, G., KECK, J., ENGELHORN, R. and TUPPY, H. (1967). The isolation and purification of an ovarian polypeptide with uterine-relaxing activity. *Biochim. Biophys. Acta* **140**, 45–54

HALL, K. (1960). Modification by relaxin of the response of the reproductive tract of mice to oestradiol and progesterone. *J. Endocr.* **20**, 355–364

HAMOLSKY, M. and SPARROW, R.C. (1945). Influence of relaxin on mammary development in sexually immature female rats. *Proc. Soc. exp. Biol. Med.* **60**, 8–9

HARKNESS, M.L.R. and HARKNESS, R.D. (1959). Changes in the physical properties of the uterine cervix of the rat during pregnancy. *J. Physiol.* **148**, 524–547

HARNESS, J.R. and ANDERSON, R.R. (1975). Effect of relaxin on mammary gland growth and lactation in the rat. *Proc. Soc. exp. Biol. Med.* **148**, 933–936

HARNEY, P.J., SNEDDON, J.M. and WILLIAMS, K.I. (1974). The influence of ovarian hormones upon the motility and prostaglandin production of the pregnant rat uterus *in vitro*. *J. Endocr.* **60**, 343–351

HISAW, F.L. (1926). Experimental relaxation of the pubic ligament of the guinea pig. *Proc. Soc. exp. Biol. Med.* **23**, 661–663

HISAW, F.L. (1929). The corpus luteum hormone. I. Experimental relaxation of the pelvic ligaments of the guinea pig. *Physiol. Zool.* **2**, 59–79

HISAW, F.L., Jr. and HISAW, F.L. (1964). Effect of relaxin on the uterus of monkeys (*Macaca mulatta*) with observations on the cervix and symphysis pubis. *Am. J. Obstet. Gynec.* **89**, 141–155

HISAW, F.L. and ZARROW, M.X. (1948). Relaxin in the ovary of the domestic sow. *Proc. Soc. exp. Biol. Med.* **69**, 395–398

HOLLINGSWORTH, M., GALLIMORE, S. and ISHERWOOD, C.N.M. (1980). Effects of prostaglandins $F_{2\alpha}$ and E_2 on cervical extensibility in the late pregnant rat. *J. Reprod. Fert.* **58**, 95–99

HOLLINGSWORTH, M., ISHERWOOD, C.N.M. and FOSTER, R.W. (1979). The effects of oestradiol benzoate, progesterone, relaxin, and ovariectomy on cervical extensibility in the late pregnant rat. *J. Reprod. Fert.* **56**, 471–477

ISAACS, N., JAMES, R., NIALL, H., BRYANT-GREENWOOD, G., DODSON, G., EVANS, A. and NORTH, A.C.T. (1978). Relaxin and its structural relationship to insulin. *Nature, Lond.* **271**, 278–281

JAMES, R., NIALL, H., KWOK, S. and BRYANT-GREENWOOD, G. (1977). Primary structure of porcine relaxin: Homology with insulin and related growth factors. *Nature, Lond.* **267**, 544–546

JUDSON, D.G., PAY, S., and BHOOLA, K.D. (1980). Modulation of cyclic AMP in isolated rat uterine tissue slices by porcine relaxin. *J. Endocr.* **87**, 153–159

KENDALL, J.Z., PLOPPER, C.G. and BRYANT-GREENWOOD, G.D. (1978). Ultrastructural immunoperoxidase demonstration of relaxin in corpora lutea from a pregnant sow. *Biol. Reprod.* **18**, 94–98

KENDALL, J.Z., DZIUK, P.J., NELSON, D., SHERWOOD, O.D., STREAT, C.N. and THURMON, J.T. (1980). Aberrant parturition in pigs with hypophysectomized foetuses. *Biol. Reprod.* **22**, Suppl. **1**, Abstr. 80

KENNEDY, T.G. (1974). Effect of relaxin on oestrogen-induced uterine luminal fluid accumulation in the ovariectomized rat. *J. Endocr.* **61**, 347–353

KENNEDY, T.G. (1976). Does prostaglandin $F_{2\alpha}$ ($PGF_{2\alpha}$) mediate the effect of relaxin on cervical tone in the rat? *Proc. Can. Fed. Biol. Soc.* **19**, 273

KERTILES, L.P. and ANDERSON, L.L. (1979). Effect of relaxin on cervical dilatation, parturition and lactation in the pig. *Biol. Reprod.* **21**, 57–68

KHALIGH, H.S. (1968). Inhibition by relaxin of spontaneous contractions of the uterus of the hamster *in vitro*. *J. Endocr.* **40**, 125–126

KILLIAN, D.B., GARVERICK, H.A. and DAY, B.N. (1973). Peripheral plasma progesterone and corticoid levels at parturition in the sow. *J. Anim. Sci.* **37**, 1371–1375

KNOX, F.S. and GRIFFITH, D.R. (1970). Effect of ovarian hormones upon milk yield in the rat. *Proc. Soc. exp. Biol. Med.* **133**, 135–137

KRANTZ, J.C., BRYANT, H.H. and CARR, C.J. (1950). The action of aqueous corpus luteum extract upon uterine activity. *Surg. Gynec. Obstet.* **90**, 372–375

KROC. R.L., STEINETZ, B.G. and BEACH, V.L. (1959). The effects of estrogens, progestagens, and relaxin in pregnant and nonpregnant laboratory rodents. *Ann. N.Y. Acad. Sci.* **75**, 942–980

KWOK, S.C.M., CHAMLEY, W.A. and BRYANT-GREENWOOD, G.D. (1978). High molecular weight forms of relaxin in pregnant sow ovaries. *Biochem. Biophys. Res. Commun.* **82**, 997–1005

LARKIN, L.H., FIELDS, P.A. and OLIVER, R.M. (1977). Production of antisera against electrophoretically separated relaxin and immunofluorescent localization of relaxin in the porcine corpus luteum. *Endocrinology* **101**, 679–685

LEPPI, T.J. (1964). A study of the uterine cervix of the mouse. *Anat. Rec.* **150**, 51–66

MCINNES, D.R., NAFTOLIN, F. and VAN DER REST, M. Cervical changes in pregnant women. In *Dilatation of the Uterine Cervix*, (F. Naftolin and P.G. Stubblefield, Eds.), pp. 181–193. New York, Raven Press

MACKENZIE, I.Z. and EMBREY, M.P. (1977). Cervical ripening with intravaginal prostaglandin E_2 gel. *Br. med. J.* **2**, 1381–1384

MACLENNAN, A.H., GREEN, R.C., BRYANT-GREENWOOD, G.D., GREENWOOD, F.C. and SEAMARK, R.F. (1980). Ripening of the human cervix and induction of labour with purified porcine relaxin. *Lancet* **1**, 220–223

MCMURTRY, J.P., KWOK, S.C.M. and BRYANT-GREENWOOD, G.D. (1978). Target tissues for relaxin identified *in vitro* with ^{125}I-labelled porcine relaxin. *J. Reprod. Fert.* **53**, 209–216

MARTIN, P.A., BEVIER, G.W. and DZIUK, P.J. (1978). The effect of disconnecting the uterus and ovary on the length of gestation in the pig. *Biol. Reprod.* **18**, 428–433

MATSUMOTO, D. and CHAMLEY, W.A. (1980). Identification of relaxins in porcine follicular fluid and in the ovary of the immature sow. *J. Reprod. Fert.* **58**, 369–375

MELTON, C.E. Jr. and SALDIVAR, J.T. Jr. (1964). Impulse velocity and conduction pathways in rat myometrium. *Am. J. Physiol.* **207**, 279–285

MERCADO-SIMMEN, R.C., BRYANT-GREENWOOD, G.D. and GREENWOOD, F.C. (1980). Characterization of the binding of ^{125}I-relaxin to rat uterus. *J. biol. Chem.* **255**, 3617–3623

MOCHIZUKI, M. and TOJO, S. (1980). Effect of dehydroepiandrosterone sulfate on softening and dilation of the uterine cervix in pregnant women. In *Dilatation of the Uterine Cervix*, (F. Naftolin and P.G. Stubblefield, Eds.), pp. 267–286. New York, Raven Press

MOLOKWU, E.C.I. and WAGNER, W.C. (1973). Endocrine physiology of the puerperal sow. *J. Anim. Sci.* **36**, 1158–1163

NAJAK, Z., HILLIER, K. and KARIM, S.M.M. (1970). The action of prostaglandins on the human isolated non pregnant cervix. *J. Obstet. Gynaec. Br. Commonw.* **77**, 701–709

NARA, B.S. and FIRST, N.L. (1981). Effect of indomethacin and prostaglandin $F_{2\alpha}$ on parturition in swine. *J. Anim. Sci.* **52**, 1360–1370

NARA, B.S., WELK, F.A., SHERWOOD, O.D. and FIRST, N.L. Effect of relaxin on parturition and live birth in swine. (Submitted for publication)

NELLOR, J.E., DANIELS, R.W., HOEFER, J.A., WILDT, D.E. and DUKELOW, W.R. (1975). Influence of induced delayed parturition on fetal survival in pigs. *Theriogenology* **4**, 23–31

NIALL, H.D., JAMES, R., JOHN, M., WALSH, J., KWOK, S., BRYANT-GREENWOOD, G.D., TREGEAR, G.W. and BRADSHAW, R.A. (1982). Chemical studies on relaxin. In *Proceedings of the 15th Midwest Conference on Endocrinology and Metabolism – Relaxin*, (R.R. Anderson, Ed.). New York, Plenum Press

O'BYRNE, E.M. and STEINETZ, B.G. (1976). Radioimmunoassay (RIA) of relaxin in sera of various species using an antiserum to porcine relaxin. *Proc. Soc. exp. Biol. Med.* **152**, 272–276

O'BYRNE, E.M., SAWYER, W.K., BUTLER, M.C. and STEINETZ, B.G. (1976). Serum immunoreactive relaxin and softening of the uterine cervix in pregnant hamsters. *Endocrinology* **99**, 1333–1335

PINTO, R.M., FISCH, L., SCHWARCZ, R.L. and MONTUORI, E. (1964). Action of estradiol 17β upon uterine contractility and the milk-ejecting effect in the pregnant woman. *Am. J. Obstet. Gynec.* **90**, 99–107

PORTER, D.G. (1971). The action of relaxin on myometrial activity in the guinea pig *in vivo. J. Reprod. Fert.* **26**, 251–253

PORTER, D.G. (1972). Myometrium of the pregnant guinea pig: the probable importance of relaxin. *Biol. Reprod.* **7**, 458–464

PORTER, D.G. (1974). Inhibition of myometrial activity in the pregnant rabbit. Evidence for a new factor. *Biol. Reprod.* **10**, 54–61

PORTER, D.G. (1979a). The myometrium and the relaxin enigma. *Anim. Reprod. Sci.* **2**, 77–96

PORTER, D.G. (1979b). Relaxin: Old hormone, new prospect. In *Oxford Reviews of Reproductive Biology*, Volume 1, (C.A. Finn, Ed.), pp. 1–57. Oxford, Clarendon Press

PORTER, D.G. and DOWNING, S.J. (1978). Evidence that a humoral factor possessing relaxin-like activity is responsible for uterine quiescence in the late pregnant rat. *J. Reprod. Fert.* **52**, 95–102

PORTER, D.G., DOWNING, S.J. and BRADSHAW, J.M.C. (1979). Relaxin inhibits spontaneous and prostaglandin-driven myometrial activity in anaesthetized rats. *J. Endocr.* **83**, 183–192

RANDALL, G.C.B. (1972). Observations on parturition in the sow. II. Factors influencing stillbirth and perinatal mortality. *Vet. Rec.* **90**, 183–186

RAWITCH, A.B., MOORE, W.V. and FRIEDEN, E.H. (1980). Relaxin-insulin homology: Predictions of secondary structure and lack of competitive binding. *Int. J. Biochem.* **11**, 357–362

SANBORN, B.M., KUO, H.S., WEISBRODT, N.W. and SHERWOOD, O.D. (1980). The interaction of relaxin with the rat uterus. I: Effect on cyclic nucleotide levels and spontaneous contractile activity. *Endocrinology* **106**, 1210–1215

SAWYER, W.H., FRIEDEN, E.H. and MARTIN, A.C. (1953). *In vitro* inhibition of spontaneous contractions of the rat uterus by relaxin-containing extracts of sow ovaries. *Am. J. Physiol.* **172**, 547–552

SCHWABE, C. and MCDONALD, J.K. (1977). Relaxin: A disulfide homolog of insulin. *Science* **197**, 914–915

SCHWABE, C., MCDONALD, J.K. and STEINETZ, B.G. (1976). Primary structure of the A-chain of porcine relaxin. *Biochem. Biophys. Res. Commun.* **70**, 397–405

SCHWABE, C., MCDONALD, J.K. and STEINETZ, B.G. (1977). Primary structure of the B-chain of porcine relaxin. *Biochem. Biophys. Res. Commun.* **75**, 503–510

SCHWABE, C., STEINETZ, B.G., WEISS, G., SEGALOFF, A., MCDONALD, J.K., O'BYRNE, E., HOCHMAN, J., CARRIERE, B. and GOLDSMITH, L. (1978). Relaxin. *Recent Prog. Horm. Res.* **34**, 123–211

SHERWOOD, O.D. (1982a). Isolation and characterization of porcine and rat relaxin. In *Proceedings of the 15th Midwest Conference on Endocrinology and Metabolism – Relaxin*, (R.R. Anderson, Ed.), New York, Plenum Press

SHERWOOD, O.D. (1982b). Radioimmunoassay of relaxin. In *Proceedings of the 15th Midwest Conference on Endocrinology and Metabolism – Relaxin*, (R.R. Anderson,, Ed.), New York, Plenum Press

SHERWOOD, O.D. and O'BYRNE, E.M. (1974). Purification and characterization of porcine relaxin. *Arch. Biochem. Biophys.* **160**, 185–196

SHERWOOD, O.D. and RUTHERFORD, J.E. (1981). Relaxin immunoactivity levels in ovarian extracts obtained from rats during various reproductive states and from adult cycling pigs. *Endocrinology* **108**, 1171–1177

SHERWOOD, O.D., ROSENTRETER, K.R. and BIRKHIMER, M.L. (1975). Development of a radioimmunoassay for porcine relaxin using ^{125}I-labeled polytyrosylrelaxin. *Endocrinology* **96**, 1106–1113

SHERWOOD, O.D., CHANG, C.C., BEVIER, G.W. and DZIUK, P.J. (1975). Radioimmunoassay of plasma relaxin levels throughout pregnancy and at parturition in the pig. *Endocrinology* **97**, 834–837

SHERWOOD, O.D., MARTIN, P.A., CHANG, C.C. and DZIUK, P.J. (1977a). Plasma relaxin levels in pigs with corpora lutea induced during late pregnancy. *Biol. Reprod.* **17**, 97–100

SHERWOOD, O.D., MARTIN, P.A., CHANG, C.C. and DZIUK, P.J. (1977b). Plasma relaxin levels during late pregnancy and at parturition in pigs with altered utero-ovarian connections. *Biol. Reprod.* **17**, 101–103

SHERWOOD, O.D., WILSON, M.E., EDGERTON, L.A. and CHANG, C.C. (1978). Serum relaxin concentrations in pigs with parturition delayed by progesterone administration. *Endocrinology* **102**, 471–475

SHERWOOD, O.D., NARA, B.S., CRNEKOVIC, V.E. and FIRST, N.L. (1979). Relaxin concentrations in pig plasma after the administration of indomethacin and prostaglandin $F_{2\alpha}$ during late pregnancy. *Endocrinology* **104**, 1716–1721

SHERWOOD, O.D., CRNEKOVIC, V.E., GORDON, W.L. and RUTHERFORD, J.E. (1980). Radioimmunoassay of relaxin throughout pregnancy and during parturition in the rat. *Endocrinology* **107**, 691–698

SHERWOOD, O.D., CHANG, C.C., BEVIER, G.W., DIEHL, J.R. and DZIUK, P.J. (1976). Relaxin concentrations in pig plasma following administration of prostaglandin $F_{2\alpha}$ during late pregnancy. *Endocrinology* **98**, 875–879

SHERWOOD, O.D., NARA, B.S., WELK, F.A., FIRST, N.L. and RUTHERFORD, J.E. (1981). Relaxin levels in the maternal plasma of pigs before, during, and after parturition and before, during, and after suckling. *Biol. Reprod.* **25**, 65–71

SMITH, J.C. and NALBANDOV, A.V. (1958). The role of hormones in the relaxation of the uterine portion of the cervix in swine. *Am. J. vet. Res.* **19**, 15–18

SMITH, T.C. (1954). The action of relaxin on mammary gland growth in the rat. *Endocrinology* **54**, 59–70

STEINETZ, B.G., BEACH, V.L. and KROC, R.L. (1959). The physiology of relaxin in laboratory animals. In *Endocrinology of Reproduction*, (C.M. Lloyd, Ed.), pp. 389–427. New York, Academic Press

STEINETZ, B.G., BEACH, V.L. AND KROC, R.L. (1969). Bioassay of relaxin. In *Methods in Hormone Research*, 2nd Ed., Vol. 2A, (R.I. Dorfman, Ed.), pp. 481–513. New York, Academic Press

STEINETZ, B.G., O'BYRNE, E.M. and KROC, R.L. (1980). The role of relaxin in cervical softening during pregnancy in mammals. In *Dilatation of the Uterine Cervix*, (F. Naftolin and P.G. Stubblefield, Eds.), pp. 157–177. New York, Raven Press

STEINETZ, B.G., BEACH, V.L., BLYE, R.P. and KROC, R.L. (1957). Changes in the composition of the rat uterus following a single injection of relaxin. *Endocrinology* **61**, 287–292

STRYKER, J.L. and DZIUK, P.J. (1975). Effects of fetal decapitation on fetal development, parturition, and lactation in pigs. *J. Anim. Sci.* **40**, 282–287

STYS, S.J., CLARK, K.E., CLEWELL, W.H. and MESCHIA, G. (1980). Hormonal effects on cervical compliance in sheep. In *Dilatation of the Uterine Cervix*, (F. Naftolin and P.G. Stubblefield, Eds.), pp. 147–156. New York, Raven Press

TAVERNE, M.A.M., NAAKTGEBOREN, C., ELSAESSER, F., FORSLING, M.L., van der WEYDEN, G.C., ELLENDORFF, F. and SMIDT, D. (1979). Myometrial electrical activity and plasma concentrations of progesterone, estrogens, and oxytocin during late pregnancy and parturition in the miniature pig. *Biol. Reprod.* **21**, 1125–1134

UYLDERT, I.E. and DEVAAL, O.M. (1947). Relaxation of the rat's uterine ostium during pregnancy. *Acta brev. neerl. Physiol.* **15**, 49–53

VASILENKO, P., FRIEDEN, E.H. and ADAMS, W.C. (1980). Effect of purified relaxin on uterine glycogen and protein in the rat. *Proc. Soc. exp. Biol. Med.* **163**, 245–248

VEIS, A. (1980). Cervical dilatation: A proteolytic mechanism for loosening the collagen fiber network. In *Dilatation of the Uterine Cervix*, (F. Naftolin and P.G. Stubblefield, Eds.), pp. 195–202. New York, Raven Press

WALSH, J.R. and NIALL, H.D. (1980). Use of an octadecylsilica purification method minimizes proteolysis during isolation of porcine and rat relaxins. *Endocrinology* **107**, 1258–1260

WIQVIST, N. (1959). The effect of prolonged administration of relaxin on some functional properties of the non-pregnant mouse and rat uterus. *Acta endocr., Copenh.* **32**, Suppl. **46**, 15–32

WRIGHT, L.C. and ANDERSON, R.R. (1982). Effect of relaxin on mammary gland development. In *Proceedings of the 15th Midwest Conference on Endocrinology and Metabolism – Relaxin*, (R.R. Anderson, Ed.). New York, Plenum Press

YOSHINAGA, K., HAWKINS, R.A. and STOCKER, J.F. (1969). Estrogen secretion by the rat ovary *in vivo* during the estrous cycle and pregnancy. *Endocrinology* **85**, 103–112

ZARROW, M.X., and ROSENBERG, B. (1953). Sources of relaxin in the rabbit. *Endocrinology* **53**, 593–598

ZARROW, M.X. and YOCHIM, J. (1961). Dilation of the uterine cervix of the rat and accompanying changes during the estrous cycle, pregnancy and following treatment with estradiol, progesterone, and relaxin. *Endocrinology* **69**, 292–304

ZARROW, M.X., SIKES, D. and NEHER, G.M. (1954). Effect of relaxin on the uterine cervix and vulva of young castrated sows and heifers. *Am. J. Physiol.* **179**, 687

ZARROW, M.X., NEHER, G.M., SIKES, D., BRENNAN, D.M. and BULLARD, J.F. (1956). Dilatation of the uterine cervix of the sow following treatment with relaxin. *Am. J. Obstet. Gynec.* **72**, 260–263

18

PHYSIOLOGY AND ENDOCRINOLOGY OF THE FOETUS IN LATE GESTATION

A.A. MACDONALD, B. COLENBRANDER and C.J.A.H.V. VAN VORSTENBOSCH
Institute of Veterinary Anatomy, State University of Utrecht, The Netherlands

Birth, the transfer from liquid space to gaseous atmosphere confronts the mammalian organism with substantial and novel challenges. If the genotype of the newborn, indeed the species, is to reproduce further, it must survive. In the case of the pig, survival is so commonplace that the miracle of coordination required to achieve it is often overlooked; after all, about eight out of every ten newborn piglets succeed in making the transition (Randall, 1978). How do they manage this remarkable feat and what do they require at birth to survive?

Clearly while *in utero* the placenta, perfused with foetal blood, abstracts from the maternal circulation the gases, metabolites and minerals necessary for growth. This supply is severed at birth and to survive the piglet must obtain oxygen and remove carbon dioxide. Failure to do so leads to death within minutes (Miller and Miller, 1965).

The lung, therefore, must be fully competent at birth to take over the role of gas exchange with the circulation from the placenta. How it develops that ability, and an examination of the factors which may contribute to its development, form part of this review.

Nutrient supply, its distribution, and metabolite clearance during gestation are designed for the buffered, aquatic environment *in utero*. Birth disrupts this, thrusting upon the neonate a gaseous environment, atmospheric pressures, an expanded pulmonary circulation and the destruction of the placental supply.

What developments during gestation contribute to piglet competence to cope with such changes? For example, how is the homoeostasis of the circulation maintained; are the nervous faculties developed; how is it that the piglet can stand, balance and move; what must be present to enable the piglet to locate the udder and its new supply of nutrients? This chapter will examine what is known of such developments, indicate a number of the many gaps in that knowledge, and finally ask whether we know when, or how this complex piece of expressed genotype acquires the wherewithal to protect itself against the cold, disease and competition awaiting it after birth.

The development of the lung

The lung acquires the function of gas exchange at birth, respiration up until that time having been performed by the placenta. At mid-gestation the lung is a somewhat glandular mass of mesenchymal tissue (*Figure 18.1(a)*) and quite unsuitable for respiratory gas exchange (Flint, 1906–07; Clements, 1938). Nevertheless by 80 days of gestation well-defined bronchi

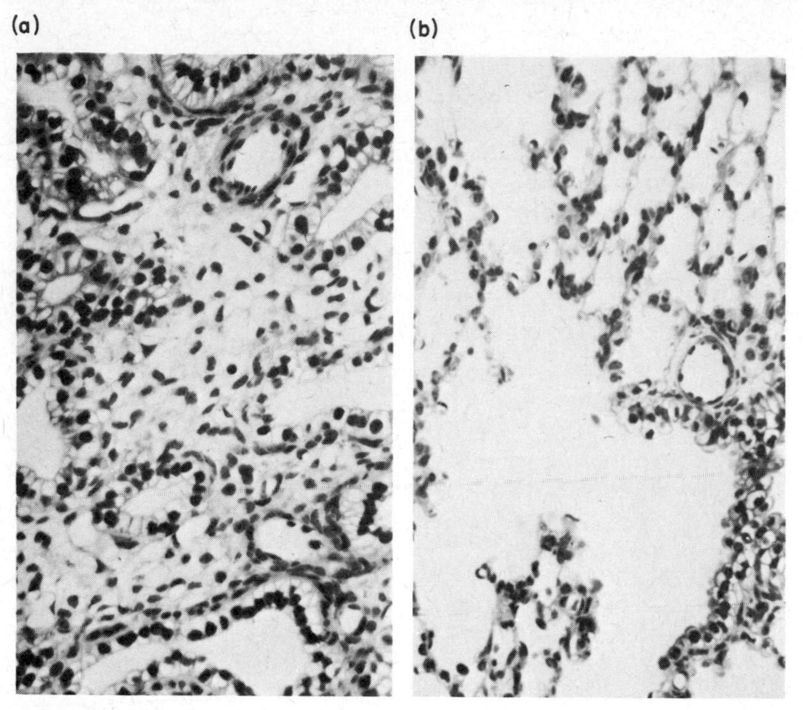

Figure 18.1 (a) Section of foetal pig lung at 70 days of gestation to demonstrate the glandlike appearance of the tissue between the mesenchyme. (b) Section of foetal pig lung at 112 days of gestation to demonstrate a terminal bronchiole leading into an alveolar duct and a cluster of alveoli. (Magn. ×290)

are present and some ten days later the epithelial cells of the bronchi are differentiated into ciliated and goblet cells (Baskerville, 1976). Likewise, whereas at 80 days the alveoli have a cuboidal or columnar epithelium, this changes to a mainly squamous type by about 90 days of gestation; differentiating type II alveolar cells may be identified by the osmophilic lamellar bodies seen in their cytoplasm.

Clusters of goblet cells with the appearance of primitive glands can be found in the bronchi at this stage of development and during the remaining three weeks of gestation they increase in number and size. Also in the bronchi, the ciliated cells increase in number such that by term they represent almost two thirds of the bronchial epithelium (Baskerville, 1976).

The shape of the alveoli changes during the last three weeks of gestation. The more or less rounded form gives way to the irregular open lattice pattern (*Figure 18.1(b)*) seen in the perinatal animal (Clements, 1938; Ham and Baldwin, 1941). The surface of the alveoli becomes covered for the most part by squamous type I alveolar cells and the type II alveolar cells increase their content of lamellar bodies (Rufer and Spitzer, 1974; Baskerville, 1976).

Studies on other species have demonstrated that it is the phospholipid material of which the lamellar bodies are composed that is responsible for forming a surfactant film over, and thereby holding the postnatal stability of, the alveoli and alveolar ducts. Between days 95 and 110 of gestation there is a sharp increase in both phospholipid content and *in vitro* lecithin production by the lung (Rufer and Spitzer, 1974). The five-fold increase in foetal pig lung compliance is consistent with these findings. Somewhat surprisingly, however, it is not the intrinsic elasticity of the phopholipid surfactant film, but rather the change in the film's surface tension that modifies lung compliance (Meban, 1980).

LUNG GLYCOGEN

The development of the foetal lung is also reflected in the decrease in its glycogen content as shown in *Figure 18.2.* Glycogen, which in the lung

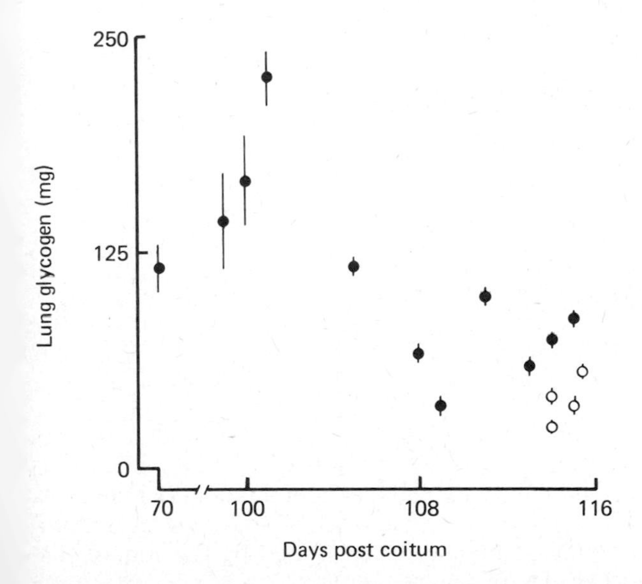

Figure 18.2 The decrease in lung glycogen ($\bar{x}$±S.E.M.) during the last two weeks of gestation. ● = prenatal; ○ = neonatal. From Macdonald (1974)

provides a local energy substrate for differentiation, is distributed throughout the epithelia of the alveoli, respiratory bronchioles and bronchi at about 100 days gestation. By the end of gestation, little of the polysaccharide remains in the alveoli, although its presence may be detected in the bronchial epithelium (Macdonald, 1974).

LUNG FLUID

The airways at term are filled with fluid (*Figure 18.1(b)*) produced by the lung tissue during the last part of gestation (Berton, 1970; Baskerville, 1976). This is removed from the airways during birth and very soon thereafter. The amount cleared from the foetal pig lung can be estimated from the sharp reduction in lung weight following birth shown in *Figure 18.3*. Studies on other species indicate that this fluid is lost as a result of

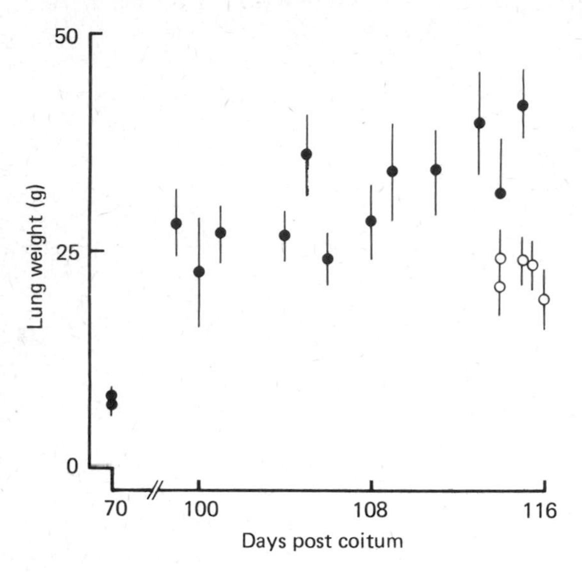

Figure 18.3 The fresh weight of lung tissue (and fluid contents) sectioned at the tracheal bifurcation: ● = prenatal; ○ = neonatal. From Macdonald (1974)

increased lymph drainage (Bland, McMillan and Bressack, 1977). A number of hormonal factors may control foetal lung development, so it is necessary to examine the foetal growth of those endocrine glands which are suspected of influencing lung maturation.

The adrenal

The role which glucocorticoids may play in foetal lung maturation has recently been reviewed (Olson, 1979), and it is clear that the interrelationship between the development of the lung and the hormones of the adrenal may be more complex than had earlier been appreciated. The adrenal gland has a definitive cortex by mid-gestation (*Figure 18.4(a)*) and, as shown by Flint (1900) and Whitehead (1903), increases in size and structural organization towards term (*Figure 18.4(b)*). Some recent studies have resulted in an array of cortical cells being termed the 'foetal cortex'; these cells, found intermingled with medullary tissue at mid-gestation, were reported to be replaced with true cortical tissue by about 80 days of gestation (Katznelson, 1965, 1966; Sedova, 1974). Preliminary results of a

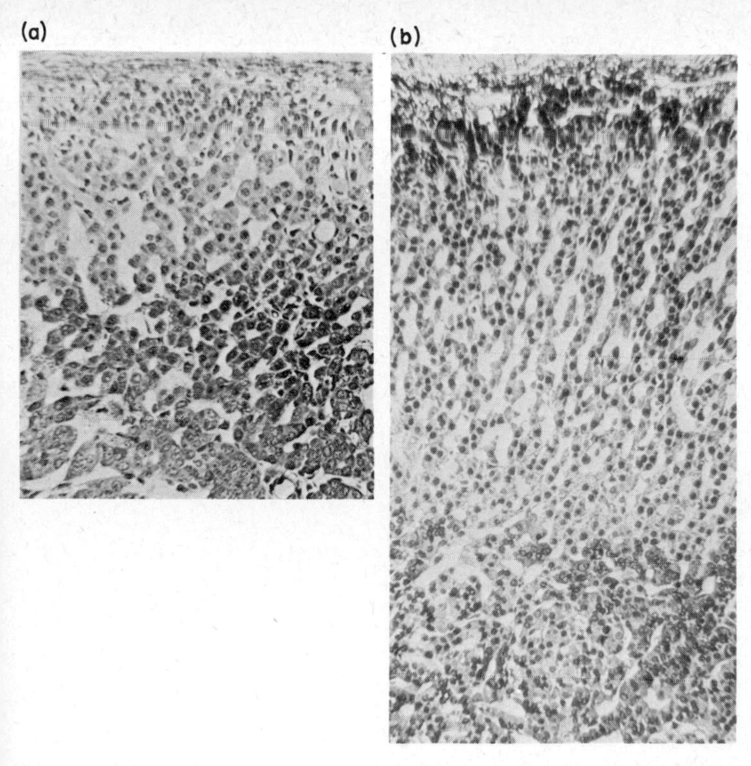

Figure 18.4 (a) Cross section of foetal pig adrenal at 70 days of gestation to illustrate the clustered (pale coloured) cortical tissue above the cells of the medulla. (b) Cross section of foetal pig adrenal at 112 days of gestation to illustrate development of cortical tissue into cords. (Magn. ×160)

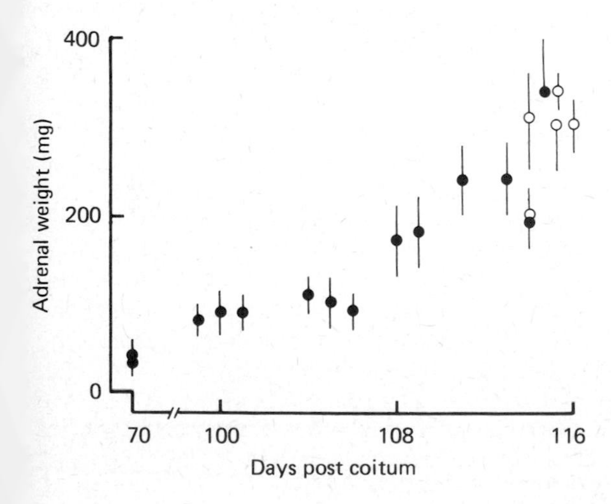

Figure 18.5 The increase in total adrenal tissue weight of pig foetuses during the last two weeks of gestation: ● = prenatal; ○ = neonatal. From Macdonald (1974)

re-examination of adrenal development could find no histological or histochemical grounds for a distinction between 'foetal' and reticularis cells of the true cortex (Colenbrander, Macdonald and Wensing, unpublished observations).

During the last two weeks before term adrenal weight increases rapidly (*Figure 18.5*) and at a rate faster than foetal body weight (Dvorak, 1972; Lohse and First, 1979). Corticosteroids are produced by the adrenal even more rapidly (Dvorak, 1972; Lohse and First, 1979) and this is reflected in circulating concentrations of 17β-hydroxy corticosteroids (Dvorak, 1972) and cortisol (Fèvre, 1975), the latter being the hormone with greatest activity (*Figure 18.6*). However, the rise in circulating corticosteroid levels

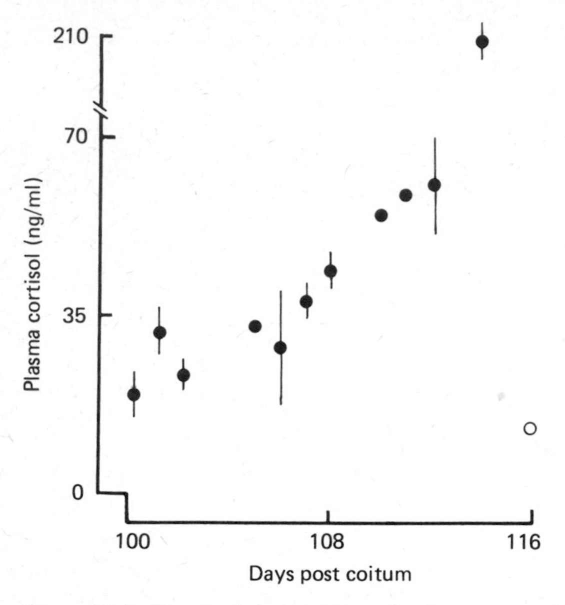

Figure 18.6 The rise in cortisol concentrations measured in foetal pig plasma during the last two weeks of gestation: ● = prenatal; ○ = neonatal. From Fèvre (1975)

is not well correlated to the timing of foetal lung maturation. It may be, therefore, that as in the foetal rabbit (Nicholas *et al.*, 1978) the lung of the pig foetus contains enzymes which can convert inactive corticosteroids into cortisol at a stage of gestation earlier than that at which cortisol is seen to increase in the general circulation. Alternatively the lung may develop as a consequence of an increased hormonal receptor population, or as a result of other hormonal action.

The thyroid

Hormones from the foetal thyroid may be associated with the development of the foetal pig lung. Anatomical studies demonstrate that thyroid follicular development has become established by about 75 days of gestation (Moody, 1906; Studzinski, Bobowiec and Rybka, 1976). However, biochemical studies reviewed recently indicate that there is little change in the glandular production or circulating concentrations of thyroid hormones between 75 days of gestation and term (Macdonald, 1979; Colenbrander *et al.*, 1980).

The pancreas

Pancreatic hormones, and in particular insulin, may play a role in type II cell maturation. Microscopic studies have demonstrated that the cell of the islets of Langerhans increase in number and size after about 95 days gestation (Aron, 1922; Comline *et al.*, 1981). The pronounced proliferation of insulin-producing cells is also reflected in the sharp increase by day 100 in the circulating concentrations of insulin (Atinmo *et al.*, 1976; Fowden, Comline and Silver, 1981; Comline *et al.*, 1981).

Thus the alveolar type II cell development takes place against an apparently stable background of adrenocorticosteroid and thyroid hormones and is possibly in parallel with changes in the pancreatic production and secretion of insulin. Do these hormones influence lung development? What role does the nervous system or prolactin play? It is clear that further study is required before the functional development of the pig's type II alveolar cells is fully understood.

The heart and circulation

The demands placed on the foetal heart are, as in the adult, those of distribution of nutrients and removal of metabolites. It is not surprising, therefore, that there is a close linear relationship between the growth of the heart and either the increase in foetal body weight or the increase in weight of the foetal body plus placental membranes, the latter being a more appropriate description of the total mass of tissue perfused by the foetal blood stream (Macdonald, 1971; Macdonald, unpublished observations).

THE CIRCULATION

The pattern of the circulation during foetal life is specifically adapted to the intrauterine environment (*Figure 18.7(a)*). It has been known for some time that some of the blood returning in the venae cavae passes through the foramen ovale (Pohlman, 1909; Kellogg, 1928). During gestation the lungs receive only a small proportion of combined ventricular output, much of the cranial vena cava return flowing via the ductus arteriosus to the aorta (Macdonald, Rudolph and Heymann, unpublished observations). About 30% of combined ventricular output flows to the placenta (Macdonald, Rudolph and Heymann, 1980). The venous return from the placenta passes through the ductus venosus in the liver (*Figure 18.7(a)*) and flows with the venous drainage from the lower body tissues back to the heart (Pohlman, 1909; Kaman, 1968a; Barnes *et al.*, 1979).

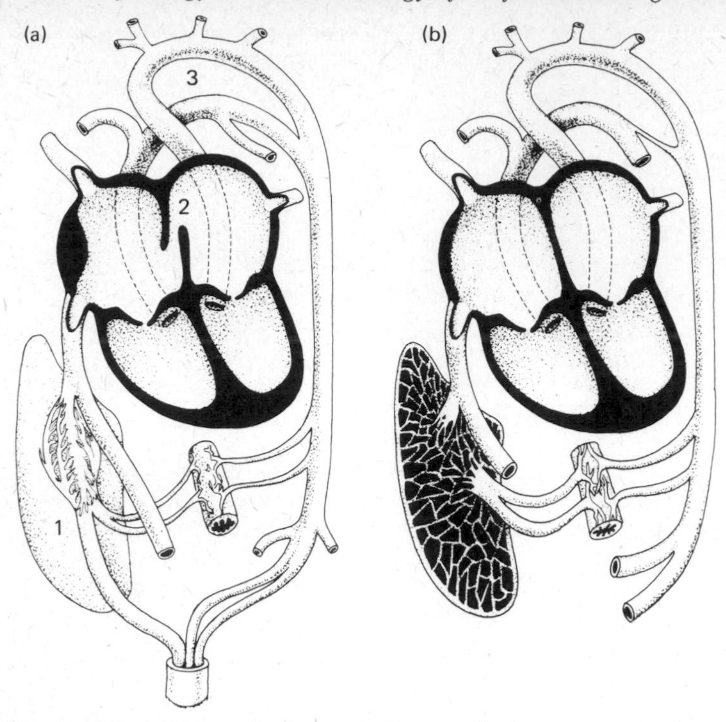

Figure 18.7 (a) Schematic diagram to illustrate the foetal circulation. The umbilical venous return from the placenta passes through the ductus venosus (1) in the liver and joins the caudal vena cava flow to enter the right atrium. The left and right atria connect through the foramen ovale (2). Blood flows from the right ventricle mainly via the ductus arteriosus (3) into the aorta. (b) Schematic diagram to indicate the developed postnatal circulation. Venous return from the intestine enters the liver. There is no flow from right to left atrium. The ductus arteriosus is closed. Blood flows from the right ventricle to the lungs, from the left ventricle to the aorta.

Following birth this circulatory path changes (*Figure 18.7(b)*). The placenta is no longer part of the piglet's general circulation. The lung takes over the respiratory function of the placenta and blood flow is directed along the pulmonary arteries as a result of the gradual closure of the ductus arteriosus (Evans *et al.*, 1963; Rowe *et al.*, 1964). Recent studies have shown that the patency of the ductus arteriosus can be maintained by administration of prostaglandins of the A and E series, or their analogues (Starling *et al.*, 1976, 1978). The disruption at birth of the endogenous production of these substances is probably responsible for the closure of this foetal vessel. As blood pressure in the systemic circulation rises above that in the pulmonary circulation the valve-like foramen ovale is more often held shut than open and the opening is gradually sealed (van Nie *et al.*, 1970; Versprille *et al.*, 1970). The ductus venosus no longer receives blood from the umbilical vein, which closes (Kaman, 1968b).

CONTROL OVER THE HEART AND CIRCULATION

At about four weeks before birth the mean pressure within the arteries of the foetus is 38 mm Hg (Macdonald *et al.*, 1981b) and, as shown in *Figure*

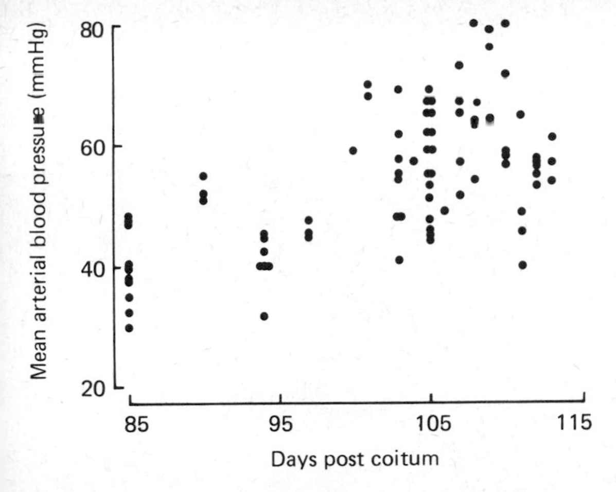

Figure 18.8 Mean arterial blood pressure measured in chronically catheterized pig foetuses during the last month of gestation.

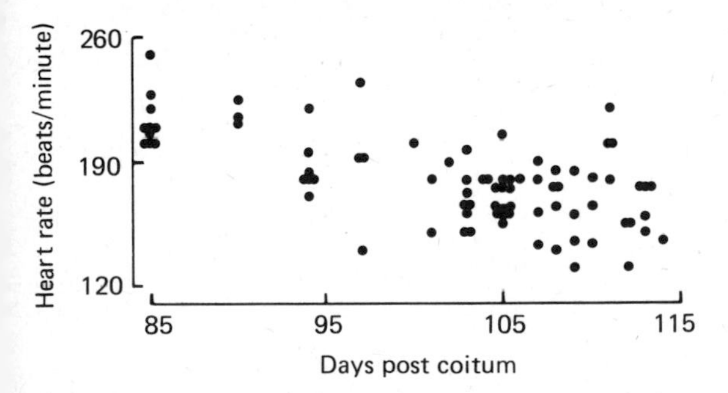

Figure 18.9 Heart rate measurements made on chronically catheterized pig foetuses during the last month of gestation.

18.8, there is a gradual increase until term, although a pronounced variability is seen between foetuses of similar age. Over the same period of time (*Figure 18.9*) foetal heart rate slows down (Fraser, Nagaratnam and Callicott, 1971; Too *et al.*, 1974; Macdonald *et al.*, 1981b).

In the adult, control over the heart and circulation is exerted by a complex array of hormonal, nervous and physical factors (Guyton, Coleman and Granger, 1972). Substantially less is known, however, about which of these mechanisms is available to the foetus for controlling its circulation, and to what extent they are used during the transition to postnatal life.

Neural control

Anatomical studies of the heart clearly demonstrate that neural elements are laid down early in gestation (Wensing, 1964) and that both catecholamine- and acetylcholinesterase-containing fibres are present and increase in number as gestation proceeds (Macdonald *et al.*, 1981a). It has also been shown recently that the foetal heart is responsive during the last month of gestation to both catecholamines (or their agonists) and acetylcholine (Comline, Fowden and Silver, 1979; Macdonald *et al.*, 1981b; Macdonald and Colenbrander, 1981). Thus the heart of the pig foetus possesses end organ receptors sensitive to neurotransmitters. In order to test whether the nerves observed anatomically are physiologically functional, the endogenous catecholamine action was blocked with propranolol and the vagal action was blocked with atropine (Macdonald *et al.*, 1981b). The heart rate slowed or accelerated respectively in response to the two drugs, indicating that the heart was under functional nervous control. The increase in responsiveness between 85 days gestation and term was interpreted as a sign of development within the system during the last month of gestation.

The neural control of the vasculature was similarly tested. Injection of methoxamine caused the blood pressure to increase as a result of peripheral vasoconstriction (Macdonald *et al.*, 1981b; Macdonald and Colenbrander, 1981). Blockade of endogenous α-sympathetic action with phentolamine resulted in a fall in blood pressure.

The enzymes monoamine oxidase and catechol-O-methyl transferase are responsible for deactivation of catecholamines and they are present in the heart during the last 10 days of gestation with high and increasing activity (Stanton *et al.*, 1975).

Adrenal hormones

Catecholamines are also produced by cells of the adrenal medulla (see *Figure 18.4(a)* and *(b)*) which histological studies have demonstrated increases in amount during the second half of gestation (Flint, 1900; Wiesel, 1901; Fenger, 1912; Weymann, 1922–23). The concentration of norepinephrine in the gland at birth is 1.65 mg/g and norepinephrine represents 72% of the gland's total catecholamine content (Stanton and Woo, 1978). Although no measurements appear to have been made of foetal gland catecholamine content, preliminary results of histological studies suggest that the number of cells containing epinephrine increases during gestation (Verhoffstad, personal communication). This is a trend consistent with that seen after birth in biochemically measured epinephrine content (Stanton and Woo, 1978).

Among the factors known to act on the development of the sympathetic system are the adrenal corticosteroids. Evidence to support this view is the poor development of epinephrine-containing cells in the medulla of pig foetuses whose cortex has been stunted by removal of ACTH support through chronic foetal decapitation (Verhofstad *et al.*, 1981).

Renin–Angiotensin

Another substance which may have an influence on foetal cardiovascular function is angiotensin II. This hormone is produced from a precursor plasma globulin, angiotensinogen, following the actions first of the proteolytic enzyme renin (to produce physiologically inactive angiotensin I)

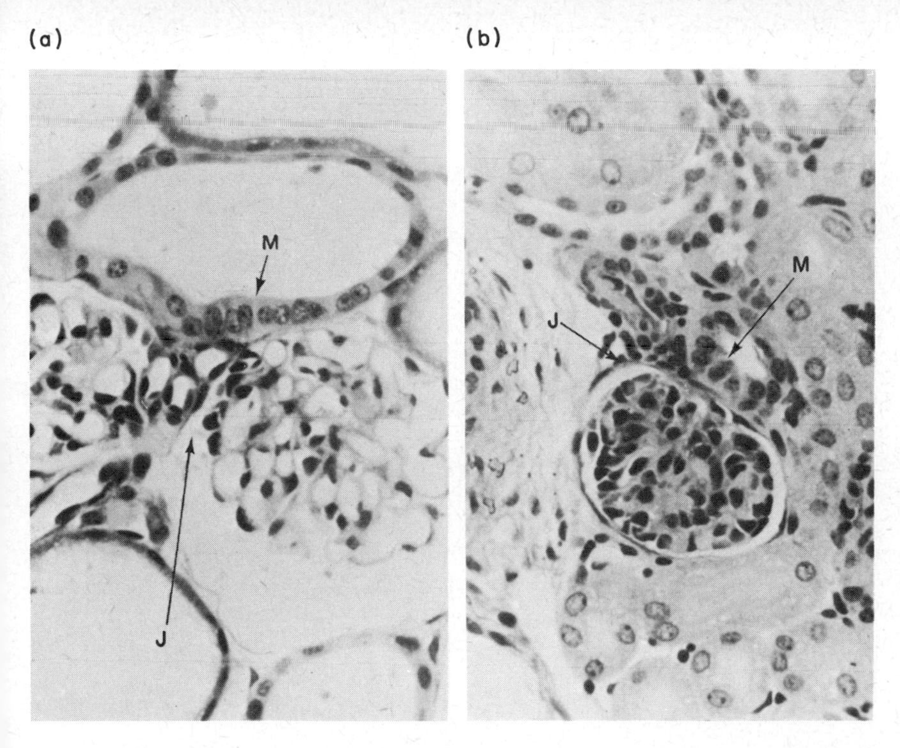

Figure 18.10 Sections of foetal pig kidney at (a) 70 days of gestation and (b) 112 days of gestation illustrating juxtaglomerular cells (J) lying between the glomerulus and the macula densa (M) of the collecting tubule. (Magn. ×400)

and then of a converting enzyme which is a protein circulating in the blood plasma. Renin is produced in response to a variety of stimuli by the juxtaglomerular cells of the kidney (*Figure 18.10*). The foetal development of these cells has not been studied in the pig, but an account of their appearance in the kidney of the newborn piglet has been given by Kazimierczak (1970). *In vitro* culture of kidney cortex from pig foetuses at approximately 65 days gestation resulted in renin production (Szalay and Gyevai, 1967). Earlier studies in which extracts were made of kidney from foetuses aged between 35 days and term demonstrated a gestational trend of increased pressor activity/gram of kidney extract particularly during the last month of gestation (Kaplan and Friedman, 1943). Therefore, although angiotensin II has not yet been measured in the circulation of the pig foetus, these studies when taken together with the observation that plasma angiotensin II concentrations are elevated in the newborn piglet (Osborn, 1979), would imply foetal competence to produce the hormone.

The role which the renin–angiotensin system may play in aldosterone secretion by the adrenal is also presently unknown in the pig foetus. Furthermore the part which aldosterone plays in conjunction with the other physiological mechanisms in the maintenance of foetal blood pressure and volume remains to be investigated.

Fluid balance

The control of fluid balance is more difficult to study in the foetus than the neonate or adult animal because of the larger number of tissues and compartments through which minerals and solutes may be exchanged. In addition to the placenta, foetal membranes and umbilical cord, fluid can transfer through the foetal skin (France, 1976). It is also produced by the foetal lung as mentioned earlier (p.380), either to be swallowed with amniotic fluid, or expelled into the amniotic cavity by intrauterine breathing, sighing and gasping movements (Randall, 1978; 1979). Urine is produced by the foetal kidneys, but is excreted not only into the amniotic cavity via the urethra, but also to the allantoic cavity via the urachus; it is this aspect of fluid balance which has been the most accessible to study.

During the second half of gestation the microscopic anatomy of the kidney changes. At about mid-gestation the glomeruli are large, have an 'open' appearance and are relatively few in number, lying surrounded by proximal tubules with wide internal diameters (*Figure 18.11(a)*). By the end of gestation the glomeruli have increased in number and are more densely packed together (*Figure 18.11(b)*); individually the glomeruli appear to have a more intricate structure and all stages of their development can be seen in the zone of growth of the outer cortex. More tubules

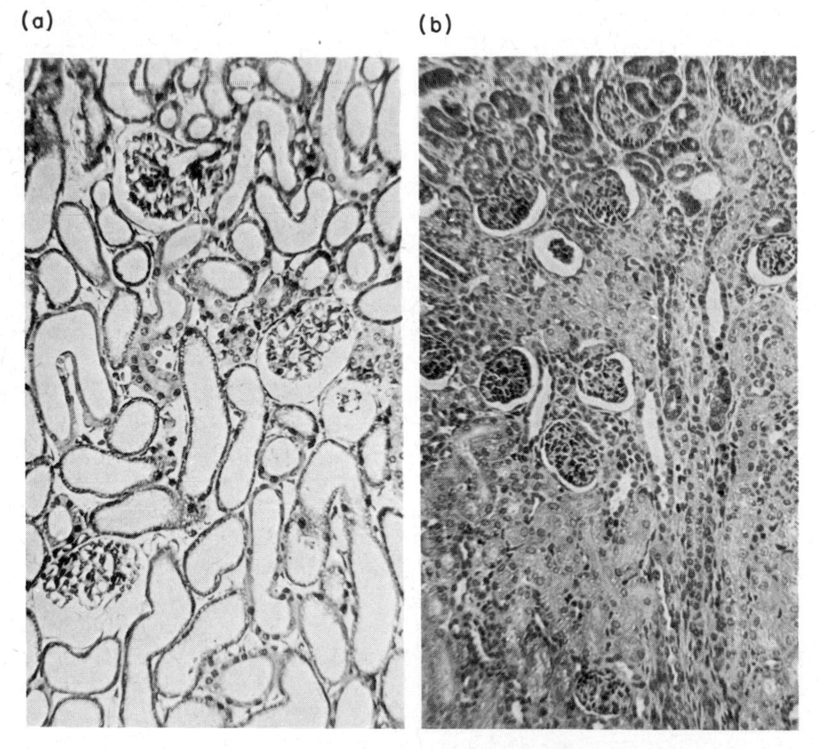

Figure 18.11 (a) Section of foetal pig kidney at 70 days of gestation demonstrating the large glomeruli and large fluid-filled proximal tubules. (b) Section of foetal pig kidney at 112 days of gestation illustrating the cortical zone of differentiating glomeruli (top) and the compact glomeruli surrounded by densely packed tubules. (Magn. ×150)

are present and their diameters are smaller (Hill, 1905; Kunska, 1971a, b; Bergelin and Karlsson, 1975).

By about 70 days of gestation urine flow is estimated to be 33 μl/minute/g kidney weight (Perry and Stanier, 1962) and the decrease to 5.4 μl/minute/g kidney weight at birth (Joppich *et al.*, 1979) reflects the development of competence by the perinatal kidney to produce a more concentrated urine (McCance and Stanier, 1960; Alt *et al.*, 1981). The studies by Alt *et al.* (1981) have demonstrated that although the kidneys of foetuses one week before birth are functionally quite mature, they are nevertheless significantly less mature than those of piglets one week after birth. The mechanism whereby this is brought about remains a matter of speculation.

The nervous system

It is necessary for the brain and peripheral nervous system to possess a large degree of functional competence at birth. The morphological development of the central nervous system in the pig has been studied qualitatively by Bradley (1903), Larsell (1954) and Welento (1960; 1961). Quantitative studies have demonstrated that the brain rapidly increases in weight through the last month of gestation and the first month after birth. Although the rate of total brain growth does not seem to change during this period, the different constituent parts of the brain do show differing rates of growth (Pomeroy, 1960; Dickerson and Dobbing, 1966; Done and Herbert, 1968; Brooks and Davis, 1969). The complexity of brain development is, however, poorly represented by gross morphological or morphometric observations alone.

Increases in total cell number are reflected in the concentration of DNA-P which in the cerebellum and brainstem is seen to rise to a peak at 95 days, the increase occurring somewhat earlier in other parts of the brain (Dickerson and Dobbing, 1966). The number of neurons, as indicated by the concentration of gangliosides, sharply increases during the last three weeks of gestation (Dickerson, Merat and Widdowson, 1971). The number of synaptic complexes isolated biochemically from the cerebellum increases with foetal maturation in line with the change seen in the morphology of the cells. At 70 days of gestation the predominant synaptic complex is associated with climbing fibre terminals whereas parallel fibre terminals predominate after birth (Kornguth *et al.*, 1972).

In the near term pig foetus electrical activity has been recorded from cells of the post-central gyrus of the cerebral cortex and the patterns of activity were very similar to those observed in prepubertal animals (Konda *et al.*, 1979). Stimulation of particular sites in the brain stem of the newborn pig produced appropriate changes in heart rate and blood pressure, indicating that the cardiovascular regulatory centres of the piglet are functional at birth (Gootman *et al.*, 1972; Marshall and Breazile, 1974a, b; Gootman, Buckley and Gootman, 1978; 1979).

The concentration of cholesterol in various parts of the brain and spinal cord (Dickerson and Dobbing, 1966) is closely correlated to myelination of the foetal nervous system during the last month of gestation (Ziolo, 1965; Majstruk-Majewska, 1966). The amount of myelin in the spinal cord, for example, increased from 33.3±1.9 mg/g fresh tissue at 100 days of gestation to 50.0±2.29 mg/g at birth (Patterson *et al.*, 1976).

THE SENSES

It is pertinent to ask in what way development proceeds with regard to the various organs which will sense the new environment. To be successful in obtaining food the newly born piglet must manage to get from the rear of the sow to the udder. To do this effectively it should be able to maintain some degree of balance. Moreover, it is conceivable that one or several of the senses—sight, touch, hearing, taste, smell—are involved in guiding the newborn to the new site of nutrient supply.

Very little is known about the development *in utero* of these senses in any species of animal (Gottlieb, 1971; Bradley and Mistretta, 1975), although a general ontogenetic sequence would appear to be touch, balance, hearing and sight. Spinal reflexes and responsiveness to touch are present by at least 40 days of gestation (Carey, 1922–23; Macdonald, unpublished observations). Anaesthetized chronically decapitated pig foetuses show greater sensitivity to touch than control foetuses (Stryker and Dziuk, 1975; Kraeling *et al.*, 1978). In view of the importance of the pig's snout as a tactile organ after birth (Adrian, 1943a, b; Woolsey and Fairman, 1946), information concerning its prenatal development would be of interest. The same may be said for the senses of balance and taste, although fragmentary information concerning their embryological development is present in the earlier literature (see Patten, 1931).

Studies of the developmental anatomy of the ear suggest that it may be responsive to sound during the second half of gestation (Prentiss, 1913; Hardesty, 1915). However, no studies on the piglet have been found in which the sensitivity of the ear to sound has been explored during gestation.

The eyes are open at birth, and it is clear from studies of their anatomical development that they already possess a high degree of structural organization by mid-gestation (Rabl, 1900; Kölliker, 1904). The extent to which the various structures are functional before birth remains unknown. However, recent studies indicate that all the adult components of visually evoked electroencephalograph responses are present at birth, though demonstrating signs of immaturity (Mattsson *et al.*, 1978).

AUTONOMIC NERVOUS SYSTEM

Rather more information is available concerning the development of the autonomic nervous system, and in particular the development of foetal competence to recognize and respond to changes in blood pressure and oxygen supply. When blood pressure is reduced by removing 15–30% of foetal blood volume, a reflex increase in heart rate and increased circulating concentrations of lysine vasopressin are observed (Forsling, Macdonald and Ellendorff, 1979; Macdonald *et al.*, 1979; Biermann *et al.*, 1979). Similarly, when the foetal supply of oxygen is reduced by lowered maternal oxygen intake (Harris and Cummings, 1973), cord clamping (Randall, 1978, 1979; Forsling, Macdonald and Ellendorff, 1979) or complete removal from the uterus of the foetus within intact placental membranes (Vesalius, 1543; see Macdonald, 1981), then the heart rate of the foetus increases; gasping movements, recognizably different from normal intra-uterine breathing movements (Randall, 1979) are also seen, and there is an increase in circulating vasopressin concentrations. The foetus becomes more responsive to these stimuli during the last month of gestation (Forsling, Macdonald and Ellendorff, 1979).

PITUITARY AND HYPOTHALAMUS

The foetal pituitary shows signs of considerable development during the second half of gestation and particularly during the last three weeks before term (Nelson, 1933; Liwska, 1975; 1978). Concentrations of the gonado-trophic hormones, LH and FSH, measured in the pituitary and general circulation rise to high levels before term (Melampy *et al.*, 1966; Elsaesser *et al.*, 1976; Colenbrander *et al.*, 1977, 1980). The amount of prolactin in the pituitary and blood tissues increases similarly (Colenbrander *et al.*, 1982).

In the foetal hypothalamus the development of neurophysin II containing elements is such that an adult-like pattern of distribution is achieved by the time of birth (Livett, Uttenthal and Hope, 1971; Ellis and Watkins, 1975). As gestation proceeds to term the amount of lysine vasopressin present in the pituitary increases (Perks and Vizsolyi, 1973) and is more readily released into the circulation following hypoxic or haemorrhagic stimuli (Forsling, Macdonald and Ellendorff, 1979; Biermann *et al.*, 1979).

The importance of the pituitary–adrenal axis in parturition is implied by a number of observations. The concentration of circulating ACTH is high at the end of gestation (Brenner, Gurtler and Reinhardt, 1978), the adrenal responsiveness to ACTH increases towards term (Dvorak, 1972) and there is a sharp increase in circulating concentrations of corticosteroids prior to term (Fèvre, 1975; Silver *et al.*, 1979). In addition, removal of the foetal pituitary, either by electrocoagulation (Bosc *et al.*, 1979) or foetal decapitation (Stryker and Dziuk, 1975; Colenbrander *et al.*, 1979) results in retarded adrenal development and delayed parturition. The precise mechanism whereby parturition is stimulated and controlled, however, remains something of a mystery in this polytocous species (Macdonald, 1979; Silver *et al.*, 1979).

The body frame

A general impression of the growth in bone and skeletal muscle may be gained from the changes seen in the shape of the foetal body as gestation progresses. The limbs grow larger, the body's length to girth ratio increases and the head becomes less rounded and takes on a more blunted conical shape (Patten, 1931).

THE SKELETON

Growth of the limb bones, seen radiographically as changes in the length of calcified diaphyses, occurs at rates which are constant for individual bones but which differ between bones (Hodges, 1953; Wenham, McDonald and Elsley, 1969; Gjesdal, 1972; Wrathall, Bailey and Hebert, 1974). However, these simple linear relationships of bone growth to age belie the complex three-dimensional processes of anabolism and catabolism occurring within the tissue itself (Patten, 1931; Vaughan, 1980). The growth observed radiographically is ossification of the bone. This is a process which involves, firstly, the degeneration of collagen cells in the diaphysis; an initial tissue calcification then follows which is in turn followed by extensive remodelling of the bone's internal structure. This latter process involves both deposition and excavation of calcified tissue (Vaughan, 1980). The actual increase in length of limb bones is produced by growth of cartilage cells which constitute the epiphyseal plates and which precede the region of ossification.

Radiography has also demonstrated that ossification centres other than those in the diaphyses appear and develop in the epiphyses during gestation. The appearance of the majority of these secondary ossification centres within the last three weeks of gestation (Wrathall, Bailey and Hebert, 1974) may be indicative of changes in the factors controlling bone growth. As mentioned earlier (p.382), cortisol, thyroxine and insulin are present in the foetal circulation during this period, as are somatomedin, calcitonin and parathyroid hormone (Phillippo, Care and Hinde, 1969; Littledike, Arnaud and Whipp, 1972; Care *et al.*, 1978; Ross *et al.*,1980; Charrier, 1980). Moreoever the latter two hormones, together with 1,25-dihydroxycholecalciferol have been shown to influence, or be influenced by, circulating concentrations of calcium in the foetus. However, the nature of the interactions between these hormones and their role in foetal bone growth remains a subject for future study.

The skeleton is the basic frame to which the skeletal muscles attach and it would seem relevant to ask whether the growth in muscle influences bone growth, and if so to what extent. Just such an approach was earlier taken by Carey (1922–23) in his analysis of the early morphogenesis of the hind limb. As indicated above, there are changes taking place in bone shape and he interpreted these to be the result, in part, of associated skeletal muscle development.

THE SKELETAL MUSCULATURE

Currently the development of foetal muscle is being studied with a view to understanding its basic biochemistry and microscopic anatomy (Swatland and Cassens, 1973; Ashmore, Addis and Doerr, 1973; Ward, 1978b). Although the presence of primary and secondary muscle fibre types has been known for some time (Schwann, 1839) it is as a consequence of recent work that we know that primary fibres form the structural framework upon which the formation and subsequent growth of secondary fibres takes place. The latter rapidly increase in number between 50 and 70 days of gestation and their histochemical differentiation into slow-contracting, fatigue-resistant Type I fibres or fast-contracting, fatigue-sensitive Type II fibres would seem to be neurally regulated (Beermann, Cassens and Hausman, 1978; Szentkuti and Cassens, 1979). After about 70 days of gestation muscle tissue grows by hypertrophy of the individual myofibrils and the primary muscle fibres assume a more developed morphology shortly before birth.

The competence of foetal limb and neck muscles to contract and relax may be seen early in the second month of gestation in response to electrical or touch stimulation (Carey, 1922–23; Macdonald, unpublished observations). Similarly, foetal breathing, gasping and sighing movements, the result of movement by the diaphragm and/or intercostal muscles are observed during gestation (Randall, 1978; 1979).

The way in which these movements develop, and the manner in which they become coordinated remains unclear. Very little appears to be known about the prenatal development of functional bone–skeletal muscle relationships (Ward, 1978a, b).

The defence of the individual

Many separate factors, acting individually or in concert, constitute the defence of the newborn piglet. It is necessary to present only one or two examples to give an impression of their range and variety.

ENERGY RESERVES

Energy reserves are required to give the piglet a measure of protection against hypoxia during the birth process and heat loss following birth; in addition, they cover the temporary break in nutrient uptake as a consequence of the move from a placental to an alimentary supply. However, the piglet is born with little or no reserves of fat (Widdowson, 1950). It relies on the glucose deposited as glycogen in the liver, heart and skeletal muscle during the last month of gestation (Padalikova, Holub and Jezkova, 1972; Randall and l'Ecuyer, 1976). Although the glycogen in the liver and heart may be mobilized by the foetus at any time during the last two or three weeks of gestation (Comline, Fowden and Silver, 1979; Randall, 1979), the largest proportion of the body's energy reserves, representing 90% of body glycogen content, is in the skeletal musculature (Macdonald, 1974; Okai *et al.*, 1978). These reserves are mobilized after birth, partly for locomotion and partly for body temperature maintenance.

AGGRESSIVENESS

Newborn piglets spend time competing aggressively with one another to establish a 'teat order' (Hartsock, Graves and Baumgardt, 1977; Fraser *et al.*, 1979), those heavier at birth tending to win more fights and successfully defend their teat. Teeth are used actively in these encounters and it is of interest to note that these weapons erupt as early as 90 days of gestation (Gjesdal, 1972) and increase in size up until birth.

IMMUNITY

Immunoresistance to infection is not transferred across the placenta during gestation in the pig (Brambell, 1970). Protection is afforded first by the non-selective absorption of colostral immunoglobulins during the 24 hours after birth, and thereafter partly by the wash of milk immunoglobulins over the piglet's gut lumen and partly as a result of the rapid activation of the piglet's own immune system (Porter, 1979).

The gastric mucosa of the stomach develops during the second half of gestation (*Figure 18.12(a)* and *(b)*) such that although the stomach contents have an acid pH (Macdonald, unpublished observations) and prochymosin may be detected about three weeks before birth, no pepsin is present until about five days after birth; this is consistent with the period of early absorption of antibodies from colostrum as pepsin has the ability to cleave immunoglobulins (Tudor, Schofield and Titchen, 1977; Foltmann *et al.*, 1981). Although the development of the intestine begins earlier than that

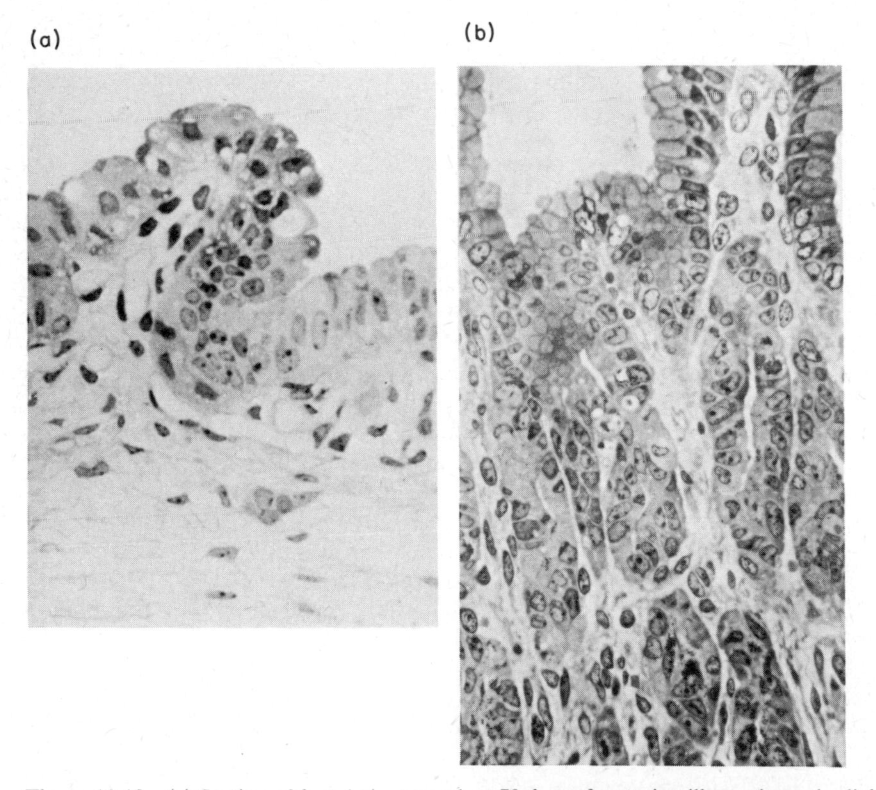

Figure 18.12 (a) Section of foetal pig stomach at 70 days of gestation illustrating only slight folding of the lining epithelium. (b) Section of foetal pig stomach at 112 days of gestation illustrating the deep invagination and glandular development of the stomach body. (Magn. ×350)

of the stomach (Lindberg and Karlsson, 1970; Hardy, Hockaday and Tapp, 1971; Karlsson, 1972) it only proceeds to a stage consistent with the passage and intact absorption of colostral gammaglobulins at birth (Lecce and Morgan, 1962; Burton and Smith, 1977)

Lymphoid differentiation occurs early in the second month of gestation (Chapman, Johnson and Cooper, 1974; Kovaru and Jaroskova, 1979) and piglets after about mid-gestation respond to injected antigens with endogenous antibody production (Binns, 1967; Bourne *et al.*, 1974). Evidence to support the view that this early immunological development may in the past have had, and may still have, functional significance was provided by a recent survey of more than eleven hundred foetal piglets (Chaniago *et al.*, 1978). Immunoglobulins representative of autologous foetal antibody production were found in five foetuses.

Conclusions

This chapter has discussed the foetal development of an assortment of those components, examples only of the many subsystems, whose competence to function at birth can drastically affect the ability of the piglet to survive. Much is known; however, there are also very large areas of ignorance. The practical relevance of such knowledge may be questioned as most piglets normally survive. The important question remains, however; why do the few fail?

A greater understanding of how and when the separate parts of the piglet's physiology and endocrinology normally achieve competence would enable us to assist and to exert control over this critical part of reproduction.

Acknowledgements

Sincere gratitude is expressed to the Meat and Livestock Commission, the Fulbright Foundation and the Deutsche Forschungsgemeinschaft for financially supporting A.A. Macdonald to carry out parts of this work.

References

ADRIAN, E.D. (1943a). Sensory areas of the brain. *Lancet* **ii**, 33–36

ADRIAN, E.D. (1943b). Afferent areas in the brain of ungulates. *Brain* **66**, 89–103

ALT, J.M., COLENBRANDER, B., MACDONALD, A.A., MAESS, B. and BIERMANN, U. (1981). Development of glomerular and tubular function in fetal and newborn pigs and their response to hypotonic saline load. In *Advances in Physiological Sciences, Vol. 11, Kidney and Body Fluids*, (L. Takács, Ed.), pp. 125–129. Oxford, Pergamon Press

ARON, M. (1922). L'evolution morphologique et fonctionelle des ilots endocrines du pancreas embryonnaire. *Archs Anat. Histol. Embryol.* **1**, 69–112

ASHMORE, C.R., ADDIS, P.B. and DOERR, L. (1973). Development of muscle fibers in the foetal pig. *J. Anim. Sci.* **36**, 1088–1093

ATINMO, T., BALDISAO, C., POND, W.G. and BARNES, R.H. (1976). Maternal protein malnutrition during gestation alone and its effect on plasma insulin levels of the pregnant pig, its foetuses and the developing offspring. *J. Nutr.* **106**, 1647–1653

BARNES, R.J., COMLINE, R.S., DOBSON, A. and SILVER, M. (1979). On the presence of a ductus venosus in the foetal pig in late gestation. *J. Devel. Physiol.* **1**, 105–110

BASKERVILLE, A. (1976). Histological and ultrastructural observations on the development of the lung of the foetal pig. *Acta Anat.* **95**, 218–233

BEERMANN, D.H., CASSENS, R.G. and HAUSMAN, G.J. (1978). A second look at fibre type differentiation in porcine skeletal muscle. *J. Anim. Sci.* **46**, 125–132

BERGELIN, I.S.S. and KARLSSON, B.W. (1975). Functional structure of the glomerular filtration barrier and the proximal tubuli in the developing foetal and neonatal pig kidney. *Anat. Embryol.* **148**, 223–234

BERTON, J.P. (1970). Augmentation du volume pulmonaire et ascites après ligature intrauterine de la trachea chez le foetus de porc miniature. *C. r. Ass. Anat.* **147**, 140–150

BIERMANN, U., FORSLING, M.L., ELLENDORFF, F. and MACDONALD, A.A. (1979). The cardiovascular responses of the chronically catheterised pig foetus to infused lysine vasopressin and to haemorrhage. *J. Physiol.* **296**, 28–29P

BINNS, R.M. (1967). Bone marrow and lymphoid cell injection of the pig fetus resulting in transplantation tolerance or immunity, and immunoglobulin production. *Nature, Lond.* **214**, 179–181

BLAND, R.D., McMILLAN, D.D. and BRESSACK, M.A. (1977). Movement of water and protein in the fetal and newborn lung. *Annls Rech. vet.* **8**, 418–427

BOSC, M., DU MESNIL DU BUISSON, F. and LOCATELLI, A. (1974). Mise en evidence d'un contrôle foetal de la parturition chez la truie. Interactions avec la fonction luteale. *C. r. hebd. Sèanc. Acad. Sci., Paris* **D 278**, 1507–1510

BOURNE, F.J., CURTIS, J., JOHNSON, R.H. and COLLINGS, D.F. (1974). Antibody formation in porcine fetuses. *Res. vet. Sci.* **16**, 223–227

BRADLEY, O.C. (1903). On the development and homology of the mammalian cerebellar fissures. *J. Anat. Physiol.* **37**, 211–240

BRADLEY, R.M. and MISTRETTA, C.M. (1975). Fetal sensory receptors. *Physiol. Rev.* **55**, 352–382

BRAMBELL, F.W.R. (1970). The transmission of passive immunity from the mother to young. In *Frontiers of Biology, Volume 18*, (A. Neuberger and E.L. Tatum, Eds.). Amsterdam, North-Holland

BRENNER, K.V., GURTLER, H. and REINHARDT, P. (1978). Zum Gehalt an ACTH im Blutplasma bei Feten und neugeborenen Ferkeln. *Endokrinologie* **71**, 154–158

BROOKS, C.C. and DAVIS, J.W. (1969). Changes in the perinatal pig. *J. Anim. Sci.* **29**, 325–329

BURTON, K.A. and SMITH, M.W. (1977). Endocytosis and immunoglobulin transport across the small intestine of the newborn pig. *J. Physiol., Lond.* **270**, 473–488

CARE, A.D., PICKARD, D.W., ROSS, R., GAREL, J.M., PAPAPOULOS, S., O'RIORDAN, J.L.H. and ROBINSON, J.S. (1978). Calcium homoeostasis in the foetus and the placental calcium pump. In *Endocrinology of Calcium Metabolism*, (D.H. Copp and R V. Talmage, Eds.), p.380. Amsterdam, Excerpta Medica

CAREY, E.J. (1922–23). Direct observations on the transformation of the mesenchyme in the thigh of the pig embryo (*Sus scrofa*) with especial reference to the genesis of the thigh muscles, of the knee and hip joints, and of the primary bone of the femur. *J. Morphol.* **37**, 1–78

CHANIAGO, T.D., WATSON, D.L., OWEN, R.A. and JOHNSON, R.H. (1978). Immunoglobulins in blood serum of foetal pigs. *Aust. vet. J.* **54**, 30–33

CHAPMAN, H.A., JOHNSON, J.S. and COOPER, M.D. (1974). Ontogeny of Peyer's patches and immunoglobulin-containing cells in the pigs. *J. Immunol.* **112**, 555–563

CHARRIER, J. (1980). Somatomedin-A (Sm-A) bioactivity in serum and amniotic fluid as related to weight in the fetal pig. *Reprod. Nutr. Dével.* **20 (1B)**, 301–310

CLEMENTS, L.P. (1938). Embryonic development of the respiratory portion of the pig's lung. *Anat. Rec.* **70**, 575–595

COLENBRANDER, B., KRUIP, Th.A.M., DIELEMAN, S.J. and WENSING, C.J.G. (1977). Changes in serum LH concentrations during normal and abnormal sexual development in the pig. *Biol. Reprod.* **17**, 506–513

COLENBRANDER, B., MACDONALD, A.A., WONG, C.C. and PARVIZI, N. (1980). Plasma TSH, T_4 and T_3 concentrations in the perinatal pig. *J. Endocr.* **85**, 38P–39P

COLENBRANDER, B., ROSSUM-KOK, C.M.J.E. van, STRAATEN, H.W.M. van, and WENSING, C.J.G. (1979). The effect of fetal decapitation on the testis and other endocrine organs in the pig. *Biol. Reprod.* **20**, 198–204

COLENBRANDER, B., MACDONALD, A.A., MEIJER, J.D., ELLENDORFF, F., VAN DE WIEL, D.F.M. and BEVERS, M.M. (1982). Prolactin in the pig fetus. *Eur. J. Obstet. Gynec. reprod. Biol.* (in press)

COMLINE, R.S., FOWDEN, A.L. and SILVER, M. (1979). Carbohydrate metabolism in the fetal pig during late gestation. *Quart. J. exp. Physiol.* **64**, 277–289

COMLINE, R.S., FOWDEN, A.L., ROBINSON, P.M. and SILVER, M. (1981). Morphological development of the endocrine tissue in the foetal pig pancreas. *J. Physiol.* **307**, 11–12P

DICKERSON, J.W.T. and DOBBING, J. (1966). Prenatal and postnatal growth and development of the central nervous system of the pig. *Proc. Roy. Soc.* **166**, 384–395

DICKERSON, J.W.T., MERAT, A. and WIDDOWSON, E.M. (1971). The effect of development on the gangliosides of pig brain. *Biochem. J.* **125**, 40–41P

DONE, J.T. and HERBERT, C.N. (1968). The growth of the cerebellum in the foetal pig. *Res. vet. Sci.* **9**, 143–148

DVORAK, M. (1972). Adrenocortical function in foetal, neonatal and young pigs. *J. Endocr.* **54**, 473–481

ELLIS, H.K. and WATKINS, W.B. (1975). Ontogeny of the pig hypothalamic neurosecretory system with particular reference to the distribution of neurophysin. *Cell Tiss. Res.* **164**, 543–557

ELSAESSER, F., ELLENDORFF, F., POMERANTZ, D.K., PARVIZI, N. and SMIDT,

D. (1976). Plasma levels of luteinizing hormone, progesterone, testosterone and 5alpha-dehydrotestosterone in male and female pigs during sexual maturation. *J. Endocr.* **68**, 347–348

EVANS, J.R., ROWE, R.D., DOWNIE, H.G. and ROWSELL, H.C. (1963). Murmurs arising from ductus arteriosus in normal newborn swine. *Circ. Res.* **12**, 85–93

FENGER, F. (1912). On the presence of active principles in the thyroid and suprarenal glands before and after birth. *J. biol. Chem.* **11**, 489–492

FEVRE, J. (1975). Corticosteroides maternels et foetaux chez la truie en fin de gestation. *C. r. hebd. Sèanc. Acad. Sci., Paris Series D* **281**, 2009–2012

FLINT, J.M. (1900). The blood vessels, angiogenesis, organogenesis, reticulum, and histology, of the adrenal. *Johns Hopkins Hosp. Rep.* **9**, 152–231

FLINT, J.M. (1906–07). The development of the lungs. *Am. J. Anat.* **6**, 1–138

FOLTMANN, B., JENSEN, A.L., LONBLAD, P., SMIDT, E. and AXELSEN, N.H. (1981). A developmental analysis of the production of chymosin and pepsin in pigs. *Comp. Biochem. Physiol.* **68B**, 9–13

FORSLING, M.L., MACDONALD, A.A. and ELLENDORFF, F. (1979). The neurohypophysial hormones. *Anim. Reprod. Sci.* **2**, 43–56

FOWDEN, A.L., COMLINE, R.S. and SILVER, M. (1981). Insulin release in the chronically catheterised sow and fetus. In *Advances in Physiological Sciences, Vol. 20: Advances in Animal and Comparative Physiology*, (G. Pethes and V.L. Frenyo, Eds.), Oxford, Pergamon Press

FRANCE, V.M. (1976). Active sodium uptake by the skin of foetal sheep and pigs. *J. Physiol.* **258**, 377–392

FRASER, A.F., NAGARATNAM, V. and CALLICOTT, R.M. (1971). The comprehensive use of Doppler ultrasound in farm animal reproduction. *Vet. Rec.* **88**, 202–205

FRASER, D., THOMPSON, B.K., FERGUSON, D.K. and DARROCH, R.L. (1979). The "teat order" of suckling pigs. 3. Relation to competition within litters. *J. agric. Sci., Camb.* **92**, 257–261

GJESDAL, F. (1972). Age determination of swine foetuses. *Acta vet. scand., Suppl.* **40**, 1–29

GOOTMAN, P.M., BUCKLEY, N.M. and GOOTMAN, N. (1978). Postnatal maturation of the central neural cardiovascular regulatory system. In *Fetal and Newborn Cardiovascular Physiology, Vol. 1: Developmental Aspects*, (L.D. Longo and D.D. Reneau, Eds.), pp. 93–152. London, Garland Stpm.

GOOTMAN, P.M. BUCKLEY, N.M. and GOOTMAN, N. (1979). Postnatal maturation of neural control of the circulation. *Rev. Perinatal Med.* **3**, 1–72

GOOTMAN, N., GOOTMAN, P.M., BUCKLEY, N.M., COHEN, M.I., LEVINE, M. and SPIELBERG, R. (1972). Central vasomotor regulation in the newborn piglet *Sus scrofa*. *Am. J. Physiol.* **222**, 994–999

GOTTLIEB, G. (1971). Ontogenesis of sensory function in birds and mammals. In *The Biopsychology of Development*, (E. Tobach, L.R. Aronson and E. Shaw, Eds.), pp. 67–128. New York, Academic Press

GUYTON, A.C., COLEMAN, T.G. and GRANGER, H.J. (1972). Circulation: overall regulation. *Ann. Rev. Physiol.* **34**, 13–46

HAM, A.W. and BALDWIN, K.W. (1941). A histological study of the development of the lung with particular reference to the nature of alveoli. *Anat. Rec.* **81**, 363–379

HARDESTY, I. (1915). On the proportions, development and attachment of the tectorial membrane. *Am. J. Anat.* **18**, 1–73

HARDY, R.N., HOCKADAY, A.R. and TAPP, R.L. (1971). Observations on the structure of the small intestine in foetal, neonatal and suckling pigs. *Phil. Trans. Roy. Soc., Lond.* **259B**, 517–531

HARRIS, W.H. and CUMMINGS, J.N. (1973). Maternal and fetal responses to varying levels of oxygen intake in swine. *J. appl. Physiol.* **34**, 584–589

HARTSOCK, T.G., GRAVES, H.B. and BAUMGARDT, B.R. (1977). Agonistic behaviour and the nursing order in suckling piglets: relationships with survival growth and body composition. *J. Anim. Sci.* **44**, 320–330

HILL, E.C. (1905). On the first appearance of the renal artery, and the relative development of the kidneys and Wolffian bodies in pig embryos. *Johns Hopkins Hosp. Bull.* **16**, 60–64

HODGES, P.C. (1953). Ossification in the fetal pig. *Anat. Rec.* **116**, 315–325

JOPPICH, R., KIEMANN, U., MAYER, G. and HABERLE, D. (1979). Effect of antidiuretic hormone upon urinary concentrating ability and medullary c-AMP formation in neonatal piglets. *Pediat. Res.* **13**, 884–888

KAMAN, J. (1968a). Der Umbau des Ductus Venosus des Schweines. 1. Pränatales Stadium. *Anat. Anz.* **122**, 252–266

KAMAN, J. (1968b). Der Umbau des Ductus Venosus beim Schwein. II. Postnatales Stadium. *Anat. Anz.* **122**, 476–486

KAPLAN, A. and FRIEDMAN, M. (1943). Studies concerning the site of renin formation in the kidney. III. The apparent site of renin formation in the tubules of the mesonephros and metanephros of the hog fetus. *J. exp. Med.* **76**, 307–316

KARLSSON, B.W. (1972). Ultrastructure of the small intestine epithelium of the developing pig foetus. *Z. Anat. EntwGesch.* **135**, 253–264

KATZNELSON, Z.S. (1965). Zur Fruhentwicklung der Nebenniere des Schweines. *Z. mikrosk.-anat. Forsch.* **73**, 187–199

KATZNELSON, Z.S. (1966). Späthistogenese der Nebenniere des Schweines. *Z. mikrosk.-anat. Forsch.* **74**, 193–208

KAZIMIERCZAK, J. (1970). Histochemical observations of the developing glomerulus and juxtaglomerular apparatus. *Acta path. microbiol. scand.* **78A**, 401–413

KELLOGG, H.B. (1928). The course of the blood flow through the fetal mammalian heart. *Am. J. Anat.* **42**, 443–465

KÖLLIKER, A. (1904). Die Entwicklung und Bedeutung des Glaskörpers. *Z. wiss. Zool.* **76**, 1–25

KONDA, N., DYER, R.G., BRUHN, T., MACDONALD, A.A. and ELLENDORFF, F. (1979). A method for recording single unit activity from the brains of foetal pigs *in utero*. *J. Neurosci. Meth.* **1**, 289–300

KORNGUTH, S.E., FLANGAS, A.L., GEISON, R.L. and SCOTT, G. (1972). Morphology, isopycnic density and lipid content of synaptic complexes isolated from developing cerebellums and different brain regions. *Brain Res.* **37**, 53–68

KOVARU, F. and JAROSKOVA, L. (1979). Development of Erosette formation in ontogeny of pigs. *Folia biol., Praha* **25**, 399–400

KRAELING, R.R., RAMPACEK, G.B., CAMPION, D.R. and RICHARDSON, R.L.

(1978). Longissimus muscle and plasma enzymes and metabolites in fetally decapitated pigs. *Growth* **42**, 457–468

KUNSKA, A. (1971a). Histological studies on the development of the kidney (metanephros) in embryos of the domestic pig. I. *Folia morph.* **30**. 1–20

KUNSKA, A. (1971b). Histochemical study on the development of the kidney (metanephros) in embryos of the domestic pig. II. *Folia morph.* **30**, 259–270

LARSELL, O. (1954). The development of the cerebellum of the pig. *Anat. Rec.* **118**, 73–107

LECCE, J.G. and MORGAN, D.O. (1962). Effect of dietary regimen on cessation of intestical absorption of large molecules (closure) in the neonatal pig and lamb. *J. Nutr.* **78**, 263–268

LINDBERG, T. and KARLSOON, B.W. (1970). Changes in intestinal dipeptidase activities during fetal and neonatal development of the pig as related to the ultrastructure of mucosal cells. *Gastroenterology* **59**, 247–256

LITTLEDIKE, E.T., ARNAUD, C.D. and WHIPP, S.C. (1972). Calcitonin secretion in ovine, porcine and bovine fetuses. *Proc. Soc. exp. Biol. Med.* **139**, 428–433

LIVETT, B.G., UTTENTHAL, L.O. and HOPE, D.B. (1971). Localisation of neurophysin-II in the hypothalamo–neurohypophysial system of the pig by immunofluorescence histochemistry. *Phil. Trans. Roy. Soc. Lond.* **261B**, 371–378

LIWSKA, J. (1975). Development of the adenohypophysis in the embryo of the domestic pig. *Folia morph.* **34**, 211–217

LIWSKA, J. (1978). Ultrastructure of the adenohypophysis in the domestic pig (*Sus scrofa domestica*). Part 1: Cells of the pars anterior. *Folia histochem. cytochem.* **16**, 307–314

LOHSE, J.K. and FIRST, N.L. (1979). Development of the porcine fetal adrenal in late gestation: steroidogenesis and histology. *Biol. Reprod.* **20**, *Suppl.* **1**, 126A

MACDONALD, A.A. (1971). The foetal and postnatal growth of *Sus domesticus* L. with a study of the foetal-maternal placental vascularisation. Thesis. Glasgow University

MACDONALD, A.A. (1974). Studies into foetal and neonatal development of the pig (*Sus scrofa* L.). Thesis. Edinburgh University

MACDONALD, A.A. (1979). Patterns of endocrine change in the pig foetus. *Anim. Reprod. Sci.* **2**, 289–304

MACDONALD, A.A. (1981). Studies on the anatomy and physiology of the pig fetus and placenta: an historical review. In *Advances in Physiological Sciences, Vol. 21: History of Physiology*, (E. Schultheisz, Ed.), pp. 53–60. Oxford, Pergamon Press

MACDONALD, A.A. and COLENBRANDER, B. (1981). Cardiovascular responses of the fetal pig to autonomic stimulation. In *Advances in Physiological Sciences, Vol. 8: Cardiovascular physiology: Heart, Peripheral circulation and Methodology*, (A.G.B. Kovach, E. Monos and G. Rubany, Eds.), pp. 319–325. Oxford, Pergamon Press

MACDONALD, A.A., RUDOLPH, A.M. and HEYMANN, M.A. (1980). The pig placenta: vascular anatomy and blood flow. *Proc. Int. Symp. Primate Non-Primate Placental Transfer, Rotterdam* (Abstract)

MACDONALD, A.A., COLENBRANDER, B., MEIJER, J.C., POOT, P. and WENSING, C.J.G. (1981a). Development of innervation to the heart of the pig fetus. *Acta morph. neerl. scand.* **19**, 257–258

MACDONALD, A.A., FORLING, M.L., WILLIAMS, H. and ELLENDORFF, F. (1979). Plasma vasopressin and oxytocin concentrations in the conscious pig foetus: response to haemorrhage. *J. Endocr.* **81**, 124P–125P

MACDONALD, A.A., LLANOS, A.J., HEYMANN, M.A. and RUDOLPH, A.M. (1981b). Cardiovascular responsiveness of the pig fetus to autonomic blockade. *Pfluger's Arch. ges. Physiol.* **390**, 262–264

McCANCE, R.A. and STANIER, M.W. (1960). The function of the metanephros of foetal rabbits and pigs. *J. Physiol.* **151**, 479–483

MAJSTRUK-MAJEWSKA, T. (1966). Histochemistry of some lipids in the course of myelinisation of the spinal cord in domestic pig. *Annls Univ. Marie Curie-Sklodowska* **21 DD**, 1–18

MARSHALL, A.E. and BREAZILE, J.E. (1974a). Localization of cardiovascular centers in myelencephalon of newborn and older pigs. *Am. J. vet. Res.* **35**, 223–229

MARSHALL, A.E. and BREAZILE, J.E. (1974b). Evidence for maturation of myelencephalic cardiovascular control in the postnatal pig. *Am. J. vet. Res.* **35**, 231–236

MATTSSON, J.L., FRY, W.N., BOWARD, C.A. and MILLER, E. (1978). Maturation of the visual evoked response in newborn miniature pigs. *Am. J. vet. Res.* **39**, 1279–1281

MEBAN, C. (1980). Surface elastic properties of surfactant from the lungs of neonatal pigs. *Biol. Neonate* **37**, 308–312

MELAMPY, R.M. HENRICKS, D.M., ANDERSON, L.L., CHEN, C.L. and SCHULTE, J.R. (1966). Pituitary follicle-stimulating hormone and luteinizing hormone concentrations in pregnant and lactating pigs. *Endocrinology* **78**, 801–804

MILLER, J.A. and MILLER, F.S. (1965). Studies on prevention of brain damage in asphyxia. *Dev. Med. Child Neurol.* **7**, 607–619

MOODY, R.O. (1906). Some features of the histogenesis of the thyroid gland in the pig. *Anat. Rec.* **4**, 429–452

NELSON, W.O. (1933). Studies on the anterior hypophysis. I. The development of the hypophysis in the pig (*Sus scrofa*). II. The cytological differentiation in the anterior hypophysis of the foetal pig. *Am. J. Anat.* **52**, 307–332

NICHOLAS, T.E., JOHNSON, R.G., LUGG, M.A. and KIM, P.A. (1978). Pulmonary phospholipid biosynthesis and the ability of the fetal rabbit lung to reduce cortisone to cortisol during the final ten days of gestation. *Life Sci.* **22**, 1517–1524

OKAI, D.B., WYLLIE, D., AHERNE, F.X. and EWAN, R.C. (1978). Glycogen reserves in the fetal and newborn pig. *J. Anim. Sci.* **46**, 391–401

OLSON, E.B. (1979). Role of glucocorticoids in lung maturation. *J. Anim. Sci.* **49**, 225–238

OSBORN, J.L. (1979). Immaturity of renal function in newborn pigs: factors affecting renal hemodynamics, sodium excretion and plasma renin activity. PhD Thesis. Michigan State University

PADALIKOVA, D., HOLUB, A. and JEZKOVA, D. (1972). Glycogen in the placenta and pig foetus tissues in the last third of intra-uterine life. *Vet. Med., Praha* **17**, 649–656

PATTEN, B.M. (1931). *The Embryology of the Pig.* Philadelphia, Blakiston

PATTERSON, D.S.P., DONE, J.T., FOULKES, J.A. and SWEASY, D. (1976). Neurochemistry of the spinal cord in congenital tremor of piglets (type

A11) a spinal dysmyelinogenesis of infectious origin. *J. Neurochem.* **26**, 481–485

PERKS, A.M. and VIZSOLYI (1973). Studies of the neurohypophysis in foetal mammals. In *Foetal and Neonatal Physiology*, (R.S. Comline, K.W. Cross, G.S. Dawes and P.W. Nathanielsz, Eds.), pp. 430–438. Cambridge, University Press

PERRY, J.S. and STANIER, M.W. (1962). The rate of flow of urine of foetal pigs. *J. Physiol.* **161**, 344–350

PHILLIPPO, M., CARE, A.D. and HINDE, F.R. (1969). The effect of thyrocalcitonin in neonatal animals. *J. Endocr.* **43**, XV–XVI

POHLMAN, A.G. (1909). The course of the blood through the heart of the fetal mammal, with a note on the reptilian and amphibian circulations. *Anat. Rec.* **3**, 75–109

POMEROY, R.W. (1960). Infertility and neonatal mortality in the sow. III. Neonatal mortality and foetal development. *J. agric. Sci., Camb.* **54**, 31–56

PORTER, P. (1979). Structural and functional characteristics of immunoglobulins of the common domestic species. *Adv. vet. Sci. comp. Med.* **23**, 1–21

PRENTISS, C.W. (1913). On the development of the mebrana tectoria with reference to its structure and attachments. *Am. J. Anat.* **14**, 425–459

RABL, C. (1900). Uber den Bau und die Entwicklung der Linse. III. Die Linse der Säugetiere. Ruckblick und Schluss. *Z. wiss. Zool.* **67**, 1–138

RANDALL, G.C.B. (1978). Perinatal mortality: some problems of adaptation at birth. *Adv. vet. Sci. comp. Med.* **22**, 53–81

RANDALL, G.C.B. (1979). Studies on the effect of acute asphyxia on the fetal pig *in utero*. *Biol. Neonate* **36**, 63–69

RANDALL, G.C.B. and l'ECUYER, C. (1976). Tissue glycogen and blood glucose and fructose level in the pig fetus during the second half of gestation. *Biol. Neonate* **28**, 74–82

ROSS, R., CARE, A.D., PICKARD, D.W., GAREL. J.M. and WEATHERLEY (1980). Placental transfer of calcium in the pig. *J. Endocr.* **85**, 53P–54P

ROWE, R.D., SINCLAIR, J.D., KERR, A.R. and GAGE, P.W. (1964). Duct flow and mitral regurgitation during changes of oxygenation in newborn swine. *J. appl. Physiol.* **29**, 1157–1163

RUFER, R. and SPITZER, H.L. (1974). Liquid ventilation in the respiratory distress syndrome. *Chest* **66**, *Suppl.* 29S–30S

SCHWANN, T. (1839). *Mikroskopische Untersuchungen über die Uebereinstimmung in der Struktur und dem Wachstume der Tiere und Pflanzen.*

SEDOVA, E.V. (1974). The fetal cortex and X-zone in the adrenal gland of some mammals. *Archs Anat. Histol. Embryol.* **66**, 77–82

SILVER, M., FOWDEN, A.L., COMLINE, R.S., CLOVER, L. and MITCHELL, M.D. (1979). Prostaglandins in the foetal pig and prepartum endocrine changes in mother and foetus. *Anim. Reprod. Sci.* **2**, 305–322

STANTON, H.C. and WOO, S.K. (1978). Development of adrenal medullary function in swine. *Am. J. Physiol.* **234**, E137–E145

STANTON, H.C., CORNEJO, R.A., MERSMANN, H.J., BROWN, L.J. AND MUELLER, R.L. (1975). Ontogenesis of monoamine oxidase and catechol-O-methyl transferase in various tissues of domestic swine. *Archs int. Pharmacodyn. Ther.* **213**, 128–144

STARLING, M.B., NEUTZE, J.M., ELLIOTT, R.L. and ELLIOT, R.B. (1976). Studies on the effects of prostaglandins E_1, E_2, A_1 and A_2 on the ductus arteriosus of swine *in vitro* using cineangiography. *Prostaglandins* **12**, 335

STARLING, M.B., NEUTZE, J.M., ELLIOTT, R.L., TAYLOR, I.M.M. and ELLIOTT, R.B. (1978). The effects of some methyl prostaglandin derivatives on the ductus arteriosus of swine *in vivo. Prostaglandins Med.* **1**, 267–281

STRYKER, J.L. AND DZIUK, P.J. (1975). Effects of fetal decapitation on fetal development parturition and lactation in pigs. *J. Anim. Sci.* **40**, 282–287

STUDZINSKI, T., BOBOWIEC, R. and RYBKA, A. (1976). Histomorphological and functional development of thyroid gland in the pig during embryonic and postnatal state. *Annls Univ. Marie Curie-Sklodowska* **31 DD**, 119–128

SWATLAND, H.J. and CASSENS, R.G. (1973). Prenatal development histochemistry and innervation of porcine muscle. *J. Anim. Sci.* **36**, 343–354

SZALAY, K.S. and GYEVAI, A. (1967). Renin production by tissue cultures of renal cortex. *Life Sci.* **6**, 925–928

SZENTKUTI, L. and CASSENS, R.G. (1979). Motor innervation of myofiber types in porcine skeletal muscle. *J. Anim. Sci.* **49**, 693–700

TOO, K., KAWATA, K., FUKUI, Y., SATO, K., KAGOTA, K. and KAWABE, K. (1974). Studies on pregnancy diagnosis in domestic animals by an ultrasonic Doppler method. 1. *Jap. J. vet. Res.* **22**, 61–71

TUDOR, E.M., SCHOFIELD, G.C. and TITCHEN, D.A. (1977). Structural and functional development of the gastric parietal cell population in the newborn pig. *Annls Recherch. Vet.* **8**, 450–459

VAN NIE, C.J., VERSPRILLE, A., GIESBERTS, M.A.H., RIEDSTRA, J.W., BENEKEN, J.E.W. and ROHMER, J. (1970). Functional behaviour of the Foramen ovale in the newborn piglet. *Pfluger's Arch. ges. Physiol.* **314**, 154

VAUGHAN, J. (1980). Bone growth and modelling. In *Growth in Animals.* (T.L.J. Lawrence, Ed.), pp. 83–99. London, Butterworths

VERHOFSTAD, A.A.J., STEINBUSCH, H.W.M., JOOSTEN, H.W.J., COLENBRANDER, B. and MACDONALD, A.A. (1981). Development of the noradrenaline and adrenaline-storing cells in the adrenal medulla and its control by the adrenal cortex. *Acta morph. neerl-scand.* **19**, 330

VERSPRILLE, A., SOETEMAN, D.W., STULE, J. and VAN NIE, C.J. (1970). Flow resistance of the foramen ovale in newborn pigs. *Pfluger's Arch. ges Physiol.* **318**, 269

VESALIUS, A. (1543). *De humani corporis fabrica.* Oporinus, Baseleae

WARD, P.S. (1978a). The splayleg syndrome in newborn pigs. A review. Part 1. *Vet. Bull.* **48**, 279–295

WARD, P.S. (1978b). The splayleg syndrome in newborn pigs. A review. Part 2. *Vet. Bull.* **48**, 381–399

WELENTO, J. (1960). Statistical investigations on the correlation between the weight of the brain and the weight and length of the foetuses of the Pulawska pig. *Annls Univ. Marie Curie-Sklodowska* **15**, 81–84

WELENTO, J. (1961). The development of the cerebrum trunk in the pig. *Annls Univ. Marie Curie-Sklodowska* **16 DD**, 87–101

WENHAM, G., MCDONALD, I. and ELSLEY, F.W.H. (1969). A radio-graphic study of the development of the skeleton of the foetal pig. *J. agric. Sci., Camb.* **72**, 123–130

WENSING, C.J.G. (1964). The conductive system and its nervous component in the pig's heart. PhD Thesis. Utrecht

WEYMANN, M.F. (1922–23). The beginning and development of function in the suprarenal medulla of pig embryos. *Anat. Rec.* **24**, 299–313

WHITEHEAD, R.H. (1903). The histogenesis of the adrenal in the pig. *Am. J. Anat.* **2**, 349–360

WIDDOWSON, E.M. (1950). Chemical composition of newly born mammals. *Nature, Lond.* **166**, 626–628

WIESEL, J. (1901). Uber die Entwicklung der Nebenniere des Schweines besonders der Marksubstanz. *Anat. Hefte* **16 S**, 117–148

WOOLSEY, C.N. and FAIRMAN, D. (1946). Contralateral, ipsilateral, and bilateral representation of cutaneous receptors in somatic areas I and II of the cerebral cortex of pig, sheep and other mammals. *Surgery* **19**, 684–702

WRATHALL, A.E., BAILEY, J. and HEBERT, C.N. (1974). A radiographic study of development of the appendicular skeleton in the fetal pig. *Res. vet. Sci.* **17**, 154–168

ZIOLO, I. (1965). Myelinization of nerve fibres of pig spinal cord. *Acta Anat.* **61**, 297–320

19

PLACENTAL STEROID METABOLISM IN LATE PREGNANCY

V.A. CRAIG
A.R.C. Institute of Animal Physiology, Babraham, Cambridge, UK

There are considerable differences between species in the contribution of the placenta to the concentration of steroids in the maternal circulation. This is particularly evident in the placental contribution to the maternal plasma concentration of progesterone, the hormone necessary for the maintenance of pregnancy. Some species depend predominantly on progesterone synthesized by the placenta for the maintenance of pregnancy, as in women (throughout the majority of gestation) and sheep (for the final two-thirds of gestation), while in other species, as in the goat and the pig, the placental contribution to maternal progesterone is relatively low and pregnancy is maintained by luteal progesterone secretion.

Changes in placental steroid metabolism have been implicated in the sequence of hormonal events in the maternal circulation which leads to the onset of labour in the goat and the sheep. In sheep these endocrine changes have been described in considerable detail (Anderson, Flint and Turnbull, 1975; Steele, Flint and Turnbull, 1976; Flint and Ricketts, 1979). However it is only in recent years that certain parallel mechanisms have become apparent in the goat. In this species, parturition occurs as a direct consequence of the cessation of luteal function which is associated with the release of uterine prostaglandin $F_{2\alpha}$ and which results in the removal of the 'progesterone block'. As in sheep, an early step in the chain of events leading to the release of prostaglandin $F_{2\alpha}$ is a rise in the concentration of cortisol in the foetal circulation which occurs over the last 4–10 days of pregnancy (Currie and Thorburn, 1977). This has been shown to induce the activity of the placental steroid metabolizing enzyme 17α-hydroxylase (Flint *et al.*, 1978) leading to the formation of increased amounts of substrate for the synthesis of oestrogens. Of the oestrogens thus formed, it has been demonstrated that oestradiol-17β will induce the release of uterine prostaglandin $F_{2\alpha}$ into the uterine vein (Currie, Cox and Thorburn, 1976) thus providing an important link between the increase in foetal plasma cortisol and regression of the corpus luteum of pregnancy.

The pig is another species in which the corpora lutea are the major source of maternal progesterone and certain similarities exist between the pig and the goat in the hormonal changes occurring in both the foetal and the maternal circulation prior to parturition. In the pig, the peripheral plasma concentration of progesterone remains steady at approximately 10 ng/ml until about 48 hours prior to parturition when it declines rapidly

(Ash *et al.*, 1973; Robertson and King, 1974; Ash and Heap, 1975; Baldwin and Stabenfeldt, 1975). The plasma concentrations of oestrone (*Figure 19.1*) and oestradiol-17β start to rise on or about 108 days of gestation and then fall rapidly once delivery is completed. The fall in the plasma concentration of progesterone is also associated with the release of uterine prostaglandin $F_{2\alpha}$ into the maternal circualtion (Nara and First, 1977; First and Bosc, 1979). Evidence has been obtained which suggests that increased foetal cortisol secretion presages the rise of oestrone and oestradiol-17β

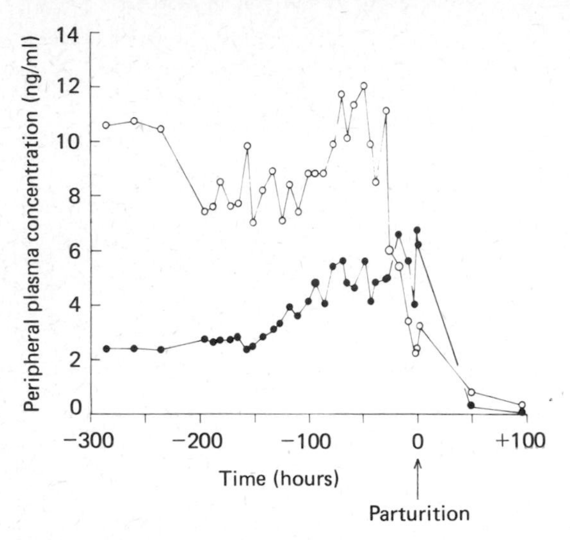

Figure 19.1 Changes in the peripheral plasma concentration of progesterone (○——○) and oestrone (●——●) over the last two weeks of gestation. The data were obtained from two animals and the peripheral plasma steroid concentration is expressed in hours relative to the delivery of the first piglet

seen in the maternal plasma over the last week of gestation; foetal cortisol rises concurrently with, or even precedes, that of oestrone and oestradiol-17β, and dexamethasone or $ACTH_{1-24}$ administered to foetuses after 100 days of gestation will induce premature delivery accompanied by the expected changes in maternal plasma progesterone and oestrogens (North, Hauser and First, 1973; Bosc, 1973).

Although there is evidence to support the view that the porcine placenta synthesizes steroids, the possibility that the endometrium also contributes to the production of steroids by the gravid uterus should not be ignored. In order to provide a more comprehensive picture of steroid synthesis by the uterus in late pregnancy, the metabolism of steroids by both the placenta and endometrium will be discussed.

Uterine production of steroids *in vivo*

Studies of the uterine secretion of progesterone and oestrogens during late gestation in the pig are relatively few although there is evidence that the umbilical venous concentration of progesterone is greater than that of the umbilical artery, which is consistent with intrauterine progesterone production (Godke and Day, 1973; Barnes, Comline and Silver, 1974; Silver *et al.*, 1979; MacDonald *et al.*, 1980). Evidence for an extra-ovarian source of oestrogens throughout gestation has been achieved by ovariectomizing sows in early pregnancy (the pregnancy being maintained by daily intramuscular injections of 300 mg of progesterone) and determining the amount of oestrogens excreted into the urine (Fèvre, Léglise and Rombauts, 1968). By using this technique it was found that the excretion of urinary oestrogens by ovariectomized animals did not differ from intact animals throughout the course of gestation.

A modification of this method has been utilized in the present study to determine the output of steroids by the uterus in late pregnancy. Four gilts were ovariectomized at 98 days of gestation and catheters were inserted into the uterine artery and vein. Pregnancy was maintained by intramuscular injection of 10 mg of medroxyprogesterone acetate every alternate day until 115 days post coitum (one day after the expected day of delivery) when the animals were killed. Daily paired samples were taken from the uterine artery and vein and the plasma concentration of progesterone, 17α-hydroxyprogesterone, androstenedione and unconjugated oestrogens determined using specific radioimmunoassays. Medroxyprogesterone acetate did not cross-react in any of the assays used.

A positive venous — arterial (V — A) difference was obtained for all the steroids assayed. There was an apparent increase in uterine progesterone

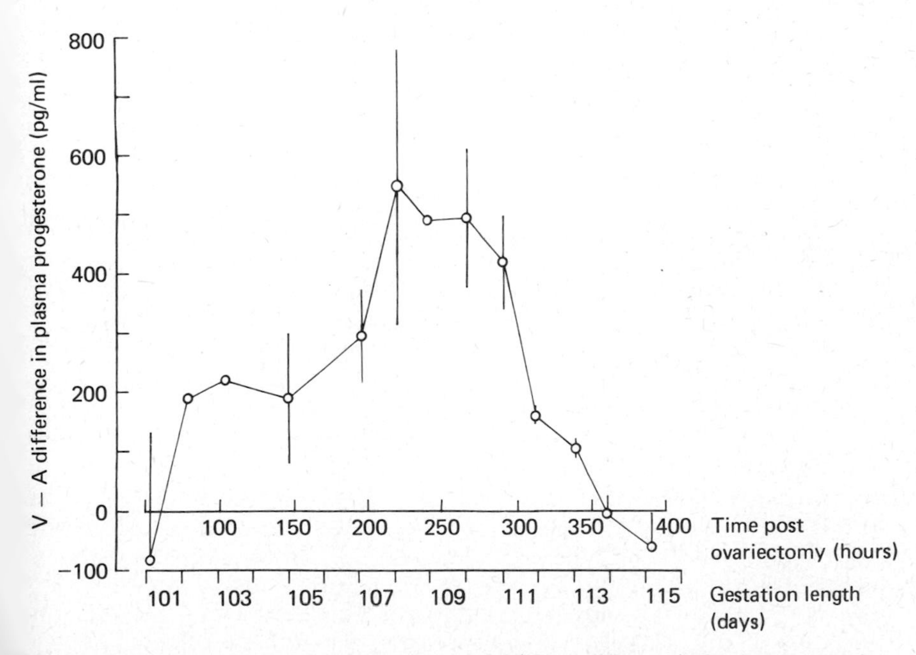

Figure 19.2 Uterine venous minus uterine arterial (V – A difference) plasma concentration of progesterone for gilts ovariectomized at 98 days post coitum in which pregnancy was maintained with medroxyprogesterone acetate. Each point represents the mean ±S.E.M. V – A difference expressed in pg/ml for up to four animals and is plotted against both the time in hours after ovariectomy and the length of gestation in days

production between 104 days and 108 days post coitum with a peak V — A difference of 500 pg/ml (*Figure 19.2*). Thereafter the venous concentration fell until, by the expected day of delivery (114 days post coitum), the V — A difference had reached zero. The results obtained also demonstrated an increase in the output of 17α-hydroxyprogesterone and androstenedione. In the case of androstenedione, this occurred from 108–110 days post coitum with a peak V — A difference of 115 pg/ml. In these animals a rise in the plasma concentration of unconjugated oestrogens occurred which was comparable to that obtained in intact animals thus indicating that the placenta or endometrium (or a combination of the two) is the major source of maternal circulating oestrogens in late pregnancy.

Placental and endometrial metabolism of steroids in the pig; studies *in vitro*

Histochemical evidence has been obtained for the presence of $\Delta^5$3β- and 17β-hydroxysteroid dehydrogenases in placental tissue from the fourth week of gestation until term (Christie, 1968; Dufour and Raeside, 1969). Earlier studies by Bloch and Newman (1966) which demonstrated the conversion of dehydroepiandrosterone to androstenedione by porcine placental tissue obtained at the end of gestation also provide evidence for the presence of an active $\Delta^5$3β-hydroxysteroid dehydrogenase. An *in vitro* study of placental tissue taken in late pregnancy (112 days post coitum) revealed the presence of an active aromatase complex (Ainsworth and Ryan, 1966). Placental preparations were shown to convert [7α-^{3}H]dehydroepiandrosterone and [4-^{14}C]androstenedione to oestrogens with oestrone being the major oestrogen formed. More recently, Choong and Raeside (1974) reported the presence of high concentrations of unconjugated and conjugated oestrogens in placental tissue and suggested that the placenta was the site of synthesis. High concentrations of unconjugated oestrogens have also been found in allantoic and amniotic fluid; Knight *et al.* (1977) showed that from 60–100 days the concentration of oestrone present in allantoic fluid increased from 0.9 ng/ml to 537.7 ng/ml, which greatly exceeds the concentration of oestrone in the maternal circulation and again indicates that the placenta is the probable site of synthesis. Although Ainsworth and Ryan (1966) demonstrated that the porcine placenta could utilize androgen substrates for oestrogen synthesis, no synthesis of oestrogens from either pregnenolone or progesterone could be detected by these authors. From these results it was concluded that the porcine placenta did not possess all the enzymes of the Δ^4 pathway, in particular C-17,20-lyase, necessary for the conversion of pregnenolone to oestrogens. An outline of the enzymes of the Δ^4 pathway which have been shown to be present in late pregnancy is given in *Figure 19.3*.

In vitro studies have demonstrated that both blastocyst (Gadsby and Heap, 1980) and endometrial tissue (Dueben *et al.*, 1980) taken in early pregnancy are capable of synthesizing oestrogens from the C-21-precursors pregnenolone or progesterone respectively. However, relatively little is known about the metabolism of steroids by the placenta and endometrium in late gestation in the pig. In view of the accumulating evidence for the

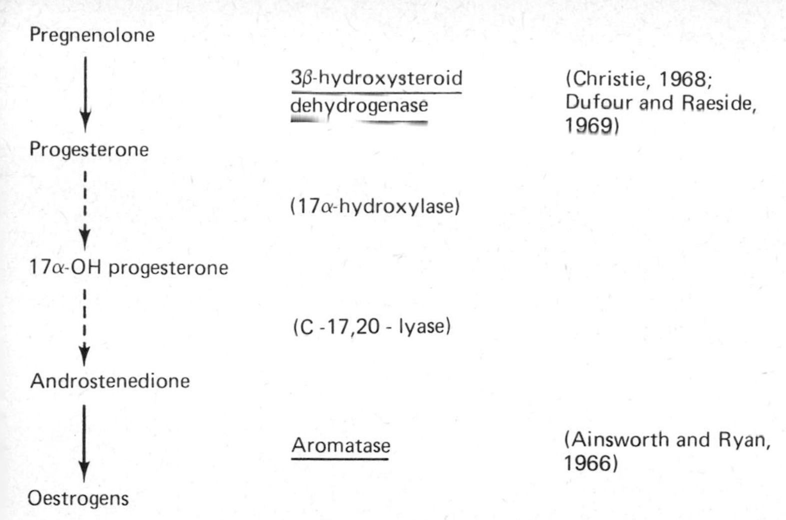

Figure 19.3 The Δ^4-3-keto pathway from pregnenolone to oestrogens. Enzymes whose presence has been confirmed by histochemical or *in vitro* incubation techniques are underlined.

placental synthesis of oestrogens, it was decided to re-examine the possibility that these tissues might be capable of synthesizing oestrogens from C-21 steroids. Furthermore, the metabolism of steroids by the endometrium in late pregnancy has not been studied extensively although there is some histochemical evidence that endometrial tissue possesses both $\Delta^5$3β- and 17β-hydroxysteroid dehydrogenase activity (Christie, 1968; Dufour and Raeside, 1969). The enzyme $\Delta^5$3β-hydroxysteroid dehydrogenase occupies a key position in the synthesis of steroid hormones and its presence in endometrial tissue gives rise to the possibility that the endometrium may be able to synthesize progesterone from its immediate precursor, pregnenolone.

To determine which enzymes of the Δ^4 pathway were present in the placenta and endometrium in late pregnancy, samples of these tissues were taken from gilts slaughtered at 100–112 days post coitum. Placental tissue obtained after both prostaglandin $F_{2\alpha}$-induced and spontaneous delivery

Table 19.1 INTERMEDIATES OF THE Δ^4 PATHWAY ISOLATED FROM INCUBATIONS OF PLACENTAL OR ENDOMETRIAL TISSUE OBTAINED AT 100 DAYS POST COITUM

Substrate	*Product*	*Enzyme*	*Tissue*	
			Placenta	*Endometrium*
Pregnenolone	progesterone	3β-hydroxysteroid dehydrogenase	√	√
Progesterone	17α-hydroxy progesterone	17α-hydroxylase	√	×
17α-hydroxy progesterone	androstenedione	C-17,20-lyase	√	×
Androstenedione	oestrone	aromatase	√	√

Tissues incubated with appropriate tritium-labelled precursor steroid in presence of excess co-factor.
√ conversion demonstrated; × conversion undetectable.

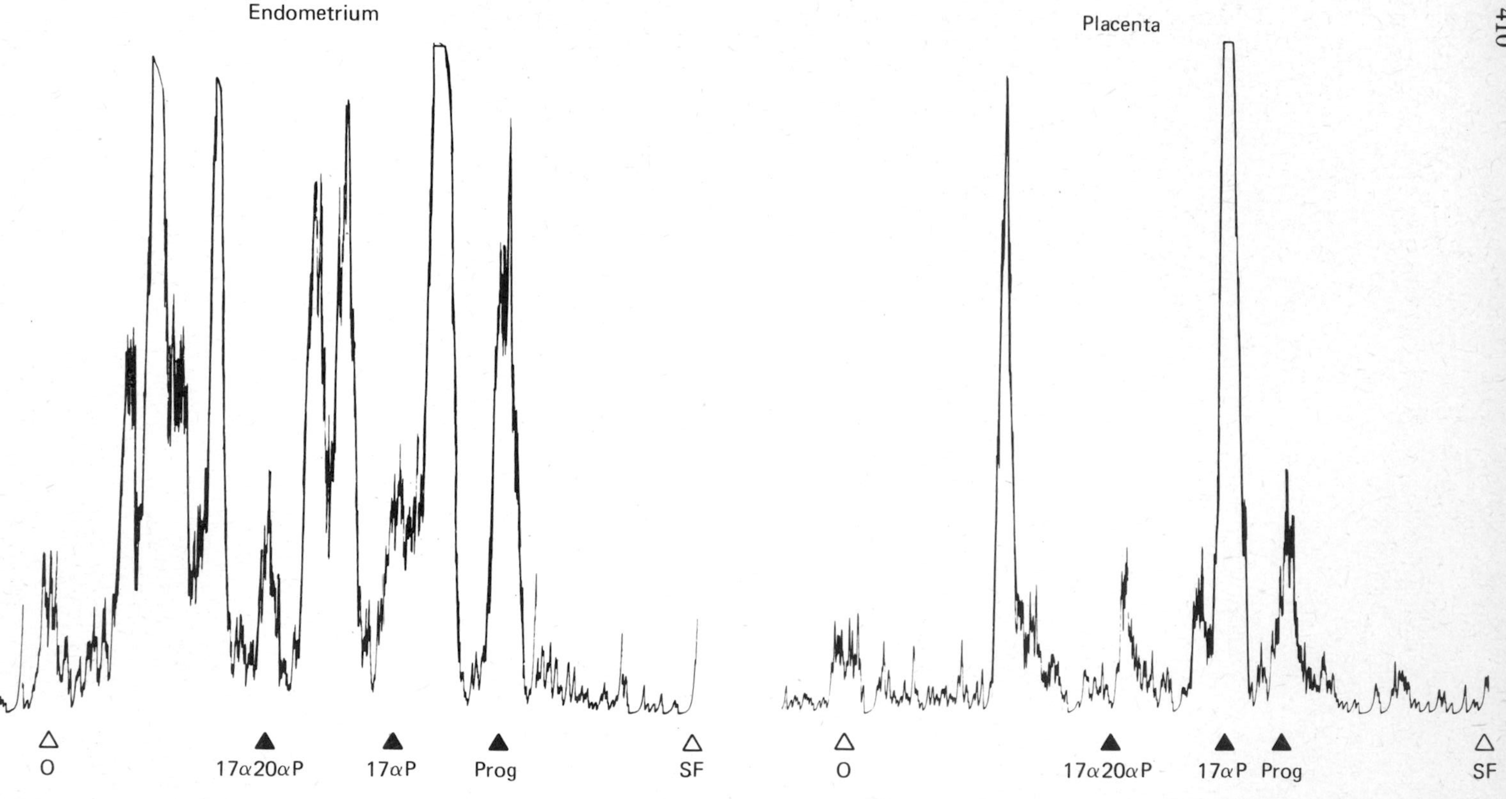

Figure 19.4 Radiochromatogram scan illustrating the metabolism of [1α,2α-^{3}H] progesterone (in the presence of excess NADPH) by homogenates of placental and endometrial tissue obtained at 100 days post coitum. Open triangles indicate the position of the origin (O) and solvent front (SF). Closed triangles indicate the position of the authentic marker steroids, progesterone (Prog), 17α-hydroxyprogesterone (17αP) and 17α,20α-dihydroprogesterone (17α20αP).

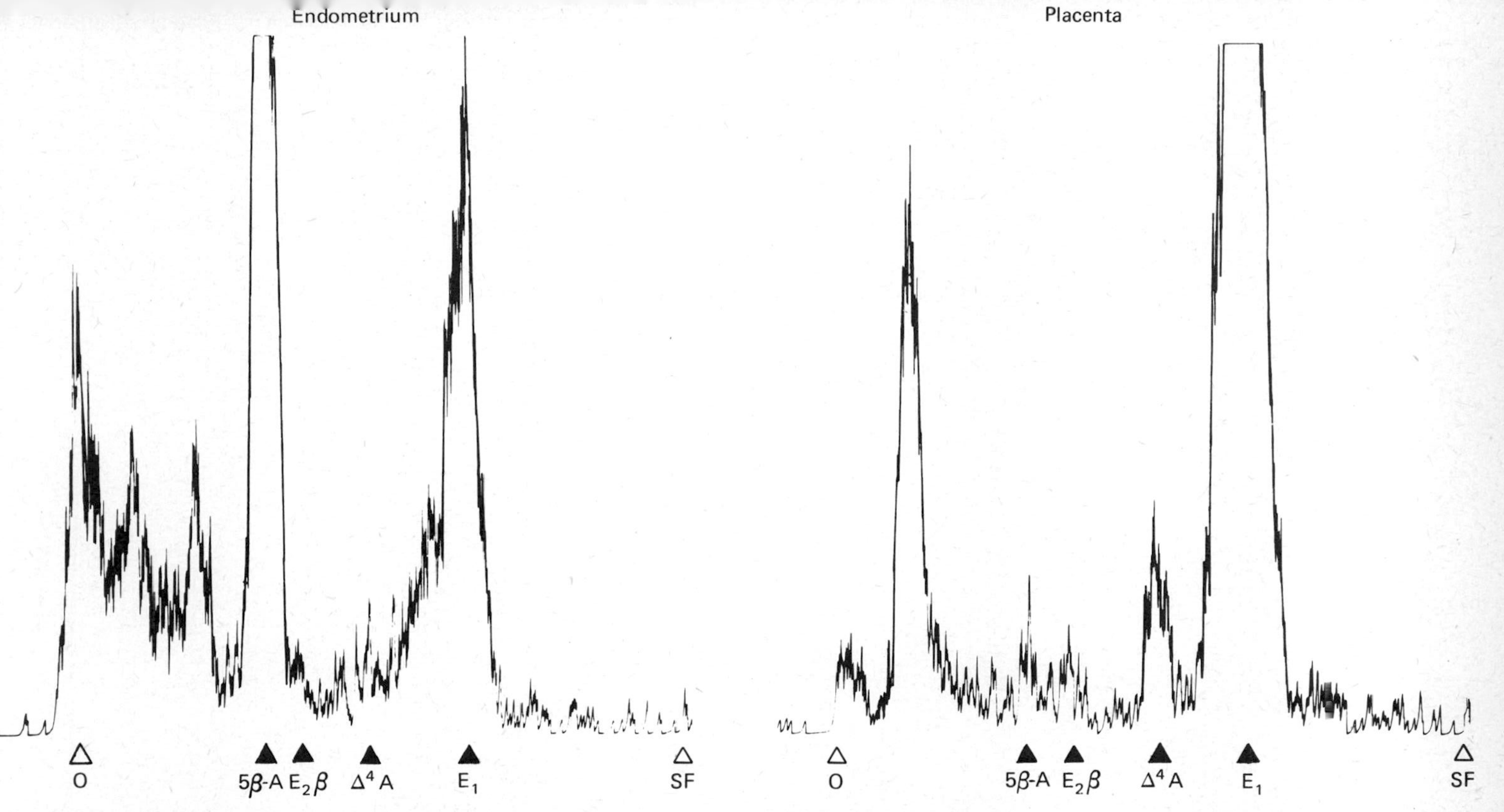

Figure 19.5 Radiochromatogram scans illustrating the metabolism of [1,2,6,7-^{3}H] androstenedione (in the presence of excess NADPH) by microsomal preparations of placental and endometrial tissue obtained at 100 days post coitum. Open triangles indicate the position of the origin (O) and solvent front (SF). Closed triangles indicate the position of the authentic marker steroids, oestrone (E_1), androstenedione (Δ^4A), oestradiol-17β ($E_2\beta$), 3α-hydroxy 5β-androstane-17-one (5β-A).

was also studied. Homogenates of the tissue were incubated with the following tritium-labelled precursors, [7α-^{3}H]pregnenolone, [1α,2α-^{3}H]progesterone and [7α-^{3}H]17α-hydroxyprogesterone in the presence of excess cofactor. Aromatase activity was determined by incubating microsomal preparations of the tissue with [1,2,6,7-^{3}H]androstenedione. At the end of the incubation period the unconjugated metabolites were extracted from the medium with diethyl ether. These ether-soluble metabolites were then separated by thin layer chromatography and detected by using a radiochromatogram scanner. Identification of the metabolites was achieved by derivative formation and recrystallization with authentic carrier steroid to constant specific activity.

Table 19.1 indicates which enzymes of the Δ^4 pathway were shown to be present in endometrial and placental tissue at 100 days post coitum. The major products of progesterone metabolism by the endometrium were 5β-reduced pregnanediols and pregnanetriols together with 5β-pregnan-3α-ol-20-one (5β-pregnanolone). Less than 10% of the metabolites were 5α-reduced. In contrast to this, the products of placental progesterone metabolism were predominantly 5α- and 5β-pregnanolones. 17α-hydroxylated products accounted for less than 5% of the metabolites isolated (see *Figure 19.4*). The major product of androstenedione metabolism by placental tissue was oestrone, while the endometrium synthesized predominantly 5β-reduced androstanolones and 5β-androstanediols (see *Figure 19.5*).

Changes in the activity of the enzymes of the Δ^4 pathway before term

$\Delta^5$3β-hydroxysteroid dehydrogenase

There was a decrease in the percentage of progesterone synthesized by placental tissue over the last two weeks of gestation while synthesis of progesterone by the endometrium showed little change. These findings contrast with the histochemical studies of Dufour and Raeside (1969) which suggested that a moderate increase in placental $\Delta^5$3β-hydroxysteroid dehydrogenase activity occurred near term and with those of Christie (1968) who found only trace activity of this enzyme from 101–112 days of pregnancy.

17α-hydroxylase

The two major products of progesterone metabolism by placental tissue obtained after spontaneous delivery at term (114 days post coitum) were 17α-hydroxyprogesterone and 17α,20α-dihydroxy-4-pregnen-3-one (17α,20α-progesterone). These two 17α-hydroxylated metabolites accounted for 42% of the substrate added (see *Table 19.2*), with 50% of the radioactivity being associated with unmetabolized progesterone. These findings may represent an increase in 17α-hydroxylase activity. An alternative explanation which needs to be investigated is the apparent loss of 5β-reductase activity, since no 5β- reduced products were obtained by

Table 19.2 PLACENTAL STEROID SYNTHESIS: CHANGES BEFORE TERM

Substrate	*Enzyme*	*Product*	*Stage of gestation* (days)		
			100	*106*	*Term (114)*
Progesterone	17α-hydroxylase	17α-hydroxyprogesterone	<5	<5	19.8±4.8 (n = 6)
		17α,20α-progesterone	<5	<5	22.5±2.9 (n = 6)
17α-hydroxyprogesterone	C-17,20-lyase	Oestrone	18.4±4.4 (n = 3)	30.6±1.8 (n = 4)	34.4±3.6 (n = 7)
Androstenedione	Aromatase	Oestrone	43.8±7.9 (n = 4)	48.8±11.8 (n = 3)	74.[illegible]±4.4 (n = 4)

Values given as mean percentage conversions (± S.E.M.)
Figures in parentheses indicate the number of animals (at least three observations, in duplicate, were made for each animal)

metabolism of progesterone by placental tissue obtained post delivery. Whether this loss of 5β-reductase activity occurred prior to delivery or was the result of anoxia arising from cessation of placental blood flow at term remains uncertain. No 17α-hydroxyprogesterone could be isolated from incubations of endometrial tissue with progesterone at any of the stages of pregnancy studied.

C-17,20-lyase

Placental C-17,20-lyase activity as determined by conversion of 17α-hydroxyprogesterone to oestrone was found to increase on or about day 105 post coitum. At 100 days post coitum conversion to oestrone was 18.5%, by 106 days 30% and by the end of pregnancy had reached 30–35% (see *Table 19.2*). Conversion of 17α-hydroxyprogesterone to oestrone by endometrial tissue remained at less than 3% throughout the last two weeks of gestation.

Aromatase

An increase in the activity of both placental (see *Table 19.2*) and endometrial aromatase was seen from 105–114 days post coitum. The increase in oestrone produced from androstenedione could, in part, be due to a decrease in the synthesis of 5β-reduced metabolites.

A composite diagram of the pathways of steroid metabolism present *in vitro* at 100 days and 114 days post coitum in both placental and endometrial tissue is given in *Figure 19.6*.

Conclusions

From the above studies, the following points may be made.

(a) Ovariectomy and medroxyprogesterone acetate replacement therapy allow the study of the uterine contribution to circulating steroids in late pregnancy. Using this technique, it has been demonstrated that the gravid uterus is capable of producing progesterone, 17α-hydroxyprogesterone and androstenedione although the plasma concentration of these steroids is 5–10% that of intact, untreated animals. In contrast, plasma concentrations of unconjugated oestrogens were unchanged by ovariectomy thus providing further evidence for an extra-ovarian source of oestrogens in late pregnancy.

(b) *In vitro* investigations have demonstrated that both endometrial and placental tissue will actively metabolize steroids of the Δ^4 pathway, although from these studies it would appear that only the placenta possesses all the enzymes necessary for the synthesis of oestrogens from pregnenolone by the Δ^4 pathway.

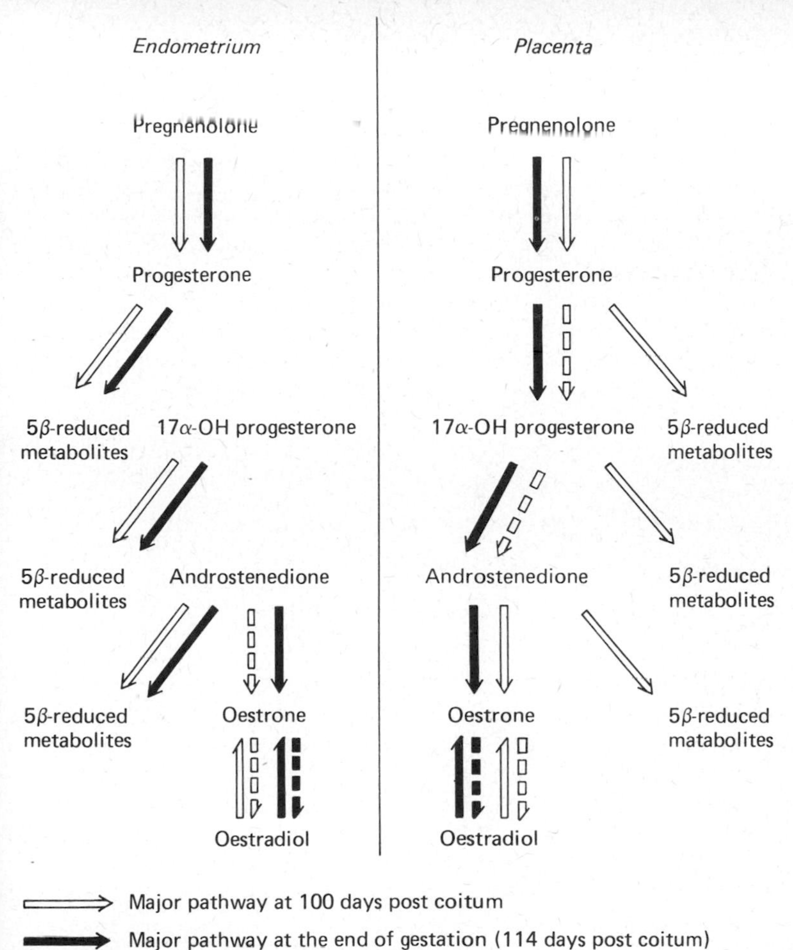

Figure 19.6 A composite diagram to illustrate the qualitative changes which occur in the metabolism of steroids of the Δ^4 pathway by both placental and endometrial tissue between 100 days of gestation and term

(c) The apparent activity of placental 17α-hydroxylase, C-17,20-lyase and aromatase increases between 100 days post coitum and term concurrent with the increase in maternal circulating oestrogens seen over this period.

Elevated foetal plasma cortisol concentrations at the end of pregnancy have been shown to influence the Δ^4-3-keto pathway in both the sheep and the goat leading to an increase in placental oestrogen synthesis. Since the plasma cortisol concentration increases during the last two weeks of gestation in the foetal pig, at a time when the maternal plasma concentration of oestrogens is rising, it is possible that a similar mechanism may also operate in this species.

There is, however, an important species difference in the role of oestradiol-17β in the control of parturition. Whereas in the goat the rising plasma concentration of oestradiol-17β plays an important role in controlling uterine prostaglandin $F_{2\alpha}$ production and is therefore indirectly luteolytic, no such role for oestradiol-17β has been demonstrated in the control of parturition in the pig. Infusion of oestradiol-17β into the maternal circulation (Flint, Ricketts and Craig, 1979) or intramuscular injection of oestradiol-17β (Coggins, 1975; First and Bosc, 1979) in late pregnancy has no effect on the length of gestation. It may be postulated however that oestrogens play a facilitatory rather than obligatory role in the pig, possibly by raising the concentration of the myometrial oxytocin receptor and thereby preparing the uterus for oxytocin release during labour.

Acknowledgements

The author acknowledges the support of an M.R.C. studentship.

References

AINSWORTH, L. and RYAN, K.J. (1966). Steroid hormone transformations by endocrine organs from pregnant mammals. I. Oestrogen biosynthesis by mammalian placental preparations *in vitro*. *Endocrinology* **79**, 875–883

ANDERSON, A.B.M., FLINT, A.P.F. and TURNBULL, A.C. (1975). Mechanism of action of glucocorticoids in induction of ovine parturition: effect on placental steroid metabolism. *J. Endocr.* **66**, 61–70

ASH, R.W. and HEAP, R.B. (1975). Oestrogen, progesterone and corticosteroid concentration in peripheral plasma of sows during pregnancy, parturition, lactation and after weaning. *J. Endocr.* **64**, 141–154

ASH, R.W., BANKS, P., BAILES, G., BROAD, S. and HEAP, R.B. (1973). Plasma oestrogen, progesterone and corticoid concentrations in the pregnant, parturient and lactating sow. *J. Reprod. Fert.* **33**, 359–360

BALDWIN, D.M. and STABENFELDT, G.H. (1975). Endocrine changes in the pig during late pregnancy, parturition and lactation. *Biol. Reprod.* **12**, 508–515

BARNES, R.J., COMLINE, R.S. and SILVER, M. (1974). Foetal and maternal progesterone concentrations in the pig. *J. Endocr.* **62**, 419–420

BLOCH, E. and NEWMAN, E. (1966). Comparative placental steroid synthesis. I. Conversion of (7-^{3}H)-dehydroepiandrosterone to (^{3}H)-androst-4-ene 3,17-dione. *Endocrinology* **79**, 524–530

BOSC, M.J. (1973). Modification de la durée de gestation de la Truie après administration d'ACTH aux foetus. *C. r. hebd. Seanc. Acad. Sci., Paris* **276**, 3183

CHOONG, C.H. and RAESIDE, J.I. (1974). Chemical determination of oestrogen distribution in the foetus and placenta of the domestic pig. *Acta endocr.* **77**, 171–185

CHRISTIE, G.A. (1968). Distribution of hydroxysteroid dehydrogenase in the placenta of the pig. *J. Endocr.* **40**, 285–291

COGGINS, E.G. (1975). Mechanisms controlling parturition in swine. PhD Thesis. University of Wisconsin, Madison, USA

CURRIE, W.B. and THORBURN, G.D. (1977). The foetal role in timing the initiation of parturition in the goat. In *The Fetus and Birth*, pp. 49–72. Amsterdam, Elsevier/Excerpta Medica/North Holland

CURRIE, W.B., COX, R.I. and THORBURN, G.D. (1976). Release of prostaglandin F, regression of corpora lutea and induction of premature parturition in goats treated with oestradiol-17β. *Prostaglandins* **12**, 1093–1103

DUEBEN, B.D., WISE, T.H., BAZER, F.W., FIELDS, M.J. and KALRA, P.S. (1980). Metabolism of H^3-progesterone to estrogens by pregnant gilt endometrium and conceptus. *J. Anim. Sci.* **49**, *Suppl.* **1**, 293(Abstract 375)

DUFOUR, J. and RAESIDE, J.I. (1969). Hydroxysteroid dehydrogenase activity in the placenta of the domestic pig. *Endocrinology* **84**, 426–431

FÈVRE, J., LÉGLISE, P.-C and ROMBAUTS, P. (1968). Du rôle de l'hypophyse et des ovaires dans la biosynthese des oestrogènes aux cours de la gestation chez la truie. *Annls Biol. anim. Biochim. Biophys.* **8**, 225–233

FIRST, N.L. and BOSC, M.J. (1979). Proposed mechanisms controlling parturition and the induction of parturition in swine. *J. Anim. Sci.* **48**, 1407–1421

FLINT, A.P.F. and RICKETTS, A.P. (1979). Control of placental endocrine function: role of enzyme activation in the onset of labour. *J. Steroid Biochem.* **11**, 493–500

FLINT, A.P.F., RICKETTS, A.P. and CRAIG, V.A. (1979). The control of placental steroid synthesis at parturition in domestic animals. *Anim. Reprod. Sci.* **2**, 239–251

FLINT, A.P.F., KINGSTON, E.J., ROBINSON, J.S. and THORBURN, G.D. (1978). Initiation of parturition in the goat: evidence for control by foetal glucocorticoid through activation of placental C_{21}-steroid 17α-hydroxylase. *J. Endocr.* **78**, 367–378

GADSBY, J.E. and HEAP, R.B. (1980). Oestrogen synthesis by embryos of the pig, cow and sheep. *J. Anim. Sci.* **49**, *Suppl.* **1**, 299 (Abstract 389)

GODKE, R.A. and DAY, B.N. (1973). Maternal and fetal plasma progestin levels in swine. *J. Anim. Sci.* **37**, 313

KNIGHT, J.W., BAZER, F.W., THATCHER, W.W., FRANK, D.E. and WALLACE, M.D. (1977). Conceptus development in intact and unilaterally hysterectomized and ovariectomized gilts: inter-relations among hormonal status, placental development, fetal fluids and fetal growth. *J. Anim. Sci.* **44**, 620–637

MACDONALD, A.A., COLENBRANDER, B., ELSAESSER, F. and HEILHECKER, A. (1980). Progesterone production by the pig fetus and the response to stimulation by adrenocorticotrophin. *J. Endocr.* **85**, 34P

NARA, B.S. and FIRST, N.L. (1977). Effect of indomethacin and $PGF_{2\alpha}$ on porcine parturition. *J. Anim. Sci.* **45**, *Suppl.* **1**, 191

NORTH, S.A., HAUSER, E.R. and FIRST, N.L. (1973). Induction of parturition in swine and rabbits with the corticosteroid dexamethasone. *J. Anim. Sci.* **36**, 1170–1174

ROBERTSON, H.A. and KING, G.J. (1974). Plasma concentrations of progesterone oestrone, oestradiol-17β and oestrone sulphate in the pig at implantation during pregnancy and at parturition. *J. Reprod. Fert.* **40**, 133–141

SILVER, M., BARNES, R.J., COMLINE, R.S., FOWDEN, A.L., CLOVER, L. and MITCHELL, M.D. (1979). Prostaglandins in the foetal pig and prepartum endocrine changes in mother and foetus. *Anim. Reprod. Sci.* **2**, 305–322

STEELE, P.A., FLINT, A.P.F. and TURNBULL, A.C. (1976). Increased utero-ovarian androstenedione production before parturition in sheep. *J. Reprod. Fert.* **46**, 443–445

20

MYOMETRIAL ACTIVITY DURING PREGNANCY AND PARTURITION IN THE PIG

M.A.M. TAVERNE
Clinic for Veterinary Obstetrics, A.I. and Reproduction, State University of Utrecht, The Netherlands

In contrast to the quite substantial number of publications on the endocrine control of parturition in the pig (First and Bosc, 1979; Ellendorff *et al.*, 1979a; First, Chapter 16), very little information is available on the actual activity of one of the most important target tissues, i.e. the myometrium. Including studies on the non-pregnant uterus, only about twenty papers have appeared on this topic during the last twenty years. Many therapeutic measures have been undertaken to reduce the rate of stillbirths in the pig (Sprecher *et al.*, 1974; Dziuk, 1979) without a documented knowledge of their ultimate effects on uterine contractions during the parturition process. One can hardly believe in a species of such economic importance and from which tissue samples are easily obtainable, that even some of the very basic questions on the morphology of the myometrium still remain to be answered. (a) Is the architecture of the porcine myometrium, where muscle fibres have been reported to traverse from the outer to the inner layer only incidently (Nagler, 1956), really different from other polytocous species, like the rat, where the two layers were found to be built up by the same bundles of muscle cells (Ludwig, 1952)? (b) What is the relative distribution of cholinergic and adrenergic nerve fibres in the myometrium during pregnancy and parturition? (c) Are changes in the density of the innervation, if they exist, under hormonal control, as reported for the myometrium of the non-pregnant pig (Colenbrander, 1974)?

This chapter will summarize some of the more recent data on other aspects of uterine physiology such as (a) the temporary changes in myometrial activity around parturition, (b) the mechanical aspects of the delivery process, and (c) the *in vivo* and/or *in vitro* measurement of the effects of progesterone, oestrogens, oxytocin, prostaglandins and catecholamines on the myometrium of the pig.

Evolution of uterine contractions

Spontaneous changes in uterine activity during late pregnancy and parturition have been investigated by *in vivo* recordings of intrauterine pressure changes (Zerobin, 1968; Zerobin and Spörri, 1972; Ngiam, 1974; 1977) and by uterine electromyography (Zerobin, 1968; Zerobin and Spörri, 1972;

Taverne, Naaktgeboren and van der Weyden, 1979; Taverne *et al.*, 1979). Although the pressure recordings made during the last weeks of gestation have revealed the existence of only feeble, local contractions that occur asynchronously with electrical activity (Zerobin and Spörri, 1972), prolonged EMG recording sessions performed during the last three weeks of pregnancy in the miniature pig have demonstrated the existence of a definite pattern of myometrial electrical activity (Taverne *et al.*, 1979). Episodes of EMG activity of several minutes' duration occur with low frequency (0.5–3.5/hour) in uterine segments that contain a piglet while empty parts of the uterus are relatively inactive.

A similar pattern of myometrial electrical activity has been found in the sheep (Naaktgeboren *et al.*, 1975; Prud'homme and Bosc, 1977), the cow (Taverne, van der Weyden and Fontijne, 1979) and the goat (Taverne and Scheerboom, unpublished data). Implantation of electrodes before a successful mating in the ewe allows EMG recordings to be performed during the entire course of pregnancy (van der Weyden *et al.*, 1981a). Episodes of electrical activity were first detected as early as the 5th week of gestation in the sheep and during the last trimester concurrent recordings of real-time ultrasound images of the ovine conceptus and of EMG activity showed that during these episodes of electrical activity the conceptus is passively displaced within the abdominal cavity of the ewe (Scheerboom and Taverne, 1981). The physiological implications of EMG activity during late pregnancy in the pig remain to be investigated. The observation that this activity sometimes appears in consecutive parts of a pregnant uterine horn (*Figure 20.1*) indicates that it can be initiated at one foetal compartment and that either a muscular or neural pathway exists for its propagation. However, in most instances no clear propagation can be detected during pregnancy and EMG activity appears more or less synchronously at different foetal locations.

The few studies that have been performed so far indicate that the typical pregnancy pattern of myometrial activity changes only 4–9 hours before expulsion of the first piglet. At this time local contractions have disappeared and bi- or triphasic increases of intrauterine pressure, lasting for 1–3 minutes, occur at regular intervals synchronized with episodes of EMG activity. Empty parts of the uterine horns also contract and the propagation of contractions along the horn in both a tubocervical and cervicotubal direction is frequently observed at this stage of the parturition process.

Further evolution of myometrial activity (i.e. an increase in the frequency and amplitude of the contractions and the gradual appearance of straining efforts of the dam) takes place during the last few hours before the birth of the first piglet but the few data that are available are difficult to compare in a quantitative way because of the differences in the experimental protocols and in the techniques used. In addition, large differences in the characteristics of the uterine contractions between individual sows have been reported (Zerobin, 1968; *Figure 20.2*). The mean frequency of uterine contractions is maximal during the delivery of the piglets and placentae (Taverne *et al.*, 1979); however, during this stage of parturition, contraction frequency, duration and amplitude may vary considerably from one hour to another (Zerobin, 1968; Ngiam, 1974).

The combination of individual foetal marking with uterine electromyography in the minipig showed that both the duration and the direction of

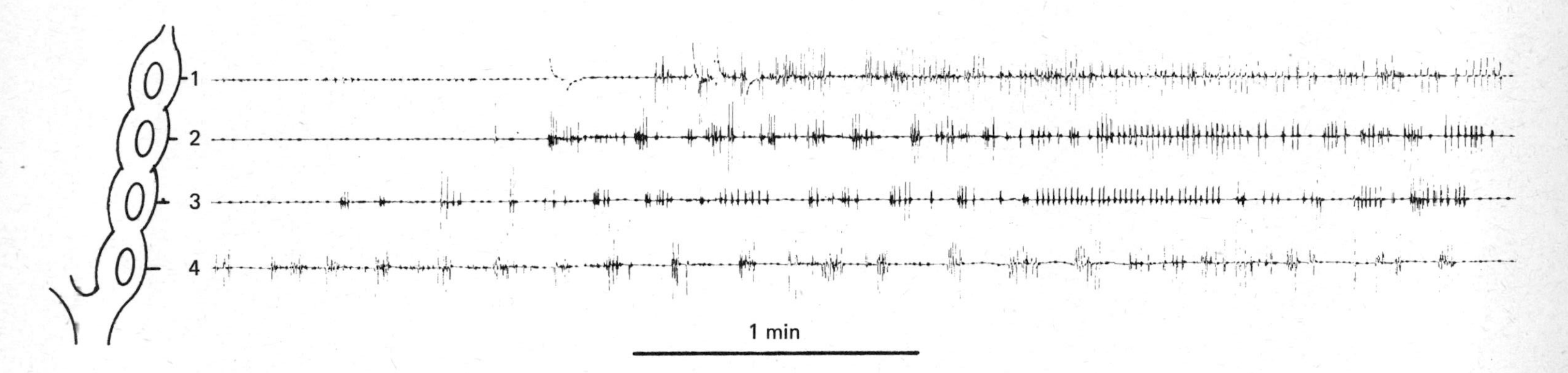

Figure 20.1 Uterine electromyograph from a miniature sow at day 105 of pregnancy, eight days before spontaneous parturition. The diagram shows the sites of implantation of the four bipolar electrodes. An episode of EMG activity appears consecutively at electrode Nos. 4, 3, 2 and 1, indicating that propagation of the electrical activity occurs in a cervicotubal direction

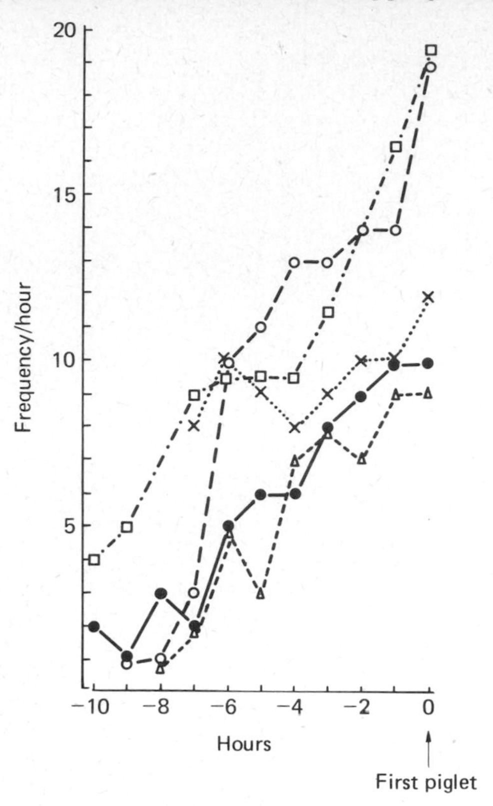

Figure 20.2 The increase in the frequency (per hour) of episodes of myometrial electrical activity in five miniature sows during the 10-hour period preceding the spontaneous delivery of the first piglet. (In each sow the surface electrode had been implanted in the middle of a foetal compartment, at least seven days before.)

the propagation of uterine contractions change upon emptying of a uterine horn (Taverne, Naaktgeboren and van der Weyden, 1979). Without knowledge of the latter, intrauterine pressure measurements should be interpreted with caution because one cannot be sure that the tip of the catheter is enclosed between a piglet and the uterine wall or that it just remains in an empty portion of the uterine lumen. For this same reason actual values for the pressures that have been measured are only useful when the physical conditions around the tip of the catheter or the balloon remain more or less stable throughout the measurements. The only place where this probably occurs seems to be close to the utero–isthmic junction. These methodological implications should be kept in mind when *in vivo* studies are performed in the future.

Mechanical aspects of delivery and uterine contractions

The birth canal along which each piglet must pass consists of that part of the uterus where the piglet is situated, the empty parts of the horn, the uterine body, the cervix, the vagina and the vulva. The caudal part of the cervix and the vagina are situated in the bony part of the birth canal that is formed by the more or less rectangular pelvic inlet, the pelvic cavity and the pelvic outlet. It is not known if the cervix and vagina actively contribute to the delivery process.

Transport of foetuses through the uterus is effected by the myometrial contractions. How this transport is achieved has not been investigated in detail in the sow. The importance of research into the nature of an unimpaired and quick transport of piglets through the uterine horns is emphasized by at least three observations:

(a) Two-thirds of all piglets that die during or shortly after delivery are expelled during the final third of the delivery process (Randall, 1972) and a majority of these piglets are born with broken umbilical cords;
(b) Nearly all the piglets which have occupied the most tubal parts of an otherwise empty uterine horn and consequently have had to traverse a long segment of previously unoccupied, undilated uterine tissue before reaching the uterine body, are stillborn (Bevier and Dziuk, 1976);
(c) Surgical inversion of the uterine horn in the sow seriously impairs the normal expulsion of piglets from that horn (Bosc *et al.*, 1976), thus once again raising the question about the existence of special morphological structures for peristalsis in the uterus.

Multiple implantation of intrauterine catheters (Zerobin, 1968) or surface electrodes (Taverne *et al.*, 1979) have demonstrated the existence of a synchronization of myometrial contractions at all segments of a uterine horn during delivery. Contractions were found to be initiated preferentially at the two ends of the uterine horn and subsequently propagated in either a tubocervical or cervicotubal direction. Even 'echo-propagation' has been documented (Taverne *et al.*, 1979), i.e. a tubocervically-directed contraction, once it reaches the end of the horn, rebounds in the opposite direction. A combination of foetal marking with uterine electromyography has shown that cervicotubally-directed contractions normally occur during the parturition process until the moment that the horn is empty. The presence of piglets close to the cervix obviously plays a role in the initiation of these cervicotubally-directed contractions during labour. However, the exact function of this type of contraction remains to be elucidated, perhaps by direct observation of the exposed uterus, as has been performed in the parturient rabbit (Carter, Naaktgeboren and van Zon- van Wagtendonk, 1971; Naaktgeboren and Carter, 1971). Nevertheless from the present data two possible functions for these cervicotubal contractions can be postulated:

(1) When contractions start at the caudal part of the uterus they cause an extra shortening of the distance to be traversed by the piglets that are still left within the horn and this additional shortening is superimposed on the retraction of those parts of the uterine horn that have already been emptied (Taverne, 1979).
(2) Cervicotubally-directed contractions retain those piglets in the horn that are not yet due to be expelled. The uterine horn is transformed

into a slippery tube by the rupture of the chorionic ends of the individual placentae and this allows piglets to move freely through the chorionic membranes of their neighbours (Perry, 1954; Ashdown and Marrable, 1970). To prevent accumulation of piglets at the caudal part of the horn, transport in the direction of the oviduct would keep the piglets more or less in place and would prevent premature rupture of umbilical cords.

Nothing is known about a possible mutual relationship between contractions in the two horns. Although expulsion of piglets from the two horns takes place by chance (Dziuk and Harmon, 1969; Taverne *et al.*, 1977), complete independence of the two sites seems very unlikely because muscle fibres from both horns fuse at the common uterine body (Leibrecht, 1953; Nagler, 1956). In this respect it would be interesting to make a comparison with the dog in which expulsion of puppies takes place alternately from the two horns in the majority of cases (van der Weyden *et al.*, 1981b).

Effects of hormones and drugs on the myometrium

PROGESTERONE AND OESTROGENS

A relationship between the morphology of the ovaries and the motility of the uterus (*in vitro* observation) of the sow was reported many years ago (Keye, 1923; King, 1927). *In vivo* recordings of myometrial activity in the sow have been performed during different stages of the oestrous cycle (as judged by teasing behaviour) in either anaesthetized (Zerobin and Spörri, 1972) or conscious sows (Döcke and Worch, 1963; Bower, 1974). From these studies and from experiments using ovariectomized sows treated with progesterone and/or oestradiol (Zerobin, 1968), one may conclude that changes in steroid hormone production by the ovaries of the cycling sow are responsible for the changes in spontaneous activity of the myometrium. However, concurrent investigations of both uterine contractility and plasma or tissue levels of progesterone and oestradiol have not been published so far for the non-pregnant pig.

The observation in the late pregnant minipig that the onset of the parturient pattern of myometrial contractions is preceded by a significant decrease of the plasma progesterone concentration and a significant increase of the oestrone and oestradiol-17β concentrations (Taverne *et al.*, 1979), raises the question of how essential these steroid changes are for the stimulation of the myometrium.

There are few direct observations to indicate that progesterone suppresses myometrial activity in the pig. Contractions are either of low amplitude or even completely absent during the greater part of the luteal phase of the cycle (Zerobin and Spörri, 1972; Bower, 1974) and in ovariectomized sows treated with progesterone (Zerobin, 1968). However, simultaneous injections of progesterone and very high doses of oestradiol can initiate powerful contractions comparable to those registered during the first and third day of oestrus (Zerobin, 1968). An experiment with one late pregnant sow provided with open intraluminal uterine catheters for the recording of myometrial contractions revealed that after six days of treatment with a progestin (CAP), typical uterine pressure waves were absent during parturition. The amplitude of the remaining atypical uterine pressure waves was reduced by about 50% but their duration was increased and eleven piglets (two of which were dead) had to be delivered manually from the vagina (Jöchle *et al.*, 1974). When a single injection of the prostaglandin $F_{2\alpha}$ analogue cloprostenol was given to a pregnant sow (provided with uterine surface electrodes) on day 110 of pregnancy, stimulation of the myometrium was not observed immediately after injection. Only 23 hours after injection of the prostaglandin, when peripheral plasma progesterone concentrations had reached values below 5 ng/ml, did the onset of the parturient pattern of myometrial activity take place (Taverne, 1979). This would be in agreement with the data of Ellicott and Dziuk (1973) which suggest that a concentration of about 4 ng/ml of progesterone in the peripheral plasma appears to be minimal for maintenance of pregnancy in the pig.

Despite the overwhelming evidence that progesterone withdrawal is an essential step in the processes that lead to the normal delivery of piglets (see review by First and Bosc, 1979; First, Chapter 16), the following observations tend to indicate that the activation of the myometrium and complete expulsion of the litter can take place even when the animal is still being treated with progesterone or a progestin, or when peripheral plasma levels of progesterone are still elevated:

(a) In pigs in which surface electrodes were surgically implanted on the uterus during late pregnancy, regular and frequent episodes of EMG activity were recorded during the first week after surgery despite the presence of high circulating progesterone levels in the blood (Taverne, unpublished data);
(b) A single subcutaneous injection of 100 mg progesterone in oil on day 105 of pregnancy in a sow provided with uterine electrodes raised the peripheral plasma progesterone concentration for many days and although an injection of the prostaglandin $F_{2\alpha}$ analogue, cloprostenol, caused a significant decrease in the plasma progesterone concentrations in this animal, a progesterone level of 14.0 ng/ml was measured during the spontaneous delivery of three living piglets 30 hours after the injection of cloprostenol (EMG activity in this animal is shown in *Figure 20.3*);
(c) In some animals continuously treated with progesterone, the delivery of piglets was not prevented (Minar and Schilling, 1970; Ellicott and Dziuk, 1973);
(d) In a recent study it was found that after subcutaneous injections of progesterone for six days, even though delivery was delayed, expulsion of living piglets still occurred while plasma progesterone levels were very high (Taverne *et al.*, 1982);
(e) Current experiments in our laboratory on the EMG activity in the cycling pig indicate that regular bursts of electrical activity occur during the very early stages of luteolysis, when plasma progesterone levels are still elevated (*Figure 20.4*).

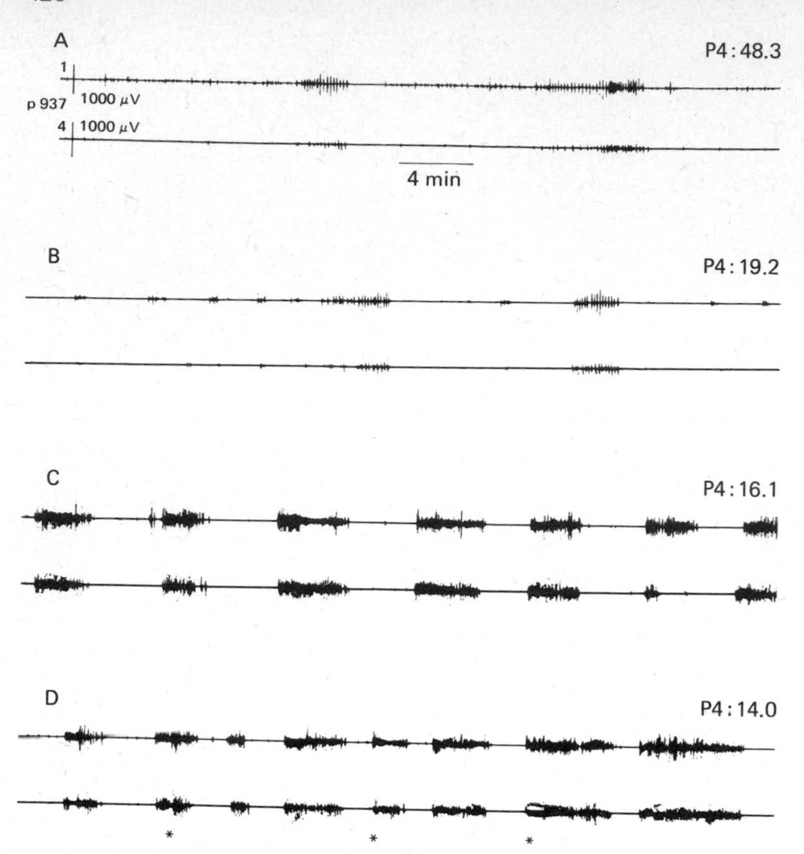

Figure 20.3 Uterine EMGs recorded from two different sites of a horn in a pregnant sow that was injected subcutaneously with 100 mg progesterone (in oil) on day 105 of gestation and in which parturition was induced by a single intramuscular injection of a prostaglandin $F_{2\alpha}$ analogue (ICI 80996, Cloprostenol) on day 110 of gestation. A: day 109 of gestation, one day before prostaglandin injection; B: 14 hours after prostaglandin injection; C: 25 hours after prostaglandin injection; D: during spontaneous delivery of three living piglets (*) 30 hours after injection of the prostaglandin analogue. The peripheral plasma progesterone levels indicated (P_4:ng/ml) were estimated according to the method of Dieleman and Schoenmakers (1979)

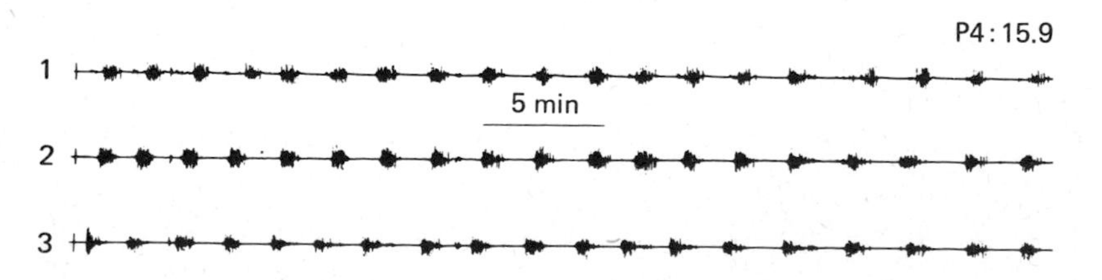

Figure 20.4 Regular bursts of EMG activity recorded from three different sites of a uterine horn in a miniature sow during late dioestrus, six days before the onset of the next oestrus. Peripheral plasma progesterone levels measured 30.9 ng/ml on the day before, 15.9 ng/ml during, and 2.6 ng/ml 16 hours after this recording

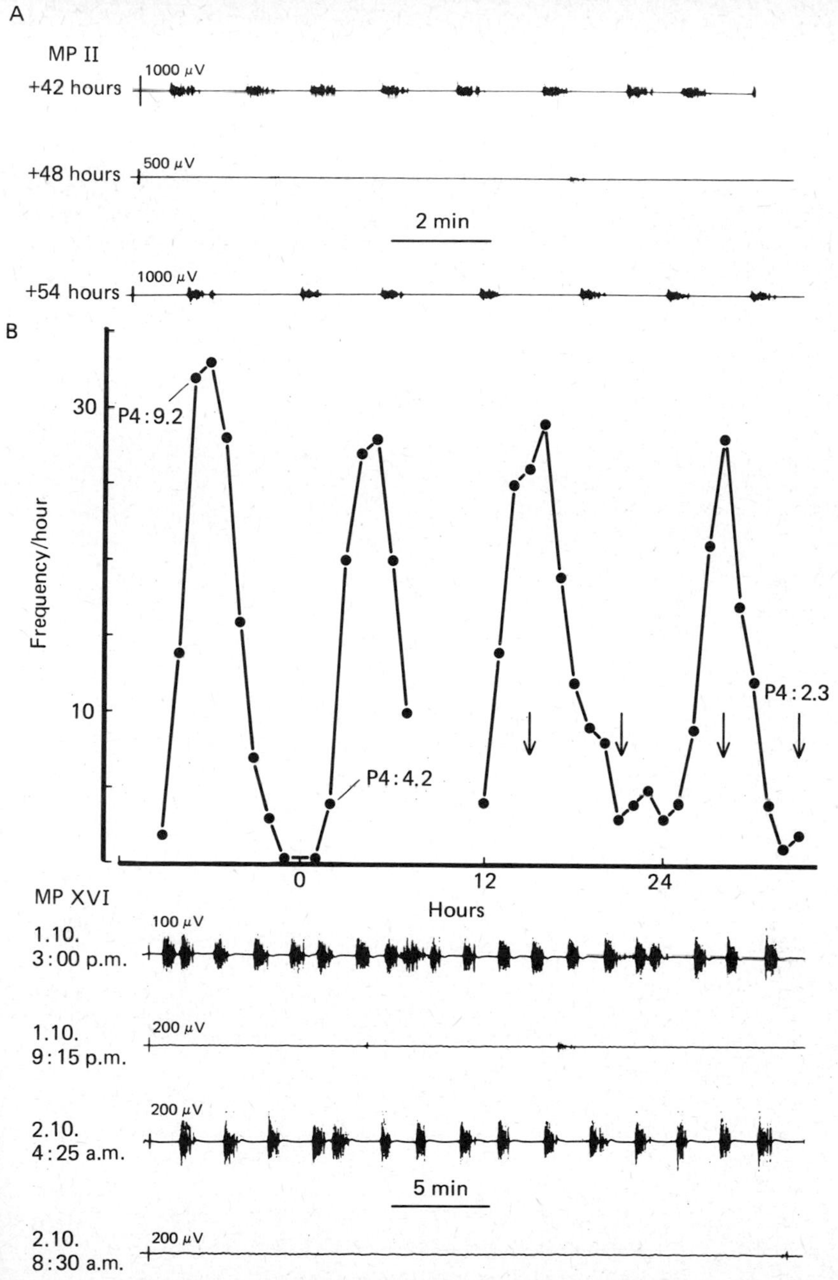

Figure 20.5 A. EMG recorded from one site of a uterine horn in an ovariectomized miniature sow 42, 48 and 54 hours after the intramuscular injection of 100 μg oestradiol-17β; (from the 8th to the 3rd day before the administration of oestradiol, the sow was injected daily with 12.5 mg progesterone). B. The frequency (per hour) of bursts of electrical activity recorded from one site of a uterine horn in a miniature sow during two days of pro-oestrus. Peripheral plasma progesterone levels were determined three times during continuous recording of EMG activity. Four segments of the electromyograph (at times indicated by the arrows in the graph) are shown at the bottom

Thus there is some doubt about the significance of plasma progesterone levels in relation to the action of progesterone on the myometrium. A similar conclusion was made by Csapo, Eskola and Ruttner (1980) after the treatment of rats with progesterone shortly before the expected time of delivery.

The role of oestrogens for the stimulation of the myometrium at parturition is even less clear. Infusion of oestradiol did not change gestation length in two late pregnant sows (Flint, Ricketts and Craig, 1979). Expulsion of the conceptuses in sows with prolonged pregnancies, in which all the piglets were mummified, has been reported after a single injection of prostaglandins (Wrathall, 1980), even when oestrogen levels can be expected to be very low.

In vitro, the motility of uterine strips obtained from sows during parturition is inhibited when oestradiol is added to the organ bath (Dias e Silva, 1979). However, *in vivo*, after pretreating ovariectomized non-pregnant sows with progesterone, a single injection of oestradiol-17β induces a cyclic pattern of myometrial electrical activity in which hours with regular and well propagated bursts of electrical activity alternate with hours of almost complete rest (*Figure 20.5*). This pattern lasts for at least two days after the injection of oestradiol and it is interesting that a similar pattern of EMG activity has been recorded during luteolysis and the greater part of the follicular phase of the oestrous cycle (*Figure 20.5*). Elevated plasma levels of prostaglandins have been reported during this stage of the cycle (Moeljono *et al.*, 1977; Shille *et al.*, 1979), so one might speculate that prostaglandins are involved in the effects of oestrogens on the myometrium of the pig. This is supported by recent data on the effects of oestrogens on the non-pregnant uterus in sheep that were treated with prostaglandin synthesis inhibitors (Lye, 1980; Prud'homme, 1980).

OXYTOCIN AND PROSTAGLANDINS

Despite the warning of Zerobin (1968) that an injection of only 10 iu of oxytocin is sufficient to restore contractions in sows with uterine inertia during delivery, oxytocin is probably still one of the hormones that is too frequently misused in farrowing barns.

Even in a recent report on the improvement of synchronization of farrowing after prostaglandin injection, a dose of 50 iu of oxytocin was given several hours before the first piglet was born (Welk and First, 1979). Although oxytocin treatment in this latter study did not cause an additional increase in the incidence of stillbirths, application of drugs to the sow during the periparturient period should be avoided as much as possible when their effects on both the myometrium and uterine vasculature are not known. The pig foetus is very susceptible to intrauterine asphyxia (Randall, 1979) and a prolonged period of intrauterine hypoxia caused by uterine hypertension might deplete the glycogen stores of the piglets that would otherwise help them to survive when they are due to be expelled.

The myometrium of the pregnant pig does not react with frequent and well propagated contractions upon a single injection (intramuscular or intravenous) of even high doses of oxytocin. After application of the drug,

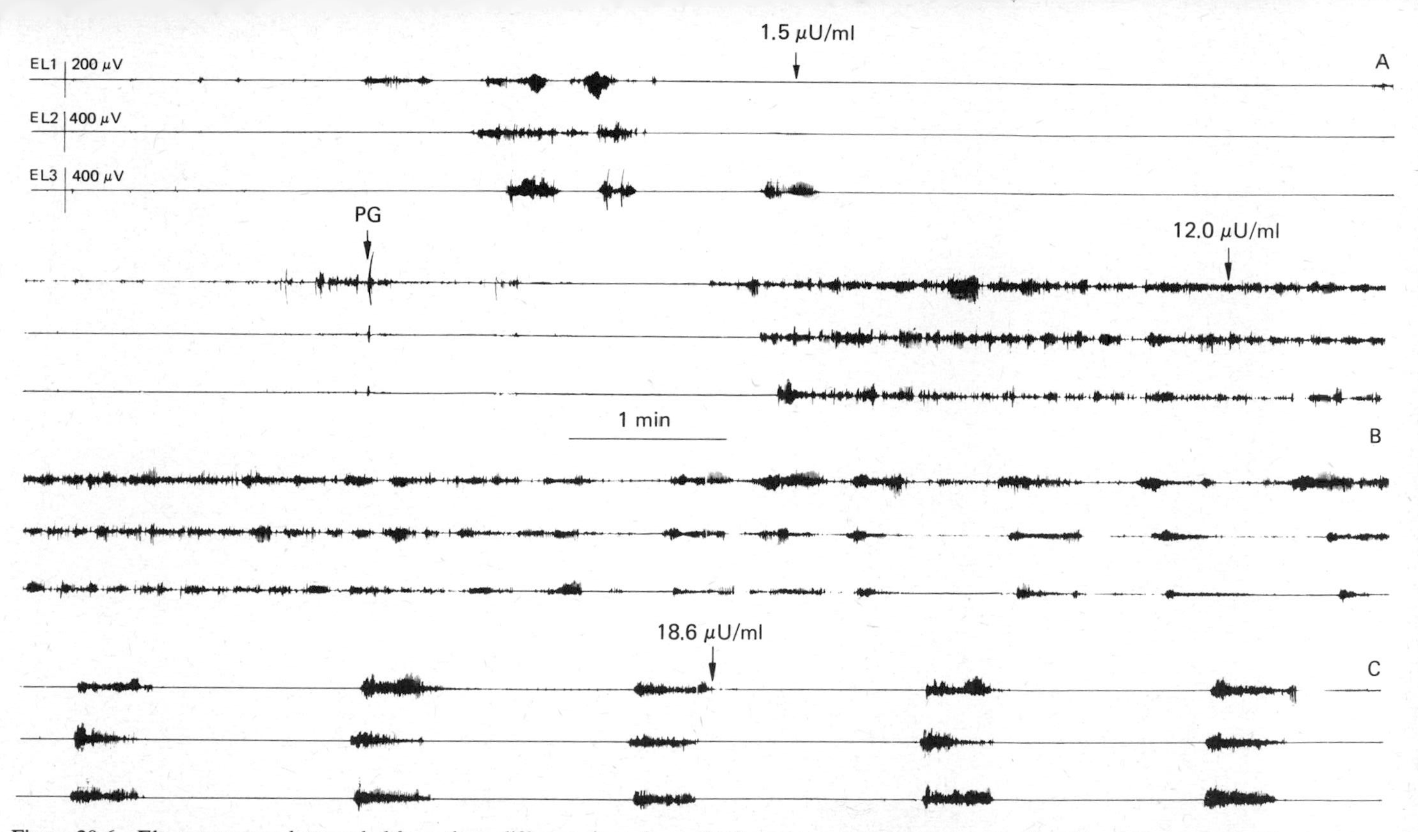

Figure 20.6 Electromyograph recorded from three different sites of a uterine horn in a miniature sow on the first day of dioestrus, at (A) 60 minutes before, (B) during and (C) at 60 minutes after the intramuscular administration of 5 mg prostaglandin $F_{2\alpha}$ (arrow PG; Dinoprost, Upjohn). Plasma oxytocin levels of three samples taken (arrow) during the illustrated parts of the EMG are indicated. Oxytocin concentrations were measured as described by Chard and Forsling (1976)

a prolonged slight increase of intrauterine pressure (Zerobin, 1968) and a prolonged episode of EMG activity (Ellendorff *et al.*, 1979b) have been recorded but these subside after several minutes and no further stimulation is seen despite the presence of still elevated peripheral oxytocin levels (Ellendorff *et al.*, 1979b). It would be interesting to know the myometrial response to a more prolonged infusion of oxytocin or to a long acting oxytocin analogue (Cort, Einarsson and Viring, 1979) during late pregnancy, since the presence of oxytocin receptors has been demonstrated in a pregnant pig, about one week before delivery (Soloff and Swartz, 1974). An increase in the density of these binding sites probably occurs shortly before delivery, as judged by the increase of myometrial sensitivity to exogenous oxytocin (Zerobin, 1968) and the elevated levels of oestrogens at this time may be responsible for this (Soloff, 1975).

Evolution of myometrial activity during parturition takes place concurrently with an elevation of the peripheral plasma oxytocin concentration (Taverne *et al.*, 1979; Forsling *et al.*, 1979, Forsling, Macdonald and Ellendorff, 1979). The relative significance of oxytocin stimulation is hard to judge during this stage because prostaglandin $F_{2\alpha}$ concentrations, as judged by measurements of the 15-keto metabolite, are elevated as well on the day of parturition (Silver *et al.*, 1979; First and Bosc, 1979; Martin, 1980). Prostaglandin $F_{2\alpha}$ can both stimulate the myometrium and cause an oxytocin release in the pig (Ellendorff *et al.*, 1979b; *Figure 20.6*). In addition *in vitro* studies with uterine strips taken from sows during the follicular stage of the cycle indicated that, depending on the dose in the organ bath, prostaglandin E_2 can either stimulate or inhibit uterine contractions and that the inhibition can be blocked by the beta-blocking agent propanolol (Rüsse, 1972). Indomethacin prevented both spontaneous and PGE_2-induced contractions in the latter study. *In vivo*, indomethacin treatment of the late pregnant sow postpones parturition (Sherwood *et al.*, 1979; Taverne *et al.*, 1982) but delivery of piglets can be achieved during indomethacin treatment by infusion of high doses of prostaglandin $F_{2\alpha}$ (Nara, 1979).

CATECHOLAMINES

In myometrial tissue obtained from pigs in oestrus, dioestrus and from pregnant sows, the presence of both α- and β-adrenoreceptors has been demonstrated by studying the response to noradrenalin and adrenalin before and after the addition of α- and β-blocking agents to the organ bath (Noreisch, 1973). This *in vitro* study explained the results obtained *in vivo* by Zerobin (1968) and Bower (1974), who showed that during dioestrus, pro-oestrus, oestrus and pregnancy a biphasic response (a stimulation followed by an inhibition of several minutes) was recorded after the application of 100 μg adrenalin; the stimulatory effect results from α-receptor stimulation and the inhibitory effect from β-receptor stimulation. Also Ngiam (1974) demonstrated *in vivo* in the post partum sow that inhibition of uterine contractions by adrenalin could be blocked by the β-blocking agent propanolol. The physiological implications of these experiments have been illustrated both during oestrus by Bower (1974) and during parturition by Naaktgeboren (1979) when inhibition of uterine contractions was recorded for a varying period of time after disturbances to the environment of the sows. The authors suggested that endogenous adrenalin was responsible for this inhibition.

Recently the highly selective β_2-receptor sympathicomimetic compound NAB 365 (Planipart®, Boehringer Ingelheim, West Germany) was found to postpone or interrupt the delivery process in sows for several hours if intravenous administration occurred either when gilts demonstrated pre-parturient milk production or after delivery of 1–3 piglets (Zerobin and

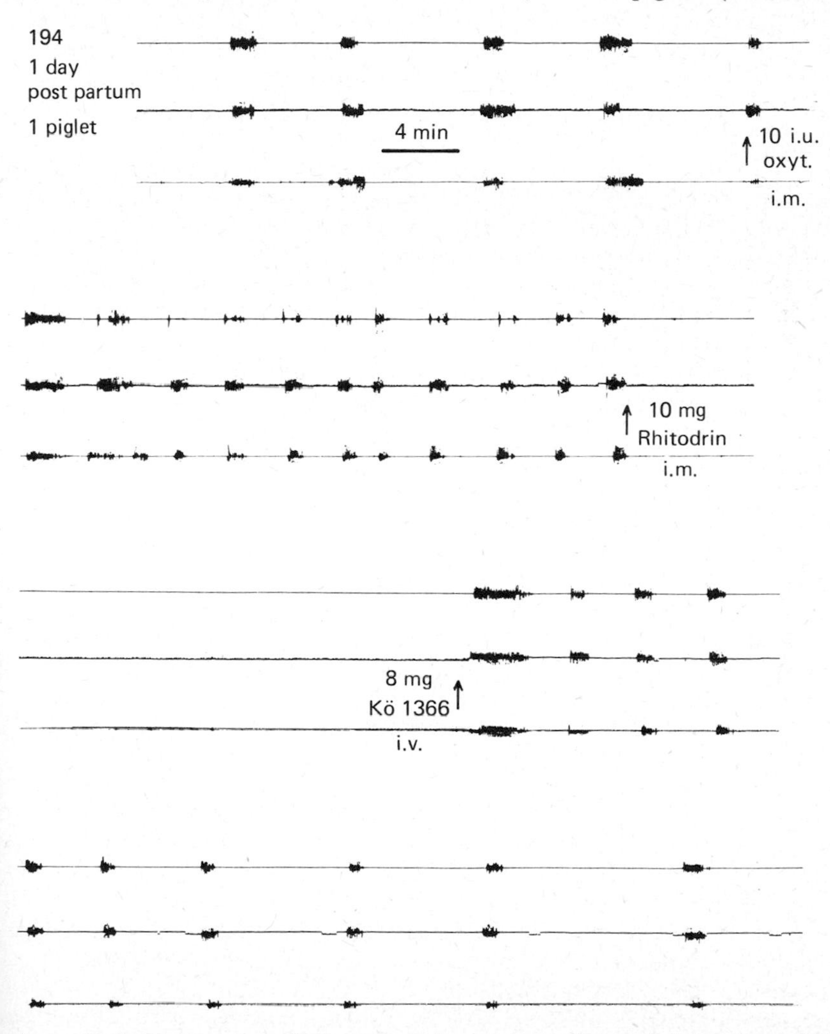

Figure 20.7 Electromyograph (in four continuous sections) recorded from three different sites of a uterine horn in a (Landrace) sow on the first day after parturition (one piglet was left with the sow). After stimulation with oxytocin (Oxytocin-S, Intervet, Boxmeer, The Netherlands), myometrial electrical activity was completely abolished by the betamimetic compound rhitodrin (Prepar, Philips Duphar, Weesp, The Netherlands) but restoration immediately occurred after the intravenous injection of the beta-blocking agent Kö 1366 (Boehringer Ingelheim, West Germany)

Kündig, 1980). No deleterious effects on mother or on offspring were observed. The tocolytic effect of the drug was demonstrated both by recordings of EMG activity and intrauterine pressure changes. The myometrium remained responsive to oxytocin during the tocolysis but surprisingly the betalytic compound Kö 1366 (Boehringer Ingelheim, West Germany; Müller-Tyl, Reinhold and Hernuss, 1974) did not antagonize the effects of Planipart (Zerobin and Kündig, 1980). This is in contrast with our own observations in the post partum sow (*Figure 20.7*) in which the β-blocking drug Kö 1366 (charge 51218) immediately restored uterine activity that had been abolished previously by the administration of the betamimetic agent rhitodrin (Prepar®, Philips Duphar, Weesp, The Netherlands), an agent frequently used in human obstetrics.

References

ASHDOWN, R.R. and MARRABLE, A.E. (1970). The development of the embryonic membranes in the pig: observations on the afterbirths. Res. *vet. Sci.* **11**, 227–231

BEVIER, G.W. and DZIUK, P.J. (1976). The effect of the number of fetuses and their location in the uterus on the incidence of stillbirth. *Proc. Int. Pig Vet. Soc.*, p.D21

BOSC, M.J., LOCATELLI, A., NICOLLE, A. and DU MESNIL DU BUISSON, F. (1976). Effect of inversion of one or both uterine horns on farrowing in sow. *Annls Biol. anim. Biochim. Biophys.* **16**, 645–648

BOWER, R.E. (1974). Factors affecting myometrial activity in the pig. PhD Thesis. University of Minnesota

CARTER, A.M., NAAKTGEBOREN, C. and VAN ZON- VAN WAGTENDONK, A.M. (1971). Parturition in the rabbit: spontaneous uterine activity during late pregnancy, parturition and the post partum period and its relation to normal behaviour. *Eur. J. Obstet. Gynec.* **2**, 37–68

CHARD, T. and FORSLING, M.L. (1976). *Hormones in Blood*, (H.N. Antoniades, Ed.). pp. 485–516. Harvard University Press.

COLENBRANDER, B. (1974). The influence of oestradiol and progesterone on the catecholamine content of the genital tract of the sow. *Acta morph neerl. scand.* **12**, 243–256

CORT, N., EINARSSON, S. and VIRING, S. (1979). Actions of oxytocin and a long acting carba oxytocin analog on the porcine myometrium *in vitro* and *in vivo*. *Am. J. vet. Res.* **40**, 430–432

CSAPO, A.J., ESKOLA, J. and RUTTNER, Z. (1980). The biological meaning of progesterone levels. *Prostaglandins* **19**, 203–211

DIAS E SILVA, U. (1979). In-vitro Untersuchungen zur Beeinfluszbarkeit isolierter Uterusstreifen von Sauen mit unterschiedlicher Geburtsdauer. *Inaugural Dissertation Tierärztliche Hochschule, Hannover.*

DIELEMAN and SCHOENMAKERS (1979). Radioimmunoassays to determine the presence of progesterone and estrone in the starfish *Asterias rubens. Gen. Comp. Endocr.* **39**, 534–542

DÖCKE, F. and WORCH, H. (1963). Untersuchungen über die Uterusmotilität und die Paarungsreaktionen der Sau. *Zuchthygiene* **7**, 169–178

DZIUK, P.J. (1979). Control and mechanics of parturition in the pig. *Anim. Reprod. Sci.* **2**, 335–342

DZIUK, P.J. and HARMON, B.G. (1969). Succession of fetuses at parturition in the pig. *Am. J. vet. Res.* **30**, 419–421

ELLENDORFF, F., TAVERNE, M., ELSAESSER, F., FORSLING, M., PARVIZI, N., NAAKTGEBOREN, C. and SMIDT, D. (1979a). Endocrinology of parturition in the pig. *Anim. Reprod. Sci.* **2**, 323–334

ELLENDORFF, F., FORSLING, M., PARVIZI, N., WILLIAMS, H., TAVERNE, M. and SMIDT, D. (1979b). Plasma oxytocin and vasopressin concentrations in response to prostaglandin injection into the pig. *J. Reprod. Fert.* **56**, 573–577

ELLICOTT, A.R. and DZIUK, P.J. (1973). Minimum daily dose of progesterone and plasma concentrations for maintenance of pregnancy in ovariectomized gilts. *Biol. Reprod.* **9**, 300–304

FIRST, N.L. and BOSC, M.J. (1979). Proposed mechanisms controlling parturition and the induction of parturition in swine. *J. Anim. Sci.* **48**, 1407–1421

FLINT, A.P.F., RICKETTS, A.P. and CRAIG, V.A. (1979). The control of placental steroid synthesis at parturition in domestic animals. *Anim. Reprod. Sci.* **2**, 239–251

FORSLING, M.L., MACDONALD, A.A. and ELLENDORFF, F. (1979). The neurohypophysial hormones. *Anim. Reprod. Sci.* **2**, 43–56

FORSLING, M.L., TAVERNE, M.A.M., PARVIZI, N., ELSAESSER, F., SMIDT, D. and ELLENDORFF, F. (1979). Plasma oxytocin and steroid concentrations during late pregnancy, parturition and lactation in the miniature pig. *J. Endocr.* **82**, 61–69

JÖCHLE, W., OROZCO, L., ZEROBIN, K., ESPARZA, H. and HIDALGO, M.A. (1974). Effects of a progestin on parturition and the post-partum period in pigs. *Theriogenology* **2**, 11–20

KEYE, J.D. (1923). Periodic variations in spontaneous contractions of uterine muscle, in relation to the oestrous cycle and early pregnancy. *Johns Hopkins Hosp. Bull.* **384**, 60–63

KING, J.L. (1927). Observations on the activity and working power of the uterine muscle of the non-pregnant sow. *Am. J. Physiol.* **81**, 725–737

LEIBRECHT, R. (1953). Uber die Struktur der Ringmuskelschicht am Uterus von Rind und Schwein and ihre funktionelle Bedeutung. *Inaugural Dissertation Tierärztliche Fakultät, Universität München.*

LUDWIG, K.S. (1952). Die Architektur der Muskelwand im Rattenuterus. *Acta Anat.* **15**, 23–41

LYE, S.J. (1980). The hormonal control of myometrial activity in the sheep and rat. PhD Thesis. University of Bristol

MARTIN, K.A. (1980). Effects of partial fetectomy in the sow. Thesis. University of Guelph

MINAR, M. and SCHILLING, E. (1970). Die Beeinflussung des Geburtstermins beim Schwein durch gestagene Hormone. *Dt. tierärztl. Wschr.* **77**, 428–431

MOELJONO, M.P.E., THATCHER, W.W., BAZER, E.W., FRANK, M., OWENS, L.J. and WILCOX, C.J. (1977). A study of prostaglandin $F_{2\alpha}$ as the luteolysin in swine: II. Characterization and comparison of prostaglandin $F_{2\alpha}$, estrogen and progestin concentrations in utero-ovarian vein plasma of non-pregnant and pregnant gilts. *Prostaglandins* **14**, 543–555

MÜLLER-TYL, E., REINHOLD, E. and HERNUSS, P. (1974). Gleichzeitige

Anwendung einer beta-mimetischen und beta-rezeptorenblockierenden Substanz bei der Wehenhemmung. *Z. Geburtsh. Perinat.* **178**, 128–134

NAAKTGEBOREN, C. (1979). Behavioural aspects of parturition. *Anim. Reprod. Sci.* **2**, 155–166

NAAKTGEBOREN, C. and CARTER, A.M. (1971). *Oryctolagus cuniculus* (Leporidae): Uterusaktivität wahrend der Geburt. In *Encyclopaedia Cinematographica*, (G. Wolf, Ed.), Film E 1649

NAAKTGEBOREN, C., POOL, C., VAN DER WEYDEN, G.C., TAVERNE, M.A.M., SCHOOF, A.G. and KROON, C.H. (1975). Elektrophysiologische Untersuchungen über die Uteruskontraktionen des Schafes während der Trächtigkeit und der Geburt. *Z. Tierzucht. Zücht Biol.* **92**, 220–243

NAGLER, M. (1956). Untersuchungen uber Struktur und Funktion des Schweine-uterus. *Inaugural Dissertation Tierärztliche Fakultät, Universität München*

NARA, B.S. (1979). Mechanisms controlling prepartum luteolysis in swine. PhD Thesis. University of Wisconsin, Madison, USA

NGIAM, T.T. (1974). A study of the involuting porcine uterus with special reference to its histology, histochemistry and its response to bacterial infection. PhD Thesis. Royal Veterinary College, University of London

NGIAM, T.T. (1977). A study of the motility of the uterus of the sow during the peri-parturient period. *Singapore vet. J.* **1**, 13–27

NOREISCH, W. (1973). Adrenozeptoren im Myometrium des Schweines. *Inaugural Dissertation Tierärztliche Fakultät, Universität München*

PERRY, J.S. (1954). Parturition in the pig. *Vet. Rec.* **66**, 706–708

PRUD'HOMME, M.J. (1980). Effect of an inhibition of prostaglandin synthesis on uterine motility in the ovariectomized ewe during induced oestrus. A preliminary report. *Theriogenology* **14**, 349–359

PRUD'HOMME, M.J. and BOSC, M.J. (1977). Motricité utérine de la brebis, avant, pendant et après la parturition spontanée ou après traitement par la dexamethasone. *Annls Biol. anim. Biochim. Biophys.* **16**, 645–648

RANDALL, G.C.B. (1972). Observations on parturition in the sow. II. Factors influencing stillbirth and perinatal mortality. *Vet. Rec.* **90**, 183–186

RANDALL, G.C.B. (1979). Studies on the effect of acute asphyxia on the fetal pig *in utero*. *Biol. Neonate* **36**, 63–69

RÜSSE, M.W. (1972). Die Bedeutung von Prostaglandin E_2 in der Steuerung der Kontraktionen des Uterus. 1. Mitteilung: In-vitro Unter-suchungen am Myometrium des Schweines. *Zuchthygiene* **7**, 162–169

SCHEERBOOM, J.E.M. and TAVERNE, M.A.M. (1981). Combined electromyography and real-time ultrasound scanning of the pregnant uterus of the ewe. *Eur. J. Obstet. Gynec.* and *reprod. Biol.*, in press

SHERWOOD, O.D., NARA, B.S., CRNEKOVIC, V.E. and FIRST, N.L. (1979). Relaxin concentrations in pig plasma after the administration of indomethacin and prostaglandin $F_{2\alpha}$ during late pregnancy. *Endocrinology* **104**, 1716–1721

SHILLE, V.M., KARLBOM, I., EINARSSON, S., LARSSON, K., KINDAHL, H. and EDQVIST, L.E. (1979). Concentrations of progesterone and 15-keto-13,14-dihydro-prostaglandin $F_{2\alpha}$ in peripheral plasma during estrus cycle and early pregnancy in gilts. *Zentbl. VetMed.* **A26**, 169–181

SILVER, M., BARNES, R.J., COMLINE, R.S., FOWDEN, A.L., CLOVER, L. and MITCHELL, M.D. (1979). Prostaglandins in the foetal pig and prepartum endocrine changes in mother and foetus. *Anim. Reprod. Sci.* **2**, 305–322

SOLOFF, M.S. (1975). Uterine receptor for oxytocin: effects of estrogen. *Biochem. Biophys. Res. Commun.* **65**, 205–212

SOLOFF, M.S. and SWARTZ, T.L. (1974). Characterization of a proposed oxytocin receptor in the uterus of the rat and sow. *J. biol. Chem.* **249**, 1376–1381

SPRECHER, D.J., LEMAN, A.D., DZIUK, P.J., CROPPER, M. and DEDECKER, M. (1974). Causes and control of swine stillbirths. *J. Am. vet. med. Ass.* **165**, 698–701

TAVERNE, M. (1979). Physiological aspects of parturition in the pig. PhD Thesis. University of Utrecht

TAVERNE, M.A.M., NAAKTGEBOREN, C. and VAN DER WEYDEN, G.C. (1979). Myometrial activity and expulsion of fetuses. *Anim. Reprod. Sci.* **2**, 117–131

TAVERNE, M.A.M., VAN DER WEYDEN, G.C. and FONTIJNE, P. (1979). Preliminary observations on myometrial electrical activity before, during and after parturition in the cow. In *Calving Problems and Early Viability of the Calf*, (B. Hoffman, I.L. Mason and J. Schmidt, Eds.). *Current Topics in Veterinary Medicine and Animal Science* **4**, 297–311

TAVERNE, M., BEVERS, M., BRADSHAW, J., DIELEMAN, S.J., WILLEMSE, A.H. and PORTER, D.G. (1982). Plasma concentrations of prolactin, progesterone, relaxin and oestradiol-17β in sows treated with either progesterone, bromoergocryptine or indomethacin during late pregnancy. *J. Reprod. Fert.* **65**, in press

TAVERNE, M.A.M., VAN DER WEYDEN, G.C., FONTIJNE, P., ELLENDORFF, F., NAAKTGEBOREN, C. and SMIDT, D. (1977). Uterine position and presentation of mini-pig fetuses and their order and presentation at birth. *Am. J. vet. Res.* **38**, 1761–1764

TAVERNE, M.A.M., NAAKTGEBOREN, C., ELSAESSER, F., FORSLING, M.L., VAN DER WEYDEN, G.C., ELLENDORFF, F. and SMIDT, D. (1979). Myometrial electrical activity and plasma concentrations of progesterone, oestrogens and oxytocin during late pregnancy and parturition in the miniature pig. *Biol. Reprod.* **21**, 1125–1134

WELK, F. and FIRST, N.L. (1979). Effect of oxytocin on the synchrony of parturition induced by $PGF_{2\alpha}$ (ICI 80996). *J. Anim. Sci.* **49** (*Suppl. 1*), 347–348

VAN DER WEYDEN, G.C., TAVERNE, M.A.M., DIELEMAN, S.J. and FONTIJNE, P. (1981a). Myometrial electrical activity throughout the entire course of pregnancy in the ewe. *Eur. J. Obstet. Gynec. reprod. Biol.* **11**, 347–354

VAN DER WEYDEN, G.C., TAVERNE, M.A.M., OKKENS, A.C. and FONTIJNE, P. (1981b). The intrauterine position of canine foetuses and their sequence of expulsion at birth. *J. Small Anim. Pract.*, **22**, 503–510

WRATHALL, A.E. (1980). Pathology of the ovary and ovarian disorders in the sow. *Proc. 9th Int. Congr. Anim. Reprod. A.I., Madrid, 1980*, Vol. I, pp. 223–244

ZEROBIN, K. (1968). Untersuchungen über die Uterusmotorik des Schweines. *Zentbl. VetMed.* **A15**, 740–798

ZEROBIN, K. and KÜNDIG, H. (1980). The control of myometrial functions during parturition with a β_2-mimetic compound, Planipart®. *Theriogenology* **14**, 21–35

ZEROBIN, K. and SPÖRRI, H. (1972). Motility of the bovine and porcine uterus and Fallopian tube. *Adv. vet. Sci. comp. Med.* **16**, 303–354

V

REPRODUCTIVE FUNCTION IN THE POST-PARTUM SOW

21

THE ENDOCRINOLOGY OF THE POST-PARTUM SOW

S. EDWARDS
Department of Physiology and Environmental Studies, University of Nottingham, UK†

A discussion on the reproductive physiology of the post-partum sow in the context of a symposium on the Control of Pig Reproduction must include a survey of the management practices currently in use during pregnancy and lactation and a consideration of the current state of knowledge relating to the endocrine events associated with lactation. The post-partum rebreeding of sows is one of the few areas of pig production which is still open to significant improvement in that losses in productivity still occur consistently at this time. However, it is not the purpose of this chapter to review the changing management practices being adopted, but more to summarize the endocrine status of the sow after parturition and through weaning to the first post-weaning oestrus.

In general there is a change in hypothalamic, pituitary and ovarian activity during pregnancy, induced by progesterone, which leads to the observed failure of ovulation in the pig and other species. If lactation does not occur, i.e. the offspring are weaned at birth or shortly afterwards, there is an interval post-partum during which reproductive cycles and ovulation are still suppressed. The suppression of reproductive activity in the absence of suckling and lactation can only be associated with a post-partum block on ovulation. In a majority of situations, however, this post-partum block on reproductive activity is complicated by the influence of suckling and lactation (Lamming, 1978). Reproduction post-partum is therefore dependent upon, and intimately concerned with, the complex interrelationships of the endocrine events controlling lactation. It is, therefore, relevant to examine the development of the axis controlling gonadotrophin secretion and to determine the stages of lactation at which its functional components become operational and also to propose ideas which may account for the lack of operational competence prior to this time.

The endocrinology of the sow during lactation

Since the two major groups of ovarian steroids, the progestagens and the oestrogens, are both capable of reducing the pituitary secretion of gonadotrophic hormones they may, by such means, directly suppress the

†Present address: An Foras Talúntais, Grange Research Station, Dunsany, Co. Meath, Ireland.

occurrence of oestrus and ovulation post-partum. However, a consideration of the literature available indicates that the corpora lutea of pregnancy are in a state of regression shortly before parturition, and for the first 2–3 days of lactation are represented by inactive corpora albicantia (Warnick, Casida and Grummer, 1950; Burger, 1952; Palmer, Teague and Venzke, 1965a,b). Associated with these morphological changes, the plasma levels of progesterone decrease sharply around the time of parturition and remain low throughout lactation (Ash and Heap, 1975; Baldwin and Stabenfeldt, 1975; Parvizi *et al.*, 1976).

The demise of the corpora lutea at farrowing is associated with an increase in both the plasma (Sasser *et al.*, 1973; Ash *et al.*, 1973; Robertson and King, 1974; Ash and Heap, 1975) and urinary (Raeside, 1963) concentrations of oestrogens, which then fall rapidly after parturition and remain low throughout lactation (Ash *et al.*, 1972; 1973; Edqvist, Einarsson and Settergren, 1974; Ash and Heap, 1975; Stevenson, Cox and Britt, 1981) as illustrated in *Figure 21.1*. Therefore, a block on pituitary

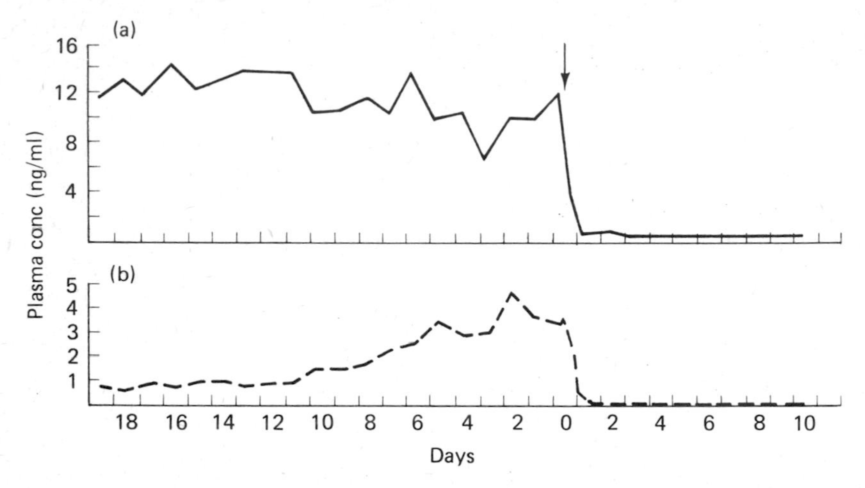

Figure 21.1 Plasma concentrations of (a) progesterone and (b) total unconjugated oestrogens in a sow during late pregnancy, parturition and early lactation. Day 0 and the arrow denote the day of parturition. From Ash and Heap (1975)

gonadotrophin secretion induced by ovarian steroids would not appear to account for the lack of ovulation seen in the lactating sow.

It has commonly been observed that a proportion of sows show behavioural oestrus at or just after farrowing (Warnick, Casida and Grummer, 1950; Burger, 1952; Baker *et al.*, 1953; Self and Grummer, 1958) and whilst the proportion of sows showing this farrowing oestrus varies, it can be as high as 100%. Since no surge of luteinizing hormone (Parvizi *et al.*, 1976) or ovulation (Burger, 1952 and others) occurs at this oestrus the overt signs of heat have been attributed to the high levels of oestrogens seen at parturition, since no subsequent rise in oestrogens occurred in sows showing such an oestrus (Holness and Hunter, 1975). Hence lactation in the sow would appear to be associated with a period of ovarian follicular quiescence resulting in anoestrus and anovulation. Studies on the morphology of the sow ovary during lactation have indicated that follicle size and number decreases significantly during the first week of lactation but thereafter increases steadily (Palmer, Teague and Venzke, 1965a,b). Information from this and other laboratories has indicated that these follicles may reach a size capable of oestrogen secretion by 21–35 days post-partum. However, oestrus and ovulation do not normally occur until after weaning (Edqvist, Einarrson and Settergren, 1974; Edwards, 1980).

The two gonadotrophic hormones secreted by the anterior pituitary, follicle stimulating hormone (FSH) and luteinizing hormone (LH), are themselves under hypothalamic control via the releasing factor(s) gonadotrophin-releasing hormone(s) (GnRH). The difficulties in assaying protein hormones, coupled with the technical difficulties in using the pig as an experimental animal, has meant that little data has been available until recently relating to the synthesis and release of gonadotrophins during lactation. This information is critical to our understanding of the factors controlling lactational anoestrus, since data previously discussed suggest that a lack of gonadotrophic stimulation may account for the depressed follicular activity associated with lactation.

The earliest data available on pituitary FSH and LH concentrations indicate that pituitary FSH is high, with little evidence of change throughout lactation (Lauderdale *et al.*, 1965; Melampy *et al.*, 1966; Crighton and Lamming, 1969). In contrast there is a sharp drop in pituitary LH activity between the end of pregnancy and day 14 of lactation (Melampy *et al.*, 1966) which is maintained even after longer periods of lactation up to 56 days (Crighton, 1967). These data were interpreted by Crighton and Lamming (1969) as evidence for a lactation and/or suckling induced block on FSH release (but not synthesis), causing a lack of follicular growth and an inhibition of LH synthesis (and consequently release). However, a re-examination of the information on follicular development post-partum suggests that the pituitary secretions gradually escape from the inhibitory effects of suckling and lactation as follicular size and number increase gradually after the first 7–10 days of lactation (Palmer, Teague and Venzke, 1965a,b). Aherne *et al.* (1976) have reported levels of FSH in the plasma of lactating sows at three weeks post-partum which were similar to those observed by Rayford *et al.* (1974) in cyclic sows and gilts. Furthermore FSH levels appear to increase as lactation progresses particularly during the latter stages of a five-week lactation (Edwards, 1980; Stevenson, Cox and Britt, 1981) as illustrated in *Figure 21.2*. Whether this elevated secretion of FSH serves any physiological function is unclear at present, but it appears to reflect the aforementioned changes in ovarian follicular development seen as lactation progresses. The factors controlling FSH secretion during lactation are still undetermined but appear to involve an ovarian secretion (probably non-steroidal) which inhibits FSH release early in lactation (Stevenson, Cox and Britt, 1981; see *Figure 21.2*). This factor is probably folliculostatin, the putative inhibitor of FSH production (Campbell and Schwarz, 1979). As lactation progresses the secretion of this substance may serve only to limit excess, but not apparently normal, FSH secretion.

LH secretion during lactation is low (Parvizi *et al.*, 1976; Stevenson, Cox and Britt, 1981), reflecting the low pituitary content of LH after the first

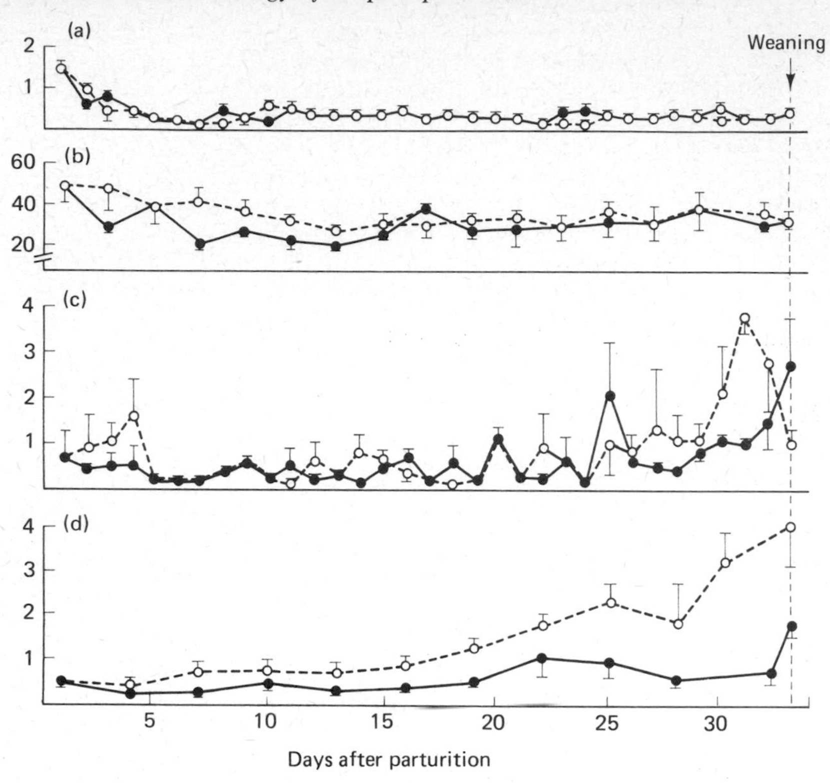

Figure 21.2 Serum concentrations of (a) progesterone (ng/ml), (b) total oestrogens (pg/ml), (c) LH (ng/ml) and (d) FSH (ng/ml) during lactation in sows ovariectomized (○----○) 2–4 days after parturition or left intact (●——●). Note that FSH concentrations are consistently higher after ovariectomy whereas LH concentrations are not different. By courtesy of Stevenson, Cox and Britt (1981)

week of lactation (*Figure 21.2*), but equivocal data exist as to whether basal LH secretion during the latter stages of lactation in particular differs from that seen throughout the oestrous cycle. Whereas Booman and van de Wiel (1980) report that basal LH is suppressed, others (Parvizi *et al.*, 1976; Edwards, 1980) have shown that basal LH secretion is not significantly depressed during lactation. In comparison, Stevenson and Britt (1981) report that whilst LH secretion is suppressed early in lactation (day 7) it gradually increases during late lactation (21 days). A detailed study of the mechanisms controlling LH secretion is required before such equivocal data can be resolved, but work to date has indicated that, unlike FSH, ovarian secretions play no part in any inhibition of LH release which may occur during lactation (*Figure 21.2*) (Crighton and Lamming, 1969; Parvizi *et al.*, 1976; Stevenson, Cox and Britt, 1981). Suckling may, however, directly suppress the release of LH releasing hormone, since hypothalamic stores of GnRH are lower at weaning in the pig than at any time during the weaning to oestrus interval (Cox and Britt, 1981) and by such means maintain depressed basal LH secretion during lactation as has been postulated in the rat (Minaguchi and Meites, 1967).

A hormone classically associated with lactation and more recently with reproduction, is prolactin. Since analytical methods have become available to study the patterns of prolactin secretion *in vivo* its role in controlling lactational anoestrus in several species, including the pig, has been re-assessed. Nursing sows produce high levels of prolactin which decline gradually as lactation progresses, but are still 20–40 times basal levels at five weeks post-partum (van Landeghem and van de Wiel, 1978; Bevers, Willemse and Kruip, 1978; Stevenson, Cox and Britt, 1981) as illustrated in *Figure 21.3*. The ability of prolactin to block the stimulatory effects of the

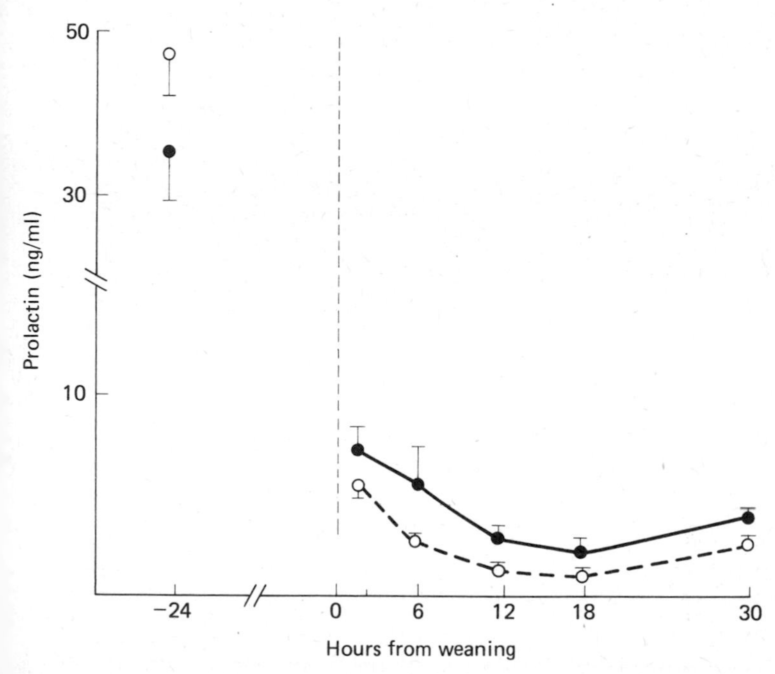

Figure 21.3 Serum prolactin concentrations in sows weaned 29–36 days post-partum. Note that a precipitous decline in prolactin concentrations occurs reaching basal values within 12 hours of weaning. By courtesy of Stevenson, Cox and Britt (1981)

gonadotrophic hormones has been demonstrated in the human female in that it disturbs the normal steroid secreting capacity of Graafian follicles cultured *in vitro* (McNatty, Sawers and McNeilly, 1974) and also prevents any increase in steroid output in response to endogenous or exogenous gonadotrophins *in vivo* (Seddon, 1970; Zarate *et al.*, 1972; Reyes, Winter and Faiman, 1972; Rolland *et al.*, 1975). In addition, evidence has accumulated in recent years for a dysfunction in the hypothalamic response to a positive oestrogen feedback signal, i.e. the induction of a preovulatory LH surge is blocked in hyperprolactinaemic women (Glass *et al.*, 1975; 1976), sheep (Kann, Martinet and Schirar, 1976) and cattle (Radford, Nancarrow and Mattner, 1976; 1978) and a similar mechanism appears to be present in the lactating sow (Elsaesser and Parvizi, 1980) particularly during early lactation. The anti-gonadotrophic effects of prolactin have not been tested directly in the lactating sow but attempts to induce ovulation with gonadotrophin preparations such as pregnant mare's serum gonadotrophin (PMSG), human chorionic gonadotrophin (HCG) and gonadotrophin releasing hormone (GnRH) have had limited success before 21 days post-partum (Cole and Hughes, 1946; Heitman and Cole, 1956; Guthrie, Pursel and Frobish, 1978) indicating a block on the responsiveness of the ovary to gonadotrophic stimulation; this may, in part, be induced by the elevated plasma levels of prolactin.

In summary the endocrine events which control lactational anoestrus in the sow are still poorly understood but available data suggests that a combination of factors results in anovulation post-partum. The elevated levels of prolactin seen during lactation and maintained by suckling, may play a multifactorial role in that they may block both the effects of normal or elevated FSH secretions whilst also preventing positive oestrogen feedback where follicular development and oestrogen synthesis are present. Suckling, possibly by a direct neural suppression of LH releasing hormone secretion, also appears to depress LH, but not FSH, secretion and may therefore disturb the delicate FSH/LH balance which is required for sustained follicular growth and development. Whilst other exteroceptive stimuli associated with piglet presence, e.g. sight, sound and smell, may have a complementary role to play in the maintenance of ovarian inactivity during lactation, this role would appear to be less important than that exerted by suckling.

The endocrinology of the sow at weaning

The total removal of the sow from her litter at weaning after 3–5 weeks of lactation normally results in an acceleration of follicular growth culminating in overt oestrus and ovulation within 4–8 days. It is the aim of this section to elucidate the factor(s) responsible for triggering and maintaining such follicular activity.

Pituitary levels of FSH change little in the immediate post-weaning period (Crighton and Lamming, 1969) and an examination of plasma FSH has indicated a similar pattern when sows are weaned after either three or five weeks of lactation, as shown in *Figure 21.4* (Edwards, 1980).

In contrast pituitary LH levels rise significantly at weaning (Crighton and Lamming, 1969) indicating that weaning may functionally re-establish the synthesis and secretion of LH. Although daily measurements have indicated that plasma LH remains generally low at weaning (Aherne *et al.*, 1976; Parvizi *et al.*, 1976), very recent information has shown that a transient increase in basal LH secretion occurs at weaning and lasts 1–2 days as illustrated in *Figure 21.5* (van de Wiel *et al.*, 1979; Edwards, 1980). Furthermore this increase in LH secretion is associated with similar increases in hypothalamic GnRH content (Cox and Britt, 1981) within 60 hours of weaning, thereby increasing the potential for LH synthesis by the pituitary in preparation for the sustained LH secretion which occurs at oestrus.

Hence the increase in LH concentrations, coupled with normal or elevated FSH and decreased prolactin secretion, may represent the combination of factors (or trigger) required to stimulate follicular growth after weaning. The relative importance of each of these factors, and particularly their relationship to the stimuli generated by the piglets, has

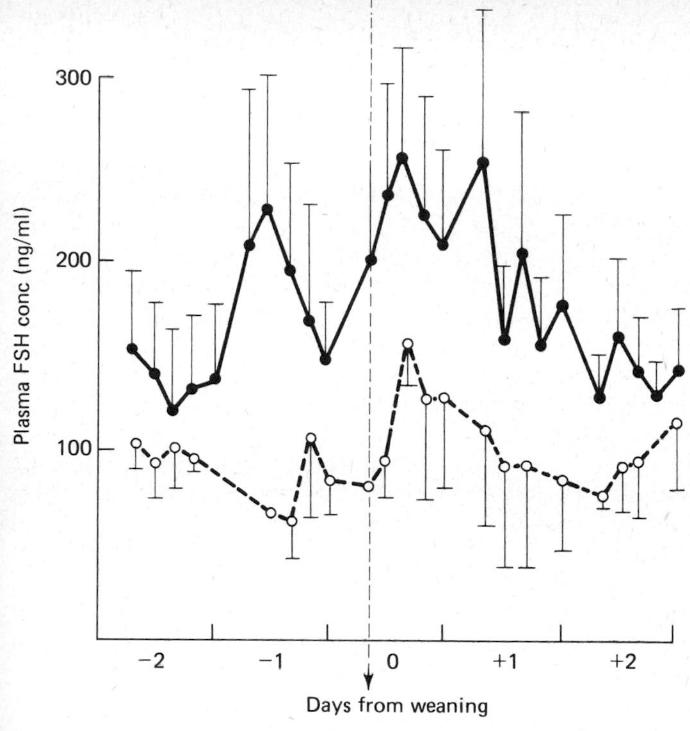

Figure 21.4 Mean plasma FSH concentrations for 3-week (○- - -○) and 5-week (●——●) weaned sows. The vertical dotted line indicates weaning. From Edwards (1980)

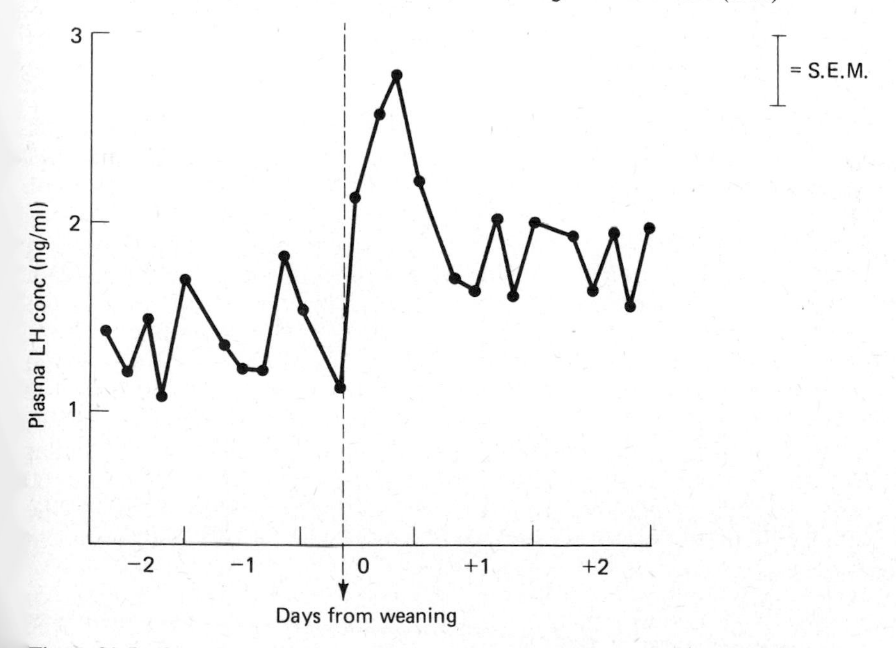

Figure 21.5 Overall mean plasma LH concentrations for 3-week and 5-week weaned sows synchronized to 08.00 hours on the day of weaning as indicated by the vertical dotted line. From Edwards (1980)

not yet been determined. However, Booman and van de Wiel (1980) have provided some useful information in this regard. This elegant series of experiments has indicated that:

(a) Separation of the sow and her litter for periods of four hours or more results in a rapid decline in plasma prolactin levels associated with an increase in basal LH secretion.
(b) Administration of exogenous prolactin does not completely prevent the increase in LH levels caused by removal of the piglets; however the rise in LH concentration is significantly reduced in sows infused with prolactin compared with those which are not.
(c) The suckling stimulus is the most potent stimulus generated by the piglets, when compared with sight, sound and smell, for increasing plasma prolactin and inhibiting LH secretion.

The consequences of the sustained follicular growth occurring after weaning will be considered in the following section.

Endocrine changes at the post-weaning oestrus

The endocrine changes associated with the post-weaning oestrus are basically similar to those seen in the oestrous cycle, but some interesting and potentially significant differences are apparent which appear to be associated with the length of the preceding lactation.

Two to six days after removal of the sow from her litter plasma oestrogens begin to rise and remain elevated for 2–3 days prior to, and partially including, oestrus (Ash and Heap, 1975; Aherne *et al.*, 1976; van de Wiel *et al.*, 1979; Edwards, 1980), as illustrated in *Figure 21.6*. Peak levels of oestrogen comparable to, or marginally lower than, those seen during the oestrous cycle are observed and the data also suggest that neither the length of lactation (Edwards, 1980) nor the interval from weaning to oestrus (Ash and Heap, 1975; van de Wiel *et al.*, 1979) has any effect on the peak levels of oestrogen attained in sows weaned after lactations ranging from 2–6 weeks, although the numbers of animals are of necessity small in such experiments.

Oestrogen secretion is rapidly terminated by the pre-ovulatory surge of LH which occurs at oestrus (Aherne *et al.*, 1976; Edwards, 1980; see *Figure 21.6*), an observation which is consistent with the reported ability of LH to terminate oestrogen secretion by cultured sheep Graafian follicles *in vitro* (Moor, 1974) and which also correlates well with observations from cyclic sows (see Chapter 8).

The characterization of this pre-ovulatory surge of LH has recently formed part of a detailed study of the reproductive physiology of the post-partum sow (Edwards, 1980). Since the pre-ovulatory surge of LH is directly responsible for ovulation, any changes in its characteristics, particularly in relation to the length of lactation, could have a major influence on the post-weaning reproductive performance of sows. A comparison of the amounts of circulating LH during the pre-ovulatory surge in 3-week or 5-week weaned sows is shown in *Figure 21.7*. A significant

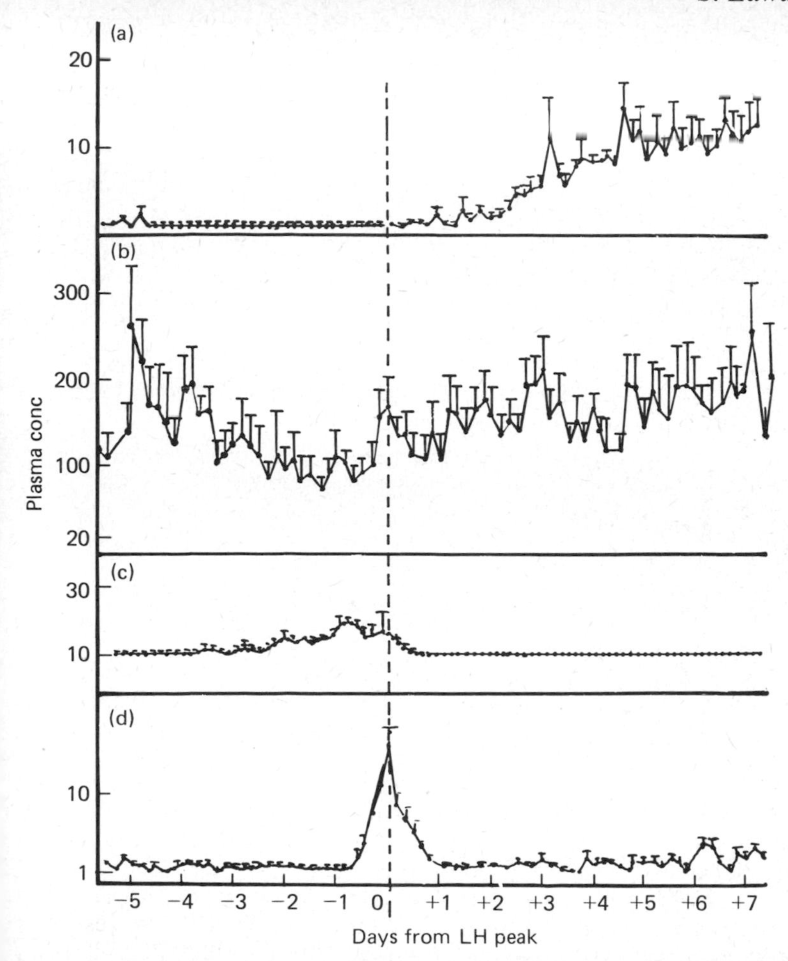

Figure 21.6 Mean plasma concentrations of (a) progesterone (ng/ml), (b) FSH (ng/ml) (c) oestradiol-17β (pg/ml) and (d) LH (ng/ml) normalized around the first pre-ovulatory LH surge after weaning. From Edwards (1980)

depression in terms of mean peak height and total LH secreted (area under the curve) is seen after a 3-week compared with a 5-week lactation (or to cyclic sows at oestrus). Although the physiological basis for this difference is not clear a decrease in the responsiveness to GnRH induced by prolonged progesterone secretion during pregnancy and carried over into the post-partum period, may be inferred from studies in other species (Jequier, Vanthuyne and Jacobs, 1973; Jenkin and Heap, 1974; Webb *et al.*, 1977). A similar recovery of the LH response to the positive feedback of oestradiol benzoate during lactation has recently been demonstrated in the pig (Elsaesser and Parvizi, 1980) and lends further support to the idea that the post-partum interval *per se* and not a reduction in the intensity of the suckling stimulus in late lactation, is responsible for the increased LH levels seen during the pre-ovulatory surges of 5-week weaned sows. The importance of these observations is obvious when

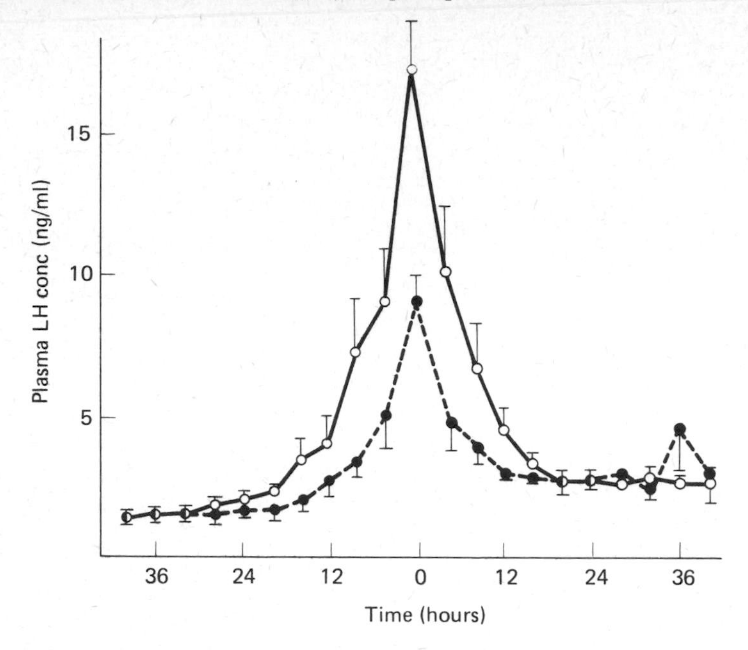

Figure 21.7 Mean plasma LH concentrations for 3-week (●----●) and 5-week (○——○) weaned sows during the first pre-ovulatory LH surge after weaning. Time 0 represents the peak LH value observed. From Edwards (1980)

considering the lowered fecundity often associated with shorter lactations; reproductive performance may be seriously affected by early weaning, particularly in herds where reproductive problems are endemic.

Since LH secretion is so significantly affected during this critical phase of the sow's breeding cycle it is pertinent to examine the changes occurring in the secretion of the second pituitary gonadotrophin, FSH. A decline in pituitary FSH levels, in concert with LH levels, at the post-weaning oestrus has been considered representative of hormone release and the trigger for follicular growth and ovulation (Crighton and Lamming, 1969). The idea of elevated FSH secretion accompanying follicular growth (and oestrogen secretion) has in recent years been modified by an analysis of the plasma concentrations of this hormone during both the follicular phase of the oestrous cycle (Wilfinger, 1974; Foxcroft and Edwards, unpublished observations) and that preceding the post-weaning oestrus (Aherne *et al.*, 1976; Edwards, 1980; Stevenson, Cox and Britt, 1981). FSH secretion is, in fact, at its lowest at the time of sustained follicular growth and maturation, when compared with either the luteal phase of the oestrous cycle, lactation or the first few days after weaning. Oestrogen is therefore capable of chronically inhibiting the secretion of FSH at a time when such a stimulus would appear to be most required and a detailed study of the receptor populations associated with follicular development in the pig is required before this anomalous situation can be resolved.

The determination of FSH levels in the plasma of cyclic sows has shown that approximately 50–60% of animals show a concomitant surge of LH and FSH (Wilfinger, 1974; Foxcroft and Edwards, unpublished observations) at oestrus. Of the studies in the weaned sow, one (Aherne *et al.*, 1976) has indicated the presence of a slight increase in FSH concentrations on the day of oestrus (when sampling ceased) following a 3-week lactation whereas a second, more detailed, study (Edwards, 1980) has indicated an effect of lactation length on the occurrence of synchronous LH/FSH surges at oestrus after weaning at 3-weeks or 5-weeks of lactation. Of sows weaned at five weeks post-partum 75% produced distinct, coincident surges of FSH and LH whilst none of the sows weaned after three weeks of lactation showed such a rise, indicating that the hypothalamic–pituitary response after a longer lactation more closely resembles that seen in the cyclic sow at oestrus in terms of both the characteristics and quantity of FSH and LH released; such decreases may not be of critical importance to the induction of ovulation or the continuance of cyclic activity however, since 40–50% of cyclic sows continue to ovulate and cycle in similar circumstances. However, the combination of significantly depressed LH secretion, coupled with lowered tonic secretion of FSH at weaning and during the pre-follicular secretion of oestrogen and the absence of an FSH surge during oestrus, indicates that the earlier weaning of sows results in a general lowering of gonadotrophin secretion (and potential ovarian stimulation) at a time when such stimulation is most needed. This supports the earlier suggestion that reduced gonadotrophic stimulation may be the cause of lowered fertility and ultimately lowered productivity often associated with severely shortened lactations.

The role of prolactin at ovulation is an obscure one, if indeed it exists. A period of elevated prolactin secretion occurs at or around oestrus in cyclic sows (Wilfinger, 1974; van Landeghem and van de Wiel, 1977), but information from weaned sows is limited and contradictory. Several authors (Bevers, Willemse and Kruip, 1978; van Landeghem and van de Wiel, 1978; Stevenson, Cox and Britt, 1981) have shown a rise in prolactin at oestrus, but Edwards (1980) failed to identify such a rise in weaned sows. Since the role of prolactin in ovulation is unclear, it is not surprising to find that blockade of prolactin secretion at this time does not interfere with the ovulation process (Niswender, 1972) and it has been suggested that the oestrogen trigger responsible for the initiation of the LH surge also initiates the rise in prolactin associated with oestrus in the pig (Bevers *et al.*, 1978; Stevenson, Cox and Britt, 1981) and other species (Chen and Meites, 1970; Schams, 1974).

What may be of greater importance to the continuance of reproductive cycles is the sustained secretion of FSH, in the absence of significant LH secretion, seen around the time of ovulation approximately 24–48 hours after the pre-ovulatory LH/FSH surge (Edwards, 1980; Stevenson, Cox and Britt, 1981). This phase of secretion is consistently observed in cyclic sows (Wilfinger, 1974; Foxcroft and Edwards, unpublished observations) and its characteristics are similar in sows weaned at 3 or 5 weeks of lactation, in contrast to the differences observed during the pre-ovulatory gonadotrophin surge. Although this phase of secretion is observed in cyclic and weaned sows after oestrus its physiological significance remains unclear. It would appear that normal negative feedback responses to exogenous oestrogen occur at this time, i.e. FSH secretion can be suppressed, but in such situations ovulation is not adversely affected (Edwards, 1980). The cumulative effects of such treatment, however, on subsequent follicular growth and ovulation are unknown; repeated treatment may decrease ovulation rate and/or follicular function and result in lowered fertility and/or anoestrus.

Following oestrus and ovulation progesterone levels rise sharply and enter a phase of secretion which is indistinguishable from the luteal phase of a normal oestrous cycle (Ash and Heap, 1975; Baldwin and Stabenfeldt, 1975; Parvizi *et al.*, 1976; Stevenson, Cox and Britt, 1981). There is no apparent effect of lactation length on the pattern or peak levels of progesterone secretion attained (Edwards, 1980) even when the litter is removed at birth (Ash and Heap, 1975) as shown in *Figure 21.8*. The

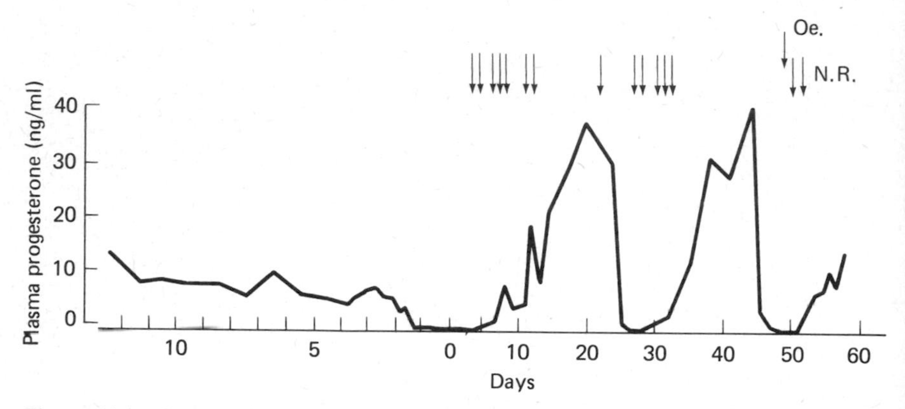

Figure 21.8 Early weaning and the plasma concentrations of progesterone. Parturition occurred on day 0, and the piglets were separated from the sow immediately after delivery. Oe (top arrow), oestrus and fertile mating: N.R. (lower arrows), non-receptive when tested with the boar. From Ash and Heap (1975)

significance of some small amounts of progesterone secreted in a minority of sows prior to the LH surge is not apparent. Since these fluctuations occurred at the time of elevated oestrogen secretion they may only reflect a metabolic by-product of such secretion (Edwards, 1980). However, no evidence for this type of secretion has been seen in the follicular phase of the normal oestrous cycle (Foxcroft and Edwards, unpublished observations).

A proportion of sows, however, fail to show oestrus and ovulation within 10–12 days of weaning and this failure of reproductive activity is associated with a lack of oestrogen production (Aherne *et al.*, 1976; van de Wiel *et al.*, 1979; Edwards, 1980) indicating a lack of ovarian responsiveness to gonadotrophin stimulation. Although the endocrine function in such sows is poorly understood due to a lack of basic data, it would appear that FSH levels are low initially and may rise within 1–3 days of weaning (Aherne *et al.*, 1976), or are normal and become elevated 3–6 days after weaning. This normal or elevated secretion of FSH may continue for at least two weeks post-weaning if oestrus is not stimulated. Basal LH levels appear to rise at weaning (Edwards, 1980) but, in contrast to sows returning to oestrus within 4–6 days, remain elevated or increase for 6–8 days after weaning. A period of depressed LH secretion may then ensue, associated with the secretion of oestrogen; conversely basal LH levels may remain elevated for periods in excess of 2–3 weeks post-weaning if oestrogen secretion is not stimulated. Similar changes in the pattern of gonadotrophin secretion have been observed in sows ovariectomized whilst lactating (Stevenson, Cox and Britt, 1981). In such sows oestrogen replacement 8–10 days post-weaning results in normal feedback responses and a pre-ovulatory LH surge on day 11, indicating that the protracted weaning to oestrus interval seen in some sows is the result of a lack of an adequate ovarian response and not due to a malfunction of the hypothalamus or pituitary. The factors controlling the ovarian response to gonadotrophins post-weaning are unclear at present but do not appear to include a carryover of the hypersecretion of prolactin from lactation into the post-weaning period (van de Wiel *et al.*, 1979; Edwards, 1980). Elegant studies in the rat indicate that the ovarian response to gonadotrophic stimulation is intimately concerned with the proper, sequential development of follicular gonadotrophin receptors (Richards, Rao and Ireland, 1978). Similar studies of the follicular receptor populations on the ovaries of lactating and weaned sows are needed to enable a clearer understanding of the factors limiting ovarian development both during lactation and after weaning.

Episodic LH secretion and ovarian function

Great emphasis in recent years has been placed on the pulsatile or episodic nature of LH secretion in the control of gonadal function in several species, e.g. sheep (Karsch *et al.*, 1978) and humans (Souvatzoglou *et al.*, 1973). The ability of gonadal steroids and in particular, oestrogen, to modulate this pulsatile LH secretion has been reported in pigs (Foxcroft, Pomerantz and Nalbandov, 1975) and Rhesus monkeys (Knobil, 1974) amongst others. Progesterone is thought to play a pivotal role in the control of the sheep oestrous cycle (Hauger, Karsch and Foster, 1977) by its modification of tonic or pulsatile LH secretion and similar changes in pulsatile LH patterns have been observed in the luteal phase of the sow oestrous cycle (Foxcroft and Edwards, unpublished observations). Whether these characteristic changes in pulsatile LH secretion are intimately concerned with ovarian function or simply reflect the changing patterns of steroids during the oestrous cycle has not yet been determined for the pig. Since lactation is associated with a block on sustained follicular development, and the period immediately after weaning represents a transitional phase in ovarian development, it would be useful to know if this transition is associated with any characteristic changes in pulsatile gonadotrophin output. Evidence has already been presented for a weaning-associated rise in basal LH secretion which resembles that seen in the follicular phase of the sheep oestrous cycle. Studies in Nottingham (Edwards, 1980) in the lactating and weaned sow have indicated that (see *Figures 21.9* and *21.10*):

(a) the length of lactation has no consistent effect on the characteristics of pulsatile LH secretion;

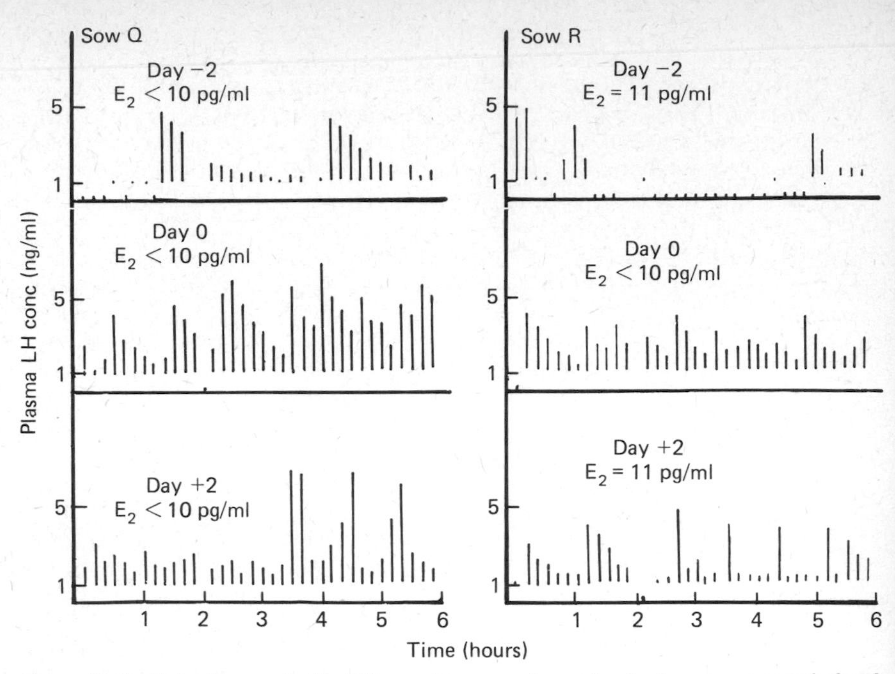

Figure 21.9 Episodic or pulsatile LH secretion around weaning for two sows weaned after 3 weeks of lactation. Weaning occurred on day 0 and plasma progesterone concentrations were basal at all times. E_2 = oestradiol concentration. Note the rise in LH baseline secretion at day 0 in the absence of any consistent effect on the characteristics of pulsatile secretion. From Edwards (1980)

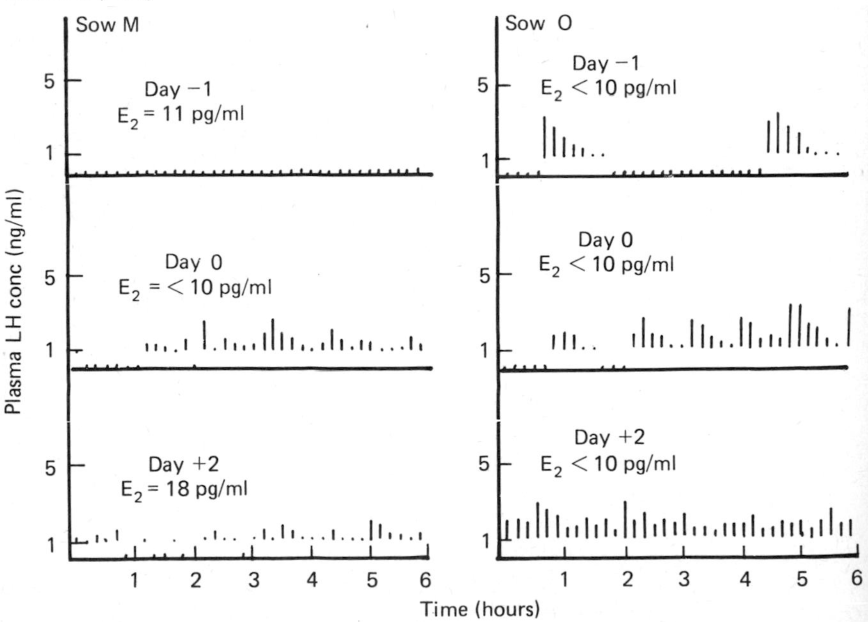

Figure 21.10 Episodic or pulsatile LH secretion around weaning for two sows weaned after 5 weeks of lactation. Weaning occurred on day 0 and plasma progesterone concentrations were basal at all times. E_2 = oestradiol concentration. Again a rise in basal LH secretion occurs at day 0 in the absence of any consistent effect on the characteristics of pulsatile LH secretion. From Edwards (1980)

(b) weaning is not associated with characteristic changes in this pulsatile pattern, although basal LH secretion is elevated;
(c) patterns of pulsatile LH secretion only become ordered during the pre-ovulatory secretion of oestradiol, when they resemble the changes seen in the follicular phase of the oestrous cycle.

Subsequent to the pre-ovulatory LH surge, pulsatile secretion is similar in all respects to that seen in the oestrous cycle (*Figure 21.11*). Since it is not possible to demonstrate a characteristic change in pulsatile LH secretion in relation to weaning, the function of such a mode of secretion

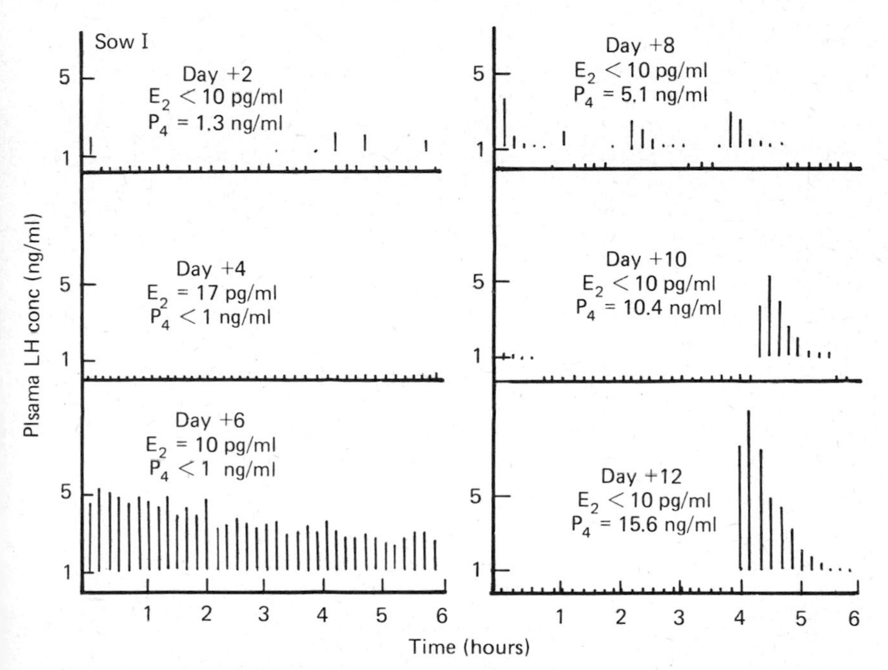

Figure 21.11 Episodic LH secretion in a sow from 2–12 days after weaning at 5 weeks post-partum. E_2 = oestradiol, P_4 = progesterone concentrations. The pre-ovulatory surge of LH occurred on days 5–6. Similar changes in pulsatile LH secretion occurred in sows weaned after 3 weeks of lactation (data not shown). From Edwards (1980)

remains unclear and may simply reflect a pulsatile release of GnRH from the hypothalamus. Until further research is undertaken the problem will remain the subject of much speculation.

Conclusions

As regards the overall control of reproductive function during lactation it appears that the hypothalamic–pituitary unit, primarily under the influence of the suckling stimulus, is responsible for the limited follicular development, anoestrus and anovulation which normally prevails. Although our knowledge of the endocrine events which accompany these morphological and behavioural changes has increased dramatically over the last five years, much more detailed information is required before a complete understanding can be had of the intricate endocrine relationships involved in the control of post-partum reproduction in the pig. In particular the changes occurring at the ovarian level in terms of follicular gonadotrophin receptors need to be elucidated and studied in relation to the endocrine events now known to occur as lactation progresses and the sow is weaned. The influence of lactation length on such phenomena is also an area which requires much greater research effort.

Whether or not it will ever be possible, or even desirable, to stimulate reproduction during lactation is debatable but a more controlled system of sow breeding must be the ultimate aim of current work, using either managemental changes and/or hormone treatment to produce the optimal number of live piglets when required.

It remains to be determined just how quickly such a system can be made practical and attractive to the producer.

References

AHERNE, F.X., CHRISTOPHERSON, R.J., THOMPSON, J.R. and HARDIN, R.T. (1976). Factors affecting the onset of puberty, post weaning oestrus and blood hormone levels of Lacombe gilts. *Can. J. Anim. Sci.* **56**, 681–692

ASH, R.W. and HEAP, R.B. (1975). Oestrogen, progesterone and corticosteroid concentrations in peripheral plasma of sows during pregnancy, parturition, lactation and after weaning. *J. Endocr.* **64**, 141–154

ASH, R.W., BANKS, P., BROAD, S. and HEAP, R.B. (1972). Techniques for studying changes in some peripheral blood components and steroid hormones in pregnant and lactating sows. *J. Physiol. (Lond.)* **224**, 40P–41P

ASH, R.W., BANKS, P., BAILES, G., BROAD, S. and HEAP, R.B. (1973). Plasma oestrogen, progesterone and cortisosteroid concentrations in the pregnant, parturient and lactating sow. *J. Reprod. Fert.* **33**, 359–360

BAKER, L.N., WOEHLING, H.L., CASIDA, L.E. and GRUMMER, R.H. (1953). Occurrence of oestrus in sows following parturition. *J. Anim. Sci.* **12**, 33–38

BALDWIN, D.M. and STABENFELDT, G.H. (1975). Endocrine changes in the pig during late pregnancy, parturition and lactation. *Biol. Reprod.* **12**, 508–515

BEVERS, M.M., WILLEMSE, A.H. and KRUIP, Th.A.M. (1978). Plasma prolactin levels in the sow during lactation and the postweaning period as measured by radioimmunoassay. *Biol. Reprod.* **19**, 628–634

BOOMAN, P. and VAN DE WIEL, D.F.M. (1980). Lactational anoestrus in the pig: possible relationship with hyperprolactinaemia. *Report B-157*. Instituut voor Veeteeltkundig Onderzoek "Schoonoord", Driebergseweg 10, Zeist, Holland

BURGER, J.F. (1952). Sex physiology of pigs. *Onderstepoort J. vet. Res. Suppl. 2*, 3–218

CAMPBELL, C.J. and SCHWARTZ, N.B. (1979). Time course of serum FSH suppression in ovariectomized rats injected with porcine follicular fluid (folliculostatin): Effect of oestradiol treatment. *Biol. Reprod.* **20**, 1093–1098

CHEN, C.L. and MEITES, J. (1970). Effects of oestrogen and progesterone on serum and pituitary prolactin levels in ovariectomized rats. *Endocrinology* **86**, 503–506

COLE, H.H. and HUGHES, E.H. (1946). Induction of oestrus in lactating sows with equine gonadotrophin. *J. Anim. Sci.* **5**, 25–29

COX, N.M. and BRITT, J.H. (1981). Relationship between endogenous GnRH and post-weaning endocrine events in sows (Abstract). *Annual Meeting of the Southern Section, American Society of Animal Science, Atlanta, Georgia*

CRIGHTON, D.B. (1967). Effects of lactation on the pituitary gonadotrophins of the sow. In *Reproduction in the Female Mammal* (G.E. Lamming and E.C. Amoroso, Eds.), pp. 223–238. London, Butterworths

CRIGHTON, D.B. and LAMMING, G.E. (1969). The lactational anoestrus of the sow; the status of the anterior pituitary–ovarian system during lactation and after weaning. *J. Endocr.* **43**, 507–519

EDQVIST, L.E., EINARSSON, S. and SETTERGREN, I. (1974). Ovarian activity and peripheral plasma levels of oestrogens and progesterone in the lactating sow. *Theriogenology* **1**, 43–49

EDWARDS, S. (1980). Reproductive physiology of the post-parturient sow. PhD Thesis. University of Nottingham

ELSAESSER, F. and PARVIZI, N. (1980). Partial recovery of the stimulatory oestrogen feedback action on LH release during late lactation in the pig. *J. Reprod. Fert.* **59**, 63–67

FOXCROFT, G.R., POMERANTZ, D.K. and NALBANDOV, A.V. (1975). Effects of oestroadiol-17-β on LHRH/FSHRH induced and spontaneous LH release in pre-pubertal female pigs. *Endocrinology* **96**, 551–557

GLASS, M.R., SHAW, R.W., BUTT, W.R., EDWARDS, R.L. and LONDON, D.R. (1975). An abnormality of oestrogen feedback in amenorrhea–galactorrhea. *Br. med. J.* **3**, 274–275

GLASS, M.R., SHAW, R.W., EDWARDS, R.L., BUTT, W.R., WILLIAMS, J. and LONDON, D.R. (1976). Modulation of gonadotrophin release by steroid hormones in women with hyperprolactinaemia. *J. Endocr.* **69**, 46P–47P

GUTHRIE, H.D., PURSEL, V.G. and FROBISH, L.T. (1978). Attempts to induce conception in lactating sows. *J. Anim. Sci.* **47**, 1145–1151

HAUGER, R.L., KARSCH, F.J. and FOSTER, D.L. (1977). A new concept for control of the oestrous cycle of the ewe based on the temporal relationships between LH, oestradiol and progesterone in peripheral serum and evidence that progesterone inhibits tonic LH secretion. *Endocrinology* **101**, 807–817

HEITMAN, H. and COLE, H.H. (1956). Further studies in the induction of oestrus in lactating sows with equine gonadotrophin. *J. Anim. Sci.* **15**, 970–977

HOLNESS, D.H. and HUNTER, R.H.F. (1975), Post-partum oestrus in the sow in relation to the concentration of plasma oestrogens. *J. Reprod. Fert.* **45**, 15–20

JENKIN, G. and HEAP, R.B. (1974). The lack of response of the sheep pituitary to LHRH stimulation in gestation and early lactation; the probable role of progesterone. *J. Endocr.* **61**, XII

JEQUIER, A.M., VANTHUYNE, C. and JACOBS, H.S. (1973). Gonadotrophin secretion in lactating women: response to LHRH/FSHRH in the puerperium. *J. Endocr.* **59**, XIV

KANN, G., MARTINET, J. and SCHIRAR, A. (1976). Impairment of luteinizing hormone release following oestrogen administration to hyperprolactinaemic ewes. *Nature, Lond.* **264**, 465–466

KARSCH, F.J., LEGAN, S.J., RYAN, K.D. and FOSTER, D.L. (1978). The feedback effects of ovarian steroids on gonadotrophin secretion. In *Control of Ovulation*, (D.B. Crighton, N.B. Haynes, G.R. Foxcroft and G.E. Lamming, Eds.), pp. 29–43. London, Butterworths

KNOBIL, E. (1974). On the control of gonadotrophin secretion in the Rhesus monkey. *Recent Prog. Horm. Res.* **30**, 1–36

LAMMING, G.E. (1978). Reproduction during lactation. In *Control of Ovulation*, (D.B. Crighton, N.B. Haynes, G.R. Foxcroft and G.E. Lamming, Eds.), pp. 335–353. London, Butterworths

LAUDERDALE, J.W., KIRKPATRICK, R.L., FIRST, N.L., HAUSER, E.R. and CASIDA, L.E. (1965). Ovarian and pituitary gland changes in periparturient sows. *J. Anim. Sci.* **24**, 1100–1103

McNATTY, K.P., SAWERS, R.S. and McNEILLY, A.S. (1974). A possible role for prolactin in control of steroid secretion by the human Graafian follicle. *Nature, Lond.* **250**, 653–655

MELAMPY, R.M., HENRICKS, D.M., ANDERSON, L.L., CHEN, C.L. and SCHULTZ, J.R. (1966). Pituitary FSH and LH concentrations in pregnant and lactating pigs. *Endocrinology* **78**, 801–804

MINAGUCHI, H. and MEITES, J. (1967). Effects of suckling on hypothalamic LH-releasing factor and prolactin inhibiting factor and on pituitary LH and prolactin. *Endocrinology* **80**, 603–607

MOOR, R.M. (1974). The ovarian follicle of the sheep: inhibition of oestrogen secretion by luteinizing hormone. *J. Endocr.* **61**, 455–463

NISWENDER, G.D. (1972). The effect of ergocornine on reproduction in sheep. *Biol. Reprod.* **7**, 138–139

PALMER, W.M., TEAGUE, H.S. and VENZKE, W.G. (1965a). Macroscopic observations on the reproductive tract of the sow during lactation and early post-weaning. *J. Anim. Sci.* **24**, 541–545

PALMER, W.M., TEAGUE, H.S. and VENZKE, W.G. (1965b). Histological changes in the reproductive tract of the sow during lactation and early post-weaning. *J. Anim. Sci.* **24**, 1117–1125

PARVIZI, N., ELSAESSER, F., SMIDT, D. and ELLENDORFF, F. (1976). Plasma LH and progesterone in the adult female pig during the oestrous cycle, late pregnancy and lactation, and after ovariectomy and pentobarbitone treatment. *J. Endocr.* **69**, 193–203

RADFORD, H.M., NANCARROW, C.D. and MATTNER, P.E. (1976). Evidence for hypothalamic dysfunction in suckled beef cows. *Theriogenology* **6**, 641

RADFORD, H.M., NANCARROW, C.D. and MATTNER, P.E. (1978). Ovarian function in suckling and non-suckling beef cows post-partum. *J. Reprod. Fert.* **54**, 49–56

RAESIDE, J.I. (1963). Urinary oestrogen excretion in the pig during pregnancy and parturition. *J. Reprod. Fert.* **6**, 427–431

RAYFORD, P.L., BRINKLEY, H.J., YOUNG, E.P. and REICHERT, L.E. (1974). Radioimmunoassay of porcine FSH. *J. Anim. Sci.* **39**, 348–354

REYES, F.I., WINTER, J.S.D. and FAIMAN, C. (1972). Pituitary–ovarian interrelationships during the puerperium. *Am. J. Obstet. Gynec.* **114**, 589–594

RICHARDS, J.S., RAO, M.C. and IRELAND, J.J. (1978). Actions of pituitary gonadotrophins on the ovary. In *Control of Ovulation*, (D.B. Crighton, N.B. Haynes, G.R. Foxcroft and G.E. Lamming, Eds.), pp. 197–216. London, Butterworths

ROBERTSON, H.A. and KING, C.J. (1974). Plasma concentrations of progesterone, oestrone, oestradiol-17-β and oestrone sulphate in the pig at implantation, during pregnancy and at parturition. *J. Reprod. Fert.* **40**, 133–141

ROLLAND, R., LEQUIN, R.M., SCHELLEKENS, L.A. and de JONG, F.H. (1975). The role of prolactin in the restoration of ovarian function during the early *post-partum* period in the human female. I. A study during physiological lactation. *Clin. Endocr.* **4**, 15–25

SASSER, R.G., HEGGE, F.N., CHRISTIAN, R.E. and FALK, D.E. (1973). Plasma oestrogens in pregnant gilts. *J. Anim. Sci.* **37**, 327

SCHAMS, D. (1974). *Untersuchungen uber Prolactin beim Rind.* Hamburg and Berlin, Verlag Parey

SEDDON, R.J. (1970). The ovarian response in women to large single injections of human menopausal gonadotrophin and human chorionic gonadotrophin. *J. Reprod. Fert.* **23**, 299–305

SELF, H.L. and GRUMMER, R.H. (1958). The rate and economy of pig gains and the reproductive behaviour in sows when litters are weaned at 10 days, 21 days or 56 days of age. *J. Anim. Sci.* **17**, 862–868

SOUVATZOGLOU, A., METZGER, I., KULLACK, W., BUTENANDT, O. and HENDERKOTT, V. (1973). The functional significance of episodic LH secretion. *Acta endocr. Suppl.* **173**, Abstract 42

STEVENSON, J.S., COX, N.M. and BRITT, J.H. (1981). Role of the ovary in controlling LH, FSH and prolactin secretion during and after lactation in pigs. *Biol. Reprod.* **24**, 341–353

STEVENSON, J.S. and BRITT, J.H. (1981). Luteinizing hormone, total oestrogens and progesterone secretion during lactation and after weaning in sows. *Theriogenology* **14**, 453–462

VAN DE WIEL, D.F.M. VAN LANDEGHEM, A.A.J., WILLEMSE, A.H. and BEVERS, M.M. (1979). Endocrine control of ovarian function after weaning in the domestic sow. *J. Endocr.* **80**, 69P

VAN LANDEGHEM, A.A.J. and VAN DE WIEL, D.F.M. (1977). Plasma prolactin levels in gilts during the oestrous cycle and at hourly intervals around the time of oestrus. *Acta endocr.* **85**, *Suppl. 212,* Abstract 233

VAN LANDEGHEM, A.A.J. and VAN DE WIEL, D.F.M. (1978). Radioimmunoassay for porcine prolactin: plasma levels during lactation, suckling and weaning and after TRH administration. *Acta endocr.* **88**, 653–667

WARNICK, A.C., CASIDA, L.E. and GRUMMER, R.H. (1950). The occurrence of oestrus and ovulation in *post-partum* sows. *J. Anim. Sci.* **9**, 66–72

WEBB, R., LAMMING, G.E., HAYNES, N.B., HAFS, H.D. and MANNS, J.G. (1977). Response of cyclic and *post-partum* suckled cows to injections of synthetic LHRH. *J. Reprod. Fert.* **50**, 203–210

WILFINGER, W.W. (1974). Plasma concentrations of LH, FSH and prolactin in ovariectomized, hysterectomized and intact swine. *Diss. Abstr.* **35**, 2985-B.

ZARATE, A., CANALES, E.S., SORIA, J., RUIZ, F. and MACGREGOR, C. (1972). Ovarian refractoriness during lactation in women: effect of gonadotrophin stimulation. *Am. J. Obstet. Gynec.* **114**, 589–594

22

THE TIME OF WEANING AND ITS EFFECTS ON REPRODUCTIVE FUNCTION

M.A. VARLEY
Department of Applied Nutrition, Rowett Research Institute, Aberdeen, UK

Conventional systems of pig production include a lactation length of six to eight weeks duration followed by a quick return to oestrus and ovulation. The reproductive life of the sow consists therefore of two complete production cycles in one twelve-month period. Throughout the 1970s, many producers dispensed with this hitherto acceptable system in favour of weaning piglets from the sow at a much earlier age. The reasons for this are many and varied. By weaning earlier the average time from one farrowing to the next is reduced and hence each sow has more than two reproductive cycles in one year. For approximately the same input of sow feed a significant increase in the number of weaners is produced per sow per year. As a result of this reasoning a number of quite different systems have evolved and these include 35-day weaning, 21-day weaning and 10-day weaning. Each of these systems is associated with markedly different reproductive function and it is the purpose of this chapter to review the considerable amount of recent work in this area and to examine how weaning age influences many of the parameters used to assess reproductive efficiency.

The interval from weaning to conception

At the end of a 6–8 week lactation the sow may have already escaped, at least partially, from the known inhibition of hypothalamic and pituitary function induced by the suckling stimulus. Gonadotrophins have been produced and stored in the anterior pituitary gland and at weaning a rapid follicular phase results in oestrus and ovulation some four or five days later. Reducing lactation length below 42 days may leave the hypothalamic–pituitary axis unable to respond in exactly the same way at weaning and therefore sows may take longer to return to oestrus and the variability of the interval from weaning to oestrus increases. This has been shown in a number of studies (Aumaitre, 1972; te Brake, 1972; Dyrendahl *et al.*, 1958; Moody and Speer, 1971; Self and Grummer, 1958; Van der Heyde, 1972; Svajgr *et al.*, 1974; Cole, Varley and Hughes, 1975; Varley and Cole, 1976a). The relationship between lactation length and the interval from weaning to oestrus is illustrated in *Figure 22.1*. It can be seen that as lactation length is reduced the interval from weaning to oestrus rises

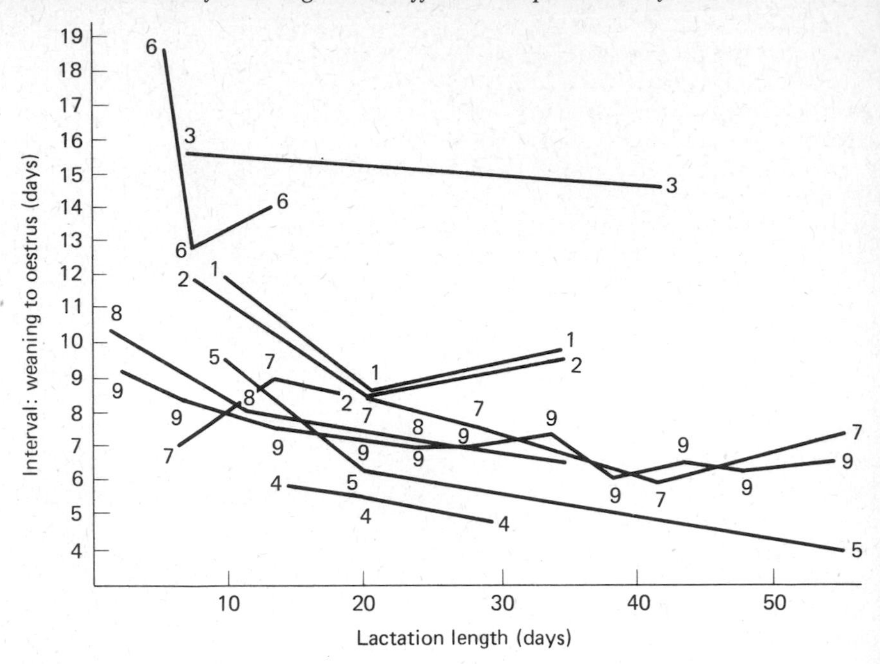

Figure 22.1 The effect of lactation length on the interval from weaning to oestrus. Numbers refer to references as follows: 1. Aumaitre (1972); 2. Aumaitre and Rettigliati (1972); 3. te Brake (1972); 4. Moody and Speer (1971); 5. Self and Grummer (1958); 6. Van der Heyde (1972); 7. Puyaoan and Castillo (1963); 8. Svajgr *et al.* (1974); 9. Smidt, Scheven and Steinbach (1965)

significantly. When weaning age is three weeks the average interval from weaning to oestrus appears to be at a value of around seven days. When weaning age is further reduced to ten days the average interval rises again to around nine days. This relationship has been established numerically by Cole, Varley and Hughes (1975) who found that:

$$\text{Log } Y = 0.931 - 0.0077X$$

where Y = the interval from weaning to oestrus in days
X = lactation length in days

In a series of experiments the same workers compared very early weaned sows with sows weaned after a 42-day lactation and found that both the interval from weaning to oestrus and its variability increased markedly as lactation length dropped below 21 days (see *Table 22.1*).

Table 22.1 THE EFFECT OF LACTATION LENGTH ON THE MEAN AND STANDARD DEVIATION OF THE INTERVAL FROM WEANING TO OESTRUS

	Lactation length (days)					
	7–10		*21*		*42*	
Authors	*Mean*	*SD*	*Mean*	*SD*	*Mean*	*SD*
Varley and Cole (1976a)	8.2	2.8	–	–	4.5	0.5
Varley and Cole (1976b)	8.2	1.6	7.2	3.3	5.0	0.5

There are many diverse factors that can affect the interval from weaning to oestrus and these have been reviewed in detail elsewhere (Varley, 1976; Hughes and Varley, 1980). Briefly they can be listed as: genotype, age, parity and season. Plane of nutrition plays a negligible role in this respect (Brooks *et al.*, 1975; Clark *et al.*, 1972; Dyck, 1972).

In sows weaned following conventional lactation lengths, pituitary FSH (follicle stimulating hormone) rises very little in the immediate post-weaning period but pituitary LH (luteinizing hormone) rises significantly (Crighton, 1967). Increased production and storage of LH must therefore characterize the sow's post-weaning endocrine status. Peripheral LH activity is not seen until just prior to ovulation (Parlow, Anderson and Melampy, 1964). FSH therefore must be undergoing very rapid production and release from the pituitary in view of the rapid growth in the number and size of follicles on the ovaries at this time. Oestrus and ovulation subsequently occur due to the surge of oestrogen produced from the follicles which induces LH release from the pituitary under the permissive influence of pulsatile GnRH (gonadotrophin releasing hormone) (Edwards and Foxcroft, personal communication; Knobil, 1980).

This sequence of events is disturbed in some way by reducing the lactation length. If the pulsatile pattern of GnRH release is the same in the pig as has been demonstrated in other species (Knobil, 1980), then the suckling stimulus in early lactation will result in an inhibition of production and release of the hormone into the pituitary portal circulation. The sow therefore remains in anoestrus completely in early lactation. As the lactation proceeds beyond the 21-day stage the inhibitory effect of the suckling stimulus on the hypothalamic production of GnRH diminishes. This is clearly indicated by the fact that many sows in the late part of a 42–56 day lactation will either show spontaneous oestrus and ovulation whilst still lactating or can be induced to show a fertile heat by a variety of husbandry techniques (Burger, 1952; Smith, 1961; Rowlinson and Bryant, 1974). Pulsatile or episodic release of GnRH must therefore begin again at some point in mid lactation. By weaning much earlier than 21 days the

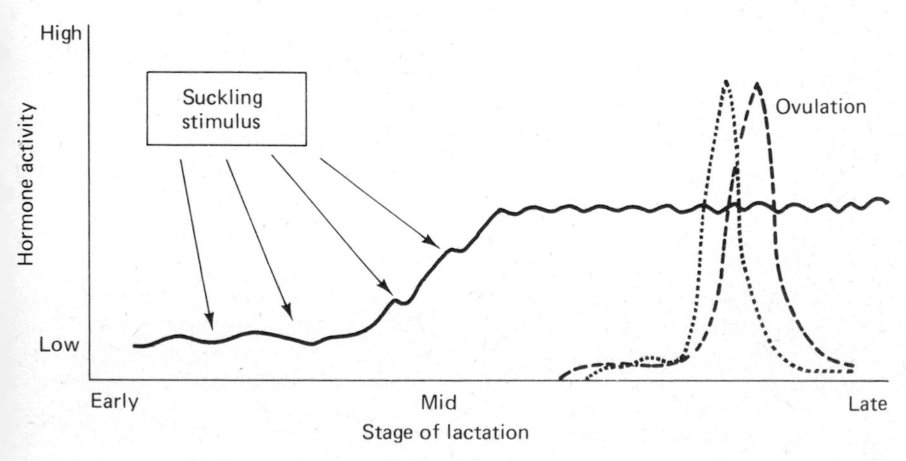

Figure 22.2. A diagrammatic representation of the sow's gradual escape from the lactational anoestrous condition. —— pulsatile GnRH; ······· oestrogen; ---- gonadotrophins

hypophyseal axis has not yet been 'primed' by both GnRH and the modulating influence of oestrogens acting via the negative feedback loop on the hypothalamus. The sow is therefore unable to show a rapid follicular phase and the return to oestrus and ovulation is significatnly delayed. *Figure 22.2* gives a diagrammatic representation of this effect. It does seem that despite this effect, sows do still show a detectable heat after a 7–10 day lactation and the percentage of sows becoming anoestrous is similar to that seen in sows weaned after conventional lactation lengths (Varley and Cole, 1976a,b; Varley and Cole, 1978).

Conception rate

There is some doubt about the effect lactation length has on the ability of the sow to conceive at the first post-weaning oestrus. Aumaitre (1972) found that sows weaned after a 35-day lactation showed a conception rate of 97.6%, whereas the figure for sows which were weaned at seven days was 86.0%. te Brake (1972) observed a larger reduction in his study where sows weaned at 42 days had an overall conception rate of 79.2% compared to 53.3% for sows weaned at seven days. In contrast, Cole, Varley and Hughes (1975) and Varley and Cole (1976a) found no differences in either conception or farrowing rates for sows weaned after lactation lengths of 42 days or 10 days. Van der Heyde (1972) concluded that there is considerable between-farm variation in conception rates and that it should be possible to achieve the same conception rate for both very early weaning and conventional systems. He added that the apparent low conception rates on some farms were often associated with incorrect frequency and timing of insemination. It might be that on farms changing to earlier weaning, stockmen are not prepared for the modification of the interval from weaning to oestrus and do not time admission of the boar correctly.

The effect of lactation length on prolificacy

The ultimate prolificacy in terms of the number of viable piglets produced at parturition per litter is a function of ovulation rate, fertilization rate and the survival of embryos and foetuses. All of these factors could be associated with the previous lactation length and will be reviewed in turn.

OVULATION RATE

In the light of the endocrine disturbances between weaning and remating of sows subjected to very short lactations, a concomitant reduction in the number of ova shed at ovulation might also be expected. This would depend on the pattern and magnitude of the release of both FSH and LH from the pituitary. To date the peripheral gonadotrophins between weaning and remating have never been measured in early weaned sows. The effect of lactation length on ovulation rate however has been measured and the results are presented in *Table 22.2*.

Table 22.2 THE EFFECT OF LACTATION LENGTH ON OVULATION RATE

	Lactation length (days)					
Authors	*2*	*7–10*	*13*	*21*	*35*	*42–56*
Self and Grummer (1958)	–	12.8	–	13.2	–	16.6
Svajgr *et al.* (1974)	15.6	–	15.0	15.2	14.4	–
Varley and Cole (1976b)	–	15.7	–	17.1	–	15.1
Varley and Cole (1978)	–	15.3	–	–	–	15.1

In 1958 Self and Grummer observed a large reduction in ovulation rate which approached significant proportions ($P<0.06$). This has not been substantiated by more recent studies (Varley and Cole, 1976, 1978; Svajgr *et al.*, 1974) where no differences were found in ovulation rate, as determined by corpora lutea counts, between sows weaned at 42 days and sows weaned at seven days.

FERTILIZATION RATE

Fertilization rate and its degree of association with the preceding lactation length has been the subject of a number of studies. Self and Grummer (1958) found that following 10, 21 or 56-day lactations, fertilization rates were 93.4%, 90.6% and 98.1% respectively and the differences were non-significant. For lactation lengths of 2, 13, 24 and 35 days, Svajgr *et al.* (1974) obtained fertilization rates of 81.9%, 86.3%, 96.5% and 98% respectively. The latter authors attributed this result to two sows in the group weaned at two days and one in the 13-day group which showed complete absence of cleaved ova at day 3 post coitum. Where conception took place, fertilization rates for sows weaned after 2–10 day lactations were comparable with fertilization rates for sows weaned after a more conventional (i.e. longer) lactation length.

EMBRYONIC SURVIVAL RATE

Two studies have shown clearly that lactation length adversely affects the ability of embryos to survive the first three weeks of gestation. Varley and Cole (1976b) evaluated the embryonic survival rate at day 20 post coitum for sows which had previously lactated for either 7, 21 or 42 days. Survival rates in these sows were 59.2%, 63.9% and 81.7% respectively and these large differences were significant ($P<0.01$). Similarly Svajgr *et al.* (1974) for lactation lengths of 2, 13, 24 and 35 days found survival rates of 54.3%, 70.7%, 71.6% and 79.5% respectively. Embryo loss is therefore considerable even for sows weaned after long lactations; one in five fertilized eggs which begin developing after mating are lost by the end of the third week of pregnancy (Wrathall, 1971). For the early weaned sow this loss increases to about two in five.

Losses can be ascribed to a large number of factors and can be separated into two parts; firstly, those losses occurring during blastulation and before implantation and secondly those occurring around the time of implantation or soon after, as a result of malfunctions in the process of attachment (Wrathall, 1971). In order to study the relative importance of these two components, Varley and Cole (1978) slaughtered two groups of sows, weaned following either a 7- or a 42-day lactation, at nine days post coitum, i.e. just before the first attachment of the trophoblast to the endometrium. Survival of embryos for sows weaned at 7 and 42 days was 74.3% and 83.5% respectively but this difference was not significant. The authors concluded that although the trend was towards larger losses for the early weaned group, in the light of the earlier work where survival at day 20 post coitum (Varley and Cole, 1976b) was measured, the bulk of embryo losses for the early weaned sow was incurred at or around implantation.

THE EFFECT OF LACTATION LENGTH ON LITTER SIZE

The product of ovulation, fertilization and embryonic survival ultimately determines the size of the litter at the next parturition. Many studies have shown that the effect of high embryo losses for early weaned sows outlined above eventually result in a significantly reduced litter size (Aumaitre, 1972; Aumaitre and Rettigliati, 1972; Moody and Speer, 1971; Van der Heyde, 1972; Smidt *et al.* 1965; te Brake, 1972; Varley and Cole, 1976a; Cole, Varley and Hughes, 1974). This is illustrated in *Figure 22.3.*

A most exhaustive study has been carried out by Smidt, Scheven and Steinbach (1965) (Reference 5 in *Figure 22.3*). No significant difference was observed in this study in litter size of sows weaned after lactation

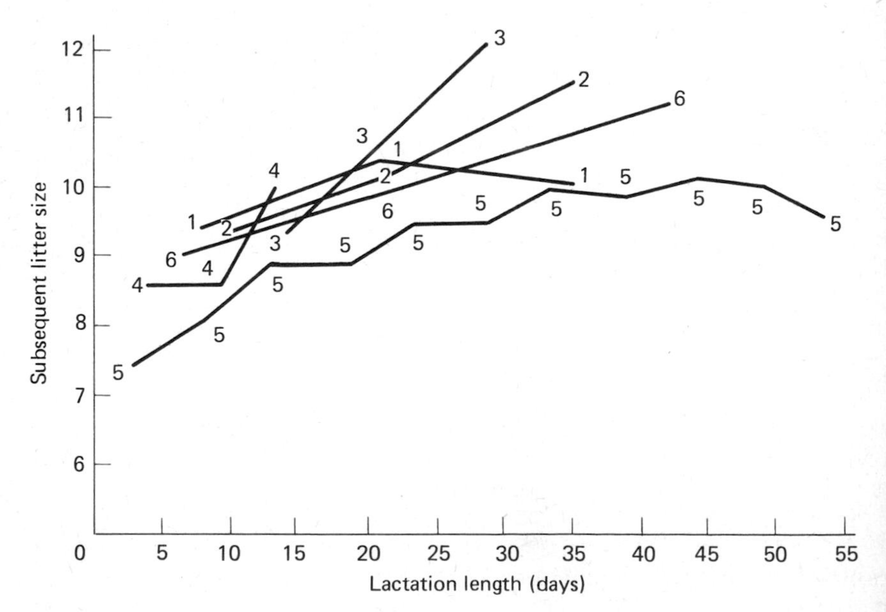

Figure 22.3 The effect of lactation length on subsequent litter size. Numbers refer to references as follows: 1. Aumaitre (1972); 2. Aumaitre and Rettigliati (1972); 3. Moody and Speer (1971); 4. Van der Heyde (1972); 5. Smidt, Scheven and Steinbach (1965); 6. te Brake (1972)

lengths of 56 days or 25 days. However, some authors (Aumaitre and Rettigliati, 1972; te Brake, 1972) have observed a difference in this respect and Varley (1979) has concluded that on average a small reduction in litter size will be observed with sows weaned after 21-day lactations compared with sows weaned after 35–42 day lactations. Furthermore Varley (1979) estimated that the magnitude of this reduction is in the order of 0.1–0.2 piglets per litter. There is however a large reduction in litter size as lactation length is reduced below three weeks and Varley and Cole (1976a) concluded that a sow weaned at 7–10 days post partum would show a decrease of between 1.5 and 2.0 piglets per litter at the next farrowing. This result is consonant with the findings on embryonic mortality (Varley and Cole, 1978).

The results of a series of experiments carried out at Nottingham University in the mid 1970s are presented schematically in *Figure 22.4* and show the number of viable individuals per litter throughout the various stages of gestation.

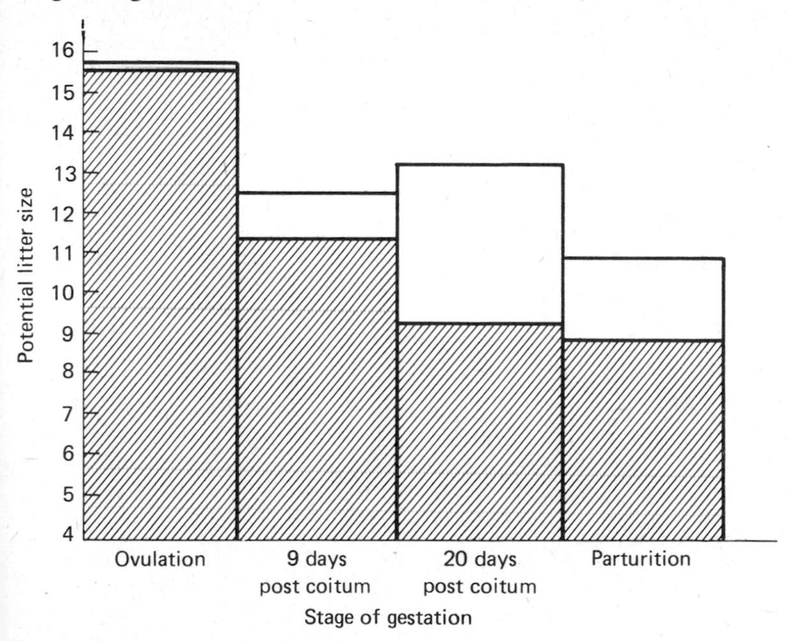

Figure 22.4 Litter size for the early weaned sow. Shaded area, 7–10 day weaned; open area, 42-day weaned

The small difference in litter size seen at day 9 post coitum is in striking contrast to the large difference seen at day 20 post coitum. This difference is then reflected in the numbers born alive at parturition.

There have been reports from commercial practice (Pay, 1973; Looker, 1974) suggesting that very early weaning does not substantially reduce prolificacy. The large number of reports to the contrary however cannot be ignored and a commercial producer embarking on a programme of weaning earlier than three weeks must expect reduced prolificacy. Some of the differences between farms could be accounted for by variation in feeding systems but Varley and Cole (1976a) have observed that despite very large differences in feed intake throughout lactation and from weaning to remating, all sows weaned after 10-day lactations exhibited reduced prolificacy.

An interesting observation was made by Walker *et al.* (1979) from the results of a large scale, long term study including 146 sows at two centres. These sows were allocated at random to one of three lactation lengths (10, 25 or 40 days) in their first parity. Each sow remained on the same treatment throughout the course of five parities and reproductive performance at each parity was recorded. The results are presented in *Table 22.3*.

Table 22.3 THE EFFECT OF LACTATION LENGTH AND PARITY ON THE NUMBER OF PIGLETS BORN ALIVE PER LITTER AND THE NUMBER OF PIGLETS WEANED PER LITTER

	Lactation length					
	10 (days)		*25* (days)		*40* (days)	
Parity	*Born alive*	*Weaned*	*Born alive*	*Weaned*	*Born alive*	*Weaned*
1	9.19	8.40	9.15	8.24	8.72	8.20
2	9.01	8.81	9.28	8.39	8.36	7.73
3	8.82	8.31	9.58	8.77	9.87	9.33
4	8.62	8.29	9.65	8.88	11.31	9.99
5	9.19	8.64	9.34	8.26	11.47	9.75

From Walker *et al.* (1979)

For the first two parities, there was little difference in either numbers born alive or numbers of piglets surviving to weaning. As the age and reproductive experience of the sow increased the sows weaned after long lactations produced significantly larger litters born alive when compared to sows weaned at 10 days. This large difference at birth however was not seen at weaning and the difference in numbers weaned between the 10-day weaned group and the 40-day group was not significant for parities 1–5.

This effect could well explain a proportion of the between-herd variation seen in practice. Few herds have identical age structure and newly established herds would not tend to notice changes in prolificacy as a result of earlier weaning, whereas long established herds particularly where culling rate had been low would experience a sharp drop in production of piglets born alive. Walker *et al.* (1979) suggested that the reason numbers were similar at weaning was that sows weaned at 40 days had very large numbers born alive which they were incapable of rearing throughout a long lactation. In contrast sows weaned at 10 days start off with fewer piglets at birth and then lose fewer in the short period of lactation. All of this strongly suggests that for all systems husbandry will be a major component influencing the final output of piglets per sow per year at weaning.

Effect of lactation length on endocrine status

Polge (1972) has speculated on possible differences in endocrine status between sows weaned after conventional lactation lengths and for very early weaned sows. He concluded from the work of Lauderdale *et al.* (1963) and Crighton (1967) that the concentration of LH in the pituitary gland increases much less in sows weaned within 1–11 days post partum than in sows weaned at 56 days post partum. In addition, FSH concentration in the pituitary dropped in sows weaned at eight weeks. This suggests that the very early weaned sow shows a much reduced pituitary response and this is in accord with the observed effect on the weaning to oestrus interval. Peters *et al.* (1969) and Kirkpatrick *et al.* (1965) both attempted to modify the endocrine balance of very early weaned sows by using exogenous applications of both FSH and pituitary extract. Sows could be induced to ovulate shortly after parturition but fertility was negligible. Moody and Speer (1971) have measured the luteal progesterone content of plasma at day 25 post coitum after lactation lengths of either 14, 21 or 28 days and have found no differences.

Varley and Cole (1976b) measured plasma progesterone concentration from weaning to day 20 post coitum and concluded that differences might exist between sows weaned after different lactation lengths but in this study sows were bled only five times over the course of the experiment and therefore a definitive relationship could not be established.

In order to provide more data in this area, Varley, Atkinson and Ross (1981) have studied the plasma concentrations of progesterone and oestradiol-17β in sows weaned after lactation lengths of either 10 or 42 days. Sows were bled every second day from mating to day 26 post coitum. The plasma steroid concentrations are shown in *Figure 22.5* and *Figure 22.6*.

The results illustrated in *Figure 22.5* indicate the existence of a different pattern of progesterone secretion in early gestation between early and late weaned sows. Plasma progesterone between days 4 and 10 post coitum

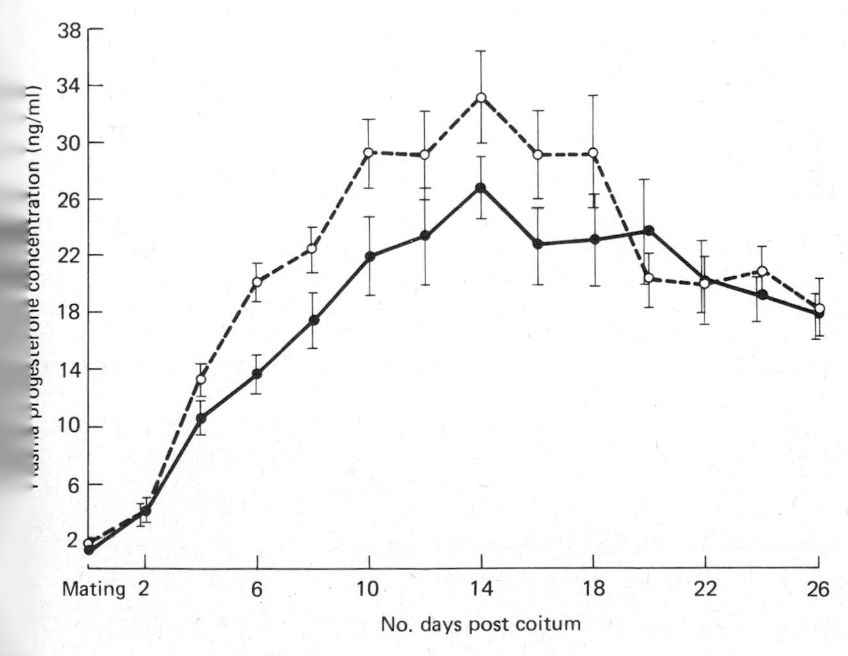

igure 22.5 Variation in plasma progesterone concentration with number of days post oitum. ○ early weaned; ● control. (N.B. Control bars offset to left for clarity in some cases)

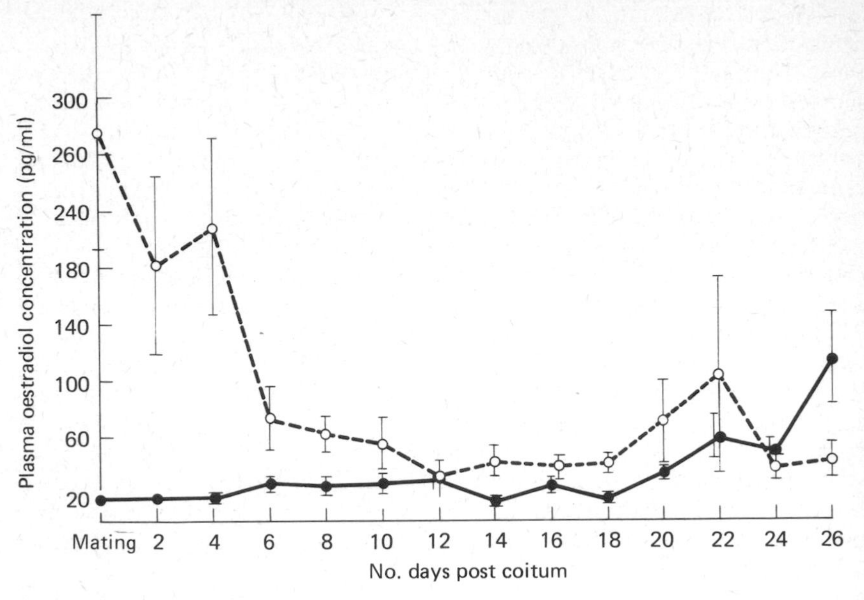

Figure 22.6 Variation in plasma oestradiol concentration with number of days post coitum. ○ early weaned; ● control

increases significantly faster for early weaned sows. Part of this difference could result from chance differences in ovulation rate and unfortunately this was not measured. The two groups of sows nevertheless were of similar genotype, age, parity and liveweight and therefore minimal systematic bias of this kind was introduced in the experiment. It is more probable that the differences in progesterone were a direct reflection of the oestradiol concentrations seen in *Figure 22.6*. Early weaned sows showed a prolonged oestrogen surge during and after mating when compared to control sows weaned at 42 days.

The ramifications of these very aberrant steroid levels in early gestation in terms of embryonic survival were not measured and are the subject of a current study at the Rowett Institute. It is a matter of conjecture to implicate the observations shown in *Figures 22.5* and *22.6* with embryonic survival but it seems likely that such steroid profiles could have a number of possible effects. Firstly, the rate of passage of fertilized eggs down the oviducts could be adversely affected (Day and Polge, 1968) and this almost certainly would result in eggs being lost prior to implantation. Varley and Cole (1978) observed mean differences of 1.3 embryos per litter at 9 days post coitum. Secondly, fertilized eggs experiencing abnormal tubal transport might subsequently be exposed to conditions within the uterine lumen in terms of secretory proteins with which they were not in synchrony. The net result would therefore be losses of embryos prior to implantation and then considerable loss during the implantation process itself.

It is now established that during the very early part of gestation in the pig the secretion of a number of different pregnancy specific proteins into both the uterine lumen and probably the maternal circulation takes place (Bazer *et al.*, 1969; Murray and Grifo, 1976; Knight *et al.*, 1974; Bashe *et al.*, 1980

Etzel *et al.*, 1978). This has also been reported in other species and comprehensive reviews on these have been published elsewhere (Beier, 1979; Klopper, 1979). The relative amounts of each of these proteins changes daily as gestation proceeds, presumably to service the rapidly changing needs of the growing embryo. This highlights the need to secure absolute synchrony in physiological status between embryo and uterine environment if maximum survival is to be attained. Some of these proteins may originate in the endometrium, others in the maternal liver and a plethora of names have been associated with them.

In the rabbit the terms *blastokinin* or *uteroglobin* have been used to name one protein fraction (Krishnan and Daniel, 1967; Beier and Maurer, 1975). *Pregnancy associated proteins* and *progesterone induced proteins* are other generic terms which have been used (Murray and Grifo, 1976; Klopper, 1979). This latter term refers to the fact that circulating progesterone activity plays a significant mediating role in the secretion of uterine proteins. Knight *et al.* (1974) established a dose response relationship between exogenous progesterone and porcine uterine protein secretion. The same group however (Knight, Bazer and Wallace, 1974) were unable to elucidate a relationship between induced uterine proteins and embryonic survival.

The function of these uterine proteins may not be one of simple egg nurture. It has been shown in many species including the pig that they have high immunoreactivity (Etzel *et al.*, 1978; McIntyre and Faulk, 1979; Roberts, 1977; Murray, Segerson and Brown, 1978). It has therefore been postulated that they act in a permissive way to allow implantation by suppressing the normal host/graft immune rejection mechanism at the interface between trophoblast and endometrium (Etzel *et al.*, 1978; Murray, Segerson and Brown, 1978). Attempts have also been made to implicate other substances capable of effecting a similar suppression of the immune mechanism for the same purpose. One of these is α-fetoprotein in the mouse (Murgita and Tomasi, 1975), the bovine (Smith *et al.*, 1979), sheep (Lai *et al.*, 1978) and woman (Hay *et al.*, 1976).

If the integrated effects of these various immune suppressive factors is functioning at the correct intensity, then implantation is allowed to take place and the embryo implants onto the endometrium. Presumably the converse is also true, namely that if insufficient immune suppression is effected (i.e. insufficient uterine protein) then embryos fail to implant and die shortly after. In the case of the pig, it is possible that this delicate mechanism of immune acceptance or rejection acts as a natural selection process to cull a proportion of the high number of embryos (15–20) originally entering the uterus after fertilization. By disturbing the peripheral steroids as a result of very early weaning we may be shifting this delicate balance towards fewer embryos being able to overcome the less favourable uterine environment. It has been suggested (Bazer *et al.*, 1969) that embryos 'compete' for an essential biochemical factor which may be uterine proteins, in order to achieve survival and further development. In this way the uterus may impose a ceiling on the number of offspring carried to term. It should be emphasized however that to date no causal relationship between progesterone induced proteins and the rate of survival of embryos has been elucidated (Knight *et al.*, 1974) and the two appear to

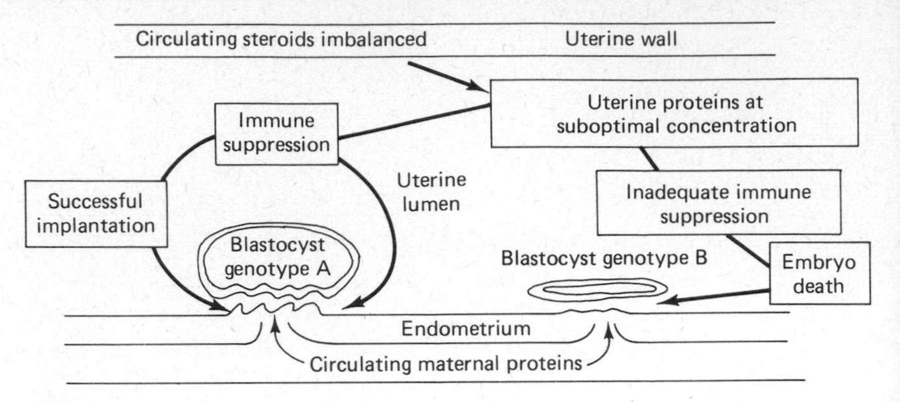

Figure 22.7 A schematic representation of embryonic death and its possible relationship with uterine environment

express themselves independently. *Figure 22.7* shows a schematic representation of embryonic death and its possible relationship with uterine protein activity.

One of the progesterone induced uterine proteins has been identified as a 32 000 dalton glycoprotein with a high iron content (Murray *et al.*, 1972; Murray, Segerson and Brown, 1978). This has been named *uteroferrin* to add to the complexity of nomenclature (Bashe *et al.*, 1980; Buhi *et al.*, 1979). Bazer *et al.* (1975) have indicated that this protein appears to function in iron transport to the conceptus. Riboflavin has also recently been shown to be present in the uterus in significant amounts under the influence of progesterone in the pre-implantational stage (Murray *et al.*, 1980). The precise role of this vitamin in embryonic nutrition, development and survival has yet to be evaluated.

The involvement of all of these substances in the expression of embryonic mortality seen in the early weaned sow would seem to be a fertile area for future studies.

Uterine involution and endometrial repair

Following parturition there is a decrease in the length and weight of the distended uterus (Palmer, Teague and Venzke, 1965a,b). This involution process is rapid the first week following farrowing and then proceeds more slowly until a minimum length and weight is attained by days 21–28 of lactation. Degeneration of the endometrium is also seen during the first week of lactation, but by the end of the week regenerative processes are also evident and complete rebuilding of the endometrium may take place by days 14–21 post partum. Hence although complete involution does not occur until about three weeks into lactation it seems that much of the important rebuilding process has occurred by as early as day 7 post partum. From the evidence available there seems little reason why the uterus should not be fully competent again by 2–3 weeks post partum (Palmer, Teague and Venzke, 1965a,b; Graves *et al.*, 1967; Svajgr *et al.*

1974). This latter point is pertinent to the success or failure of very early weaning systems. If a sow is weaned at day 7 post partum and is remated thereafter at day 15 post partum, then fertilized eggs will appear in the uterus at about day 18 post partum. They may then be subject to an adverse uterine environment and will suffer accordingly.

Suckled sows show a faster rate of involution than sows which have had their litters removed early in lactation (Graves *et al.*, 1967). Smidt, Thume and Jochle (1969), however, have noted that a quicker regeneration of the uterine gland itself takes place in sows which have been weaned compared with sows which have been suckled.

Although the uterine glandular capacity appears histologically competent within about 15 days post partum (Svagjr *et al.*, 1974), the question not yet resolved is: how responsive are the uterine target cells and receptor systems to the steroid hormones for the production of the proteins mentioned above?

It might be that the interactions between plasma steroids, endometrial competence and uterine protein production may be responsible for the degree of embryonic survival observed in the very early weaned sow. They may also explain why in sows weaned after conventional lactation lengths embryonic mortality is still, on average, 25–30% (Wrathall, 1971).

Weaning at birth

Systems of weaning at or near to the moment of birth have been investigated by a number of authors (Alexander, 1969; Braude *et al.*, 1970; Robertson *et al.*, 1971; Sharman *et al.*, 1971; Jones, 1972; Tate, 1974). The advantage of weaning at birth rather than later at 7–10 days is that the high percentage of neonatal death which occurs in any system involving piglets suckling the sow is avoided. Much of the early work on weaning at birth has explored the nutritional requirements of the young piglets (Alexander, 1969; Braude and Newport, 1973; Braude, Keal and Newport, 1976, 1977; Tate, 1974; Newport, Storry and Tuckley, 1979). Difficulty has always been experienced in avoiding excessive scouring. Alexander (1969) observed clinical enteritis in 50% of piglets weaned at two days of age and 20% of all piglets scoured and 18% died following weaning at between two and four days. Williams (1976) has attributed this magnitude of loss to primary infection of piglets with rotavirus followed by colibacillosis. Considering the difficulties encountered with the young piglets it is not surprising that little attention has been paid to the reproduction of the sow following weaning at birth.

However, a recent paper (Elliot, King and Robertson, 1980) has described the subsequent reproductive performance of sows weaned at birth. Based on plasma progesterone analysis, these sows ovulated on average at 17.0±6.6 days post partum but showed a significantly increased incidence of cystic ovaries compared to control sows weaned at 30 days post partum. Most of the sows weaned at birth exhibited nymphomania for the first two weeks post partum. The farrowing to effective service interval was 34.9±4.0 days for control sows and 42.4±26.2 days for sows weaned at birth. Weaning at birth was not associated with improved sow productivity and the overall farrowing interval for sows weaned at birth was 159 days compared to 148 days for the sows weaned at 30 days post partum. Clearly the sow weaned at birth experiences very serious endocrine disturbances.

In the spring of 1980 a unit was commissioned at the Rowett Research Institute to examine in detail both the sow's reproductive potential and the nutrition and immune development of piglets weaned at birth. The facility includes incubators for 96 piglets and flat deck accommodation to house piglets up to eight weeks of age as well as space for a self-contained herd of 24 sows and two boars. Under these conditions where the unit has been conceived, designed and built specifically for the purpose it is used, early results have shown that piglet mortality up to eight weeks can be reduced to under 5% with a low incidence of enteric disorders. The philosophy and concept behind this project has been described by Fowler and Varley (1981).

Currently in this unit the use of a new oral progestagen (Regumate; allyl trenbolone, Hoechst, UK) is being examined as a hypophyseal blocking agent for sows weaned at birth. This steroid is given for seven days from parturition and early results indicate that oestrus and ovulation will occur at day 14 post partum. This implies the possibility of a high order of sow reproductive performance. Fertility levels however have yet to be assessed.

Practical considerations

The sow weaned following a conventional long lactation length will show oestrus and ovulation rapidly and consistently after weaning. Although some losses of fertilized eggs occur, the eventual litter size will be large, particularly for sows in the third parity or above. By imposing a much reduced lactation period on the sow there is significant gain in terms of the number of litters produced per sow per year but there is also a loss in terms of the number of piglets per litter at each farrowing due to the effects reviewed in this Chapter. How much is gained or lost for each given lactation length has been assessed by Varley and Cole (1978) who concluded that as lactation length is reduced from the conventional 6–8 weeks down to about three weeks, there is a general increase in the annual sow productivity due to improving farrowing interval.

For lactation lengths much below three weeks any further improvement in farrowing interval is far outweighed by the reduced litter size and hence annual sow productivity falls. This conclusion is substantiated by the reports of Smidt, Scheven and Steinbach (1965) and Aumaitre, Perez and Chauvel (1975) illustrated in *Figure 22.8*.

Producers aiming to maximize sow productivity should consider the 3-week weaning system carefully. Many still take the view that 5-week weaning offers the maximum potential, although survey work (MLC, 1979; Ridgeon, 1979) consistently comes out in favour of 3-week weaning. The biological relationships described in this chapter also point emphatically to the conclusion that although the 3-week weaning system represents the interface between conventional and very early weaning systems at present it has more potential than any other system. The 3-week weaned sow can

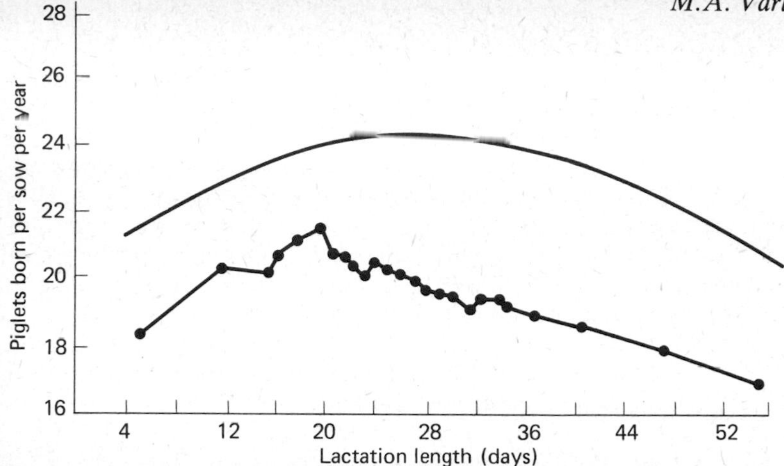

Figure 22.8 The effect of lactation length on annual sow productivity. Upper curve from Smidt, Scheven and Steinbach (1965), lower curve from Aumaitre, Perez and Chauvel (1975)

experience slightly increased embryonic mortality and Varley (1979) who has reviewed survey data on the prolificacy of these sows concluded that despite an apparent drop in average litter size of 0.2 piglets, the annual sow productivity from 3-week weaned sows can still be greater than other systems. te Brake (1976) has also carried out a comprehensive analysis of available data on this point. He concluded that 'weaning of piglets at 21–25 days of age is the most profitable way of producing piglets of 20 kg body weight'.

Part of the discrepancy in performance between farms may result from the age/parity effects reported by Walker *et al.* (1979). In a herd operating quite rapid generation turnover (for genetic improvement), the average age and parity of sows would be low. Little difference would therefore be detectable between 3-week weaning and 5-week weaning. On the other hand if a herd had a slow generation turnover and low sow culling rates the mean age and parity would be greater and therefore the effects of higher embryo mortality in the 3-week weaned sows would manifest itself as a very distinct difference in litter size compared with sows weaned at five weeks.

Looking to the future, with further work on the understanding of factors controlling embryonic mortality a highly productive 10-day weaning system can be envisaged producing around 26–28 piglets per sow per year. Alternatively by using the existing knowledge on the early induction of puberty and gestation in once bred gilts destined for meat production, coupled with weaning at birth, we may circumvent any problems of the sow's future reproductvve capacity and at the same time reduce neonatal death to a minimum.

References

ALEXANDER, V.A. (1969). Studies on the nutrition of the neonatal pig. PhD Thesis. University of Edinburgh

AUMAITRE, A. (1972). Influence de mode de sevrage sur la productivité des truies. *23rd Annual Meeting E.A.A.P. Pig Commission, Verona*

AUMAITRE, A. and RETTIGLIATI, J. (1972). Age au sevrage chez le porcelet: Repercussion sur la productivite des truies et influence sur les jeunes animaux. *Annls Zootech.* **21**, 634–635

AUMAITRE, A., PEREZ, J.M. and CHAUVEL, J. (1975). Effet de l'habitat et de l'age au sevrage sur les composentres de la productivité des truies en France. *Journées de la recherche porcine en France*, pp. 52–67. Paris, L'Institut Technique du Porc

BASHE, S.M., BAZER, F.W., GEISERT, R.D. and ROBERTS, R.M. (1980). Progesterone-induced uterine secretions in pigs. Recovery from pseudopregnant and unilaterally pregnant gilts. *J. Anim. Sci.* **50**, 113–123

BAZER, F.W., ROBINSON, O.W., CLAWSON, A.J. and ULBERG, L.C. (1969). Uterine capacity at two stages of gestation in gilts following embryo superinduction. *J. Anim. Sci.* **29**, 30–34

BAZER, F.W., CHEN, T.T., KNIGHT, J.W., SCHLOSNAGLE, D.C., BALDWIN, N.J. and ROBERTS, R.M. (1975). Presence of a progesterone-induced, uterine specific, acid phosphatase in allantoic fluid of gilts. *J. Anim. Sci.* **41**, 112

BEIER, H.M. (1969). Endometrial secretion proteins—biochemistry and biological significance. In *The Biology of the Fluids of the Female Genital Tract*, (F.K. Beller and G.F.B. Schumacher, Eds.), pp. 89–113. Amsterdam, Elsevier

BEIER, H.M. and MAURER, R.R. (1975). Uteroglobin and other proteins in rabbit blastocyst fluid after development *in vivo* and *in vitro*. *Cell Tiss. Res.* **159**, 1–10

TE BRAKE, J.H.A. (1972). Pig rearing in cages—the extra early weaning of piglets and the fertility of sows. *23rd Annual Meeting E.A.A.P. Commission on Pig Production, Verona*

TE BRAKE, J.H.A. (1976). An assessment of the most profitable lactation length for producing piglets of 20 kg body weight. *27th Annual Meeting E.A.A.P. Pig Commission, Zurich*

BRAUDE, R. and NEWPORT, M.J. (1973). Artificial rearing of pigs. 4. The replacement of butterfat in a whole-milk diet by either beef tallow, coconut oil or soya-bean oil. *Br. J. Nutr.* **29**, 447–455

BRAUDE, R., KEAL, H.D. and NEWPORT, M.J. (1976). Artificial rearing of pigs. 5. The effect of different proportions of beef tallow or soya bean oil and dried skim milk in the diet on growth, feed utilisation, apparent digestibility and carcass composition. *Br. J. Nutr.* **35**, 253–258

BRAUDE, R., KEAL, H.D. and NEWPORT, M.J. (1977). Artificial rearing of pigs. 6. The effect of different levels of fat, protein and methionine in a milk substitute diet containing skim milk and soya-bean oil. *Br. J. Nutr.* **37**, 187–194

BRAUDE, R., MITCHELL, K.G., NEWPORT, M.J. and PORTER, J.W.G. (1970). Artificial rearing of pigs. 1. Effect of frequency and level of feeding on performance and digestion of milk proteins. *Br. J. Nutr.* **24**, 501

BROOKS, P.H., COLE, D.J.A., ROWLINSON, P., CROXON, V.J. and LUSCOMBE, J.R. (1975). Studies in sow reproduction. 3. The effect of nutrition between weaning and remating on the reproductive performance of multiparous sows. *Anim. Prod.* **20**, 407–412

BUHI, W., BAZER, F.W., DVCSAY, C.D., CHUN, P.W. and ROBERTS, R.M. (1979). Iron content, molecular weight and possible function of the progesterone-induced purple glycoprotein of the porcine uterus. *Fed. Proc.* **38**, 733

CLARK, J.R., DAILEY, R.A., FIRST, N.L., CHAPMAN, A.B. and CASIDA, L.E.

(1972). Effect of feed level and parity on ovulation rate in three genetic groups of swine. *J. Anim. Sci.* **35**, 1216

COLE, D.J.A., VARLEY, M.A. and HUGHES, P.E. (1975). Studies in sow reproduction. 2. The effect of lactation length on the subsequent reproductive performance of the sow. *Anim. Prod.* **20**, 401–406

CRIGHTON, D.B. (1967). Effects of lactation on the pituitary gonadotrophins of the sow. In *Reproduction in the Female Mammal*, (G.E. Lamming and E.C. Amoroso, Eds.), pp. 228–238. London, Butterworths

DAY, B.M. and POLGE, C. (1968). Effects of progesterone on fertilisation and egg transport in the pig. *J. Reprod. Fert.* **17**, 227–230

DYRENDAHL, S., OLSSON, B., FJORCK, G. and EHLERS, T. (1958). Artificial rearing of baby pigs. Part II: Additional experiments including the effect of early weaning on the fertility of sows. *Acta agric. scand.* **8**, 3–19

ELLIOT, J.I. KING, G.J. and ROBERTSON, H.A. (1980). Reproductive performance of the sow subsequent to weaning piglets at birth. *Can. J. Anim. Sci.* **60**, 65–71

ETZEL, B.J., MURRAY, F.A., GRIFO, A. P. Jr. and KINDER, J.E. (1978). Partial purification of uterine secretory protein capable of suppressing lymphocyte reactivity *in vitro*. *Theriogenology* **10**, 469–480

FOWLER, V.R. and VARLEY, M.A. (1981). Recent development in weaning at birth. *Proc. Pig Vet. Soc.* (in press).

GRAVES, W.E. LAUDERDALE, J.W., KIRKPATRICK, R.L., FIRST, N.L. and CASIDA, L.E. (1967). Tissue changes in the involuting uterus of the post partum sow. *J. Anim. Sci.* **26**, 365–369

HAY, D.M., FORRESTER, P.I., HANCOCK, R.L. and LORSCHEIDER, F.L. (1976). Maternal serum alpha-fetoprotein in normal pregnancy. *Br. J. Obstet. Gynaec.* **83**, 534–538

HUGHES, P.E. and VARLEY, M.A. (1980). *Reproduction in the Pig*. London, Butterworths

JONES, A.S. (1972). Problems of nutrition and management of early weaned piglets. *Proc. Br. Soc. Anim. Prod.* 19–32

KIRKPATRICK, R.L. LAUDERDALE, J.W., FIRST, N.L., HAUSER, E.R. and CASIDA, L.E. (1965). Ovarian and pituitary gland changes in post partum sows treated with FSH. *J. Anim. Sci.* **24**, 1104–1106

KLOPPER, A. (1979). The new placental proteins. *Biol. Med.* **1**, 89–104

KNIGHT, J.W., BAZER, F.W. and WALLACE, H.D. (1974). Effect of progesterone induced increase in uterine secretory activity on development of the porcine conceptus. *J. Anim. Sci.* **39**, 743–746

KNIGHT, J.W., BAZER, F.W., WALLACE, H.D. and WILCOX, C.J. (1974). Dose response relationships between exogenous progesterone and oestradiol and porcine uterine protein secretions. *J. Anim. Sci.* **39**, 747–751

KNOBIL, E. (1980). The neuroendocrine control of the menstrual cycle. *Recent Prog. Horm. Res.* **36**, 53–88

KRISHNAN, R.S. and DANIEL, J.C. Jr. (1967). "Blastokinin": Inducer and regulator of blastocyst development in the rabbit uterus. *Science* **158**, 490–498

LAI, P.C.W., MEARS, G.J., VAN PETTEN, G.R., HAY, D.M. and LORSCHEIDER, F.L. (1978). Fetal–maternal distribution of ovine alpha-fetoprotein. *Am. J. Physiol.* **235**, E27–31

LAUDERDALE, J.W., KIRKPATRICK, R.L., FIRST, N.L., HAUSER, E.R. and CASIDA, L.E. (1963). Some changes in the reproductive organs of the periparturient sow. *J. Anim. Sci.* **22**, 1138

LOOKER, M. (1974). Cage rearing on a large scale. *Pig Fmg* **22**, 66–67

McINTYRE, J.A. and FAULK, W.P. (1979). Trophoblast modulation of maternal allogeneic recognition. *Proc. natn. Acad. Sci. USA* **76**, 4029–4032

MLC (1979). *Pig Improvement Services.* Meat and Livestock Commission, Pig Feed Recording Services, 1979

MOODY, N.W. and SPEER, V.C. (1971). Factors affecting sow farrowing interval. *J. Anim. Sci.* **32**, 510–514

MURGITA, R.A. and TOMASI, T.B. Jr. (1975). Suppression of the immune response by α-fetoprotein. 1. The effect of mouse α-fetoprotein on the primary and secondary antibody response. *J. exp. Med.* **114**, 269–286

MURRAY, F.A. and GRIFO, A.P. Jr. (1976). Development of the capacity to secrete progesterone-induced protein by the porcine uterus. *Biol. Reprod.* **15**, 620–625

MURRAY, F.A., MOFFATT, R.J. and GRIFO, A.P. Jnr. (1980). Secretion of riboflavin by the porcine uterus. *J. Anim. Sci.* **50**, 926–927

MURRAY, F.A., SEGERSON, E.C. and BROWN, F.T. (1978). Suppression of lymphocytes *in vitro* by porcine uterine secretory protein. *Biol. Reprod.* **19**, 15–25

MURRAY, F.A., BAZER, F.W., WALLACE, H.D. and WARNICK, A.C. (1972). Quantitative and qualitative variation in the secretion of protein by the porcine uterus during the estrous cycle. *Biol. Reprod.* **7**, 314–320

NEWPORT, M.J., STORRY, J.E. and TUCKLEY, B. (1979). Artificial rearing of pigs. 7. Medium chain triglycerides as a dietary source of energy and their effect on liveweight gain, feed:gain ratio, carcass composition and blood lipids. *Br. J. Nutr.* **41**, 85–93

PALMER, W.M., TEAGUE, H.S. and VENZKE, W.G. (1965a). Macroscopic observations on the reproductive tract of the sow during lactation and early post weaning. *J. Anim. Sci.* **24**, 541–545

PALMER, W.M., TEAGUE, H.S. and VENZKE, W.G. (1965b). Histological changes in the reproductive tract of the sow during lactation and early post weaning. *J. Anim. Sci.* **24**, 1117–1125

PARLOW, A.F., ANDERSON, L.L. and MELAMPY, R.M. (1964). Pituitary follicle stimulating hormone and luteinising hormone concentrations in relation to reproductive stages in the pig. *Endocrinology* **75**, 365–376

PAY, M.G. (1973). The effect of short lactations on the productivity of sows. *Vet. Rec.* **92**, 255–259

PETERS, J.B., SHORT, R.E., FIRST, N.L. and CASIDA, L.E. (1969). Attempts to induce fertility in post partum sows. *J. Anim. Sci.* **29**, 20–24

POLGE, C. (1972). Reproductive physiology in the pig with special reference to early weaning. *Proc. Br. Soc. Anim. Prod.* 1972, pp. 5–18

PUY AOAN, R.B. and CASTILLO, L.S. (1963). A study on the effects of weaning pigs at different ages and the subsequent reproductive performance of their dams. *Philipp. Agric.* **47**, 32–44

RIDGEON, R.F. (1979). *Pig Management Scheme, Results for 1979*, Agricultural Economics Unit, University of Cambridge

ROBERTS, G.P. (1977). Inhibition of lymphocyte stimulation by bovine uterine proteins. *J. Reprod. Fert.* **50**, 337–339

ROBERTSON, V.A.W., JONES, A.S., FULLER, M.F. and ELSLEY, F.W.H. (1971)

A pig herd established by hysterectomy. 1. The techniques for rearing hysterectomy derived piglets to 5 weeks of age. *Res. vet. Sci.* **12**, 59–64

ROWLINSON, P. and BRYANT, M.J. (1974). Sows mated during lactation: Observations from a commercial unit. *Proc. Br. Soc. Anim. Prod.* **3**, 93

SELF, H.L. and GRUMMER, R.H. (1958). The rate and economy of pig gains and the reproductive behaviour in sows when litters are weaned at 10 days, 21 days or 56 days of age. *J. Anim. Sci.* **17**, 862–868

SHARMAN, G.A.M., JONES, A.S., DENERLEY, H. and ELSLEY, F.W.H. (1971). A pig herd established by hysterectomy. II. Health and performance. *Res. vet. Sci.* **12**, 65–73

SMIDT, D., SCHEVEN, B. and STEINBACH, J. (1965). The influence of lactation on the sexual function of sows. *Zuchtungskunde* **37**, 23–36

SMIDT, D., THUME, O. and JOCHLE, W. (1969). Investigations on post partum sexual regeneration in suckling and non lactating sows. *Zuchtungskunde* **41**, 36–45

SMITH, D.M. (1961). The effect of daily separation of sows from their litters upon milk yield, creep intake and energetic efficiency. *N.Z. J. agric. Res.* **4**, 232–245

SMITH, K.M., LAI, P.C.W., ROBERTSON, H.A., CHURCH, R.B. and LORSCHEIDER, F.L. (1979). Distribution of alpha-fetoprotein in fetal plasma, allantoic fluid, amniotic fluid and maternal plasma of cows. *J. Reprod. Fert.* **57**, 235–238

SVAJGR, A.J., HAYS, V.W., CROMWELL, G.L. and DUTT, R.H. (1974). Effect of lactation duration on reproductive performance of sows. *J. Anim. Sci.* **38**, 100–105

TATE, M. (1974). Studies on the utilisation of amino acids by the neonatal pig. PhD Thesis. University of Aberdeen

VAN DER HEYDE, H. (1972). A practical assessment of early weaning. *Proc. Br. Soc. Anim. Prod.* 33–36

VARLEY, M.A. (1976). Reproductive performance of the early weaned sow. PhD Thesis. University of Nottingham

VARLEY, M.A. (1979). Three week weaning and prolificacy. *Pig Fmg* **27**, 34–35

VARLEY, M.A. and COLE, D.J.A. (1976a). Studies in sow reproduction. 4. The effect of level of feeding in lactation and during the interval from weaning to remating on the subsequent reproductive performance of the early weaned sow. *Anim. Prod.* **22**, 71–77

VARLEY, M.A. and COLE, D.J.A. (1976b). Studies in sow reproduction. 5. The effect of lactation length of the sow on the subsequent embryonic development. *Anim. Prod.* **22**, 79–85

VARLEY, M.A. and COLE, D.J.A. (1978). Studies in sow reproduction. 6. The effect of lactation length on pre-implantation losses. *Anim. Prod.* **27**, 209–214

VARLEY, M.A., ATKINSON, T. and ROSS, L.N. (1981). The effect of lactation length on the circulating concentrations of progesterone and oestradiol in the early weaned sow. *Theriogenology* **16**, 179–184

WALKER, N., WATT, D., MacLEOD, A.S., JOHNSON, C.L., BOAZ, T.G. and CALDER, A.F.C. (1979). The effect of weaning at 10, 25 or 40 days on the reproductive performance of sows from the first to the fifth parity. *J. agric. Sci., Camb.* **42**, 449–456

WILLIAMS, I.H. (1976). Nutrition of the young pig in relation to body composition. PhD Thesis. University of Melbourne
WRATHALL, A.E. (1971). *Prenatal survival in pigs. Part I.* Slough, England, Commonwealth Agricultural Bureaux

23

MANAGEMENT OF THE SOW AND LITTER IN LATE PREGNANCY AND LACTATION IN RELATION TO PIGLET SURVIVAL AND GROWTH

P.R. ENGLISH and V. WILKINSON
School of Agriculture, University of Aberdeen, UK

The objectives of management in late pregnancy and lactation are to minimize the incidence of stillbirths, to maximize the number of livebirths at parturition and to improve the viability of these pigs. Thereafter, management must strive to minimize mortality and to produce healthy and well grown pigs at weaning. Treatment imposed in late pregnancy and lactation should be orientated towards achieving prompt conception following weaning and to providing the potential for establishing a large litter of viable piglets in the subsequent gestation.

Aspects relating to piglet mortality and its prevention will form the first part of this chapter followed by a consideration of supplementary feeding of the litter.

Piglet mortality

EXTENT OF MORTALITY

Losses from stillbirths range from 4–8% of all pigs born in most studies (*Table 23.1*), although considerably higher levels of stillbirths have been reported in isolated cases (Moore, Redmond and Livingston, 1965).

Table 23.1 ESTIMATES OF STILLBIRTH LOSSES

Source	*Country*	*Total born*	*Stillbirths (%)*
Braude *et al.* (1954)	Britain	12.10	5.72
Hutchinson *et al.* (1954)	U.S.A.	9.62	5.15
Gracey (1955)	Ireland	11.08	4.9
Anon (1959)	Britain	11.00	6.04
Bauman *et al.* (1966)	U.S.A.	10.07	5.67
		10.70	5.26
Sharpe (1966)	Britain	11.21	4.21
English (1969)	Britain	11.69	7.9
Randall and Penny (1970)	Britain	10.99	6.33
Randall (1972b)	Britain	10.65	6.83
Leman *et al.* (1972)	Illinois, USA	9.82	7.09
Nielsen *et al.* (1974)	Denmark	10.16	5.9
Glastonbury (1976)	New South Wales, Australia	10.4	5.8
Meat and Livestock Commission (1980)	Britain	11.1	5.4

Table 23.2 PRE-WEANING MORTALITY

Source	*Country*	*Weaning age* (weeks)	*Born alive*	*Mortality* (% of livebirths)
Braude *et al.* (1954)	Britain	8	11.4	29.5
Fraser (1966)	Jamaica		9.6	20.8
Sharpe (1966)	Britain	6	10.75	21.4
Ilančić *et al.* (1968)	Yugoslavia	6	10.30	14.5
English (1969)	Britain	8	10.77	24.4
Leman *et al.* (1972)	Illinois, USA		9.12	19.5
Nielsen *et al.* (1974)	Denmark		9.56	17.8
Glastonbury (1976)	New South Wales, Australia		9.8	14.3
Meat and Livestock Commission (1980)	Britain	<19 days	10.2	11.1
		19–25 days	10.3	12.3
		26–32 days	10.4	12.0
		33–39 days	10.5	13.5
		>39 days	10.5	14.3
		Overall	10.4	12.8

The extent of pre-weaning losses has also been very variable ranging from 12 to almost 30% (*Table 23.2*). However, even the lowest levels of mortality reported constitute considerable losses. For example, a 12% mortality is equivalent to the loss of over 250 piglets/year in a 100-sow herd.

It is very important to relate percentage losses to numbers born. In commercial terms, 10% pre-weaning mortality with 10 born alive (leaving 9 reared per sow) is considerably more serious than 20% mortality with 15 pigs born (leaving 12 reared per sow).

CAUSES OF STILLBIRTH

Stillbirths have been classified into Type I or pre-partum deaths (dead before start of parturition) and Type II or intrapartum deaths (dying during the parturition process) (Randall and Penny, 1967). Among other factors from which pre-partum deaths can result, infection during pregnancy such as that associated with the Smedi syndrome can be responsible (Dunne and Leman, 1975). In isolated cases, iron deficiency in the sow has been associated with a higher incidence of stillbirths (Moore, Redmond and Livingston, 1965) and, in such cases, administration of supplementary iron has helped to reduce the problem.

Of the total stillbirths, between 70 and 90% are of Type II or intrapartum deaths (Randall, 1972b; English and Smith, 1975) and foetal anoxia during parturition is recognized as being the major cause (Randall and Penny, 1967). Such anoxia is induced by decreased placental blood flow associated with uterine contractions or with occlusion or premature rupture of the umbilical cord or premature detachment of the placenta (Curtis, 1974).

Later born piglets are likely to suffer to a greater degree from anoxia because of the cumulative effects of successive contractions in reducing the oxygenation of the unborn piglets and also because of the greater risk of premature rupture of the umbilical cord or of detachment of the placenta as parturition progresses (English, 1969; Randall, 1972a,b). Thus, the incidence of intrapartum deaths tends to be concentrated in the second half of the birth order with about 70% being among the last three piglets born (Randall, 1972a,b; English, Smith and MacLean, 1977)

The incidence of intrapartum deaths increases markedly in older sows, probably because of poorer uterine muscle tone, and is also higher in larger litters (Anon, 1959; Sharpe, 1966; English, 1969; Randall and Penny, 1970; Bille *et al.*, 1974; English *et al.*, 1977). A higher percentage of stillbirths is also experienced in very small litters (Carmichael and Rice, 1920; Anon, 1959; Sharpe, 1966).

The same factors that increase the incidence of intrapartum deaths in later born pigs also render the later born survivors more anoxic at birth and probably less viable (English *et al.*, 1977). The blood lactate level in newborn piglets provides a useful measure of the degree of anoxia suffered during parturition (Dawes *et al.*, 1963) and the relationship of lactate level to birth order is illustrated in *Figure 23.1.*

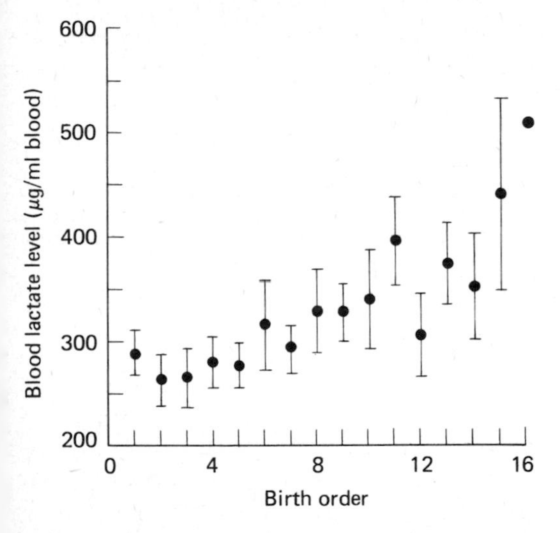

Figure 23.1 Blood lactate level in liveborn pigs at birth according to birth order (mean blood lactate ± standard error)

The relationship between the level of anoxia in the liveborn piglet at birth and subsequent survival has not been clearly established. However, it is known that severe anoxia *in utero* may lead to depression of the central nervous system and irreversible brain damage (Stanton and Carroll, 1974). Miller and Miller (1965) considered that the unborn piglet during parturition was likely to suffer from irreversible brain damage within approximately five minutes following restriction of umbilical blood flow. *Table 23.3* contains details of a comparison of blood lactate levels at birth of liveborn piglets which died before three weeks of age and those which survived after this time. Liveborn piglets dying before three weeks of age had a 26.5% higher blood lactate level at birth than those surviving after this stage, the difference between the two groups being significant ($P<0.01$). Thus, it appeared that the liveborn pigs which died suffered

Table 23.3 BLOOD LACTATE LEVELS AT BIRTH IN LIVEBORN PIGS (a) DYING BEFORE AND (b) SURVIVING TO 3 WEEKS OF AGE

	Dying before 3 weeks	*Surviving to 3 weeks*	*Difference*	*S.E. of difference between means*	*Level of significance of difference*
Number of piglets	58	252			
Mean blood lactate (μg lactate/ml blood)	383.3	303.0	80.3	25.3	**

**Significant at $P<0.01$.
From English and Smith (1975)

from a higher degree of prenatal anoxia than those which survived. It seems probable that this may have contributed to their death.

ATTEMPTS TO IMPROVE THE EFFICIENCY OF PARTURITION

It has been established that inefficiencies associated with parturition result in increased incidence of intrapartum deaths and may reduce the viability of liveborn piglets. Various attempts have been made to increase the efficiency of the parturition process. One simple approach in commercial practice is to cull sows timeously since older sows take longer to farrow and have a higher incidence of intrapartum deaths (Friend, Cunningham and Nicholson, 1962; English, 1969; English *et al.*, 1977). Other approaches to improving efficiency of parturition have included the administration of posterior pituitary extract which contains oxytocin, of oxytocin itself and of parasympathomimetic drugs. The main findings of such work are summarized in *Table 23.4*.

While pituitary extract and oxytocin are used widely to stimulate uterine muscle contractions in cases of dystocia, their routine use in an attempt to increase the efficiency of parturition and reduce the incidence of intrapartum deaths has produced inconsistent and generally disappointing results (Asdell and Willman, 1941; Lee, 1977). The explanation is likely to be connected with the very short period over which oxytocin appears to be effective in stimulating uterine muscle contractions; its biological half life appears to be no more than five minutes (Denamur, 1965; Hayes and Van Demark, 1952). There is also the risk that if oxytocin is administered in excess it could produce uterine hypotonus (tetany) as it does in the human (Lloyd D.J., personal communication). In cases of dystocia in the human, administration of oxytocin is carried out by continuous intravenous infusion, this being regulated according to the rate and intensity of the contractions (Turnbull and Anderson, 1968). Brenner, Schulze and Gurtler (1978) appear to be the only workers who have used this approach in the pig but their findings provide no indications of the effect on the incidence of intrapartum deaths.

In work conducted to date, the parasympathomimetic drugs have given most promise in relation to possibilities for improving the efficiency of the parturition process. These synthetic compounds act on the parasympathetic nervous system, resulting in stimulation of smooth muscle contraction and acceleration of the parturition process. Parasympathomimetic drugs which have been evaluated include neostigmine, bethanecol, pilocarpine and carbachol. Sprecher *et al.* (1974) reported the effects of a 2 or 3 mg injection of carbachol given after the birth of the first pig in two trials. While the treatment reduced birth interval by over 30% in both trials relative to untreated sows, only in the second trial was some reduction in the incidence of stillbirths evident. However, in later trials by Sprecher, Leman and Carlisle (1975) using carbachol, and by Lee (1977) using both bethanecol and pilocarpine, these drugs were administered midway through parturition in an attempt to accelerate the delivery of the last pigs in the litter, that is, those which are most at risk from intrapartum death through anoxia. In these trials stillbirth rate was reduced by over 0.5 pig/litter relative to controls. While these drugs were found to be equally effective in reducing the incidence of stillbirths, neostigmine was preferred because of the comparative absence of any unpleasant side effects. McInnes (1977) evaluated neostigmine in a large scale on-farm trial and claimed similar reductions in stillbirth rate to Sprecher, Leman and Carlisle (1975) and Lee (1977). However, neostigmine has not produced the same response in all trials since Hendrix *et al.* (1978) found no reduction in either birth interval, or in stillbirths relative to controls, in a trial in which piglets were removed from the sow as they were born. Such removal of piglets as they were born is likely to have reduced the stimulus for oxytocin release from the pituitary gland and may thus have affected the response to neostigmine (Murdoch, 1980).

In trials in which neostigmine has reduced the incidence of intrapartum deaths, it might be expected to decrease the degree of anoxia suffered by liveborn piglets and thus to increase their viability. However, this aspect has not yet been investigated in any of the studies reported.

BASIC REQUIREMENTS OF THE NEWBORN PIGLET

The newborn piglet has several basic requirements if it is to survive and thrive. The herd must have a good health status, the piglet must be anatomically normal, it must be provided with an adequate thermal environment so as to conserve scarce energy reserves, it requires adequate and regular nutrition and it must be protected from being overlain by its dam.

The causes of piglet mortality determined in various studies reflect the extent to which these basic requirements of the newborn piglet are, or are not, provided in practice.

TIMING OF MORTALITY

Over 50% of the losses of liveborn piglets occur within the first 2–3 days of life (Anon, 1959; Pomeroy, 1960; Bauman, Kadlec and Powlen, 1966; Sharpe, 1966; English, 1969; Fahmy and Bernard, 1971; English and Smith, 1975; Glastonbury, 1976). Such data, of course, underestimate the importance of this very early period in relation to losses, for many of the later deaths are triggered off by events in the first few hours of life

Table 23.4 RESULTS OF EXPERIMENTS INVESTIGATING THE USE OF INFLUENCE PARTURITION

Workers	*Compound*		*No. animals in study Controls*	*treated*	*Dose*	*Method of administration*
Asdell and Willman (1941)	Pituitary extract	(1) (2)			1 or 2 ml 1 or 2 ml	s.c.[a] s.c.
Muhrer *et al.* (1955)	Oxytocin		43	36	Not less than 20 iu	i.m.[b]
Lee (1977)	Oxytocin		31	33	40 iu	i.m.
Brenner *et al.* (1978)	Oxytocin	(1) (2) (3)	13 13 13	8 6 8	25 iu 0.125 iu/min 25 iu and 0.125 iu/min	i.m. i.m. slow drip infusion i.m. injection and i.m. slow drip infusion
Sprecher *et al.* (1975)	Neostigmine bromide		17	33	5 mg	s.c.
Lee (1977)	Neostigmine methylsulphate		31	35	5 mg	i.m.
McInnes (1977)	Neostigmine methylsulphate		141	100	5 mg	s.c.
Hendrix *et al.* (1978)[c]	Neostigmine methylsulphate		20	19	5 mg	s.c.
Sprecher *et al.* (1974)	Carbachol	(1) (2)	20 55	16 52	3 mg 2 mg	s.c. s.c.
Sprecher *et al.* (1975)	Carbachol		17	34	2 mg	s.c.
Lee (1977)	Bethanecol		31	45	12 mg	s.c.
Lee (1977)	Pilocarpine		31	69	40 mg	s.c.

[a]s.c. = subcutaneous injection [b]i.m. = intramuscular injection
[c]These results may have been influenced by the fact that all piglets were removed from the sow as they were born
ns = not significantly different from controls

(English, 1969). The fact that such a high proportion of losses occurs in very early life must indicate either that many piglets are very weak at birth and/or that the conditions provided for the newborn piglet are, in general, very inadequate.

PITUITARY EXTRACT, OXYTOCIN AND PARASYMPATHOMIMETIC DRUGS TO

Timing of treatment	*Decreases (–) or increases (+) relative to controls (percentage changes in brackets)*			*Other effects of treatment*
	Interval between births (min)	*Farrowing duration (first to last pig)* (min)	*Stillbirths per litter*	
After first pig			(–56.8%)	
After first pig			(+78.2%)	
(1) As soon as milk was available		–78 (–37.1%)		
(2) After farrowing had started		–60 (–28.6%)		
After fifth pig	–0.2 (–0.7%) ns		–0.01 (+0.8%) ns	
After first pig		–89 (–37.2%) ns		
From first to last pig		–19 (–7.9%) ns		
Injection after first pig. Infusion between first and last pig		–150 (–62.8%) *P*<0.01		
After third or fourth pig			–0.67 (–76.1%) *P*<0.001	
After fifth pig	–8.5 (–29.9%) ns		–0.54 (–73.0%) *P*<0.01	Mild salivation in small animals
After first or second pig		–135 (–51.2%)	–0.33 (–54.1%)	
After fifth pig	–2.3 (–6.2%) ns	+ 135 (+52.5%) ns	+0.16 ns	
After first pig	–6.0 (–36.4%)		(+0.32%) ns	
Ater first pig	–6.7 (–31.6%) *P*<0.05		–0.05 (–10.75%) ns	
After fourth pig			–0.65 (–73.5%) *P*<0.001	Severe salivation and vomiting
After fifth pig	–4.9 (–17.3%) ns		–0.61 (–81.9%) *P*<0.01	Moderate to severe salivation, some vomiting
After fifth pig	–2.8 (–9.9%) ns		–0.55 (–75.5%) *P*<0.01	Very severe salivation and vomiting

CAUSES OF MORTALITY OF LIVEBIRTHS

Deaths of liveborn piglets fall into two categories, namely, whole litter loss and the insidious loss of one or a few piglets from most litters. Most attention will be given to the latter category which is by far the most serious source of loss.

Congenital and genetic abnormalities

Congenital and genetic abnormalities account for an average of 5% of losses in all studies ranging from 2.1% (Sharpe, 1966) to 12.3% (English and Smith, 1975). The most common causes of death in this category are atresia ani, congenital splay leg and cardiac abnormality.

Disease

While disease conditions such as agalactia in sows and scour in piglets are responsible for some losses, most studies have shown that disease as a primary factor in death accounts for only about 6% of deaths (Braude, Clarke and Mitchell, 1954; Hutchinson *et al.*, 1954; Gracey, 1955; Sharpe, 1966; English and Smith, 1975). Taking essential precautions such as careful selection of the source of replacement stock, maintaining a high level of hygiene especially in the farrowing quarters, providing comfortable conditions for sows and piglets, arranging for adequate nutrition and adopting prophylactic or prompt therapeutic treatment to combat problems such as piglet scour, are usually effective in keeping piglet deaths from primary disease down to low levels.

Starvation and overlying by the sow

The major causes of death of baby piglets in most studies have been starvation (leading to hypoglycaemia) and overlying by the sow. Together, these two factors accounted for 74.8%, 79.0%, 50.1%, 73.7% and 75.9% of the deaths respectively in the studies of Braude, Clarke and Mitchell (1954), Gracey (1955), Anon (1959), Bauman, Kadlec and Powlen (1966) and English and Smith (1975). In attempts to reduce the extremely high losses suffered through starvation and overlying, efforts are required on various fronts.

Table 23.5 LIVEBIRTH LOSSES IN RELATION TO BIRTHWEIGHT

Birthweight (g)	*% of pigs*	*Mortality* (%)	*Contribution to losses* (%)
Under 800	7.7	56.5	28.3
800–1000	9.4	26.8	16.4
1000–1200	16.2	15.5	16.3
Over 1200	66.7	9.0	39.0
			100.0

From English *et al.* (1977)

A very close relationship exists between birthweight and mortality (*Table 23.5*). It can be seen that while over 70% of the total losses are suffered by pigs of 800 g and over, the small proportion of pigs weighing less than 800 g at birth suffer extremely high losses. Piglets of low birthweight and vigour are very liable to succumb to starvation and overlying (Carroll, Krider and Andrews, 1962; English, 1969; England, 1974; English and Smith, 1975). Thus, one approach to reducing piglet losses from starvation and overlying is to attempt to increase the birthweight and vigour of the newborn piglet.

The poor insulation and inadequate temperature regulating mechanism of the newborn pig (Newland, MacMillen and Reineke, 1952; Mount, 1972) renders it very vulnerable to chilling unless an adequate thermal environment is provided for it. In addition, because of the very low energy reserves of the baby pig (Elsley, 1964; Elliot and Lodge, 1977) and its small stomach capacity, there is a pressing need to ensure adequate and regular nutrition. Further, since the newborn piglet constitutes only about 1% of its dam's weight and because sows can be very restless during and immediately after parturition, there is a need to ensure adequate protection of the newborn piglet from overlying.

Thus, the approaches to attempting to reduce losses from starvation and overlying should include the following:

(a) Improving the birthweight and vigour of the newborn piglet.
(b) Ensuring adequate and regular nutrition.
(c) Provision of an adequate thermal environment and protection from overlying.

These three objectives are interdependent in that weaker piglets at birth are more prone to chilling, starvation and overlying. An underfed piglet is also more liable to chilling and to being overlain by the sow, while a chilled piglet is less capable of competing for a suckling position and is therefore more liable to starvation and to being overlain.

ATTEMPTS TO IMPROVE THE BIRTHWEIGHT AND VIGOUR OF THE NEWBORN PIGLET

Increasing birthweight

As already indicated, there is an inverse relationship between birthweight and losses of liveborn pigs. This is because piglets of higher birthweight lose less heat to their surroundings because of a lower surface area to weight ratio (Stanton and Carroll, 1974). Thus, they are less liable to chilling than piglets of lower birthweight. Higher birthweight pigs are also likely to have larger reserves of glycogen which is the source of energy which the piglet can utilize most effectively in early life (Dale, 1975).

Piglet birthweight can be improved by higher feed or energy intakes in pregnancy (Clawson *et al.*, 1963; Lodge, Elsley and MacPherson, 1966a,b; O'Grady, 1967; Elsley *et al.*, 1969; Baker *et al.*, 1969; Buitrago, Maner and Gallo, 1970; Lodge, 1972; Elsley and Shirlaw, 1976). The response in terms of increased birthweight is greater when daily energy intake of the sow is increased up to 30 MJ DE than that obtained by increasing daily energy intake above this level (Elsley and Shirlaw, 1976). However, considerable increases in food or energy intake are required to bring about very modest improvements in birthweight. It has been estimated from the results of the studes cited above that to improve average piglet birthweight by only 0.1 kg, an extra 100 kg (approximately 11 MJ DE) of a conventional UK sow diet would have to be given in pregnancy (Elsley and Shirlaw, 1976). The cost/benefit of increasing feed intake in pregnancy to improve mean piglet birthweight and piglet survival will depend on the cost of the extra food required and the value of the extra piglets likely to be saved.

Attempts to improve piglet birthweight and vigour by marked increases in feed or energy intake in very late pregnancy have produced conflicting

Table 23.6 INFLUENCE OF LEVEL AND SOURCE OF ENERGY INTAKE IN LATE PREGNANCY ON FACTORS RELATED TO PIGLET VIABILITY AND ON SURVIVAL

Workers	*Stage of pregnancy studied (days)*	*Energy intake in earlier pregnancy (MJ DE/day)*	*Sows per treatment*	*Treatments*		*Differences in piglets at birth relative to controls*						
				Daily energy intake (MJ DE/day)	*Source of supplementary energy*	*Birth weight*	*Glycogen in: Liver*	*Glycogen in: Skeletal muscle*	*Blood glucose*	*Carcass lipids*	*Piglet survival to weaning*	*Fat in colostrum (%)*
Kotarbinska	100–110	33.5	75	33.5		+15.5%					+4.4%	
(unpublished)				50.2								
Seerley *et al.*	109–	22.4	19–23	16		ns					ns	
(1974)	farrowing			36	10% Corn oil				+	+		+
				36	10% Corn starch		+	+				
Friend (1974)	109–		11	27.2	10% Corn oil					ns		+
	farrowing			23.4	10% Corn starch	+						
Elliot and	100–	31.0	7	5.9		ns		ns		ns		
Lodge (1977)	farrowing			30.5			+					
Okai *et al.*	100–	25.3	19–21	25.3		ns	ns			ns	ns	ns
(1977)	farrowing			53.2								
				58.4	10% Sucrose							
				58.1	10% Tallow							
Okai *et al.*	100–	28.1	4–5	25.6		ns		ns	ns			
(1978)	farrowing			67.7								
				59.7	10% Sucrose							
				71.9	10% Tallow		Lower than control ($P<0.01$)					
Boyd *et al.*	100–	25.3	4	25.3		ns	ns			ns		
(1978a) I	farrowing			40.9	20% Tallow							
				40.9	32.4% Corn starch							
II	109–	37.9	8–9	37.9					ns			
	farrowing			37.9	15% Tallow							
Boyd *et al.*	100–	25.3	22–30	25.3		ns					ns	
(1978b)	farrowing			40.9	20% Tallow							+
				40.9	32.4% Corn starch							
English and	100–	28.8	106	28.8		ns						
Dias (1978)	farrowing			45.0								
Hillyer and	90–	30	150	30.0		ns						
Phillips (1980)	farrowing			42.5								

results and the relevant work is summarized in *Table 23.6.* Kotarbinska (unpublished data) compared normal (33.5 MJ DE) and high (50.2 MJ DE) daily energy intakes from day 100 to day 110 of pregnancy and claimed that a 15.5% increase in birthweight and a 1.4% improvement in piglet survival was associated with the higher energy intake. However, Elliot and Lodge (1977), who compared normal energy intake (30.5 MJ DE/day) in the last 15 days of gestation with very low energy intake (5.9 MJ DE/day) in the same period, did not find any significant reduction in birthweight associated with the lower energy intake. However, they did find a significantly lower liver glycogen level ($P<0.05$) to be associated with low energy intake. Other workers who have applied differential energy intakes in the last 5–15 days of gestation also failed to bring about a significant improvement in birthweight (Seerley *et al.*, 1974; Boyd *et al.*, 1978a,b; Okai *et al.*, 1977, 1978; English and Dias, 1978; Hillyer and Phillips, 1980).

Other attempts to improve piglet birthweight have included the use of food additives and selection. England (1974) found that the inclusion in the sow diet for the last 30 days of gestation of 2,2-dichlorovinyl dimethyl phosphate (dichlorvos) increased birthweight in some trials but not in others. Since the heritability of birthweight is near zero (Craig, Norton and Terrill, 1956) there is little hope of effecting improvement in this trait by selection. Thus, it appears to be very difficult to bring about worthwhile improvements in piglet birthweight.

Variation in piglet birthweight is just as important as birthweight *per se*, if not even more so, in relation to chances of survival (English, 1969; Fahmy and Bernard, 1971; English and Smith, 1975) as indicated in *Table 23.7.* It can be seen that litters with high mean birthweight but with high variation in birthweight within litters suffered as high mortality as both low birthweight litter groups. The lowest mortality rate (13.7%) was in the high birthweight group with low within-litter variation in birthweight.

The competition for possession of a suckling position, and therefore for nutrition, can be severe and small piglets in an uneven litter (in terms of birthweight) are at a severe physical disadvantage. Within-litter birthweight variation tends to be reduced with crossbreeding and to increase with advancing parity but, apart from these relationships, there is no known way of influencing this character so as to obtain more uniform

Table 23.7 VARIATION IN BIRTHWEIGHT WITHIN LITTERS IN RELATION TO MORTALITY OF LIVEBIRTHS

Litter groups	*High birthweight*			*Low birthweight*		
	Mean birthweight (g)	*Within-litter standard deviation in birthweight*[a]	*Mortality* (%)	*Mean birthweight* (g)	*Within-litter standard deviation in birthweight*[a]	*Mortality* (%)
High standard deviation	1397	277	20.1	1148	268	19.9
Low standard deviation	1374	154	13.7	1129	177	19.3

[a]Overall average of means for within-litter standard deviation in birthweight for each litter size group.
Data in each of four subgroups based on 78 litters and 895 livebirths.
From English and Smith (1975)

litters at birth. The only effective way to achieve more uniform birthweights within litters is to artificially create such litters very soon after birth by crossfostering between simultaneously farrowed litters. This approach will be discussed in a later section.

Improving energy reserves and piglet vigour at birth

Several attempts have been made to increase the energy reserves of piglets at birth by manipulations of nutrition in very late pregnancy and the results of these trials are summarized in *Table 23.6.* It has already been mentioned that Elliot and Lodge (1977) found significantly lower glycogen levels ($P<0.05$) at birth in piglets born to sows on a low energy intake in the last 15 days of gestation. When Seerley *et al.* (1974) supplemented a basal diet with either corn starch or corn oil from the 109th day of gestation so as to more than double the sows' energy intake in this period, significant increases ($P<0.05$) in glycogen content of both the liver and the longissimus dorsi muscle of newborn piglets was associated with the corn starch supplemented treatment. Supplementation of the basal diet with corn oil resulted in significant increases in both body fat and blood glucose at birth and there were some indications (although not statistically significant) of improved piglet survival.

Supplementation of a basal diet (25.3 MJ DE/day) with either corn starch or stabilized tallow to provide 40.9 MJ DE/day from day 100 of pregnancy to parturition did not result in any apparent differences in piglets at birth (Boyd *et al.*, 1978b), although the fat content of the colostrum of sows supplemented with tallow was significantly higher than in controls ($P<0.05$). In another experiment Boyd *et al.* (1978a) failed to demonstrate any significant differences in liver glycogen, or in carcass fat of newborn piglets from sows supplemented with either corn starch or tallow from day 100 of pregnancy relative to piglets from sows receiving a much lower energy intake from a control diet in this period.

Other studies by Okai *et al.* (1977), Okai *et al.* (1978), Boyd *et al.* (1978b), Curtis, Heidenreich and Foley (1965), and Reedy *et al.* (1966) involving various manipulations of nutrition from day 100 of gestation to parturition failed to produce a consistent response in terms of increased energy reserves or indices of improved viability in piglets at birth.

Thus, while increased energy intake in the last 1–2 weeks of pregnancy has given promise in some trials of more viable pigs at birth, responses obtained in general have been very inconsistent and, on the basis of existing evidence, there would appear to be no sound basis for making any recommendations for commercial practice regarding manipulations of nutrition in late pregnancy likely to improve viability of newborn piglets.

However, many of the trials which have examined the effects of higher energy intake or of particular energy supplements in late pregnancy have been carried out with relatively small numbers of animals (see *Table 23.6*). Other trials which have involved larger numbers have merely examined the effect of higher energy levels in late pregnancy on piglet birthweight (English and Dias, 1978; Hillyer and Phillips, 1980), and have failed to monitor other indices of piglet viability or piglet survival. Thus, there is a need for a large scale comprehensive experiment to examine the effects of differential energy intakes in late pregnancy on all important indices of piglet viability at birth and on piglet survival to weaning.

One important provision for improving the vigour of the newborn piglet which is implemented widely in practice is that of crossbreeding. It is well established that for commercial production a policy of crossbreeding should be operated because of the important influence of heterosis on piglet vigour at birth and on survival (Fredeen, 1957; Smith and King, 1964).

Improving vigour at birth: general position

As far as the problems of low and variable piglet birthweights and inadequate vigour at birth are concerned, it is clear from the foregoing reviews that these problems can be resolved only to a small degree through breeding strategy and manipulations of prenatal nutrition. Research work related to increasing piglet vigour at birth must be continued and given adequate support. However, since the problems of inadequate piglet vigour at birth and variation in birthweight cannot, as yet, be resolved effectively by amendments of prenatal management, the present onus in commercial practice is on postnatal management to recognize piglets at risk from these factors and to take all necessary precautions, within the limits of economic and practical feasibility, to maximize the survival chances of these disadvantaged and under-privileged piglets.

IMPROVING THE CHANCES OF ADEQUATE AND REGULAR NUTRITION

Much research and development effort has been devoted to improving the opportunities for adequate and regular nutrition of each viable piglet born. One overriding consideration in these studies must be the fact that most piglets which ultimately die from malnutrition around 2–3 days of age are actually moribund almost from the moment of birth because of their inability, or reduced ability, to obtain adequate and regular nutrition mainly because of the very competitive situation which prevails at suckling (English, 1969). Thus, measures to improve survival of such disadvantaged and under-privileged piglets must be implemented at parturition or immediately afterwards. Studies related to improving opportunities for adequate and regular nutrition have involved the following approaches.

(a) Conservation of the newborn piglet's scarce energy reserves.
(b) Studies on sow rearing capacity and fostering strategy.
(c) Studies related to minimizing the effects of agalactia.
(d) Provision of supplementary feeding to newborn piglets.
(e) Artificial rearing of piglets surplus to rearing capacity.
(f) Synchronized farrowing and its supervision.

Conservation of the newborn piglet's scarce energy reserves

Reference has already been made to the comparatively low energy reserves (Elsley 1964; Elliott and Lodge, 1977) and the poor insulation and temperature regulating mechanism of the newborn pig (Newland, MacMillen and Reineke, 1952; Mount, 1972). Estimates of the environmental temperature requirement of the newborn pig vary from 32–36 °C, the smaller pigs requiring a higher temperature (Mount, 1972).

Provision of a perfect thermal environment in the farrowing pen is vitally important in minimizing heat loss so as to prevent chilling and to ensure sufficient energy is available to help each piglet to obtain and retain a functional feeding place at the udder under the stress of severe competition from litter mates. It is important for supplementary heating, without draughts, to be provided in the farrowing pen at the site of birth and adjacent to the udder to ensure, as far as possible, that the newborn piglet will have an adequate thermal environment from birth. Piglets have a natural inclination to spend a high proportion of their first 24 hours of life adjacent to the udder (Titterington and Fraser, 1976; Wilkinson and English, 1981), so that supplementary heating should be concentrated in the side creep areas outside the confines of the farrowing crate at least in the first 24 hours of life.

Studies on sow rearing capacity and fostering strategy

Sow rearing capacity refers to the number of functional teats exposed to piglets at nursing. This aspect has been reviewed extensively by English, Smith and MacLean (1977). To increase the opportunities for adequate and regular nutrition from birth and to cater for a large number of piglets per litter it is very important to ensure that selected gilts have adequate 'rearing capacity'. Gilts should have at least 14 well developed, evenly spaced teats and the lateral distance between equivalent pairs of teats should not be excessive so as to increase the chances that all teats on the lower row will be adequately exposed to piglets when the sow is in the recumbent position at nursing. Older sows, with more pendulous udders, experience greater difficulty in exposing the more posterior teats on the lower row at nursing (English, Smith and MacLean, 1977). It is important that such problems are detected promptly so as to guide both fostering and culling strategy.

Fostering strategy should cater not only for piglets which are surplus to 'rearing capacity' (supernumerary piglets) but should also have the objective of reducing variation in birthweight within litters. Crossfostering between simultaneously farrowed litters so as to have all the small piglets on one sow and the larger ones on another sow has been shown to be effective in effecting significant reductions in piglet losses (English, Smith and MacLean, 1977). It is important to correct problems of variation in birthweight within litters and of supernumerary piglets very soon after piglets have had the opportunity to obtain adequate colostrum from their own dam.

Studies related to minimizing the effects of agalactia

Agalactia refers to complete lactational failure while hypogalactia is the term used to describe partial lactational failure. It is part of the complex condition of MMA (M = mastitis or inflammation of the udder, M = metritis or inflammation of the womb and A = agalactia). The MMA syndrome can involve metabolic, bacterial and hormonal factors with stress playing a part (Ringarp, 1960; English, Smith and Maclean, 1977).

Since its main effect is loss of milk in the first three days after farrowing, the condition contributes to piglet losses from starvation and, while research findings do not appear to have provided a basis for effective preventative measures, some positive steps can be taken in practice to minimize the worst effects of the condition.

In many cases, elevated body temperature is associated with the condition (English, 1969; Bugeac, 1971) so that regular monitoring of sow rectal temperature in the first 2–3 days after farrowing can be effective in helping to detect some cases of agalactia promptly. Once detected, the course of treatment found to be effective in resolving similar cases in the past must be applied promptly to ensure that full milk production is restored as quickly as possible before piglets begin to suffer unduly from malnutrition. This usually takes the form of antibiotic therapy sometimes accompanied by oxytocin (Martin and Threfall, 1970). If agalactia is not detected until other signs, such as loss of appetite by the sow or restlessness or loss of body condition in piglets, become evident then it is usually much more difficult to resolve the condition quickly so as to minimize the degree of malnutrition suffered by piglets. In situations where the condition is difficult to resolve, alternative prompt provision for piglets such as fostering or artificial rearing must be made so as to minimize losses from starvation.

It is possible that a form of agalactia can be induced by heat stress through having a high ambient temperature in the farrowing house as suggested by Fraser (1970). Thus, while it is extremely important to provide a very warm environment for newborn piglets in the farrowing house, the lactating sow is likely to benefit from much cooler conditions.

Provision of supplementary feeding to newborn piglets

The provision of supplementary milk substitute to suckling piglets may be justified to improve the nutrition of individual piglets in larger litters which experience difficulty in commanding a suckling position regularly and such provision may also be justified to provide for piglets on sows suffering from agalactia or hypogalactia.

However, there appears to be no published work on such a practice. English, Dawson and Ritchie (1980, unpublished data) evaluated the use of such liquid supplements of piglet milk substitute for litters in the first three days of life. Although considerable quantities of liquid milk substitute were consumed by some litters, no improvement in piglet survival was observed although supplemented piglets were significantly heavier ($P<0.05$) at 7 days of age. However, piglet mortality in this study was very low, being 5.5%. There appears to be a need to evaluate this practice on a wider basis. Progressive commercial units are using it, especially for very large litters, and claim to be obtaining a useful response. Such an approach may well be preferable to artificial rearing of supernumerary piglets in situations where such piglets cannot be fostered to other sows with spare rearing capacity.

Other approaches to improving the energy stores and nutrition of suckling pigs have involved the feeding of colostrum or milk substitute to weakly pigs by stomach tube and the administration of glucose to newborn pigs either orally or by peritoneal injection. MacPherson and Jones (1976) reported a slight improvement in survival following the intraperitoneal administration of 2 g glucose (10 ml of a 20% solution) to newborn piglets.

Artificial rearing of piglets surplus to rearing capacity

Many attempts have been made to develop suitable artificial rearing systems for piglets. Some systems were designed for rearing hysterectomy-derived pigs as a basis for producing specific pathogen-free (SPF) pigs or for rearing entire litters of pigs from birth for other specialized reasons (Young and Underdahl, 1953; Lecce and Matrone, 1960; Betts, Lamont and Littlewort, 1960). Other artificial rearing systems were designed to rear piglets surplus to sow rearing capacity from within a few hours of birth with the remainder of the litter being reared naturally (Dyrendahl *et al.*, 1953; Braude, Clarke and Mitchell, 1954; English and Smith, 1976; Scott and Pringle, 1976; Scott, 1981). The ease of operation and efficiency of artificial rearing systems has varied, some being fairly successful. However, to be effective and useful in commercial practice, artificial rearing systems must be almost fully automated so as to reduce labour requirement. In turn, automated systems must be very reliable so as to ensure that piglets receive adequate and regular nutrition. As yet, no automated artificial rearing system has been developed which is suitable and reliable for use in commercial practice. Until a suitable system is developed, provision for supernumerary piglets must be made by fostering or by supplementary feeding within the litter. In fact, if these latter approaches can be exploited to cater adequately for supernumerary piglets then this is likely to be very advantageous in terms of ease and cost of operation relative to an artificial rearing system.

Synchronized farrowing and its supervision

It is clear that the basic factors which predispose to malnutrition and ultimate death from this cause in the great majority of cases operate almost from the moment of birth. Thus, it is very important that strategic measures are taken to increase the chances of regular nutrition for each viable piglet born from very soon after birth. In addition, sows can be very restless during the parturition process and during this time, as piglets explore the area of their pen almost at random, they are very vulnerable to death from overlying by the sow (English, 1969).

It is widely accepted that supervision of farrowing and provision of strategic assistance to vulnerable piglets is very cost effective in increasing piglet survival (England, 1966). Supervision of farrowing can be better justified, in economic terms, if a batch system of farrowing is practised (English, 1978) and the use of chemicals such as analogues of prostaglandin $F_{2\alpha}$ ($PGF_{2\alpha}$) have been shown to be fairly effective in inducing parturition at specific times to facilitate supervision of a batch of farrowing sows (Ash and Heap, 1973; Downey *et al.*, 1976; Hammond and Carlyle, 1976; English *et al.*, 1977; Walker, 1977). The latter three groups of workers used an injection of 175 μg of a $PGF_{2\alpha}$ analogue (Cloprostenol, ICI 80996) between days 111 and 116 of pregnancy and found that between 66% and 80% of farrowings took place within 24–34 hours following the injection.

While analogues of $PGF_{2\alpha}$ are very effective in inducing and synchronizing farrowing of a group of sows, certain precautions must be exercised in their use. Inducement of farrowing implies premature birth to a varying degree. It appears that liver glycogen content increases 15-fold during the last 15 days of foetal life (Elliot and Lodge, 1977) and inducement of farrowing 1–3 days early is likely to be associated with lower energy reserves in newborn piglets. Both Downey *et al.* (1976) and Walker (1977) found that litters induced 1–3 days early relative to controls had lower birthweight and lower survival rates. English *et al.* (1977), despite applying very intensive measures of care to the piglets from induced farrowings, failed to improve survival relative to controls. Piglets induced to farrow on day 115 (average gestation for the herd) had better survival than controls whereas those induced to farrow one and two days early had progressively poorer survival. Hammond and Carlyle (1976) have also drawn attention to the dangers of premature farrowing.

Thus, while induced farrowing using analogues of $PGF_{2\alpha}$ provides a very useful means of achieving batch farrowing, the risks of inducing farrowing too prematurely must be recognized. However, the technique is being applied in commercial practice successfully, particularly to induce parturition in those sows which have failed to farrow by the end of the average gestation period for the herd.

Research work is in progress examining the possibility of achieving synchronized farrowing by prolonging gestation slightly through the use of progestagens (Gooneratne *et al.*, 1979; Varley and Brooking, 1981).

When farrowings are being supervised, then practices known to be effective in improving piglet survival include:

(a) placing weaker piglets under the heat source and assisting them to suckle;
(b) feeding very weak piglets with sow or cow colostrum by stomach tube;
(c) cross-fostering piglets between simultaneously farrowed litters so as to equalize, as far as possible, birthweights within litters;
(d) artificial rearing of piglets which are surplus to the rearing capacity of their dam after the technique of fostering has been exploited to the fullest extent;
(e) regular monitoring of sow health and prompt detection and treatment of any condition likely to lead to agalactia.

These practices have been part of the systems operated by England, Chapman and Bertun (1961), and by English, Smith and MacLean (1977) which were effective in achieving very high piglet survival rates.

An essential part of any system which sets out to achieve high piglet survival rate is availability of skilled, knowledgeable and well motivated stockmen who are given sufficient time and opportunity to exercize their skills.

REDUCING LOSSES FROM OVERLYING

Various approaches have been adopted for reducing losses from overlying (English, Smith and MacLean, 1977). The risk of overlying is influenced by the relative activity of the dam and her piglets and by the wellbeing and relative contentment of the latter. Factors which help to reduce the restlessness of piglets include measures to ensure their adequate nutrition as discussed previously (p.491). Associated with this objective is the provision of a very comfortable creep area within the farrowing pen, preferably situated at each side adjacent to the udder at least for the first 24 hours of life. The more adequate the nutrition of the piglets and the more comfortable they are, the less restless they tend to be.

Sow activity tends to be particularly high during parturition and around the regular feeding periods (English, 1969) with risks of overlying being correspondingly high at these times. In commercial units where these high risk periods are recognized and piglets are enclosed in the creep areas during these periods, the incidence of overlying appears to have been reduced.

In view of the peaks of sow restlessness which prevail in a farrowing house around the regular twice daily sow feeding times, alternative feeding strategies such as once daily feeding are worthy of study. In some situations adoption of *ad libitum* feeding during at least the early part of lactation has been shown to remove the peaks of sow activity, reduce its degree and result in improved piglet survival (English, 1969).

The work routine in the farrowing house is also worthy of study. Where sows are fed on a regular basis it is likely to be desirable to carry out all piglet management tasks while the sows are feeding. Such a measure would be likely to result in a much more peaceful situation for the remainder of the day in what is, in effect, a nursery area.

Most research and development work in relation to reducing losses from overlying has centred on the farrowing pen and crate. Robertson *et al.* (1966) demonstrated the benefits of providing a well designed farrowing pen incorporating a farrowing crate. Further work has been devoted to particular aspects of design of the farrowing pen and crate in relation to making the whole arrangement as 'fail safe' as possible in relation to risk of overlying, to ensuring maximum comfort for piglets and for the sow and to effecting labour economy in its operation.

The efforts towards development of more effective farrowing accommodation have been reviewed by English, Smith and MacLean (1977). Aspects of particular importance in relation to minimizing losses from overlying include the following:

(a) Arranging an atractive and comfortable creep area for the piglets.
(b) Providing an adequate floor surface to allow good mobility of the piglets and ensure good foothold by the sow.
(c) Provision of a farrowing crate designed in such a way as to help control the descent of the sow as she lies from the standing position. It is very important that the design obliges the sow to lie down carefully on to her belly first before rolling over on to either side (English, Smith and MacLean, 1977).

OVERALL STRATEGY FOR REDUCING PIGLET LOSSES IN PRACTICE

The information resulting from research and development work related to piglet viability and prevention of losses has been synthesized into an 'advisory package' and this has been applied in practice to a varying degree in commercial breeding units. This 'package' incorporates advice on the provision of suitable farrowing accommodation, the maintenance of good herd health status, good hygiene, and intensive care strategy in catering for the needs of the farrowing sow and her newborn piglets and on the basis of sound sow and litter management in lactation. In commercial units where such advice is applied rigorously, losses of liveborn pigs have been reduced to around 5% (England Chapman and Bertun, 1961; English, Smith and Maclean, 1977). In view of the difficulty of eliminating the incidence of genetic and congenital abnormalities, it is probably unrealistic in practice to expect losses of livebirths to be reduced much below this level.

FAILURE TO APPLY EXISTING KNOWLEDGE IN PRACTICE

Although some commercial units are achieving high piglet survival rates of 95% or over, even with 11 livebirths or more per litter (English, Smith and Maclean, 1977), and about 20% of recorded herds in the UK are achieving survival rates of between 90 and 95% with 10.5 or over livebirths per litter, the majority of pig units are still experiencing between 10 and 20% losses of livebirths in practice (Meat and Livestock Commission, 1980). This is illustrated in *Table 23.8* which summarizes the performance of herds recorded by the Meat and Livestock Commission in the UK.

Table 23.8 BETWEEN-HERD DISTRIBUTION IN PRE-WEANING MORTALITY OF LIVEBIRTHS. (HERDS AVERAGING 10.5 OR OVER LIVEBIRTHS PER LITTER)

Livebirth losses (%)	*Herds*	
	Number	*Per Cent*
0–5	3	1
5–10	42	19
10–15	102	47
15–20	58	27
Over 20	13	6
	218	100

From Meat and Livestock Commission (1980)

It is likely that on the great majority of units high piglet losses are still being experienced because existing knowledge is not being applied adequately in practice. Achieving high piglet survival is dependent on many factors and these must be catered for in establishing a sound farrowing and rearing system as illustrated in *Figure 23.2.* Failure to make adequate provision for all requirements will result in a less reliable system and higher piglet losses.

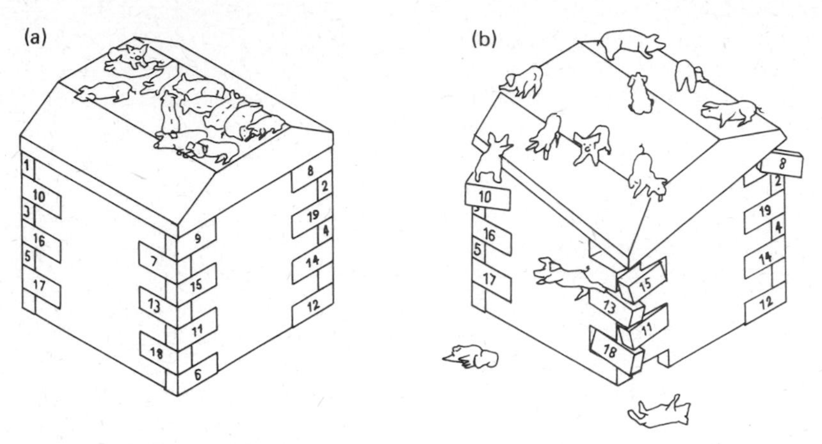

Figure 23.2 An analogy of contrasting piglet rearing systems. (a) A sound system with all vital components provided and all piglets firmly established. (b) An unsound system with one or two vital components missing or inadequately provided. This puts more piglets at risk and increases piglet losses

On the majority of pig units, farrowing accommodation combined with the farrowing and lactation management system, tends to cater only for the best 85% or so of liveborn piglets. They fail to cater adequately for the remaining 15% of the smaller and weaker piglets. Important components of such a system likely to be most defective are faulty design of the farrowing pen and crate, inadequate thermal environment and the failure to apply appropriate intensive care strategy to the weaker and more vulnerable piglets at farrowing Most pig keepers have failed to realize that the exact stage at which piglets die is not so important as is the point at which they 'start to die'. The point has already been made that most piglets dying from malnutrition will die on the second or third day of life but most become moribund almost from the moment of birth. There is a general failure to assist weakly piglets at birth at a sufficiently early stage. Other crucial problems which tend not to be detected and treated with sufficient promptness include dystocia and agalactia.

NEED FOR ROUTINE MONITORING OF EVENTS DURING AND IMMEDIATELY AFTER PARTURITION

The basic reason why there is general failure to detect and attempt to rectify problems associated with dystocia, weakness at birth and agalactia promptly may be related to the non-routine incidence of such events. It is likely that stockpersons deal much more competently with routine tasks than they do with events that are unpredictable and occur in a random manner. If such an hypothesis is correct, it calls for a strategy of much more routine monitoring of events during and immediately after parturition being put into operation on commercial breeding units.

Supplementary feeding of the litter

Aspects of supplementary feeding in the immediate postnatal period to increase energy reserves and to improve nutrition of the underprivileged and disadvantaged piglets have already been reviewed (p.493).

'Creep' feeding refers to the provision of supplementary dry feed to the suckling litter and it has two main purposes. The first is to augment the sow's milk supply which is inadequate for maximum growth especially after three weeks of age. It is also required to accustom suckling piglets to dry feed so as to encourage increased consumption by weaning in an attempt to reduce growth checks and digestive disturbances at this stage.

The practice of creep feeding was initiated when weaning took place at 6–8 weeks of age. With the sow's milk yield declining after three weeks of lactation, sound creep feeding practice was found to be essential to supplement the declining milk yield after three weeks so as to ensure reasonable growth to weaning and acceptable weaning weight. With the advent of very early weaning at 2–3 weeks of age it is perhaps more difficult to justify the practice of creep feeding. However, the justification of the practice of creep feeding on early weaning systems would be increased if it could be demonstrated to stimulate an earlier development of the mature digestive enzyme system. This may help to reduce the effects of the malabsorption syndrome common at weaning (Kenworthy, 1967) and to reduce growth checks (Okai, Aherne and Hardin, 1976) and post-weaning mortality (English, 1980).

The digestive system of the baby pig is equipped to deal only with lactose, casein and easily digested fats. The ability of the digestive system to deal with more complex carbohydrates, non-milk sugars and with proteins other than casein develops slowly, but there is evidence that earlier development of sucrase, maltase, amylase and trypsin activity can be induced by encouraging earlier consumption of non-milk sugars, starch and protein (Aumaitre and Rerat, 1966; Aumaitre, 1972).

The practice of creep feeding, using a diet that would be both acceptable and digestible to young pigs (Fowler, 1980), might therefore be expected to stimulate an earlier development of the mature digestive enzyme system and result in a reduced growth check following early weaning. The work of Friend, Gorrill and MacIntyre (1970) and that of Okai, Aherne and Hardin (1976) did not produce strong evidence of such an effect but the work of English, Robb and Dias (1980) indicated that a sound creep feeding system using a highly digestible diet, not only resulted in improvements in piglet growth during the suckling period, but also led to improved feed intake and performance following weaning. The reason for the failure of creep feeding to improve post-weaning performance in the work of Friend, Gorrill and MacIntyre (1970) and of Okai, Aherne and Hardin (1976), in which weaning took place at either three or five weeks of age, might be associated with the relatively small creep feed intakes achieved in their work relative to that achieved by piglets in the work of English, Robb and Dias (1980). In the latter work, the highly digestible diet used was in meal form and was offered from one week of age to pigs which were to be weaned at four weeks of age. The creep feed was made readily available in shallow troughs and uneaten feed was discarded and fresh feed added twice daily. Although this work demonstrated that a soundly based and well managed creep feeding strategy could be beneficial for pigs to be weaned at four weeks of age, the benefits of creep feeding for pigs to be weaned at earlier ages have not yet been demonstrated.

The essential features of a sound creep feeding system would appear to be as follows:

(a) An acceptable and digestible diet.
(b) A suitable form of diet (a flaky meal or small pellet in preference to finely ground material).
(c) A suitable feeder which makes the creep feed readily visible and available.
(d) Suitable placement of the creep feeder within the pen to minimize risk of fouling.
(e) Feeding of creep on a 'little and often' basis with strict attention to cleanliness of the system, frequent and regular removal of uneaten creep and replacement with fresh material.

Need for more research

The point has already been made that much of the inefficiency associated with piglet rearing is due to failure to apply existing knowledge in practice. However, there is considerable scope for further research on many aspects relating to piglet survival and growth, and the most crucial area in which further research effort should be concentrated is that of improvement of viability of the newborn pig. Piglets of low birthweight and those otherwise weak at birth are most demanding in terms of the conditions which must be provided and the skill and effort involved in the intensive care strategy necessary to ensure their survival.

References

ANON (1959). A survey of the incidence and causes of mortality in pigs. 1. Sow survey. *Vet. Rec.* **71**, 777–786

ASDELL, S.A. and WILLMAN, J.P. (1941). The causes of stillbirth in swine and an attempt to control it. *J. agric. Res.* **63**, 345–353

ASH, R.W. and HEAP, R.B. (1973). The induction and synchronisation of parturition in sows treated with I.C.I.79,939, an analogue of $PGF_{2\alpha}$. *J. agric. Sci., Camb.* **81**, 365–368

AUMAITRE, A. (1972). Development of enzyme activity in the digestive tract of the suckling pig: nutritional significance and implications for weaning. *Wld Rev. Anim. Prod.* **8**, 54–68

AUMAITRE, A. and RERAT, A. (1966). Unpublished data cited by Aumaitre, A. (1972). *Wld. Rev. Anim. Prod.* **8**, 54–68

BAKER, D.H., BECKER, D.E., NORTON, H.W., SASSE, C.E., JENSEN, A.H. and HARMON, B.G. (1969). Reproductive performance and progeny development in swine as influenced by feed intake during pregnancy. *J. Nutr.* **97**, 489–495

BAUMAN, R.M., KADLEC, J.E. and POWLEN, P.A. (1966). Some factors affecting death loss in baby pigs. *Purdue Univ. Agric. Exp. Stn. Res. Bull.* **810**

BETTS, A.O., LAMONT, P.H. and LITTLEWORT, M.C.G. (1960). The production by hysterectomy of pathogen-free, colostrum-deprived pigs and the foundation of a minimal-disease herd. *Vet. Rec.* **72**, 461–468

BILLE, N., NIELSEN, N.C., LARSEN, J.L. and SVENDSEN, J. (1974). Preweaning mortality in pigs. 2. The perinatal period. *Nord. VetMed.* **26**, 294–313

BOYD, R.D., MOSER, B.D., PEO, E.R. and CUNNINGHAM, P.J. (1978a). Effect of energy source prior to parturition and during lactation on tissue lipid, liver glycogen and plasma levels of some metabolites in the newborn pig. *J. Anim. Sci.* **47**, 874–882

BOYD, R.D., MOSER, B.D., PEO, E.R. and CUNNINGHAM, P.J. (1978b). Effect of energy source prior to parturition and during lactation on piglet survival and growth and on milk lipids. *J. Anim. Sci.* **47**, 883–892

BRAUDE, R., CLARKE, P.M. and MITCHELL, K.G. (1954). Analysis of the breeding records of a herd of pigs. *J. agric. Sci.* **45**, 19–27

BRENNER, K.V., SCHULZE, H. and GURTLER, H. (1978). Application of oxytocin during parturition-impact on the processes of birth in the sow and on glucose and lactate levels in blood plasma or blood of newborn piglets. *Mh. VetMed.* **33**, 304–308

BUGEAC, T. (1971). Observations and research on the metritis–mastitis–agalactia syndrome in sows and diarrhoea in newborn piglets. *Revta Zootech. Med. Vet.* **21**, 50–58

BUITRAGO, J., MANER, J.H. and GALLO, J.T. (1970). Effect of gestation energy level on reproductive performance. *J. Anim. Sci.* **31**, 197 (Abstract)

CARMICHAEL, W.J. and RICE, J.B. (1920). Variations in farrow: with special references to the birthweight of pigs. *Illinois Agric. Exp. Stn Bull.* **226**, 67–95

CARROLL, W.E., KRIDER, J.L. and ANDREWS, F.N. (1962). *Swine Production.* Third Edition. p.137. New York, McGraw-Hill Book Company

CLAWSON, A.J., RICHARDS, H.L., MATRONE, G. and BARRICK, E.R. (1963). Influence of level of total nutrient and protein intake on reproductive performance in swine. *J. Anim. Sci.* **22**, 662

CRAIG, J.V., NORTON, H.W. and TERRILL, S.W. (1956). A genetic study of weight at five ages in Hampshire swine. *J. Anim. Sci.* **15**, 242–256

CURTIS, S.E. (1974). Responses of the piglet to perinatal stressors. *J. Anim. Sci.* **38**, 1031–1036

CURTIS, S.E., HEIDENREICH, C.J. and FOLEY, C.W. (1965). Effect of late prepartum glucose loading of sows on birthweight in pigs. *J. Anim. Sci.* **24**, 915 (Abstract)

DALE, H.E. (1975). Energy metabolism. In *Duke's Physiology of Domestic Animals.* 8th Edition. Ithaca and London, Comstock Publishing Association

DAWES, G.S., JACOBSON, H.N., MOTT, J.C., SHELLEY, H.J. and STAFFORD, A. (1963). The treatment of asphyxiated, mature foetal lambs and rhesus monkeys with intravenous glucose and sodium carbonate. *J. Physiol. Lond.* **169**, 167–184
DENAMUR, D. (1965). The hypothalamo–neurohypophyseal system and the milk ejection reflex. *Dairy Sci. Abstr.* **27**, 193–244
DOWNEY, B.R., CONLON, P.D., IRVINE, D.S. and BAKER, R.D. (1976). Controlled farrowing program using a prostaglandin analogue, AY24,655. *Can. J. Anim. Sci.* **56**, 655–659
DUNNE, H.W. and LEMAN, A.D. (1975). *Diseases of Swine.* 4th Edition. Ames, Iowa, Iowa State University Press
DYRENDAHL, S., SWAHN, O., BJÖRCK, G. and HELLVING, L. (1953). Artificial raising of baby pigs. *Acta Agric. Scand.* **3**, 334–354
ELLIOT, J.I. and LODGE, G.A. (1977). Body composition and glycogen reserves in the neonatal pig during the first 96 hours postpartum. *Can. J. Anim. Sci.* **57**, 141–150
ELSLEY, F.W.H. (1964). The physiological development of the young pig. *Annls Zootech.* **13**, 75–84
ELSLEY, F.W.H. and SHIRLAW, D.W.G. (1976). Aspects of the energy nutrition of sows. *27th Annual Meeting of the E.A.A.P., Zurich*
ELSLEY, F.W.H., BANNERMAN, M., BATHURST, E.V.J., BRACEWELL, A.G., CUNNINGHAM, J.M.M., DODSWORTH, T.L., DODDS, P.A., FORBES, T.J. and LAIRD, R. (1969). The effect of level of feed intake in pregnancy and in lactation upon the productivity of sows. *Anim. Prod.* **11**, 225–241
ENGLAND, D.C. (1966). Saving the pig crop. Summary Article. Oregon State University
ENGLAND, D.C. (1974). Husbandry components in prenatal and perinatal development in swine. *J. Anim. Sci.* **38**, 1045–1049
ENGLAND, D.C., CHAPMAN, V.M. and BERTUN, P.L. (1961). Relationship between birthweight and volume of milk consumed by artificially reared baby pigs. *Proceedings of the Western Section, American Society of Animal Production,* **12**, 1–5
ENGLISH, P.R. (1969). Mortality and variation in growth of piglets:A study of predisposing factors with particular reference to sow and piglet behaviour. PhD Thesis. University of Aberdeen
ENGLISH, P.R. (1978). Benefits of batch farrowing. *Pig Fmg.* **26**, 35–37
ENGLISH, P.R. (1980). Establishing the early weaned pig. *Proc. Pig Vet. Soc.,* **7**, 29–37
ENGLISH, P.R. and DIAS, M.F.M. (1978). Effect of feeding level of the sow in late pregnancy on piglet birthweight. *Research, Investigation and Field Trials, North of Scotland College of Agriculture, 1977–78,* 57–58
ENGLISH, P.R. and SMITH, W.J. (1975). Some causes of death in neonatal piglets. *Vet. Ann.* **15**, 95–104
ENGLISH, P.R. and SMITH, W.J. (1976). Experiments on complementary artificial rearing of piglets. 1. The need for artificial rearing. *Fm. Building Prog.* **45**, 5–7
ENGLISH, P.R., DAWSON, I. and RITCHIE, R.M. (1980). Unpublished data
ENGLISH, P.R., ROBB, C.M. and DIAS, M.F.M. (1980). Evaluation of creep feeding using a highly digestible diet for litters weaned at 4 weeks of age *Anim. Prod.* **30**, 496 (Abstract)

ENGLISH, P.R., SMITH, W.J. and MACLEAN, A. (1977). *The Sow:Improving her Efficiency*. 311 pages. Ipswich, Suffolk, Farming Press Ltd.

ENGLISH, P.R., HAMMOND, D., DAVIDSON, F.M., SMITH, W.J., SILVER, C.L., DIAS, M.F.M. and MACPHERSON, R.M. (1977). Evaluation of an induced farrowing system using cloprostenol, (I.C.I. 80996) a synthetic analogue of prostaglandin $F_{2\alpha}$. *Anim. Prod.* **24**, 139–140 (Abstract)

FAHMY, M.H. and BERNARD, C. (1971). Causes of mortality in Yorkshire pigs from birth to 20 weeks of age. *Can. J. Anim. Sci.* **51**, 351–359

FOWLER, V.R. (1980). The nutrition of weaner pigs. *Pig News and Information* **1**, 11–15

FRASER, A.F. (1966). Studies of piglet husbandry in Jamaica. 2. Principal causes of loss between birth and weaning. *Br. vet. J.* **112**, 325–332

FRASER, A.F. (1970). Field observations in Jamaica on thermal agalactia in the sow. *Trop. Anim. Hlth. Prod.* **2**, 175–181

FREDEEN, H.T. (1957). Crossbreeding and swine production. *Anim. Breed. Abstr.* **25**, 339–347

FRIEND, D.W. (1974). Effect on the performance of pigs from birth to market weight of adding fat to the lactation diet of their dams. *J. Anim. Sci.* **39**, 1073–1081

FRIEND, D.W., GORRILL, A.D.L. and MACINTYRE, T.M. (1970). Performance and proteolytic enzyme activity of the suckling piglet creep-fed at 1 or 3 weeks of age. *Can. J. Anim. Sci.* **50**, 349–354

FRIEND, D.W., CUNNINGHAM, H.M. and NICHOLSON, J.W.G. (1962). The duration of farrowing in relation to the reproductive performance of Yorkshire sows. *Can. J. comp. Med.* **26**, 127–130

GLASTONBURY, J.R.W. (1976). A survey of preweaning mortality in the pig. *Aust. Vet. J.* **52**, 272–276

GOONERATNE, A., HARTMANN, P.E., McCAULEY, I. and MARTIN, C.E. (1979). Control of parturition in the sow using progesterone and prostaglandin. *Aust. J. Biol. Sci.* **32**, 587–595

GRACEY, J.F. (1955). Survey of pig losses. *Vet. Rec.* **67**, 984–990

HAKKARAINEN, J. (1975). Developmental changes of protein, RNA, DNA, lipid and glycogen in the liver, skeletal muscle and brain of the piglet. *Acta vet. scand. (Suppl. 59)* **59**, 1–198

HAMMOND, D. and CARLYLE, W.W. (1976). Controlled farrowing on commercial pig breeding units using cloprostenol, a synthetic analogue of prostaglandin $F_{2\alpha}$. *8th Int. Congr. Anim. Reprod. A. I., Krakow, 102* (Abstract)

HAYES, R.L. and VAN DEMARK, N.L. (1952). Effects of hormones on uterine motility and sperm transport in the perfused genital tract of the cow. *J. Dairy Sci.* **35**, 499–500 (Abstract)

HENDRIX, W.F., KELLEY, K.W., GASKINS, C.T. and HINRICHS, D.J. (1978). Porcine neonatal survival and serum gamma globulins. *J. Anim. Sci.* **47**, 1281–1286

HILLYER, G.M. and PHILLIPS, P. (1980). The effect of increasing feed level to sows and gilts in late pregnancy on subsequent litter size, litter weight and maternal body weight change. *Anim. Prod.* **30**, 469 (Abstract)

HUTCHINSON, H.D., TERRILL, S.W., MORRILL, C.C., NORTON, H.W., MEADE, R.J., JENSEN, A.H. and BECKER, D.E. (1954). Causes of baby pig mortality. *J. Anim. Sci.* **13**, 1023 (Abstract)

ILANČIĆ, D., NIKOLIĆ, P. and PAVLOVIĆ, D. (1968). Analysis of farrowing and mortality during suckling in a herd of white meat pigs. *Vet. Glasn.* **22**, 601–607

KENWORTHY, R. (1967). Intestinal malabsorption and diarrhoea in the newly weaned pig. M.V.Sc. Thesis. University of Liverpool.

KOTARBINSKA, M. Level of feeding in late pregnancy of the sow. Unpublished data

LECCE, J.G. and MATRONE, G. (1960). Porcine neonatal nutrition:the effect of diet on blood serum proteins and performance of the baby pig. *J. Nutr.* **70**, 13–20

LEE, CHANG WOO (1977). Effects of oxytocin and parasympathomimetic drugs on porcine stillbirths. *Korean J. vet. Res.* **17**, 9–12

LEMAN, A.D., KNUDSON, C., RODEFFER, H.E. and MUELLER, A.G. (1972). Reproductive performance of swine on 76 Illinois farms. *J. Am. vet. med. Ass.* **161**, 1248–1250

LODGE, G.A. (1972). Quantitative aspects of nutrition in pregnancy and lactation. In *Pig Production* (D.J.A. Cole, Ed.), pp. 399–416. London, Butterworths

LODGE, G.A., ELSLEY, F.W.H. and MACPHERSON, R.M. (1966a). The effects of level of feeding of sows during pregnancy. 1. Reproductive performance. *Anim. Prod.* **8**, 29–38

LODGE, G.A., ELSLEY, F.W.H. and MACPHERSON, R.M. (1966b). The effects of level of feeding of sows during pregnancy. II. Changes in body weight. *Anim. Prod.* **8**, 499–506

McINNES, A.A. (1977). Unpublished data – personal communication

MACPHERSON, R.M. and JONES, A.S. (1976). The effect of administration of glucose on survival of the neonatal pig. *Anim. Prod.* **22**, 153 (Abstract)

MARTIN, C.E. and THREFALL, W.R. (1970). Clinical evaluation of hormone therapy in the agalactia syndrome of sows. *Vet. Rec.* **87**, 768–771

MEAT and LIVESTOCK COMMISSION (1980). MLC Commercial Pig Yearbook 1980

MILLER, J.A. and MILLER, F.S. (1965). Studies on prevention of brain damage in asphyxia. *Dev. Med. Child Neurol.* **7**, 607–619

MOORE, R.W., REDMOND, H.E. and LIVINGSTON, C.W. (1965). Iron deficiency anaemia as a cause of stillbirths in swine. *J. Am. vet. med. Ass.* **147**, 746–748

MOUNT, L.E. (1972). Environmental physiology in relation to pig production. In *Pig Production* (D.J.A. Cole, Ed.), pp. 71–90. London, Butterworths

MUHRER, M.E., SHIPPEN, O.F. and LASLEY, J.F. (1955). The use of oxytocin for initiating parturition and reducing farrowing time in sows. *J. Anim. Sci.* **14**, 1250 (Abstract)

MURDOCH, Y.P. (1980). Effects of isolating newborn piglets from the sow for varying periods on the duration of farrowing. BSc (Hons) Dissertation. University of Aberdeen

NEWLAND, H.W., MacMILLEN, W.N. and REINEKE, E.P. (1952). Temperature adaptation in the baby pig. *J. Anim. Sci.* **11**, 118–133

NIELSEN, N.C., CHRISTENSEN, K., BILLE, N. AND LARSEN, J.L. (1974). Preweaning mortality in pigs. 1. Herd investigations. *Nord. VetMed.* **26**, 137–150

O'GRADY, J.F. (1967). Effect of level and pattern of feeding during pregnancy on weight change and reproductive performance of sows. *Ir. J. agric. Res.* **6**, 57–71

OKAI, D.B., AHERNE, F.X. and HARDIN, R.T. (1976). Effects of creep and starter composition on feed intake and performance of young pigs. *Can. J. anim. Sci.* **56**, 573–586

OKAI, D.B., AHERNE, F.X. and HARDIN, R.T. (1977). Effects of sow nutrition in late gestation on the body composition and survival of the neonatal pig. *Can. J. anim. Sci.* **57**, 439–448

OKAI, D.B., WYLLIE, D., AHERNE, F.X. and EWAN, R.C. (1978). Glycogen reserves in the foetal and newborn pig. *J. Anim. Sci.* **46**, 391–401

POMEROY, R.W. (1960). Infertility and neonatal mortality in the sow. III. Neonatal mortality and foetal development. *J. agric. Sci., Camb.* **54**, 31–56

RANDALL, G.C.B. (1972a). Observations on parturition in the sow. 1. Factors associated with the delivery of the piglets and their subsequent behaviour. *Vet. Rec.* **90**, 178–182

RANDALL, G.C.B. (1972b). Observations on parturition in the sow. 2. Factors influencing stillbirth and perinatal mortality. *Vet. Rec.* **90**, 183–186

RANDALL, G.C.B. and PENNY, R.H.C. (1967). Stillbirth in pigs: the possible role of anoxia. *Vet. Rec.* **81**, 359–361

RANDALL, G.C.B. and PENNY, R.H.C. (1970). Stillbirth in the pig: An analysis of the breeding records of five herds. *Br. vet. J.* **126**, 593–603

REEDY, L.P., HEIDENREICH, C.J., FOLEY, C.W. and CURTIS, S.E. (1966). Orally-administered glucose effects on swine. *J. Anim. Sci.* **25**, 1267 (Abstract)

RINGARP, N. (1960). A post-parturient syndrome with agalactia in sows. *Acta Agric. Scand., Suppl.*, **7**,

ROBERTSON, J.B., LAIRD, R., HALL, J.K.S., FORSYTH, R.J., THOMPSON, J.M. and WALKER-LOVE, J. (1966). A comparison of two indoor farrowing systems for sows. *Anim. Prod.* **8**, 171–178

SCOTT, G.T. (1981). Development of a system of artificial rearing for piglets surplus to the sow's rearing capacity and evaluation of its repercussions. PhD Thesis. University of Aberdeen

SCOTT, G.T. and PRINGLE, R.T. (1976). Experiments on complementary artificial rearing of piglets. 3. Feeding system and cages. *Fm. Building Prog.* **45**, 23–28

SEERLEY, R.W., PACE, T.A., FOLEY, C.W. and SCARTH, R.D. (1974). Effect of energy intake prior to parturition on milk lipids and survival rate, thermostability and carcass composition of piglets. *J. Anim. Sci.* **38**, 64–70

SHARPE, H.B.A. (1966). Preweaning mortality in a herd of Large White pigs. *Br. vet. J.* **122**, 99–111

SMITH, C. and KING, J.W.B. (1964). Crossbreeding and litter production in British pigs. *Anim. Prod.* **6**, 265–271

SPRECHER, D.J. LEMAN, A.D. and CARLISLE, S. (1975). Effects of parasympathomimetics on porcine stillbirth. *Am. J. vet. Res.* **36**, 1331–1333

SPRECHER, D.J., LEMAN, A.D., DZUIK, P.D., CROPPER, M. and DEDECKER, M. (1974). Causes and control of swine stillbirth. *J. Am. vet. med. Ass.* **165**, 698–701

STANTON, H.C. and CARROLL, J.K. (1974). Potential mechanisms responsible for prenatal and perinatal mortality or low viability of swine. *J. Anim. Sci.* **38**, 1037–1044

TITTERINGTON, R.W. and FRASER, D. (1976). The lying behaviour of sows and piglets during early lactation in relation to the position of the creep heater. *Appl. Anim. Ethol.* **2**, 47–53

TURNBULL, A.C. and ANDERSON, A.B.M. (1968). Induction of labour. Part II. Intravenous oxytocin infusion. *J. Obstet. Gynaec. Br. Commonw.* **75**, 24–31

VARLEY, M.A. and BROOKING, P. (1981). The control of parturition in the sow using an oral progestagen. *Anim. Prod.* **32**, 369 (Abstract)

WALKER, N. (1977). The effects of induction of parturition in sows using an analogue of prostaglandin $F_{2\alpha}$. *J. agric. Sci., Camb.* **89**, 267–271

WILKINSON, V. and ENGLISH, P.R. (1981). Unpublished data

WRATHALL, A.E. (1971). Prenatal survival in pigs. Part 1. Ovulation rate and its influence on prenatal survival and litter size in pigs. *Commonw. Bur. Anim. Hlth Rev. Ser.* **9**, Slough, UK, Commonwealth Agricultural Bureaux

YOUNG, G.A. and UNDERDAHL, N.R. (1953). Isolation units for growing baby pigs without colostrum. *Am. J. vet. Res.* **14**, 571–574

VI

FACTORS THAT INTERACT IN THE CONTROL OF REPRODUCTION

24

SEASONALITY OF REPRODUCTION IN THE WILD BOAR

R. MAUGET
C.N.R.S. Centre d'Etudes Biologiques des Animaux Sauvages, Villiers-en-Bois, 79360 Beauvoir-sur-Niort, France

In the management of large breeding units of domestic pig, particular attention is currently paid to seasonal fluctuations in reproductive performance. Although the domestic pig is known to reproduce throughout the year, there is a seasonal decrease in breeding performance. Considerable reproductive inefficiency has been reported during the summer and autumn months in the domestic sow in France (Corteel, Signoret and du Mesnil du Buisson, 1964), England (Stork, 1979), Italy (Enne, Beccaro and Tarocco, 1979), USSR, (Radev, Andreev and Kostov, 1976), USA (Hurtgen, 1976; Hurtgen, Leman and Crabo, 1980) and Australia (Paterson, Barker and Lindsay, 1978). Although not exhaustive, this list shows the general trend. The European wild boar represents the wild form from which the modern breeds of domestic pig have been derived by intensive selection directed towards growth and productivity criteria. However, most studies of the European wild boar have been conducted from an ecological or hunting viewpoint (Oloff, 1951; Haber, 1969; Snethlage, 1974). The aim of this chapter is to review our knowledge of the reproductive biology of this wild species as a basis for comparison with the domestic pig.

Most of the data presented here were obtained from an indigenous population of wild boar living in a natural environment, the Chizé forest in midwestern France. They form a part of a study dealing with the ecological, behavioural and physiological (reproductive) aspects of the adaptation of the wild boar to its environment (Mauget, 1980).

Reproductive performance

Compared with other ungulates of similar body size, the wild boar appears to be the species having the highest reproductive capacity. This characteristic, associated with a great faculty of adaptation, seems likely to be at the origin of domestication, the result of which has been to optimize productivity.

FREQUENCY OF BREEDING

Generally one litter is produced each year with farrowing occurring in late winter and early spring. However, under certain conditions, which will be considered later, a second farrowing in the year may occur.

LITTER SIZE

Information has been obtained from the reproductive tracts of 57 slaughtered females in which the foetuses were examined. There were, on average, 4.60±0.18 foetuses per sow. Variations from one year to another were not statistically significant. However, variations were observed in the litter size in relation to the weight and subsequently to the age and parity of the female. Average litter size varied from 2.50±0.51 for young primiparous females (weighing 30–39 kg and 9–15 months of age) to 5.43±0.26 for older females (weighing more than 80 kg and over 3 years of age).

OVULATION RATE AND FOETAL SURVIVAL

The number of ova shed per female was determined by corpora lutea count. Ovulation rate for the combined mature female age group was 5.26±0.25 (n = 31), with a 12.5% intrauterine loss.

CONCEPTION RATE

Conception rate (the percentage of mature females which are bred) varied considerably among the weight–age classes. Thus in the primiparous females (30–39 kg, 9–15 months) the maximum conception rate was 68.8% (n = 52 females). It reached 97.9% (n = 48) for animals weighing 40–59 kg and aged 16–24 months and 100% (n = 53) for pigs of 60 kg and more and over two years of age.

GESTATION LENGTH

The gestation period of wild females reared in enclosures was 119±0.7 days (n = 18) with a range of 112–126 days.

Table 24.1 gives a comparison of the reproductive performance established in the wild boar with that of the domestic pig and feral pigs (i.e. domesticated animals returned to a wild status). While modern breeds of domestic pig may have a litter size as high as 20, with a mean value of about 12, the wild boar averages only 4.6 young. However, the intrauterine mortality which reaches 30% in the domestic sow, is significantly lower in the wild pig. In feral pigs, ovulation rate is intermediate between the other two but intrauterine losses still remain high. Mean gestation length appears slightly higher in the wild pig than in the domestic sow. While it is clear that the wild boar differs in reproductive performance from domestic forms, the essential characteristic of its reproduction is its seasonal pattern.

Season of breeding

In pure breeds of wild boars having no history of hybridization with domesticated animals, as assessed by chromosomal studies (McFee, Banner and Rary, 1966; Mauget *et al.*, 1977), reproduction is clearly seasonal.

Table 24.1 REPRODUCTIVE PERFORMANCE OF THE DOMESTIC, FERAL AND WILD FORMS OF *SUS SCROFA*

	No. of corpora lutea	*Intrauterine losses (%)*	*Gestation length (days)*	*Average litter size*	*Farrowing*		*References*
					Frequency	*Period*	
Domestic pig		30	114	12	→2.5	Jan.→Dec.	Asdell (1964)
Feral pig (U.S.)	8.5	34	116–118	5–7	2	Jan.→Dec.	Barrett (1978)
	8.7±0.3	29.1		6.2±0.5			Hagen and Kephart (1980)
Feral pig (Corsica, France)				5–7 (1–11)	→2	Jan.→Dec.	Molénat and Casabianca (1979)
Wild boar × feral pig (U.S.)				4.2	1	Jan.→Dec.	Pine and Gerdes (1973)
	→6			3.2	1	Jan.→Dec.	Duncan (1974)
European wild boar			120	4.6			Henry (1968)
				5	1	Spring–Summer	Briederman (1971)
	4.6→5.7	13→23		3.5→5.0	1	Spring–Summer	Aumaitre (1979, personal communication)
	5.26±0.25	12.5	119±0.7	4.6±0.18	1→2	Jan.→Sept.	Mauget (1972)

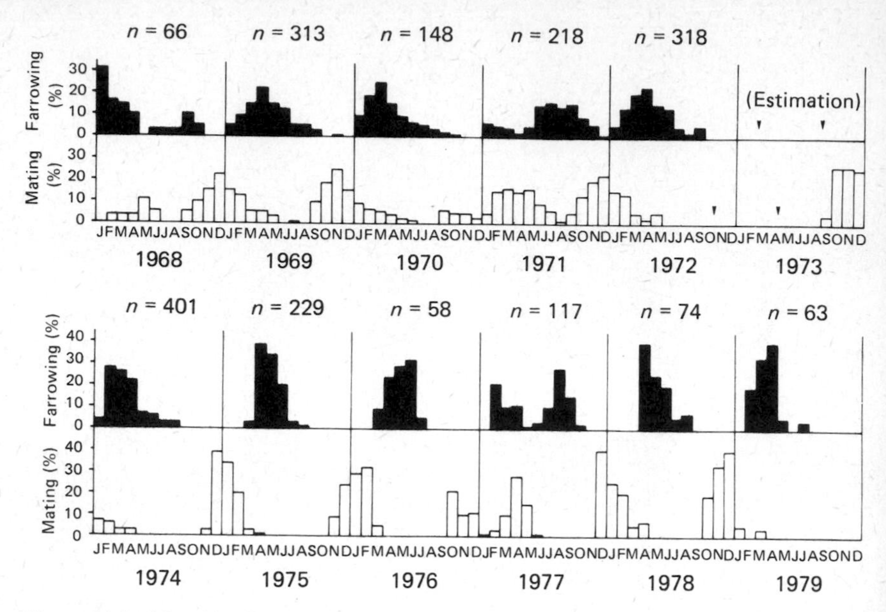

Figure 24.1 Monthly distribution of farrowings and matings recorded in the Chizé forest from 1968 to 1979, expressed as the percentage of total annual number (n) of observations

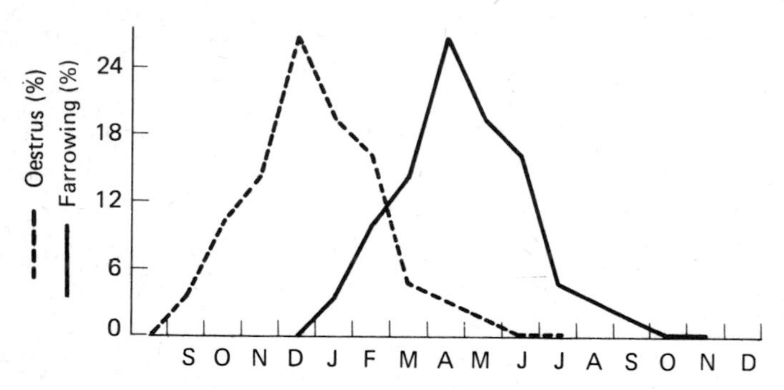

Figure 24.2 Unimodal distribution of farrowings and corresponding oestrus (pooled data from seven years, estimated by back-aging dates of birth of 1584 young).

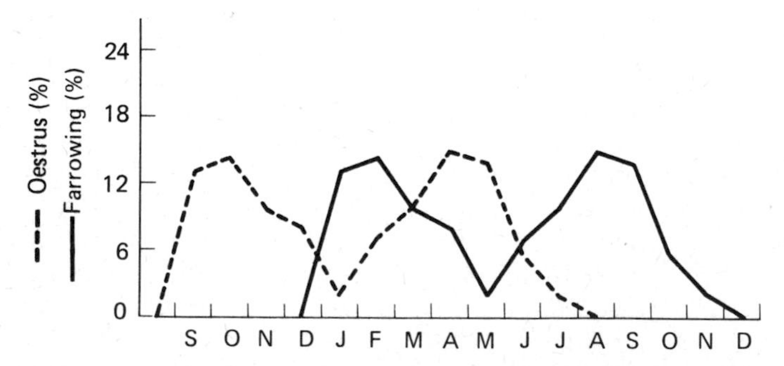

Figure 24.3 Bimodal distribution of farrowings and corresponding oestrus (pooled data from three years, estimated by back-aging dates of birth of 401 young).

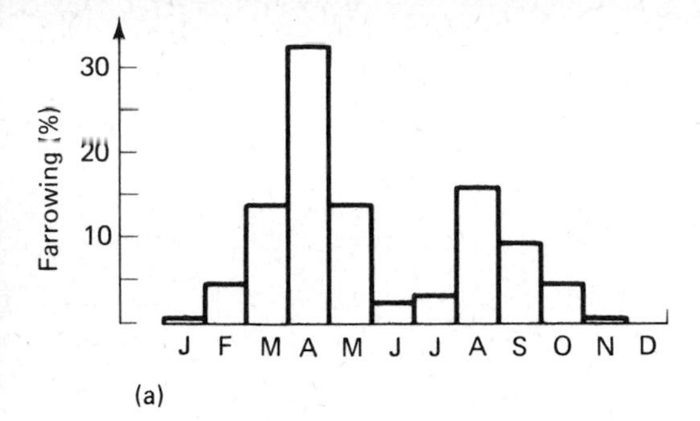

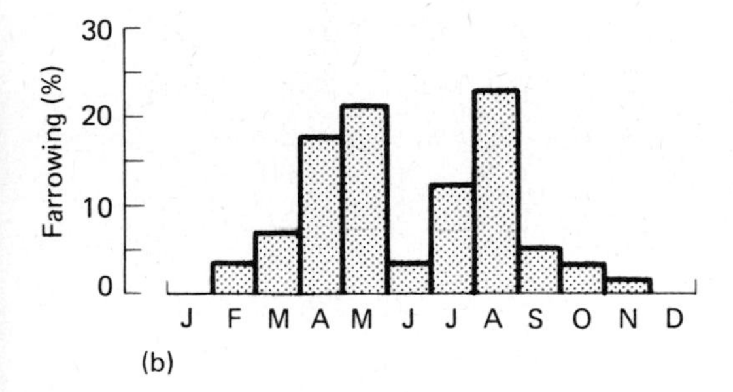

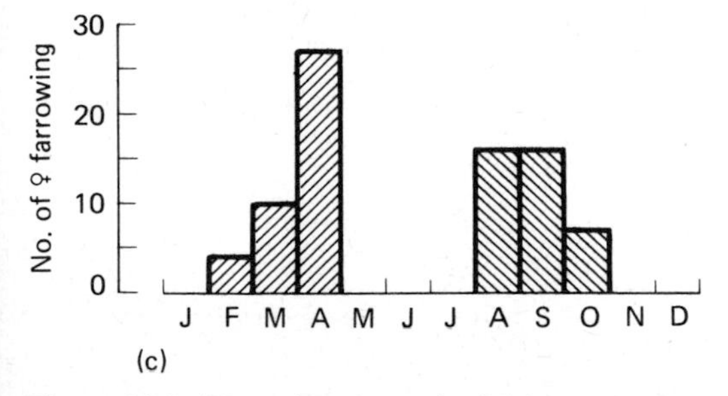

Figure 24.4 Farrowing data of wild boars in the commercial rearing unit. (a) Whole population (n = 226); (b) females reaching puberty (n = 56); (c) females farrowing twice a year, ▨ first litter; ▧ second litter

Analysis of farrowing data recorded both in forest and in enclosures supports the evidence of such seasonal breeding (Mauget, 1978).

The monthly distribution of farrowings and corresponding oestrus observed in the population of the Chizé forest from 1968 to 1979 are presented in *Figure 24.1* and show considerable variation. There appear to be two types of farrowing distribution. One is unimodal (*Figure 24.2*) with the peak of farrowing occurring in April and May and a wide distribution ranging from January to September. In the bimodal type (*Figure 24.3*), two peaks are seen. The first peak occurs in January and February and the second one in August and September. It is interesting to note that in years with two periods of birth the peak occurs before the unimodal one.

The animals involved in each breeding period can be determined from data collected in enclosures on individually marked animals (*Figure 24.4*).

Table 24.2 INFLUENCE OF THE TIME OF YEAR ON REPRODUCTIVE STATE (FEMALES CAPTURED IN THE WILD)

Reproductive state	*Monthly frequency* (no. of animals)											
	Jan.	*Feb.*	*March*	*Apr.*	*May*	*June*	*July*	*Aug.*	*Sept.*	*Oct.*	*Nov.*	*Dec.*
Cyclic	**7**	**7**	**5**	–	**2**	**2**	–	–	1	–	–	–
Pregnant	**5**	**10**	**43**	–	–	–	–	–	1	–	–	–
Lactating	–	–	–	–	–	**13**	**20**	**3**	–	2	1	–
Anoestrus	–	–	–	–	–	**4**	**10**	**7**	**11**	**10**	**9**	**5**
Immature	5	17	17	–	4	1	–	2	1	19	11	15

The histogram of the monthly distribution of births recorded in a rearing unit is typically bimodal as shown in panel (a); the first peak is in April followed by a second in August. The animals involved in these two main periods are young females reaching puberty at one of the peaks (panel (b)) and adult females farrowing twice a year (panel (c)). The second litter in the year is obtained as a result of the separation of the sow from her young three weeks after parturition.

Annual cycle of ovarian activity

Farrowing distribution results from seasonal ovarian activity and on the basis of samples collected all year round in the forest, an anoestrous period occurs during the summer and autumn months.

Reproductive tracts from female wild pigs were examined to determine their physiological state. Five reproductive states could be distinguished: cycling, pregnancy, lactation, anoestrus and immature. The histological characteristics of ovaries and vaginal epithelium corresponding to each stage agree with the morphological descriptions classically reported in the domestic sow (Corner, 1921; Bal, Wensing and Getty, 1969; Steinbach and Schmidt, 1970). Reproductive states varied at different times of the year (*Table 24.2*). The occurrence of anoestrus in females was seen as early as June and as late as December.

Table 24.3 PLASMA PROGESTERONE LEVELS (MEAN ±2 S.E.M.: 95% CONFIDENCE LIMITS) ESTIMATED ON FEMALE WILD BOARS IN DIFFERENT REPRODUCTIVE STATES

Reproductive state	*No. of females*	*Plasma progesterone* (ng/ml)
Cyclic	11	12.47±2.07
Pregnant	17	13.45±0.73
Lactating	30	3.95±0.51
Anoestrus	37	2.19±0.25
Immature	13	3.20±0.65

Plasma progesterone levels related to the various reproductive states are shown in *Table 24.3*. Both cyclic and pregnant animals had high mean progesterone levels. In animals with no functional corpora lutea, progesterone levels were somewhat higher when compared with the basal levels generally reported in domestic sows (about 1 ng/ml). A stress-induced adrenal secretion of progesterone could be envisaged in such trapped wild animals.

In order to determine precisely the timing of ovarian function, longitudinal analysis was performed on adult females older than eighteen months of age which had been captured in the wild and maintained in enclosures without males. From weekly bleeding, the evolution of plasma progesterone concentration as a reflection of ovarian activity was determined. Marked seasonal variations in weekly mean values were observed (*Figure 24.5*). From December and January to June and July, progesterone levels were high. The weekly mean values were associated with standard errors

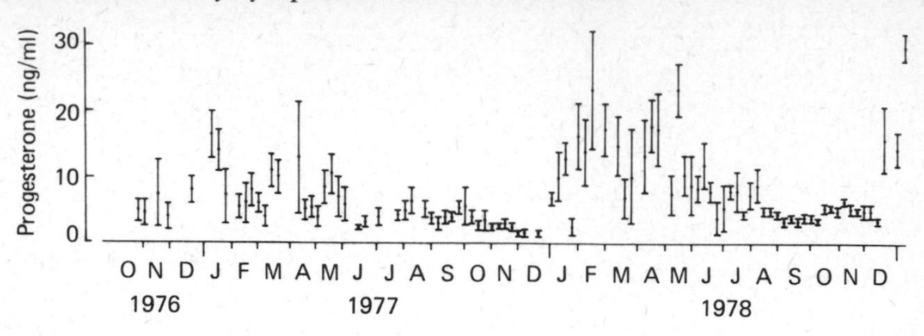

Figure 24.5 Seasonal variations of plasma progesterone levels in unbred adult female wild pigs (mean ±S.E.M. based on weekly values from six animals)

which reflected a wide range in individual progesterone levels in relation to cyclic ovarian activity. They varied from 3–4 ng/ml to 20–30 ng/ml with a mean of 13.45 ng/ml. These values are in agreement with those that have been reported during the oestrous cycle in the domestic sow (Stabenfeldt *et al.*, 1969; Tillson, Erb and Niswender, 1970; Edqvist and Lamm, 1971; Henricks, Guthrie and Handlin, 1972; Shearer *et al.*, 1972). However, the basal levels appear somewhat higher and as discussed above, the adrenal cortex might contribute to this. Time series analysis (the method of Halberg *et al.*, 1972) has revealed a cyclicity of progesterone level ranging from 19–23 days. This periodicity is comparable to the length of oestrous cycle determined by Henry (1968) in penned female wild pigs monitored for oestrus and is within the range of values for the domestic sow. During the summer and autumn months progesterone concentrations remain at a low level characterizing an anoestrous stage. Thus, a regular alternation of ovarian cyclic activity and anoestrous sequences appeared over the 28 months of this study.

Environmental involvement

It is a general feature of wild mammals that environmental factors may influence the seasonal breeding activity (see reviews by Perry and Rowlands, 1973; Assenmacher and Farner, 1978; Mauget, Boissin-Agasse and Boissin, 1981). The seasonal synchronization of sexual activity might result from the adjustment of an 'endogenous oscillatory system' (Assenmacher, 1974) by cyclic environmental factors. It is established that for numerous species natural photoperiodism is the main synchronizer (Menaker, Takahashi and Eskin, 1978). However, other environmental factors (climatic, nutritional and social) might be involved in the timing of seasonal breeding.

NUTRITIONAL FACTORS

Under the controlled conditions of rearing, food is exclusively supplied by man. From data recorded on 53 wild boar rearing units it appeared that although plane of nutrition varied between rearings, there was always an

Table 24.4 FOOD LEVELS AND THE ONSET OF OVARIAN ACTIVITY

Plane of nutrition	*Weight gain* (kg)	*Onset of cyclic activity* (no. of animals)				
	Oct. → *Dec.*	*Dec. 11–17*	*Dec. 18–24*	*Dec. 25–31*	*Jan. 1–7*	*Jan. 8–14*
I	2.2±0.9	–	–	6	3	
II	7.1±1.4	6				

Group I: 9 animals receiving a control diet of 1 kg of a commercial pelleted food and 0.5 kg of barley or corn/animal/day.
Group II: 6 animals receiving 1.5 × the amount given to the control group.

anoestrous period whose minimum duration extended from July to September. The onset of sexual activity was never observed before the beginning of October. To define nutritional effects more precisely, two planes of nutrition were tested in females kept in enclosures (*Table 24.4*). The onset of sexual activity occurred earlier in the females which were fed liberally (i.e. 1.5 × the control level).

In the forest, availability of food fluctuates throughout the year and the onset of the breeding season occurs between the months of October and January. Data on the first incidence of fertile matings, collected over eight years, are presented in *Table 24.5*. It appears that earliness or delay in the onset of the breeding season (October or January) is related to the level of mast production in the forest in the autumn.

Table 24.5 ONSET OF BREEDING SEASON IN THE WILD BOAR POPULATION IN RELATION TO THE PRODUCTIVITY OF THE FOREST IN AUTUMN

Year	*Mast index*[(a)]	*Onset of breeding*[(b)]	*No. animals*[(c)]
1972	2	Nov.–Dec.	16
1973	4	Oct.–Nov.	401
1974	2	Dec.–Jan.	229
1975	1	Dec.–Jan.	58
1976	3	Oct.–Nov.	117
1977	1	Dec.–Jan.	74
1978	4	Oct.–Nov.	63

[(a)] 1–4: Autumn mast estimates from minimum to maximum productivity.
[(b)] Month during which more than 10% of females were bred.
[(c)] Total number of trapped animals under observation each year.

It seems likely that nutritional factors may, in part, influence the timing of reproduction in the wild boar. Nutritional problems have been widely studied in the sow (e.g. reviews in Cole, 1972). Food levels mainly influence reproductive performance. In wild animals, low food availability has been reported often to cause a delay in puberty attainment or in the onset of sexual season (Sadleir, 1969). For example, this has been shown in the bear (Rogers, 1976) and in the white-tailed deer (Verme, 1965).

SOCIAL FACTORS

As reported above (p.513), births in the population of the forest can be spread over several months. However, within each social group, constituted by the association of a small number of females (about 4), a close synchronization of births is found generally within 10–15 days. This reflects

Table 24.6 FIRST FERTILE MATINGS AFTER THE ANOESTROUS PERIOD FOR THREE REARING GROUPS

Group	*No. of females*	*Mating dates* (day of the year) *No. of first mating*								
I	19	311	312	315	319	321		341	342	344
		2	2	2	1	1		3	3	5
II	19	349	351	352	353	356	357			
		5	3	3	2	4	2			
III	14	337	338	340	357					
		3	4	2	5					

a synchronization of the onset of the breeding season for females of the same social group.

In groups of females in rearing units, a similar synchronization of the onset of cyclic activity after the anoestrous period is also found. Thus, from the data presented in *Table 24.6* it appears that the return to oestrus in each rearing group occurs within about a week.

PHOTOPERIOD

The cessation of breeding seems to be independent of plane of nutrition as it is even observed under stable feeding conditions. Furthermore, the cessation of ovarian activity begins in mid-April, when environmental temperatures are still low (mean: 10.2 °C). This suggests the involvement of a reliable yearly environmental factor which might be photoperiod.

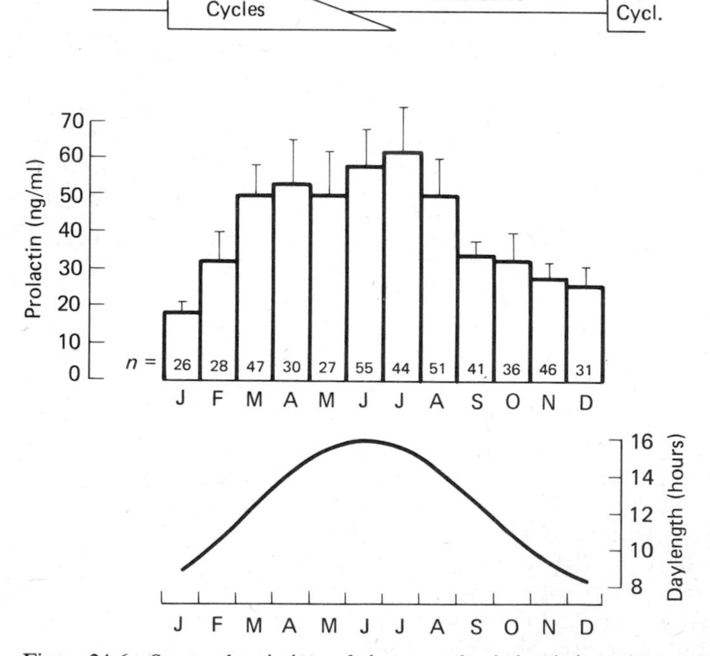

Figure 24.6 Seasonal variations of plasma prolactin levels (monthly pooled data of *n* samples obtained from 16 animals) related to daylength variations and timing of reproduction. The 95% confidence intervals are indicated by the vertical lines

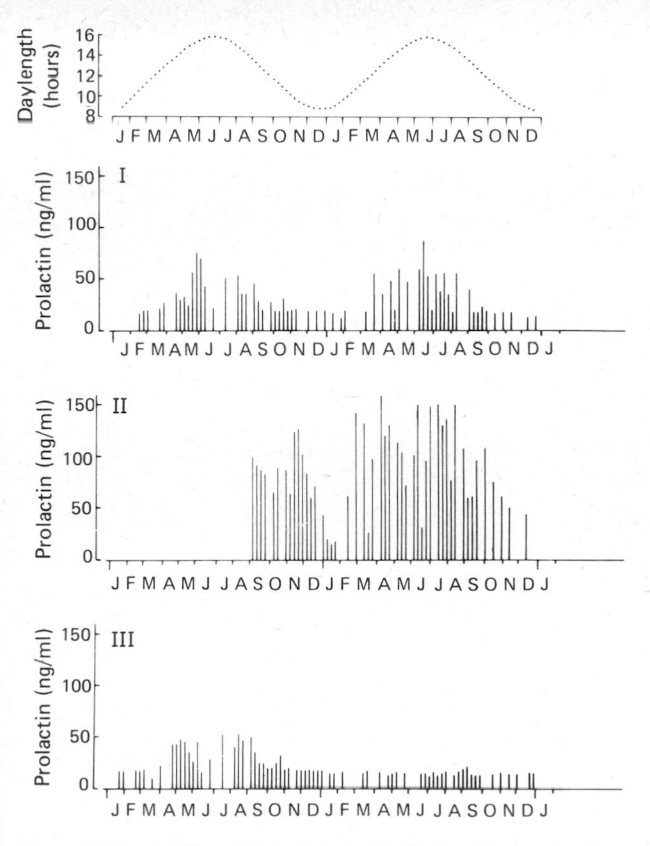

Figure 24.7 An illustration of marked individual fluctuations in prolactin profiles. (Frequency of each type: I = 11 animals, II = 3 animals, III = 2 animals)

The sensitivity of the wild boar to the annual variations of photoperiod has been indirectly investigated by studying seasonal changes in plasma prolactin concentrations. The results of monthly determinations of plasma prolactin levels are shown in *Figure 24.6*. A seasonal rhythm appears with peak values occurring in summer and minimum values in winter following the seasonal changes in natural daylength. Individual prolactin profiles are given in *Figure 24.7*, which shows marked individual fluctuations. While most females exhibit a pattern similar to the one shown for the pooled data, some deviations are observed. They relate to peak values reaching 180 ng/ml in some animals, while others have an annual prolactin profile that remains flat and at a low level. Stressful effects related to handling and venepuncture might be involved in increased prolactin levels, as has been reported in cattle (Raud, Kiddy and Odell, 1971; Leining, Bourne and Tucker, 1979). Nevertheless, it may be considered that over the two years of the study, the animals had adjusted to the sampling routine.

A relationship between the photoperiod and prolactin has been reported in many domestic ungulates e.g. cattle (Karg and Schams, 1974; Leining, Bourne and Tucker, 1979), goats (Buttle, 1974; Hart, 1975) and sheep (Pelletier, 1973; Lamming, Moseley and McNeilly, 1974; Ravault, 1976;

Walton *et al.*, 1977; Thimonier, Ravault and Ortavant, 1978). Manipulations of natural daylength have clearly demonstrated that the seasonal prolactin rhythm is primarily determined by photoperiod. Recent studies have shown similar prolactin rhythms in wild mammals such as the white-tailed deer (Mirarchi *et al.*, 1978; Schulte *et al.*, 1980), the badger, the red fox (Maurel, 1981) and the roe deer (Sempéré, 1982). In the domestic pig, values during different reproductive stages (e.g. oestrus and lactation) have been established (Brinkley, Wiltinger and Young, 1973; Van Landeghem and Van de Wiel, 1978; Bevers, Willemse and Kruip, 1978). The only study of long duration is that conducted by Ravault *et al.*, 1981. It has been shown that both in the cyclic and spayed sows mean plasma prolactin concentrations were low throughout the year. However, a trend to seasonal variations has been shown with prolactin levels being slightly higher during the spring and summer months and the lowest values occurring from September to December. This feature might be interpreted as a 'vestige' of an ancestral photoperiodic rhythm which could reflect the seasonal prolactin rhythm reported in the wild species.

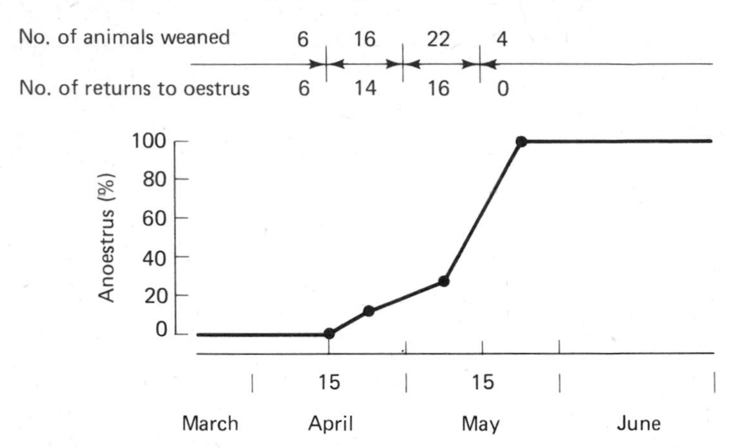

Figure 24.8 Occurrence of anoestrus in relation to time of weaning

In order to appreciate the seasonal influence, the natural timing of gestation and lactation were altered and the return to cyclic ovarian activity recorded (*Figure 24.8*). When abortion or weaning occur before mid-April, they are always followed by a return to oestrus. When weaning occurs from 15th April to 15th May, a maximum of 75% of females exhibit subsequent ovarian cyclic activity. A post-weaning oestrus does not occur in later weaned females. The change to cessation of ovarian cyclic activity in

Table 24.7 THE CHANGE TO CESSATION OF OVARIAN CYCLIC ACTIVITY IN UNBRED FEMALES (NUMBER OF FEMALES)

	Period of the year							
	April		*May*		*June*		*July*	
	1–15	*16–30*	*1–15*	*16–31*	*1–15*	*16–30*	*1–15*	*16–31*
Cycling	16	15	11	8	6	3	1	0
In anoestrus	0	1	5	8	10	13	15	16
Cumulative (%)		(6)	(31)	(50)	(62)	(81)	(94)	(100)

unbred females exhibits a similar progressive pattern (*Table 24.7*). The cumulative percentage of animals in anoestrus increases slightly from mid-April and reaches 100% in July.

If the present data suggest that the wild boar is photoperiodic with respect to reproduction, then further investigations of the physiological mechanisms responsible for the seasonality of breeding are necessary. It is not known whether a relationship exists between increased prolactin levels and cessation of ovarian activity. From the literature, the effects of prolactin remain controversial. Hyperprolactinaemia has been correlated with a reduction of both gonadotrophin secretion and ovarian steroidogenesis in rats (Beck *et al.*, 1977), women (Rolland *et al.*, 1975; McNatty, Sawers and McNeilly, 1974) and sheep (Kann, Martinet and Schirar, 1978; Munro, McNatty and Renshaw, 1980). Other studies have failed to provide evidence of antigonadotrophic effects of prolactin in ewes (Niswender, 1974; Schanbacher, 1980) and bovine heifers (Williams and Ray, 1980). As stated by Walton *et al.* (1980) it seems unlikely that prolactin levels *per se* are solely responsible for seasonal effects on ovarian activity. The internal mechanisms involved in the seasonal pattern of reproduction in the wild boar might be similar to those developed in other seasonal breeders. Present theories (Turek and Campbell, 1979; Karsch, Goodman and Legan, 1980) suggest that the feedback interplay between gonadal hormones and gonadotrophin is seasonally modified by a photoperiodic alteration of hypothalamic–pituitary activity.

Conclusions

The data presented here support the evidence of an environmental control of reproductive activity in the wild boar. As schematically represented in *Figure 24.9*, a number of environmental factors, whose relative importance

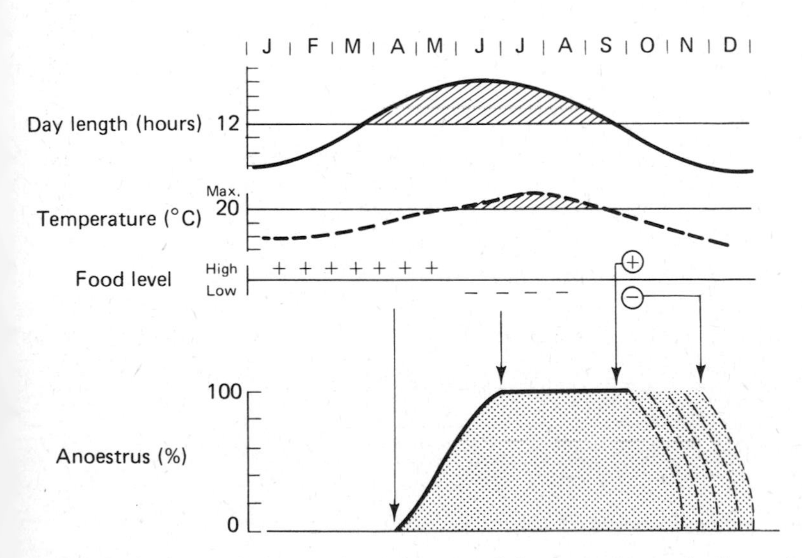

Figure 24.9 Seasonal changes in environmental factors and reproductive activity in the European wild boar

is different may be involved at the cessation and at the onset of the sexual season.

From April to June, the frequency of anoestrus increases progressively. It begins prior to the onset of high environmental temperatures and during a period of relatively high food availablity. This led to the consideration of the influence of the factor showing greatest change, namely, daylight increasing from 12 to 16 hours. The progressive nature of the response may be related to individual variations of sensitivity.

From July to September, the whole population is anoestrous. All the external stimuli may act as limiting factors during this period, i.e. more than 12 hours daylight, external temperature over 20 °C and low summer food availability.

From October, the onset of the breeding season can be regulated by the availability of food. Depending on annual variations in mast availability in the autumn, cyclic activity occurs early (October) or is delayed (December). Subsequently, there can be either one or two litters per year.

Thus, the sensitivity of the European wild boar to its environment allows the synchronization of reproduction at the most advantageous time of the year. The seasonal anoestrus, the length of which determines the frequency of breeding, appears to be the major mechanism controlling the dynamics of the natural population. In modern breeds of domestic pig, this adaptative quality has disappeared although a trend to a summer reproductive inefficiency does persist and may reflect the dependence of reproduction on environmental factors that have been described for the wild boar.

References

ASDELL S.A. (1964). *Patterns of Mammalian Reproduction.* New York, Cornell University Press

ASSENMACHER I. (1974). External and internal components of the mechanism controlling reproductive cycles in drakes. *Circannual Clocks* (E.T. Pingelley, Ed.), pp. 197–239. New York, Academic Press

ASSENMACHER I. and FARNER D.S. (Eds.) (1978). *Environmental Endocrinology.* Berlin, Heidelberg and New York, Springer Verlag

BAL H.S., WENSING C.J.G. and GETTY R. (1969). Morphological changes in the vaginal epithelium and ovary of swine of various ages as a mean of evaluating cyclic phases. *Iowa State J. Sci.* **43**, 341–358

BARRETT R. (1978). The feral hog on the Dye Creek Ranch, California. *Hilgardia* **46**, 283–355

BECK W., ENGELBART S., GELATO M. and WUTTKE W. (1977). Antigonadotrophic effects of prolactin in adult castrated and in immature female rats. *Acta endocr.* **84**, 62–71

BEVERS M.M., WILLEMSE A.H. and KRUIP T.A.M. (1978). Plasma prolactin levels in the sow during lactation and the postweaning period as measured by radioimmunoassay. *Biol. Reprod.* **19**, 628–634

BRIEDERMAN I. (1971). Zur Reproduction des Schwarzwildes in der Deutschen Demokratischen Republik. *TagBer. dt. Akad. LandwWiss. Berl.* **113**, 169–186

BRINKLEY H.J., WILFINGER, W.W. and YOUNG E.P. (1973). Plasma prolactin in the estrous cycle of the pig. *J. Anim. Sci.* **37**, 303

BUTTLE H.L. (1974). Seasonal variation of prolactin of male goats. *J. Reprod. Fert.* **37**, 95

COLE D.J.A. (1972). *Pig Production*. London, Butterworths

CORNER G.W. (1921). Cyclic changes in the ovaries and uterus of the sow and their relation to mechanism of implantation. *Contr. Embryol.* **13**, 117–146

CORTEEL, J.M., SIGNORET J.P. and DU MESNIL DU BUISSON F. (1964). Variations saisonnières de la reproduction de la Truie et facteurs favorisant l'anoestrus temporel. *5th Congr. Int. Reprod. Anim. Insem. Artif., Trente* **3**, 536–546

DUNCAN R.W. (1974). Reproductive biology of the european wild hog in the Great Smoky Mountains National Park. M.S. Thesis. University of Tennessee

EDQVIST L.E. and LAMM A.M. (1971). Progesterone levels in plasma during the oestrous cycle of the sow, measured by a rapid competitive protein binding technique. *J. Reprod. Fert.* **25**, 447–449

ENNE G., BECCARO P.V. and TAROCCO C. (1979). A note on the effect of climate on fertility in pigs in the Padana valley of Italy. *Anim. Prod.* **28**, 115–117

HABER A. (1969). *Dzik. Panstw. Wyd. Roln i Lesn., Warsaw,* 1–215

HAGEN D.R. and KEPHART K.B. (1980). Reproduction in domestic and feral swine. I. Comparison of ovulatory rate and litter size. *Biol. Reprod.* **22**, 550–552

HALBERG F., JOHNSON E.A., NELSON W., RUNGE W. and SOTHERN R. (1972). Autorhythmometry: Procedures for physiologic self-measurements and their analysis. *Physiol. Teacher* **1**, 3–11

HART I.C. (1975). Concentrations of prolactin in serial blood samples from goats before, during and after milking throughout lactation. *J. Endocr.* **64**, 305–312

HENRICKS D.M., GUTHRIE H.D. and HANDLIN D.L. (1972). Plasma estrogen, progesterone and luteinizing hormone levels during the estrous cycle in pigs. *Biol. Reprod.* **6**, 210–218

HENRY V.G. (1968). Length of estrous cycle and gestation in European wild hogs. *J. Wildl. Mgmt.* **32**, 406–408

HURTGEN J.P. (1976). Seasonal anestrus in a Minnesota swine breeding herd. *Proc. 4th Int. Pig Vet. Soc.*, D 22

HURTGEN J.P., LEMAN A.D. and CRABO B. (1980). Seasonal influence on estrous activity in sows and gilts. *J. Am. vet. med. Ass.* **176,** 119–123

KANN G., MARTINET J. and SCHIRAR A. (1978). Hypothalamic-pituitary control during lactation in sheep. In *Control of Ovulation,* (D.B. Crighton, N.B. Hayes, G.R. Foxcroft and G.E. Lamming, Eds.), pp. 319–333. London, Butterworths

KARG H. and SCHAMS D. (1974). Prolactin release in cattle. *J. Reprod. Fert.* **39**, 463–472

KARSCH F.J., GOODMAN R.L. and LEGAN S.J. (1980). Feedback basis of seasonal breeding : test of an hypothesis. *J. Reprod. Fert.* **58**, 521–535

LAMMING G.E., MOSELEY S.R. and McNEILLY J.R. (1974). Prolactin release in the sheep. *J. Reprod. Fert.* **40**, 151–168

LEINING K.B., BOURNE R.A. and TUCKER H.A. (1979). Prolactin response to duration and wavelength of light in prepubertal bulls. *Endocrinology* **104**, 289–294

McFEE A.F., BANNER H.W. and RARY J.H. (1966). Variation in chromosome number among European wild pigs. *Cytogenetics* **5**, 75–81

McNATTY K.P., SAWERS R.S. and McNEILLY A.S. (1974). A possible role for prolactin in control of steroid secretion by the human graafian follicle. *Nature, Lond.* **250**, 653–655

MAUGET R. (1972). Observations sur la reproduction du Sanglier (*Sus scrofa L.*) à l'état sauvage. *Ann. Biol. anim. Biochem. Biophys.* **12**, 195–202

MAUGET R. (1978). Seasonal reproductive activity of the european wild boar; comparison with the domestic sow. In *Environmental Endocrinology*, (I. Assenmacher and D.S. Farner, Eds.), pp. 79–80. Berlin, Heidelberg and New York, Springer Verlag

MAUGET R. (1980). Régulations écologiques, comportementales et physiologiques (reproduction) de l'adaptation du Sanglier, *Sus scrofa*, au milieu. Doct. Thesis. University of Tours

MAUGET R., BOISSIN-AGASSE L. and BOISSIN J. (1981). Ecorégulatin du cycle de la fonctin de reproduction chez les Mammifères sauvages. *Bull. Soc. Zool. France* **106**, 431–443

MAUGET R., CASTET M.C., MARAUD C. and CANIVENC R. (1977). Etude dynamique et caryotypique d'une population de sangliers à robe claire. *C.r. Soc. Biol.* **171**, 592–596

MAUREL D. (1981). Variations saisonnières des fonctions testiculaire et thyroidienne en relation avec l'utilisation de l'espace et du temps chez le Blaireau européen (*Meles meles*) et le Renard roux (*Vulpes vulpes*). Doct. Thesis. University of Montpellier

MENAKER M., TAKAHASHI, J.S. and ESKIN A. (1978). The physiology of circadian pace makers. *Ann. Rev. Physiol.* **40**, 501–526

MIRARCHI R.E., HOWLANDS B.E., SCANLON P.F., KIRKPATRICK R.L. and SANFORD L.M. (1978). Seasonal variation in LH, FSH, prolactin, and testosterone concentrations in adult male white tailed deer. *Can. J. Zool.* **56**, 121–127

MOLENAT M. and CASABIANCA F. (1979). Contribution à la maîtrise de l'élevage porcin extensif en Corse, *I.N.R.A., Bull. Tech. Dept. Gent. Anim.* **32**, 72 pp

MUNRO C.J., McNATTY K.P. and RENSHAW L. (1980). Circa-annual rhythms of prolactin secretion in ewes and the effect of pinealectomy. *J. Endocr.* **84**, 83–89

NISWENDER G.D. (1974). Influence of 2-Br-α-ergocryptine on serum levels of prolactin and the estrous cycle in sheep. *Endocrinology* **94**, 612–615

OLOFF H.B. (1951). Zur biologie und ökologie des wildschweines. *Beitr. Tierk. Tierz.* **2**, 1–95

PATERSON A.M., BARKER I. and LINDSAY D.R. (1978). Summer infertility in pigs: its incidence and characteristics in an Australian commercial piggery. *Aust. J. exp. Agric. Anim. Husb.* **18**, 698–701

PELLETIER J. (1973). Evidence for photoperiodic control of prolactin release in rams. *J. Reprod. Fert.* **35**, 143–147

PERRY J.S. and ROWLANDS I.W. (1973). The environment and reproduction in mammals and birds. *J. Reprod. Fert., Suppl.* **19**

PINE D.S. and GERDES G.L. (1973). Wild pigs in Monterey County California. *Calif. Fish Game* **59**, 126–137

RADEV G., ANDREEV A. and KOSTOV L. (1976). The influence of age and season on the weaning to oestrus period in sows. *Proc. Intern. Congr. Anim. Reprod. Artif. Insem., Krakow*, **1**, *Communication Abstracts*, 208

RAUD H.R., KIDDY C.A. and ODELL W.D. (1971). The effect of stress upon the determination of serum prolactin by radioimmunoassay. *Proc. Soc. exp. Biol. Med.* **136**, 689

RAVAULT J.P. (1976). Prolactin in the ram: seasonal variations in the concentration of blood plasma from birth until three years old. *Acta endocr. (Kbh)* **83**, 720–725

RAVAULT, J.P. (1981). (to be published)

ROGERS L.L. (1976). Effects of mast and berry crop failures on survival, growth and reproductive success of black bears. *Trans. N. Am. Wildl. Conf.* **41**, 431–437

ROLLAND R., LEQUIN R.M., SCHELLEKENS L.A. and DE JONG F.H. (1975). The role of prolactin in restoration of ovarian function during the early post-partum period in the human female. I. *Clin. Endocr.* **4**, 15–25

SADLEIR R.M.F.S. (1969). The role of nutrition in the reproduction of wild mammals. *J. Reprod. Fert., Suppl.* **6**, 39–48

SCHANBACHER B. (1980). Relationship of daylength and prolactin to resumption of reproductive activity in anestrous ewes. *J. Anim. Sci.* **50**, 293–297

SCHULTE B.A., PARSONS J.A., SEAL U.S., PLOTKA E.D., VERME L.J. and OZOGA J.J. (1980). Heterologous radioimmunoassay for deer prolactin. *Gen. Comp. Endocrinol.* **40**, 59–68

SEMPERE, A. (1982). Fonction de reproduction et caractères sexuels secondaires chez la chevreuil (*Capreolus capreolus* L.): variations saisonnières et incidences sur l'utilisation du budget temps–espace. Doct. Thesis, University of Tours

SHEARER I.J., PURVIS K., JENKIN G. and HAYNES N.B. (1972). Peripheral plasma progesterone and estradiol-17β levels before and after puberty in gilts. *J. Reprod. Fert.* **30**, 347–360

SNETHLAGE K. (1974). *Das Schwarzwild.* Berlin, Paul Parey Verlag

STABENFELDT G.H., AKINS E.L., EWING L.L. and MORRISSETTE M.C. (1969). Peripheral plasma progesterone levels in pigs during the oestrus cycle. *J. Reprod. Fert.* **20**, 443–449

STEINBACH J. and SCHMIDT D. (1970). Cyclical phenomena in the female genital tract of swine. Histological observations. *J. Anim. Sci.* **30**, 573–577

STORK M.G. (1979). Seasonal reproductive inefficiency in large pig breeding units in Britain. *Vet. Rec.* **104**, 49–52

THIMONIER J., RAVAULT J.P. and ORTAVANT R. (1978). Plasma prolactin variations and cyclic ovarian activity in ewes submitted to different light regimens. *Ann. Biol. animal. Biochem. Biophys.* **18**, 1229–1235

TILLSON S.A., ERB R.E. and NISWENDER G.D. (1970). Comparison of luteinizing hormone and progesterone in blood and metabolites of progesterone in urine of domestic sows during the oestrous cycle and early pregnancy. *J. Anim. Sci.* **30**, 795

TURECK F.W. and CAMPBELL (1979). Photoperiodic regulation of neuroendocrine-gonadal activity. *Biol. Reprod.* **20**, 32–50

VAN LANDEGHEM, A.A.J. and VAN DE WIEL, D.F.M. (1978). Radioimmunoassay for porcine prolactin: plasma levels during lactation, suckling and weaning and after TRH administration. *Acta endocr.* **88**, 653–667

VERME L.J. (1965). Reproductive studies in penned white tailed deer. *J. Wildl. Mgmt.* **29**, 74–79

WALTON J.S., EVINS J.D., FITZGERALD B.P. and CUNNINGHAM F.J. (1980). Abrupt decrease in daylength and short-term changes in the plasma concentrations of FSH, LH and prolactin in anoestrous ewes. *J. Reprod. Fert.* **59**, 163–171

WALTON J.S., McNEILLY J.R., McNEILLY A.S. and CUNNINGHAM F.J. (1977). Changes in concentrations of FSH, LH, prolactin and progesterone in the plasma of ewes during the transition from anoestrus to breeding activity. *J. Endocr.* **75**, 127–136

WILLIAMS G.L. and RAY D.E. (1980). Hormonal and reproductive profiles of early postpartum beef heifers after prolactin suppression or steroid-induced luteal function. *J. Anim. Sci.* **50**, 906–918

25

FACTORS AFFECTING REPRODUCTIVE EFFICIENCY OF THE BREEDING HERD

G.J. TOMES
Muresk Agricultural College, Western Australia

and

H.E. NIELSEN
National Institute of Animal Science, Copenhagen, Denmark

The great variation that exists in the productivity of sow herds is well illustrated by a Danish survey based on a large number of commercial breeding herds where the number of piglets weaned/sow/year ranged from 10 to 22 (Aagaard and Studstrup, 1977). The financial consequences are clearly demonstrated in *Figure 25.1* where overall profitability of the breeding unit rises as the number of pigs weaned/sow/year increases. It is impossible to make a general assessment of financial implications of reproductive efficiency for various countries but such a trend is general.

In herds producing small numbers of pigs/sow/year there are often many unproductive sows with the attendant penalties of extra feed, labour and

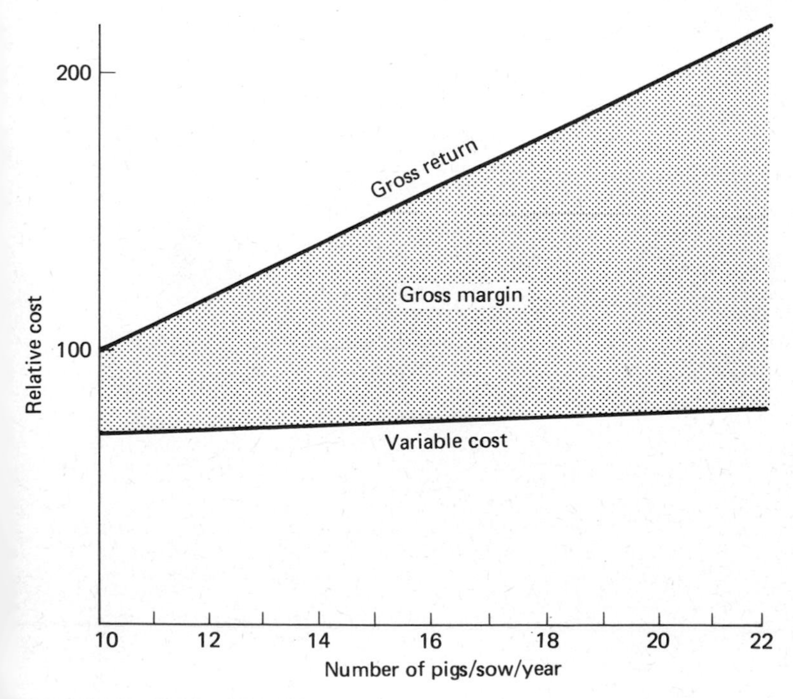

Figure 25.1 Profitability in breeding herds. From Aagaard and Studstrup (1977)

Table 25.1 NUMBER OF SOWS NEEDED TO PRODUCE 1000 PIGS PER YEAR AT VARIOUS PRODUCTION LEVELS

No. of weaned pigs/sow/year	*No. of sows needed*
12	83
16	62
20	50
24	42

Table 25.2 UNPRODUCTIVE DAYS OF SOWS AT DIFFERENT LEVELS OF EFFICIENCY BASED ON A 5-WEEK WEANING SYSTEM

No. of litters/ sow/year	*Sow empty days*	*Percentage of the year* (%)	*Equivalent herd sizes to achieve same total margin over feed costs* (£)
1.6	127	35	76
2.2	37	10	50

From Aherne (1980, personal communication)

under-utilized production facilities. Some production parameters for herds at different levels of efficiency are given in *Tables 25.1* and *25.2*. Pig producers are well aware that large litters and short intervals between farrowings are essential to maintain high annual output per sow (Nielsen 1981a). However, they often overlook the impact of herd age structure and culling patterns on reproductive efficiency. The major factors affecting reproductive efficiency in sow herds are given in *Figure 25.2* and are discussed in this chapter.

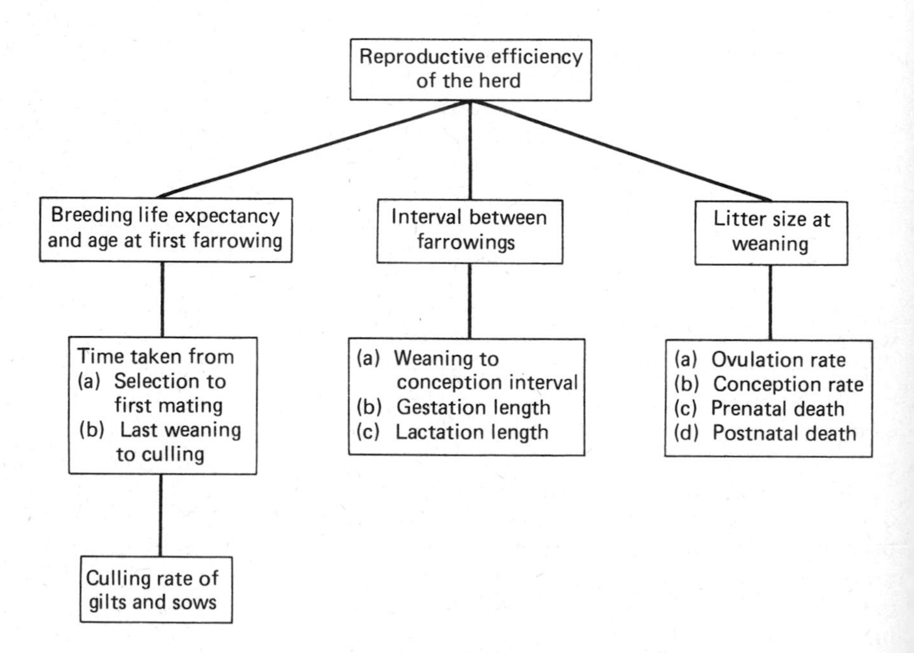

Figure 25.2 The major factors affecting reproductive efficiency of the breeding herd

Breeding life expectancy and age at first farrowing

BREEDING LIFE EXPECTANCY

The definition of 'breeding sows' varies in the literature and this causes problems when results of various studies are interpreted. An acceptable definition has been offered recently by Sundgren *et al.* (1980) who included all female pigs selected for breeding and older than 200 days.

Under commercial conditions there is a high wastage rate in sows, the main causes of culling being reproductive failure, leg problems and small litters (Tomes, Nielsen and Jacobsen, 1977). Reproductive failure may be connected with genotype, nutrition, housing conditions, herd health and management practices. A number of studies have shown that approximately 40% of sows are culled or die each year in European and Australian production units (Jones, 1967; 1968; Einarsson and Settergren, 1974; English, Smith and MacLean, 1977; Dagorn and Aumaitre, 1979; Tomes and Nielsen, 1979).

Culling rate has a great impact on the composition of sow herds and their productivity. Kroes and van Male (1979) have shown that herds with a low culling rate (31.3%) produced on average 1.5 weaners/sow/year more than herds where 55.4% sows were replaced. The difference in profitability between units with low and high culling rates was even more pronounced. The average age of culled sows was low (24–35 months) and 35–45% produced less than three litters.

Seasonal variations in the percentage of culled sows have been reported in Europe by Tomes, Nielsen and Jacobsen (1977) and by Dagorn and Aumaitre (1979). The large increase in the number of sows culled during summer and early autumn is caused by breeding failure. It may be due to heat, but is probably caused by other factors, as the temperatures seldom are critical in most parts of Europe. Results recorded in Australia (Tomes and Nielsen, 1979) show culling rates similar to those found in Europe, but the proportion of animals dying reached 8% (cf. approximately 3% of the sow population recorded in Denmark). Most of these losses occurred during hot summers when the temperature in the piggery was in the vicinity of 40 °C. Sows at late pregnancy or shortly after farrowing are particularly susceptible to high temperatures. Speer (1981) has confirmed increased sow mortality during hot summer months; for sows kept under non-confinement he also observed marked increases in mortality during harsh winters.

Reports in the literature suggest that only about 30% of sows produce more than five litters, i.e. reach the period when the largest litters can be expected (Dagorn and Aumaitre, 1979). However, in many herds sows can produce six or more litters before culling (Dagorn and Aumaitre, 1979).

Individual attention, efficient oestrous detection and the close supervision of mating are the prerequisites for long breeding life and increased productivity in sow herds. Under most conditions the standard of management and stockmanship is the most important factor determining the breeding life expectancy. Several devices exist for detecting pregnancy in pigs; however, Aumaitre and Etienne (1981) have claimed that a boar detecting heat gives as good a result as any of the equipment available.

It is essential that non-productive animals are identified and excluded from the herd as soon as possible. Failure to do so has a substantial effect on herd output. Long delays in culling sows from commercial herds were reported by Pomeroy (1960), Dagorn and Aumaitre (1979) and Pattison, Cook and Mackenzie (1980a,b). Kroes and van Male (1979) calculated that the time lost as a result of culling increased the average farrowing interval by six to eight days. Efficient culling must be accompanied by a ready supply of replacement gilts to avoid a suboptimal sow population and the corresponding economic penalties.

AGE AT FIRST FARROWING

The selection and pre-service management of gilts is an area of major change and development. Not so long ago a pig often took over 200 days to reach 90 kg liveweight and gilts were first mated between 8–10 months of age (third oestrus). This practice was based on studies showing that gilts served at third oestrus produced larger litters than those served at first oestrus (Pay and Davies, 1973).

Under good production conditions gilts can reach 90 kg in less than 150 days (Barker, 1979). Many such animals are capable of reaching puberty when exposed to mature boars at this stage (Brooks and Cole, 1970). Examples of the early onset of puberty in gilts are presented in *Figures 25.3* and *25.4*. There is also substantial evidence that gilts served at second or third oestrus by the time they reach market weight will produce as large a litter as those served at 130 kg or 8 months (Brooks and Cole, 1973; Pay and Davies, 1973). The subsequent production life of these early mated animals appears to be similar to those mated later (Brooks and Smith,

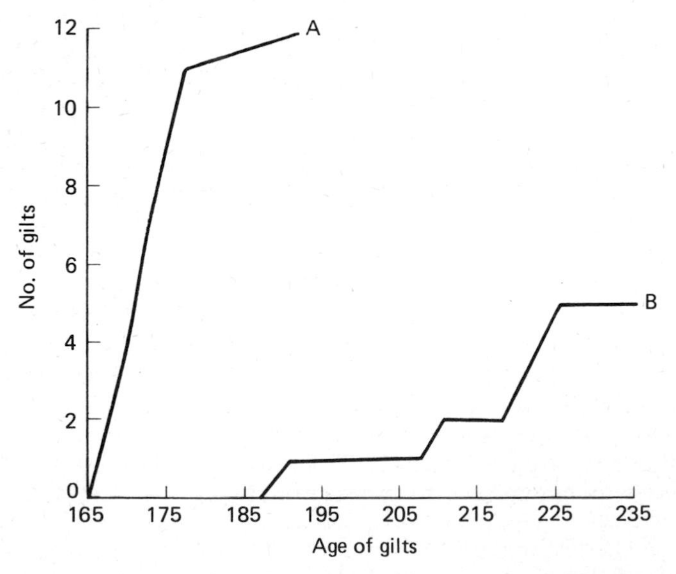

Figure 25.3 The effect of boar presence or absence on attainment of puberty in gilts. A: boars introduced at 165 days; B: no boars present. From Brooks and Cole (1970)

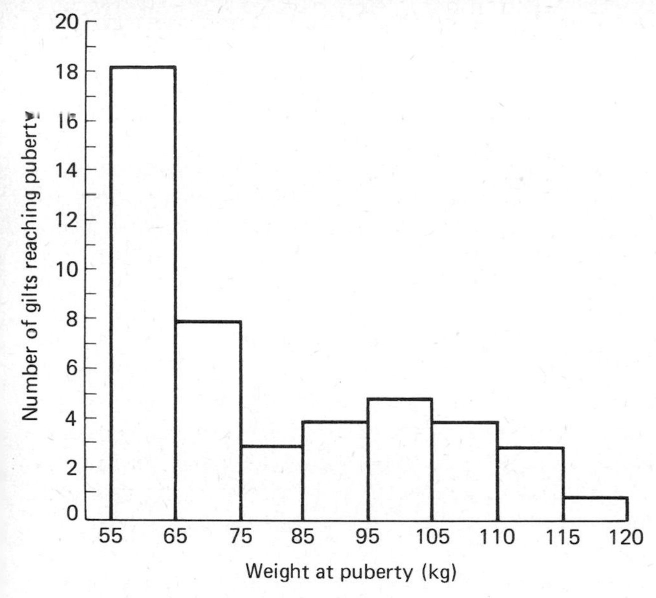

Figure 25.4 The distribution of weight at puberty in Lacombe gilts following exposure to boars. From Aherne *et al.* (1976)

1980). Often conflicting opinions exist about the optimum age for mating and about replacing gilts in commercial herds. O'Grady (1979) has recommended not to mate before the gilts reach 240 days of age; Brooks and Smith (1980) have advocated mating at second heat but as early in life as possible.

Gilts produce smaller litters than older sows and the individual birthweights of first litter pigs are also lower than of those from older sows (Nielsen, 1968). There is some evidence to suggest that piglets born in first litters are more susceptible to infection than piglets from older sows. However, when the gilts are properly introduced into the herd, protection via colostrum and milk is adequate (Kruse, 1981).

A practical system benefiting producers in terms of reduced food consumption and earlier production from gilts has been described by Aherne (1979, unpublished data) and is shown in *Figure 25.5*. Gilts are housed separately after weaning until they reach a weight of approximately 65 kg. Mature boars are then introduced for a period of 10 days and all gilts showing strong and consistent heats within three weeks are considered for breeding. At third heat the selected animals are served. The number of gilts served should be considerably higher than the required number of breeding animals to allow for the final selection among served gilts. Gilts returning to oestrus, some pregnant animals above the requirement for replacements and those rejected for various other reasons are marketed.

This method is likely to increase the consistency of heats by selection. It also provides a ready pool of replacements essential to avoid seasonal changes caused by delayed puberty (Bane, Einarsson and Larsson, 1976; Tomes and Nielsen, 1979) or increased culling rate (Tomes, Nielsen and Jacobsen, 1977; Dagorn and Aumaitre, 1979). Nielsen (1981b) reported

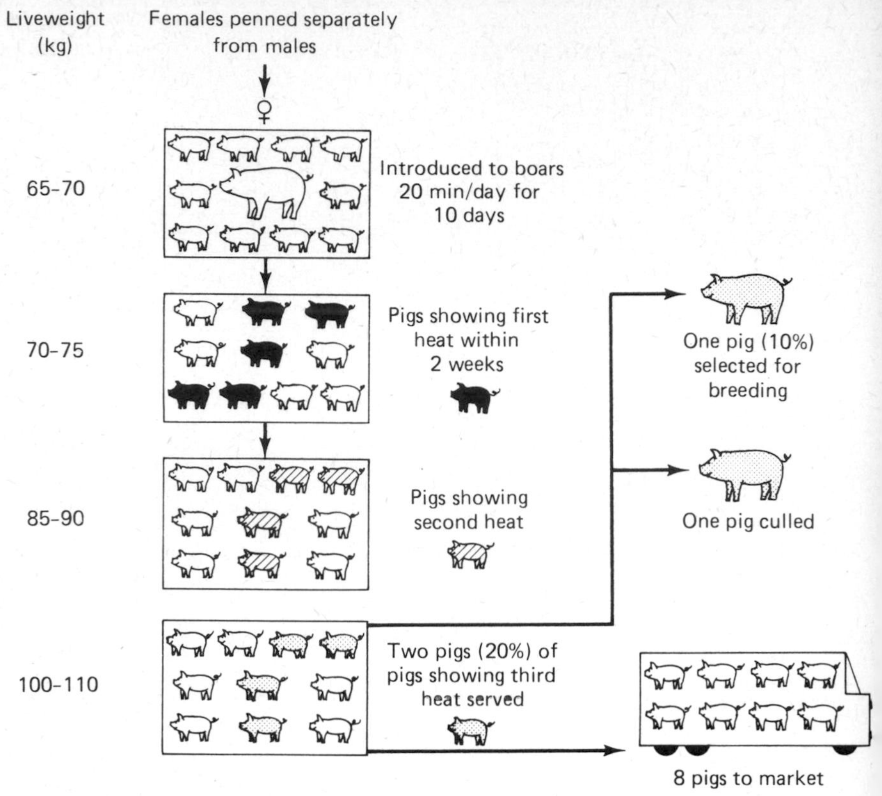

Figure 25.5 Selection of female pigs for breeding

the use of Aherne's method in a large Danish unit and recommended it for application in batch farrowing systems as suggested by Nielsen and Danielsen (1976).

Interval between farrowings

The number of piglets produced/sow/year can be increased by reducing the interval between farrowings. This can be done by reducing lactation length and the period between weaning and conception.

Normal gestation can range from 112–118 days and although a negative correlation exists between gestation period and the number of pigs born, little can be done under commercial conditions to shorten gestation. Prostaglandin analogues are increasingly used to induce parturition in commercial herds. Average gestation length is reduced by 1–3 days but there may be even greater potential benefits resulting from reduced perinatal losses and more efficient use of farm labour (King, 1981).

In theory the number of litters/sow/year may be increased by approximately 0.1 if lactation is shortened by one week (*Figure 25.6*). Practical results show a smaller improvement, particularly when very short lactation periods are adopted.

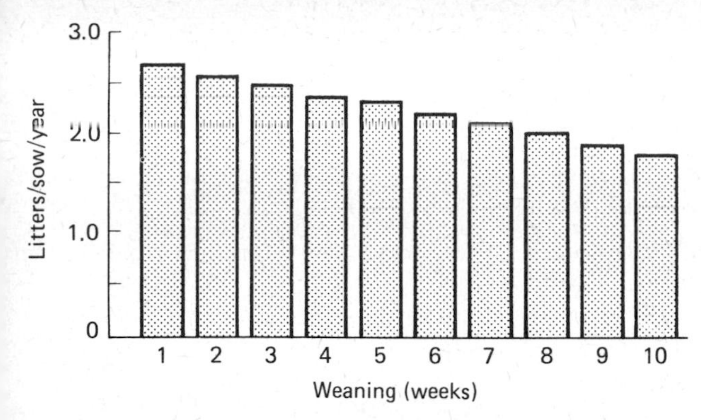

Figure 25.6 Theoretical number of litters/sow/year with various lactation lengths

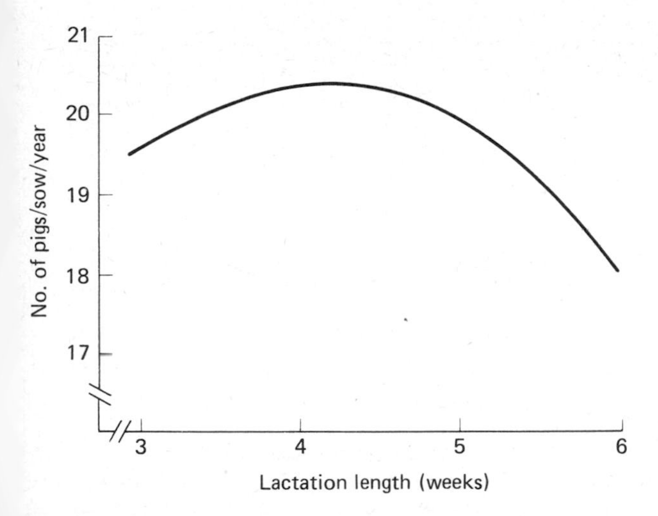

Figure 25.7 The influence of lactation length on the number of pigs/sow/year. From Danielsen and Ruby (1979)

In a recent comprehensive study, Danielsen (1980b) has stated that weaning before four weeks is detrimental to overall productivity. Weanings before three weeks of age may result in more litters/sow/year but will also cause a significant reduction in subsequent litter size (Varley and Cole, 1976; te Brake, 1978). Other reproductive parameters such as the incidence of post-weaning oestrus (Allrich *et al.*, 1979; Danielsen, 1980a) and conception rate (Varley and Cole, 1976) are also negatively affected by shorter lactation length. *Figure 25.7* shows the effect of lactation length on the annual production of piglets/sow recorded in a comprehensive Danish study. Results recorded in the literature generally suggest lactation lengths of 21–30 days for maximum production/sow/year. The characteristics of individual herds are likely to be important in determining where within this interval highest productivity can be achieved.

The period between weaning and conception varies considerably between herds. Results recorded in research studies indicate that the normal period between weaning and oestrus is 4–7 days. In commercial units,

Table 25.3 PATTERN OF RETURN TO SERVICE ON FRENCH FARMS

Days after weaning	*Sows mated* (%)
1–14	58.5
15–20	5.5
21–30	9.5
31–40	8.5
41–50	6.8
Over 50	11.2

From Legault, Dagorn and Taster (1975)

Table 25.4 EFFECT OF PARITY ON THE INTERVAL FROM WEANING TO MATING EXPRESSED AS A PERCENTAGE OF SOWS RETURNING IN THE TIME SPECIFIED

Parity	*Days after weaning*			
	0–10 (%)	*11–23* (%)	*24–30* (%)	*30* (%)
1 (*n* = 2143)	70.6	14.8	8.1	6.5
2–9 (*n* = 5939)	92.4	4.8	1.9	0.9

From Paterson, Barker and Lindsay (1980)

under excellent conditions, this period may be shorter than 10 days (Paterson, Barker and Lindsay, 1980) but it is generally far longer (Dagorn and Aumaitre, 1979). *Table 25.3* shows results from a French survey (Legault, Dagorn and Taster, 1975) where less than 60% of sows were served within two weeks after weaning. The onset of oestrus is delayed in primiparous sows, and Paterson, Barker and Lindsay (1980) have reported that primiparous sows are mated on average five days later after weaning than older sows. Some of their results are presented in *Table 25.4*. They also found that crossbred sows exhibited oestrus on average two days earlier than purebred Landrace sows and nearly one day before purebred Large White sows.

Tomes and Nielsen (1979) compared reproductive performance in Australia and Denmark and found that seasonal anoestrus and low conception rates occurred in both countries despite the great difference in seasonal temperature.

Delayed oestrus is usually associated with low conception rates (Einarsson, 1977; Miskovic, Cerne and Jancic, 1977) and with shorter periods of standing heat (Tomes and Smith, 1975). Several factors may be responsible for the lack of normal oestrus in sows. The presence of mature boars has a marked stimulating effect on breeding activity in sows (Hughes and Cole, 1976). Newly weaned sows and gilts approaching maturity should be housed in close proximity to mature boars (English, Smith and MacLean, 1977). Increased intake of feed during lactation and prior to oestrus may stimulate early post-weaning oestrus in primiparous sows (Tomes, 1978a). The level of feeding of young sows is important; Whittemore, Franklin and Pearce (1980) have shown that the amount of body fat is reduced considerably during the first two parities even when the sows are fed up to 2 kg feed/day.

Litter size

The main factors determining litter size at weaning are shown in *Figure 25.2*. The first limit imposed on the potential litter size is ovulation rate. The number of ova is higher for sows than for gilts, but in gilts there is a pronounced increase in ovulation after the first oestrus. MacPherson, Hovell and Jones (1977) noted that gilts mated at first, second and third oestrus produced 7.8, 9.8 and 10.2 piglets, respectively, when they first farrowed.

When gilts are fed a low plane of nutrition, an increased energy intake for 10–14 days before mating usually results in increased ovulation rate (Zimmerman, Spies and Self, 1960) but its effect on live litter size is inconsistent (Brooks and Cole, 1973). Larger litters are usually recorded only with young, lean sows. Tomes (1978a) found no response to increased energy intake after second parity. There is also evidence suggesting that the administration of antibiotics at parturition may increase ovulation rate (Hays *et al.*, 1978). Some producers practise starvation of the sows after weaning. Shearer and Adam (1973) have shown that starvation can have an adverse effect on the onset of oestrus and on conception rates. It is recommended that high levels of feeding are used during the first week after weaning.

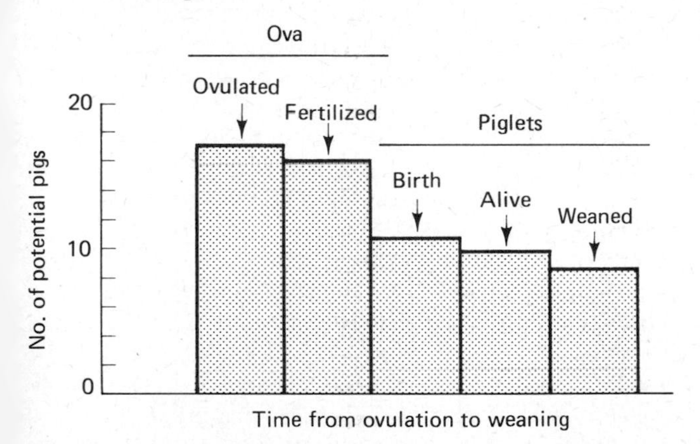

Figure 25.8 Loss of potential pigs from the time of ovulation until weaning. From Schofield (1972)

Under normal conditions nearly all eggs are fertilized after mating (*Figure 25.8*). However, the chances of conception are reduced if the sow or gilt is mated too early or too late during oestrus. The timing of mating also affects litter size. This phenomenon has been studied by Polge (1974) and his results are presented in *Figure 25.9*. It should be noted, however, that ovulation does not occur at a specific time after the onset of oestrus (boar acceptance). This may explain the usual increase in conception rate and even in litter size when sows and gilts are mated twice or three times.

Changes in ovulation patterns caused by the administration of high levels of exogenous gonadotrophins may be used to explain reduced conception rates after superovulation (Tomes, 1978b). Relatively low levels of exogenous gonadotrophins have not been found to depress conception and are

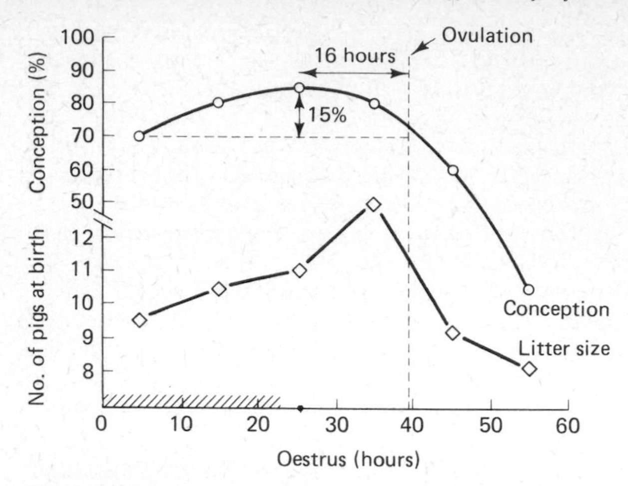

Figure 25.9 Conception and litter size in relation to time of mating. From Polge (personal communication)

commonly used to stimulate and synchronize oestrus in herds with a previous history of reproductive disturbances.

Conception rate and infertility depend on environmental conditions e.g. climate (*Table 25.5*) and on the health of the sows and fertility of the boars. Several workers have reported big differences in conception rate and litter size when individual boars have been evaluated (Skjervold, 1963; Nielsen, 1968; Garnett, 1981). These differences often exceed 23% in conception rate and two piglets/litter. Significant improvements in production can be made if young boars are tested with a relatively small number of sows (30–40) and then only considered for further breeding. This is particularly relevant in large units and with artificial insemination.

Table 25.5 SUMMER INFERTILITY IN SOWS IN AUSTRALIA

Temperature range (°C)	*Duration* (weeks/year)	*Total no. of services*	*Returns* (%)
20–23.9	7.8	1116	12.2
24–27.9	15.1	2584	11.4
28–31.9	13.1	2392	14.4
⩽32	16.0	2919	19.7

From Paterson (1978)

The role of crossbreeding as an effective method of increasing the number of pigs produced/sow/year is well recognized by pig breeders and is further discussed, in addition to other genetic means of improving productivity, in Chapter 26.

Prenatal and postnatal mortality cause serious problems in many herds. The economical implications are often underestimated by producers. This wide ranging area is beyond the scope of this chapter and detailed information can be found in Chapters 12, 13 and 23.

More emphasis should be placed on monitoring the performance of individual boars where large savings can be achieved by early culling of those with low libido, poor conception rate or small litters. In addition, recent results (Tomes and Nielsen, 1979) show seasonal variation in the birthweights of piglets and also differences resulting from the use of various sires (Tomes, unpublished).

There is a clear need for more information about individual boars, and about the subsequent performance and survival of piglets sired by them. In the meantime a significant improvement in litter size may be achieved by double mating using two different boars during the same oestrus instead of the usual two matings with one boar (Meding, 1979 unpublished). Although this system is suitable only for the production of slaughter pigs it can increase the annual production per sow by 0.6–0.8 at no extra cost.

References

AAGAARD, R. and STUDSTRUP, N. (1977). *Svineholds okonomi. Landsudvalget for driftsøkonomi.* Det Fagligte landscenter, Viby, Denmark

AHERNE, F.X., CHRISTOPHERSON, R.J. THOMPSON, J.R. and HARDIN, R.T. (1976). Factors affecting the onset of puberty, post-weaning estrus and blood hormone levels of Lacombe gilts. *Can. J. Anim. Sci.* **56**, 681–692

ALLRICH, R.D., TILTON, J.E., JOHNSON, J.N., SLANGER, W.D. and MARCHELLO, M.J. (1979). Effect of lactation length and fasting on various reproductive phenomena of sows. *J. Anim. Sci.* **48**, 359–362

AUMAITRE, A. and ETIENNE, M. (1981). *Proceedings Guelph Pork Symposium. Management of the Sow and Young Pig.* University of Guelph, pp. 1–16

BANE, A., EINARSSON, S. and LARSSON, K. (1976). Effect of season on age at puberty in crossbred gilts in Sweden. *VIIth Int. Congr. Anim. Reprod., Krakow,* pp. 117–119

BARKER, I. (1979). *Australian Pig Manual*, p. 191. Australian Pig Industry Research Committee

TE BRAKE, J.H.A. (1978). An assessment of the most profitable length of lactation on producing piglets of 20 kg body weight. *Livest. Prod. Sci.* **5**, 81–94

BROOKS, P.H. and COLE, D.J.A. (1970). The effect of presence of a boar on the attainment of puberty in gilts. *J. Reprod. Fert.* **23**, 435–440

BROOKS, P.H. and COLE, D.J.A. (1973). Meat production from pigs which have farrowed. 1. Reproductive performance and food conversion efficiency. *Anim. Prod.* **17**, 305–315

BROOKS, P.H. and SMITH, D.A. (1980). The effect of mating age on the reproductive performance, food utilization and liveweight change of the female pig. *Livest. Prod. Sci.* **7**, 67–78

DAGORN, J. and AUMAITRE, A. (1979). Sow culling: reason for and effect on productivity. *Livest. Prod. Sci.* **6**, 167–178

DANIELSEN, V. (1980a). Results favour 4-week weaning. *Pig International* **10**, 6.50

DANIELSEN, V. (1980b). The effect of weaning age on reproductive performance in the pig. *E.A.A.P. Commission on Pig Production, Munich*

DANIELSEN, V. and RUBY, V. (1979). Evaluation of data from experiments and field test on reproductive performance in sows. *E.A.A.P. Commission on Pig Production, Harrogate*

EINARSSON, S. (1977). Disturbances in reproduction and measures for reduction of these phenomena. *E.A.A.P. Commission on Pig Production, Brussels*

EINARSSON, S. and SETTERGREN, I. (1974). Fertility and culling in some pig breeding herds in Sweden. *Nord. VetMed.* **26**, 576–584

ENGLISH, P., SMITH, W. and MACLEAN, A. (1977). *The sow – improving her efficiency*. Farming Press, Suffolk. p. 305

GARNETT, I. (1981). *Important Factors to be considered in selecting your Breeding Sow. Proceedings of Alberta Pork Seminar*, University of Alberta, pp. 160–167

HAYS, V.W., KRUG, J., CROMWELL, L., DUTT, G.L. and KRATZER, D.D. (1978). Effect of lactation length and dietary antibiotics on reproductive performance of sows. *J. Anim. Sci.* **46**, 884–891

HUGHES, P.E. and COLE, D.J.A. (1976). Reproduction in the gilt. 2. The influence of gilt age at boar introduction on the attainment of puberty. *Anim. Prod.* **23**, 89–94

JONES, J.E.T. (1967). An investigation of the causes of mortality and morbidity in sows in a commercial herd. *Br. vet. J.* **123**, 327–339

JONES, J.E.T. (1968). The cause of death in sows. A one year survey of 106 herds in Essex. *Br. vet. J.* **124**, 45–55

KING, G.J. (1981). *Proceedings Guelph Pork Symposium. Management of the Sow and Young Pig*. University of Guelph, pp. 43–47

KROES, Y. and VAN MALE, J.P. (1979). Reproductive lifetime of sows in relation to economy of production. *Livest. Prod. Sci.* **6**, 179–183

KRUSE, P.E. (1981). Immunity – what you should know. *Proceedings Alberta Pork Seminar*, University of Alberta, pp. 49–65

LEGAULT, C., DAGORN, J. and TASTER, D. (1975). *Journées de la recherche porcine en France*, pp. 43–51. Paris, L'Institut Technique du Porc

MACPHERSON, R.M., HOVELL, F.D. and JONES, A.S. (1977). Performance of sows first mated at puberty or second or third oestrus, and carcass assessment of once-bred gilts. *Anim. Prod.* **24**, 333–343

MISKOVIC, M., CERNE, F. and JANCIC, S. (1977). Losses in gilts and sows and their principal causes in large units. *E.A.A.P. Commission on Pig Production, Brussels*

NIELSEN, H.E. (1968). Growth rate, fertility and longevity of boars on different gilts during rearing, and some results concerning fertility in sows. *Report from National Institute of Animal Science, Copenhagen*, p.375

NIELSEN, H.E. (1981a). Sow productivity. *Proceedings Alberta Pork Seminar,* University of Alberta, pp. 1–17

NIELSEN, H.E. (1981b). Gilt selection and management to farrowing. *Proceedings Guelph Pork Symposium. Management of the Sow and Young Pig*. University of Guelph, pp. 1–16

NIELSEN, H.E. and DANIELSEN, V. (1976). Maximising reproduction in pigs. *Proc. Pig Vet. Soc.* **1**, 76–89

O'GRADY, J.R. (1979). *Proceedings Pork Industry Conference*, University of Illinois, pp. 59–66

PATERSON, A.M. (1978). *J. exp. Agric. Anim. Husb.* **18**, 691–698

PATERSON, A.M., BARKER, I. and LINDSAY, D.R. (1980). *Proc. aust. Soc. Anim. Prod.* **13**, 389–392

PATTISON, H.D., COOK, G.L. and MACKENZIE, S. (1980a). A study of natural service farrowing rates and associated fertility parameters. *Anim. Prod.* **30**, 452 (Abstract)

PATTISON, H.D., COOK, G.L. and MACKENZIE, S. (1980b). A study of culling patterns in commercial pig breeding herds. *Anim. Prod.* **30**, 462–463 (Abstract)

PAY, M.G. and DAVIES, T.E. (1973). Growth, food consumption and litter production of female pigs mated at puberty and at low body weight. *Anim. Prod.* **17**, 85–91

POLGE, C. (1974). Meeting of the European Association of Animal Production, Copenhagen, 1974, Unpublished data

POMEROY, R.W. (1960). Infertility and neonatal mortality in the sow. *J. agric. Sci., Camb.* **54**, 18–31

SCHOFIELD, A.M. (1972). Embryonic mortality. In *Pig Production*, (D.J.A. Cole, Ed.), pp. 367–383. London, Butterworths

SHEARER, I.J. and ADAM, J.L. (1973). Nutritional and physiological development in reproduction of pigs. *Proc. N.Z. Soc. Anim. Prod.* **33**, 62–76

SKJERVOLD, H. (1963). To what extent do boars affect the litter size? The Agricultural University of Norway, Report 166

SPEER, V.C. (1981). Feeding sows during lactation. *Proceedings Guelph Pork Symposium. Management of the Sow and Young Pig.* University of Guelph, pp. 19–29

SUNDGREN, P.E., VAN MALE, J.P., AUMAITRE, A., KALM, E. and NIELSEN, H.E. (1980). Sow and litter recording procedures. Report of a working party of the EAAP Commission of Pig Production. *Livest. Prod. Sci.* **7**, 393–401

TOMES, G.J. (1978a). Effects of post-weaning feeding levels and pregnant mares serum gonadotrophin (PMSG) on sow reproductive performance. *Proc. aust. Soc. Anim. Prod.* **12**, 255

TOMES, G.J. (1978b). Influence of short term feed intake variation and hormonal supplements on sow reproductive performance. PhD Thesis. University of Western Australia

TOMES, G.J. and NIELSEN, H.E. (1979). Seasonal variation in the reproductive performance of sows under different climatic conditions. *Wld Rev. Anim. Prod. XV* **1**, 9–20

TOMES, G.J. and SMITH, C.A. (1975). Effect of energy intake and hormonal supplements on litter size. *Proc. Aust. Pig Prod. Rev. Conf.* **2**, 101–103

TOMES, C.J., NIELSEN, H.E. and JACOBSEN, K.A. (1977). Review of 15 years recording of culling sows in Danish production units. *E.A.A.P. Commission on Pig Production, Brussels*

VARLEY, M.A. and COLE, D.J.A. (1976). Studies in sow reproduction. 5. The effect of lactation length of the sow on subsequent embryonic development. *Anim. Prod.* **22**, 79–85

WHITTEMORE, C.T., FRANKLIN, M.F. and PEARCE, B.S. (1980). Fat change in breeding sows. *Anim. Prod.* **31**, 181–190

ZIMMERMAN, D.R., SPIES, H.G. and SELF, H.J. (1960). Ovulation rate in swine as affected by increased energy intake just prior to ovulation. *J. Anim. Sci.* **19**, 295–301

26

GENETICS OF REPRODUCTION IN THE PIG

W.G. HILL
Institute of Animal Genetics, University of Edinburgh, UK
and
A.J. WEBB
A.R.C. Animal Breeding Research Organisation, Edinburgh, UK

Genetic principles

Reproductive performance depends both on the genotype of the pig and the environment which it encounters. Some aspects of this environment, for example the feeding regime, can be specified; other random factors affecting individual pigs cannot. Similarly, there are genetic differences among breeds and animals of the same breed such that there is both genetic and non-genetic variability between pigs of the same breed under the same management system. For traits of reproduction, such as litter size, and its components, such as ovulation rate and embryonic survival, genetic variability is due to segregation of many genes which have, presumably, different effects on each trait and also exist with different frequencies in the population.

Whatever type of management system, feeding regime and disease control the pig producer adopts, he can still benefit from the best choice of breeds and crosses among them, and from genetic changes within them. Genetic and non-genetic improvement are not alternatives for they can, and should, proceed together. An awareness of genetic differences may also enable the research worker in other disciplines to use them to advantage in his experiments or, at least, not confound genetic and other effects.

Firstly in this chapter some of the evidence on genetic variability in reproductive performance is reviewed and then consideration is given to how it is being used in genetic improvement programmes and how it might be used in the future. Most of the information refers to simple measures such as litter size because data on some aspects, e.g. reproductive performance, are scanty.

It is necessary to distinguish between genetic differences in performance observed among and within identifiable populations. These populations can be different breeds, or lines of the same breeds in different herds, as long as they can be identified so that stock can be repeatably drawn from them. For example, breed differences in litter size reflect deviations in the frequency of genes affecting litter size among the populations, and can be utilized by selection between them. However, once this selection is made and populations substituted, the gains are maintained without further cost, but cannot be repeated. Variation within a population arises as a consequence of Mendelian segregation and much or all of it is reconstituted each generation, depending on the size of the population and selection intensity (Falconer, 1960a). Therefore, selection can be repeated over many generations and genetic progress made continually. It is not usually possible to get precise estimates of genetic variation within any population and, within the margins of statistical sampling error, estimates tend to be similar for different populations. Thus, although an estimate of variance strictly applies to one population, for example a breed at one particular time, it is customary and often necessary to extrapolate results outside this set.

Similarly, the effects of heterosis between breeds will be considered, as will the converse of inbreeding effects within breeds and their extrapolation to other populations. The primary genetic mechanism used to explain heterosis and inbreeding depression is that of dominance at individual loci, so that the heterozygote is superior to the mean of the corresponding homozygotes. Thus, if the dominant alleles differ in frequency between two breeds, the frequency of heterozygotes and therefore performance will be higher in the cross than the parental average. Similarly inbreeding causes a reduction in heterozygosity and therefore performance. Sheridan (1981) has argued, however, that interaction between effects of genes at different loci is often an important cause of heterosis, but there is not adequate evidence of this in the pig.

Reproductive traits show low values of heritability and high values of inbreeding depression compared with growth and carcass traits, which is a measure of the amount of variation that can be utilized by selection and is typical of all animal species.

Variation among breeds

DIFFERENCES BETWEEN BREEDS

Breeds of the world exhibit considerable variation in reproductive performance. For example, litter sizes at birth range from around six in wild breeds up to a reputed 22 in some Chinese breeds. This range is much narrower for breeds used in Western pig production, as shown in *Table 26.1* which summarizes results from two recent breed comparisons in Europe and the USA. In Europe the Large White strains consistently appear to wean slightly larger litters than Landrace, although differences in total litter weight at weaning are relatively small. In the Dutch study, the Pietrain and Belgian Landrace breeds, both noted for their stress susceptibility and extreme ham conformation, weaned at least 1.6 pigs less than Dutch Large Whites (*Table 26.1*). Part of this difference may be explained by the higher frequency of the gene for halothane sensitivity in the two breeds, which may cause a reduction in litter size (Webb, 1981). Differences in litter size between American dam breeds arise from differences in both ovulation rate and embryo survival (*Table 26.2*).

Direct comparisons of up-to-date samples of European and American breeds in the same environment are scarce. In the Dutch evaluation of American breeds several generations after importation into Europe, litter

Table 26.1 SUMMARY OF SOME PUBLISHED WITHIN-COUNTRY BREED COMPARISONS OF LITTER PRODUCTIVITY[a]

Breed	*No. females*	*At birth*			*At weaning*			
		Litter size	*Piglet weight* (kg)	*Litter weight* (kg)	*Litter size*	*Mortality from birth* (%)	*Piglet weight* (kg)	*Litter weight* (kg)
(1) Holland, weaning 49 days[b]								
Dutch Large White	118	9.9	1.41	13.9	8.4	15	14.5	121.8
Dutch Landrace	122	9.4	1.54	14.5	8.1	14	14.3	135.0
Belgian Landrace	124	7.6	1.44	10.9	6.2	18	13.2	81.8
Pietrain	98	8.5	1.34	11.4	6.8	20	12.2	103.7
American Duroc	109	9.8	1.51	14.7	8.6	12	13.3	114.0
American Hampshire	107	8.6	1.36	11.7	7.5	13	13.4	100.8
(2) Oklahoma, USA, weaning 21 days[b]								
American Duroc	55	8.9	1.30	11.4	5.6	37	4.5	25.3
American Hampshire	56	8.4	1.24	10.2	5.4	36	5.0	26.7
American Yorkshire	50	10.2	1.08	10.7	7.7	25	4.7	35.4

[a]References: (1) Brascamp, Cöp and Buiting (1979); (2) Young, Johnson and Omtvedt (1976)
[b](1) Average of parities 1–4; (2) all parties adjusted to gilt basis

size at weaning in the Duroc was comparable to that of the Large White, but for the Hampshire it was intermediate between the Large White and Pietrain. Mortality from birth to weaning in the Duroc and Hampshire was three times higher in the American than in the Dutch study, and it is not clear how far this results from management differences between the two countries. However, in a cross-fostering experiment between American Hampshires and British Saddlebacks (King, 1975) the lower pre-weaning

Table 26.2 DIFFERENCES IN LITTER PRODUCTIVITY BETWEEN DUROC AND YORKSHIRE (D–Y) AND HAMPSHIRE AND YORKSHIRE (H–Y) BREEDS WHEN USED AS SIRES AND DAMS IN THE OKLAHOMA CROSSBREEDING EXPERIMENT

Trait	*Dam breeds*		*Sire breeds*	
	D–Y	*H–Y*	*D–Y*	*H–Y*
30 days after mating (n = 212)[a]				
No. corpora lutea	0.2	–1.5*	–	–
No. live embryos	–0.2	–2.2**	0.0	–0.1
Corpora lutea/live embryos (%)	–3.2	–7.1	0.8	–2.7
At birth (n = 450)				
Litter size	–0.8*	–2.0**	–0.5	–0.7*
Average piglet weight (kg)	0.24**	0.22**	0.0	0.0
Litter weight (kg)	1.4**	–0.6	–0.7	–0.7
42 days after birth (n = 437)				
Litter size	–1.2**	–1.4**	–0.6*	–1.2**
Survival from birth (%)	–6.5*	–0.5	–4.8	–9.0**
Average piglet weight (kg)	–0.03	0.18	–0.03	–0.06
Litter weight (kg)	–11.0**	–11.8**	–7.1*	–12.5**

From Young, Johnson and Omtvedt (1976)
[a]Total number of females in comparison
*P<0.05
**P<0.01

growth rate of Hampshire piglets could not be improved when nursed by Saddleback dams with greater milk production.

The effect of the breed of boar, other than through heterosis (see following section), on the litter productivity of the dam to which he is mated, is less clear. In the Oklahoma experiment (*Table 26.2*), sire breed significantly affected litter sizes at birth and weaning, as well as piglet survival and total 49-day litter weight. Within Large White strains, sires have been shown to influence significantly the litter sizes of their mates (Legault, 1970; Strang, 1970), but the proportion of variation due to sires amounted to less than 1% of the total. Martin and Dziuk (1977) also claimed to demonstrate differences between and within breeds in the conception rates and litter sizes of individual Duroc and Yorkshire boars. Although significant breed of sire effects may exist, it seems unlikely that they will be of major economic importance.

CROSSBREEDING AND HETEROSIS

Crossbreeding is used to bring together the desirable characteristics from two or more breeds, and to exploit hybrid vigour or heterosis. Heterosis may be defined as the amount by which the performance of the offspring exceeds the mean of its parents. As the parental breeds become more genetically distinct giving a higher level of heterozygosity in the cross, the amount of heterosis is expected to increase. In most circumstances heterosis will only be economically useful if the progeny outperform the better of the two parental breeds.

In pigs useful heterosis can be demonstrated in the genotypes of both the crossbred dam and her crossbred progeny. Published estimates of heterosis have been comprehensively reviewed by Sellier (1976), and are reproduced in *Table 26.3*. For litter size at weaning, heterosis estimates for the dam and her progeny amount to 11% and 6% respectively, giving a total of 17% or 1.3 pigs weaned per litter. For litter weight at weaning the total advantage from crossbreeding is 22%. These may slightly overestimate the benefits of crossbreeding in Europe for the figures include American studies in which average litter sizes and environmental conditions may be poorer. Nevertheless heterosis in litter productivity has been the main justification for commercial crossbreeding in pigs for the last 10–15 years It seems likely that heterosis may also be expressed in age at sexua maturity and conception rate in maiden gilts (e.g. Hutchens *et al.*, 1978) but these effects may result partly from the improved growth rate o crossbreds.

The value of crossbreeding in the male is less well established. Heterosi might be expected in fitness-related traits. For example, Neely, Johnson and Robison (1980) demonstrated 34% heterosis in total number of sperm but this fell to 14% after adjustment for the greater body weight o crossbreds. Other studies in US and British breeds (Wilson, Johnson an Wetteman, 1977; Lishman *et al.*, 1975) have shown no clear advantage fo crossbred boars on litter productivity. In a British experiment, F_1 Pietrai × Hampshire boars increased the litter size of their mates by 0.78 pigs a birth when compared with other types of boars, mainly Large White. Th

Table 26.3 ESTIMATED LEVELS OF HETEROSIS, EXPRESSED IN UNITS OF EACH TRAIT AND AS A PERCENTAGE OF THE MID-PARENT VALUE, FOR SOME TRAITS OF ECONOMIC IMPORTANCE

Trait	*Genotype showing heterosis*			
	Progeny		*Dam*	
	Units of trait	*% of mid-parent*	*Units of trait*	*% of mid-parent*
Litter size at birth (pigs)	+0.30	3	+0.75	8
Litter size at weaning (pigs)	+0.45	6	+0.85	11
Individual piglet weight at weaning (kg)[a]	+0.5	5	0	0
Litter weight at weaning (kg)[a]	+9	12	+8	10
Post-weaning growth rate (kg/day)	+0.04	6	0	0
Age at slaughter (days)	−10	5	0	0
Food conversion ratio (kg feed/kg liveweight gain)	−0.08	3	0	0
Body composition and meat quality	0	0	0	0

From Sellier (1976)
[a]Estimated for weaning at 6 weeks of age

authors (King and Thorpe, 1973) speculated that the Pietrain × Hampshire advantage could stem from improved mating behaviour under the particular conditions of paddock mating adopted, and might not be detected with hand-mating systems. In future experiments, crossbred boars need to be evaluated under the exact conditions in which they would be used commercially.

In general the performance of a particular cross may be adequately predicted from the mean of its parental breeds together with published estimates of heterosis. For example, recent trials in Holland have shown higher litter sizes at weaning for Dutch Landrace × Duroc females than Dutch Landrace × Dutch Large White (Brascamp and Buiting, 1980), and this might have been predicted from the purebred litter sizes in *Table 26.1*. As in the case of the Pietrain × Hampshire boars, there will be exceptions. A further complication could be the existence of differences between reciprocal crosses of the form *A*♂ × *B*♀ or *B*♂ × *A*♀, arising mainly from maternal effects of the dam breed. The Oklahoma crossbreeding experiment showed significant reciprocal differences for Duroc × Yorkshire crosses in piglet weights at birth and 21 days, but not in litter size (Johnson, Omtvedt and Walters, 1978).

SYSTEMS OF CROSSBREEDING

A variety of different systems of crossbreeding can be used to produce different commercial end products. The systems differ in the amounts of heterosis expressed, and some of the most common are listed in *Table 26.4*. The systems may be divided into two categories: continuous and discontinuous. In discontinuous systems the crossbred end product is slaughtered, and replacement male and female breeding stock must be specially bred on the farm or purchased from outside. In continuous systems replacement females are selected from the slaughter generation, and only

Table 26.4 EXPECTED LEVELS OF HETEROSIS IN GENOTYPES OF PARENTS AND OFFSPRING FROM DIFFERENT CROSSING SYSTEMS (EXPRESSED AS PERCENTAGE OF HETEROSIS IN F_1)

Crossing system		*Genotype showing heterosis*		
		Maternal	*Paternal*	*Individual*
Discontinuous				
F_1	*A*♂ × *B*♀	0	0	100
F_2	(*A* × *B*)♂ × (*A* × *B*)♀	100	100	50
Backcross	*A*♂ × (*A* × *B*)♀	100	0	50
3-breed cross	*C*♂ × (*A* × *B*)♀	100	0	100
4-breed cross	(*C* × *D*)♂ × (*A* × *B*)♀	100	100	100
Continuous[a]				
2-breed rotation	...(*B* × (*A* × (*B* × (*A* × *B*)	67	0	67
3-breed rotation	...(*B* × (*A* × (*C* × (*A* × *B*)	86	0	86
4-breed rotation	...(*A* × (*D* × (*C* × (*A* × *B*)	94	0	94
Purebred male and 2-breed rotation female:	*C*♂ × (...(*A* × (*B* × (*A* × *B*))♀	67	0	100

[a]Heterosis levels shown are those attained at equilibrium, roughly five generations after starting the rotation crosses

males are brought in. Continuous systems therefore offer less disease risk from incoming stock, lower replacement costs, but also less of the important maternal heterosis than discontinuous systems.

In the UK with only two main breed types, Large White and Landrace, the backcrossing system predominates, with breeding companies offering 'package deals' of replacement F_1 gilts and purebred boars. In recent years the loss of 50% heterosis in individual pig performance from a backcross (*Table 26.4*) has not been thought sufficient to justify the widespread use of a third 'sire' breed, with possibly inferior conversion of food to lean. Rotation crossing schemes are commonplace in the USA.

As the differences in litter productivity between crossing systems are likely to be small, very large numbers of litters are required to test theoretical expectations in practice. In 1968 a long-term experiment was started at ABRO (King, 1978) to compare the litter productivity of F_1 and F_2 Large White × Landrace females with 2- and 3-breed rotation crosses involving Large White, Landrace and Saddleback. Both types of rotation cross were conducted using either outbred boars or boars which were of inbreeding at least that of two consecutive brother × sister matings. The preliminary results (*Table 26.5*) show the expected advantage in litter size

Table 26.5 LITTER PERFORMANCE OF VARIOUS CROSSBREDS EXPRESSED AS DEVIATIONS FROM LARGE WHITE GILTS

Type of cross	*No. of litters*		*Average no. born alive*	*Average no. weaned per litter*	*Average weaning weight (kg)*
	Birth	*Weaning*[a]			
Purebred Large White (absolute values)	97	90	9.36	7.48	13.6
F_1 generation	118	114	0.84	1.01	0.4
F_2 generation	78	76	0.28	0.87	0.0
2-breed rotation with outbred boars	88	80	–0.03	0.53	0.5
2-breed rotation with inbred boars	124	121	0.94	1.32	–0.1
3-breed rotation with outbred boars	91	88	0.68	0.99	0.9
3-breed rotation with inbred boars	92	87	1.08	1.19	0.5
Average standard error of deviation of cross from Large White	–	–	0.42	0.36	0.3

From King (1978)
[a]Litters were weaned at 50 days of age

weaned for the F_1 over purebred Large White, although heterosis and breed effects are confounded. There is the expected reduction in the F_2 and 2-breed rotation. Prior inbreeding of the boars used in both rotations appears to give a boost in litter size, perhaps due to natural selection during inbreeding, but this deserves further investigation.

A theoretical full economic comparison of crossing systems of the American breeds has been undertaken using actual results from the Oklahoma experiment (Wilson and Johnson, 1979). Productivity was expressed as the number of pigs produced from each system per 10 000 total females farrowing either as purebreds or crossbreds. Numbers of pigs were adjusted to compensate for differences in production costs and carcass values arising from differences in growth rate, feed efficiency and backfat. The results are summarized in *Figure 26.1*. The most productive crossing system was a backcross of a Yorkshire boar to an F_1 Duroc ×

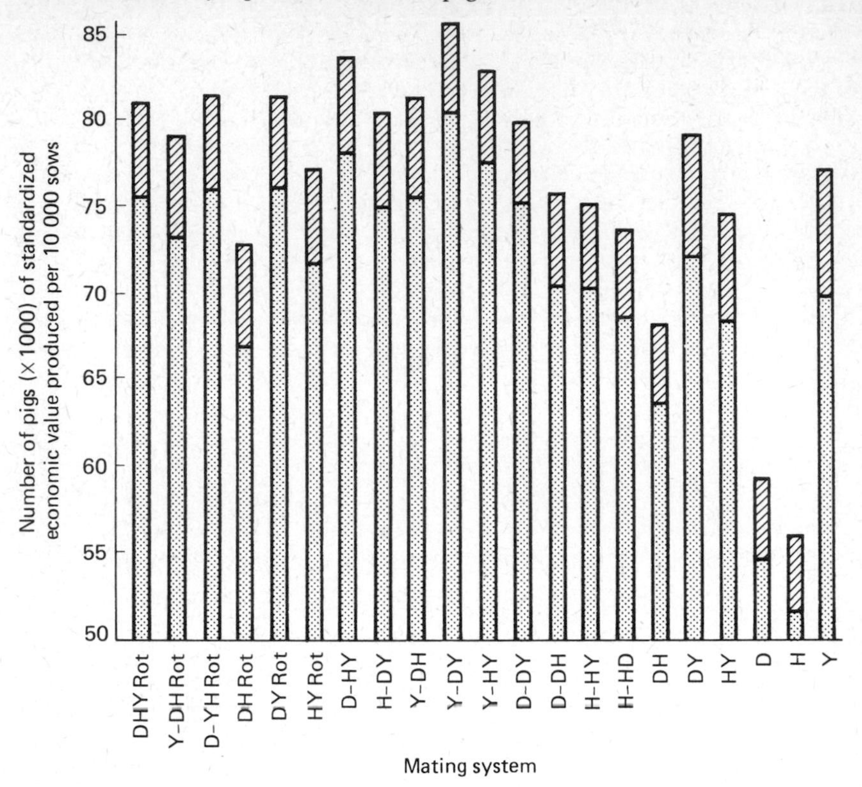

Figure 26.1 Predicted total number of pigs produced per 10 000 sows from different crossing systems involving Duroc (D), Hampshire (H) and Yorkshire (Y) breeds. DHY Rot = 3-breed rotation of D, H and Y; Y–DH Rot = Y terminal sire mated to 2-breed rotation dam containing D and H; Y–DH = Y terminal sire mated to F_1 D × H dam; DH = F_1 cross of D sire mated to purebred H dam. ▨ Replacement gilts; ▩ market hogs. From Wilson and Johnson (1979)

Yorkshire female (Y–DY), and the least was the purebred Hampshire (H) with 34% lower productivity. Both the 3-breed rotation (DHY Rot) and the best 2-breed rotation (DY) were 3% less productive than the best 3-breed discontinuous cross (D–HY). The 'winning' backcross (Y–DY) was, however, only 2% more productive than the D–HY discontinuous cross. Another comparison of systems obtained by pooling all published crossbreeding results from the whole of North America led to broadly similar, though not identical, conclusions (Quintana, 1979).

Variation within populations

REPEATABILITY AND HERITABILITY

There have been extensive studies of genetic variability in reproductive traits based on field data and therefore over a range of management schemes (for a review of earlier work see Legault, 1970). As a basis, figures for means, standard deviations and coefficients of variation averaged over litters from the extensive study on British Large Whites by Strang (1970) are given in *Table 26.6*. Notable are the high coefficients of variation, around 25% for numbers and weights and almost 100% for mortality, which are much higher than for growth and carcass traits.

Correlations of reproductive rate over successive litters of sows, customarily referred to as repeatabilities, are also shown in *Table 26.6* for the

Table 26.6 MEANS, STANDARD DEVIATIONS (SD), COEFFICIENTS OF VARIATION (CV) AND REPEATABILITY OF PERFORMANCE IN SUCCESSIVE LITTERS (r) IN BRITISH LARGE WHITE PIGS

Trait	*Mean*[a]	*SD*[a]	*CV (%)*[a]	*r(%)*[b]
Number of pigs alive				
Birth	10.9	2.8	26	15
3 weeks	9.0	2.3	26	14
8 weeks	8.8	2.3	26	14
Percentage mortality				
0–3 weeks	16.5	15.6	95	10
0–8 weeks	18.4	16.2	88	12
Average piglet weight (kg)				
3 weeks	6.0	1.1	18	17
8 weeks	16.8	4.6	27	4
Litter weight (kg)				
3 weeks	53	15	28	15
8 weeks	144	50	25	4

[a]From Strang (1970)
[b]From Strang and King (1970)

same data set (Strang and King, 1970). These figures are typical of others in the literature, which are fairly consistent (summarized by Strang and Smith, 1979), and around 15% for litter size. By weaning, piglet and litter weights have a lower repeatability, presumably through the influence of different sires. Providing there is no negative environmental correlation of performance in successive farrowings, the repeatability can be regarded as an upper limit to heritability of traits dependent solely on the sow's genotype, because the correlation includes non-additive genetic and environmental components common to all litters.

Despite the extensive data sets, several involving over 30 000 litters, published estimates of heritability (the ratio of additive, i.e. transmissable, genetic variance to total or phenotypic variance) differ markedly (*Table 26.7*). Some of these differences may reflect real genetic differences among the populations analysed, but many probably reflect sampling error and the difficulties of correcting for identifiable sources of environmental variability, such as herds and seasons.

An effect of the size of the litter in which the pig is born on her own subsequent productivity has been found in two studies. Nelson and Robison (1976) reared piglets from 24 hours in litters of 6 or 14, and observed differences in their subsequent first litters in favour of those reared in the smaller litters of 0.88 in numbers born, 1.18 in numbers born alive and 0.70 kg in litter weight at birth, although all effects were non-significant ($P>0.05$). Similarly, Rutledge (1980) compared the productivity of pigs reared in reduced litters of mean size 5.8 with those reared

Table 26.7 HERITABILITY (%) OF REPRODUCTIVE PERFORMANCE AS ESTIMATED BY REGRESSION OF DAUGHTER ON DAM (DD), DAUGHTER ON GRANDDAM (DG), AND CORRELATION OF HALF SIBS (HS)

	(1a)[a]	*(1b)*	*(2)*		*(3)*		*(4)*	*(5)*		*(6)*		*Average*
	DD	*HS*	*DD*	*HS*	*DD*	*HS*	*HS*	*DD*	*HS*	*DD*	*DG*	
Number of pigs alive												
Birth	7±2	4±4	9[b]	7±3	11±3	7±2	19±7	12(7)	25(6)	7±8	29±19	12
3 weeks	7±2	3±4	10	−2±4	–	1±2	16±7	9(1)	6(1)	1±10	28±20	8
Weaning	9±3	5±5	6	0±6	8±4	1±2	–	10(5)	23(2)	0±10	28±21	9
Weight of piglet												
3 weeks	7±2	11±4	11	9±4	–	–	28±7	23(1)	–	25±11	8±23	15
Weaning	11±2	39±5	5	12±6	–	4±2	–	21(1)	–	24±12	<0	14
Weight of litter												
3 weeks	8±2	9±4	11	7±4	–	–	34±7	17(2)	9(1)	–	–	14
Weaning	3±2	18±5	3	12±6	5±5	3±2	–	27(3)	28(2)	–	–	12

Modified from Strang and Smith (1979)
[a]Details of studies:
(1a) Strang and King (1970); (1b) Strang and Smith (1979): British Large Whites (38 000 litters).
(2) Strang and Smith (1979): British Landrace (35 000 litters)
(3) Legault (1970): French Large White (11 000 litters).
(4) Eikje (1973): Norwegian Landrace (38 000 litters).
(5) Average of other studies, based on Strang and Smith (1979) (no. of studies, each with over 1000 litters).
(6) Vangen (1980): Norwegian Landrace (821 DD and 711 DG pairs).
[b]SE stated as less than 3% in this study, but values not given.

in unaltered litters of mean size 10.1, and found an increase in the former of 0.81 in size of their first litter, although the estimate is confounded with selection effects.

A negative environmental correlation between the litter size of the dam and her progeny has also been demonstrated for mice (Falconer, 1960b). Presumably the effects are mediated through the influence of litter size on subsequent bodyweight. In view of the negative environmental correlation, Revelle and Robison (1973) have argued that heritability estimates from the regression of daughter on dam are biased downwards, and are likely to be smaller than those calculated from the correlation of half-sibs. The two kinds of heritability estimates are shown in *Table 26.7* and although the estimates are erratic, there seems to be no consistent difference between them. Vangen (1980) recently obtained higher heritabilities of litter size from daughter–granddam than daughter–dam regressions, although with large standard errors, and argued that this resulted from the negative maternal correlation between generations.

A typical figure for heritability is about 10% for litter size with somewhat higher values for litter weights. Higher values would not, however, be consistent with the repeatability estimates shown in *Table 26.6.* In contrast, typically quoted figures for the heritability of growth rate and feed conversion ratio are about 30%, and for fat depths and other carcass quality measures they are about 50%.

The heritability of the size of individual litters appears to decline somewhat with increasing parity in Strang and King (1970) and Strang and Smith (1979)'s data, presumably as environmental effects accumulate. Vangen (1980) found the reverse in a smaller data set and argued that negative dam–daughter environmental covariances were declining. The heritability of mean litter size, however, increases as more records are included.

Different measures of reproduction rate in a single litter are quite highly correlated both genetically and phenotypically, not surprisingly since they are usually part-whole relationships. For example, estimates of the genetic correlation between numbers born alive and numbers at eight weeks are 0.9 by Strang and King (1970) and 0.7 by Legault (1970), and those between number at eight weeks and litter weight at the same age are 0.3 and 1.0, respectively.

SELECTION EXPERIMENTS

Selection experiments can be practised in a population both to estimate genetic variability within it and to assess the short-term and long-term responses likely in commercial practice. Little experimental work has been done on selection for litter size in pigs; much more has been done in mice which seems a relevant model because of their litter size has a similar mean and variability.

Pigs

A selection experiment for increased litter size has been conducted in France. Although the response after the first five generations was greater than expected at 0.29±0.45 piglets/litter, there was no significant total response in numbers born in the first two litters at generation 10 (Ollivier and Bolet, 1981). Rutledge (1980) reported on the first two generations of selection for litter size in two selected lines, one with litter size reduced at birth, the other unaltered (see p.549). There is an indication, as yet inconclusive, of a response in the former.

Mice, as a model for pigs

There have been several experiments selecting directly for litter size in mice (Falconer, 1960b; Bateman, 1966; Bradford, 1979; Bakker, Wallinga and Politiek, 1976; Joakimsen and Baker, 1977; Eisen, 1978). All showed responses for increased litter size over periods of selection from 11–20 generations, ranging from 1.6–4.6 mice born/litter with a mean of over 3 mice/litter, equivalent to 21–50% with a mean near 35% of the initial litter size. The realized heritabilities (i.e. the heritability achieved during selection) ranged from 13–22% with a mean of 16%. These mice data suggest that responses could be obtained to selection for litter size in pigs, but in view of the rather lower heritability estimates in pigs of about 10% (*Table 26.7*), responses of the same rate are unlikely.

INDIRECT SELECTION

An alternative to selecting on litter size alone is to devote some or all selection effort to traits which are correlated with it. Of these the obvious one is ovulation rate which can be measured, albeit laboriously, by laparoscopy and repeat measurements can be made over successive oestrous cycles before breeding decisions are made. Cunningham *et al.* (1979) obtained a substantial response from selection on ovulation rate, but almost no response in litter size because all the gains in ovulation rate were wiped out by a corresponding increase in embryo mortality. This same result was obtained with mice by Land and Falconer (1969) in selection for natural ovulation rate. In contrast, selection for litter size in the French experiment, although unsuccessful, appears to have slightly increased ovulation rate by about 1.2 eggs (Ollivier and Bolet, 1981). A more indirect measure proposed by Land (1973) as a correlate of litter size is testis size. This has been selected for in mice, but the correlated response in ovulation rate was not accompanied by a response in litter size (Islam, Hill and Land, 1976). On the other hand, Bradford (1979) was able to demonstrate improvements in both ovulation rate and prenatal survival from long-term selection for litter size in mice.

ASSOCIATION WITH GROWTH TRAITS

Extensive data are needed to enable estimation of the genetic correlations between reproductive performance and traits of the growing animal. A summary of two analyses available is given in *Table 26.8*. Although the correlation between measures of reproductive rate and total index score

Table 26.8 GENETIC CORRELATIONS (%) BETWEEN TRAITS OF THE GROWING ANIMAL AND REPRODUCTIVE TRAITS (LS = LITTER SIZE AT BIRTH; LW3 AND LW8 = LITTER WEIGHT AT 3 AND 8 WEEKS)

	British[a]				*French*[b]	
Breed	*Large White*		*Landrace*		*Large White*	
Trait	*LS*	*LW3*	*LS*	*LW3*	*LS*	*LW8*
Daily gain	6	13	44	77	–8	6
Food conv. ratio	–15	–9	–21	–40	8	9
Killing out %	–63	–45	–49	–65	–	–
Backfat	–18	–15	–36	–21	11	–11
Hindquarters %	–41	–18	–12	–20	–	–
Ham and loin %	–	–	–	–	2	49
Total index points	1	–1	–4	2	–	–

[a]Morris (1975): SE's 30% for Large White, 45% for Landrace
[b]Legault (1971): SE's not given; correlation of LW8 with ham and loin percentage significant ($P<0.05$); remainder non-significant

for growth/carcass traits was small in the study of Morris (1975) and for most traits in the study conducted by Legault (1971), there is some indication of a positive correlation between litter size and daily liveweight gain and negative correlations between litter size and killing out percentage (an unfavourable correlation) and between litter size and backfat depth (a favourable correlation). These latter observations accord with the view that genetically fast growing lean pigs at a fixed (slaughter) weight are immature, and tend to have a higher mature body size. There is good evidence from mice, for example, that adult body size and litter size are positively correlated (Falconer, 1973). Lines of pigs selected for high and low fat depths at Beltsville, however, showed very little change in litter size in either of two breeds for either direction of selection (Hetzer and Miller, 1970). In summary, litter and growth traits in pigs seem to be weakly correlated, but the data are not very conclusive.

SINGLE GENES

The majority of reproductive characters can be assumed to be controlled by a large number of genes each of small effect. However, a small number of associations have been demonstrated between reproductive rate and single loci with two or more segregating alleles. The loci most readily detected in the past have been those controlling the red cell antigens, and polymorphic enzymes which can be distinguished by electrophoresis. The associations with performance can arise either from direct effects from the loci themselves, or from close linkage with other loci or groups of loci with major effects. For example, matings involving heterozygotes for the 'C' allele at the serum transferrin locus (T_f^C) have been shown to reduce litter size at birth by 1.4 pigs, possibly due to linkage with a lethal gene (Imlah, 1970). More recent studies with the halothane test have shown a reduction in litter size of about 1.1 pigs weaned for recessive homozygotes at the halothane locus (HAL^{nn}) (Webb, 1981). In turn, HAL is closely linked to the loci for serum phosphohexose isomerase (PHI) and the 'H' red cell antigen system (Andresen, 1979). In the Duroc and Yorkshire breeds, in which the HAL^n allele is assumed to be absent or at a very low frequency, the difference between sires of phenotypes $H^{a/}$ and $H^{-/-}$ in their mate's litter size at weaning averaged 0.9 pigs (Rasmusen and Hagen, 1973). The apparent effect of HAL on litter size could therefore result from linkage with H.

As reviewed recently by Smith and Webb (1980), an association with a single gene may be a potentially useful aid to selection, provided its effects are adequately estimated. Although normal methods of selection will automatically make use of the locus, for a trait such as litter size which has a low heritability and is expressed only in females, a single locus may add significantly to the rate of progress.

EFFECT OF INBREEDING

Inbreeding is the mating together of individuals more closely related to each other than are members of the same population chosen at random. It is measured as the coefficient of inbreeding, on a scale from 0 to 100%, and is defined as the probability that the two genes at any locus in an individual are identical by descent. Thus, for example a full-sib or offspring × parent mating results in progeny with an inbreeding coefficient of 25%, and a half-sib mating gives a coefficient of 12.5%. In a closed random mating population the rate of inbreeding, or annual increase in inbreeding coefficient, depends on the number of breeding animals, and will be greater in small populations, particularly if few males are used. Litter size and weight are expressions of both the dam's genotype (e.g. ovulation rate, milk production) and her progeny's genotype (e.g. viability). Since inbred dams can have non-inbred litters and vice versa, the inbreeding effects have to be partitioned into dam and litter components.

The effects of inbreeding on litter productivity are summarized in *Table 26.9*, updated from Sellier (1970) by including results of Mikami, Fredeen and Sather (1977). These inbreeding effects were computed by assuming a linear relation between performance and inbreeding coefficient, as predicted by simple genetic models. The studies involve relatively low average levels of inbreeding (below 50%), and at higher levels the effects may be different.

On average, litter size at weaning appears to be reduced by about one third of a pig per 10% increase in the inbreeding coefficient of the litter, and by one quarter of a pig per 10% increase in the dam's inbreeding. Individual pig and whole litter weights at weaning are also reduced by

Table 26.9 EFFECTS OF A 10% INCREASE IN INBREEDING ON LITTER PRODUCTIVITY[a]

Inbred genotype	*Litter size*		*Weight at 56 days* (kg)	
	Born alive	*56 days*	*Piglet*	*Litter*
Litter	–0.13	–0.34	–0.36	–5.96
Dam	–0.23	–0.23	–0.26	–3.31

After Sellier (1970) with additional results from Mikami, Fredeen and Sather (1977)
[a]Based on over 13 000 litters for each of litter and dam effects

inbreeding, with greater reductions for inbreeding of the litter than the dam. In an analysis of 10 000 litters in the USA (Bereskin, Shelby and Cox, 1973), survival from birth to weaning was reduced by 1.2% of piglets per 10% increase in inbreeding of the litter, and by only 0.1% per 10% increase in dam's inbreeding.

Estimated rates of inbreeding in British breeds ranged from 0.35–0.70% per generation in the early 1960s, and currently average about 0.52% per generation or 0.30% per year (Smith *et al.*, 1978). A similar increase of around 0.50% per generation has been reported in Danish Landrace from 1934 onwards (Jonsson, 1971), and 0.61% per generation in the Poland China from 1885 to 1929 (Lush and Anderson, 1939). Other estimated rates in European breeds range from 0.3% to 0.8% per generation (Langholz, 1968; Hanset, 1973). A rate of around 0.5% per generation therefore seems typical for many pig populations, and seems unlikely to lead to a noticeable decline in reproductive performance. Indeed, because natural selection would be opposing the inbreeding effects, it is possible that with low rates of inbreeding the effects are smaller than shown in *Table 26.9*. The use of crossbreeding eliminates any cumulative inbreeding effects.

In maize the production and subsequent crossing of inbred lines has proved a highly successful method of genetic improvement. In pigs this technique is ruled out by the length of time required to reach high inbreeding coefficients, the loss of selection pressure for growth traits and the loss of lines. For example, of 146 inbred Large White lines started in Britain in the fifties by mating a boar to all his full-sisters, only 18 (12%) survived to inbreeding coefficients of 40% or more (King, 1967). The main reasons for loss were small litter size or piglet weights at birth and weaning. Losses from genetic defects, such as monorchidism or intersexes, resulting from increased homozygosity of harmful recessive genes occurred in only 6% of lines.

Genetic improvement

REPRODUCTION VERSUS GROWTH

The pig breeder has to effect genetic improvement in the overall efficiency of the pig enterprise from breeding to fattening, and cannot consider reproductive rate alone. Indeed the greater part of the total cost, particularly of food, is incurred by the slaughter animal after weaning. Methods for comparing the economic importance of litter size and growth (these terms are used to imply overall reproductive performance and the overall composite of gain, food efficiency and carcass value) were formalized by Moav (1966) in terms of profit equations. These have been revised using more recent cost and return figures by Clarke and Smith (1979) and their equation for bacon pigs in the UK is:

$$P = 47.6 - 10.2\,Y - 240/X \qquad (1)$$

where P = profit in pence/kg liveweight at slaughter, Y = food conversion ratio (food/liveweight gain), and X = number of pigs sold per sow per year.

This is illustrated as iso-profit contours in *Figure 26.2.* In these formulae, returns depend on the inverse of litter size, on the basis that the annual sow cost is spread over the *X* piglets in an integrated operation. These figures do not necessarily apply to any single producer who can, for example, market the additional weaner pig at a price independent of his mean litter size. A breeder, however, has to take an industry-wide view, so the inverse relationship is more relevant, at least in a fixed market.

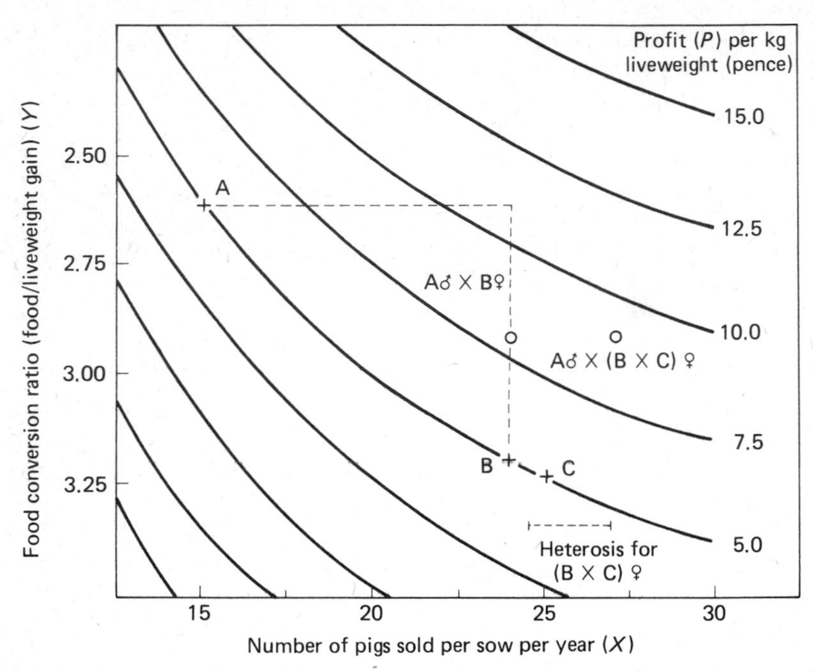

Figure 26.2 Iso-profit contours from production of breeds A, B and C, and their crosses. Profit (*P*), expressed in pence/kg liveweight, is calculated from Equation (1). Based on figures of Clarke and Smith (1979)

The relative advantages of alternative potential purebreds (A, B or C) are illustrated in *Figure 26.2* which shows that at commercial levels of 20 pigs/sow/year, little worsening (0.056) in lean tissue feed conversion can be tolerated if a breed B producing 21 pigs/sow/year is to be substituted. The figure also illustrates the benefits of crossbreeding lines differing in growth and litter size, using the more prolific breed (B) as the dam, but since the dam breed contributes one half of the genes for growth, it cannot be too inferior for growth rate. Finally, the gains from heterosis in using the crossbred B × C sow are illustrated.

BREED EVALUATION

In an earlier section the relative merits of different crossing systems were discussed in relation to breed evaluation (see p.546). In principle, breed and cross evaluation for reproductive rate and for total economic merit is quite straightforward, but there are major practical problems, particularly for reproductive rate. Litter size is highly variable, so, for example, to obtain a standard error of less than 0.25 pigs/litter in a breed comparison requires nearly 150 unrelated sows of each breed each having two litters. If related animals are used, the numbers must be higher. Any assessment of lifetime performance would involve further expense and facilities. Furthermore, care must be taken to eliminate any environmental effects at their source on the sows' subsequent performance. This can be avoided if comparisons are required and made through crossbreds by use of semen, but the results cannot be extrapolated to purebred performance because of possible differential heterosis between alternative breed combinations.

It is for reasons such as these that our knowledge of differences in reproductive rate among pig populations in Britain, for example, among commercial breeding companies' stocks is inadequate. Similarly there are no good recent estimates of the amount of heterosis between British Large Whites and Landrace which may have diverged further since the early 1960s as inbreeding has progressed.

SYNTHETIC LINES

An alternative to breed replacement or maintenance of a fixed crossing structure is to develop new synthetic breeds. The aim may be to combine the best features of each component breed, but inevitably it must represent close to mean performance for each trait, together with one half of the heterosis (i.e. as for an F_2, see *Table 26.4*). The synthetic breed also has a potential benefit in restoring variation if the parental breeds are inbred. Nevertheless the synthetic pig breeds established many years ago in North America have not been very successful, perhaps largely because their foundation populations were too small (Lopez-Fanjul, 1974).

IMPROVEMENT WITHIN BREEDS

Most pig breeders have devoted little selection effort to reproductive performance in recent years, and have concentrated on performance testing for growth and carcass traits. This largely derives from calculations of responses from alternate selection schemes (Smith, 1964; Moav and Hill, 1966) which showed that the responses in litter size were likely to be slow. This resulted from its low heritability and expression solely in one sex and at later ages than growth characteristics, and consequently monetary returns are small in view of the economic computations illustrated in *Figure 26.2*. There are increased benefits from developing specialized sire and dam lines in which reproductive performance is ignored in the former and can get doubled weight in the latter. However, using Equation (1), Clarke and Smith (1979) agreed with earlier work in concluding that the extra benefits were small.

The obvious and simplest method of attempting to improve litter size is by selecting replacement stock from among animals born in the largest litters. For example, selecting boars from the top 2.5% of litters and gilts from the top 25%, assuming a heritability of 10%, a phenotypic standard deviation of 2.8 pigs, and a generation interval of one year, a predicted annual rate of progress of 0.25 piglets/litter would result. This theoretical rate can be improved in several ways such as keeping sows for two or more litters and selecting on their total record, and including litter records on full- and half-sisters (i.e. the young pigs' aunts) in a selection index, which can double the predicted rate of improvement. Even so, the economic benefit from selecting on growth and carcass traits alone is likely to be four or more times greater.

A problem is that selection between litters, particularly when it includes records on relatives, leads to much more rapid rates of inbreeding than selection on individual performance. To overcome this, the size of herd in which selection takes place would have to be rather large. Progeny testing programmes are also restricted by inbreeding, but in any event are likely to be less effective than programmes using contemporary relatives' records, since the interval between generations would be longer.

A method of obtaining an initial boost in litter size is to screen the national herd for sows which have proved highly prolific over several litters and bring these together to found a 'hyperprolific' population by mating the sows themselves to the sons of others in the same group. It should, in theory, be feasible to gain about one pig/litter in this way, but the wider the screening net, the lower the average genetic merit of the sows for growth and carcass traits. This approach has been tried in France (Legault and Gruand, 1976), where an improvement of 2.6 eggs in ovulation rate in a Large White hyperprolific line was offset by a 69% increase in embryonic mortality to give no change in size of first litters at birth. However, evidence is now emerging (Legault, Gruand and Bolet, 1981) that the superior ovulation rate of the hyperprolific line may be expressed as a significant increase in litter size when measured in the improved uterine environment either of crossbred or second litter, purebred, hyperprolific sows. In Norway, national Landrace litter records were screened to give founder hyperprolific dams with a phenotypic superiority of 4.0 piglets born alive over their first three litters (Skjervold, 1979), but no estimate of the genetic gain has yet been published.

Future developments

AREAS FOR RESEARCH

It was shown earlier (*Table 26.8*) that the available evidence suggests that genetic associations between reproductive performance and growth and carcass traits are weak. Continued intense selection from traits of the growing pig is not therefore expected to lead to substantial changes in reproductive rate. However, the studies were rather inconclusive and data were collected over 10 years ago, so genetic trends in reproduction should continue to be monitored in national populations. Ideally, proper experiments should be designed and set up to measure the correlated responses in reproductive and fitness traits resulting from direct selection for growth and carcass traits. Such experiments would be large and expensive but could, perhaps, be combined with planned or existing selection studies on growth and efficiency. In their absence, the maintenance of unselected control populations (e.g. Smith, 1977) should give warning of any adverse trends in reproductive traits.

One of the consequences of present selection programmes may be an increase in the mature size of the sow, as a result of the reduction in the age and fatness of pigs at a fixed slaughter weight. In turn this could have the effect of increasing sow food costs and possibly raising age at sexual maturity. Other reproductive traits such as length of productive life, ease of mating, conception rate and teat number could also be influenced. In a selection programme for growth, it could be possible to select unintentionally for high or low litter size if maternal effects were not properly eliminated, for example if pigs in smaller litters grew faster. Recent evidence has come to light of small negative selection pressures being applied to litter size in some British nucleus populations (Guy and Steane, 1978), and of selection differentials below the values expected from natural selection in Norwegian Landrance nucleus herds (Skjervold, 1979).

Although selection for improved litter size would have been expected to produce only a slow improvement in the past, two new techniques which could lead to faster rates of improvement will deserve further investigation. The first, simple standardization of litter size to, say, eight pigs/litter, would remove the possible negative maternal correlation between successive generations, and therefore increase the heritability. At present it is unclear just how much the heritability would be expected to increase and, although selection experiments using the technique are under way in American breeds, it will be important that they are repeated in European breeds and under management conditions with larger average litter sizes.

The second new technique deserving continued attention will be the measurement of ovulation rate by laparoscopy. Genetic changes in ovulation rate have so far been accompanied by very small changes in litter size, presumably as a result of the large negative correlation between ovulation rate and embryonic survival. The realized genetic association has been observed mainly in first parity purebred gilts, however, and it is at least possible that improvements in ovulation rate may be better able to be expressed as extra piglets in the improved uterine environment either of a crossbred or of a second or later parity purebred sow (Legault, Gruand and Bolet, 1981). If this is so then selection on some combination of ovulation rate and litter size could prove to be more efficient than selection on litter size alone. In any case further investigation of the genetics and physiology of embryonic survival as the limiting factor would seem to be indicated. In particular, it would be useful to find some method of predicting genetic potential for embryo survival, either soon after first mating or preferably even before puberty.

One of the most neglected aspects of genetic improvement in reproductive potential has been the value of the crossbred boar in libido and semen quality, as affecting conception rates and litter sizes of his mates. A switch to crossbred rather than purebred boars would be an operationally simple, though costly step, and experiments are needed to show whether sufficient heterosis exists in male reproductive characteristics to make this worthwhile. Another subject for investigation will be the inheritance of freezability of boar semen, since the semen from roughly 25% of British boars cannot be frozen (H.C.B. Reed, personal communication). In storing frozen semen for experimental or commercial purposes, any adverse association between freezability and any other aspect of performance would be highly undesirable.

Comparisons of the performance of reciprocal crosses (Johnson, Omtvedt and Walters, 1973) and of gilts reared in litters of different sizes (Nelson and Robison, 1976) have shown that the 'maternal environment' can influence growth, carcass and reproductive traits. The control of the maternal environment through genetic, nutritional or managemental measures could therefore greatly affect the profitability of pig production in the future. New techniques such as ovum transplantation should allow, for example, separation of effects resulting from the pre- and postnatal environments, and a further understanding of the factors which can influence subsequent lifetime performance.

SELECTION PROGRAMMES

How much emphasis should practical pig breeders place on reproductive performance in present and future genetic improvement programmes? The answer to this question depends on the relative economic benefits of improvement in reproductive versus growth and carcass traits, and on the methods available for predicting an individual pig's genetic potential for reproduction. For the present, selection for reproductive performance would not be justified in terminal sire breeds, since their reproductive rate has little influence on total profitability. Any selection should be confined to the breeds used as constituents of the dam of the slaughter generation, and the emphasis given to reproduction can be greater in these breeds if a terminal sire breed is used than if a backcrossing scheme is used. With present economic values and low heritabilities for litter productivity, there are only small benefits from inclusion of reproduction in a selection index, particularly for the European Large White and Landrace breeds in which litter sizes are already relatively high. More selection for reproduction is justified in the USA where average litter sizes appear to be lower (*Table 26.1*), and feed costs and carcass grading differentials are lower (Clarke and Smith, 1979).

In the future, as pigs become leaner and variation in fat declines, the relative economic value of a genetic change in all aspects of reproduction will increase. There are already indications that the fat cover on entire male carcasses is approaching an optimum for some processing requirements. In addition, the relative value of an extra live piglet at birth might be greatly increased by the introduction of successful artificial rearing systems. Similarly the relative importance of libido in both sexes might be increased by the need to save labour at mating. The new methods now being investigated for improvement of litter size, such as standardization of litter size at birth or direct measurement of ovulation rate, could raise the effective heritability of the trait to the point where selection would be worthwhile even at present economic values.

It therefore seems likely that the importance of reproduction in genetic improvement programmes will increase in the medium or long term. If so, the challenge to find inexpensive and accurate methods of predicting reproductive potential in the young pig will be greater than ever.

References

ANDRESEN, E. (1979). Evidence indicating the sequence *Phi, Hal, H* of three closely linked loci in pigs. *Nord. VetMed.* **31**, 443–444

BAKKER, H., WALLINGA, J.H. and POLITIEK, R.D. (1976). Reproduction and body weight of mice after longterm selection for litter size. *Proc. 27th Ann. Mtg. EAAP, Zurich*

BATEMAN, N. (1966). Ovulation and post-ovulational losses in strains of mice selected from large and small litters. *Genet. Res.* **8**, 229–241

BERESKIN, B., SHELBY, C.E. and COX, D.F. (1973). Some factors affecting pig survival. *J. Anim. Sci.* **36**, 321–327

BRADFORD, G.E. (1979). Genetic variation in prenatal survival and litter size. *J. Anim. Sci.* **49**, *Supp. II*, 66–74

BRASCAMP, E.W. and BUITING, G.A.J. (1980). Preliminary results of Duroc as maternal grandsire of fattening pigs. *Proc. 31st Ann. Mtg. EAAP, Munich*

BRASCAMP, E.W., COP, W.A.G. and BUITING, G.A.J. (1979). Evaluation of six lines of pigs for crossing. 1. Reproduction and fattening in pure breeding. *Z. Tierzücht. ZüchtBiol.* **96**, 160–169

CLARKE, P.K. and SMITH, C. (1979). The value of litter size as an objective. *Proc. European Pig Testing Conf., Harrogate*

CUNNINGHAM, P.J., ENGLAND, M.E., YOUNG, L.D. and ZIMMERMAN, D.R. (1979). Selection for ovulation rate in swine : correlated response in litter size and weight. *J. Anim. Sci.* **48**, 509–516

EIKJE, E.D. (1973). [Phenotypic and genetic parameters for litter size in pigs]. *Meld. Norg. LandbrHøgsk.* **53** (38), 23pp

EISEN, E.J. (1978). Single trait and antagonistic index selection for litter size and body weight in mice. *Genetics* **88**, 781–811

FALCONER, D.S. (1960a). *Introduction to Quantitative Genetics.* Edinburgh, Oliver and Boyd

FALCONER, D.S. (1960b). The genetics of litter size in mice. *J. cell. comp. Physiol.* **56**, 153–167

FALCONER, D.S. (1973). Replicated selection for body weight in mice. *Genet. Res.* **22**, 291–321

GUY, D.R. and STEANE, D.E. (1978). Correlated selection differential for litter size in Meat and Livestock Commission Pig Improvement Scheme herds. *Anim. Prod.* **26**, 372 (Abstract)

HANSET, R. (1973). [Inbreeding and relationships in the Pietrain pig]. *Ann. Génét. Sél. Anim.* **5**, 177–188

HETZER, H.O. and MILLER, R.H. (1970). Influence of selection for high and low fatness on reproductive performance of swine. *J. Anim. Sci.* **30**, 481–495

HUTCHENS, L.K., JOHNSON, R.K., WELTY, S.D. and SCHOOLEY, J. (1978). Age and weight at puberty for purebred and crossbred gilts of four breeds. *Rep. Okla. agric. Exp. Stn.* **MP-103**, 126–130

IMLAH, P. (1970). Evidence for the T_f locus being associated with an early lethal factor in a strain of pigs. *Anim. Bld. Grp. Biochem. Genet.* **1**, 5–13
ISLAM, A.B.M.M., HILL, W.G. and LAND, R.B. (1976). Ovulation rate in lines of mice selected for testis weight. *Genet. Res.* **27**, 23–32
JOAKIMSEN, Ø. and BAKER, R.L. (1977). Selection for litter size in mice. *Acta Agric. scand.* **27**, 301–318
JOHNSON, R.K., OMTVEDT, I.T. and WALTERS, L.E. (1973). Evaluation of purebreds and two breed crosses in swine : feedlot performance and carcase merit *J. Anim. Sci.* **37**, 18–26
JOHNSON, R.K., OMTVEDT, I.T. and WALTERS, L.E. (1978). Comparison of productivity and performance for two-breed and three-breed crosses in swine. *J. Anim. Sci.* **46**, 69–82
JONSSON, P. (1971). Population parameter estimates of the Danish Landrace pig. *Acta Agric. scand.* **21**, 11–16
KING, J.W.B. (1967). Pig breeding research. *Proc. 9th Int. Congr. Anim. Prod., Edinburgh,* pp. 9–16. Edinburgh, Oliver and Boyd
KING, J.W.B. (1975). Retarded growth in Hampshire piglets. *Livestock Prod. Sci.* **2**, 69–77
KING, J.W.B. (1978). *Alternative crossing systems for pigs.* ABRO Report 1978, pp. 15–18. London, Agricultural Research Council
KING, J.W.B. and THORPE, W. (1973). Experiments with Pietrain/Hampshire crossbred boars. *Proc. 24th Ann. Mtg. EAAP, Vienna*
LAND, R.B. (1973). The expression of female, sex-limited characters in the male. *Nature (Lond.)* **241**, 208–209
LAND, R.B. and FALCONER, D.S. (1969). Genetic studies of ovulation rate in the mouse. *Genet. Res.* **13**, 25–46
LANGHOLZ, H.-J. (1968). [Inbreeding and relationships in Norwegian Landrace pigs]. *Meld. Norg. LandbrHøisk.* **47**, (2), 10pp
LEGAULT, C. (1970). [Statistical and genetical study of the performance of Large White sows. II. Direct effect of the boar, heritability, repeatability, correlations]. *Ann. Génét. Sél. Anim.* **2**, 209–227
LEGAULT, C. (1971). [Relationship between reproductive performance and fattening and carcase characters in the pig]. *Ann. Génét. Sél. Anim.* **3**, 153–160
LEGAULT, C. and GRUAND, J. (1976). [Improvement of the reproductive performance of sows by creation of a "hyperprolific" line and use of artificial insemination : principle and preliminary results]. *Journées de la recherche porcine en France*, pp. 201–206. Paris, L'Institut Technique du Porc
LEGAULT, C., GRUAND, J. and BOLET, G. (1981). [Results from the purebred and crossbred use of a "hyperprolific" line]. *Journées de la recherche porcine en France*, pp. 255–260. Paris, L'Institut Technique du Porc
LISHMAN, W.B., SMITH, W.C., BICHARD, M. and THOMPSON, R. (1975). The comparative performance of purebred and crossbred boars in commercial pig production. *Anim. Prod.* **21**, 69–75
LOPEZ-FANJUL, C. (1974). Selection from crossbred populations. *Anim. Breed. Abstr.* **42**, 403–416
LUSH, J.L. and ANDERSON, A.L. (1939). A genetic history of Poland China swine. II. Founders of the breed, prominent individuals, length of generation. *J. Hered.* **30**, 219–224

MARTIN, P.A. and DZIUK, P.J. (1977). Assessment of relative fertility of males (cockerels and boars) by competitive mating. *J. Reprod. Fert.* **49**, 323–329

MIKAMI, H., FREDEEN, H.T. and SATHER, A.P. (1977). Mass selection in a pig population. 2. The effects of inbreeding within the selected populations. *Can. J. Anim. Sci.* **57**, 627–634

MOAV, R. (1966). Specialised sire and dam lines. I. Economic evaluation of crossbreds. *Anim. Prod.* **8**, 193–202

MOAV, R. and HILL, W.G. (1966). Specialised sire and dam lines. IV. Selection within lines. *Anim. Prod.* **8**, 375–390

MORRIS, C.A. (1975). Genetic relationships of reproduction with growth and with carcase traits in British pigs. *Anim. Prod.* **20**, 31–44

NEELY, J.D., JOHNSON, B.H. and ROBISON, O.W. (1980). Heterosis estimates for measures of reproductive traits in crossbred boars. *J. Anim. Sci.* **51**, 1070–1077

NELSON, R.E. and ROBISON, O.W. (1976). Effects of postnatal maternal environment on reproduction of gilts. *J. Anim. Sci.* **43**, 71–77

OLLIVIER, L. and BOLET, G. (1981). [Selection for prolificacy in the pig : results after ten generations of selection]. *Journées de la recherche porcine en France*, pp. 261–268. Paris, L'Institut Technique du Porc

QUINTANA, F.G. (1979). Crossbreeding in swine, an evaluation of systems. PhD Thesis. North Carolina State University

RASMUSEN, B.A. and HAGEN, K.L. (1973). The H blood-group system and reproduction in pigs. *J. Anim. Sci.* **37**, 568–573

REVELLE, T.J. and ROBISON, O.W. (1973). An explanation for the low heritability of litter size in swine. *J. Anim. Sci.* **37**, 668–675

RUTLEDGE, J.J. (1980). Fraternity size and swine reproduction. 1. Effect on fecundity of gilts. *J. Anim. Sci.* **51**, 868–870

SCHNEIDER, A., SCHMIDLIN, J. and GERWIG, C. (1980). Investigation of reproductive performances of purebred and crossbred sows of a pig production organisation. *Proc. 31st Ann. Mtg. EAAP, Munich*

SELLIER, P. (1970). [Heterosis and crossbreeding in swine]. *Ann. Génét. Sél. Anim.* **2**, 145–207

SELLIER, P. (1976). The basis of crossbreeding in pigs; a review. *Livestock Prod. Sci.* **3**, 203–226

SHERIDAN, A.K. (1981). Crossbreeding and heterosis. *Anim. Breed. Abstr.* **49**, 131–144

SKJERVOLD, H. (1979). What about the genetic improvement of litter size? *Acta Agric. scand. Supplement* **21**, 176–184

SMITH, C. (1964). The use of specialised sire and dam lines in selection for meat production. *Anim. Prod.* **6**, 337–344

SMITH, C. (1977). Use of stored frozen semen and embryos to measure genetic trends in farm livestock. *Z. Tierzücht ZuchtBiol.* **94**, 119–127

SMITH, C., JORDAN, C.H.C., STEANE, D.E. and SWEENEY, M.B. (1978). A note on inbreeding and genetic relationships among British tested pigs. *Anim. Prod.* **27**, 125–128

SMITH, C. and WEBB, A.J. (1980). Effects of major genes on animal breeding strategies. *Proc. 31st Ann. Mtg. EAAP, Munich*

STRANG, G.S. (1970). Litter productivity in Large White pigs. 1. The relative importance of some sources of variation. *Anim. Prod.* **12**, 225–233

STRANG, G.S. and KING, J.W.B. (1970). Litter productivity in Large White pigs. 2. Heritability and repeatability estimates. *Anim. Prod.* **12**, 235–243

STRANG, G.S. and SMITH, C. (1979). A note on the heritability of litter traits in pigs. *Anim. Prod.* **28**, 403–406

VANGEN, O. (1980). Studies on a two trait selection experiment in pigs. VI. Heritability estimates of reproductive traits. Influence of maternal effects. *Acta Agric. scand.* **30**, 320–326

WEBB, A.J. (1981). The halothane sensitivity test. *Proc. Symp. Porcine Stress and Meat Quality, Moss, Norway,* pp. 105–124. Ås, Norwegian Agricultural Food Research Society

WILSON, E.R. and JOHNSON, R.K. (1979). A comparison of mating systems utilizing Duroc, Hampshire and Yorkshire breeds for swine production. *Rep. Okla. agric. Exp. Stn.* **MP-104**, 178–183

WILSON, E.R., JOHNSON, R.K. and WETTEMAN, R.P. (1977). Reproductive and testicular characteristics of purebred and crossbred boars. *J. Anim. Sci.* **44**, 939–947

YOUNG, L.D., JOHNSON, R.K. and OMTVEDT, I.T. (1976). Reproductive performance of swine bred to produce purebred and two-breed cross litters. *J. Anim. Sci.* **42**, 1133–1149

27

INVESTIGATION AND CONTROL OF REPRODUCTIVE DISORDERS IN THE BREEDING HERD

A.E. WRATHALL
Central Veterinary Laboratory, Ministry of Agriculture, Fisheries and Food, Weybridge, Surrey, UK

At a national level there are unceasing efforts through research, technological innovation and education of producers to raise productivity of the pig industry to levels above the contemporary norm. Within this pace-setting framework the individual pig producer must set his own herd performance standard or target, and then try to maintain herd output as close as possible to that level with the greatest efficiency.

Performance standards in commercial herds include both physical and financial ones. However the latter, because they are so prone to market fluctuations beyond the producer's control, are less useful as criteria of efficiency. Some physical criteria which are extremely precise e.g. feed used per kilogram of pigmeat produced, or weaners (of defined age) per sow *place* per year, do emphasize important aspects of herd management and are dealt with in other chapters. In this Chapter attention is confined to the area of reproductive failure, i.e. failure by the breeding herd to reproduce adequate numbers of healthy newborn piglets within a specified time period. The term 'control' is used in the broad sense, i.e. any action which is taken to prevent, correct or reduce reproductive failure to a level where it no longer affects herd profitability.

It will be apparent from the foregoing remarks that, in the context of modern pig production, any distinction between control of overt diseases of reproduction (the traditional province of the veterinarian) and the control of reproductive failures arising from managemental inefficiency is very blurred. In the past the veterinary role tended to be of the 'fire-brigade' type, with intermittent responses to outbreaks of frank reproductive disease. This role is now changing rapidly with emphasis on closer and more regular involvement, not only with disease matters, but also with target setting, monitoring, stockmanship and many other aspects of herd management. Control of reproductive failure should, in fact, be seen as an integral part of the management process, the aim of which is to achieve and maintain a high level of herd health, performance and profitability. The principal steps involved in control of reproductive failure are shown in *Figure 27.1* and it is these steps which are the main objects of discussion in this chapter.

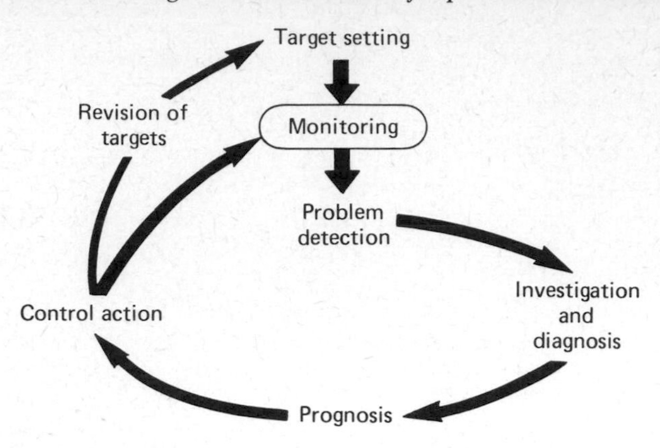

Figure 27.1 Flow diagram showing the principal steps involved in control of reproductive failure

Target setting, recording and analysis of records

The setting and periodic adjustment of herd performance targets are important preliminaries for effective control of reproductive failure. The overall target should be realistic and set with due regard to past performance of the herd as well as any aspirations based on published data from other herds of the same type. A guide to overall production standards is given in *Table 27.1*.

Although 'pigs reared per sow per unit time period' is the commonly used parameter for gauging overall herd performance, a useful criterion which relates more specifically to reproduction *per se* is 'liveborn piglets per sow per unit time' (LPSUT). Composite parameters like these are adequate by themselves only whilst the targets are being met or exceeded. When problems occur, and all herds do experience them from time to time, retrospective parameters like LPSUT are not suitable for early detection of trouble and are too general to pin-point exactly where in the reproductive cycle the failure lies. LPSUT, however, is composed of numerous sub-components, each of which contributes positively or negatively to the

Table 27.1 OVERALL REPRODUCTIVE PERFORMANCE FIGURES EXPECTED FROM AN AVERAGE HERD PRACTISING 4-WEEK WEANING AND CULLING FEMALES AFTER AN AVERAGE OF FOUR LITTERS

Output per female	*Reference figure or standard*		
	Gilts[a]	*Sows*	*All females*
Litters per year	1.89	2.34	2.23
Live piglets born per litter	10.0	11.0	10.75
Live piglets born per year	18.90	25.74	24.00
Live piglets born per month	1.57	2.14	2.00

[a]The term 'gilt' in this table refers to females (mated or unmated) which are over 6 months old but which have not yet weaned a litter. In some recording systems *all* females, once mated, are referred to as 'sows', and virgin females intended for breeding are termed 'maiden gilts'. Under the latter system the number of live piglets born per 'sow' per year would be 25.47 (2.12 per month).

Table 27.2 A GUIDE TO NORMAL REFERENCE FIGURES ('STANDARDS') FOR REPRODUCTION PARAMETERS, AND 'DECISION BOUNDARIES' ABOVE OR BELOW WHICH ACTIVE CONTROL MEASURES MAY BE NECESSARY

Reproduction parameter	*Reference figure or 'standard'*	*Decision boundary*
Age at first service	225 days	>240 days
Weaning-to-service interval	8 days	>12 days
Regular returns (21±3 days)	10%	>20%
Irregular returns (>24 days)	3%	>6%
Abortions	1%	>2.5%
Failures to farrow (detected at term)	1%	>2%
Farrowing rate	85%	80%
Piglets born alive per litter (gilts)	9.5–10.0	<9.0
Piglets born alive per litter (sows)	10.5–11.0	<10.0
Piglets born dead (stillborn)	5%	>7.5%
Piglets born mummified	1.5%	>3.0
Smaller litter index[a]	12%	>25%

[a]Small litter index = percentage of litters which contain ≤8 piglets (alive and dead).

overall figure (see column 3 of *Table 27.2*). Recording of these sub-components or derivation of their values by the analysis of records is, therefore, very important for the effective detection of reproductive failure.

The mechanics of collecting records in the piggery is a complex topic and the systems used for it vary greatly from farm to farm depending on herd size and type, producer preference and other factors. Two important prerequisites for a good recording system are the standardization of terms and definitions, and the permanent and legible identification of breeding stock.

A recent booklet published by the Agricultural Development and Advisory Service in the UK (MAFF, 1979) gives a comprehensive list of terms and emphasizes the need for standardization, especially for computer-based recording systems. The booklet also describes pig identification methods. Practical aspects of recording are also discussed by Rodeffer, Leman and Mueller (1975), Muirhead (1978a) and Sundgren *et al.* (1980).

For obvious reasons the amount and types of data which are recorded routinely, and the level of sophistication in analysing them, tend to be proportional to herd size. For example, small herds with 20–30 sows need relatively little recorded information compared with those containing 200–300 sows where it is impossible for stockmen to remember many details about the performance of individual sows. Whilst most recording systems are designed to give information about whole-herd performance there is sometimes a need for more complex systems, such as systems which can carry the records of health and diseases of individual sows and boars as well as conventional reproductive data. These more complex individual-animal record systems, as well as whole-herd systems for the very large herd, really need to be computer-based if they are to be effective. Computers may appear at first sight to be an unnecessary extravagance, but when the time and labour costs of a manual system are taken into account this may not be the case. The real costs of computer equipment ('hardware') are declining, and good programmes ('software') specifically designed for herd recording purposes are rapidly becoming available (MAFF, 1979; Pepper, Boyd and Rosenberg, 1977; Pepper, 1980; Walton, Martin and Ward, 1980; Wilson, McMillan and Swaminathan, 1980). Operating costs for computerized pig data systems in the UK are currently about £5 per month for whole-herd records, and £2–3 per sow per year for the individual-animal systems. It is not difficult to justify expenditure of this scale in many large herds because, as part of the herd health programme, it facilitates the raising of sow performance and consequently of herd profit margins.

Computer systems for herd recording utilize either the large 'mainframe' type of computer which is normally shared with many other users, and which is situated some distance away from the farm, or the 'mini' or 'micro' computer which is installed in the farm office or local veterinary practice. Transfer of data between the farm and mainframe computer can be by post, telephone, or by a terminal on the farm which is linked by telephone line directly to the computer. The postal system is, naturally, subject to delays but is less demanding in technological expertise than a terminal. The farm terminal system, however, like the mini or micro system, has the great advantage of allowing the herd manager or veterinarian to interrogate the computer himself, and to get immediate answers. Minicomputers suitable for most small businesses (like farms) now cost approximately £3000–6000, depending on requirements, and these figures do allow for necessary attachments such as a visual display unit (VDU) and a printer. Simple 'off-the-shelf' programmes may be included in the price but programmes specifically designed for analysis of pig herd records are still expensive and extra consultancy costs will probably be involved initially to ensure proper installation and operation.

An important but often overlooked component of herd recording is the inventory. To obtain this simply involves recording the numbers of adult males and females, and possibly the young stock, which are on the site at a particular moment. Ideally it is done regularly at the same time each week. Computers enable the inventory to be done extremely quickly and automatically, although sometimes a visual head count is worthwhile for validation purposes. The inventory contains essential raw data for calculating some of the important herd performance parameters, including LPSUT, and it will also show whether the farm is carrying its full complement of sows. The latter, of course, is vital if the return on capital expenditures such as buildings, land and labour, is to be maximized. If, as shown in *Table 27.3*, females in different reproductive categories are

Table 27.3 INVENTORY OF BREEDING FEMALES IN AN AVERAGE HERD PRACTISING 4-WEEK WEANING AND CULLING FEMALES AFTER AN AVERAGE OF FOUR LITTERS

	Gilts[(a)]	*Sows*	*All females*
Unserved	24%	5%	10%
Served, not yet farrowed	61%	75%	72%
Farrowed, not yet weaned	15%	18%	17%
Weaned, awaiting culling[(b)]	–	2%	1%
All classes	29%	71%	–

[(a)]Females aged over 6 months but not having weaned a litter (but see *Table 27.1*).
[(b)]Interval to culling should be minimal but market factors may make the percentage in this class higher than indicated

enumerated and categorized, the inventory will rapidly show up bottlenecks in the breeding cycle.

There are four important aspects of the actual analysis of herd records to emphasize whether it is done manually or by computer:-

(a) Frequency of analysis, when done routinely, should be inversely proportional to herd size, with at least 25 female cycles (to smooth out random variations) but not more than 100 cycles included on each occasion. For example, records from 100 sow herds ought to be analysed at least every 2–3 months, 200 sow herds every month, and herds with more than 600 sows each week.
(b) The depth or detail to which analysis is carried out depends very much on herd circumstances, including the type as well as the size of the unit, and its past performance record. For example, a nucleus breeding herd with several sire and dam lines, and a large commercial herd with a poor reproductive performance record, may both benefit from frequent and very detailed record analyses. In the latter instance, however, once the problems have been brought under control the depth of analysis may be reduced. Sometimes, when characterizing a specific problem area by data analysis, or when looking for relationships between suspected causal factors and the problem, it is helpful to display the information visually in the form of histograms or charts. An example might be breakdown of returns to service into regular and irregular ones, as shown in *Figure 27.2*. Statistical methods will also be necessary in some instances, for example when conception rates achieved by different boars have to be examined for statistically significant differences.
(c) Speed of analysis (turn-round time) should be rapid, within 3–4 days if possible. In the past some farming organizations offered a

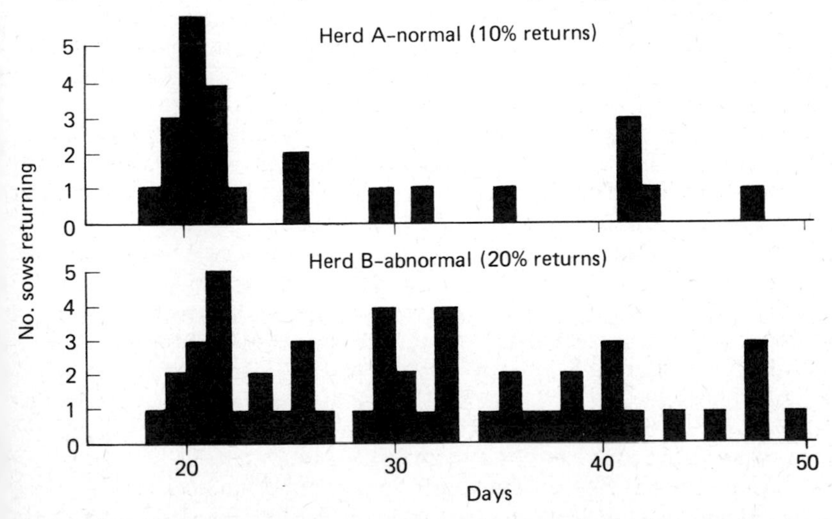

Figure 27.2 Histograms showing returns to service in two herds of comparable size. In the normal herd (A) returns occurred mainly at three weeks (19–23 days) or multiples thereof, whereas those in herd B were much more irregular. Irregular returns are often indicative of a high level of embryonic mortality (i.e. with whole litter loss)

computerized analysis service for herd records but the completion and return of results (by post) often took several weeks, by which time the value for herd health control was largely lost. It is helpful if routine veterinary visits can be arranged to coincide with arrival of the results of analysis; of course, this is no difficulty when private computing facilities are available.

(d) The manner in which the final analysis is displayed is of great importance; results should, above all, be readable and easily understood. Ideally, all the basic information relating to reproduction will be summarized on a single table, and any other data, such as those concerned with feed utilization and herd finance, will be put on other sheets. In addition to the performance parameters for the latest time period the values obtained in the previous one or two periods, together with the herd standards or targets for those parameters, should be easily visible for comparison. Some of the important parameters may also be displayed in graphic form, the graphs being prepared by hand or (as is now possible with some machines) by the computer's VDU or printer. Such graphs greatly facilitate monitoring of reproductive parameters.

Monitoring reproductive parameters

Monitoring is the regular observation of quantitative data emanating sequentially from a continuous production process. The purpose of monitoring is to detect any significant negative or undesirable deviations of the values from the predefined standard or acceptable range and, whenever such deviations occur, to carry out some form of remedial action.

In the manufacturing industries monitoring is not a new concept and is utilized routinely in many instances for quality control. In agriculture, however, the deliberate use of monitoring to supervise livestock performance is quite a recent innovation; appreciation of its value and a more scientific usage has been stimulated by growth in herd sizes and increasing remoteness between managers and their livestock. In the veterinary context effective monitoring of selected, key parameters undoubtedly constitutes a very logical basis for herd health control.

Probably the simplest and most useful monitoring system for use in the pig breeding herd involves use of control charts of the 'Shewhart' type (Shewhart, 1931). These consist of lengths of graph paper upon which successive results for a specific parameter are plotted against time, as they become available (see *Figure 27.3a*). Before results are actually plotted the standard or reference figure (derived from the average of at least 100 previous observations of the parameter, or from a knowledge of values in similar types of herd in the national population—see *Table 27.2*, column 2) is ruled as a horizontal line on the chart. 'Decision boundaries', sometimes referred to as 'action' or 'tolerance limits' (see *Table 27.2*, column 3), are also ruled on the paper, their purpose being to enable the observer to distinguish between significant changes in value of the monitored parameter and natural biological variations.

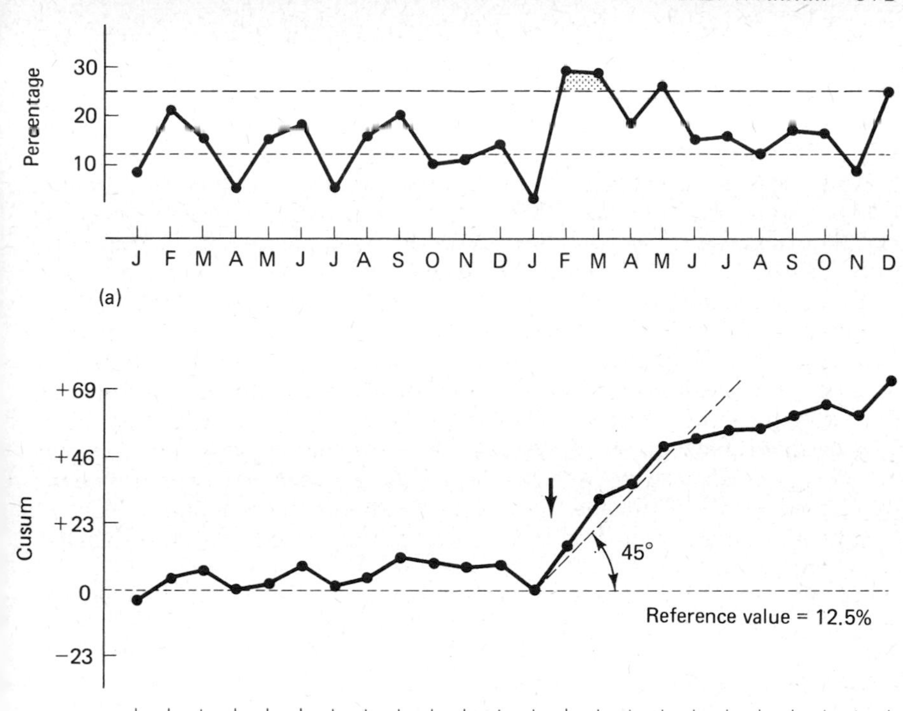

Figure 27.3 Examples of control charts used for monitoring reproductive parameters. (a) Shewhart chart showing a plot against time of the small litter index (litters with ⩽ 8 piglets born alive and dead). ---- Decision boundary; .... Reference value. Note that the values in Feb–March 1977 fall slightly outside the decision boundary. (b) Cusum chart showing a plot of the cumulative sums of the small litter index. Note how the plot changes direction sharply in February 1977 and continues through the year indicating that a new trend, i.e. towards an increased percentage of smaller litters, has occurred. Because the change in direction, at least for the first four months, exceeds 45°, it is probably a highly significant one.

Decision boundaries may be set empirically, i.e. based on past experience, or they can be derived mathematically. The latter method can, however, be rather a complex business (see Page, 1961; Pepper, 1980; Sard, 1979). It requires a knowledge, firstly, of the standard deviation (SD) of the reference figure, and secondly, because averages of samples (e.g. those of monthly batches of sows which get mated or farrow) are scattered less widely about the overall population mean than are the individual values, the average sample size (n) must be taken into account as well. A mathematically-derived decision boundary is, in fact, usually based on the standard error of the mean (SEM) of the average sample i.e.

$$\text{SEM} = \text{SD}/\sqrt{n}$$

An important question to be asked when setting decision boundaries is "how often are we prepared to investigate false alarms?". If decision boundaries based on two SEMs are used then, on average, we must expect a false alarm (with values falling on only one side—the 'problem' side—of the mean) once in every 40 times. For important problems, such as abortions or returns to service, it might be sensible to set decision boundaries closer to the reference figure than this, e.g. at one SEM, in which case an unnecessary alarm might occur once in every six occasions. In practice, although knowledge of the SEM is a very valuable guide, rigid adherence to mathematical boundaries is not always wise. For one thing the size of the 'samples' (i.e. batch sizes) in a breeding herd varies from week to week and month to month, so an SEM-based boundary is seldom completely valid. Adjustment can, of course, be made according to experience and according to the sensitivity required for that particular parameter. Sometimes it is also useful to set 'warning limits' inside the decision boundaries on the Shewhart chart, in which case a warning on two or three consecutive occasions should be taken to constitute an alarm.

A method often used by farmers and others for 'smoothing out' unwanted random variations on control chart graphs is the 'moving' or 'rolling average', in which the values for the latest batch are, before averaging, weighted by adding values for a number of previous batches as well. The number of previous batches included in the rolling average varies but often 3, 6 or 12 consecutive months are taken. Although of some help, particularly for monitoring trends, rolling averages are inevitably retrospective, insensitive and rather cumbersome.

Whereas traditional Shewhart control charts are very effective for detecting large, abrupt changes in the value of a reproductive parameter, they are less so when changes are small, resulting from 'drift' in the value over a period of time. An alternative and more sensitive method for detecting such trends is the cumulative sum technique, and control charts incorporating this method are known as 'Cusum charts' (see *Figure 27.3b*). In practice the operation of Cusum charts is similar to that for Shewhart charts, the main difference being in the criteria for taking action. Essentially they consist of a continual plot against time of deviations of the cumulative value of the monitored parameter from its reference value (k), the latter having been previously established in a manner similar to that used for the Shewhart chart. In the example shown in *Figure 27.3b* the reference value (k) has been set at 12.5%. As the values (x_1, x_2 and so on) for the batches in each successive time period become known the cumulative sums (C_1, C_2 and so on) are calculated as follows:

Period 1 $C_1 = x_1 - \mathrm{k}$
Period 2 $C_2 = (x_1 - \mathrm{k}) + (x_2 - \mathrm{k}) = C_1 + (x_2 - \mathrm{k})$
Period 3 $C_3 = C_2 + (x_3 - \mathrm{k})$, and so on.

On the Cusum chart it is changes in direction and slope of the plotted line which are important, rather than the position of the Cusum value above or below the reference value. In fact, even a small, non-significant but consistent drift in the values of x may mean that the plotted C line strays far from the zero point, and eventually falls off the graph paper. The average slope of the line over a succession of plotted Cusum values is therefore, very important, because any significant, persisting change in the mean value of the parameter under observation in relation to the set reference value (irrespective of the current position of *C* on the chart) will appear as a significant change in the direction of the plot. Plateaux, on the other hand, indicate periods of stability.

The scale of the vertical in relation to the horizontal (time) axis on a Cusum chart is very important because it determines at what angle a change in slope of the Cusum plot is a significant one. This question is discussed in depth elsewhere (Page, 1961; Sard, 1979; Davies and Goldsmith, 1972) but basically, as with the Shewhart chart, scaling should be done with reference to the SEM (i.e. SD of $k/\sqrt{n}$). If, in drawing the Cusum chart, one time period equals one unit on the horizontal axis, and two SEMs of Cusum equal one unit on the vertical axis, then any change of 45° or more will be highly significant, false alarms only occurring once in every 40 times. In practice smaller angle changes would probably be of considerable interest to the manager as well, and sophisticated methods (such as use of perspex V-masks) for gauging importance of various angle changes are described in the references cited above.

A general question which needs careful consideration in relation to monitoring is which of the many possible reproductive parameters should be monitored by control chart procedures. Some lend themselves to it readily. Total litter size (piglets born alive and dead) is one obvious example, but inevitably it is retrospective, depending largely on numbers of ovulations and embryo deaths, both of which may have occurred months before the litter is born. By the time significantly low litter size values, especially short-term, non-repetitive ones (such as those caused by parvovirus-induced embryonic death) appear on the control chart it may be too late to take any effective action. Parameters more closely related to the actual determinative event include puberty (age at first mating), weaning-to-service interval, abortions and stillbirths. Stillbirths may include some early neonatal deaths if farrowings are unattended, and this should be borne in mind when monitoring because the mechanisms of loss are often different. Although it is one of the more retrospective parameters, monitoring the small litter index (see *Table 27.2* and *Figure 27.3*) is sometimes useful to gauge specifically litter size variability.

Investigation and diagnosis of reproductive disorders

Although monitoring is the key to detection of reproductive failure it does not, in itself, identify causes or bring about solutions. It must be considered, therefore, how a system to carry out these functions (when it is necessary) can be integrated into the herd management operations.

The veterinary role in target setting and monitoring has already been discussed and the logical extension to this is for services of the same veterinarian to be used routinely for investigation and problem solving as well. This not only enables reproductive and other disease problems to be dealt with promptly but it also provides a powerful means for overall improvement of herd health, performance and profitability. It does not, of course, eliminate the need for the services of specialists in other disciplines (e.g. nutrition, genetics, ventilation), or for occasional emergency veterinary visits. By virtue of training and historical association with the farm the practising veterinarian is usually the best qualified to give broad advice on matters of health and disease; he may have undergone species specialization and further training for this. Obviously such a service is not cheap and must give value for money, but studies on the economic effects of disease, including reproductive disorders, in pig herds, and on the benefits of effective control measures leave no doubt that veterinary involvement of the right calibre can be financially beneficial to most pig farmers (Ellis and James, 1979; Govier, 1978; Muirhead, 1978b, 1980).

HERD VISITS AND CLINICAL EXAMINATIONS

Veterinary involvement includes routine herd visits, the frequency and duration of these depending on herd type, size and expected economic benefit. Ideally visits should be timed to coincide with arrival of the herd record analysis and up-dating of the control charts. Muirhead (1980) gives detailed recommendations for the conduct of such visits, and his paper also provides examples of check lists (to ensure that important points are not overlooked), advice on educational programmes for the stockmen, and guidance on preparing written reports to maximize the impact of the visit.

Clinical signs of ill health include changes in appetite, body temperature, pulse rate and so on, but conventional manifestations like these are seldom apparent in the case of reproductive disorders; this is one reason why monitoring reproductive parameters is so important. If temporal correlations between changes in the various monitored parameters are evident they may provide valuable clues as to the nature of the problem (Wrathall, 1977). However, examination of the animals themselves and their environment is usually a necessary requisite for diagnosis once a problem has been detected.

A systematic tour of the herd, commencing with maiden gilts and recently weaned sows, and progressing stage by stage to lactating sows with their litters, is often helpful because problems manifesting at one stage of reproduction often have their origins in other stages. For example, inadequate feed levels during pregnancy and/or lactation in early parities (particularly the first) may lead to anoestrus or low conception rate in future cycles (Elsley, 1971; Akkermans, 1980). Recently attention has focused increasingly on changes in weight, body condition score, and fatness of sows because depletion of body fat in successive reproductive cycles can have important adverse effects on long-term reproductive performance (MAFF, 1978; Elsley, 1972; Deering, 1977; Whittemore, Franklin and Pearce, 1980). The problem of low energy reserves in sows may have been aggravated by progress over recent years in selecting for leanness in growing pigs; it probably means that more gilts now enter their first pregnancy with low levels of body fat, levels which are insufficient to tide them through several reproductive cycles, especially if they are in negative energy balance for extended periods.

Negative energy balance can arise not only from inadequate dietary energy but also from adverse climatic conditions, lack of bedding and immobility (e.g. confinement in stalls). Parasites (e.g. mange mites and helminth infestations) can exacerbate the effects. In addition to body condition and nutrition of the animals, therefore, an appraisal of environmental influences should be made; they include climatic and social factors and the construction and layout of buildings. High ambient temperatures, for example in summer, can seriously depress fertility in both boars and sows (Tomes and Nielsen, 1979; Wetteman, Wells and Johnson, 1979; Wrathall, 1975). Reproductive photoperiodicity associated with declining daylength is also being recognized increasingly as being responsible for depressed reproductive performance in the autumn (Hurtgen and Leman, 1980; 1981; Karlberg, 1980; Stork, 1979).

Social factors include the effects of pheromones and other male sexual stimuli on the onset of puberty, weaning-to-oestrus interval, intensity of oestrus, and conception rate in females (Brooks, 1978; Signoret, 1971; Kirkwood and Hughes, 1979; Hillyer, 1976; Hemsworth, Beilharz and Findlay, 1978). It is important, therefore, to assess how freely and effectively these stimuli operate in the herd. Recent work also suggests that the social environment during the rearing period as well as after puberty, is an influential determinant of normal sexual behaviour in boars (Hemsworth, Findlay and Beilharz, 1978; Hemsworth, 1980).

Building structure and design can have multiple influences on reproductive performance. They are sometimes effected via the climate or social environment, sometimes by effects on hygiene and spread of infections and sometimes by affecting the ease with which animals can be handled and supervised by the stockman. Matters of some concern, too, are the effects on sow performance of group size (e.g. at weaning) and of confining sows in stalls. The latter, in particular, is prone to depress manifestation of oestrus, making detection of sows which fail to conceive rather difficult (Jensen *et al.*, 1970; England and Spurr, 1969; Laird and Walker-Love, 1972; Baxter, 1978).

The herd inspection provides opportunity for dialogue between the veterinarian and stockman regarding routine matters such as oestrus detection, feeding patterns, pregnancy diagnosis and farrowing procedures. Where a definite problem has been identified, however, attention has to be focused specifically onto that stage of reproduction in which the problem is manifested. For example, if there is an unacceptably high rate of regular, 3-week returns to service then a particularly careful examination of boars and mating procedures will be necessary. Similarly if the total litter size is normal but stillbirth rate is above the decision boundary then attention will focus on health and management of sows in late pregnancy and at farrowing.

In contrast to most other domestic animals special gynaecological examinations are used little in pigs, though they can be of value, particularly for older sows where the pelvic diameter is large enough for manual palpation of the genital organs *per rectum*. Ovarian abnormalities and pregnancy can be diagnosed in this way (Cameron, 1977; Meredith, 1977), and use of a speculum enables the vaginal lumen and cervix to be examined for inflammation, discharges, etc. Ultrasonic apparatus is quite widely used now for pregnancy diagnosis in pigs, and with Doppler types it is sometimes possible to make an assessment of foetal viability as well (Fraser, Nagaratnam and Callicott, 1971).

PATHOLOGICAL EXAMINATIONS

There has been an unfortunate tendency to overlook pathology in the investigation of reproductive disorders. However, because it is concerned with detecting and describing disease lesions and with elucidating mechanisms (pathogenesis) of disease, it is a valuable discipline particularly for discerning links between causal factors and the presenting problems (Hall, 1976). In addition to the classical techniques of gross and histopathology used to study morphological lesions it is important to remember the very wide and increasing array of biochemical and endocrinological methods by which the functional pathology of reproduction can now be investigated.

Obviously selection of tissues or fluids for pathology must be guided by information gained at earlier stages of the investigation. If, for example, mated sows are returning to oestrus irregularly, and some actually reach full term but fail to farrow, then pathological (as well as clinical) investigations could be useful to elucidate, from among the following possibilities, what type(s) of problem it really is:

(a) failure by the stockman to observe sows in oestrus
(b) 'silent' (undetectable) oestrous cycles
(c) true anoestrus, with quiescent ovaries
(d) undetected abortions
(e) cystic ovarian disease
(f) embryonic death
  (i) early death of the whole litter with irregular return or pseudopregnancy
  (ii) late death of the whole litter and retention of mummified foetuses *in utero*.

The pathological component of an investigation of this type of problem might include blood hormone assays as well as post-mortem studies of affected sows (Williamson, Hennessy and Cutler, 1980).

With regard to pregnancy disorders, investigations may be hindered by the fact that many disease processes go on *in utero* without visible external manifestations until, perhaps several weeks later, the diseased litter is aborted or farrowed. Pathological lesions of embryos and foetuses tend, moreover, to be rather different to those of postnatal life, and sequelae to death of the conceptus also vary considerably according to the stage of development (Wrathall, 1980a). Radiography is a useful technique for ascertaining age of death and skeletal lesions which preceded death in mummified and aborted foetuses. Where several foetuses are available from a litter it may also be possible to differentiate between simultaneous disease insults (e.g. maternal fever or transplacental toxic effects) and sequential ones where the disease spreads from foetus to foetus *in utero*, e.g. intrauterine virus infections (Wrathall, Bailey and Hebert, 1974).

Aborted foetuses are sometimes expelled without any obvious abnormalities other than subcutaneous congestion and haemolysis. On other occasions, however, examination shows that foetal disease and death preceded the abortion. Careful pathological study may, therefore, enable a distinction to be made between abortions of the maternal failure type and those of the embryonic failure type (Wrathall, 1977; 1980b). When freshly dead foetuses, stillbirths, or early neonatal deaths constitute a major component of a problem these may be subjected to more thorough examinations. First, gestational age should be established (from the service date), then body size and possibly organ weights may be taken for comparison with the appropriate standards for normal foetuses or piglets of the same gestational age. Such standards are especially valuable in foetal pathology because abnormalities of growth are relatively common in prenatal diseases (Langley, 1971; Wrathall, 1977).

LABORATORY TESTS

Innumerable special tests are applicable to reproductive failure investigations, but it would be inappropriate to discuss them in detail here. Their main purpose is to finally identify specific causal agents or aetiological factors, and they should, ideally, be used only when, on the basis of previous investigations, a direct causal relationship is strongly suspected, or when active measures to control a problem would be unjustified without greater certainty as to its aetiology. It will be clear, therefore, that laboratory tests are not invariably necessary (or appropriate) for the control of reproductive disorders. For example, when climatic or social factors are suspected, valid confirmation may only come after corrective action has been applied and a satisfactory response obtained. Too much reliance on specific tests, e.g. those for infectious organisms, toxins or nutrient deficiencies, is to be avoided because even a positive result may actually be the wrong answer as far as the real herd problem is concerned. This is especially liable to happen if preliminary investigations have been too superficial.

Prognosis

Before deciding upon active control measures it is important first to make a forecast as to the probable duration and chances of recurrence of the problem, and also the degree to which overall herd performance is liable to be affected in the future. Of course such forecasts do rely heavily upon a knowledge of the performance deficits so far, and also upon a correct diagnosis. However, assuming these, the prognosis should ideally be converted into monetary terms to show how the problem might affect future profitability. This can be done relatively simply in terms of 'opportunity piglet' values ('lost opportunity of production'—Govier, 1978) which assumes that overheads are approximately the same however many piglets are produced, i.e.

> Cost of problem = deficit of liveborn piglets × value of a piglet (about £20 at the time of writing).

Although such forecasts inevitably entail some guesswork, they do provide a good foundation for deciding upon the right kind of control action to take.

Table 27.4 MEASURES FOR CONTROL OF REPRODUCTIVE FAILURE

Causal factors	*Control measures*		
	Prevention	*Mitigation*	*Protection*
Genetic Factors			
1. Parental reproductive defects 2. Progeny developmental defects	Difficult to prevent parental or progeny defects completely.	Monitor reproductive performance and remove offenders quickly. Monitor for progeny defects (including test matings) and remove carriers.	Planned cross-breeding policy.
Nutrition			
1. Macronutrition (feed levels)	Generous maintenance rations (wasteful?) Check feed levels and nutrient content regularly.	Individual feeding/prevent competition. Avoid climatic extremes, allow free movement. Use straw bedding.	Check body condition regularly. Provide comfortable environment. Control parasites. Reduce stressors.
2. Micronutrients (vitamins, minerals, etc.)	Keep above recommended levels for all known micronutrients.	Vary sources of ingredients. Don't overheat feed.	
Environment			
1. Climate (temperature, light etc) 2. Social influences 3. Structure (building design, confinement, etc)	Difficult to prevent completely but design of buildings is fundamental.	Managerial vigilance. Flexible housing structure and layout. Ventilation. Straw bedding. Facilitate necessary social interactions.	Adequate nutrition plus body condition checks. Use genetically adapted breeds/strains of pig.

Toxic substances			
1. Toxic food ingredients and contaminants (e.g. mycotoxins)	Precautions re food ingredients and storage.	Avoid exposure to possible toxic factors at highly vulnerable stages of breeding cycle.	Seldom possible to promote any resistance.
2. Chemicals (e.g. pesticides, preservatives, gases)	Precautions re chemical hazards in environment.		
3. Drugs (e.g. side effects of hormones, vaccines)	Precautions re use of drugs (especially hormones)		
Infections			
1. Commensals—opportunist pathogens (mainly bacteria)	Rarely possible to prevent completely.	Hygiene, disinfection, strategic use of antibiotics	Boost immunity by exposure to infection at times of low risk (some vaccines, e.g. erysipelas). Use resistant breeds/strains of pig. Maintain body condition.
2. Common contagious transplacental pathogens (viruses)	Keep self-contained herd and maintain strict disease security. New breeding stock by hysterectomy, AI, or by embryo transplant.	Difficult and may be inadvisable except for very vulnerable stages (e.g. early pregnancy) where isolation, hygiene and disinfection may help.	Promote immunity by exposure to virus at times of low vulnerability: Use virus excretors, contaminated environment, or infected foetuses and placenta. Vaccines?
3. Specific reproductive pathogens (e.g. *Brucella suis, Leptospira* and many others)	Eradicate (national policy) or eliminate from herd by testing and culling. Strict disease security, testing of new entrants and quarantine.	Seldom practical in long term. In short term hygiene, antibiotic treatments, culling of clinical cases and special measures for individual diseases may help.	Vaccination or exposure to infection at times of low vulnerability. Use resistant breeds/strains of pig.

Principles for the application of control measures

Measures used to control reproductive disorders fall into three main categories:

(a) Prevention: elimination of the causal factor from the herd, or prevention of its entry.
(b) Mitigation: reduction of the amount (weight) of the causal factor which is encountered by vulnerable animals in the herd.
(c) Protection: promotion of resistance to the causal factor in vulnerable animals.

In view of the great multiplicity of causal factors and types of control action for reproductive disorders, it would be inappropriate to discuss these in detail here. *Table 27.4* attempts to set out in a simplified manner some of the measures which may be used for control once the aetiology has been established. More detailed coverage may be found in other texts (e.g. Dunne and Leman, 1975; English, Smith and MacLean, 1977; Wrathall, 1975).

It is worth mentioning that positive control action is not invariably needed. For example, with certain infectious disorders which act early in pregnancy and lead to early embryonic death, the problem may not become apparent for many weeks and by the time it is detected a strong immunity may have already developed in all the adult breeding stock. If serological tests confirm this then further action may be superfluous. Active control measures may be ruled out in other instances for economic reasons, the cost of applying them being liable to exceed the effect the problem is having on profitability. Ultimately, of course, this concept applies to every aspect of control, whether it is specific action for a particular disorder or the overall attempts by management to improve herd performance. The aim is to achieve a right balance where the input costs of control measures approximate to, but do not exceed, the level where financial return from the action disappears. With regard to the overall control of reproductive failure, as outlined in this chapter, most herds have a long way to go before the law of diminishing returns will even start to operate.

References

AKKERMANS, J.P.W.M. (1980). Recent findings on fertility disorders in the pig. *Fortschr. Vet. Med.* **30**, 95–98

BAXTER, S.H. (1978). Some general principles of housing for reproductive efficiency in pigs. In *The Veterinary Annual*, 18th Edition. (C.S. Grunsell and F.W.G. Hill, Eds.), pp. 126–133. Bristol, John Wright and Sons

BROOKS, P.H. (1978). Early sexual maturity and mating of gilts. *ADAS Quart. Rev.* **30**, 139–152

CAMERON, R.D.A. (1977). Pregnancy diagnosis in the sow by rectal examination. *Aust. Vet. J.* **53**, 432–435

DAVIES, O.L. and GOLDSMITH, P.L. (1972). *Statistical Methods in Research and Production.* 4th Edition. London and New York, Longman Group Ltd.

DEERING, J. (1977). Condition scoring sows. *Pig. Fmg.* **25**, 57–58

DUNNE, H.W. and LEMAN, A.D. (1975). *Diseases of Swine.* 4th Edition. Ames, Iowa, The Iowa State University

ELLIS, P.R. and JAMES, A.D. (1979). The economics of animal health: (2) Economics in farm practice. *Vet. Rec.* **105**, 523–526

ELSLEY, F.W.H. (1971). Recent advances in sow nutrition and their application in practice. *ADAS Quart. Rev.* **1**, 30–38

ELSLEY, F.W.H. (1972). Some aspects of productivity in the sow. In *The Improvement of Sow Productivity* (A.S. Jones, V.R. Fowler and J.C.R. Yeats, Eds.), pp. 71–87. Aberdeen, Rowett Research Institute

ENGLAND, D.C. and SPURR, D.T. (1969). Litter size of swine confined during gestation. *J. Anim. Sci.* **28**, 220–223

ENGLISH, P.R., SMITH, W. and MACLEAN, A. (1977). *The Sow – Improving her Efficiency.* Ipswich, Farming Press Ltd.

FRASER, A.F., NAGARATNAM, V. and CALLICOTT, R.B. (1971). The comprehensive use of Doppler ultra-sound in farm animal reproduction. *Vet. Rec.* **88**, 202–205

GOVIER, R.J. (1978). The economics of pigs disease: A method of calculating the on farm cost. *Pig Vet. Soc. Proc.* **2**, 101–111

HALL, S.A. (1976). Problems, diseases and diagnoses: A personal view. *Vet. Rec.* **98**, 379–381

HEMSWORTH, P.H. (1980). The social environment and the sexual behaviour of the domestic boar. *Appl. Anim. Ethol.* **6**, 306

HEMSWORTH, P.H., BEILHARZ, R.G. and FINDLAY, J.K. (1978). The importance of the courting behaviour of the boar on the success of natural and artificial matings. *Proc. Aust. Soc. Anim. Prod.* **12**, 247

HEMSWORTH, P.H., FINDLAY, J.K. and BEILHARZ, R.G. (1978). The importance of physical contact with other pigs during rearing on the sexual behaviour of the male domestic pig. *Anim. Prod.* **27**, 201–207

HILLYER, G.M. (1976). An investigation using a synthetic porcine pheromone and the effect on days from weaning to conception. *Vet. Rec.* **98**, 93–94

HURTGEN, J.P. and LEMAN, A.D. (1980). Seasonal influence on estrous activity in sows and gilts. *J. Am. vet. med. Ass.* **176**, 119–123

HURTGEN, J.P. and LEMAN, A.D. (1981). Effect of parity and season of farrowing on the subsequent farrowing interval of sows. *Vet. Rec.* **108**, 32–34

JENSEN, A.H., YEN, J.T., GEHRING, M.M., BAKER, D.H., BECKER, D.E. and HARMON, B.G. (1970). Effects of space restriction and management on pre- and postpubertal response of female swine. *J. Anim. Sci.* **31**, 745–750

KARLBERG, K. (1980). Factors affecting postweaning oestrus in the sow. *Nord. Vet. Med.* **32**, 185–193

KIRKWOOD, R.N. and HUGHES, P.E. (1979). The influence of age at first boar contact on puberty attainment in the gilt. *Anim. Prod.* **29**, 541–548

LAIRD, R. and WALKER-LOVE, J. (1972). A comparison of the performance of sows housed in sow stalls or in yards during pregnancy. *Proc. 54th Meet. Br. Soc. Anim. Prod.*, 147

LANGLEY, F.A. (1971). The perinatal postmortem examination. *J. clin. Path.* **24**, 159–169

MEREDITH, M.J. (1977). Clinical examination of the ovaries and cervix of the sow. *Vet. Rec.* **101**, 70–74

MINISTRY OF AGRICULTURE, FISHERIES AND FOOD (1978). *Nutrient allowances for pigs.* Advisory Paper No. 7

MINISTRY OF AGRICULTURE, FISHERIES AND FOOD (1979). *Pig health and production recording.* Booklet 2075

MUIRHEAD, M.R. (1978a). Constraints on productivity in the pig herd. *Vet. Rec.* **102**, 228–231

MUIRHEAD, M.R. (1978b). The economics of pig disease: The veterinary surgeon's angle. *Pig Vet. Soc. Proc.* **2**, 113–122

MUIRHEAD, M.R. (1980). The pig advisory visit in preventive medicine. *Vet. Rec.* **106**, 170–173

PAGE, E.S. (1961). Cumulative Sum Charts. *Technometrics* **3**, 1–9

PEPPER, T.A. (1980). Observations on the use of a computer program for analysing pig breeding and performance records. *Pig Vet. Soc. Proc.* **6**, 57–63

PEPPER, T.A., BOYD, H.W. and ROSENBERG, P. (1977). Breeding record analysis in pig herds and its veterinary applications. 1: Development of a program to monitor reproductive efficiency and weaner production. *Vet. Rec.* **101**, 177–180

RODEFFER, H.E., LEMAN, A.D. and MUELLER, A.G. (1975). Development of a record system for measuring swine breeding herd efficiency. *J. Anim. Sci.* **40**, 13–18

SARD, D.M. (1979). Dealing with data: The practical use of numerical information – (14) Monitoring changes. *Vet. Rec.* **105**, 323–328

SHEWHART, W.A. (1931). *The Economic Control of the Quality of the Manufactured Product.* New York, Macmillan

SIGNORET, J.P. (1971). The reproductive behaviour of pigs in relation to fertility. *Vet. Rec.* **88**, 34–38

STORK, M.G. (1979). Seasonal reproductive inefficiency in large pig breeding units in Britain. *Vet. Rec.* **104**, 49–52

SUNDGREN, P.E. VAN MALE, J.P., AUMAITRE, A., KALM, E. and NIELSEN, H.E. (1980). Sow and litter recording procedures. Report of a working party of the EAAP Commission on pig production. *Livest. Prod. Sci.* **7**, 394–401

TOMES, G.J. and NIELSEN, H.E. (1979). Seasonal variations in the reproductive performance of sows under different climatic conditions. *Wld. Rev. Anim. Prod.* **15**, 9–19

WALTON, J.R., MARTIN, J.W. and WARD, W.R. (1980). The collection and use of data on a pig farm. *Pig Vet. Soc. Proc.* **6**, 23–28

WETTEMANN, R.P., WELLS, M.E. and JOHNSON, R.K. (1979). Reproductive characteristics of boars during and after exposure to increased ambient temperature. *J. Anim. Sci.* **49**, 1501–1505

WHITTEMORE, C.T., FRANKLIN, M.F. and PEARCE, B.S. (1980). Fat changes in breeding sows. *Anim. Prod.* **31**, 183–190

WILLIAMSON, P., HENNESSY, D.P. and CUTLER, R. (1980). The use of progesterone and oestrogen concentrations in the diagnosis of pregnancy, and in the study of seasonal infertility in sows. *Aust. J. agric. Res.* **31**, 233–238

WILSON, M.R., McMILLAN, I. and SWAMINATHAN, S.S. (1980). Computerized health monitoring in swine health management. *Pig Vet. Soc. Proc.* **6**, 64–71

WRATHALL, A.E. (1975). *Reproductive Disorders in Pigs*. Slough, UK, Commonwealth Agricultural Bureaux

WRATHALL, A.E. (1977). Reproductive failure in the pig: Diagnosis and control. *Vet. Rec.* **100**, 230–237

WRATHALL, A.E. (1980a). Mechanisms of porcine reproductive failure. In *The Veterinary Annual*, 20th Edition. (C.S.G. Grunsell and F.W.G. Hill, Eds.), pp. 265–274. Bristol, John Wright and Sons

WRATHALL, A.E. (1980b). Ovarian disorders in the sow. *Vet. Bull.* **50**, 253–272

WRATHALL, A.E., BAILEY, J. and HEBERT, C.N. (1974). A radiographic study of development of the appendicular skeleton in the foetal pig. *Res. Vet. Sci.* **17**, 154–168

28

SOCIAL ENVIRONMENT AND REPRODUCTION

P.H. HEMSWORTH
Department of Agriculture, Animal Research Institute, Werribee, Victoria, Australia

Social environment is generally considered to consist of visual, tactile, olfactory, auditory and other stimuli which together form the means of communication between individual animals of a species. The role of the social environment on reproduction of domestic animals is a relatively recent topic for research but with recent developments in animal production it is assuming greater importance. Intensification of animal production has and will continue to produce marked changes in the social environment of the animal and thus it is essential that scientists, advisers and producers are aware of its role in reproduction. In a review of reproductive disorders of pigs, Wrathall (1975) suggested that many of the disorders encountered in intensive pig production originate from neglect or ignorance of the importance of social factors. Thus, failure to provide an environment in which reproduction can proceed naturally and efficiently will result in reproductive failure. It is the object of this chapter to review the current state of knowledge of the role of the social environment on reproduction of the boar, gilt and sow.

Sexual stimuli and immediate sexual behaviour

Courtship and copulation consist of a sequence of specific responses to specific stimuli from the partners which lead to the successful insemination of the female. The stimuli involved in sexual behaviour have been classified by Schein and Hale (1965) into three functional types i.e. broadcast, identification and synchronizing stimuli. It is useful to consider briefly the sexual stimuli of the pig in these three classes.

BROADCAST STIMULI

According to Schein and Hale (1965), these serve to advertize the availability of an individual and thereby, elicit approach responses on the part of potential partners to bring them into spatial proximity. During oestrus the exploratory activity of the sow greatly increases (Signoret, 1970a,b) and sensory signals from the boar provide strong attraction for the sow (Signoret, 1972). However, in T-maze experiments boars were observed to show similar attraction to oestrous and dioestrous sows (Signoret, 1970b, 1971). Therefore, since the sow's exploratory activity appears to be largely oriented by stimuli from the boar and the boar is less sensitive to broadcast-type stimuli from the oestrous sow, the oestrous sow plays the active role in bringing the two sexual partners together.

IDENTIFICATION STIMULI

These are stimuli that signal the sexual receptivity of the individual and are released in response to cues from the potential partner (Schein and Hale, 1965). The initial courtship display of the boar is an identification stimulus and the 'standing response' or immobility response (Signoret, 1970a) by the sow reveals her receptivity. Pressure on the back, odour of the boar and frequency and rhythm of the courting grunts of the boar have been identified as the main stimuli eliciting the standing response in the sow (Signoret, 1970a,b). The important role of olfactory and auditory stimuli explains the significance of the high level of naso-nasal contact between the boar and sow during courtship (Signoret, 1970a,b).

Since the boar is attracted to and courts an oestrous or dioestrous sow in a similar manner (Signoret, 1970b), it appears that the boar detects that the sow is in oestrus by her behavioural response to his stimulation.

SYNCHRONIZING STIMULI

These govern the coordination of the motor patterns leading to the successful insemination of the female (Schein and Hale, 1965). The signals are emitted once the partner shows receptivity via the identification stimuli. The main stimulus which releases mounting behaviour in the boar appears to be the visual cue(s) from the shape and immobility of the sow diplaying the standing response (Fraser, 1968a; Signoret, 1970b). The pressure of the cervix on the glans penis is the essential stimulus for ejaculation (Walton, 1960).

In addition to the role of sexual stimuli on the achievement of insemination, these stimuli may also have an immediate influence on the partner's physiological mechanisms and influence the chances of fertilization. Fraser (1968b) has suggested that genital stimulation by the male will accelerate uterine motility and thus sperm transport in the female. Similarly, Signoret (1972) proposed that sexual stimuli from the boar during mating may influence reproductive efficiency of the sow. He has suggested that the presence of the courting boar may explain the difference in farrowing rates to natural and artificial insemination. This proposal is supported by recent observations at Werribee on the sexual behaviour of 24 commercial boars. Those which displayed a high frequency of nosing the flanks of the sow during courtship had a higher farrowing rate (percentage of mated sows that farrowed) than those of lower 'nosing activity' (Hemsworth, Beilharz and Brown, 1978). Döcke and Worch (1963) reported that the presence and behaviour of the boar during courting led to an increase in uterine contractions in the sow, and Pitkjanen (1964) has shown that the physical presence of the boar increases the transit of spermatozoa along the uterine horns. Thus increased nosing activity by the boar may stimulate uterine contractions in the sow and thereby increase sperm transport, the number of spermatozoa in the oviduct and consequently the chances of fertilization. An alternative explanation arises from the work of Signoret, du Mesnil du Buisson and Mauleon (1972) who reported that a stimulus or stimuli associated with mating advanced the timing of ovulation and possibly improved the timing of fertilization.

Sexual stimuli may also influence physiological mechanisms involved in reproduction of the boar. As with bulls, sexual stimulation before semen collection, achieved by brief restraint after mounting or observation of a mating, will increase the number of spermatozoa in the ejaculate of the boar (Hemsworth and Galloway, 1979).

Thus, sexual stimuli from the partners have an important role in the achievement of insemination of the sow and, perhaps, the success of the insemination.

Prepubertal social environment of the boar

Until recently, little was known of the influence of the social environment during rearing on reproduction of the pig. This is surprising since there is considerable variation in the amount and type of social contact that the young breeding pig may receive in modern pig production. Studies with a number of laboratory species including the guinea pig, rat and rhesus monkey, have demonstrated the importance of social contact during rearing on sexual behaviour (Young, 1957; Young, Goy and Phoenix, 1968). Males that had been socially deprived from an early age showed a serious inability to properly orient towards, mount and clasp the receptive female. Similarly, females reared with limited social contact displayed poor oestrous behaviour, as indicated by a reduction in the duration of the standing response.

In view of these reports a series of studies was conducted to investigate the influence of the prepubertal social environment on the development of the sexual behaviour of the boar (Hemsworth, Beilharz and Galloway, 1977; Hemsworth, Findlay and Beilharz, 1978; Hemsworth and Beilharz, 1979). Boars reared from 3–30 weeks of age without visual or physical contact with pigs ('social restriction') achieved fewer copulations and displayed less courting behaviour in a series of mating tests than boars reared in either an all-male or mixed sex group (*Table 28.1*). Since the mating dexterity of the socially-restricted boars appeared satisfactory, it was likely that they were of low sexual motivation. The rearing treatments had no effect on size of the testicles or semen quality.

A subsequent study indicated that lack of physical contact with pigs *per se* was predominantly responsible for the depression in the level of sexual behaviour caused by lack of both physical and visual contact with pigs during rearing (*Table 28.2*). Comparison of the sexual behaviour of boars reared in social restriction from either 3 or 12 weeks of age shows that the age of the boar when first deprived of social contact influences the extent to which the level of sexual behaviour is depressed (*Table 28.2*).

Table 28.1 THE SEXUAL BEHAVIOUR OVER 12 MATING TESTS OF BOARS REARED IN SOCIAL RESTRICTION (VISUAL AND PHYSICAL ISOLATION FROM PIGS, GROUP A). AN ALL-MALE GROUP (B) AND A MIXED SEX GROUP (C)

	Rearing treatment		
	A	*B*	*C*
Number of copulations/boar (mean)	2.1^{y}	10.6^{x}	8.5^{x}
Sum of all courting behaviour activities/boar (mean)	320.0^{y}	572.0^{x}	690.6^{x}

^{x,y}: means in the same row with a different superscript differ significantly ($P<0.01$)
From Hemsworth, Beilharz and Galloway (1977)

Table 28.2 SEXUAL BEHAVIOUR OVER 20 MATING TESTS OF BOARS REARED TO 32 WEEKS OF AGE IN SOCIAL RESTRICTION (VISUAL AND PHYSICAL ISOLATION FROM PIGS) FROM 3 WEEKS (A) AND FROM 12 WEEKS (B), PHYSICAL ISOLATION FROM PIGS FROM 3 WEEKS (C) AND IN AN ALL-MALE GROUP (D)

	Rearing treatment			
	A	*B*	*C*	*D*
Number of copulations/boar (mean)	8.8^{b}	13.8^{ab}	13.3^{ab}	24.0^{a}
Sum of all courting behaviour activities/boar (mean)	316.0^{b}	603.8^{ab}	381.0^{b}	733.0^{a}

^{a,b}: means in the same row with a different superscript differ significantly ($P<0.05$)
From Hemsworth, Findlay and Beilharz (1978)

A final study examined the effects on sexual behaviour of rearing boars in individual wiremesh pens which allowed restricted physical contact with neighbouring boars. Rearing from either 3 or 12 weeks of age in these individual pens did not adversely affect the copulatory performance of the boars (Hemsworth and Beilharz, 1979). However, boars reared in these pens from three weeks of age displayed low levels of nosing activity which is apparently an important component of courtship. As previously discussed, the level of nosing activity by the boar during courting may affect the proportion of mated females that farrow.

These studies did not investigate the influence of olfactory contact with females. According to Booth and Baldwin (1980), the removal of the olfactory lobes at 10–12 weeks of age does not significantly affect the sexual behaviour of the boar. This suggests that olfactory communication with female pigs from this early age is not critical in the development of the sexual behaviour of the boar. However, the test used to assess sexual behaviour was qualitative and hence it is unlikely that it could discriminate between boars of varying levels of sexual behaviour. The test would assess mating competency and it appears that olfactory bulbectomy does not adversely affect the ability and capacity of the boar to display the motor patterns leading to copulation.

Prepubertal social environment also appears to influence the age when the boar first displays a fully coordinated mating response to an oestrous female. Boars reared in groups displayed the mating response at an earlier age then those reared in individual pens (Thomas *et al.*, 1979). The age of attainment of the mating response was not significantly different for boars housed in groups with or without contemporary gilts. However, all boars were exposed weekly to an oestrous female, presumably from an early age, to observe the attainment of the mating response. Further, whether the boars had olfactory contact with females outside the rearing pens was not reported. Thus, contact with female pigs during rearing may have been effectively similar between the two rearing treatments.

Thus, social environment during rearing will exert a major influence on the level of sexual behaviour of the boar. Physical contact with pigs appears to be a critical aspect of the social environment in this regard. Failure to provide the young boar with physical contact with pigs will severely impair his subsequent level of sexual behaviour.

Postpubertal social environment of the boar

There is little information in the literature on the influence of the presence of the female on reproduction of the male. Observations on adult male rhesus monkeys during the non-breeding season showed that the normally sexually quiescent males can be returned to a sexually active state by exposure to oestrogen-treated females (Vandenbergh, 1969). The female rhesus monkey appears to signal her receptivity to the male by way of behavioural displays and pheromones (Michael and Zumpe, 1970; Michael *et al.*, 1972). During the breeding season rams continually exposed to sexually receptive ewes achieved a greater number of copulations in a single mating test than rams isolated from ewes (Illius, Haynes and Lamming, 1976). Recent studies at our laboratory also suggest that the presence of the female pig will influence the level of sexual behaviour of the mature boar.

Isolation of mature boars from female pigs severely reduces their level of sexual behaviour (Hemsworth *et al.*, 1977). However, housing them near to sexually receptive females successfully restores their level of sexual behaviour. A recent study revealed that the presence of sexually receptive or sexually non-receptive females was equally effective in stimulating the sexual behaviour of the mature boar (*Table 28.3*). Thus, the mature boar requires the stimulation from the presence of female pigs, regardless of their oestrous status, to maintain his potential level of sexual behaviour.

The type of stimulus (or stimuli) responsible for this effect has not been identified. However, olfactory and, to a lesser extent, auditory stimuli appear to be the most likely, since postpubertal boars housed 8 or 15 m

Table 28.3 SEXUAL BEHAVIOUR OVER 9 MATING TESTS OF MATURE BOARS HOUSED IN ISOLATION FROM FEMALES (A), NEAR SEXUALLY NON-RECEPTIVE FEMALES (B) AND NEAR SEXUALLY RECEPTIVE FEMALES (C)

	Treatment		
	A	*B*	*C*
Number of copulations/boar (mean)	4.4^{a}	6.1^{b}	6.3^{b}
Sum of individual courting behaviour activities/boar (mean)	196.4^{y}	313.7^{x}	284.3^{x}

^{a,b}: Means in same row with a different superscript differ significantly ($P<0.05$)
^{x,y}: Means in same row with a different superscript differ significantly ($P<0.01$)
From Hemsworth, Winfield and Chamley (1981)

from female pigs have little visual or gustatory contact with females but display similar levels of sexual behaviour to those housed 3 m from females (*Table 28.4*).

Table 28.4 THE SEXUAL BEHAVIOUR OVER 6 MATING TESTS OF 15 MATURE BOARS HOUSED 3, 8 AND 15 m FROM FEMALE PIGS

	Distance of boar from female pigs		
	3 m	*8 m*	*15 m*
Number of copulations/boar (mean)	5.0	5.4	5.2
Sum of individual courting behaviour activities/boar (mean)	200.4	239.4	171.4

The results of a study by Booth and Baldwin (1980) may be useful in this discussion. If an olfactory stimulus or stimuli from the female is important in maintaining the level of sexual behaviour of the mature boar, then olfactory bulbectomy, irrespective of when it occurred, would be expected to reduce the level of sexual behaviour of the mature boar. However in this study sexual behaviour was not affected. As discussed earlier, it is possible that the behavioural test used may not have been able to clearly discriminate between boars of varying levels of sexual behaviour.

Thus, as with the social environment during rearing, the social environment after puberty has a marked influence on the level of sexual behaviour of the boar. A stimulus or stimuli from the female pig appears to be necessary to maintain the potential level of sexual behaviour of the mature boar.

Prepubertal social environment of the gilt

In contrast to the boar, considerably more research has been conducted on the effect of the prepubertal social environment on reproduction of the gilt, possibly because the practical implications are more apparent. It is now well established that the introduction of a mature boar to immature gilts induces the precocious attainment of puberty (Brooks and Cole, 1970; Hughes and Cole, 1976, 1978). This male effect on puberty in the gilt is discussed in detail in Chapters 6 and 7.

Initial studies on male effect were predominantly concerned with attainment of puberty and little attention was paid to reproductive efficiency of the gilt. However, Hughes and Cole (1978) noted that gilts not previously exposed to boars exhibited agitation, an inadequate standing response and a marked reduction in mating rate (defined as the percentage of gilts exhibiting a standing heat that were successfully mated) when first introduced to a boar at their second oestrus (*Table 28.5*). Kirkwood and Hughes (1980a) also reported that gilts having had no boar contact before second oestrus and having changed pens daily from 165 days of age, showed an increased tendency not to stand for the boar, with a result that mating rate was reduced. However, the control gilts (those that did not receive boar stimulation or change pens) had a similar mating rate to those exposed to boars from 165 days of age.

Table 28.5 AGE AT PUBERTY AND MATING RATE AT SECOND OESTRUS OF CONTROL (C) GILTS, AND GILTS GIVEN OESTRADIOL BENZOATE (E) AND BOAR CONTACT (B)

	Treatment groups			
	E + B	*E*	*C + B*	*C*
Mean age at puberty (days)	153.2^{x}	160.1^{y}	159.5^{z}	176.0^{xyz}
Mating rate (%)	100	50	80	77.8
Pregnancy rate at 20 days (%)	100	100	100	85.7

^{x,y,z}: means in same row with a different superscript differ significantly ($P<0.01$)
From Hughes and Cole (1978)

Observations from a recent study at our laboratory also indicated a marked reduction in mating rate (defined as the percentage of oestruses in which a successful copulation occurred) at the pubertal and subsequent hormone-induced oestruses of gilts reared in isolation from boars (treatment groups A and B in *Table 28.6*). The main reasons for reduced mating rate were failure to stand for the boar and failure of duration of ejaculation to exceed the 1.5 minutes that has been defined as necessary for a successful copulation. At the pubertal oestrus, ovulation without overt oestrus also contributed to mating failure. In addition, gilts isolated from boars during rearing took longer to display the standing response after the boar had mounted and the gilts terminated a higher proportion of copulations in a series of mating tests. Thus, both oestrus expression and sexual receptivity of the gilt appear to be adversely affected by rearing in isolation from boars.

When gilts were reared near mature boars, those that received regular boar introduction during rearing had a higher mating rate (*Table 28.6*).

Table 28.6 OESTROUS BEHAVIOUR AT THE PUBERTAL AND THREE SUBSEQUENT INDUCED OESTRUSES OF 40 GILTS REARED FROM FOUR WEEKS OF AGE IN SOCIAL RESTRICTION (NO VISUAL OR PHYSICAL CONTACT WITH PIGS) AND ISOLATED FROM BOARS (A), A GROUP ISOLATED FROM BOARS (B), A GROUP WITHIN 3 m OF MATURE BOARS (C) AND A GROUP NEAR MATURE BOARS AND INTRODUCED TO A MATURE BOAR THREE TIMES PER WEEK (D)

	Rearing treatment groups			
	A	*B*	*C*	*D*
Pubertal oestrus				
Mating rate (%)	30.0^{x}	50.0^{y}	87.5^{z}	90.0^{z}
Matings in which the gilt terminated the copulation (%)	70.0^{z}	30.0^{y}	10.0^{x}	0.0^{x}
Pubertal and induced oestruses				
Mating rate (%)	72.5^{x}	73.7^{x}	78.9^{y}	92.5^{z}
Average time from when the boar mounted until the gilt displayed the standing response (s)	24.1^{b}	32.7^{b}	16.0^{a}	5.0^{a}
Matings in which the gilt terminated the copulation (%)	40.0^{y}	73.5^{y}	17.6^{x}	10.5^{x}

^{a,b}: Means in same row with different superscript differ significantly ($P<0.05$)
^{x,y}: Means in same row with different superscript differ significantly ($P<0.01$)
From Hemsworth, Cronin and Hansen (1982)

Increased boar contact of this group D compared with gilts housed 3 m from boars (group C) may have been responsible for this difference in mating rate and it appears that a progressive increase in mating rate was associated with the increased boar contact during rearing.

On the basis of observations on the attraction by a male and the standing response to a back pressure test of gilts reared either in the absence or presence of boars, Signoret (1970c) concluded that the sexual behaviour of the gilt was not influenced by prepubertal social environment. However, Signoret did not observe the level of the sexual response of the gilts when introduced to a boar.

Further research is needed to identify the nature of the stimulus from the mature boar which is responsible for the effect on oestrous behaviour of the gilt. The timing and duration of mature boar contact also requires investigation. It is interesting that Hughes and Kirkwood (1980) reported that the stimulus from the mature boar responsible for early attainment of puberty in the gilt is olfactory in nature and that the production of this pheromone is age dependent. Mature boar contact during rearing does not appear to significantly influence conception rate, ovulation rate and litter size of the gilt (Hughes and Cole, 1976, 1978; Hughes and Kirkwood, 1980; Kirkwood and Hughes, 1979, 1980a,b).

In comparison with group housing during rearing, the housing of gilts either in tethers or in individual pens appears to delay the onset of puberty (Jensen *et al.*, 1970; Robison, 1974; Mavrogenis and Robison, 1976). In addition, type of housing may also affect the mating rate of gilts. Gilts reared in social restriction and isolated from boars (group A) had a lower mating rate at the pubertal oestrus than gilts reared in groups but isolated from boars (group B) (*Table 28.6*). Difficulties with oestrus expression, such as a reduction in the duration of overt oestrus and even a failure to display overt oestrus, were the main reasons for the reduced mating rate in the group A gilts. Jensen *et al.* (1970) also reported that gilts reared in tethers and in groups in pens had a 68% and 92% mating rate, respectively. Further, England and Spurr (1969) reported 83% and 94% mating rates in gilts housed in individual and group pens respectively, from approximately 7.5 months of age. In both studies a higher frequency of ovulation without overt oestrus was observed in the individually-housed gilts. However, in these two studies the gilts remained on their rearing treatments during oestrus and this, rather than the rearing treatment itself, may have been responsible for the reduction in oestrus expression.

Thus, in addition to the adverse influence on the attainment of puberty, the absence of mature boar contact during rearing appears to adversely affect the reproductive efficiency of the gilt by influencing oestrus expression and sexual receptivity. There is some evidence to suggest that individual housing, either in tethers or in individual pens during rearing, will also reduce the expression of oestrus in the gilt.

Postpubertal social environment of the sow

There is considerable evidence in the literature that the presence of the male can modify or regulate the sexual activity of the female. Synchronization and induction of oestrus occurs with the introduction of a mature male mouse into a colony of female mice (Whitten, 1956; 1959). A similar phenomenon occurs in the ewe. Introduction of a ram to a group of ewes prior to the commencement of the breeding season synchronizes and induces oestrus much earlier than otherwise would occur (Schinckel, 1954; Watson and Radford, 1960). Post-weaning anoestrus is one of the commonest reproductive disorders in sows, particularly first-litter sows, (Wrathall, 1975; Meredith, 1979) and yet, surprisingly, the influence of the social environment on the induction of oestrus following parturition or weaning has received little attention from research workers.

Although the period of lactation in the sow is characterized by a suppression of oestrus, probably due to the inhibiting influence of the suckling stimulus (Peters, First and Casida, 1969), a number of studies have demonstrated that the combination of several factors, including social factors, will induce a fertile oestrus during lactation (Rowlinson, Boughton and Bryant, 1975; Rowlinson and Bryant, 1976). Attempts to repeat these studies have produced equivocal results, for which variation in the quality and quantity of these factors may have been responsible.

Ad libitum feeding, grouping and the continuous presence of a boar within the group after day 20 or 21 of lactation were associated with the successful induction of oestrus in lactating sows (Rowlinson, Boughton and Bryant, 1975; Rowlinson and Bryant, 1976). An acceptable conception rate was achieved with mating at this oestrus in the first study. According to Brooks (1974), Rowlinson found that the occurrence of lactational oestrus was reduced from 80% to 40% when a boar was housed in an adjacent pen rather than with the sows.

In contrast, several studies have reported a low incidence of lactational oestrus. Lactating sows fed *ad libitum,* individually housed and introduced daily to a boar for 10–15 minutes from day 10 of lactation failed to show oestrus during lactation (Cole, Brooks and Kay, 1972). A small percentage of lactating sows separated from boars exhibited oestrus following restricted feeding and housing in either individual pens (1.3%) or group pens (2.6%) from day 10 of lactation (Petchey, Dodsworth and English, 1978). Housing sows in groups with the majority adjacent to, rather than with a boar, from day 14 of lactation and feeding the sows 6.4 kg of feed/day during lactation resulted in 48% of the lactating sows exhibiting oestrus (Petchey and Jolly, 1979). It is of interest that in this study sows mated during lactation had a higher liveweight at weaning than those mated after weaning.

Therefore, in combination with *ad libitum* feeding, there is evidence to suggest that lactational oestrus can be successfully induced by housing sows in groups and housing a boar within the group. However, it is obvious that further research is required in which the quality and quantity of these factors is adequately controlled and described. For example, in all the above-mentioned studies the age of the boar was not reported and yet this may be an important factor particularly in the light of the study by Hughes and Kirkwood (1980) on the importance of the age of the boar on puberty induction in the gilt.

Even less research has been conducted on the influence of the social environment on the onset of oestrus in the weaned sow. Hillyer (1976) reported that the weaning to conception interval was reduced from 27 to 9 days by spraying sows shortly after weaning with a synthetic boar odour containing the steroid 5α-androst-16-en-3-one. It appears that an aerosol spray containing this steroid will induce the standing response in the oestrous female (Melrose, Reed and Patterson, 1971) and thus, it is not clear whether the synthetic boar odour used by Hillyer stimulated the onset of oestrus or aided oestrous detection.

A survey of 75 Norwegian piggeries found that weaning to mating interval was not significantly affected by housing weaned sows in either the presence or absence of boars (Karlberg, 1980). However, little information on the social environment, apart from the presence or absence of a boar, was provided by the author. In contrast, a recent study conducted on 1100 sows in the Netherlands indicated that the factors of intense boar stimulation, achieved by daily introduction to a mature boar, and group housing were significantly associated ($P<0.005$ and $P<0.05$, respectively) with an earlier oestrus after weaning in the sow (*Table 28.7*). Examination of the

Table 28.7 THE INFLUENCE OF POST-WEANING SOCIAL TREATMENTS OF INDIVIDUAL (I) OR GROUP (G) HOUSING AND PRESENCE (+ ♂) OR ABSENCE (−♂) OF DAILY BOAR INTRODUCTION ON THE WEANING TO MATING INTERVAL

	Treatment groups			
	I−♂	*I*+♂	*G*−♂	*G*+♂
Average weaning to mating interval (days)	16.5 (23.9)[a]	14.4 (20.8)	15.7 (22.0)	11.7 (16.4)
Sows mated by day 10 after weaning (%)	46.8 (9.7)	54.3 (24.1)	49.1 (13.3)	64.3 (48.1)

[a]Figures in parentheses are data for first-litter sows.
From Hemsworth, Salden and Hoogerbrugge (1982)

ovaries of unmated sows and observations on the response of unmated sows to pregnant mare's serum gonadotrophin/human chorionic gonadotrophin (PMSG/HCG) injection suggested that these social treatments were stimulating the onset of oestrus rather than affecting the efficiency of oestrus detection. Surprisingly, daily introduction of a mature boar for 10 minutes into an area adjacent to weaned sows failed to significantly affect the onset of oestrus. Therefore, an element(s) involved with daily movement and introduction to a mature boar appears to stimulate the onset of oestrus in the weaned sow. Research presently being conducted in the Netherlands supports this finding. The proximity of mature boars was not significantly associated with the weaning to mating interval, while daily introduction to a mature boar for the first seven days after weaning significantly ($P<0.05$) advanced the onset of oestrus (*Table 28.8*). Further research is required to identify the element(s) involved in movement and introduction to a mature boar that is influencing the onset of oestrus. Olfactory bulbectomy disrupted ovarian activity in postpubertal gilts (Signoret and Mauleon, 1962) and thus, an olfactory stimulus, or stimuli, from the courting boar is likely to be involved in the stimulatory effect on oestrus. The study by Hughes and Kirkwood (1980) on the importance of the olfactory stimulus from boars on puberty induction of the gilt, tends to support this view.

The results of studies on the influence of the penning system on

Table 28.8 THE INFLUENCE OF THE POST-WEANING SOCIAL TREATMENTS OF HOUSING WITHIN 3 m OR GREATER THAN 9 m OF MATURE BOARS AND THE PRESENCE (+♂) OR ABSENCE (−♂) OF DAILY BOAR INTRODUCTION ON THE WEANING TO MATING INTERVAL

	Treatment groups			
	9m−♂	*9m+♂*	*3m−♂*	*3m+♂*
Weaning to mating interval (days)	12.9 (18.7)[a]	11.2 (17.0)	11.9 (17.6)	10.0 (11.9)
Sows mated by day 10 post-weaning (%)	54.0 (20.0)	61.7 (21.4)	61.0 (16.7)	69.7 (53.3)

[a]Figures in parentheses are data for first-litter sows.
From Salden and Hoogerbrugge (unpublished data)

reproduction of the sow are contradictory. Variation in such factors as group size, space allowance, tethering or loose housing, pen design, contact with pigs and feeding system may be predominantly responsible for this conflict. Sommer (1980) reported a shorter weaning to mating interval for sows housed in groups rather than in individual pens (7.9 and 23.0 days, respectively) and Hemsworth, Salden and Hoogerbrugge (1982) found that group housing from weaning to mating was associated with a significantly earlier oestrus (*Table 28.7*). No significant difference in this interval was reported in sows housed either in groups or individual pens (England and Spurr, 1969; Fahmy and Dufour, 1976; Karlberg, 1980). However, Maclean (1969) reported a shorter interval for sows housed individually rather than in groups (9.0 and 11.5 days, respectively).

While the above studies on the weaning to mating interval provide inconsistent results, studies on the influence of the post-weaning penning system suggest that individual housing may be associated with a reduction in fertility. A higher conception rate was reported for sows housed in groups of 5–6 after weaning than for sows individually housed (87.2% and 82.4%, respectively) (Knap, 1969). Fahmy and Dufour (1976), in an attempt to demonstrate the adverse effects of stress caused by grouping, found that sows housed after weaning in groups of 8–10 animals had a higher pregnancy rate at 42 days than sows individually housed (81.1% and 66.5%, respectively). Similarly, Hemsworth, Beilharz and Brown (1978) observed that sows housed in pairs after weaning had a higher farrowing rate and litter size (born alive) than those housed in individual pens (farrowing rates were 86.7% and 50.0%, respectively; litter sizes were 10.31 and 6.67, respectively). All sows were briefly courted by a boar prior to artificial insemination. A higher litter size (total piglets born) was also found for sows housed in groups of four from weaning to mating than for those individually housed during this period (10.91 and 10.69, respectively) (Hemsworth, Salden and Hoogerbrugge, 1982).

The problems associated with research on the influence of the penning system on reproduction of the sow are further highlighted in an examination of the literature on the role of the penning system during gestation. The results of a number of studies on this subject are given in *Table 28.9*. Clearly the results of studies with the gilt are inconclusive; however, those with the sow indicate a tendency for a reduction in conception rate or

Table 28.9 THE INFLUENCE OF THE PENNING SYSTEM AT MATING AND DURING GESTATION ON THE FERTILITY OF THE GILT AND SOW

Study	*Conception or pregnancy rate with housing*	
	Individually	*In groups*
Knap (1969)[(a)]		
325 sows	82.6[a]	87.2[b]
England and Spurr (1969)[(a)]		
140 gilts	73.0	67.0
153 sows	71.0	82.0
Jensen *et al.* (1970)[(a)]		
240 gilts	80.0	82.0
Schlegel and Sklenar (1972)		
809 gilts	58.8	62.2
1765 sows	69.4[a]	73.5[b]
Klatt and Schlisske (1975)		
23 sows	90.0	81.9
Fahmy and Dufour (1976)[(a)]		
177 sows	66.5[a]	81.1[b]

[(a)]Housing treatment imposed prior to mating and during gestation.
[a,b]: Figures with a different superscript are significantly different ($P<0.05$)

pregnancy rate when sows are individually housed. The results of studies on the influence of the penning system during gestation on the fecundity of the gilt and sow are contradictory (England and Spurr, 1969; Jensen *et al.*, 1970; Nygaard *et al.*, 1970; Gustafsson, 1972; Schlegel and Sklenar, 1972; Bäckström, 1973; Klatt and Schlisske, 1975). Thus, further research is required to study the influence of the penning system prior to and after mating on the fertility and fecundity of the gilt and sow and it is essential that this research considers the role of variables such as space allowance, group size, pen design, tethering and loose housing and feeding system. For instance, Yakimchuk (1980) reported a reduction in fertility and fecundity of sows and gilts housed in groups of 20, 40 and 50 compared with those housed in groups of 10 animals.

The mechanism by which individual housing may influence reproduction is not clear, but two recent studies have investigated the influence of the penning system on the physiology and behaviour of the sow. Barnett, Cronin and Winfield (1981) concluded that individually-housed sows were exhibiting a chronic stress response as evidenced by changes in adrenal function. Due to increased behavioural disturbances, Sambraus, Sommer and Kräusslich (1978) concluded that the 'well-being' of sows is adversely affected by individual housing. These two studies provide support for the view that one or more of the physiological mechanisms involved in reproduction of the female pig may be disrupted or inhibited by individual housing.

Thus, there is considerable evidence suggesting that the social environment of the sow may have a marked influence on her reproduction. Beneficial aspects of the social environment appear to be the presence of mature boars prior to mating and group housing prior to and after mating. However, further research is urgently required to define the manner in which these social factors positively influence reproduction of the sow.

Conclusions

A review of the literature on the role of the social environment on reproduction of the pig clearly shows that the social environment is a major determinant of the sexual behaviour. Physical contact during rearing, achieved either within a group or through a wire-mesh wall, is required for the development of high levels of sexual behaviour of the boar. As far as the prepubertal gilt is concerned, contact with, and especially introduction to, mature boars during rearing positively influences oestrus expression and sexual receptivity, which in turn improves mating rate.

After puberty social environment has a further influence on the stimulation and maintenance of sexual behaviour and reproduction of the pig. Mature boars require stimulation from the presence of female pigs to maintain their potential level of sexual behaviour, while oestrus in the weaned sow appears to be stimulated by introduction to mature boars. Thus, when the two sexual partners are in proximity there appears to be coordination of sexual responsiveness, ensuring efficient reproduction. At present research has not identified the stimuli from the partners responsible for these effects and, obviously, this is the next main objective of research in this area.

In addition, penning system appears to influence reproduction of the female pig. There is evidence to indicate that individual housing at various stages of the reproductive cycle will adversely affect reproduction of the gilt and sow. Research is required to identify the factor(s) responsible and so define the penning system in terms of space allowance, group size, amount and type of contact with pigs available, etc. to facilitate efficient reproduction of the female.

Therefore, it is obvious that the social environment exerts a major influence on reproduction of the boar, gilt and sow. Many of the reproductive disorders encountered in modern pig production may be a result of a suboptimum social environment and thus marked improvements in reproductive efficiency are likely to occur by using and improving our knowledge of the role of the social environment on reproduction of the pig.

Acknowledgements

The research reported in this chapter was supported by the Australian Pig Industry Research Committee. The assistance of Mr C.G. Winfield in the preparation of this chapter is gratefully acknowledged.

References

BÄCKSTRÖM, L. (1973). Environment and animal health in piglet production. A field study of incidences and correlations. *Acta vet. scand., Suppl.* **41**, 1–240

BARNETT, J.L., CRONIN, G.M. and WINFIELD, C.G. (1981). The effects of individual and group penning of pigs on total and free plasma corticosteroids and the maximum corticosteroid binding capacity. *Gen. Comp. Endocr.* **44**, 219–225

BOOTH, W.D. and BALDWIN, B.A. (1980). Lack of effect on sexual behaviour or the development of testicular function after removal of olfactory bulbs in prepubertal boars. *J. Reprod. Fert.* **58**, 173–182

BROOKS, P.H. (1974). Oestrus detection and synchronisation in gilts and sows. In *The Detection and Control of Breeding Activity in Farm Animals*, (J.B. Owen, Ed.), pp. 73–83

BROOKS, P.H. and COLE, D.J.A. (1970). The effect of boar presence on the attainment of puberty in gilts. *J. Reprod. Fert.* **23**, 435–440

COLE, D.J.A., BROOKS, P.H. and KAY, R.M. (1972). Lactational anoestrus in the sow. *Vet. Rec.* **90**, 681–683

DÖCKE, V.F. and WORCH, H. (1963). Investigations into uterine motility and mating reactions of sows. *Zuchthygiene* **7**, 169–178

ENGLAND, D.C. and SPURR, D.T. (1969). Litter size of swine confined during gestation. *J. Anim. Sci.* **28**, 220–223

FAHMY, M.H. and DUFOUR, J.J. (1976). Effects of post-weaning stress and feeding management on return to oestrus and reproductive traits during early pregnancy in swine. *J. Anim. Prod.* **23**, 103–110

FRASER, A.F. (1968a). *Reproductive Behaviour in Ungulates.* London and New York, Academic Press

FRASER, A.F. (1968b). The 'male effect' on reproductive responses in female farm animals. *VIth Congr. Int. Reprod. A.I., Paris,* **II**, 1661–1662

GUSTAFSSON, B. (1972). Environment and results in piglet production. Agricultural College of Sweden, November 13–14, 1972, pp. 6 (Abstract).

HEMSWORTH, P.H. and BEILHARZ, R.G. (1979). The influence of restricted physical contact with pigs during rearing on the sexual behaviour of the male domestic pig. *Anim. Prod.* **29**, 311–314

HEMSWORTH, P.H. and GALLOWAY, D.B. (1979). The effect of sexual stimulation on the sperm output of the domestic boar. *Anim. Reprod. Sci.* **2**, 387–394

HEMSWORTH, P.H., BEILHARZ, R.G. and BROWN, W.J. (1978). The importance of the courting behaviour of the boar on the success of natural and artificial matings. *Appl. Anim. Ethol.* **4**, 341–347

HEMSWORTH, P.H., BEILHARZ, R.G. and GALLOWAY, D.B. (1977). Influence of social conditions during rearing on the sexual behaviour of the domestic boar. *Anim. Prod.* **24**, 245–251

HEMSWORTH, P.H., CRONIN, G.M. and HANSEN, C. (1982). The influence of social restriction during rearing on the sexual behaviour of the gilt. *Anim. Prod.* (in press)

HEMSWORTH, P.H., FINDLAY, J.K. and BEILHARZ, R.G. (1978). The importance of physical contact with other pigs during rearing on the sexual behaviour of the male domestic pig. *Anim. Prod.* **27**, 201–207

HEMSWORTH, P.H., SALDEN, N.T.C.J. and HOOGERBRUGGE, A. (1982). The influence of the post-weaning social environment on the weaning to mating interval of the sow. *Anim. Prod.* (in press)

HEMSWORTH, P.H., WINFIELD, C.G. and CHAMLEY, W.A. (1981). The influence of the presence of the female on the sexual behaviour and plasma testosterone levels of the mature male pig. *Anim. Prod.* **32**, 61–65

HEMSWORTH, P.H., WINFIELD, C.G., BEILHARZ, R.G. and GALLOWAY, D.B. (1977). The influence of social conditions post-puberty on the sexual behaviour of the domestic male pig. *Anim. Prod.* **25**, 305–309

HILLYER, G.M. (1976). An investigation using a synthetic porcine pheromone and the effect on days from weaning to conception. *Vet. Rec.* **98**, 93–94

HUGHES, P.E. and COLE, D.J.A. (1976). Reproduction in the gilt. 2. The influence of gilt age at boar introduction on the attainment of puberty. *Anim. Prod.* **23**, 89–94

HUGHES, P.E. and COLE, D.J.A. (1978). Reproduction in the gilt 3. The effect of exogenous oestrogen on the attainment of puberty and subsequent reproductive performance. *Anim. Prod.* **27**, 11–20

HUGHES, P.E. and KIRKWOOD, R.N. (1980). Boar-induced precocious puberty in the gilt. *31st Annual Meeting of EAAP, Munich*, P 5/6. 39

ILLIUS, A.W., HAYNES, N.B. and LAMMING, G.E. (1976). Effects of ewe proximity on peripheral plasma testosterone levels and behaviour in the ram. *J. Reprod. Fert.* **48**, 25–32

JENSEN, A.H., YEN, J.T., GEHRING, M.M., BAKER, D.H., BECKER, D.E. and HARMON, B.G. (1970). Effects of space restriction and management of pre- and post-pubertal response of female swine. *J. Anim. Sci.* **31**, 745–750

KARLBERG, K. (1980). Factors affecting post-weaning oestrus in the sow. *Nord. VetMed.* **32**, 183–193

KIRKWOOD, R.N. and HUGHES, P.E. (1979). The influence of age at first boar contact on puberty attainment in the gilt. *Anim. Prod.* **29**, 231–238

KIRKWOOD, R.N. and HUGHES, P.E. (1980a). A note on the influence of 'boar effect' component stimuli on puberty attainment in the gilt. *Anim. Prod.* **31**, 209–211

KIRKWOOD, R.N. and HUGHES, P.E. (1980b). A note on the efficacy of continuous versus limited boar exposure on puberty attainment in the gilt. *Anim. Prod.* **31**, 205–207

KLATT, G. and SCHLISSKE, W. (1975). The effects on performance of minimal exercise of sows pregnant after an extremely short suckling period. *Anim. Breed. Abstr.* **43**, 466–467 (Abstract)

KNAP, J. (1969). Effect of group and individual housing of sows after weaning on length of the interval to the first mating and conception rate. *Anim. Breed. Abstr.* **38**, 641–642 (Abstract)

MACLEAN, C.W. (1969). Observations on non-infectious infertility in sows. *Vet. Rec.* **85**, 675–682

MAVROGENIS, A.P. and ROBISON, O.W. (1976). Factors affecting puberty in swine. *J. Anim. Sci.* **42**, 1251–1255

MELROSE, D.R., REED, H.C.B. and PATTERSON, R.L.S. (1971). Androgen steroids associated with boar odour as an aid to the detection of oestrus in pig artificial insemination. *Br. vet. J.* **127**, 497–501

MEREDITH, M.J. (1979). The treatment of anoestrus in the pig : A review. *Vet. Rec.* **104**, 25–27

MICHAEL, R.P. and ZUMPE, D. (1970). Sexual initiating behaviour by female rhesus monkeys (*Macaca mulatta*) under laboratory conditions. *Behaviour* **36**, 168–186

MICHAEL, R.P., ZUMPE, D., KEVERNE, E.B. and BONSALL, R.W. (1972). Neuroendocrine factors in the control of primate behaviour. *Recent Prog. Horm. Res.* **28**, 665–706

NYGAARD, A., AULSTAD, D., LYSO, A., KRAGGERUD, H. and STANDAL, N.

(1970). Housing for pregnant sows. *Anim. Breed. Abstr.* **40**, 744 (Abstract)

PETCHEY, A.M. and JOLLY, G.M. (1979). Sow service in lactation : An analysis of data from one herd. *Anim. Prod.* **29**, 183–191

PETCHEY, A.M., DODSWORTH, T.L. and ENGLISH, P.R. (1978). The performance of sows and litters penned individually or grouped in late lactation. *Anim. Prod.* **27**, 215–221

PETERS, J.B., FIRST, N.L. and CASIDA, L.E. (1969). Effects of pig removal and oxytocin injections on ovarian and pituitary changes in mammillectomized post-partum sows. *J. Anim. Sci.* **28**, 537–541

PITKJANEN, I.G. (1964). Investigations on reproductive biology and A.I. in swine. *Vth Congr. Int. Anim. Reprod. A.I., Trento,* Vol. 6, pp. 25–30

ROBISON, O.W. (1974). Effects of boar presence and group size on age at puberty in gilts. *J. Anim. Sci.* **39**, 224 (Abstract)

ROWLINSON, P. and BRYANT, M.J. (1976). The effect of lactating management on the incidence and timing of oestrus in lactating sows. *Anim. Prod.* **22**, 139 (Abstract)

ROWLINSON, P., BOUGHTON, H.G. and BRYANT, M.J. (1975). Mating of sows during lactation : Observations from a commercial unit. *Anim. Prod.* **21**, 233–241

SAMBRAUS, H.H., SOMMER, B. and KRÄUSSLICH, H. (1978). The behaviour of sows in different systems of husbandry. *1st World Congress on Ethology Applied to Zootechnics, Madrid*, Vol. 1, Plenary session, pp. 99–102

SCHEIN, M.W. and HALE, E.B. (1965). Stimuli eliciting sexual behaviour. In *Sex and Behaviour*, (F.A. Beach, Ed.), pp. 440–482. New York, J. Wiley and Sons, Inc.

SCHLEGEL, W. and SKLENAR, V. (1972). The effect of different management systems on reproductive performance in sows. *Anim. Breed. Abstr.* **41**, 551 (Abstract)

SCHINCKEL, P.G. (1954). The effect of the presence of the ram on the ovarian activity of the ewe. *Aust. J. agric. Res.* **5**, 465–479

SIGNORET, J.P. (1970a). Swine behaviour in reproduction. In *Effect of Disease and Stress on Reproductive Efficiency in Swine*, pp. 28–45.

SIGNORET, J.P. (1970b). Reproductive behaviour of pigs. *J. Reprod. Fert., Suppl.* **11**, 105–107

SIGNORET, J.P. (1970c). Sexual behaviour patterns in female domestic pigs (*Sus scrofa* L.) reared in isolation from males. *Anim. Behav.* **18**, 165–168

SIGNORET, J.P. (1971). The reproductive behaviour of pigs in relation to fertility. *Vet. Rec.* **88**, 34–38

SIGNORET, J.P. (1972). The mating behaviour of the sow. In *Pig Production,* (D.J.A. Cole, Ed.), pp. 295–313. London, Butterworths

SIGNORET, J.P. and MAULEON, P. (1962). The effect of surgical removal of the olfactory bulbs on the sexual cycle and the genital tract of sows. *Annls Biol. anim. Biochim. Biophys.* **2**, 167–174

SIGNORET, J.P., DU MESNIL DU BUISSON, F. and MAULEON, P. (1972). Effect of mating on the onset and duration of ovulation in the sow. *J. Reprod. Fert.* **31**, 327–330

SOMMER, B. (1980). Sows in individual pens and group housing – oestrous behaviour, parturition, fertility and damage to limbs. *Anim. Breed. Abstr.* **48**, 619 (Abstract)

THOMAS, H.R., KATTESH, H.G., KNIGHT, J.W., GWAZDAUSKAS, F.C., MEACHAM, T.N. and KORNEGAY, E.T. (1979). Effects of housing and rearing on age and puberty and libido in boars. *Anim. Prod.* **28**, 231–234

VANDENBERGH, J.G. (1969). Endocrine coordination in monkeys : Male sexual responses to the female. *Physiol. Behav.* **4**, 261–264

WALTON, A. (1960). Copulation and natural insemination. In *Marshall's Physiology of Reproduction*, 3rd Edition, (A.S. Parkes, Ed.), pp. 130–160. London, Longman, Green and Company

WATSON, R.H. and RADFORD, H.M. (1960). The influence of rams on the onset of oestrus in Merino ewes in spring. *Aust. J. agric. Res.* **11**, 65–71

WHITTEN, W.K. (1956). Modification of the oestrus cycle of the mouse by external stimuli association with the male. *J. Endocr.* **13**, 399–404

WHITTEN, W.K. (1959). Occurrence of anoestrus in mice caged in groups. *J. Endocr.* **18**, 102–107

WRATHALL, A.E. (1975). *Reproductive Disorders in Pigs,* Commonwealth Bureau of Animal Health, Review Series 11, Commonwealth Agricultural Bureau, Slough, England.

YAKIMCHUK, N.V. (1980). Conception rate and litter size of sows in relation to group size. *Anim. Breed. Abstr.* **48**, 370 (Abstract)

YOUNG, W.C. (1957). Genetic and psychological determinants of sexual behaviour patterns. In *Hormones, Brain Function and Behaviour*, (H. Hoaglan, Ed.), pp.75–98. New York, Academic Press

YOUNG, W.C., GOY, R.W. and PHOENIX, C.H. (1968). Hormones and sexual behaviour. *Science* **143**, 212–218

29

NUTRITION AND REPRODUCTION

D.J.A. COLE
Department of Agriculture and Horticulture, University of Nottingham School of Agriculture, Sutton Bonington, Loughborough, Leicestershire, UK

It is well known that the establishment of nutrient requirements of the sow is made difficult by the considerable variation that exists in reproductive characteristics. Much less success has been achieved than with the growing pig where performance is usually measured by some aspect of the efficient production of lean meat.

Thus, it is necessary to establish the production objectives of the breeding sow. For example, it is generally considered that the production of the maximum number of piglets per unit time (such as, per year) is important. At the same time it is important that the piglets are of adequate size and viability at birth and at whatever age they are weaned. Consequently the influence of nutrition during pregnancy and lactation needs to be well established. As the sow does not reach reproductive maturity until about the fourth parity, it is desirable that it is kept well beyond this stage to take advantage of the most prolific period of its life. Thus, the influence of nutrition is of interest in more than just the short term.

While the measurement of the response of the sow to its nutrition is not simple, the following can be identified as areas of importance :

1. *Short-term reproductive performance* e.g. how nutrition in pregnancy affects subsequent litter size or how nutrition in lactation affects piglet and sow weight at weaning.
2. *Medium-term reproductive performance* e.g. the effect of nutrition in lactation on performance in the next pregnancy.
3. *Long-term reproductive performance* e.g. the effect of nutrition in the early stages of reproductive life on the whole breeding lifetime.

While these have been listed as three separate items they are, of course, all connected. In order to reconcile these different objectives an approach is often taken of considering the influence of nutrition on immediate reproductive performance and also on the sow's body condition as a measure of its likely reproductive performance in the medium and long term. Weight change of the sow has often been used as a measure of body condition. However, it is recognized that the role of changes in both body condition and liveweight of the sow need further study.

A strategy of maximum conservation

A number of feeding regimens have been based on the marked depletion of body reserves during lactation with a high degree of restoration through high level feeding in the subsequent pregnancy. Such changes have usually been monitored through weight change and it is unlikely that loss and gain of weight reflect identical loss and gain in body tissues. It is proposed that the strategy needed for the long-term nutrition of the sow is one based on the maximum conservation of body condition in lactation with the minimum loss of weight. Such a strategy would then rely on minimum restoration of weight and condition in the following pregnancy. The

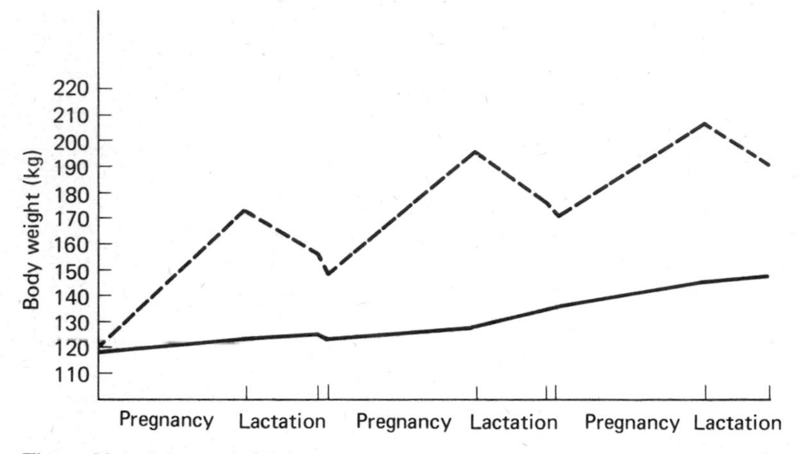

Figure 29.1 Maternal weight change of sows fed 1.8 kg + 0.36 kg/piglet suckled/day during lactation and 2.8 kg/day (---) or 1.4 kg/day (—) during pregnancy. From Lodge, Elsley and MacPherson (1966b)

benefits of this are continuous as it is well known that the greater the weight gains in pregnancy the greater the weight loss in lactation. An extreme example of this is given in *Figure 29.1*. In addition the sow would be liberally fed during the period of maximum production (i.e. lactation). Thus, the aim is a strategy of carefully controlled and limited weight gains in pregnancy with the maximum conservation of weight and condition in lactation.

The gilt

The influence of nutrition during the rearing phase has been dealt with in Chapter 11. However, nutrition before the oestrus of first mating is of consequence because of its well-established effect on ovulation rate at this stage. For example, in an extensive review of the subject Anderson and Melampy (1972) suggested that the most effective time to start high level feeding was about 14 days before oestrus; such an effect may account for the type of response given in *Table 29.1*.

Table 29.1 INFLUENCE OF FEED LEVEL DURING THE REARING PERIOD ON THE LITTER SIZE OF GILTS

Number of piglets born Maximum 2.7 kg/day			*Ad libitum*
1st heat	*2nd heat*	*3rd heat*	*1st heat*
8.4	9.8	10.4	11.0

From MacPherson, Hovell and Jones (1977)

Pregnancy

ENERGY AND FEED INTAKE

The response to energy intake in pregnancy is generally reflected in response to total feed intake. During pregnancy the sow needs food to meet the demands of the developing litter and to achieve some weight gain, either as true growth or pregnancy anabolism. At some stage pregnancy anabolism gives way to the catabolism of lactation and the point at which this occurs appears to be in late pregnancy but has not been well established. However, it has already been suggested that large weight gains in pregnancy are not desirable in maintaining optimum weight change and body condition in the long-term reproduction of the sow.

A well known characteristic of the pregnant sow is its considerably greater efficiency of feed utilization than the non-pregnant female. This was well illustrated by Salmon-Legagneur and Rerat (1962) who showed that the sow was able to produce a litter of pigs on a diet comprising little more than a maintenance ration for a non-pregnant animal (*Table 29.2*). Consequently the establishment of energy requirements during pregnancy does not lend itself to the use of factorial estimates.

Table 29.2 WEIGHT GAINS OF SOWS DURING PREGNANCY ON LOW (0.87 kg/100 kg LIVEWEIGHT/DAY) AND HIGH (1.8 kg/100 kg LIVEWEIGHT/DAY) PLANES OF NUTRITION IN PREGNANCY

	Total gestation food intake (kg)	*Weight at mating* (kg)	*Weight before mating* (kg)	*Weight after parturition* (kg)
Low plane				
Pregnant sows	225	229.7	273.9	249.8
Non-pregnant sows	224	230.7	235.0	235.0
High plane				
Pregnant sows	418	230.2	308.2	284.1
Non-pregnant sows	419	231.0	270.0	270.0

Gain of pregnant sows[(a)]	*Low plane* (kg)	*High plane* (kg)
Foetuses	15.4	13.8
Loss of parturition (placenta, fluids)	8.7	10.3
True anabolism	15.8	14.9
Growth	4.3	39.0
Total	44.2	78.0

[(a)]Difference between weight before mating and weight at mating
From Salmon-Legagneur and Rerat (1962)

In terms of energy level in pregnancy it has been suggested by Brooks and Cole (1971) that digestible energy (DE) intake should not fall below 25 MJ/day. This should be regarded as a minimum figure and takes no account of variation in age, weight or environmental conditions. While there is likely to be some increase in birthweight when energy intake exceeds 25 MJ DE/day, it is only of practical importance where a birth-weight problem exists. The major response to increased energy intake will be as maternal weight gain. Up to intakes of about 40 MJ DE/day, this is illustrated by the equation established by Van Schoukbroek and Van Spaendonck (1973):

$$\text{Increase in sow weight (kg)} = -11.7 + 2.631x - 0.018x^2 \pm 12.1 \ (r = 0.71)$$

where x = energy intake (MJ ME/day). It was further illustrated by the work of Salmon-Legagneur and Rerat (1962) which also shows the influence of weight gain in pregnancy on overall weight change in the reproductive cycle (*Table 29.3*). Sows that had been fed to gain 20.1 kg in pregnancy lost 7.4 kg in lactation, whereas sows fed at a higher level to gain 53.9 kg in pregnancy had a weight loss of 48.3 kg in lactation. Thus, those sows fed to make a large weight gain in pregnancy had a lower net weight gain over the whole reproductive cycle (5.6 kg) than those fed to make only modest weight gains (a net gain of 12.7 kg).

Table 29.3 WEIGHT CHANGES IN PREGNANCY AND LACTATION OF SOWS FED LOW (0.87 kg/100 kg LIVEWEIGHT/DAY) AND HIGH (1.8 kg/100 kg LIVEWEIGHT/DAY) PLANES OF NUTRITION IN PREGNANCY

Plane of nutrition	*Weight at mating* (kg)	*Weight after farrowing* (kg)	*Gain during pregnancy* (kg)	*Weight at weaning* (kg)	*Loss during lactation* (kg)	*Total weight change* (kg)
Low	229.7	249.8	20.1	242.4	7.4	+12.7
High	230.2	284.1	53.9	235.8	48.3	+5.6

From Salmon-Legagneur and Rerat (1962)

In addition to considering the overall feeding of pregnancy, it is necessary to consider if any one part of it needs a different nutritional regimen to another.

Early pregnancy

Controversy has existed over nutrition in early pregnancy. In a number of cases comparisons of feed level have shown no influence on embryo survival but the data are difficult to interpret because of the variation in duration of treatments. However, several workers have reported improved embryo survival with gilts given lower feed levels. For example, Dutt and

Table 29.4 THE INFLUENCE OF FEED INTAKE IN EARLY PREGNANCY

Duration of treatment in pregnancy			
Days 0–10		*Days 10–20*	
Feed intake (kg/day)	*Embryo survival* (%)	*Feed intake* (kg/day)	*Embryo survival* (%)
4.1	66.0	4.1	67.3
2.5	72.1	2.5	72.0
1.25	78.4	1.25	71.9

From Dutt and Chaney (1968)

Table 29.5 THE INFLUENCE OF FEED INTAKE IN EARLY PREGNANCY ON EMBRYO SURVIVAL TO DAY 30–35

Feed intake (kg/day)			
Days 0–10	*Days 10–30*	*Conception rate* (%)	*Embryo survival* (%)
2.5	2.5	87.1	75.8
2.5	1.5	86.2	76.9
1.5	2.5	87.1	85.4
1.5	1.5	64.3	86.7

From Dyck and Strain (1980)

Chaney (1968) showed a small benefit when feed intake was reduced from the time of implantation onwards but a much bigger benefit when the period of feed restriction was from the day of mating onwards (*Table 29.4*). A similar trend was shown by Dyck and Strain (1980) with quite marked improvements for reduced feed intake from mating to day 10 of pregnancy (*Table 29.5*). However, low feed levels (1.5 kg/day) from mating to the end of the experiment resulted in a lower conception rate. It is important that any trends are considered in relation to the range of dietary treatment levels and their relevance to practical feeding situations. For example, the range of feed intakes in *Table 29.4* should be outside that encountered in practice.

Late pregnancy

Generally energy levels in late pregnancy have had little effect other than to increase piglet birthweight when this might have been expected to be low. However, a similar benefit might also have been achieved by spreading the extra feed over the whole of the pregnancy (Lodge, Elsley and MacPherson, 1966a,b). More recently there have been reports of much shorter term nutrition in late pregnancy influencing pig birthweight. For example, an increase of feed intake from 2.8 to 4.0 kg/day between days 100 and 110 of pregnancy increased piglet birthweight from 1.42 kg to 1.64 kg (Kotarbinska, unpublished). However, other workers (e.g. Hillyer and Phillips, 1980) have failed to obtain these gains.

PROTEIN

There has been considerable variation both in the suggested requirements for protein in pregnancy and in the practical application of research findings. Generally, pregnant sows have a greater nitrogen retention than non-pregnant sows. For example, up to 10% increases were suggested by Salmon-Legagneur (1965) and Heap and Lodge (1967). Although this is well recognized now, it was not accounted for in early estimates of protein requirements.

Reproductive characteristics such as breeding regularity, litter size and piglet birthweight and composition show little response above 140 g crude protein/day (see *Figure 29.2*). There has been little work with protein

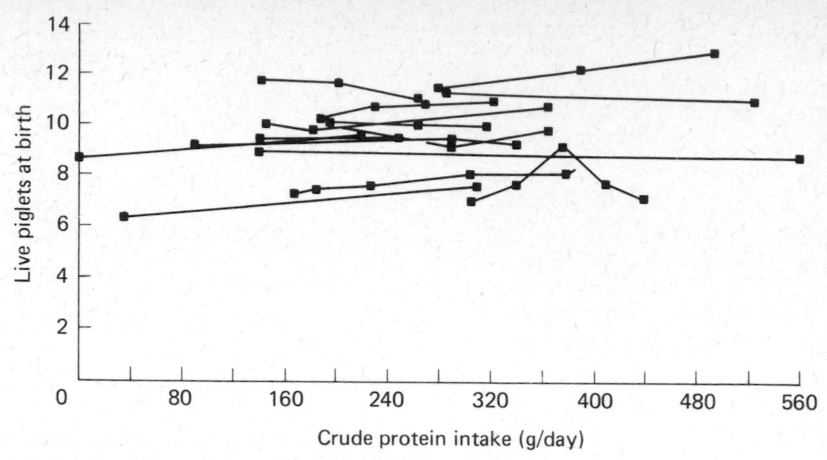

Figure 29.2 Influence of daily crude protein intake in pregnancy on number of piglets at birth. From data of Baker *et al.* (1970a); Boaz (1962); Clawson *et al.* (1963); Elsley and MacPherson (1972); Frobish *et al.* (1966); De Geeter *et al.* (1970a,b); Greenhalgh *et al.* (1977); Hawton and Meade (1971); Hesby *et al.* (1970a); Holden *et al.* (1968); Kemm and Pieterse (1968); Pike and Boaz (1969); Pond *et al.* (1968); Rippel *et al.* (1965a)

intakes lower than this although pigs have been kept on protein-free diets for long periods of pregnancy without adverse effect (Pond *et al.*, 1968). In contrast to reproductive characteristics, maternal weight gain has responded to higher levels, up to about 300 g crude protein/day (*Figure 29.3*).

The study of protein quality as reflected by the requirements for individual amino acids during pregnancy has received little attention. This could reflect both lack of limitation of dietary protein and the difficulty of measuring responses in the breeding animal. As a consequence of this

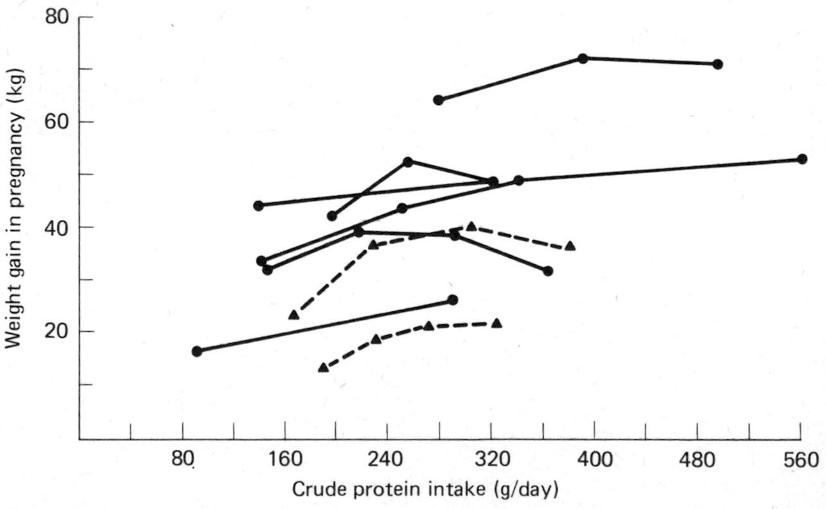

Figure 29.3 Influence of daily crude protein intake in pregnancy on gross (●——●) and net (▲---▲) weight gain of the sow. From data of Baker *et al.* (1970a); Boaz (1962); Clawson *et al.* (1963); Hesby *et al.* (1970a); Holden *et al.* (1968); Rippel *et al.* (1965a)

latter point most of the work has involved the use of indirect techniques to establish requirement values and measurements of nitrogen retention, blood urea and plasma amino acids have been used. The basis for the use of blood urea is that it is the principal excretory product in the pig and represents the difference between protein supply and protein requirements. Blood urea will increase when the dietary supply of protein is raised and fall when it is reduced. However, if the dietary protein level is kept constant, blood urea can be used to monitor the effect of dietary protein quality, and the requirement for an individual amino acid. Thus, the addition of a single amino acid to a deficient diet can be expected to reduce the level of blood urea (due to the improved efficiency of use of the protein) until the 'requirement' level is reached. Further additions of the dietary amino acid would then increase the level of blood urea. The responses of individual amino acids in the blood also have been used to measure dietary adequacy but generally with less success than with blood urea.

Recently there has been a move to describe an ideal protein for particular productive functions (Cole, 1978). For example, with growing pigs it has been suggested that the difference between the requirements of pigs of different sex, breed and liveweight for the deposition of 1 g lean is likely to be in the quantity rather than the quality of protein. Consequently, it should be possible to identify an optimum balance of amino acids which, when supplied with sufficient nitrogen for the synthesis of non-essential nitrogen, would constitute the 'ideal protein'. The balance of essential amino acids in the ideal protein can be described as their ratio to lysine which is usually first limiting in pig diets. A similar approach could be taken for the breeding pig but it is recognized that few response data exist.

The dietary requirement for lysine probably does not exceed 10 g/day (*Table 29.6*). On the basis of the small amount of work reported, the

Table 29.6 SOME ESTIMATES OF LYSINE REQUIREMENTS DURING PREGNANCY

	Feed intake (kg/day)	*Lysine* (% diet)	*Lysine* (g/day)	*Criteria used for estimating requirement*
Woerman and Speer (1976)		0.41	7.5	N retention and piglet performance
Baker *et al.* (1970b)		0.42	8.0	Sow weight change and piglet performance
Rippel *et al.* (1965c)		0.42	7.6	N retention and piglet performance
Duee and Rerat (1974; 1975)		0.43	8.6	Blood urea, sow weight change and piglet performance
Salmon-Legagneur and Duee (1972)	1.90	0.44	8.4	N retention
Hesby *et al.* (1970a,b)	2.22	0.49	10.8	Reproductive performance
Allee and Baker (1970)	2.00	0.49	9.8	N retention
Sohail *et al.* (1978a)	1.82	0.64	10.0	Blood urea and plasma amino acids
Miller *et al.* (1969)		0.66	12.16	N retention

Table 29.7 SOME ESTIMATES OF AMINO ACID REQUIREMENTS DURING PREGNANCY

	Rippel et al. *(1965d)*		*Lucas* et al. *(1969)*		*Miller* et al. *(1969)*	
	(% of diet)	(% of lysine)	(% of diet)	(% of lysine)	(% of diet)	(% of lysine)
Lysine	0.42	100	0.43	100	0.64	100
Methionine + cystine	0.29	69	0.30	70	0.50	78
Tryptophan	0.07	17	0.10	23	0.13	20
Threonine	0.34	81	0.44	102	0.53	83
Leucine	0.56	133	0.72	167	1.42	221
Crude protein	12.5		8.0		15.0	

requirements for other amino acids are given as a percentage of lysine in the diet and also as a balance relative to lysine (*Table 29.7*). It has been suggested that the requirement for non-essential nitrogen is low, of the order of 4.3 g/day (Allee and Baker, 1970).

Lactation

The sow is capable of considerable milk production to meet the needs of the suckling litter and the demands of milk production for energy and nutrients result in very high requirement values relative to those of pregnancy. However, as with other high yielding domestic livestock, allowances established from nutrient requirements need to be within the appetite limits of the animal. As a result of practical problems the level of voluntary feed intake in the lactating sow has received greater attention in recent years.

Feeding during lactation is further complicated by the differences in weaning age that are used. A considerable volume of information exists on requirement values established with lactation lengths of 42–56 days. Less is known about requirements with weaning ages much shorter than this and about the relationships between lactation, reproduction and weight change in the earlier weaned sows.

REQUIREMENTS FOR LACTATION

Energy

Requirements during lactation can be regarded as the sum of the needs of maintenance and production. The equation is completed by the sparing effect on requirements of energy contributed by the weight loss of the sow. However, the desirability and extent of weight loss in relation to the short and long term objectives of reproduction must be considered carefully.

A major factor determining energy requirements during lactation will be variation in milk yield as a result of variation in litter size. Also it is well known that the stage of lactation will influence the energy requirement. The largest changes in milk composition are in early lactation when the change from colostrum to milk results in a fall in milk protein over the first

Table 29.8 CALCULATION OF ENERGY REQUIREMENTS OF SOWS AT DIFFERENT STAGES OF LACTATION

Lactation and requirements	*Weeks of lactation*		
	1 + 2	*3 + 4*	*5 + 6*
Milk yield (kg/day)	5.80	7.15	6.77
Milk energy (MJ/day)[(a)]	26.22	32.32	30.60
Feed required/day (kg)[(b)]	6.6	8.15	7.62

[(a)]Assumed to contain 4.52 MJ/kg
[(b)]Assumed to contain 12.55 MJ DE/kg
From O'Grady (1980)

10–14 days and a rise in milk fat during the first three days of lactation. Thereafter there is a small but gradual rise in milk protein and a decline in fat. Typically, sow's milk contains about 5.1–5.3 MJ/kg which is produced at an efficiency of about 60%.

Although the stage of lactation can influence energy requirements, it has been suggested that the plane of nutrition is unlikely to influence the level of milk production during the first three weeks (Lodge, 1972). At the other extreme, lactation lengths of greater than six weeks are uncommon. Thus, in calculating the influence of length of lactation on daily energy requirement, it is likely that the greatest emphasis should be placed on weeks 3–6. Examples of suggested energy requirements at different stages of lactation (O'Grady, 1980) are given in *Table 29.8*.

Protein

The quality and quantity of milk produced in lactation are the biggest factors in the determination of protein requirements at this stage. However, there is considerable variation in the suggested requirement values and it is difficult to give recommended values for lactation. The gross efficiency of milk production has been calculated to be of the order of 33% (Lodge, 1959) to 43% (Elsley and MacPherson, 1966). However, responses are normally measured in terms of weight gains of piglets. Generally, low levels of protein have resulted in a slower piglet growth rate, e.g. Greenhalgh *et al.* (1977) who used 9–13% crude protein in the diet and DeGeeter *et al.* (1972) who used 45–309 g crude protein/day, but benefits have not been achieved in raising crude protein intake above about 800 g/day.

Of the essential amino acids, lysine has received most attention in studies of protein quality. The various estimates are mostly based on indirect measures of adequacy and show considerable variation (*Table 29.9*). However, a large number of values fall between 30 and 40 g lysine/day. Protein quality in lactation could lend itself to expression as the requirement for an 'ideal protein'. In the absence of a large amount of experimental evidence the composition of sow's milk might be used as a guideline to the balance of essential amino acids needed in the diet, an approach taken by Speer (1975). However, the validity of such an approach relies on the requirements for milk production being quantitatively much greater than the requirements for maintenance. Wilkinson (1978) calculated that maintenance had little effect on balance when

Table 29.9 SOME ESTIMATES OF LYSINE REQUIREMENTS OF LACTATING SOWS

	Feed intake (kg/day)	*Lysine* (% of diet)	*Lysine intake* (g/day)	*Criteria used for estimating requirement*
Baker *et al.* (1970a,b)	4.0	0.81	32.4	Reproductive performance
Boomgaardt *et al.* (1972)	*ad libitum*	0.60	20.0	Reproductive performance and blood urea
Salmon-Legagneur and Duee (1972)	5.42	0.69	37.4	Piglet performance
Lewis and Speer (1973)	5.45	0.56	30.5	N retention and milk protein
Sohail, Cole and Lewis (1978b)	4.5	0.85	38.4	Plasma amino acids and blood urea
Wilkinson, Cole and Lewis (unpublished)			49.5	Plasma amino acids and blood urea
Wilkinson, Cole and Lewis (unpublished)			40.0	Plasma urea

Table 29.10 THE BALANCE OF DIETARY AMINO ACIDS (LYSINE = 100) BASED ON VALUES FOR MAINTENANCE AND TWO LEVELS OF MILK YIELD

Amino acid	*Maintenance + 250g milk protein*	*Maintenance + 400g milk protein*
Lysine	100	100
Isoleucine	59	57
Leucine	112	113
Methionine	32	30
Threonine	62	60
Tryptophan	17	17
Phenylalanine	55	55
Valine	80	80

From Wilkinson (1978)

estimated from published values for maintenance and the composition of sow's milk. The amino acid most affected was methionine but not to a large extent (*Table 29.10*). However, it should be pointed out that there is little information on which to base the maintenance requirements for amino acids and the values used suggest that they are low (e.g. Baker *et al.*, 1966; Baker and Allee, 1970).

Some estimates of the requirements of other essential amino acids are presented in *Table 29.11*, together with the composition of sow's milk which has been used as a basis for establishing requirement values in some cases (e.g. Speer, 1975). Values also have been expressed as a balance using lysine, the most likely limiting amino acid, as the reference.

Table 29.11 AMINO ACID REQUIREMENTS FOR LACTATION AND AMINOACID COMPOSITION OF SOW'S MILK

	Requirement						*Sow's milk*	
	Baker et al. *(1970a) (gilt)*			*Speer (1975) (sow)*			*Elliot* et al. *(1971)*	
	(% of diet)	(g/day)	(% of lysine)	(% of diet)	(g/day)	(% of lysine)	(% of protein)	(% of lysine)
Arginine	0.34	13.6	42	0.41	22.4	67	4.7	67
Histidine	0.26	10.5	32	0.32	12.5	38	3.6	51
Isoleucine	0.67	27.0	83	0.35	19.1	58	3.7	53
Leucine	0.99	39.6	121	0.68	37.1	112	8.1	116
Lysine	0.81	32.6	100	0.61	33.2	100	7.0	100
Methionine + cystine	0.38	14.2	44	0.27	14.7	44	3.1	44
Phenylalanine + tyrosine	1.00	40.1	123	0.71	38.7	116	8.7	124
Threonine	0.51	20.4	63	0.37	20.2	61	4.4	63
Tryptophan	0.13	5.4	17	0.11	6.0	18	1.3	19
Valine	0.68	27.2	83	0.43	23.5	71	4.4	63

Interrelationships between pregnancy and lactation

Generally in establishing nutrient requirements each part of the reproductive cycle has been treated separately. However, it is important to question the extent to which different parts of the reproductive cycle (e.g. pregnancy and lactation) influence each other, not only in terms of requirements, but also in terms of appetite and body condition. This also reinforces the need to examine the short, medium and long term consequences of current feeding strategy.

Recently one of the major interests in these relationships has been the way in which they might influence feed intake in lactation. It has been known for some time that there is a relationship between feed intake in pregnancy and feed intake in lactation. This is well illustrated by the classical work of Salmon-Legagneur and Rerat (1962). *Table 29.12* shows

Table 29.12 THE RELATIONSHIP BETWEEN FEED INTAKE DURING PREGNANCY AND LACTATION IN THE SOW

Feed intake during pregnancy	*Daily feed intake during pregnancy* (kg)	*Daily feed intake during lactation* (kg)
High	3.68	4.95
Low	1.95	6.23

From Salmon-Legagneur and Rerat (1962)

that doubling of feed intake in pregnancy results in a 20% lower feed intake in lactation. Generally sows fed liberally in gestation have lower feed intakes, greater weight losses and higher milk yields in lactation. However, a more efficient system is not to overfeed in pregnancy but rather to give any extra food necessary in lactation. Such a system is also likely to be beneficial in terms of the long-term condition of the sow.

Feed intake in lactation may also be influenced by protein nutrition in pregnancy. For example, Mahan and Mangan (1975) reported that sows given 12% crude protein in lactation ate more when they had received high levels in pregnancy. However, the feed intakes of sows given 18% crude

Table 29.13 THE RELATIONSHIP BETWEEN FEED INTAKE (KG/DAY) IN LACTATION AND LEVELS OF DIETARY CRUDE PROTEIN IN PREGNANCY AND LACTATION

Protein in pregnancy (%)	*Protein in lactation*	
	12%	*18%*
9	4.2	6.2
13	4.8	6.5
17	5.9	6.2

From Mahan and Mangan (1975)

protein in lactation were unaffected by the protein level during pregnancy (*Table 29.13*). Only one reproductive cycle was involved but work over three and four reproductive cycles has supported their findings (O'Grady, 1971; O'Grady and Hanrahan, 1975). Some experiments have failed to show such relationships (Elsley and MacPherson, 1972; Greenhalgh *et al.*, 1977).

Weaning to remating

Although a lot of attention has been paid to nutrient requirements in pregnancy and lactation the period from weaning to remating is often neglected. It is interesting to consider the results of two experiments conducted at the University of Nottingham. The first of these examined the feed level from weaning to remating of sows which had just weaned their first litter. The results indicated a marked response to feed intake (*Table 29.14*) but this was not supported by the results of an experiment in which third parity sows were used (*Table 29.15*).

Table 29.14 EFFECT OF FEED LEVEL FROM WEANING TO REMATING ON REPRODUCTIVE PERFORMANCE OF GILTS HAVING JUST WEANED THEIR FIRST LITTER

	Feed intake (kg/day)		
	1.8	*2.7*	*3.6*
Interval from weaning to first oestrus (days)	21.6	12.0	9.3
Litter size	9.4	10.1	11.6
Conception rate (%)	58.3	75.0	100

From Brooks and Cole (1972)

Table 29.15 EFFECT OF FEED LEVEL FROM WEANING TO REMATING ON THE REPRODUCTIVE PERFORMANCE OF SOWS HAVING WEANED THEIR THIRD LITTER

	Feed intake (kg/day)			
	1.8	*2.7*	*3.6*	'*semi*-ad libitum'
Interval from weaning to first oestrus	4.92	4.69	5.0	5.0
Litter size	12.6	11.8	12.2	12.3
Conception rate (%)	100	100	100	100

From Brooks *et al.* (1975)

The explanation might be that sows having just weaned their first litter are particularly susceptible to reproductive failure and the extra energy intake would have a beneficial effect, whereas sows having just weaned their third litter would be at the peak of their reproductive performance. In this work there was also a greater weight loss in the first experiment than the second. The period from weaning to remating is short and it is useful insurance to feed high levels from weaning to remating particularly after the first litter. Short term 'flushing' (e.g. for a single day) has generally not been successful (Brooks and Cole, 1974). High protein levels can also be beneficial in improving the return to oestrus (Svajgr *et al.*, 1972).

Conclusions

In conclusion it can be said that while our knowledge on sow nutrition has served us well in the past, it is necessary to ensure that it continues to meet the needs of changes in the future. For example, it is important that we establish the changes that are necessary in nutrition as a result of changes in weaning age. It is also important that experiments are of sufficient duration to enable responses to be measured over longer periods.

References

ALLEE, G.L. and BAKER, D.H. (1970). Limiting nitrogenous factors in corn protein for adult female swine. *J. Anim. Sci.* **30**, 748–752

ANDERSON, L.L. and MELAMPY, R.M. (1972). Factors affecting ovulation rate in the pig. In *Pig Production* (D.J.A. Cole, Ed.), pp. 329–366. London, Butterworths

BAKER, D.H. and ALLEE, G.L. (1970). Effect of dietary carbohydrate on assessment of the leucine need for maintenance of adult swine. *J. Nutr.* **100**, 277–280

BAKER, D.H., BECKER, D.E., NORTON, H.W., JENSEN, A.H. and HARMON, B.G. (1966). Some qualitative amino acid needs of adult swine for maintenance. *J. Nutr.* **88**, 391–396

BAKER, D.H., BECKER, D.E., JENSEN, A.H. and HARMON, B.G. (1970a). Reproductive performance and progeny development as influenced by nutrition during pregnancy and lactation. *Illinois Pork Industry Day Report*, University of Illinois, Urbana, As-655a, p.15

BAKER, D.H., BECKER, D.E., JENSEN, A.H. and HARMON, B.G. (1970b). Protein source and level for pregnant gilts: A comparison of corn, Opaque-2 corn and corn soybean meal diets. *J. Anim. Sci.* **30**, 364–367

BOOMGAARDT, J., BAKER, D.H., JENSEN, A.H. and HARMON, B.G. (1972). Effect of dietary lysine levels on 21 day lactation performance. *J. Anim. Sci.* **34**, 408–410

BROOKS, P.H. and COLE, D.J.A. (1971). Effect of nutrition on reproductive performance in the pig. In *Proceedings of Nutrition Conference for Feed Manufacturers, University of Nottingham* (H. Swan and D. Lewis, Eds.), pp. 21–37. London, Churchill Livingstone

BROOKS, P.H. and COLE, D.J.A. (1972). Studies in sow reproduction. 1. The effect of nutrition between weaning and remating on the reproductive performance of primiparous sows. *Anim. Prod.* **15**, 259–264

BROOKS, P.H. and COLE, D.J.A. (1974). The effect of nutrition during the growing period and the oestrous cycle on the reproductive performance of the pig. *Livest. Prod. Sci.* **1**, 7–20

BROOKS, P.H., COLE, D.J.A., ROWLINSON, P., CROXSON, V.J. and LUSCOMBE, J.R. (1975). Studies in sow reproduction. 3. The effects of nutrition between weaning and remating on the reproductive performance of multiparous sows. *Anim. Prod.* **20**, 407–412

CLAWSON, A.J., RICHARDS, H.L., MATRONE, G. and BARRICK, E.R. (1963). Influence of level of total nutrient and protein intake on reproductive performance in swine. *J. Anim. Sci.* **22**, 662–669

COLE, D.J.A. (1978). Animo acid nutrition of the pig. In *Recent Advances in Animal Nutrition – 1978* (W. Haresign and D. Lewis, Eds.), pp. 59–72. London, Butterworths

DEGEETER, M.J., HAYS, V.W., KRATZER, D.D. and CROMWELL, G.L. (1970a). Reproductive and progeny performance of protein restricted gilts. *J. Anim. Sci.* **31**, 199 (Abstract)

DEGEETER, M.J., HAYS, V.W., CROMWELL, G.L. and KRATZER, D.D. (1970b). Reproductive and progeny performance of protein restricted gilts. *J. Anim. Sci.* **31**, 1020 (Abstract)

DEGEETER, M.J., HAYS, V.W., KRATZER, D.D. and CROMWELL, G.L. (1972). Reproductive performance of gilts fed diets low in protein during gestation and lactation. *J. Anim. Sci.* **35**, 772–777

DUEE, P.H. and RERAT, A. (1974). Etude du besoin en lysine de la truie gestante nullipare. *Journées de la recherche porcine en France,* pp. 49–56. Paris L'Institut Technique du Porc

DUEE, P.H. and RERAT, A. (1975). Etude de besoin en lysine de la truie gestante nullipare. *Annls Zootech.* **24**, 447–464

DUTT, R.H. and CHANEY, C.H. (1968). Feed intake and embryo survival in gilts. *Prog. Rep. Ky agric. Exp. Stn* **No.176**, 33–35

DYCK, G.W. and STRAIN, J.H. (1980). Post-mating feed consumption and reproductive performance in gilts. *Can. J. Anim. Sci.* **60**, 1060 (Abstract)

ELLIOTT, R.F., VAN DER NOOT, G.W., GILBREATH, R.L. and FISHER, H. (1971). Effect of dietary protein level on composition changes in sow colostrum and milk. *J. Anim. Sci.* **32**, 1128–1137

ELSLEY, F.W.H. and MACPHERSON, R.M. (1966). *9th International Congress on Animal Production, Edinburgh,* Science Progress p. 104 (Abstract)

ELSLEY, F.W.H. and MACPHERSON, R.M. (1972). Protein and amino acid requirements in pregnancy and lactation. In *Pig Production* (D.J.A. Cole, Ed.), pp. 417–434. London, Butterworths

FROBISH, L.T., SPEER, V.C. and HAYS, V.W. (1966). Effect of protein and energy intake on reproductive performance in swine. *J. Anim. Sci.* **25**, 729–733

GREENHALGH, J.F.D., ELSLEY, F.W.H., GRUBB, D.A., LIGHTFOOT, A.L., SAUL, D.W., SMITH, P., WALKER, N., WILLIAMS, D. and YEO, M.L. (1977). Comparison of four levels of dietary protein in gestation and two in lactation. *Anim. Prod.* **24**, 307–321

HAWTON, J.D. and MEADE, R.J. (1971). Influence of quantity and quality of protein fed the gravid female on reproductive performance and development of offspring in swine. *J. Anim. Sci.* **32**, 88–95

HEAP, F.C. and LODGE, G.A. (1967). Changes in body composition of the sow during pregnancy. *Anim. Prod.* **9**, 237–245

HESBY, J.H., CONRAD, J.N., PLUMLEE, M.P. and MARTIN, T.G. (1970a). Opaque-2 corn, normal corn and corn-soybean meal gestation diets for swine reproduction. *J. Anim. Sci.* **31**, 474–480

HESBY, J.H., CONRAD, J.N., PLUMLEE, M.P. and HARRINGTON, R.B. (1970b). Nitrogen balance and serum protein response of gestating swine fed Opaque-2 corn, normal corn and corn-soybean diets. *J. Anim. Sci.* **31**, 481–485

HILLYER, G.M. and PHILIPS, P. (1980). The effect of increasing feed level to sows and gilts in late pregnancy on subsequent litter size, litter weight and maternal body-weight change. *Anim. Prod.* **30**, 469 (Abstract)

HOLDEN, P.J., LUCAS, E.W., SPEER, V.C. and HAYS, V.W. (1968). Effect of protein level during pregnancy and lactation on reproductive performance in swine. *J. Anim. Sci.* **27**, 1587–1590

KEMM, E.H. and PIETERSE, P.J.S. (1968). The influence of protein on the productivity of Large White gilts. *Proc. S. Afr. Soc. Anim. Prod.* **7**, 133–135

LEWIS, A.J. and SPEER, V.C. (1973). Lysine requirement of the lactating sow. *J. Anim. Sci.* **37**, 104–110

LODGE, G.A. (1959). Nitrogen metabolism in the lactating sow. *J. agric. Sci., Camb.* **53**, 172–176

LODGE, G.A. (1972). Quantitative aspects of nutrition in pregnancy and lactation. In *Pig Production* (D.J.A. Cole, Ed.), pp. 399–416. London, Butterworths

LODGE, G.A., ELSLEY, F.W.H. and MACPHERSON, R.M. (1966a). The effects of level of feeding of sows during pregnancy. 1. Reproductive performance. *Anim. Prod.* **8**, 29–38

LODGE, G.A., ELSLEY, F.W.H. and MACPHERSON, R.M. (1966b). The effects of level of feeding of sows during pregnancy. 2. Changes in body weight. *Anim. Prod.* **8**, 499–506

LUCAS, E.W., HOLDEN, P.J., SPEER, V.C. and HAYS, V.W. (1969). Effect of protein level during pregnancy and lactation on plasma amino acid profile in swine. *J. Anim. Sci.* **29**, 429–432

MACPHERSON, R.M., HOVELL, F.D.DeB. and JONES, A.S. (1977). Performance of sows first mated at puberty or second or third oestrus and carcass assessment of once-bred gilts. *Anim. Prod.* **24**, 333–342

MAHAN, D.C. and MANGAN, L.T. (1975). Evaluation of various sequences on the nutritional carry-over from gestation to lactation with first-litter sows. *J. Nutr.* **105**, 1291–1298

MILLER, G.M., BECKER, D.E., JENSEN, A.W., HARMON, B.G. and NORTON, H.W. (1969). Effects of protein intake on nitrogen retention by swine during late pregnancy *J. Anim. Sci.* **28**, 204–207

O'GRADY, J.F. (1971). Level and source of protein in the diets of lactating sows. *Ir. J. agric. Res.* **10**, 17–30

O'GRADY, J.F. (1980). Energy and protein nutrition of the sow. In *Recent Advances in Animal Nutrition – 1980* (W. Haresign, Ed.), pp. 121–131. London, Butterworths

O'GRADY, J.F. and HANRAHAN, T.J. (1975). Influence of protein level and amino acid supplementation of diets fed in lactation on the performance of sows and their litters. 1. Sow and litter performance. *Ir. J. agric. Res.* **14**, 127–135

PIKE, I.H. and BOAZ, T.G. (1969). The effect on the reproductive performance of sows of dietary protein concentration and pattern of feeding in pregnancy. *J. agric. Sci., Camb.* **73**, 301–309

POND, W.G., DUNN, J.A., WELLINGTON, G.H., STOUFFER, J.R. and VAN VLECK, L.D. (1968). Weight gain and carcass measurements of pigs from gilts fed adequate vs protein-free diets during gestation. *J. Anim. Sci.* **27**, 1583–1586

RIPPEL, R.H., RASMUSSEN, A.H., JENSEN, A.H., NORTON, H.W. and BECKER, D.E. (1965a). Effect of level and source of protein on reproductive performance of swine. *J. Anim. Sci.* **24**, 203–208

RIPPEL, R.H., HARMON, E.G., JENSEN, A.H., NORTON, H.W. and BECKER, D.E. (1965b). Response of the gravid gilt to levels of protein as determined by nitrogen balance. *J. Anim. Sci.* **24**, 209–215

RIPPEL, R.H., HARMON, B.G., JENSEN, A.H., NORTON, H.W. and BECKER, D.E. (1965c). Essential amino acid supplementation of intact proteins fed to the gravid gilt. *J. Anim. Sci.* **24**, 373–377

RIPPEL, R.H., HARMON, B.G., JENSEN, A.H., NORTON, H.W. and BECKER, D.E. (1965d). Some amino acid requirements of the gravid gilt fed a purified diet. *J. Anim. Sci.* **24**, 378–382

SALMON-LEGAGNEUR, E. (1962). Effect of changes at different times in the plane of nutrition of pregnant sows. *Annls Zootech.* **11**, 173–180

SALMON-LEGAGNEUR, E. (1965). Some aspects of the nutritional relationships between pregnancy and lactation in the sow. *Annls Zootech.* **14**, Special Serial No. 1, 137

SALMON-LEGAGNEUR, E. and DUEE, P.H. (1972). Lysine supplementation of a cereal basal diet in pregnant and lactating sows. *Journées de la recherche porcine en France,* pp. 157–161. Paris, L'Institut Technique du Porc

SALMON-LEGAGNEUR, E. and RERAT, A. (1962). Nutrition of the sow during pregnancy. In *Nutrition of Pigs and Poultry* (J.T. Morgan and D. Lewis, Eds.), pp. 207–223. London, Butterworths

SOHAIL, M.A., COLE, D.J.A. and LEWIS, D. (1978a). Amino acid requirements of the breeding sow: the dietary lysine requirement during pregnancy. *Br. J. Nutr.* **39**, 463–468

SOHAIL, M.A., COLE, D.J.A. and LEWIS, D. (1978b). Amino acid requirements of the breeding sow. 2. The dietary lysine requirement of the lactating sow. *Br. J. Nutr.* **40**, 369–376

SPEER, V.C. (1975). Amino acid requirements for the lactating sow. (Calculated requirements and research on levels of the essential amino acids for lactation: An update). *Feedstuffs* **47**, 21–22

SVAJGR, A.J., HAMMELL, D.L., DEGEETER, M.J., HAYS, V.W., CROMWELL, G.L. and DUTT, R.H. (1972). Reproductive performance of sows on a protein restricted diet. *J. Reprod. Fert.* **30**, 455–458

VAN SCHOUBROEK, F. and VAN SPAENDONCK, R. (1973). Faktorieller Aufban des Energiebedorfs tragender Zuchtsauen. *Z. Tierphysiol. Tierernähr. Futtermittelk.* **31**, 1–21

WILKINSON, R. (1978). Amino acid nutrition of the lactating sow. PhD Thesis. University of Nottingham

WOERMAN, R.L. and SPEER, V.C. (1976). Lysine requirement for reproduction in swine. *J. Anim. Sci.* **42**, 114–120

30

MICRONUTRIENTS AND REPRODUCTION

B. HARDY and D. FRAPE,
Dalgety Spillers Ltd., Bristol, UK

All micronutrients are required for reproduction in a general sense for maintenance, cell enlargement and multiplication and for various secretions including milk, but there is little published evidence on the interactions between micronutrients and control of the endocrine system. An interaction between diet and reproduction is likely to occur where the supply of nutrients is only marginally adequate and variable. Evidence of marginality in traditional diets has been found for several nutrients in recent years, largely as a result of changes in two other environmental factors: (1) the confinement of sows and intensification of pig production and (2) developments in crop husbandry and production.

It is very difficult to establish the micronutrient requirements of the breeding pig due to the complex nature of pregnancy, the ability of the animal to mobilize stored nutrients at different stages of the reproductive cycle and the interactions that occur between the micronutrients. Many estimates of their requirement are based on the minimum level required to prevent deficiency symptoms and therefore bear no relation to the amount

Table 30.1 MICRONUTRIENT REQUIREMENTS OF BREEDING PIGS (DRY MATTER BASIS)

	Current UK estimate	*NRC (1979)*	
	All breeding pigs	*Bred gilts and sows Young and adult boars*	*Lactating gilts and sows*
Fat-soluble vitamins			
Retinol (mg/kg) or	0.7	1.3	0.7
β-Carotene (mg/kg)	8.4	17.8	8.9
Cholecalciferol (mg/kg) or Ergocalciferol	–	5.5	5.5
α-Tocopherol (mg/kg)	10.2	11.1	11.1
Water-soluble vitamins			
Riboflavin (mg/kg)	3.0	3.3	3.3
Pantothenic acid (mg/kg)	10.0	13.3	13.3
Cyanocobalamin (μg/kg)	15.0	16.7	16.7
Pyridoxine (mg/kg)	1.5	1.1	1.1
Choline (mg/kg)	1000–1900	1400	1400
Minerals and trace elements			
Calcium (%)	0.90	0.83	0.83
Phosphorus (%)	0.70	0.67	0.55
Manganese (mg/kg)	10	11	11
Iodine (mg/kg)	0.50	0.15	0.15

of the micronutrient required for maximization of reproductive performance. The best current UK estimates of micronutrient requirements for breeding pigs have been compared with those given by the NRC (1979) in *Table 30.1*. There are insufficient data on which to base an estimate of the requirement for those vitamins, minerals and trace elements not shown in *Table 30.1*.

There is a marked difference in the total quantity of a micronutrient supplied to the animal in its diet and the 'requirement' as currently known. The relationship between the requirement, the supply from natural sources and that normally added to the diet by means of a vitamin–mineral supplement is shown in *Table 30.2*. In most cases the average vitamin supplementation rate more than covers the level of minimum requirement and the total cost of this is currently £1.15/tonne of finished feed. The addition of minerals and trace elements to the diet will add approximately 85p/tonne of finished feed. The total micronutrient supplementation represents 1–1.5% of the total cost of the feed.

Fat-soluble vitamins

RETINOL

Where sows are kept on pasture the likelihood of reproductive problems occurring as a result of deficiencies in any of the fat-soluble vitamins is very slight. The vitamin A requirements are provided as the precursor β-carotene which has a potency in the pig of approximately 9% that of retinol. A deficiency of retinol leads not only to a decline in general health but more specifically to a decrease in ovarian size and to testicular atrophy. These symptoms are associated with a lowering of the concentration of retinol in the colostrum and milk and a diminution of the hepatic reserves in the neonate. Ocular lesions and other abnormalities are also apparent in the offspring. Where the diet contains between 7 and 10 mg β-carotene/kg dietary dry matter, normality is achieved. Spring and summer pasture would normally provide adequate amounts, including sufficient hepatic reserves to sustain the sow during the winter period. Cereal-based diets given to confined sows should contain 700 μg preformed retinol/kg dietary dry matter (14 μg/kg liveweight daily).

Although the needs of gestation and lactation are likely to differ it is not possible to consider the states of gestation and lactation independently because hepatic reserves in the dam provide a source of retinol for the milk. The amount of retinol transferred from the mother to the foetus is closely related to the amount available during pregnancy (Hjarde *et al.*, 1961) both from liver reserves and from feed. Nevertheless it has been calculated (Frape *et al.*, 1969) that total placental transfer is only 1–2% of the amount given during gestation. Large doses of retinol given during the later stages of pregnancy (Thomas, Loosli and Willman, 1947; Whiting, Loosli and Willman, 1949) have increased the concentration in the colostrum and in the livers of the offspring but the amount of retinol ingested by the piglet in the colostrum under normal conditions greatly exceeds that found in the livers at birth and exceeds the concentration found in the milk (Braude *et al.*, 1947; Hjarde *et al.*, 1961).

Table 30.2 COMPARISON OF MINIMUM REQUIREMENT, AMOUNT NATURALLY PRESENT AND AVERAGE SUPPLEMENTATION RATES AND THE AVERAGE COST OF SUPPLEMENTS OF SOW DIETS IN THE UK

	Minimum requirement (mg/kg DM)	*Naturally present in diet* (mg/kg DM)	*Average supplementation rate* (mg/kg DM)	*Average cost of supplementation* (p/tonne feed)	*Total dietary vitamin content/minimum requirement*
Vitamin A as retinol	0.7	Nil	4.7	21.6	6.7
Vitamin D_3 as cholecalciferol	(0.003)	Nil	0.057	2.0	19.2
DL-α tocopheryl acetate	10.2	8	13.2	12.6	2.1
Menaphthone salts	(0.2)	1	4.6	2.4	28.0
Riboflavin	3.0	1.1	5.7	9.0	2.3
Pantothenic acid	10.0	6	13.8	7.2	2.0
Pyridoxine	1.5	4	2.6	4.0	4.4
Cyanocobalamin	0.015	Nil	0.019	2.8	1.3
Choline	1000–1900	1000	121	12.7	0.6–1.1
Thiamin	(1.5)	4	1.1	1.5	3.4
Available Nicotinic acid	(14.0)	2	17.2	3.0	1.4
Biotin	(0.15)	0.1	0.10	27.0	1.4
Folic acid	(0.4)	0.4	0.25	1.2	1.6
Ascorbic acid	Nil	Nil	7.5	7.5	

Figures in parentheses are based on limited data and cannot be considered as an estimate of requirement.

Little objective evidence is available on a positive effect of retinol on the reproductive processes of pigs under practical conditions. A direct effect on conception or implantation may exist (Sevkovic *et al.*, 1973; Saryceva, 1968). As a result, some interest has been shown in the parenteral use of retinol in large doses given by injection at the time of mating. Evidence from Eastern Europe (Gondos *et al.*, 1970, Jugina, 1966) suggests increases in the number and weight of foetuses in gilts at the 60th day of pregnancy and an increase in litter size at birth as a result of such injections. The injection of 30 mg retinol (Sevkovic *et al.*, 1969) before mating has been shown to increase litter size and weight at birth. Even better results have been obtained with the injection of between 75 and 300 mg before mating, at mid-pregnancy and at the end of pregnancy. Similar injections (Jancic *et al.*, 1970; Stumpf, 1968; Petrenko and Zirnov, 1968) have been claimed to reduce the numbers stillborn and to lower neonatal mortality. However, responses of this nature have not been produced in western countries under practical conditions.

VITAMIN D_3

Two forms of vitamin D are of importance, vitamin D_2 (ergocalciferol) and vitamin D_3 (cholecalciferol). There is some evidence that pigs do not absorb ergocalciferol from feedstuffs and therefore cholecalciferol is the main precursor of vitamin D. In the absence of any meaningful data relating to the vitamin D requirement of sows, no further information is given.

α-TOCOPHEROL

The concentration of tocopherols per unit dry matter in fresh herbage is between five and ten times as great as that in some cereals or their products. Tocopherols are labile and the preservation of cereals by ensilage has been demonstrated to cause almost the complete loss of vitamin E activity. Symptoms of a deficiency of the vitamin have, therefore, been described amongst sows receiving ensiled cereals. It is uncertain whether the requirements/unit of bodyweight for reproduction are higher or lower than those for growth. Several reports suggest that the requirement of young pigs is greater than that of their dams. Sows maintained on a diet deficient in vitamin E and selenium for five reproductive cycles (Glienke and Ewan, 1974) produced piglets which died between 30 and 48 days of age in the fifth parity. The symptoms in the young include muscular dystrophy (Adamstone, Krider and James, 1949; Aydin, Pond and Kirtland, 1973) not generally seen in their dams, low piglet survival and depressed growth rate. Considerable reserves of α-tocopherol normally occur in the sow which meet the demands of the neonatal pig for several weeks by transference across the placenta and by secretion in the colostrum. Thus, sows maintained on a diet deficient in vitamin E and selenium have been shown to produce normal piglets during the first reproductive cycle of the deficiency and symptoms occurred only after five such cycles (Glienke and Ewan, 1974). The requirement for milk secretion could be considered to be greater than that for foetal growth as the milk provides an important source for the offspring (Cline, Mahan and Moxon 1974), although a deficient gestation diet has been shown to depress milk yield (Nielsen *et al.*, 1973). This effect could be ascribed to the vitamin E potency of the diet, although it could be ascribed more directly to the oxidized herring oil also present in the sow's diet. In the absence of the herring oil (Nielsen, 1971) additional α-tocopherol had no effect on litter size at birth or at three weeks, suggesting that the vitamin E potency of the basal diet may have been higher than in the more recent study.

The tissue aberrations appear to have a more damaging effect on the neonate than on the sow. They include abnormalities of cardiac and striated muscle, blood and the liver (Eggert *et al.*, 1957; Michel, 1968; Reid, 1968; Reid *et al.*, 1968; Money, 1970; Piper *et al.*, 1975). The effects of deficiency in these animals are exacerbated by some forms of injectable complexed iron used as a prevention of anaemia in piglets. The consequences are severe bruising, a rise in serum aspartate-amino-transferase and increased mortality as ferric iron imposes a severe strain on the redox system of the pig (Tollerz and Lannek, 1964; Pedersen, 1966; Tollerz, 1973; Miller *et al.*, 1973). Protection from iron intoxication has been achieved by the intramuscular injection, not only of α-tocopherol, but also of several synthetic antioxidants. A number of such synthetic compounds, including methylene blue, butylated hydroxyanisole (BHA), butylated hydroxytoluene (BHT) and ethoxyquin, apparently possess some biological activity comparable with that of the tocopherols, substituting for them in some functions provided they penetrate to and are retained by the tissue. These functions include suppression of iron intoxication/bruising (Tollerz, 1973), prevention of steatitis, nutritional muscular dystrophy and liver necrosis (Michel, 1968) and stimulation of erythropoiesis.

Selenium and vitamin E

It is appropriate to consider the role of selenium in vitamin E metabolism at this stage as the symptoms of vitamin E deficiency in the field can frequently be overcome by the dietary provision, or the injection, of selenium as sodium selenite. In areas where hepatic necrosis and mulberry heart disease have been endemic, probably associated with soils deficient in selenium, sows treated by intramuscular injection have a lowered incidence of stillbirths, neonatal mortality and hepatic necrosis (Van Vleet, Meyer and Olander, 1973). Similar responses in deficient sows have been obtained by providing diets containing 0.13 mg selenium/kg (Mahan *et al.*, 1974). The concentration required in the diet depends to some extent upon the source of the selenium as some natural sources have a greater availability than sodium selenite (Ku *et al.*, 1972). Supplementation of the sow results in the production of milk with a higher selenium content (Mahan, Moxon and Hubbard, 1975; 1977), but the growing pig is unlikely to obtain sufficient selenium from this source to meet its needs (Mahan *et al.*, 1974; Mahan, Moxon and Hubbard, 1975; 1977). A risk attached to the injection of selenium compounds results from the narrow limits of dose within which a response can be obtained, but which avoids toxicity. Experimental evidence has demonstrated that 1.5 mg selenium/kg liveweight daily is apparently not toxic (Van Vleet, Meyer and Olander, 1974) but similar evidence (Diehl, Mahan and Moxon, 1975) has suggested that an injection at the rate of 1.65 mg/kg liveweight is toxic. A toxic level in the diet of sows lies between 5 and 10 mg/kg diet. Storage of selenium occurs in the blood and other tissues and gilts reared for breeding on a selenium-deficient diet either die before breeding or suffer reproductive failure. An adequate level in the diet from a mixture of natural and synthetic sources appears to be in the region of 0.15 mg/kg for breeding pigs.

The results of numerous experiments conducted with pigs deficient of both α-tocopherol and selenium suggest that the one nutrient can substitute for the other. This outcome undoubtedly results from a sparing effect rather than a substitution as more recent experiments entailing extended depletion periods have indicated that both are required. Such experiments (Ewan *et al.*, 1969; Wastell *et al.*, 1972) have demonstrated the requirements for both nutrients and have shown that the requirements of the young pig appear to exceed those of their dams. A major function of α-tocopherol is suggested as being a protector of the essential fatty acids of cell membranes, whereas selenium acts in glutathione peroxidase and functions to mop up toxic tissue peroxides, production of which the vitamin has failed to inhibit.

Essential fatty acids

No information of note is available on the needs of the breeding pig for essential fatty acids (EFA). In view of the role of EFAs in prostaglandin synthesis, and the evidence in breeding rats, it is speculated that intensively managed breeding boars and sows have a particularly important requirement for EFA. The germ of cereal grains is a source of oil which is a rich source of linoleic acid (C18:2ω6)—a precursor of arachidonic acid (C20:4ω6) and an essential component of cellular lipids. Relatively few unsaturated fatty acids are precursors of this C20 acid and these must be provided by the diet. Certain prostaglandins, for which a role has been demonstrated in reproduction are synthesized from arachidonic acid. Whereas the oil from cereal grains is a rich source of the ω6 acid-linoleic, that oil is a relatively poor source of the ω3 fatty acid α-linolenic (C18:3) and timnodonic acid (C20:5). These unsaturated fatty acids are precursors of another group of prostaglandins and are found abundantly in herbage lipids and fish oils. It remains to be seen whether pasture species can confer any breeding advantage in pigs through this medium.

Water-soluble vitamins

A dietary requirement in breeding stock for many of the water-soluble vitamins has not been demonstrated. This situation may arise as a consequence of the relatively voluminous hind gut of the mature pig in which a vast population of microorganisms synthesize a number of the vitamins of this group. The absorption from the intestines of some members of the B vitamin group synthesized by the gut microflora has been demonstrated in several mammalian species. The extent, however, to which this acts as a biologically important source for direct absorption remains to be determined, although undoubtedly a portion of this source is made available to the pig through the mechanism of coprophagy. Several experiments in pigs have demonstrated a better reproductive performance in sows having access to their faeces, but whether this results from an enhanced supply of B vitamins and vitamin K, or from other nutrients, or indeed from a general reduction in environmental stress, has not been determined.

RIBOFLAVIN

An indirect effect on breeding performance of sows might be anticipated from the known enhanced antibody response in the young pig to injected antigens when deficient diets are supplemented with either pantothenic acid, pyridoxine or riboflavin (Harmon *et al.*, 1963). Whether the requirements for normality of this function differ from those for optimum growth is, however, not known. Fresh pasture herbage is a relatively rich source of riboflavin, but supplementation of cereal-based diets is generally required. During absorption riboflavin is phosphorylated in the wall of the small intestine and in its phosphorylated form acts as a coenzyme in many reactions. In a deficient state sows lose their appetite and the sows either do not conceive or a high neonatal mortality occurs amongst piglets. The dietary level found adequate for growing pigs (Miller and Ellis, 1951) is inadequate for satisfactory reproduction and lactation in the sow. Breeding sows appear to require a minimum of 3.0 mg riboflavin/kg dietary dry matter (Miller and Ellis 1951; Pochernnyaeva, 1976).

NICOTINIC ACID AND PANTOTHENIC ACID

Nicotinic acid is present in cereals in a bound form which is completely, or largely, unavailable to the pig. Diets based upon maize which are low in tryptophan have been demonstrated repeatedly to cause nicotinic acid deficiency in growing pigs. On the other hand, the only experiment in which it was attempted to produce a deficiency in sows (Ensminger, Colby and Cunha, 1951) failed to produce definite symptoms, despite an absence of gut synthesis. Symptoms of pantothenic acid deficiency have, on the other hand, been produced in gilts (Ensminger, Colby and Cunha, 1951). Diets containing up to only 7.8 mg/kg dietary dry matter (Teague, Palmer and Grifo, 1970; Teague, Grifo and Palmer, 1971) caused a marginally reduced reproductive performance and symptoms in the progeny of second generation gilts. As a part of the coenzyme A molecule the vitamin has a key role in many metabolic reactions, not confined to any particular reproductive function. Supplementation of breeding diets is recommended and some evidence suggests that the breeding performance of boars, particularly under intensive conditions, can be affected by an inadequacy in natural diets causing abnormalities in gait.

PYRIDOXINE AND CYANOCOBALAMIN

Experiments on the response of gilts and sows to pyridoxine have been both small in scale and short in duration. The supplementation of diets containing 1.0–2.3 mg pyridoxine/kg dietary dry matter was associated with a non-significant increase in the number of piglets weaned per litter (Ritchie *et al.*, 1960; Wöhlbier and Siegel, 1967). Supplementation gave an increase in the pyridoxine content of the sow's milk.

Pyridoxine is widely distributed in feed ingredients and is synthesized by gut bacteria. Errors may, however, be made in estimating the potency of diets as the standard test organism *S. uvarum* underestimates the potency of pyridoxal and pyridoxamine for the pig. Pyridoxine functions biologically in the phosphorylated form as a coenzyme in transaminase and decarboxylase activities. Therefore it functions in the synthesis of porphyrin rings in haemoglobin formation and in lymphocyte production and hence in antibody production. More specifically for the breeding pig its role in the synthesis of endocrine secretions awaits investigation. Studies with rats have demonstrated such an association with the thyroid gland, adrenal glands and the testes. (The intensification of sow production with their maintenance in stalls lends itself to such investigation.)

Cyanocobalamin is also synthesized by gut bacteria and therefore sows maintained in deep litter yards may be less susceptible to deficiency than those housed in stalls. The vitamin is, however, stored in the blood and liver so that symptoms of deficiency only develop gradually. Conversely supplementation of the diet of sows is of more benefit than the supplementation of gilt diets (Frederick and Brisson, 1961) and successive litters produced by sows given a diet low in the vitamin are progressively weaker (Kralovanszky, Eöri and Kallai, 1954; Teague and Grifo, 1966). Cyanocobalamin is involved in the transfer of the so-called C1 fragment in association with folic acid. The two vitamins tend to partially spare the requirement for each other and the need for them is reduced if the diet contains a good supply of compounds acting as a source of C1 groups such as choline, betaine and methionine. An increasing scarcity of animal proteins may imply that supplementation of breeding pig diets is increasingly based on vegetable products, less reliance being placed on animal proteins, and therefore supplementation with cyanocobalamin is becoming more necessary.

Cyanocobalamin in milk is strongly bound to minor whey proteins which are present in excess in sow's milk (Gregory and Holdsworth, 1955) and it has been postulated that this increases the amount of the vitamin available to the piglet directly, but also indirectly, by preventing its uptake by intestinal microorganisms (Ford, 1974). The much higher binding capacity of sow's milk, compared with cow's milk, may have implications from the point of view of the relative requirements for the vitamin of piglets reared on sow's milk and those reared on diets based on cow's milk. The general conclusion concerning the requirements for the vitamin in sows is that they should be provided with 15 μg/kg dietary dry matter, but in the light of evidence with older pigs (Frederick, 1962; Teague and Grifo, 1966) the needs may be increased considerably for older sows.

BIOTIN

Biotin is again synthesized by gut bacteria, but there appears to be a definite need for the vitamin in the diet of sows. Much of this is undoubtedly met by natural feed sources, but the relative availability to the pig of different sources is unknown. The availability of microbiologically assessed biotin to the chick is low for barley, wheat, milo and some protein feeds of animal origin but is high for maize and oilseed meals. It is widely involved metabolically, including its involvement in the synthesis of fatty acids found in healthy skin. There is considerable storage in the liver and probably similar arguments apply to it as apply to cyanocobalamin and the age of the sow. Supplementation either by injection (Cunha, Adams and Richardson, 1968) or in the feed of sows (Brooks, Smith and Irwin, 1977) receiving a basal diet containing just over 100 μg biotin/kg according to microbiological assay, reduced the incidence of foot lesions and dry or eczematous skin. Sows given supplements (Brooks, Smith and Irwin, 1977) produced more piglets and were remated significantly sooner after weaning than were control sows.

FOLIC ACID

Folic acid is another vitamin synthesized by gut bacteria and although normally no dietary requirement occurs for the growing pig, there may be a need in sows under some conditions, particularly where vitamin B_{12} intake is limited. Folate in sow's milk is strongly bound to minor whey proteins and it has been postulated that this increases the amount of the vitamin available to piglets (Ford and Scott, 1975). The authors suggested that in view of the low concentration of folate in sow's milk and the rapid postnatal increase in liver folate, folate-synthesizing bacteria in the intestine may make a significant contribution to the folic acid nutrition of the piglet. The possibility is suggested that in acting to sequester the milk folate the unsaturated binder protein in milk may reduce the growth of folate-dependent bacteria in the gut, and hence encourage the growth of bacteria that contribute to the folate needs of the piglet.

CHOLINE

The requirement for dietary choline depends upon the quantitative presence of other methyl donors and the dietary content of vitamin B_{12} (Dyer and Krider, 1950). Earlier evidence has suggested that choline deficiency in gilts is manifested by a splay-leg condition in their neonatal progeny but this claim could not be clearly substantiated by more recent work. Experiments covering six parities (Kornegay and Meacham, 1973) have shown that sows receiving a maize–soyabean meal diet containing 150 g crude protein/kg produced fewer piglets in the fifth and sixth litters when no supplementary choline was given. An experiment in which similar diets were involved, but conducted at nine centres, showed that the addition of 770 mg choline/kg diet increased the total number of live pigs born per litter and the number of pigs alive at two weeks post-partum. A further experiment (Stockland and Blaylock, 1974) in which maize–soya diets were supplemented with 410 mg and 820 mg choline/kg showed that conception rate, farrowing rate and total and live piglets farrowed were increased by the choline addition.

The evidence in total suggests that sow diets containing 150 g of crude protein, including 2 g methionine, also require the diet to contain between 1400 mg and 1900 mg choline/kg dry matter.

Trace minerals

IRON

A number of attempts has been made to increase the haemoglobin level in the blood and the hepatic iron reserves of piglets at birth by administering various iron salts in chelated and free forms to the sow. Evidence on the success of these measures is conflicting, although iron from more complex molecules may be transferred across the placenta (Agapitova, 1970; Brady *et al.*, 1975). Higher than normal concentrations of copper in pregnancy diets have also been shown to increase the amount of iron in piglets at birth (Hemingway, Brown and Luscombe, 1974). The piglet at birth is relatively well endowed with iron (blood haemoglobin is approximately 80 g/l) and anaemia arises principally as a consequence of rapid growth rate on a diet of mother's milk deficient in iron. Regardless of the scale of iron storage at birth, it is probable that pigs denied access to iron other than that supplied by sow's milk will develop anaemia within a few weeks when blood haemoglobin may fall to 40 g/l. Various sources of iron given to the lactating sow may bring about an increase in the iron content of the milk but owing to the low concentration this can have only a small effect on the piglet. If the piglet has access to iron provided by the sow's feed and faeces, then performance may be improved even if the piglet is denied iron from other sources. It is very unlikely that an iron deficiency anaemia would develop in breeding sows or boars given conventional diets as the iron content of most natural ingredients would normally meet their needs. If the diet is, on the other hand, imbalanced in respect of some other minerals, viz. calcium, copper, manganese or phosphorus, then interference with iron metabolism may occur in the sow. For example, feeds containing growth promotion levels of copper could induce hypochromic microcytic anaemia through impairment of iron absorption (Gipp *et al.*, 1974).

ZINC

Zinc deficiency problems in pigs have arisen largely as a consequence of interactions with other mineral elements. Evidence in the growing pig has indicated that where copper is used as a growth promoter this has led to the development of parakeratosis in pigs fed vegetable protein-based diets unless the zinc supplementation has been adequate. Increased dietary concentrations of calcium and phytic acid also increase the requirement for zinc in that phytic acid adversely affects the zinc availability and a rise in the dietary content of calcium increases urinary excretion of zinc (Beardsley and Forbes, 1957). Whether high levels of dietary calcium also influence absorption from the intestines is not clear. Dry diets are more likely to produce parakeratosis than the same diets fed wet (Lewis, Grummer and Hoekstra, 1957) possibly due to the hydrolysis of phytates in diets which are soaked (Frape, Wayman and Tuck, 1979). Furthermore, iron may act as an antagonist to zinc in relieving parakeratosis (Hoefer *et al.*, 1960) and high levels of dietary zinc may depress the utilization of dietary iron (Cox and Hale, 1962). No evidence of these deficiencies and interactions has been demonstrated in sows in practice.

Observations made in a range of species other than the pig have shown that spermatogenesis and the development of the primary and secondary sex organs in the male and all phases of the reproductive process in the female from oestrus to parturition and lactation can be adversely affected by zinc deficiency. The impaired development of the secondary sex glands may be subsidiary to the inanition of zinc deficiency which could result in a reduced gonadotrophin output and consequential fall in androgen production. Testicular atrophy and failure of spermatogenesis, on the other hand, is due directly to lack of zinc. In the female congenital abnormalities and failure to suckle at birth have been observed, together with an impairment of lactation. Zinc levels in the milk are reduced by a deficiency of this element so that the offspring may suffer further. Zinc-deficient females deliver their offspring with extreme difficulty and suffer excessive bleeding.

Limited evidence in sows shows that suboptimal dietary levels of zinc reduce the size of litters and the zinc content of some of the tissues in the young. No abnormalities in foetal development or in maternal behaviour have been shown (Hoekstra *et al.*, 1967). Skeletal abnormalities have been demonstrated in foetuses of several species resulting from both zinc and manganese deficiency and a reduction in the size and strength of the femur in zinc-deficient baby pigs has been reported by Miller *et al.* (1968); but the effect appeared to be largely a consequence of reduced food intake. However, in another study with weanling pigs, the reduced skeletal growth associated with zinc deficiency was most apparent in the low activity of the epiphyseal growth plate and at other points of osteoblastic prominence (Norrdin *et al.*, 1973). Zinc is involved primarily in nucleic acid and protein metabolism and hence in the fundamental processes of cell replication. The utilization of amino acids in the synthesis of protein is impaired in zinc deficiency and the general consensus is that growing boars and gilts have a higher requirement for zinc than do castrates, probably resulting from their higher protein demands. Two papers, however, suggest that zinc requirements of breeding sows do not exceed 40 mg/kg diet containing 1.4% calcium (Pond and Jones, 1964; Hennig, 1965). One of these experiments entailed the use of fishmeal as a source of protein and as other work has shown that the availability of zinc in soyabean meal is less than that in milk

(Smith, Plumlee and Beeson, 1962), it is generally considered that the diet of breeding pigs should contain 50 mg zinc/kg, particularly where it is based on vegetable proteins. This would, in addition, allow for some body reserves in the offspring (Palludan and Wegger, 1972; Hedges, Kornegay and Thomas, 1976). The recommendation assumes normal levels of other mineral elements for breeding stock. Zinc supplementation of a maize–soyabean meal diet containing 30–34 mg Zn/kg and 1.6% calcium significantly increased the number of live pigs per litter without affecting the birth or weaning weights (Hoekstra *et al.*, 1967).

MANGANESE

The requirement of pigs for satisfactory reproduction may be higher than that required for body growth. In fact satisfactory growth in young pigs has been reported with diets supplying only 1 mg/kg (Johnson, 1943; Plumlee *et al.*, 1956). This level incurs a marked tissue manganese depletion and when such diets are fed throughout gestation and lactation, skeletal abnormalities and impaired reproduction become apparent. Other studies have shown that diets containing less than 3 mg/kg manganese cause a depression in reproductive performance (Johnson 1943) but concentrations of 6 or 12 mg/kg appear to be adequate (Johnson, 1943; Grummer *et al.*, 1950). The precise mode of action of manganese in preventing reproductive defects in both males and females has not yet been established. It has been suggested (Doisy, 1974) that lack of manganese inhibits the synthesis of cholesterol, which in turn limits the synthesis of sex hormones. In laboratory species a deficiency of manganese is associated with defective ovulation, testicular degeneration and infant mortality.

COPPER

Very little work has been conducted on the copper requirements of the breeding pig. One report (Carpenter, 1946/47) showed that by increasing the copper content of the diet from 7 to 35 mg/kg an increase in litter size of two to three pigs occurred. In the female rat and guinea pig a copper deficiency results in reproductive failure due to foetal death and resorption (Dutt and Mills, 1960; Hall and Howell, 1969; Howell and Hall, 1969; 1970). The oestrous cycle remains unaffected and conception appears to be uninhibited. Normal foetal development in copper-deficient rats was shown to cease on the 13th day of pregnancy (Howell and Hall, 1969) and necrosis of the placenta became apparent on the 15th day.

IODINE

Little work has been conducted on the iodine requirements of breeding pigs. Growing pigs fed maize–soyabean meal diets have a requirement of iodine in the region of 0.1 mg/kg (Sihombing, Cromwell and Hays, 1974) and the results of other experiments suggest that the requirement does not exceed 0.14 mg/kg. Iodine requirements are influenced by the presence of goitrogens and certain other elements in the diet. Commonly found goitrogens are the thiocyanates. Some elements including rubidium and arsenic can interfere with iodine uptake and induce goitre.

Changes in thyroxine secretion are reflected in changes in the adrenal cortex and a relationship between the thyroid and the gonads is apparent in all male and female mammals. Reproductive failure is often the outstanding manifestation of iodine deficiency with the birth of weak, dead or hairless young. Foetal development may be arrested at any stage leading to early death and resorption, abortion and stillbirth. The live birth of weak young frequently associated with prolonged gestation often occurs together with retention of foetal membranes at parturition. The occurrence of hairless piglets has been prevented by adding potassium iodide, equivalent to 0.17 mg iodine/kg diet given to pregnant sows (Hart and Steenbock, 1918) and intrauterine death has been reduced by feeding sows 0.77 mg iodine/100 kg liveweight daily from three months before service (Varganov, 1965). Supplements are frequently given to breeding females by including iodized salt which provides 0.35 mg iodine/kg diet (Andrews *et al.*, 1948). A roughly similar quantity of 0.4 mg in diets containing 12% of extracted rapeseed meal was sufficient to prevent intrauterine and neonatal deaths in piglets if fed to females throughout pregnancy (Devilat and Skoknic, 1971). Suckling piglets should receive sufficient quantities from their dams as it has been demonstrated (Iwarsson, Bengtsson and Ekman, 1973) that between 20% and 45% of the iodine intake of sows is secreted in the milk.

No scientific evidence has been produced in breeding pigs to establish their dietary requirement for vitamins D_3, K, thiamine and ascorbic acid. There appears not to be a need for the supplementation of practical diets with any of them, although the claims that navel bleeding is caused by a dietary deficiency of vitamin K and ascorbic acid may require further investigation.

Conclusion

The amount of evidence on which to base requirement figures for micronutrients for breeding pigs is sparse. The high cost of purified or semi-purified diets for sows, plus the difficulty in measuring the potency and controlling the potency of micronutrients in natural diets, has been restrictive to experimentation. Developments in instrumentation and assay techniques, coupled with the individual housing and management of sows, should lend themselves to easier research in the future.

More information is required on the total micronutrient content and availability values in raw materials and on the effect of processing techniques used in feed manufacture on micronutrient stability. This, linked to better understanding of interactions between macro and micronutrients, should enable the breeding pig's requirement, once known, to be satisfied within a given daily feed allowance, thereby ensuring the maximization of economic production of weaned pigs.

References

ADAMSTONE, F.B., KRIDER, J.L. and JAMES, M.F. (1949). Response of swine to vitamin E deficient rations. *Ann. N.Y. Acad. Sci.* **52**, 260

AGAPITOVA, G.N. (1970). Iron glycerophosphate in rations for pregnant sows. *Sb. nauch. Rab. vses. nauchno-issled. Inst. Zhivotn.* **20**, 55–57 (*Nutr. Abstr. Rev.* **42**, 769)

ANDREWS, F.N., SHREWSBURY, C.L., HARPER, C., VESTAL, C.M. and DOYLE L.P. (1948). Iodine deficiency in newborn sheep and swine. *J. Anim. Sci.* **7**, 298–310

AYDIN, A., POND, W.G. and KIRTLAND, D. (1973). Dietary PUFA – Vitamin E: Dam–progeny effects in pigs. *J. Anim. Sci.* **37**, 274 (Abstract)

BEARDSLEY, D.W. and FORBES, R.M. (1957). Growth and chemical studies on zinc deficiency in the baby pig. *J. Anim. Sci.* **16**, 1038 (Abstract)

BRADY, P.S., KU, P.K., GREEN, F.F., ULLREY, D.E. and MILLER, E.R. (1975). Evaluation of an amino acid–iron chelate hematinic. *J. Anim. Sci.* **41**, 308 (Abstract)

BRAUDE, R., COATES, M.E., HENRY, K.M., KON, S.K., ROWLAND, S.J., THOMPSON, S.Y. and WALKER, D.M. (1947). A study of the composition of sow's milk. *Br. J. Nutr.* **1**, 64–77

BROOKS, P.H. SMITH, D.A. and IRWIN, V.C.R. (1977). Biotin-supplementation of diets; the incidence of foot lesions and the reproductive performance of sows. *Vet. Rec.* **101**, 46–50

CARPENTER, L.E. (1946/47). *Rep. Hormel Inst. Univ. Minn.* p.19

CLINE, J.H., MAHAN, D.C. and MOXON, A.L. (1974). Progeny effects of supplemental vitamin E in sow diets. *J. Anim. Sci.* **39**, 974 (Abstract)

COX, D.H. and HALE, O.M. (1962). Liver iron depletion without copper loss in swine fed excess zinc. *J. Nutr.* **77**, 225–228

CUNHA, T.J., ADAMS, C.R. and RICHARDSON, C.E. (1968). Observations on biotin needs of the pig. *Feedstuffs, Minneap.* **40**, 22

DEVILAT, J. and SKOKNIC, A. (1971). Feeding high levels of rapeseed meal to pregnant gilts. *Can. J. Anim. Sci.* **51**, 715–719

DIEHL, J.S., MAHAN, D.C. and MOXON, A.L. (1975). Effects of single intramuscular injections of selenium at various levels to young swine. *J. Anim. Sci.* **40**, 844–850

DOISY, E.A. JR. (1974). Effects of deficiency in manganese upon plasma levels and clotting protein and cholesterol in man. In *Trace Element Metabolism in Animals Vol. 2*, (W.G. Hoekstra *et al.*, Eds.), pp. 664–667. Baltimore, Univ. Park Press

DUTT, B. and MILLS, C.F. (1960). Reproductive failure in rats due to copper deficiency. *J. comp. Path. Ther.* **70**, 120–125

DYER, I.A. and KRIDER, J.L. (1950). Choline versus betaine and expeller versus solvent soyabean meal for weanling pigs. *J. Anim. Sci.* **9**, 176–179

EGGERT, R.G., PATTERSON, E., AKERS, W.T. and STOKSTAD, E.L.R. (1957). The role of vitamin E and selenium in the nutrition of the pig. *J. Anim. Sci.* **16**, 1037 (Abstract)

ENSMINGER, M.E., COLBY, R.W. and CUNHA T.J. (1951). Effect of certain B-complex vitamins on gestation and lactation in swine. *Stn Circ. For. Facts agric. Exp. Stn Wash. St. Coll.*, No. 134

EWAN, R.C., WASTELL, M.E., BICKNEL, E.J. and SPEER, V.C. (1969). Performance and deficiency symptoms of young pigs fed diets low in vitamin E and selenium. *J. Anim. Sci.* **29**, 912–915

FORD, J.E. (1974). Some observations on the possible nutritional significance of vitamin B_{12} and folate-binding proteins in milk. *Br. J. Nutr.* **31**, 243–257

FORD, J.E. and SCOTT, K.J. (1975). Uptake of vitamin B_{12} in the piglet. *Rep. natn. Inst. Res. Dairy.* 1973–74, 75

FRAPE, D.L., WAYMAN, B.J. and TUCK, M.G. (1979). The utilisation of phosphorus and nitrogen in wheat offal by growing pigs. *J. agric. Sci., Camb.* **93**, 133–146

FRAPE, D.L., WOLF, K.L., WILKINSON, J. and CHUBB, L.G. (1969). Liver weight and its N and vitamin A contents in piglets from sows fed two levels of protein and food. *J. agric. Sci., Camb.* **73**, 33–40

FREDERICK, G.L. (1962). Practical and physiological studies of the relationship between vitamin B_{12} and reproduction in swine. In *Vitamin B_{12} and Intrinsic Factor: 2. Europaisches Symposium*, (H.C. Heinrich, Ed.), p. 580. Stuttgart, Ferdinand Enke Verlag

FREDERICK, G.L. and BRISSON, G.J. (1961). Some observations on the relationship between vitamin B_{12} and reproduction in swine. *Can. J. Anim. Sci.* **41**, 212–219

GIPP, W.F., POND, W.G., KALLFELZ, F.A., TASKERN, J.B., VAN CAMPEN, D.R. KROOK, L. and VISEK, W.J. (1974). Effect of dietary copper, iron and ascorbic acid levels on hematology, blood and tissue copper, iron and zinc concentrations and ^{64}Cu and ^{59}Fe metabolism in young pigs. *J. Nutr.* **104**, 532–541

GLIENKE, L.R. and EWAN, R.C. (1974). Selenium in the nutrition of the young pig. *J. Anim. Sci.* **39**, 975 (Abstract)

GONDOS, M., PALAMARU, E., HARSIAN, A., HARSIAN, E., NICHITIN, A., MAXIM, V. and NICOLOF, E. (1970). Optimum amounts of vitamin A for pigs. *Lucr. stiint. Inst. Cerc. zooteh.* **27**, 551–564

GREGORY, M.E. and HOLDSWORTH, E.S. (1955). The occurrence of a cyanocobalamin-binding protein in milk and the isolation of a cyanocobalamin–protein complex from sow's milk. *Biochem. J.* **59**, 329–334

GRUMMER, R.H., BENTLEY, O.G., PHILLIPS, P.H. and BOHSTEDT, G. (1950). The role of manganese in growth, reproduction and lactation of swine. *J. Anim. Sci.* **9**, 170–175

HALL, G.A. and HOWELL, J. McC. (1969). The effect of copper deficiency on reproduction in the female rat. *Br. J. Nutr.* **23**, 41–45

HARMON, B.G., MILLER, E.R., HOEFER, J.A., ULREY, D.E. and LUECKE, R.W. (1963). Relationship of specific nutrient deficiencies to antibody production in swine. II. Pantothenic acid, pyridoxine or riboflavin. *J. Nutr.* **79**, 269–275

HART, E.B. and STEENBOCK, H. (1918). Thyroid hyperplasia and the relation of iodine to the hairless pig malady. *J. biol. Chem.* **33**, 313–323

HEDGES, J.D., KORNEGAY, E.T. and THOMAS, H.R. (1976). Comparison of dietary zinc levels for reproducing sows and the effect of dietary zinc and calcium on the subsequent performance of their progeny. *J. Anim. Sci.* **43**, 453–463

HEMINGWAY, R.G., BROWN, N.A. and LUSCOMBE, J. (1974). The effect of iron and copper supplementation of the diet of sows during pregnancy and lactation on the iron and copper status of their piglets. In *Trace Element Metabolism in Animals Vol. 2*, (W.G. Hoekstra *et al.*, Eds.), pp. 601–604. Baltimore, Univ. Park Press

HENNIG, A. (1965). Effect of supplementing mother's ration with Ca and Zn on composition of the piglet and of sow's milk. *Arch. Tierernähr.* **15**, 331–383

HJARDE, W., NEIMANN-SORENSEN, A., PALLUDAN, B. and HAVSKOV SORENSEN, P. (1961). Investigations concerning vitamin A requirement utilisation and deficiency symptoms in pigs. *Acta Agric. scand.* **11**, 13–53

HOEFER, J.A., MILLER, E.R., ULLREY, D.E, RITCHIE, H.D. and LUECKE, R.W. (1960). Interrelationships between calcium, zinc, iron and copper in swine feeding. *J. Anim. Sci.* **19**, 249–259

HOEKSTRA, W.G., FALTIN, E.C., LIN, C.W., ROBERTS, H.F. and GRUMMER, R.H. (1967). Zinc deficiency in reproducing gilts fed a diet high in calcium and its effect on tissue zinc and blood serum alkaline phosphatase. *J. Anim. Sci.* **26**, 1348–1357

HOWELL, J.McC. and HALL, G.A. (1969). Histological observations on foetal resorption in copper deficient rats. *Br. J. Nutr.* **23**, 47–50

HOWELL, J.McC. and HALL, G.A. (1970). Infertility associated with experimental copper deficiency in sheep, guinea-pigs and rats. In *Trace Element Metabolism in Animals Vol. 1*, (C.F. Mills, Ed.), pp. 106–109. Edinburgh, Livingstone

IWARSSON, K. BENGTSSON, G. and EKMAN, L. (1973). Iodine content in colostrum and milk of cows and sows. *Acta vet. scand.* **14**, 254–262

JANCIC, S., PESUT, M., CRNOJEVIC, Z., CRNOJEVIC, T., and COSIC, H. (1970). Effect of large doses of vitamin A and lucerne carotene on reproduction in sows and vitamin A in blood of the sows and blood and liver of their piglets. *Symp. Pig Prod. Anim. Nutr., Zagreb,* pp. 72–89

JOHNSON, S.R. (1943). Studies with swine on rations extremely low in manganese. *J. Anim. Sci.* **2**, 14–22

JUGINA, A.D. (1966). Effect of a mixture of vitamins on the fertility of gilts. *Svinovodstvo* **20**, 30–32

KORNEGAY, E.T. and MEACHAM, T.N. (1973). Evaluation of supplemental choline for reproducing sows housed in total confinement on concrete or in dirt lots. *J. Anim. Sci.* **37**, 506–509

KRALOVANSZKY, U.P., EORI, E. and KALLAI, L. (1954). The influence of vitamin B_{12} on the reproduction of Mangalica sows. *Allattenyesztes* **3**, 331

KU, P.K., MILLER, E.R., WAHLSTROM, R.C. GROCE, A.W. and ULLREY, D.E. (1972). Supplementation of naturally high selenium diets. *J. Anim. Sci.* **35**, 218 (Abstract No. 207)

LEWIS, P.K. Jr., GRUMMER, R.H. and HOEKSTRA, W.G. (1957). The effect of method of feeding upon the susceptibility of the pig for parakeratosis. *J. Anim. Sci.* **16**, 927–936

MAHAN, D.C., MOXON, A.L. and HUBBARD, M.D. (1975). Efficiency of supplemental Se on sow and progeny tissue Se. *J. Anim. Sci.* **41**, 319 (Abstract No. 297)

MAHAN, D.C., MOXON, A.L. and HUBBARD, M.D. (1977). Efficacy of inorganic selenium supplementation to sow diets on resulting carry-over to their progeny. *J. Anim. Sci.* **45**, 738–746

MAHAN, D.C., PENHALE, L.H., CLINE, J.H., MOXON, A.L., FETTER, A.W. and YARRINGTON, J.T. (1974). Efficacy of supplemental selenium in reproductive diets on sow and progeny performance. *J. Anim. Sci.* **39**, 536–543

MICHEL, R.L. (1968). The role of vitamin E, selenium and methionine in dietary liver necrosis and nutritional muscular dystrophy in the pig. *Diss. Abstr.* **28**, 3764B

MILLER, C.O. and ELLIS, N.R. (1951). The riboflavin requirements of growing swine. *J. Anim. Sci.* **10**, 807–812

MILLER, E.R., LUECKE, R.W., ULLREY, D.E., BALTZER, B.V., BRADLEY, B.L. and HOEFER, J.A. (1968). Biochemical, skeletal and allometric changes due to zinc deficiency in baby pigs. *J. Nutr.* **95**, 278–286

MILLER, E.R., HITCHCOCK, J.P., KUAN, K.K., KU, P.K., ULLREY, D.E. and KEAHEY, K.K. (1973). Iron tolerance and E–Se status of young swine. *J. Anim. Sci.* **37**, 287 (Abstract)

MONEY, D.F.L. (1970). Vitamin E and selenium deficiencies and their possible aetiological role in the sudden death in infants syndrome. *N.Z. med. J.* **70**, 32

NATIONAL RESEARCH COUNCIL (1979). *Nutrient Requirements of Domestic Animals, No. 2 Nutrient Requirements of Swine,* 8th Revised Edition. Washington, National Research Council

NIELSEN, H.E. (1971). The influence of nutrition on the reproduction performance of boars and sows. *Svind Symp. Zagreb* **1**,

NIELSEN, H.E., HØJGAARD-OLSEN, N.J., HJARDE, W. and LEERBECK, E. (1973). Vitamin E content in colostrum and sow's milk and sow milk yield at two levels of dietary fat. *Acta Agric. scand., Suppl.* **19**, 35

NORRDIN, R.W., KROOK, L., POND, W.G. and WALKER, E.F. (1973). Experimental zinc deficiency in weanling pigs on high and low calcium diets. *Cornell Vet.* **63**, 264–290

PALLUDAN, B. and WEGGER, I. (1972). Zinc metabolism in pigs III. Placental transfer of zinc in normal and zinc deficient gilts and its influence on fetal development. *Arsberetn. Inst. Sterilitetsforsk.* p.27

PEDERSEN, J.G.A. (1966). Om nedsat tolerance for jern hos grise. *Nord. VetMed.* **18**, 1–18

PETRENKO, G.G. and ZIRNOV, I.I. (1968). Injection of vitamin A for pregnant sows. *Svinovodstvo* **22**, 36 (*Nutr. Abstr. Rev.* **39**, 3985)

PIPER, R.C., FROSETH, J.A., McDOWELL, L.R., KROENING, G.H. and DYER, I.A. (1975). Selenium–Vitamin E deficiency in swine fed peas (*Pisum sativum*). *Am. J. vet. Res.* **36**, 273–281

PLUMLEE, M.P., THRASHER, D.M., BEESON, W.M. ANDREWS, F.N. and PARKER, H.E. (1953). The effect of manganese deficiency on the growth and development of swine. *J. Anim. Sci.* **14**, 996 (Abstr.)

PLUMLEE, M.P., THRASHER, D.M., BEESON, W.M., ANDREWS, F.N. and PARKER, H.E. (1956). The effect of a manganese deficiency upon the growth development and reproduction in swine. *J. Anim. Sci.* **15**, 352–367

POCHERNNYAEVA, G. (1976). Riboflavin in the diet and reproductive capacity of replacement gilts. *Svinovodstvo* **4**, 42–43. (*Nutr. Abstr. Rev.* **47**, 5542)

POND, W.G. and JONES, J.R. (1964). Effect of level of zinc in high calcium diets on pigs from weaning through reproductive cycle and on subsequent growth of their offspring. *J. Anim. Sci.* **23**, 1057–1060

REID, I.M. (1968). Chemical and histologic pathology of protein deficiency alone and complicated by vitamin E deficiency in the young pig. *Diss. Abstr.* **28**, 3962B

REID, I.M., BARNES, R.H., POND, W.G. and KROOK, L. (1968). Methionine-responsive liver damage in young pigs fed a diet low in protein and vitamin E. *J. Nutr.* **95**, 499–508

RITCHIE, H.D., MILLER, E.R., ULLREY, D.E., HOEFER, J.A. and LUECKE, R.W. (1960). Supplementation of the swine gestation diet with pyridoxine. *J. Nutr.* **70**, 491–496

SARYCEVA, M.M. (1968). Effect of vitamin A on reproduction in sows. *Svinovodstvo* **22**, 39 (*Nutr. Abstr. Rev.* **38**, 8313)

SEVKOVIC, N., PUJIN, D., TESARZ, I. and VUKOVIC, S. (1969). Effect of different amounts of vitamin A on reproduction in sows. *Vet. Glasn.* **23**, 21–27 (*Nutr. Abstr. Rev.* **39**, 5962)

SEVKOVIC, N., RAJIC, I., VUJOSEVIC, J. and PETRIC, M. (1973). The effect of vitamin A on embryonic mortality in gilts. *Acta vet. Beogr.* **23**, 197–202 (*Nutr. Abstr. Rev.* **44**, 6875)

SIHOMBING, D.T.H., CROMWELL, G.L. and HAYS, V.W. (1974). Effect of protein source, goitrogens and iodine level on performance and thyroid status of pigs. *J. Anim. Sci.* **39**, 1106–1112

SMITH, W.H., PLUMLEE, M.P. and BEESON, W.M. (1962). Effect of source of protein on zinc requirement of the growing pig. *J. Anim. Sci.* **21**, 399–405

STOCKLAND, W.L. and BLAYLOCK, L.G. (1974). Choline requirement of pregnant sows and gilts under restricted feeding conditions. *J. Anim. Sci.* **39**, 1113–1116

STUMPF, J. (1968). Effect of vitamin A on production of piglets. *Veterinaria, Spofa* **10**, 127–133 (*Nutr. Abstr. Rev.* **39**, 3984)

TEAGUE, H.S. and GRIFO, A.P., Jr. (1966). Vitamin B_{12} supplementation of sow rations. *J. Anim. Sci.* **25**, 895 (Abstract)

TEAGUE, H.S., GRIFO, A.P., Jr. and PALMER, W.M. (1971). Pantothenic acid deficiency in the sow. *J. Anim. Sci.* **33**, 239 (Abstract)

TEAGUE, H.S., PALMER, W.M. and GRIFO, A.P., Jr. (1970). Pantothenic acid deficiency in the reproducing sow. *Ohio agric. Res. Dev. Center Anim. Sci.* Mimeo No. 200, pp. 1–5

THOMAS, J.W., LOOSLI, J.K. and WILLMAN, J.P. (1947). Placental and mammary transfer of vitamin A in swine and goats as affected by the prepartum diet. *J. Anim. Sci.* **6**, 141–145

TOLLERZ, G. (1973). Vitamin, selenium (and some related compounds) and iron-tolerance in piglets. *Acta Agric. scand., Suppl.* **19**, 184

TOLLERZ, G. and LANNEK, N. (1964). Protection against iron toxicity in vitamin E deficient piglets and mice by vitamin E and synthetic antioxidants. *Nature, Lond.* **201**, 846–847

VAN VLEET, J.F., MEYER, K.B. and OLANDER, H.J. (1973). Control of selenium–vitamin E deficiency in growing swine by parenteral administration of selenium–vitamin E preparations to baby pigs or to pregnant sows and their baby pigs. *J. Am. vet. med Ass.* **163**, 452–456

VAN VLEET, J.F., MEYER, K.B. AND OLANDER, H.J. (1974). Acute selenium toxicosis induced in baby pigs by parenteral administration of selenium–vitamin E preparations. *J. Am. vet. med. Ass.* **165**, 543–547

VARGANOV, A.I. (1965). Fertility of pigs in conditions of I deficiency. *Svinovodstvo* 4, 31–32 (*Nutr. Abstr. Rev.* **35**, 7109)

WASTELL, M.E., EWAN, R.C., VORHIES, M.W. and SPEER, V.C. (1972). Vitamin E and selenium for growing and finishing pigs. *J. Anim. Sci.* **34**, 969–973

WHITING, F., LOOSLI, J.K. and WILLMAN, J.P. (1949). The influence of tocopherols upon the mammary and placental transfer of vitamin A in the sheep, goat and pig. *J. Anim. Sci.* **8**, 35–40

WÖHLBIER, W. and SIEGEL, A. (1967). Vitamin B_6 supply of sucking pigs. *Arch. Tierernähr.* **17**, 257–262

LIST OF PARTICIPANTS

Allen, Dr W.R.	TBA Equine Fertility Unit, Animal Research Station, 307 Huntingdon Road, Cambridge, CB3 0JQ
Almlid, Mr T.	Department of Animal Genetics and Breeding, Agricultural University of Norway, Box 24, 1432 As-NLH, Norway
Amoroso, Prof. E.C.	ARC Institute of Animal Physiology, Babraham, Cambridge
Andersson, Dr Anne-M.	Swedish University of Agricultural Sciences, College of Veterinary Medicine, S 75707 Uppsala, Sweden
Arven, Miss K.M.	Department of Animal Breeding and Genetics, Swedish University of Agricultural Sciences, 75007 Uppsala, Sweden
Aubrey, Miss S.	Nitrovit Limited, Belle Eau Park, Bilsthorpe, Nottinghamshire
Aumaitre, Dr A.L.	Pig Husbandry Department, Centre de Rennes St. Gilles, 35590 L'Hermitage, France
Ayliffe, Dr T.R.	Northumberland Hall, Hawkshead Lane, North Mymms, Hatfield, Hertfordshire
Baars, Dr J.C.	Intervet International B.V., P.O. Box 31, 5830 AA Boxmeer, The Netherlands
Baker, Miss D.J.	Hoechst UK Ltd., Animal Health Division, Walton Manor, Walton, Milton Keynes, MK7 7AJ
Ball, Dr P.D.H.	Department of Physiology and Environmental Studies, University of Nottingham School of Agriculture
Baxter, Mr M.R.	Scottish Farm Buildings Investigation Unit, Craibstone Estate, Bucksburn, Aberdeen
Bazer, Dr F.W.	Department of Animal Science, University of Florida, Gainesville, FL 32610, USA
Beccaro Dr P.	Instituto di Zootecnia Generale, Facolta di Agraria, Via Celoria, 2 20133 Milano, Italy
Best, Mr P.	Pig International, 18 Chapel Street, Petersfield, Hampshire
Binder, Dr M.	Institute for Animal Reproduction, The Royal Veterinary and Agricultural University, Bülowsrej 13, DK-1870 Copenhagen V, Denmark
Blair, Prof. R.	Prairie Swine Centre, University of Saskatchewan, Saskatoon, Sasketchewan, Canada S7N 0W0
Blichfeldt, Mr T.	Department of Animal Genetics and Breeding, Agricultural University of Norway, Box 24, 1432 As-NLH, Norway

Boland, Dr M.	Department of Agriculture, University College, Lyons Estate, Newcastle, Co. Dublin, Eire
Bonte, Dr P.	Faculty of Veterinary Medicine, Casinoplein 24, 9000 Ghent, Belgium
Booman, Dr P.	Research Institute for Animal Husbandry "Schoonoord", Driebergseweg 10D, Zeist, The Netherlands
Booth, Dr W.D.	ARC Institute of Animal Physiology, 307 Huntingdon Road, Cambridge, CB3 0JQ
Botté, Mrs F.	Institut Technique du Porc, 149 rue de Bercy 75595, Paris Cedex 12, France
Bouffault, Dr J.C.	Roussel-Uclaf, 163 Avenue Gambetta, 75020 Paris, France
Boyd, Dr J.	Unilever Research Laboratory, Colworth House, Sharnbrook, Bedfordshire
te Brake, Dr J.H.A.	Research Institute for Animal Husbandry "Schoonoord", PO Box 501, 3700 AM Zeist, The Netherlands
Braude, Dr R.	National Institute for Research in Dairying, Shinfield, Reading RG2 9AT
Britt, Dr J.H.	Department Animal Science, N.C. State University, Raleigh, North Caroline 27650 USA
Brooks, Dr P.H.	Seale Hayne College, Newton Abbot, Devon
Brown, Mr G.	Breckland Farms Ltd., Cranwich Road, Mundford, Thetford, Norfolk
Bunker, Mr L.A.	Icknield Way Farm Ltd., Tring Road, Dunstable, Bedfordshire
Buttle, Dr H.	National Institute for Research in Dairying, Shinfield, Reading RG2 9AT
Cameron, Mr D.	Huntingdon Research Centre, Huntingdon, Cambridgeshire
Campbell, Mr I.W.	United Pig Breeders Ltd., UPB House, 42 High Street, Somersham, Huntingdon, Cambridgeshire
Chang, Prof. M.C.	Worcester Foundation, Shrewsbury, Mass. 01545, USA
Clent, Mr E.	Department of Agriculture and Horticulture, University of Nottingham School of Agriculture
Cole, Dr D.J.A.	Department of Agriculture and Horticulture, University of Nottingham School of Agriculture
Colenbrander, Dr B.	Vakgroep Functionele Morfologie, Yalelaan 1, Postbus 80157, 3508 TD Utrecht, The Netherlands
Craig, Dr V.	ARC Institute of Animal Physiology, Babraham, Cambridge
Crettenand, Dr J.	Swiss Main Office for Pig, Sheep and Goat Production, Belpstrasse 16, 3000 Bern 14, Switzerland
Crighton, Dr D.B.	Department of Physiology and Environmental Studies, University of Nottingham School of Agriculture
Crofts, Mr T.	British Farmer & Stockbreeder, Surrey House, Throwley Way, Sutton, Surrey SM1 4QQ
Davidson, Mr F.McL.	Mains of Bogfechel, Whiterashes, Aberdeenshire
Davis, Mr C.	Elanco Products Ltd., Kingsclere Road, Basingstoke, Hampshire
Day, Prof. B.N.	University of Missouri, Animal Science Department, 159 Animal Science Research Center, Columbia, Mo 65211, USA

Deeley, Miss S.M.	Butterworth & Co (Publishers) Ltd., Borough Green, Sevenoaks, Kent TN15 8PH
Deligeorgis, Mr G.S.	School of Agriculture, Animal Production Department, 581 King Street, Aberdeen AB9 1UD
Dobson, Mrs H.	Department of Veterinary Clinical Studies, University of Liverpool, Neston, S. Wirral, Cheshire
Dufour, Mr J.	Research Station, Agriculture Canada, PO Box 90, Lennoxville, Quebec, Canada J1M 173
Dyck, Dr G.W.	Agriculture Canada, Research Station, Box 610, Brandon, Manitoba, Canada, R7A 5Z7
Dziuk, Dr P.J.	University of Illinois, Dept. of Animal Science, Animal Genetics Laboratory, Urbana, Illinois 61801, USA
Eastham, Miss P.	Department of Agriculture and Horticulture, University of Nottingham School of Agriculture
Edwards, Dr D.F.	Department of Physiology, The University, Sheffield, S10 2TN
Edwards, Dr S.	An Foras Taluntais, Grange Research Station, Dunsany, Co. Meath, Eire
Egbunike, Dr G.N.	Animal Physiology Laboratory, Department of Animal Science, University of Ibadan, Nigeria
Ellendorff Dr F.	Institut fur Tierzucht und Tierverhalten, FAL, Mariensee, 3057 Neustadt 1, West Germany
Elsaesser, Dr F.	Institut fur Tierzucht und Tierverhalten, FAL, Mariensee, 3057 Neustadt 1, West Germany
English, Dr P.R.	University of Aberdeen, Aberdeen AB9 1UD
Enne Dr G.	Instituto di Zootecnia Generale, Facolta di Agraria, via Celoria 2 20133 Milano, Italy
Etienne, Mr M.	Station de Recherche sur l'Elevage des Porcs, Centre de Rennes-St. Gilles, 35590 L'Hermitage, France
First, Prof. N.L.	University of Wisconsin-Madison, Department of Meat and Animal Science, Animal Sciences Building, 1675 Obervatory Drive, Madison, Wisconsin 53706, USA
Flint, Dr A.P.F.	ARC Institute of Animal Physiology, Babraham, Cambridge CB2 4AT
Ford, Dr J.J.	US Meat Animal Research Center, Clay Center, NE 68933 USA
Foxcroft, Dr G.R.	Department of Physiology and Environmental Studies, University of Nottingham School of Agriculture
Frape, Mr D.	Dalgety Spillers Ltd., Dalgety House, The Promenade, Clifton, Bristol BS8 3NJ
Garnsworthy, Dr P.	Department of Agriculture and Horticulture, University of Nottingham School of Agriculture
Garverick, Prof. H.A.	111 Animal Science Research Centre, University of Missouri, Columbia, Missouri 65211, USA
Gayerie, Miss F.	Department of Physiology and Environmental Studies, University of Nottingham School of Agriculture
Green, Mr S.	Department of Agriculture and Horticulture, University of Nottingham School of Agriculture
Hale, Dr D.H.	University of Zimbabwe, PO Box MP167, Mount Pleasant, Gansbury, Zimbabwe

Hardy, Dr B.	Dalgety Spillers Ltd., Dalgety House, The Promenade, Clifton, Bristol BS8 3NJ
Haresign, Dr W.	Department of Agriculture and Horticulture, University of Nottingham School of Agriculture
Harvey, Dr M.J.A.	Department of Veterinary Reproduction, University of Glasgow Veterinary School, Bearsden Road, Glasgow G61
Haynes, Dr N.B.	Department of Physiology and Environmental Studies, University of Nottingham School of Agriculture
Heap, Dr R.B.	ARC Institute of Animal Physiology, Babraham, Cambridge
Hemsworth, Dr P.	Animal Research Institute, Princes Highway, Werribee, Victoria 3030, Australia
Hiley, Dr P.	Cotswold Pig Development Co., Rothwell, Lincolnshire
Hill, Dr W.G.	Department of Genetics, University of Edinburgh, Institute of Animal Genetics, King's Buildings, West Mains Road, Edinburgh EH9 3JN
Holzman, Dr A.	Veterinarmedizinsche Universitat Wien, Klinik F. Geburtshilfe, Gynakologie u. Andrologie, Linke Bahngasse 11, 1-1030 Wien, Austria
Hoogerbrugge, Prof. A.	Institute of Zootechnic, Yalelaan 17, Utrecht, The Netherlands
Hook, Mr J.	Department of Physiology and Environmental Studies, University of Nottingham School of Agriculture
Howles, Mr C.	Department of Physiology and Environmental Studies, University of Nottingham School of Agriculture
Hudson, Mr K.W.	Beecham Animal Health, Broadmead Lane, Keynsham, Bristol BS18 1ST
Hughes, Dr P.E.	Department of Animal Physiology and Nutrition, University of Leeds, Leeds LS2 9JT
Humke, Dr R.	Hoechst Aktiengesellschaft, Med. Abt/Vet. H 840, Postfach 80 03 20, d-6230 Frankfurt (M) 80, West Germany
Hunter, Dr R.H.F.	School of Agriculture, University of Edinburgh, West Mains Road, Edinburgh EH9 3JG
Jackson, Mr P.S.	ICI Limited, Pharmaceuticals Div., Animal Health Dept. Alderley House, Alderley Park, Macclesfield, Cheshire
Jochle, Dr W.	Wolfgang Jochle Associates, Old Boonton Road, Denville TWP., N.J. 07834, USA
Karlberg, Mr K.	Departmemt of Reproductive Physiology and Pathology, Veterinary College of Norway, PO Box 8146, Oslo 1, Norway
Kenyon, Mr P.J.	BOCM Silcock Ltd., Basing View, Basingstoke, Hampshire
Kenyon, Mr P.W.	Harbro Farm Sales Ltd., 62/64 Fife Street, Turriff, Aberdeenshire
Kerr, Mrs O.M.	Veterinary Research Labs., Stoney Road, Stormont, Belfast BT4 3SD
Kirby, Mr P.S.	MAFF, Government Buildings, Kenton Bar, Newcastle-upon-Tyne, NE1 2YA
Kovac, Miss M.	Univeria Edvarda Kardelja V Ljubljani, Biotehniska Fakulteta Vtozd za Zivinorejo, Geoblje 3, 61230 Domzale, Yugoslavia
Kraeling, Dr R.R.	Richard Russell Agricultural Research Center, SEA, USDA, Athens, Georgia, USA

Ladewig, Dr J.	Institut fur Tierzucht und Tierverhalten – FAL, Mariensee, 3067 Neustadt 1, West Germany
Laird, Mr R.	Animal Husbandry Department, West of Scotland Agricultural College, Auchincruive, Ayr KA6 5HW
Lamming, Prof. E.	Department of Physiology and Environmental Studies, University of Nottingham School of Agriculture
Lowe, Mr J.A.	Pig Products Manager, Simmons Watts Ltd., (RHM Agriculture) PO Box 15, Edward Street, Banbury, Oxon OX16 8SB
Lynch, Mr P.B.	The Agricultural Institute, Moorepark Research Centre, Fermoy, Co. Cork, Eire
Macdonald, Dr A.A.	Vakgroep Functionele Morfologie, Yalelaan 1, Postbus 80157, 3508 TD Utrecht University, The Netherlands
Madsen, Mr P.	Veterinary Department, ESS-Food, Axelborg, Axeltorv 3, 1609 Copenhagen V. DK-Denmark
Major, Mr R.	The Major Pig Consultancy, 48 Kenilworth Avenue, Reading. RG3 3DN
Martin, Mr M.A.	ACOT, Western Region Office, Athenry, Co. Galway, Eire
Mauget, Dr A.	Centre National de la Recherche Scientifique, Centre D'Etudes Biologiques des Animaux Sauvages, Villiers-en-Bois, 79360 Beauvoir-sur-Niort, France
Melrose, Dr D.R.	Meat and Livestock Commission, PO Box 44, Queensway House, Bletchley, MK2 2EF
Mills, Mr C.	Department of Agriculture and Horticulture, University of Nottingham School of Agriculture
Mitchell, Dr K.G.	Applied Pig Nutrition Department, National Institute for Research in Dairying, Shinfield, Reading RG2 9AT
Morton, Dr H.	Department of Surgery, University of Queensland, Queensland, Australia
Ogink, Dr W.D.	Consulentschap i.a.d. voor Varkenshoudery, Willensplantsoen 6, 3511 LA Utrecht, The Netherlands
O'Reilly, Mr P.J.	Senior Research Officer, Department of Agriculture, Agriculture House, Dublin 2, Eire
Owers, Dr M.J.	Pauls & Whites Foods, New Cut West, Ipswich, Suffolk
Page, Mr R.	Allen and Page Ltd., Quayside Mills, Norwich, Norfolk
Palludan, Dr B.	Department of Physiology, Endocrinology & Bloodgrouping, Royal Veterinary & Agricultural University, DK-Bulowsvej 13, 1870 Copenhagen V, Denmark
Parry, Mrs M.A.	Harper Adams Agricultural College, Newport, Salop
Parvizi, Dr N.	Institute of Tierzucht und Tierverhalten, Mariensee, 3057 Neustadt 1, West Germany
Paterson, Dr A.M.	Department of Physiology and Environmental Studies, University of Nottingham School of Agriculture
Petchey, Dr A.M.	North of Scotland College of Agriculture, 581, King Street, Aberdeen
Peters, Dr A.R.	Meat and Livestock Commission, P.O. Box 44, Queensway House, Bletchley, MK2 2EF
Pierantoni, Dr R.	Instituo e Museo di Zoologia, Via Mezzocannone 8 80134 Napoli, Italy

Pittman, Mr R.J.	Applied Pig Nutrition Department, National Institute for Research in Dairying, Shinfield, Reading RG2 9AT
Polge, Dr C.	ARC Institute of Animal Physiology, Animal Research Station, 307 Huntingdon Road, Cambridge CB3 0JQ
Prunier, Miss A.	INRA, Laboratoire de Physiologie de la Reproduction, 37380 Nouzilly, France
Pugh, Mr O.L.	Dorset College of Agriculture, Kingston Maurward, Dorchester, Dorset
Rampacek, Dr G.B.	Animal and Dairy Science Department, Livestock-Poultry Buildings, University of Georgia, Athens, Georgia, USA
Reed, Dr H.C.B.	Meat and Livestock Commission, Pig Breeding Centre, Leeds Road, Thorpe Willoughby, Selby, N. Yorks YO8 9HL
Riley, Miss G.M.	Department of Physiology and Environmental Studies, University of Nottingham School of Agriculture
Riley, Mr J.	ADAS, Government Buildings, Brooklands Avenue, Cambridge
Robbins, Mr S.	J. Bibby Agriculture Ltd., Feeds & Seeds Division, Nr. Banbury, Oxon
Robertson, Dr H.A.	Animal Research Centre, Agriculture Canada, Ottawa, Canada K1A 0C6
Saunders, Miss P.	ARC Institute of Animal Physiology, Babraham, Cambridge
Shaw, Miss H.J.	Department of Physiology and Environmental Studies, University of Nottingham School of Agriculture
Sherwood, Prof. O.D.	Department of Physiology and Biophysics, 524 Burrill Hall, Urbana, Illinois 61801, USA
Smith, Mr D.	Meat and Livestock Commission, PO Box 44, Queensway House, Bletchley, Milton Keynes, MK2 2EF
Smyth, Mr D.J.	Department of Agriculture, Agriculture House, Dublin 2, Eire
Sparkes, Mr G.	Department of Agriculture and Horticulture, University of Nottingham School of Agriculture
Speight, Mr D.	Nitrovit Limited, Nitrovit House, Dalton, Thirsk, North Yorkshire
Van der Steen, Dr H.	Agriculture University, Animal Breeding Department, Postbus 338, 6700 AH Wageningen, The Netherlands
Stickney, Mr K.	BP Nutrition (UK) Ltd., Wincham, Northwich, Cheshire CW9 6DF
Tagwerker, Dr F.J.	Department VT, F. Hoffmann-LA Roche & Co. Limited, CH-4002 Basle, Switzerland
Tait, Dr A.J.	Glaxo Group Research Ltd., Breakspear Road South, Harefield, Uxbridge, Middlesex
Tardivon, Miss F.	UFAC, 95450 Viegy, France
Taverne, Dr M.	State University of Utrecht, Yalelaan 7, De Uithof, 3068 TD Utrecht, The Netherlands
Thomas, Mr P.	MAFF, Government Buildings, Burghill Road, Westbury-on-Trym, Bristol
Tilton, Dr J.E.	Animal Science Department, North Dakota State University, Fargo, N.D.58105, USA

Tomes, Dr G.	Muresk Agricultural College, Muresk, W.A. 6401, Australia
Toplis, Mr P.	RHM Animal Feed Services Ltd., Deans Grove House, Colehill, Wimborne, Dorset
Tuck, Mr J.P.	W.F. Tuck & Sons Ltd., The Mills, Burston, Diss, Norfolk
van der Valk, Dr P.C.	Kliniek voor Inwendige Ziekten, Yalelaan 16, 3584 CM Utrecht, Belgium
Varley, Dr M.A.	The Rowett Research Institute, Applied Nutrition Department, Greenburn Road, Bucksburn, Aberdeen AB2 9SB
Vrieze, Dr J.D.	Intervet International B.V., PO Box 31, 5830 AA Boxmeer, The Netherlands
Walach-Janiak, Dr M.	Polish Academy of Sciences, Institute of Animal Physiology and Nutrition, 05-110 Jablonna, near Warsaw, Poland
Walters, Mr R.	Masterbreeders (Livestock Development) Ltd., Basing View, Basingstoke, Hampshire
Walton, Dr J.S.	Department of Animal & Poultry Science, University of Guelph, Guelph, Ontario, Canada N1G 2W1
Ware, Mr M.	Butterworth & Co. (Publishers) Ltd. Borough Green, Sevenoaks, Kent TN15 8PH
Waterworth, Mr D.G.	ICI Ltd., Jealotts Hill Research Station, Bracknell, Berkshire
Webb, Dr A.J.	ARC Animal Breeding Research Organisation, West Mains Road, Edinburgh EH9 3JQ
Webel, Dr S.K.	Reproductive Consulting Service, 5303 Wilmot Road, McHenry, Illinois 60050, USA
Welch, Prof. J.A.	Division of Animal & Veterinary Sciences, West Virginia University, Morgantown, WV 2650, USA
Wilby, Mr D.T.	W.F. Tuck and Son Ltd., The Mills, Burston, Diss, Norfolk
Wilkinson, Miss V.	Department of Animal Production and Health, School of Agriculture, 581 King Street, Aberdeen AB9 1UD
Will, Mr I.	ABRO, Mountmarle, Roslin, Midlothian
Willemse, Prof. A.H.	Clinic of Veterinary Obstetrics, A.I. and Reproduction, Vet. Faculty University of Utrecht, Yalelaan 7, 3508 TD Utrecht, The Netherlands
Williamson, Miss E.D.	Meat Research Institute, Langford, Bristol, BS18 7DY
Wiseman, Dr J.	Department of Agriculture and Horticulture, University of Nottingham School of Agriculture
Wrathall, Dr A.E.	MAFF, Central Veterinary Laboratory, New Haw, Weybridge, Surrey
Yamada, Dr Y.	National Institut of Animal Industry, Tsukuba Norindanchi, P.O. Box 5, Ibaraki-ken, 305, Japan
Young, Mr I.M.	Upjohn Ltd., Fleming Way, Crawley, Sussex
Ziecik, Dr A.J.	Institute of Animal Physiology and Biochemistry, University of Agriculture and Technology, 10-718 Olsityn, Poland

INDEX

References to Figures are in italics.